Lecture Notes in Computer Science   9949

Commenced Publication in 1973
Founding and Former Series Editors:
Gerhard Goos, Juris Hartmanis, and Jan van Leeuwen

More information about this series at http://www.springer.com/series/7407

Akira Hirose · Seiichi Ozawa
Kenji Doya · Kazushi Ikeda
Minho Lee · Derong Liu (Eds.)

Neural
Information Processing

23rd International Conference, ICONIP 2016
Kyoto, Japan, October 16–21, 2016
Proceedings, Part III

 Springer

Editors
Akira Hirose
The University of Tokyo
Tokyo
Japan

Kazushi Ikeda
Nara Institute of Science and Technology
Ikoma
Japan

Seiichi Ozawa
Kobe University
Kobe
Japan

Minho Lee
Kyungpook National University
Daegu
Korea (Republic of)

Kenji Doya
Okinawa Institute of Science and
  Technology Graduate University
Onna
Japan

Derong Liu
Chinese Academy of Sciences
Beijing
China

ISSN 0302-9743   ISSN 1611-3349   (electronic)
Lecture Notes in Computer Science
ISBN 978-3-319-46674-3   ISBN 978-3-319-46675-0   (eBook)
DOI 10.1007/978-3-319-46675-0

Library of Congress Control Number: 2016953319

LNCS Sublibrary: SL1 – Theoretical Computer Science and General Issues

Printed on acid-free paper

This Springer imprint is published by Springer Nature
The registered company is Springer International Publishing AG
The registered company address is: Gewerbestrasse 11, 6330 Cham, Switzerland

Preface

This volume is part of the four-volume proceedings of the 23rd International Conference on Neural Information Processing (ICONIP 2016) held in Kyoto, Japan, during October 16–21, 2016, which was organized by the Asia-Pacific Neural Network Society (APNNS, http://www.apnns.org/) and the Japanese Neural Network Society (JNNS, http://www.jnns.org/). ICONIP 2016 Kyoto was the first annual conference of APNNS, which started in January 2016 as a new society succeeding the Asia-Pacific Neural Network Assembly (APNNA). APNNS aims at the local and global promotion of neural network research and education with an emphasis on diversity in members and cultures, transparency in its operation, and continuity in event organization. The ICONIP 2016 Organizing Committee consists of JNNS board members and international researchers, who plan and run the conference.

Currently, neural networks are attracting the attention of many people, not only from scientific and technological communities but also the general public in relation to the so-called Big Data, TrueNorth (IBM), Deep Learning, AlphaGo (Google DeepMind), as well as major projects such as the SyNAPSE Project (USA, 2008), the Human Brain Project (EU, 2012), and the AIP Project (Japan, 2016). The APNNS's predecessor, APNNA, promoted fields that were active but also others that were leveling off. APNNS has taken over this function, and further enhances the aim of holding technical and scientific events for interaction where even those who have extended the continuing fields and moved into new/neighboring areas rejoin and participate in lively discussions to generate and cultivate novel ideas in neural networks and related fields.

The ICONIP 2016 Kyoto Organizing Committee received 431 submissions from 38 countries and regions worldwide. Among them, 296 (68.7 %) were accepted for presentation. The first authors of papers that were presented came from Japan (100), China (78), Australia (22), India (13), Korea (12), France (7), Hong Kong (7), Taiwan (7), Malaysia (6), United Kingdom (6), Germany (5), New Zealand (5) and other countries/regions worldwide.

Besides the papers published in these four volumes of the Proceedings, the conference technical program includes

- Four plenary talks by Kunihiko Fukushima, Mitsuo Kawato, Irwin King, and Sebastian Seung
- Four tutorials by Aapo Hyvarinen, Nikola Kazabov, Stephen Scott, and Okito Yamashita,
- One Student Best Paper Award evaluation session
- Five special sessions, namely, bio-inspired/energy-efficient information processing, whole-brain architecture, data-driven approach for extracting latent features from multidimensional data, topological and graph-based clustering methods, and deep and reinforcement learning
- Two workshops: Data Mining and Cybersecurity Workshop 2016 and Workshop on Novel Approaches of Systems Neuroscience to Sports and Rehabilitation

The event also included exhibitions and a technical tour.

Kyoto is located in the central part of Honshu, the main island of Japan. Kyoto formerly flourished as the imperial capital of Japan for 1,000 years after 794 A.D., and is presently known as "The City of Ten Thousand Shrines." There are 17 sites (13 temples, three shrines, and one castle) in Kyoto that form part of the UNESCO World Heritage Listing, named the "Historic Monuments of Ancient Kyoto (Kyoto, Uji and Otsu Cities)." In addition, there are three popular, major festivals (Matsuri) in Kyoto, one of which, "Jidai Matsuri" (The Festival of Ages), was held on October 22, just after ICONIP 2016.

We, the general chair, co-chair, and Program Committee co-chairs, would like to express our sincere gratitude to everyone involved in making the conference a success. We wish to acknowledge the support of all the sponsors and supporters of ICONIP 2016, namely, APNNS, JNNS, KDDI, NICT, Ogasawara Foundation, SCAT, as well as Kyoto Prefecture, Kyoto Convention and Visitors Bureau, and Springer. We also thank the keynote, plenary, and invited speakers, the exhibitors, the student paper award evaluation committee members, the special session and workshop organizers, as well as all the Organizing Committee members, the reviewers, the conference particⁱpants, and the contributing authors.

October 2016

Akira Hirose
Seiichi Ozawa
Kenji Doya
Kazushi Ikeda
Minho Lee
Derong Liu

Organization

General Organizing Board

JNNS Board Members

Honorary Chairs

Shun-ichi Amari   RIKEN
Kunihiko Fukushima   Fuzzy Logic Systems Institute

Organizing Committee

General Chair

Akira Hirose   The University of Tokyo, Japan

General Co-chair

Seiichi Ozawa   Kobe University, Japan

Program Committee Chairs

Kenji Doya   OIST, Japan
Kazushi Ikeda   NAIST, Japan
Minho Lee   Kyungpook National University, Korea
Derong Liu   Chinese Academy of Science, China

Local Arrangements Chairs

Hiroaki Nakanishi   Kyoto University, Japan
Ikuko Nishikawa   Ritsumeikan University, Japan

Members

Toshio Aoyagi   Kyoto University, Japan
Naoki Honda   Kyoto University, Japan
Kazushi Ikeda   NAIST, Japan

Shin Ishii   Kyoto University, Japan
Katsunori Kitano   Ritsumeikan University, Japan
Hiroaki Mizuhara   Kyoto University, Japan
Yoshio Sakurai   Doshisha University, Japan
Yasuhiro Tsubo   Ritsumeikan University, Japan

Financial Chair

Seiichi Ozawa   Kobe University, Japan

Member

Toshiaki Omori   Kobe University, Japan

Special Session Chair

Kazushi Ikeda   NAIST, Japan

Workshop/Tutorial Chair

Hiroaki Gomi   NTT Communication Science Laboratories, Japan

Publication Chair

Koichiro Yamauchi   Chubu University, Japan

Members

Yutaka Hirata   Chubu University, Japan
Kay Inagaki   Chubu University, Japan
Akito Ishihara   Chukyo University, Japan

Exhibition Chair

Tomohiro Shibata   Kyushu Institute of Technology, Japan

Members

Hiroshi Kage   Mitsubishi Electric Corporation, Japan
Daiju Nakano   IBM Research - Tokyo, Japan
Takashi Shinozaki   NICT, Japan

Publicity Chair

Yutaka Sakai   Tamagawa University, Japan

Industry Relations

Ken-ichi Tanaka   Mitsubishi Electric Corporation, Japan
Toshiyuki Yamane   IBM Research - Tokyo, Japan

Sponsorship Chair

Ko Sakai   University of Tsukuba, Japan

Member

Susumu Kuroyanagi   Nagoya Institute of Technology, Japan

General Secretaries

Hiroaki Mizuhara   Kyoto University, Japan
Gouhei Tanaka   The University of Tokyo, Japan

International Advisory Committee

Igor Aizenberg   Texas A&M University-Texarkana, USA
Sabri Arik   Istanbul University, Turkey
P. Balasubramaniam   Gandhigram Rural Institute, India
Eduardo Bayro-Corrochano   CINVESTAV, Mexico
Jinde Cao   Southeast University, China
Jonathan Chan   King Mongkut's University of Technology, Thailand
Sung-Bae Cho   Yonsei University, Korea
Wlodzislaw Duch   Nicolaus Copernicus University, Poland
Tom Gedeon   Australian National University, Australia
Tingwen Huang   Texas A&M University at Qatar, Qatar
Nik Kasabov   Auckland University of Technology, New Zealand
Rhee Man Kil   Sungkyunkwan University (SKKU), Korea
Irwin King   Chinese University of Hong Kong, SAR China
James Kwok   Hong Kong University of Science and Technology,
                                 SAR China
Weng Kin Lai   Tunku Abdul Rahman University College, Malaysia
James Lam   The University of Hong Kong, SAR Hong Kong
Kittichai Lavangnananda   King Mongkut's University of Technology, Thailand
Min-Ho Lee   Kyungpoor National University, Korea
Soo-Young Lee   Korea Advanced Institute of Science and Technology,
                                 Korea
Andrew Chi-Sing Leung   City University of Hong Kong, SAR China
Chee Peng Lim   University Sains Malaysia, Malaysia
Chin-Teng Lin   National Chiao Tung University, Taiwan
Derong Liu   The Institute of Automation of the Chinese Academy of
                                 Sciences (CASIA), China

Chu Kiong Loo	University of Malaya, Malaysia
Bao-Liang Lu	Shanghai Jiao Tong University, China
Aamir Saeed Malik	Petronas University of Technology, Malaysia
Danilo P. Mandic	Imperial College London, UK
Nikhil R. Pal	Indian Statistical Institute, India
Hyeyoung Park	Kyungpook National University, Korea
Ju. H. Park	Yeungnam University, Republic of Korea
John Sum	National Chung Hsing University, Taiwan
DeLiang Wang	Ohio State University, USA
Jun Wang	Chinese University of Hong Kong, SAR Hong Kong
Lipo Wang	Nanyang Technological University, Singapore
Zidong Wang	Brunel University, UK
Kevin Wong	Murdoch University, Australia
Xin Yao	University of Birmingham, UK
Li-Qing Zhang	Shanghai Jiao Tong University, China

Advisory Committee Members

Masumi Ishikawa	Kyushu Institute of Technology
Noboru Ohnishi	Nagoya University
Shiro Usui	Toyohashi University of Technology
Takeshi Yamakawa	Fuzzy Logic Systems Institute

Technical Program Committee

Abdulrahman Altahhan	Tetsuo Furukawa
Sabri Arik	Kuntal Ghosh
Sang-Woo Ban	Anupriya Gogna
Tao Ban	Hiroaki Gomi
Matei Basarab	Shanqing Guo
Younes Bennani	Masafumi Hagiwara
Ivo Bukovsky	Isao Hayashi
Bin Cao	Shan He
Jonathan Chan	Akira Hirose
Rohitash Chandra	Jin Hu
Chung-Cheng Chen	Jinglu Hu
Gang Chen	Kaizhu Huang
Jun Cheng	Jun Igarashi
Long Cheng	Kazushi Ikeda
Zunshui Cheng	Ryoichi Isawa
Sung-Bae Cho	Shin Ishii
Justin Dauwels	Teijiro Isokawa
Mingcong Deng	Wisnu Jatmiko
Kenji Doya	Sungmoon Jeong
Issam Falih	Youki Kadobayashi

Keisuke Kameyama
Joarder Kamruzzaman
Rhee-Man Kil
DaeEun Kim
Jun Kitazono
Yasuharu Koike
Markus Koskela
Takio Kurita
Shuichi Kurogi
Susumu Kuroyanagi
Minho Lee
Nung Kion Lee
Benkai Li
Bin Li
Chao Li
Chengdong Li
Tieshan Li
Yangming Li
Yueheng Li
Mingming Liang
Qiao Lin
Derong Liu
Jiangjiang Liu
Weifeng Liu
Weiqiang Liu
Bao-Liang Lu
Shiqian Luo
Hongwen Ma
Angshul Majumdar
Eric Matson
Nobuyuki Matsui
Masanobu Miyashita
Takashi Morie
Jun Morimoto
Chaoxu Mu
Hiroyuki Nakahara
Kiyohisa Natsume
Michael Kwok-Po Ng
Vinh Nguyen
Jun Nishii
Ikuko Nishikawa
Haruhiko Nishimura
Tohru Nitta
Homma Noriyasu
Anto Satriyo Nugroho
Noboru Ohnishi

Takashi Omori
Toshiaki Omori
Sid-Ali Ouadfeul
Seiichi Ozawa
Paul Pang
Hyung-Min Park
Kitsuchart Pasupa
Geong Sen Poh
Santitham Prom-on
Dianwei Qian
Jagath C. Rajapakse
Mallipeddi Rammohan
Alexander Rast
Yutaka Sakaguchi
Ko Sakai
Yutaka Sakai
Naoyuki Sato
Shigeo Sato
Shunji Satoh
Chunping Shi
Guang Shi
Katsunari Shibata
Hayaru Shouno
Jeremie Sublime
Davor Svetinovic
Takeshi Takahashi
Gouhei Tanaka
Kenichi Tanaka
Toshihisa Tanaka
Jun Tani
Katsumi Tateno
Takashi Tateno
Dat Tran
Jan Treur
Eiji Uchibe
Eiji Uchino
Yoji Uno
Kalyana C. Veluvolu
Michel Verleysen
Ding Wang
Jian Wang
Ning Wang
Ziyang Wang
Yoshikazu Washizawa
Kazuho Watanabe
Bunthit Watanapa

Juyang Weng

Bin Xu

Tetsuya Yagi

Nobuhiko Yamaguchi

Hiroshi Yamakawa

Toshiyuki Yamane

Koichiro Yamauchi

Tadashi Yamazaki

Pengfei Yan

Qinmin Yang

Xiong Yang

Zhanyu Yang

Junichiro Yoshimoto

Zhigang Zeng

Dehua Zhang

Li Zhang

Nian Zhang

Ruibin Zhang

Bo Zhao

Jinghui Zhong

Ding-Xuan Zhou

Lei Zhu

Contents – Part III

Topological and Graph Based Clustering Methods

Reinforcement Learning

Computational Intelligence

Data Mining

Deep Neural Networks

Time Series Analysis

Chaotic Feature Selection and Reconstruction in Time Series Prediction

Shamina Hussein[1] and Rohitash Chandra[2(✉)]

[1] School of Mathematical and Computing Sciences,
Fiji National University, Suva, Fiji
shamina.hussein@gmail.com
[2] Artificial Intelligence and Cybernetics Research Group Software Foundation,
Nausori, Fiji
c.rohitash@gmail.com
http://aicrg.softwarefoundationfiji.org

Abstract. The challenge in feature selection for time series lies in achieving similar prediction performance when compared with the original dataset. The method has to ensure that important information has not been lost by with feature selection for data reduction. We present a chaotic feature selection and reconstruction method based on statistical analysis for time series prediction. The method can also be viewed as a way for reduction of data through selection of most relevant features with the hope of reducing training time for learning algorithms. We employ cooperative neuro-evolution as a machine learning tool to evaluate the performance of the proposed method. The results show that our method gives a data reduction of up to 42 % with a similar performance when compared to the literature.

1 Introduction

Time series prediction can be cumbersome for big data related problems. The volume of data and the associated computational complexity can yield higher prediction inaccuracies due to learning irrelevant information [1]. The volume of data and the associated computational complexity can make the application very challenging and therefore, robust feature selection methods are important for removing redundant features [2].

Feature selection methods identify features that are most relevant for a time series in order to achieve faster training performance and with the hope of improving prediction performance as noisy features can be eliminated [1]. The major categories of feature selection methods include the wrapper [3], filter [4] and embedded [4] methods. In a wrapper method, the selection criterion is dependent on the learning algorithm as a part of the fitness function. The selection criterion of filtering methods are independent of the learning algorithm and the selection of feature relies on the relevance score of the feature. The embedded method is specific to a learning algorithm and searches for an optimal subset of features by estimating changes in the objective function value incurred by

© Springer International Publishing AG 2016
A. Hirose et al. (Eds.): ICONIP 2016, Part III, LNCS 9949, pp. 3–11, 2016.
DOI: 10.1007/978-3-319-46675-0_1

making moves in the variable subset. Although wrapper and filter methods are the most commonly used methods for feature selection, their drawbacks initiate a need for a simpler and less expensive method. Wrapper methods have reported superior performance but it is computationally expensive when compared to filter methods [5]. A drawback of filters is that they cannot scale for high dimensional data [5]. Some commonly used feature extraction techniques include statistical methods that feature mean and standard deviation [6], frequency count summations, KarhunenLoeve transformations, Fourier transformations and wavelet transformations [7].

Recently, big data related problems have gained much attention that highlighted further challenges in learning algorithms as they have to deal with enormous amounts of data. However, there has not been much focus on time series problems. The increase in the implementation of technologies such as internet-of-things [8] would result in enhanced data collection and time series analysis will become more difficult as it would need to deal with big data challenges. Hence, we aim to present an approach to address this for upcoming challenges in big data related time series problems [9].

We present a chaotic feature selection and reconstruction method for time series prediction with the hope to reduce the size of the original time series while retaining important information. We employ cooperative neuro-evolution to evaluate the performance of the proposed method. In principle, any machine learning method could be used, however, we selected cooperative neuro-evolution due to its promising performance for time series prediction in previous work [10].

The paper is organized as follows. Section 2 presents the proposed method and Sect. 3 presents the experiments with results and discussion. Section 4 concludes the paper with insights for future work.

2   Chaotic Feature Selection and Reconstruction

We present the details of chaotic feature selection and reconstruction (CFSR) method for chaotic time series. It essentially eliminates the smooth regions of the time series and selects the noisy and chaotic regions. We first divide the time series into subsets known as feature windows and employ simple statistical evaluations to determine if the feature window contains smooth or noisy data points. Note that statistical measurements such as the mean and the standard deviation have been used in the past in feature extraction methods [6]. In our case, they are used to identify the chaotic and noisy regions in the feature window.

In Algorithm 1, the *length* of the *feature window* define the subsets in the time series. The feature window length must be determined experimentally to find the optimal value for best prediction performance. In Step 1, feature window is used to partition the entire time series (Step 2). For each feature window until the entire time series has been considered, the *upper boundary* (Eq. 3) and *lower boundary* (Eq. 4) is defined using the *standard deviation* (Eq. 1) and the *mean* (Eq. 2). The values which falls between the boundaries are selected as the features for reconstruction.

Algorithm 1. Chaotic Feature Selection and Reconstruction

Step 1: Define chaotic feature window length
Step 2: Partition the time series into n feature windows according to its *length*
foreach *feature window* **do**
  i. Calculate the Mean *(Eq. 2)*
  ii. Calculate the Standard Deviation *(Eq. 1)*
  iii. Identify the Upper-boundary *(Eq. 3)*
  iv. Identify the Lower-boundary *(Eq. 4)*
  v. Select features that fall within Upper and Lower boundaries
end

$$\sigma = \sqrt{\frac{1}{N-1}\sum_{i=1}^{N}(x_i - \bar{x})^2} \quad (1)$$

$$B_u = \mu + \sigma \quad (3)$$

$$\mu = \frac{1}{n}\sum_{k=1}^{n} x_k \quad (2)$$

$$B_l = \mu - \sigma \quad (4)$$

We then apply Takens' embedding theorem [11] to the selected chaotic features in order to reconstruct the dataset for a one-step prediction. Given an observed time series $x(t)$, an embedded phase space $Y(t) = [x(t), x(t - T), ..., x(t(D-1)T)]$ can be generated. T is the time delay, D is the embedding dimension and N is the actual length of the observed time series [11]. This resulting dataset is then used as the input vector for training the model, which in our case is the feedforward neural network.

2.1   Cooperative Neuro-Evolution

Cooperative coevolution(CC) that was initially proposed for function optimization [12], has gained success in neuro-evolution for time series prediction [10]. CC decomposes a problem into subcomponents that are implemented as sub-populations. Much work has been done in the past that focus on problem decomposition that are based on architectural properties of the network [13].

We employ cooperative neuro-evolution (CNE) to demonstrate the effectiveness of proposed feature selection method and it has shown promising results in chaotic time series prediction [10]. CNE used for training feedforward neural networks is given in Algorithm 2. It employs neuron-level decomposition for decomposing the neural network into k subcomponents [13]. The number of subcomponents k is determined by the total number of hidden and output neurons.

In the initialization stage, each sub-population is assigned random numbers in a range and evaluated cooperatively. This is implemented by concatenating the current individual that needs to be evaluated with the fittest individual from the rest of the sub-populations. The concatenated individual is then encoded into the neural network which returns the fitness defined by the root-mean-squared-error.

The main part of the algorithm begins by evolving each of the sub-populations in a *round-robin* fashion for a certain number of generations called the *depth of search*. Any evolutionary algorithm can be chosen for evolution of the sub-populations that feature operations such as crossover, selection and mutation. However, fitness evaluation of each individual of a sub-population is evaluated cooperatively as implemented in the initialisation stage. The procedure is repeated until the termination condition has been reached which is defined by the maximum number of fitness evaluations or a fitness value.

Algorithm 2. Cooperative Coevolution

Step 1: Employ neuron level decomposition and attain k subcomponents
Step 2: Initialize and cooperatively evaluate each subcomponent implemented as a sub-population
foreach *until termination* **do**
    foreach *each Sub-population* **do**
        foreach *n Generations* **do**
            Select and create new off-springs
            Cooperatively evaluate the new off-springs
            Update sub-population
        end
    end
end

---

3 Experiments and Results

This section presents the experimental evaluation of the proposed chaotic feature selection and reconstruction (CFSR) method for time series problems. We use cooperative neuro-evolution (CNE) as the designated learning algorithm for feedforward neural network (FNN).

3.1 Problem Description

The benchmark time series data employed are Mackey-Glass times series [14] and Lorenz time series [15], the two simulated time series while the real-world time series are the Sunspot time series [16], Laser time series [17] and Astrophysics time series [18]. Takens' embedding theorem [11] is applied to the selected features to reconstruct the data set. The values for the embedding dimension (D) and the time delay (T) has been set as follows. $D = 5$ and $T = 3$ for the Astrophysics and Sunspot time series. $D = 3$ and $T = 2$ for Lorenz and Mackey Glass time series. $D = 7$ and $T = 2$ for Laser time series.

These reconstructed vectors are then used to train the feedforward neural network. The prediction performance of the feedforward neural network is measured using the root mean squared error(RMSE) (Eq. 5) and the normalized mean squared error(NMSE) (Eq. 6)

$$RMSE = \sqrt{\frac{1}{N}\sum_{i=1}^{N}(y_i - \hat{y}_i)^2} \quad (5) \qquad NMSE = \left(\frac{\sum_{i=1}^{N}(y_i - \hat{y}_i)^2}{\sum_{i=1}^{N}(y_i - \bar{y}_i)^2}\right) \quad (6)$$

where $y_i, \hat{y}_i$ and $\bar{y}_i$ are the observed data, predicted data and average of observed data, respectively. N is the length of the observed data. These results are also compared with related methods from the literature.

3.2 Experimental Design

We use a feedforward neural network with the sigmoid units in the hidden layer. In the output layer, sigmoid unit is employed for the Mackey Glass, Sunspot and Laser time series while a hyperbolic tangent unit is employed for the Lorenz and Astrophysics time series. A set of 50 independent experimental runs are executed for 3, 5 and 7 hidden neurons. Each sub-population in CNE is evolved a fixed number of generations in a round-robin fashion. This depth of search was set to 1 generation as it has shown to be suitable for neuro-evolution [10]. The G3-PCX algorithm was used to evolve all the sub-populations of CNE. A population size of 300 is used. We used 15000 as the maximum number of function evaluations for the termination condition for all the problems.

3.3 Results and Discussion

The results of 50 experimental runs with 95 % confidence interval for different number of hidden neurons are given in Table 1. We evaluate the results by comparing the different feature windows with the number of hidden neurons (H). The lowest values for the RMSE indicates the best performance.

In the Sunspot problem, the best performance was given by 5 hidden neurons on feature window size of 100. In this case, the proposed method reduced the original dataset by 42 % which has been the greatest reduction when compared to other feature windows, while achieving the best performance. The Laser and the Astrophysics problems achieved the best generalization performance. This was through the dataset generated on feature window of 50, with 7 and 5 hidden neurons, respectively. Hence, the proposed method reduced the original data set by 25 % for Laser problem, and 34 % for the Astrophysics problem.

The proposed method has been able to cope up with noise in the real world problems such as Sunspot, Astrophysics, and Laser. It can be observed from the results that large data sets get reduced greatly and also yields very comparable results. However, there is not a large reduction for smaller datasets of size 500. It is also observed that for the simulated time series, the best generalisation performance is consistently displayed by the same feature window that gives the best results. As for the real world time series, the results were not as consistent when we consider the generalization performance which could have been a result of the presence of noise.

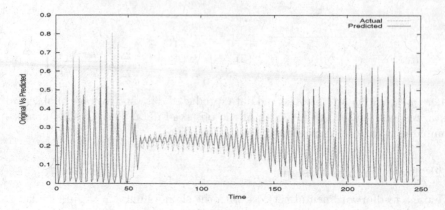

Fig. 1. Typical prediction performance of the proposed method on the test data set of Laser times series

In the Mackey-Glass and Lorenz problems, the best generalization performance was given for the reduced dataset achieved on the feature window of size 10. The proposed method reduced the Mackey-Glass data set by 35 % and the Lorenz dataset by 37 %. Figure 1 shows a typical prediction performance using the proposed data reduction method for the Laser times series on the test set.

Table 2 shows that the proposed method has been successful in reducing the training time for the featured training method when compared to the original dataset. It can be seen that the training time taken has been greatly reduced by the proposed method with the reduction in size of the original training dataset. The maximum reduction in time is of 68.39 % for the Sunspot time series data while the Astrophysics problem achieved a 61.69 % reduction, followed by Laser, Lorenz and Mackey Glass problems.

Table 1. Training and generalization performance (RMSE)

Problem	Feature window	H	Training	Generalization	Best
Sunspot	20	5	0.0243 ± 0.0005	0.0263 ± 0.0025	0.0141
	100	7	0.0108 ± 0.0005	0.0439 ± 0.0067	0.0152
Laser	50	7	0.0501 ± 0.0012	0.1450 ± 0.0068	0.0972
	100	7	0.0513 ± 0.0014	0.1555 ± 0.0043	0.1164
Astrophysics	50	5	0.0498 ± 0.0012	0.0656 ± 0.0032	0.0507
	100	5	0.0459 ± 0.0012	0.0665 ± 0.0038	0.0530
Mackey-Glass	10	7	0.0251 ± 0.0001	0.0087 ± 0.0002	0.0071
	100	5	0.0077 ± 0.0002	0.0130 ± 0.0012	0.0075
Lorenz	10	5	0.0462 ± 0.0059	0.0388 ± 0.0074	0.0103
	100	5	0.0405 ± 0.0070	0.1047 ± 0.0144	0.0133

Table 2. Training time and data reduction

Time series	Feature window	Reduced data set			Original data set			T-Test
		H	Best	Time (sec)	H	Best	Time (sec)	p value
Sunspot	100	7	0.0117	370.11	5	0.0110	1170.87	2.86E-06
Mackey Glass	10	7	0.0071	561.48	5	0.0031	746.78	3.28E-17
Laser	50	3	0.0972	387.99	5	0.0321	760.23	1.34E-42
Lorenz	10	5	0.0103	298.12	5	0.0037	575.99	5.70E-03
Astrophysics	50	5	0.0507	788.46	5	0.0507	2058.11	4.99E-06

Table 3. Comparison with the literature

Time series	Prediction method	RMSE	NMSE
Sunspot	Wavelet packet multilayer perceptron [22]		1.25E-01
	Co-evolutionary recurrent neural networks [10]	2.60E-02	3.62E-03
	Competitive co-evolutionary recurrent neural networks [23]	1.57E-02	1.31E-03
	Proposed CFSR	1.17E-02	8.15E-04
Lorenz	Recurrent neural networks [24]		1.85E-03
	Back-propagation and genetic algorithm with residual analysis [20]	2.96E-02	
	Co-evolutionary recurrent neural networks [10]	8.20E-03	1.28E-03
	Competitive co-evolutionary recurrent neural networks [23]	3.55E-03	2.41E-04
	Proposed CFSR	1.03E-02	5.58E-04
Mackey	Neural fuzzy network and meta-heuristics [19]	8.45E-03	
	Back-propagation and genetic algorithm with residual analysis [20]	1.30E-03	
	Co-evolutionary recurrent neural networks [10]	8.28E-03	4.77E-04
	Competitive co-evolutionary recurrent neural networks [23]	3.99E-03	1.11E-04
	Proposed CFSR	7.05E-03	2.93E-04
Laser	Multilayer Perceptron(MLP) [21]		1.72E-01
	Elman [21]		3.40E-02
	Non-linear Autoregressive model process with exogenous input (NARX) [21]		3.39E-02
	Proposed CFSR	9.72E-02	1.18E-01

Table 3 provides a comparison between the best results from Table 1 with related methods from literature. The RMSE and the NMSE for the best results is used for comparison. We note that the Astrophysics problem has not been used in literature. The proposed method has given better results when compared to related methods in literature such as evolutionary algorithms for training neural fuzzy networks [19] and co-evolutionary recurrent neural networks [10] for Mackey-Glass and Sunspot time series.

The proposed method performs better than back-propagation and genetic algorithm with residual analysis [20] for Lorenz time series. It also performs better than and multilayer-perceptron [21] for Laser time series. The reduced and reconstructed data set is able to eliminate irrelevant data, hence, reducing the prediction error and improving the overall efficiency of the neural network. The results also indicate that larger datasets are more favourable for the proposed method as seen with the real world problems that include Sunspot, Laser and Astrophysics time series.

In the literature, the prediction methods used the entire dataset without any feature selection. The goal for this paper was to achieve similar level of prediction performance with reduced dataset that is computationally less expensive for training. However, in some cases, the proposed method has achieved better prediction performance. This indicates that feature selection has been able to help further in generalisation performance of neural networks.

4   Conclusions and Future Work

We presented a chaotic feature selection and reconstruction method based on statistical analysis for time series prediction. It essentially implements data reduction by capturing most relevant features that are either noisy or chaotic in nature. The results show that the proposed method has been able to retain the prediction performance with a smaller dataset while reducing the training time. The results further show that the proposed method performs similar to the selected methods in the literature. Moreover, the proposed method has been able to reduce the size of the original dataset up to 42 % and the prediction time by up to 68 %.

In future work, it would be interesting to evaluate the feature selection method with other machine learning tools. The proposed method can also be extended to multi-variate time series and applied to problems that deal with very large time series datasets that include areas of astronomy and climate change.

References

1. Crone, S.F., Kourentzes, N.: Feature selection for time series prediction-a combined filter and wrapper approach for neural networks. Neurocomputing **73**(10), 1923–1936 (2010)
2. Chen, M., Mao, S., Liu, Y.: Big data: a survey. Mob. Netw. Appl. **19**(2), 171–209 (2014)
3. Kohavi, R., John, G.H.: Wrappers for feature subset selection. Artif. Intell. **97**(1), 273–324 (1997)
4. Guyon, I., Elisseeff, A.: An introduction to variable and feature selection. J. Mach. Learn. Res. **3**, 1157–1182 (2003)
5. Yu, L., Liu, H.: Feature selection for high-dimensional data: a fast correlation-based filter solution. In: ICML, vol. 3, pp. 856–863 (2003)
6. Sandya, H.B., Hemanth Kumar, P., Patil, S.B.: Feature extraction, classification and forecasting of time series signal using fuzzy and garch techniques. In: National Conference on Challenges in Research and Technology in the Coming Decades (CRT 2013), IET, pp. 1–7 (2013)
7. Olszewski, R.T.: Generalized feature extraction for structural pattern recognition in time-series data. DTIC Document, Technical report (2001)
8. Cavalcante, E., Pereira, J., Alves, M.P., Maia, P., Moura, R., Batista, T., Delicato, F.C., Pires, P.F.: On the interplay of internet of things and cloud computing: a systematic mapping study. Comput. Commun. **8990**, 17–33 (2016)
9. Chen, C.P., Zhang, C.-Y.: Data-intensive applications, challenges, techniques and technologies: a survey on big data. Inf. Sci. **275**, 314–347 (2014)

10. Chandra, R., Zhang, M.: Cooperative coevolution of Elman recurrent neural networks for chaotic time series prediction. Neurocomputing **186**, 116–123 (2012)
11. Takens, F.: On the Numerical Determination of the Dimension of an Attractor. Springer, Heidelberg (1985)
12. Potter, M.A., Jong, K.A.: A cooperative coevolutionary approach to function optimization. In: Davidor, Y., Schwefel, H.-P., Männer, R. (eds.) PPSN 1994. LNCS, vol. 866, pp. 249–257. Springer, Heidelberg (1994). doi:10.1007/3-540-58484-6_269
13. Chandra, R., Frean, M., Zhang, M.: On the issue of separability for problem decomposition in cooperative neuro-evolution. Neurocomputing **87**, 33–40 (2012)
14. Mackey, M.C., Glass, L.: Oscillation and chaos in physiological control systems. Science **197**, 287–289 (1977)
15. Lorenz, E.: Deterministic non-periodic flows. J. Atmos. Sci. **20**, 267–285 (1963)
16. SILSO World Data Center: The International Sunspot Number (1834-2001), International Sunspot Number Monthly Bulletin and Online Catalogue, Royal Observatory of Belgium, Avenue Circulaire 3, 1180 Brussels, Belgium. http://www.sidc.be/silso/. Accessed 02 Feb 2015
17. Hübner, U., Abraham, N.B., Weiss, C.O.: Dimensions and entropies of chaotic intensity pulsations in a single-mode far-infrared NH3 laser. Phys. Rev. A **40**(11), 6354–6365 (1989)
18. The Santa Fe Time Series Competition Data. http://www-psych.stanford.edu/~andreas/Time-Series/SantaFe.html. Accessed 01 May 2015
19. Lin, C.-J., Chen, C.-H., Lin, C.-T.: A hybrid of cooperative particle swarm optimization and cultural algorithm for neural fuzzy networks and its prediction applications. IEEE Trans. Syst. Man Cybern. Part C Appl. Rev. **39**(1), 55–68 (2009)
20. Ardalani-Farsa, M., Zolfaghari, S.: Residual analysis and combination of embedding theorem and artificial intelligence in chaotic time series forecasting. Appl. Artif. Intell. **25**, 45–73 (2011)
21. Menezes, J.M.P., Barreto, G.A.: Long-term time series prediction with the NARX network: an empirical evaluation. Neurocomputing **71**(16), 3335–3343 (2008)
22. Teo, K.K., Wang, L., Lin, Z.: Wavelet packet multi-layer perceptron for chaotic time series prediction: effects of weight initialization. In: Alexandrov, V.N., Dongarra, J., Juliano, B.A., Renner, R.S., Tan, C.J.K. (eds.) ICCS-ComputSci 2001. LNCS, vol. 2074, pp. 310–317. Springer, Heidelberg (2001)
23. Chandra, R.: Competition and collaboration in cooperative coevolution of Elman recurrent neural networks for time-series prediction. IEEE Trans. Neural Netw. Learn. Syst. **26**(12), 3123–3136 (2015)
24. Mirikitani, D., Nikolaev, N.: Recursive bayesian recurrent neural networks for time-series modeling. IEEE Trans. Neural Netw. **21**(2), 262–274 (2010)

$L_{1/2}$ Norm Regularized Echo State Network for Chaotic Time Series Prediction

Meiling Xu[1], Min Han[1(✉)], and Shunshoku Kanae[2]

[1] Faculty of Electronic Information and Electrical Engineering,
Dalian University of Technology, Dalian, China
minhan@dlut.edu.cn
[2] Department of Electrical, Electronic and Computer Engineering,
Fukui University of Technology, Fukui, Japan

Abstract. Echo state network contains a randomly connected hidden layer and an adaptable output layer. It can overcome the problems associated with the complex computation and local optima. But there may be ill-posed problem when large reservoir state matrix is used to calculate the output weights by least square estimation. In this study, we use $L_{1/2}$ regularization to calculate the output weights to get a sparse solution in order to solve the ill-posed problem and improve the generalized performance. In addition, an operation of iterated prediction is conducted to test the effectiveness of the proposed $L_{1/2}$ESN for capturing the dynamics of the chaotic time series. Experimental results illustrate that the predictor has been designed properly. It outperforms other modified ESN models in both sparsity and accuracy.

Keywords: Echo state networks · $L_{1/2}$ norm regularization · Chaotic time series · Prediction

1 Introduction

The echo state network (ESN) is a novel kind of recurrent neural networks. Only the output weights are modified by a simple and efficient linear regression algorithm [1]. It can overcome the local minima and vanishing gradient problems associated with traditional neural networks training algorithms. Owing to the above merits, ESNs have been extensively studied in time series prediction [2–4].

However, when least square method is used to compute the readout weights, the large reservoir state matrix may be ill-posed [5], which adversely affect the generalization of the model. To solve this problem, a series of regularization techniques have been applied in the training process of echo state networks [2, 6–8]. They are computational efficient and not prone to over fitting. For example, J.J. Steil proposed a modified L_2 norm regularized echo state network to reduce the risk of error amplification and boost model's generalization [6]. Han added an L_1 norm penalty term in the objective function to control the model's complexity [7]. But the L_2 norm regularization is a biased estimation and the L_1 norm does not satisfy oracle property [8]. Recently, $L_{1/2}$ penalty which has many promising properties, such as unbiasness, sparsity and oracle property, has been proposed and attracted growing attention [9, 10].

A. Hirose et al. (Eds.): ICONIP 2016, Part III, LNCS 9949, pp. 12–19, 2016.
DOI: 10.1007/978-3-319-46675-0_2

In this paper, we combine the L$_{1/2}$ norm regularization method with ESNs, termed L$_{1/2}$ESN, to improve the model's generalization ability.

For a chaotic time series, two trajectories in the same attractor will diverge significantly after a period of sample-by-sample iteration. Hence, minimizing the root mean square value of the prediction error is necessary, but not sufficient, for a successful mapping. The short-term predictability of the proposed model is considered herein. It is realized by iterated prediction which is to feed the output back to the input and form an autonomous system [11].

This paper is organized as follows. In Sect. 2, we give a brief introduction to general ESNs. Then in Sect. 3, we propose the L$_{1/2}$ ESN model and the iterated prediction. Afterwards, experimental results are given in Sect. 4. Finally, in Sect. 5, we draw the conclusions.

2   Echo State Networks

An echo state network consists in an input layer, a hidden layer and an output layer. The hidden layer, called dynamic reservoir, contains a large number of neurons and is regarded as a supplier of interesting dynamics [1]. The input-to-reservoir weight matrix W_{in} and the recurrent reservoir weight matrix W_x are generated randomly, whereas the reservoir-to-output weight matrix W_{out} is adapted via supervised learning [2].

Denote $m(i) = [m_1(i), m_2(i), \cdots, m_M(i)]^T \in R^{M \times 1}$ as the collected time series, $u(i) = m(i)$ and $y(i) = m(i+1)$ as the input and output signals at time step i. The basic state equation is defined by

$$s(i) = \tanh[W_{in}u(i) + W_x s(i-1)] \tag{1}$$

where $s(i) \in \mathbf{R}^{N \times 1}$ is the state of the network at time step i. $\tanh(\cdot)$ is applied element-wise, and the initial state of the reservoir is a zero vector. The dependence on the initial state gradually loses as i goes to infinity [2]. The linear output layer is defined by

$$y(i) = [s(i) : u(i)]^T \beta_1 + \beta_0 = x^T(i)\beta \tag{2}$$

where $[\cdot : \cdot]$ stands for a vertical vector concatenation, $x(i) = [s(i) : u(i) : 1]$, $\beta = [\beta_1 : \beta_0]$ are the coefficients that to be estimated. Collect $x(i)$ column-wise into a matrix $X \in \mathbf{R}^{N \times S}$, and the corresponding $y(i)$ row-wise into a matrix $Y \in \mathbf{R}^{S \times M}$, where S is the number of the samples. Then the readout weights β are computed by a linear regression method which minimizes the mean square error between the network output and the training target output [4]:

$$\min \left\| Y - X^T \beta \right\|_2^2 \tag{3}$$

where $\|\cdot\|_2$ stands for L$_2$ norm. The least square solution of (3) is

$$\hat{\beta} = X(X^T X)^{-1} Y \tag{4}$$

Sometimes, the above solution (4) is ill-posed because of the approximate collinear components in the high dimensional reservoir state matrix X. This implies a bad generalization performance [7]. A solution to this problem is to use regularization, which has the general form:

$$\min \left\| Y - X^T \beta \right\|_2^2 + \lambda \| \beta \|_k^k \tag{5}$$

where λ is a regularization parameter that balances the two objective terms, and it is chosen by five-fold cross-validation. $\| \beta \|_k$ is taken as the k norm of β.

3 $L_{1/2}$ Regularized Echo State Network

In this part, we introduce the $L_{1/2}$ penalty into echo state networks to improve the prediction performance of echo state networks, i.e., k in (5) is equal to 1/2. For a fixed non-negative λ, the cost function (5) thus can be rewritten as

$$f(\beta) = \left\| Y - X^T \beta \right\|_2^2 + \lambda \| \beta \|_{\frac{1}{2}}^{\frac{1}{2}} = \left\{ \sum_{i=1}^M \left\| y_m - x^T \beta_m \right\|_2^2 + \lambda \sum_{m=1}^M \| \beta_m \|_{\frac{1}{2}}^{\frac{1}{2}} \right\} \tag{6}$$

$$\| \beta_m \|_{\frac{1}{2}}^{\frac{1}{2}} = \sum_{j=1}^P \sqrt{\beta_{mj}}, p = N + M + 1$$

β is estimated by minimizing the following target function:

$$\min_{\beta} \sum_{m=1}^M \left(\sum_{i=1}^S \left(y_m(i) - x^T(i)\beta_m \right)^2 + \lambda \| \beta_m \|_{\frac{1}{2}}^{\frac{1}{2}} \right)$$

$$= \sum_{i=1}^S \left(y_1(i) - x^T(i)\beta_1 \right)^2 + \lambda \| \beta_1 \|_{\frac{1}{2}}^{\frac{1}{2}} + \sum_{i=1}^S \left(y_2(i) - x^T(i)\beta_2 \right)^2 + \lambda \| \beta_2 \|_{\frac{1}{2}}^{\frac{1}{2}} \tag{7}$$

$$+ \cdots + \sum_{i=1}^S \left(y_M(i) - x^T(i)\beta_M \right)^2 + \lambda \| \beta_M \|_{\frac{1}{2}}^{\frac{1}{2}}$$

This can be decomposed as M independent optimization problems:

$$\min_{\beta_m} \left\{ \left\| Y_m - X^T \beta_m \right\|_2^2 + \lambda \| \beta_m \|_{\frac{1}{2}}^{\frac{1}{2}} \right\}, \ m = 1, 2, \cdots, M \tag{8}$$

We use coordinate descent algorithm, a "one-at-a-time" approach, to obtain the optimal β_m [10]. For each coefficient, the target function is partially optimized with respect to β_{mk}, $k = 1, 2, \cdots, p$ while other elements of β_m are fixed at their recently update values.

Equation (9) can be rewritten as

$$f_k(\beta_{mk}) = \sum_{i=1}^{S}\left[y_m(i) - x_k(i)\beta_{mk} - \sum_{j\neq k}^{p} x_j(i)\bar{\beta}_{mj}\right]^2 + \lambda\left(|\beta_{mk}|^{\frac{1}{2}} + \sum_{j\neq k}^{p}|\bar{\beta}_{mj}|^{\frac{1}{2}}\right) \quad (9)$$

where $\bar{\beta}_{mj}$ represent the fixed parameters. The details of the coordinate descent algorithm [10] for L$_{1/2}$ regularized echo state network are described as follows:

Input: Set $\beta_m^{int} = 0$, $m = 1, 2, \cdots, M$, and give a nonnegative constant λ.

Output: $\beta_m = \left[\beta_{m1}, \beta_{m2}, \cdots, \beta_{mp}\right]^T$.

Step 1: $\beta_m = \beta_m^{int}$;

Step 2: Calculate $\beta_{mk}, k = 1, 2, \cdots, p,$

$$\text{if } |C_{mk}| \geq \frac{3}{4}\lambda_{mk}^{2/3}, \beta_{mk} = \frac{2}{3}C_{mk}(1 + \cos(2/3\pi - 2/3\varphi)), \text{ else } \beta_{mk} = 0,$$

where $C_{mk} = \sum_{i=1}^{S}\left([y_m(i) - \sum_{j\neq k}^{p} x_j(i)\bar{\beta}_{mj}]x_k(i)\right)/\sum_{i=1}^{S}[x_k(i)]^2,$

$$\lambda_{mk} = \lambda/\sum_{i=1}^{S}[x_k(i)]^2, \text{ and } \varphi = \arccos\left((\lambda_{mk}/8)|C_{mk}/3|^{-\frac{3}{2}}\right).$$

Step 3: if $\sum_{k=1}^{p}|\beta_{mk} - \beta_{mk}^{int}| < 10^{-4}$, the algorithm stops, else set $\beta_m^{int} = \beta_m$, go back to Step 1.

The coordinate descent algorithm for the L$_{1/2}$ regularized echo state network works well for sparsity problems, as it is not necessary to change many irrelevant parameters and recompute partial residuals for each update step.

Chaotic systems are highly sensitive to initial conditions. Small difference may yields significant diverging, rendering long-term prediction impossible [11]. Hence, a pragmatic approach for testing the short-term predictability of the L$_{1/2}$ regularized echo

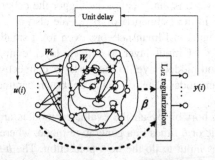

Fig. 1. Iterated prediction by L$_{1/2}$ regularized echo state network

state network model is to feed the output back to its input, forming an autonomous system, which is a realization of iterated prediction as illustrated in Fig. 1.

4 Simulations

In this section, we evaluate the performance of the proposed $L_{1/2}$ norm regularized echo state network ($L_{1/2}$ESN) on a typical chaotic system-Lorenz system. Some other models are conducted to compare with our proposed model: Elman network, echo state networks with L_1 penalty (L_1ESN) [7], echo state networks with L_2 penalty (L_2ESN) [2], and echo state networks with elastic net penalty (EESN) [8]. The accuracy of the prediction models is assessed by the root mean square error (RMSE), normalized root mean square error (NRMSE) and the symmetric mean absolute percentage error (SMAPE). The results provided herein are averages over 20 different random reservoir initializations. The Lorenz equations are defined by

$$dx/dt = \sigma(y - x), \ dy/dt = x(\rho - z) - y, \ dz/dt = xy - \beta z \qquad (10)$$

When $\sigma = 10, \beta = 3/8$ and $\rho = 28$, the system exhibits chaotic behavior. We use Runge-Kutta method to generate 5000 points with time step 0.01 from the initialized point (1, 1, 1). The Lorenz series is then reconstructed into the phase space with delay times 8, 8, 12, and embedding dimensions 2, 1, 7 for x, y, z series respectively. Afterwards, the first 80 % samples are used to train the model with 100 samples warming up the reservoir, and the remaining samples are used to test the model. The spectral radius of W_x is set as 0.9, the sparse connectivity of W_x is set as 0.05, and the input scaling parameter of W_{in} is set as 0.01. Some other parameters are given in Table 1, where the regularization parameter λ is chosen by five-fold cross-validation.

The obtained results of all the evaluated models are provided in Table 1, and the one-step-ahead prediction curves for Lorenz-$x(t)$ produced by $L_{1/2}$ESN are plotted in Fig. 2. As can be seen, the $L_{1/2}$ESN with λ equal to $10^{-7.5}$ performs much better than other models, being capable of obtaining much sparser output weights and lower RMSE, NRMSE, and SMAPE values, but when λ is chosen as $10^{-7.5}$, the effect of L_1 norm is very little as all the weights are nonzero. When λ increases to 5×10^{-5}, the L_1 norm makes sense since a large proportion of the output weights are zero. We can make the comment that $L_{1/2}$ESN is more prone to get a sparse solution than L_1ESN. One point needs to be emphasized is that λ is not the bigger the better, because a big λ will generate a big deviation in the estimation of W_{out}. We also note that there are a large number of unknown weights in Elman network even for a small hidden layer. In this experiment, the numbers of input layer to hidden layer connections, hidden layer recurrent connections, and hidden layer to output layer connections are 350, 125, 350 respectively for 35 hidden nodes. The too many unknown parameters make the Elman network underfitting.

Prediction on h-step horizon by all the evaluated models are conducted by iteratively applying the predictor h times in a generative mode, where on each step it takes its own last prediction as input to do the next prediction. The h-step-ahead prediction performance produced by $L_{1/2}$ESN in terms of RMSE is depicted in Fig. 3. In addition,

Table 1. Main parameters and performance of one-step-ahead prediction of evaluated models

Model	Num. of hidden nodes	λ	Num. of nonzero output weights	RMSE	NRMSE	SMAPE
Elman	35	–	[350 125 350]	0.2017	0.0256	0.2397
L$_1$ESN	100	5×10^{-5}	[53 18 56]	0.0153	0.0019	0.0072
L$_1$ESN	100	$10^{-7.5}$	[110 110 110]	0.0457	0.0058	0.0214
L$_2$ESN	100	5×10^{-5}	[110 110 110]	0.0201	0.0025	0.0088
EESN	100	5×10^{-5}	[51 35 59]	0.0178	0.0022	0.0245
L$_{1/2}$ESN	100	$10^{-7.5}$	**[34 14 39]**	**0.0057**	**0.0007**	**0.0025**

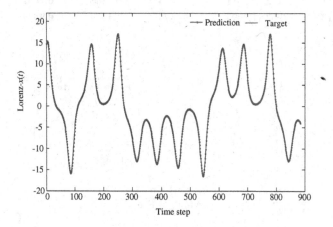

Fig. 2. One-step-ahead prediction curves produced by L$_{1/2}$ESN

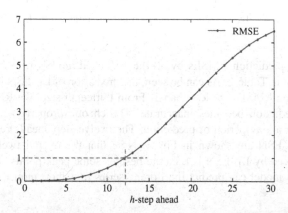

Fig. 3. RMSE of h-step-ahead prediction produced by L$_{1/2}$ESN

Table 2. RMSEs of h-step-ahead prediction of all the evaluated models

h-step	Elamn	L_1ESN $\lambda = 5 \times 10^{-5}$	L_2ESN $\lambda = 5 \times 10^{-5}$	EESN $\lambda = 5 \times 10^{-5}$	$L_{1/2}$ESN $\lambda = 10^{-7.5}$
1	0.2017	0.0153	0.0201	0.0178	**0.0057**
2	0.2957	0.0170	0.0389	0.0259	**0.0103**
3	0.4617	0.0512	0.0605	0.0377	**0.0192**
4	0.6751	0.0767	0.1014	0.0617	**0.0378**
5	0.9240	0.1458	0.1678	0.1033	**0.0709**
6	1.2027	0.2159	0.2551	0.1656	**0.1224**
7	1.5070	0.3363	0.3645	0.2588	**0.1960**
8	1.8336	0.4260	0.5129	0.3869	**0.2946**
9	2.1790	0.6374	0.6973	0.5443	**0.4213**
10	2.5400	0.8288	0.9066	0.7206	**0.5777**
11	2.9132	1.0611	1.1474	0.9148	**0.7652**
12	3.2952	1.3095	1.4174	1.1377	**0.9839**

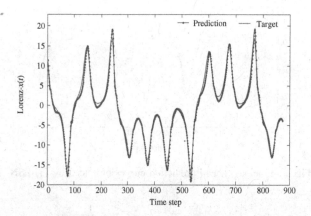

Fig. 4. Twelve-step-ahead prediction curves produced by $L_{1/2}$ESN

the h-step-ahead prediction RMSEs by all the evaluated models with h ranging from 1 to 12 are shown in Table 2. As can be seen, the prediction of $L_{1/2}$ESN is effective and accurate, as all the RMSEs are less than 1. From thirteenth step, the RMSE increases fast, and the prediction becomes inaccurate. The chaotic property of Lorenz series makes it hard for a long period of prediction. The twelve-step-ahead prediction curves produced by $L_{1/2}$ESN are shown in Fig. 4. Note that the overall twelve-step-ahead prediction produced by $L_{1/2}$ESN is accurate besides some peak points. This illustrates that the $L_{1/2}$ESN model can predict the Lorenz series in a short term.

5   Conclusions

In this paper, we propose a new model for forecasting multivariate time series, called $L_{1/2}$ESN. It applies $L_{1/2}$ norm regularization to calculate the output weights of an ESN to solve the ill-posed problem and improve the prediction performance. The $L_{1/2}$ penalty imposed on the output weights outweighs L_1 penalty in terms of sparasity. Short-term prediction is realized by iterated prediction. Experiments results of Lorenz time series show that our proposed model obtain a higher accuracy for both one-step-ahead prediction and multiple-step-ahead prediction. The short-term pre-dictability of the chaotic time series demonstrates the effectiveness of the proposed prediction model.

Acknowledgement. This work was supported by National Natural Science Foundation of China under Grant 61374154.

References

1. Jaeger, H., Haas, H.: Harnessing nonlinearity: predicting chaotic systems and saving energy in wireless communication. Science **304**(5667), 78–80 (2004)
2. Lukoševičius, M., Jaeger, H., Schrauwen, B.: Reservoir computing trends. KI-Künstliche Intelligenz **26**(4), 365–371 (2012)
3. Soh, H., Demiris, Y.: Spatio-temporal learning with the online finite and infinite echo-state gaussian processes. IEEE Trans. Neural Netw. Learn. Syst. **26**(3), 522–536 (2015)
4. Yuenyong, S., Nishihara, A.: Evolutionary pre-training for CRJ-type reservoir of echo state networks. Neurocomputing **149**, 1324–1329 (2015)
5. Chatzis, S.P., Demiris, Y.: Echo state gaussian process. IEEE Trans. Neural Networks **22**(9), 1435–1445 (2011)
6. Reinhart, R.F., Steil, J.J.: Regularization and stability in reservoir networks with output feedback. Neurocomputing **90**, 96–105 (2012)
7. Han, M., Ren, W.J., Xu, M.L.: An improved echo state network via L_1-norm regularization (in Chinese). Acta Automatica Sin. **40**(11), 2428–2435 (2014)
8. Zou, H., Hastie, T.: Regularization and variable selection via the elastic net. J. Royal Stat. Soc. Ser. B (Stat. Methodol.) **67**(2), 301–320 (2005)
9. Xu, Z.B., Chang, X.Y., Xu, F.M., Zhang, H.: $L_{1/2}$ regularization: a thresholding representation theory and a fast solver. IEEE Trans. Neural Netw. Learn. Syst. **23**(7), 1013–1027 (2012)
10. Liang, Y., Liu, C., Luan, X.Z., Leung, L.S., Chan, T.M., Xu, Z.B., Zhang, H.: Sparse logistic regression with a $L_{1/2}$ penalty for gene selection in cancer classification. BMC Bioinform. **14**(1), 198 (2013)
11. Haykin, S.S.: Neural Networks and Learning Machines, 3rd edn., pp. 711–722. Pearson Education, Prentice Hall, Upper Saddle River (2009)

SVD and Text Mining Integrated Approach to Measure Effects of Disasters on Japanese Economics

Effects of the Thai Flooding in 2011

Yuriko Yano and Yukari Shirota[✉]

Department of Management, Faculty of Economics,
Gakushuin University, Tokyo, Japan
{yuriko.yano,yukari.shirota}@gakushuin.ac.jp

Abstract. In this paper, we analyzed effects of the 2011 Thai flooding on Japanese economics. In the paper, we propose, as a new time series economics data analysis method, an integrated approach of Singular Value Decomposition on stock data and news article text mining. There we first find the correlations among companies' stock data and then in order to find the latent logical reasons of the associations, we conduct text mining. The paper shows the two-stage approach's advantages to refine the logical reasoning. Concerning the Thai flooding effects on the Japan's economy, as unexpected moves, we have found the serious harms on the Japanese food and drink companies and its quick recoveries.

Keywords: Singular value decomposition · Topic extraction · Dirichlet allocation model · Thai flooding · Disaster effects · Stock data analysis

1 Introduction

In this paper, we will analyze effects of the 2011 Thai flooding on Japanese economics. Many Japanese companies were devastated by the floods. To analyze the damages, we propose, as a new time series economics data analysis method, an integrated approach of Singular Value Decomposition (SVD) on stock data and news article text mining. There we first find the correlations among companies' stock data and then in order to find the latent logical reasons of the correlations, we conduct text mining. As the stock price data, we used Nikkei 225 that expresses the Japanese major companies' economical climates. In the paper, we focus on the time series changes after the flooding. The period is from Sept. to Dec. in 2011.

The proposed methods is conducted as follows: First, we conduct SVD on the time series data matrix X of stock price return values, so that we can obtain the matrices U, W, and V^T. Then, we have gotten two kinds of eigenvectors that are (1) Brand-eigenvector, obtained by multiplying U with W, and (2) Dailymotion-eigenvector, obtained by multiplying S with V^T. By both eigenvectors, we can identify the correlated clusters/classes. We will illustrate the monthly changes to clarify the integration and division on

© Springer International Publishing AG 2016
A. Hirose et al. (Eds.): ICONIP 2016, Part III, LNCS 9949, pp. 20–29, 2016.
DOI: 10.1007/978-3-319-46675-0_3

the classification. However, only the SVD on the stock data is not enough. To find the latent and logical reasons of the damages and changes, in addition, an analysis by text mining is required. In the paper, we propose the two-stage approach as the new method to analyze effects of a natural disaster on Japan's economy. Although our approach is not limited to on Japan's economy and we can use this method on any country, we would like to analyze the Japan's economics attributes.

In the next section, we shall explain our proposed methodologies. Then we describe the analysis results on the 2011 Thai flooding effects, by the SVD analysis in Sect. 3 and then in Sect. 4, we refine the correlations by the text mining results. Finally, we conclude the paper.

2 SVD and Text Mining Integrated Approach

In the section, we will explain our proposed two-stage methods. Our research objective is measurement of natural disaster effects on Japan's economy conditions. The disasters include Japan's earthquakes and other countries' earthquakes and floods. In the methods, first we analyze Japan's stock price data such as Nikkei 225. The analysis method we adopted is SVD (Singular Value Decomposition) [3, 4]. The SVD is used in various kinds of applications; For example, in text mining, LSA (Latent Semantic Analysis) uses the SVD. Concerning the SVD math process, Shirota et al. visually explain the intrinsic meanings and interpretation of the eigenvectors/principal components, so that readers can easily and deeply understand SVD [1, 2].

We conduct the SVD on the standardized return values of stock price data. The return value is the ratio between today's price and the previous day's one and defined as follows: $G_{i,j} = \ln(S_{i,j}/S_{i,j-1})$ where $S_{i,j}$ is the ith company's stock price on j-th day and $G_{i,j}$ is the return value on j-th day. Because different stock values have varying levels of volatility, the return value should be standardized. In addition, in the stock data analysis, each company's data during the period is standardized, so the mean value becomes 0 and the standard deviation becomes 1.

In our previous researches on the Thai 2011 flooding effects, we had found the damaged industory classes by using the SVD methods [5, 6]. Figure 1 shows the damaged company class that we found as the eigenvector #9. In other words, the ninth principal component among many principal components expresses the damaged class. Mathematically, from SVD, we can obtain two kinds of eigenvalues. In this analysis, we title them (1) Brand-Eigenvector and (2) Dailymotion-eigenvector. A SVD is a kind of Principal Component Analysis (PCA). The Brand-eigenvector covers similar movement companies. The Dailymotion-Eigenvector covers the class's average time series fluctuation. To find the damaged company class, we used the well-known fact that Japanese digital camera companies Nikon was severely devastated. We searched Nikon as the mark to find the damaged Japanese company classes such as Pioneer. The element value may be positive or negative, although in the Fig. 1 example, the damaged company is a big negative element. Its positive/negative is only up to the eigenvector direction. If the principal component has many damaged companies, the principal component can be interpreted as the damaged class.

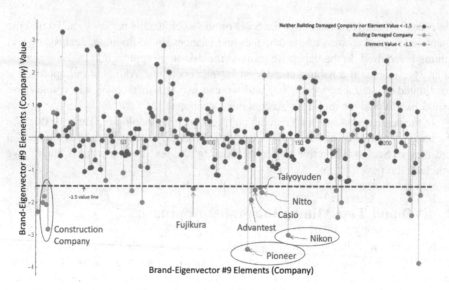

Fig. 1. The extracted damaged company Brand-Eigenvector #9 (cited from [5]).

In general, the SVD method is, in a financial analysis, called the random matrix approach and it is utilized to find the stable company classes [7–9]. Using the extracted stable classes, they make an excellent performance portfolio [10, 11]. Our usage of eigenvalues/eigenvectors in the SVD is identical to one by Plerou's proposed cross correlation analysis [10, 11]. The conversion method between both was described in our paper [5]. Our research goal is, however, completely different from theirs and we would like to find and investigate the time series effects of the disaster on stock prices. A disaster triggers a stock downfall and one industry's breakdown inflicts harm and transmit on others like a supply chain breakdown. The effects are dynamic and not stable; Some industries will recover soon and the damage will gradually diminish and others are not. We would like to investigate the time series changes from a viewpoint of time series data analysis; we are interested in the effect duration time and its magnitude.

Another research objective is extraction of logical relationships among the similar movement companies that were obtained as the same class members. The SVD offeres a hypothesis that company A and company B are damaged because they have a same attribute or that company C's decline may be caused by the disaster. The hypothesis is just a hypothesis then. We need to verify the hypothesis. Let us explain the background on this. When we consider a big data mining, we should spot the differences between Internet big data analysis and Industrious big data analysis such as in Industory 4.0. As Jay Lee says in [12, 13], Internet big data mining includes sales prediction, user relationship mining and clustering, recommendation systems, opinion mining, etc. [14, 15] and this Internet big data research focuses on 'human-generated or human-related data' instead of 'machine-generated data or industrial data', which are generated by machine controllers, sensors, manufacturing systems, etc. In other words, the latters are technologies related data for the Internet of Things (IoT). Concerning IoT, see [16]. We also think that Internet big data mining focuses on the facts and does not mind the reasons of the relationshops. The Internet big data analysis focuses on only the statistical significance level. On the other hand, in the Industrious big data mining, we need plausible logic of the relationship because they are hidden/latent logics or the intrinsic/latent meanings. Our target is stock price data mining and we would like to extract the hidden logics of extracted relationships, especially when we found the unexpected industry connection. Therefore we can conceive this disaster effect analysis as one of Industrious data mining. Of course some Internet data mining would extract the hidden logics.

3   Time Series Stock Data Analysis

In the section, we describe the result of the SVD. The data used is the Nikkei 225 during three months from October to December 2011. We have conducted SVD on each month. The matrix size on each month is 225 (companies) times 20 (days). Therefore, we can obtain 20 = Min (225, 20) principal components each month. Among the Oct. 20 principal components, we did select #2, # 3, #5, #6, #7, #8, and #13 as the damaged classes. The October was the most severely damaged month. Figure 2 shows the selected five damaged Brand-Eigenvectors and their damaged industry names in October. The selection criteria is 1.2 there. The industry elements with bigger than 1.2 are extracted and drawn there. The edge length has no meaning in the graph. In Fig. 2, the original Japanese industry names are written in Japanese, and only key company names are written in English such as NIKON and AJINOMOTO. The class, #3, can be interpreted as a food and drink industry class because they include AJI-NOMOTO, KIKKOMAN, MEIJI, KIRIN, and ASAHI. The #5 class include many financial industries such as Bank of YOKOHAMA, Bank of SHIZUOKA, Bank of CHIBA, RESONA, and SOMPO JAPAN. The #7 class include many spinning/textile companies such as TORAY, TEIJIN.

Let us consider the time series change on the classes. Figure 3 shows monthly movements on the numbers of the extracted industries with bigger than 1.2 values. For example, the number of #3 has changed as 20 – 28 – 21. Figure 4 also shows the time series changes that focuses on the eigenvalue magnitudes. The eigenvalues express the class effect impact; the bigger it is, the more harms exist. Among classes, we can see the class integrations and divisions. In October and November, # 2 could be interpreted as an electronics industry class, from the member lists. As the both values in Figs. 3 and 4 are the biggest among the classes, we can interpret that the class was the most damaged class. This is consistent with the fact that the electronics companies had severely damaged. However, in Dec., the electronics industry category has been divided into #5, #7, and #8. It is possible to assume that the electronics industry had many damages, and we can guess that they consumed a plenty of time to recover from the damages. The damages must have prolonged; therefore, the class was divided to three classes depending on the damage recovery features.

On the other hand, the drink and food industry must have been recovered quickly. In October, #3 could be interpreted as a food and drink industry class. The eigenvalue of #3 in Fig. 4 Oct. data, 14.2, is the second largest one in October; however, in November, there is no food and drink industry anymore, and in December, we can find

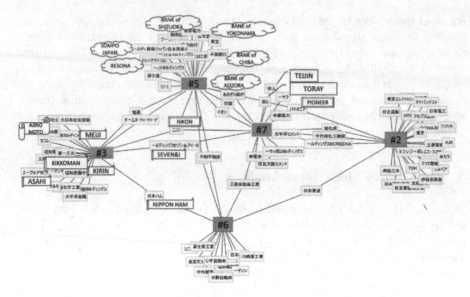

Fig. 2. The damaged Brand-Eigenvectors #2, #3, #5, #6, and #7 and the damaged industry names.

October	November	December
♯13 8 NUMEROUS kinds	♯13 8 NUMEROUS kinds	♯13 4 (Food) and Drink
♯8 14 Pharmaceutical Printing/Food	♯8 7 Chemical	♯8 11 Electronics
♯7 17 Spinning Food and Drink	♯7 18 Pharmaceutical	♯7 14 Electronics Automobile
♯6 15 Heavy industry Automobile Food/Pharmaceutical	♯6 14 Construction	♯6 14 Gas Electricity
♯5 23 Financial Electronics	♯5 19 Railroad Electricity	♯5 20 Construction Electronics
♯3 20 Food and Drink	♯3 28 Construction Financial Commercial Spinning	♯3 21 Construction Electricity Spinning
♯2 27 Electronics	♯2 27 Heavy industry Electronics	♯2 38 NUMEROUS kinds

Fig. 3. Monthly movements on the numbers of damaged industries involved in each damaged Brand-Eigenvector. From the left, Oct., Nov., and Dec.

a few beverage companies in #13. The ratio is the smallest in the month. It means that food and drink industry recovered from the flood more quickly than electronics industry.

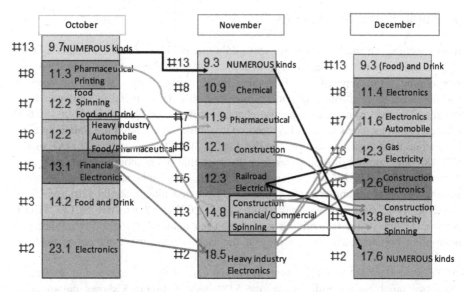

Fig. 4. Monthly movements on the eigenvalues of each damaged Brand-Eigenvector. From the left, Oct., Nov., and Dec.

Let us consider the financial industry class. The class, #5, in October can be interpreted as a financial industry, and there are lots of Japanese general insurance companies in the class. These companies themselves did not have physical direct damages. Nonetheless, the level of damage is the third largest in October. The reason must be that there were bunches of Japanese companies in Thailand and that the insurance companies had a huge amount of insurance expenses of them. Later, we could confirm the cause and effect relationship from web articles or so.

Spinning industry also does change. In October, spinning industry class is #7 (the eigenvalue is 12.2), and it changes into #3 in November. After the month, the class number is stable as #3. Absolutely, spinning industry was suffering from the flood for a while because their factories were flooded. For instance, TORAY is a maker of fibers, textiles, resins, plastics, films, chemicals, ceramics, composite materials, medical products, and electronics, according to its official website, and this company has some factories in Thailand. TTS, Thai Toray Synthetics, has three factories in Bangkok, Ayutthaya, and Nakhon Pathom. TTTM, Thai Toray Textile Mills, has one factory in Nakhon Pathom, too. The factory in Ayutthaya had the worst damages, and it needed more time to operate the factory. However, the others started to resume their operation in the year. In addition, Toray carried out alternative production immediately using its global networks. Toray could manage and make a quick recovery from the flood; hence, Toray did not have as terrible troubles as an electronics industry. It is quite possible to assume that other spinning companies had the similar actions, and as a result, spinning industry itself was not as a seriously damaged industry as an electronics industry.

4 Topic Extraction Results

In this section, we shall describe our text mining method and extracted results by that.

We analyzed the Thai early reports/news articles on the web site named "www. newsclip.be" from September to December in 2011. The news is written in Japanese and offers reliable contents to Japanese. We used only the financial related news from the site. The total number of Japanese characters of the text source is about 120,000 characters that include Chinese characters so it is informative. Input the source text, we have conducted topic extraction by the Latent Dirichlet Allocation (LDA) model [17, 18]. The number of topics was four, because that offers more clear classification than five or six. The MCMC algorithm we used is the Gibbs sampler [3, 19]. Concerning the Gibbs sampler math process, see our teaching material [20] (which is available on http://www-cc.gakushuin.ac.jp/~20010570/mathABC/SELECTED/). Another feature of the text mining is that we used noun-noun bigram terms because noun-noun bigrams analysis can prevent the lack of meaning connections between words.

To make the LDA model, at first, the whole data of the through period was input. Then, we input each month data to the LDA model. Figure 5 shows the monthly movement of the topic frequencies. We made the title of the extracted four topics; they are (1) Overall economy in Thai, (2) Damage by the Thai flooding, (3) Transport in Thai, and (4) Finance in Thai [21]. As we expected, the frequency of Topic 1 is much larger compared to others. The graph also clarifies that the transportation and finance in Thai had relationships with the flooding damages. The frequencies of Topic 1 (overall economy) keeps constantly the same level, which can be considered reasonable. Then we concluded the topic 2 is the damage-featured topic. As another data, we used the item list report concerning Thai export decrease [22]. The damaged items include (a) PCs, (b) automobiles, (c) ICs, (d) cars, (e) digital cameras/TV cameras, (f) radio receivers, (g) optical lenses, (h) printing circuits, (i) polyacetal, (j) semi-conductor devices such as diodes, (k) aluminum products, and (l) electric motors. We found the damaged items terms and the company names in our topic extraction results. However, we found other additional industry groups and related terms appeared in Topic 2. They are (A) electrical parts, (B) food (the extracted terms are "processed goods", "poultry processing", "freezing food", "food factory", and "food processing"), (C) beverage, (D) rubber commodity, (E) polyester yarn, (F) valve, (G) medical chemicals, (H) medical devices, and (I) machine processing. The extracted terms and companies names confirms correlations by the SVD analysis. In conclusion, we found that in the damaged Brand Eigenvectors, not only electronics and automobiles but also many food & beverage companies and spinning/textile companies were involved. The electronics and automobiles were expected but the latter had been unexpected as far as we concerned. Then, we conducted the text mining, and so we found that food & beverage companies and spinning/textile companies also were severely damaged.

From the results of the analysis, we found the following things: (1) the number of news articles on the damages by flooding was the largest in Oct. and then diminished gradually. Because the return data was standardized, we cannot see such a differences clearly in Figs. 3 and 4. (2) From the eigenvectors, we can spot the differences between

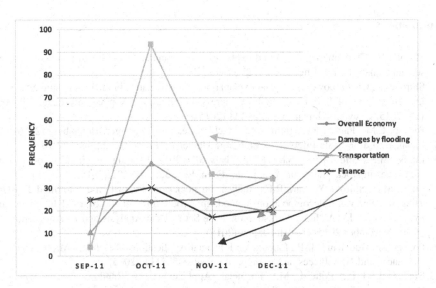

Fig. 5. Monthly movement of the topic frequencies

the electronics industry class and the food and drink one. The latter had recovered more quickly. (3) Although the news mainly focused the damages on PCs, automobile manufactures and HDD industries, contrary to my expectations, the textile and food companies were severely damaged. From the text mining results, we can confirm the damages in the SVD results.

5 Conclusion

In the paper, we propose, as a new time series economics data analysis method, an integrated approach of Singular Value Decomposition on stock data and news article text mining. There we first find the correlations among companies' stock data and then in order to find the latent logical reasons of the associations, we conduct text mining. The paper shows the two-stage approach's advantages to refine the logical reasoning. Concerning the Thai flooding effects on the Japan's economy, as unexpected moves, we have found the serious harms on the Japanese food and drink companies and its quick recoveries. Through the analysis, we found the eigenvectors that could express the categories of industries features. The paper showed the correspondence between the eigenvectors and the news and so on web data.

Acknowledgement. We thank Prof Takako Hashimoto (Chiba University of Commerce) for her wide range of knowledge about Thai economy that helps our research. This research was partly supported by funds from the Telecommunications Advancement Foundation research project in 2015 to 2016. In addition, this study was partly supported by a grant from the Japanese Society for the Promotion of Science from 2015-2017 (15K03619). We sincerely express our gratitude to the Society for its support.

References

1. Shirota, Y., Chakraborty, B.: Visual explanation of eigenvalues and math process in latent semantic analysis. Inf. Eng. Express **2**, 87–96 (2016)
2. Shirota, Y., Chakraborty, B.: Visual explanation of mathematics in latent semantic analysis. In: Proceedings of IIAI International Congress on Advanced Applied Informatics, 12–16 July 2015, Okayama, Japan, pp. 423–428 (2015)
3. Bishop, C.M.: Pattern Recognition and Machine Learning. Springer, Heidelberg (2006)
4. Efron, B.: Large-Scale Inference. Cambridge University Press, Cambridge (2010)
5. Lubis, M.F., Shirota, Y., Sari, R.F.: Thailand's 2011 flooding: its impacts on Japan companies in stock price data. Gakushuin Econ. Pap. **52**, 101–121 (2015)
6. Lubis, M.F., Shirota, Y., Sari, R.F.: Analysis on stock price fluctuation due to flood disaster using singular value decomposition method. In: Proceedings of the of JSAI International Symposia on AI, TADDA (Workshop on Time Series Data Analysis and its Applications), 16–18 November 2015, Hiyoshi, Japan (2015)
7. Potters, M., Bouchaud, J.-P., Laloux, L.: Financial Applications of Random Matrix Theory: Old Laces and New Pieces. arXiv preprint physics/0507111 (2005)
8. Bouchaud, J.-P., Potters, M.: Financial Applications of Random Matrix Theory. Oxford University Press, Oxford (2011)
9. Anderson, G.W., Guionnet, A., Zeitouni, O.: An Introduction to Random Matrices. Cambridge Studies in Advanced Mathematics. Cambridge University Press, Cambridge (2009)
10. Plerou, V., Gopikrishnan, P., Rosenow, B., Amaral, L.A.N., Stanley, H.E.: A random matrix theory approach to financial cross-correlations. Physica A: Stat. Mech. Appl. **287**, 374–382 (2000)
11. Plerou, V., Gopikrishnan, P., Rosenow, B., Amaral, L.A.N., Guhr, T., Stanley, H.E.: Random matrix approach to cross correlations in financial data. Phys. Rev. E **65**, 066126 (2002)
12. Lee, J., Bagheri, B., Kao, H.-A.: A cyber-physical systems architecture for industry 4.0-based manufacturing systems. Manuf. Lett. **3**, 18–23 (2015)
13. Lee, J., Kao, H.-A., Yang, S.: Service innovation and smart analytics for industry 4.0 and big data environment. Procedia CIRP **16**, 3–8 (2014). Product Services Systems and Value Creation. Proceedings of the 6th CIRP Conference on Industrial Product-Service Systems Edited By Hoda ElMaraghy
14. Provost, F., Fawcett, T.: Data science and its relationship to big data and data-driven decision making. Big Data **1**, 51–59 (2013)
15. McAfee, A., Brynjolfsson, E.: Big data: the management revolution. Harvard Bus. Rev. **90**, 60–68 (2012)
16. Lee, K.: How the internet of things will change your world. IdeaBook **2016**, 50–56 (2016)
17. Blei, D.M., Ng, A.Y., Jordan, M.I.: Latent dirichlet allocation. J. Mach. Learn. Res. **3**, 993–1022 (2003)
18. Griffiths, T.L., Steyvers, M.: Finding scientific topics. Proc. Natl. Acad. Sci. **101**(Suppl. 1), 5228–5235 (2004)
19. Koller, D., Friedman, N.: Probabilistic Graphical Models: Principles and Techniques. The MIT Press, Cambridge (2009)
20. Shirota, Y., Hashimoto, T., Chakraborty, B.: Visual materials to teach gibbs sampler. In: International Conference on Knowledge (ICOK 2016), 7–8 May 2016, London, UK. (in printing) (2016)

21. Shirota, Y., Hashimoto, T., Tamaki, S.: Monetary policy topic extraction by using LDA: Japanese monetary policy of the second ABE cabinet term. In: Proceedings of IIAI International Congress on Advanced Applied Informatics 2015, 12–16 July 2015, Okayama, Japan, pp. 8–13 (2015)
22. Sukegawa, S.: Chapter 3: Effects of the 2011 Thai Flooding on Industries and Companies in Thai 2011 Flooding: IDE-JETRO (2013)

Deep Belief Network Using Reinforcement Learning and Its Applications to Time Series Forecasting

Takaomi Hirata[1], Takashi Kuremoto[1(✉)], Masanao Obayashi[1],
Shingo Mabu[1], and Kunikazu Kobayashi[2]

[1] Graduate School of Science and Engineering, Yamaguchi University,
Tokiwadai 2-16-1, Ube, Yamaguchi 755-8611, Japan
{v003we,wu,m.obayas,mabu}@yamaguchi-u.ac.jp
[2] School of Information Science and Technology, Aichi Prefectural University,
1522-3 Ibaragabasama, Nagakute, Aichi 480-1198, Japan
kobayashi@ist.aichi-pu.ac.jp

Abstract. Artificial neural networks (ANNs) typified by deep learning (DL) is one of the artificial intelligence technology which is attracting the most attention of researchers recently. However, the learning algorithm used in DL is usually with the famous error-backpropagation (BP) method. In this paper, we adopt a reinforcement learning (RL) algorithm "Stochastic Gradient Ascent (SGA)" proposed by Kimura and Kobayashi into a Deep Belief Net (DBN) with multiple restricted Boltzmann machines (RBMs) instead of BP learning method. A long-term prediction experiment, which used a benchmark of time series forecasting competition, was performed to verify the effectiveness of the proposed method.

Keywords: Deep learning · Restricted boltzmann machine · Stochastic gradient ascent · Reinforcement learning · Error-backpropagation

1 Introduction

A time series is a data string to be observed in a temporal change in a certain phenomenon. For example, foreign currency exchange rate, stock prices, the amount of rainfall, the change of sunspot, etc. There is a long history of time series analysis and forecasting [1], however, the prediction study of time-series data in the real world is still on the way because of the nonlinearity and the noise affection.

Artificial neural networks (ANNs) have been widely used in pattern recognition, function approximation, nonlinear control, time series forecasting, and so on, since 1940 s [2–11]. After a learning algorithm named error-backpropagation (BP) was proposed by Rumelhart, Hinton and William for multi-layer perceptrons (MLPs) [2], ANNs had its heyday from 1980 s to 1990s. As a successful application, ANNs are utilized as time series predictors today [3–11]. Especially, a deep belief net (DBN) composed by restricted Boltzmann machines (RBMs) [6] and Multi-Layer Perceptron (MLP) [2] was proposed in [8–10] recently.

© Springer International Publishing AG 2016
A. Hirose et al. (Eds.): ICONIP 2016, Part III, LNCS 9949, pp. 30–37, 2016.
DOI: 10.1007/978-3-319-46675-0_4

Deep learning (DL), which is a kind of novel ANNs' training method, is attracting the most attention of artificial intelligent (AI) researchers. By a layer-by-layer training algorithm and stack structures, big data in the high dimensional space are available to be classified, recognized, or sparsely modeled [6]. However, although there are various kinds of deep networks such as auto-encoder, deep Boltzmann machine (DBM), convolutional neural network (CNN), etc., the learning rule used in DL is usually with the famous BP method.

In this paper, a reinforcement learning method named stochastic gradient ascent (SGA) proposed by Kimura and Kobayashi [13] is introduced to the DBN with RBMs as the fine-tuning method instead of the conventional BP learning method. The error between the output of the DBN and the sample is used as reward/punishment to modify the weights of connections of units between different layers. Using a benchmark named CATS data used in the prediction competition [4, 5], the effectiveness of the proposed deep learning method was confirmed by the time series forecasting results.

2 DBN with BP Learning (The Conventional Method)

In [8], Kuremoto et al. firstly applied Hinton and Slakhutdinov's deep belief net (DBN) with restricted Boltzmann machines (RBMs) to the field of time series forecasting. In [9, 10], Kuremoto, Hirata, et al. constructed a DBN with RBMs and a multi-layer perceptron (MLP) to improved the previous time series predictor with RBMs only. In this section, these conventional methods are introduced.

2.1 DBN with RBM and MLP

Restricted Boltzmann machine (RBM)

RBM is a Boltzmann machine consisting of two layers of the visible layer and the hidden layer, no connections between units on the same layer. It is possible to extract features to compress the high-dimensional data in low-dimensional data by performing an unsupervised learning algorithm. Each unit of the visible layer has a symmetric weighted connection to the units of hidden layer. In another word, coupling between the units are bi-direction. Now, let unit i of the visible layer has a bias b_i, unit j of the hidden layer has a bias b_j, all units in the visible layer and the hidden layer stochastically output 1 or 0, according to probabilities with sigmoid function (Eqs. (1) and (2)).

$$p(h_j = 1|v) = \frac{1}{1 + \exp(-b_j - \sum_i w_{ij} v_i)} \tag{1}$$

$$p(v_i = 1|h) = \frac{1}{1 + \exp(-b_i - \sum_j w_{ij} h_j)} \tag{2}$$

Using a learning rule which modifies the weights of connections, RBM network can reach a convergent state by observing its energy function:

$$E(v,h) = -\sum_i b_i v_i - -\sum_j b_j h_j - \sum_{i,j} w_{ij} v_i h_j \qquad (3)$$

Details of the learning algorithm of RBM can be found in [8].

DBN with RBM and MLP

A multi-layer neural network composed by multiple RBMs is proposed in [6] as a well-known deep belief net (DBN). It can extract the feature of features of high-dimensional data according to a learning algorithm which trains the stacked RBMs layer by layer, so the training method is also called "deep learning". Recently, DBN with various neural networks, such as auto-encoder, convolutional neural network (CNN), and so on, have been proposed and applied to many fields such as dimensionality reduction, image compression, pattern recognition, time series forecasting, etc.

A DBN prediction system with RBMs is proposed in [8], and another DBN prediction system using RBMs and MLP is proposed in [9] (see Fig. 1). However, all of these DBNs used the learning algorithm proposed in [6], i.e. RBM's unsupervised learning and BP learning which is a supervised learning as fine-tuning.

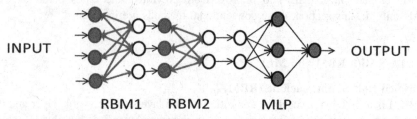

Fig. 1. A structure of DBN composed by RBM and MLP

3 DBN with SGA (The Proposed Method)

BP is a powerful learning method for the feed-forward neural network, however, it modifies the model strictly according to the teacher signal. So the robustness of the ANNs built by BP is restricted. Meanwhile, noises usually exist in the real time-series data. Therefore, we consider that prediction performance may be improved by using a reinforcement learning (RL) algorithm to modify the ANN predictors according to the rewards/punishment corresponding to a "good/bad" output. That is, if the absolute error between the output and the sample is small enough (e.g. using a threshold), then it is considered as a "good" output, and a positive reward is used to modify parameters of ANNs. In [7], MLP and a self-organized fuzzy neural network (SOFNN) with a RL algorithm called stochastic gradient ascent (SGA) [13] were shown their priority to the conventional BP learning algorithm.

Here, we intend to investigate the case when SGA is used in DBN instead of BP.

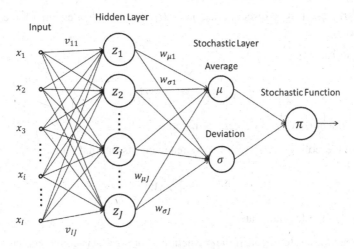

Fig. 2. The structure of MLP designed for SGA

3.1   The Structure of ANNs with SGA

In Fig. 2, a MLP type ANN is designed for SGA [7]. The main difference to the conventional MLP is that the output of network is given by two units which are the parameters of a stochastic policy which is a Gaussian distribution function. SGA Learning algorithm for ANNs

The SGA algorithm is given as follows [7, 13].

1. Observe an input $x(t)$ on time t.

2. Predict a future data $y(t)$ as $\hat{y}(t)$ according to a probability $\pi(\hat{y}(t)|W, x(t)) = \frac{1}{\sqrt{2\pi}\sigma}\exp(-\frac{(\hat{y}(t)-\mu)^2}{2\sigma^2})$ with ANN models which are constructed by parameters $W \equiv \{\mu, \sigma, w_\mu, w_\sigma, v_{ij}\}$.

3. Receive a scalar reward/punishment r_t by calculating the prediction error.

$$r_t = \begin{cases} 1 \; if \, (y(t) - \hat{y}(t))^2 \leq \varepsilon \\ -1 \; else \end{cases} \tag{4}$$

Where $\varepsilon = \beta MSE(t), MSE(t) = \frac{\sum_t (y(t)-\hat{y}(t))^2}{t}$, β is a positive constant. Note that $MSE(t)$ is calculated by the data of last training result.

4. Calculate characteristic eligibility $e_i(t)$ and eligibility trace $\bar{D}_i(t)$.

$$e_i(t) = \frac{\partial}{\partial w_i}\ln\{\pi(W)\} \tag{5}$$

$$\bar{D}_i(t) = e(t) + \gamma \bar{D}_i(t-1) \tag{6}$$

Where $0 \leq \gamma < 1$ is a discount factor, $w_i(t)$ denotes ith element of the internal variable vector W.

5. Calculate $\Delta w_i(t)$:

$$\Delta w_i(t) = (r_t - b)\bar{D}_i(t) \tag{7}$$

Where b denotes the reinforcement baseline.

6. Improve the policy $\pi(\hat{y}(t)|x(t), W)$ by renewing its internal variable W.

$$W \leftarrow W + \alpha \Delta W(t) \tag{8}$$

Where α is the learning rate.

7. For the next time step $t + 1$, if the prediction error $MSE(t)$ is converged enough then end the learning process, else, return to step 1 (Fig. 3).

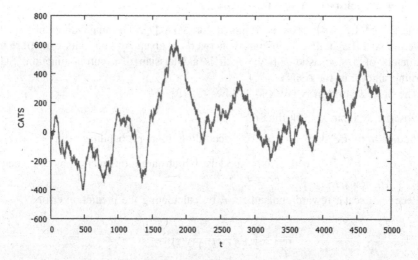

Fig. 3. Time series data of CATS

4 Prediction Experiments and Results

4.1 CATS Benchmark Time Series Data

CATS time series data is the artificial benchmark data for forecasting competition with ANN methods [4, 5]. This artificial time series is given with 5,000 data, among which 100 are missed (hidden by competition the organizers). The missed data exist in 5 blocks: elements 981 to 1,000;elements 1,981 to 2,000;elements 2,981 to 3,000; elements 3,981 to 4,000; elements 4,981 to 5,000.

The mean square error E_1 is used as the prediction precision in the competition, and it is computed by the 100 missing data and their predicted values as following:

$$E_1 = \frac{1}{100}\{ \sum_{t=981}^{1000} (y(t) - \hat{y}(t))^2 + \sum_{t=1981}^{2000} (y(t) - \hat{y}(t))^2 + \sum_{t=2981}^{3000} (y(t) - \hat{y}(t))^2 + \sum_{t=3981}^{4000} (y(t) - \hat{y}(t))^2 + \sum_{t=4981}^{5000} (y(t) - \hat{y}(t))^2\}$$

(9)

where $\hat{y}(t)$ is the long term prediction result of the missed data.

4.2 Optimization of Meta Parameters

The number of RBM that constitute the DBN and the number of neurons of each layer affect prediction performance seriously. In this paper, these meta parameters are optimized using random search method [12]. In the optimization of the ANN structure, random search is known to exhibit higher performance than the grid search. The meta parameters of DBN and their exploration limits are shown as following: the number of RBMs: 0–3; the number of neurons in each layers: 2–20; learning rate of each RBM: 10^{-5}–10^{-1}; learning rate of SGA: 10^{-5}–10^{-1}; discount factor: 10^{-5}–10^{-1}; constant β in Eq. (4) 0.5 − 2.0

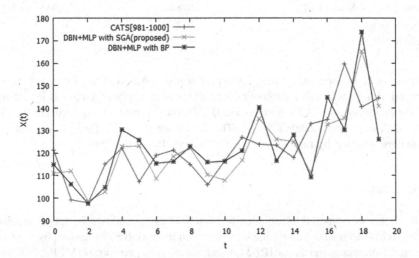

Fig. 4. The prediction results of different methods for CATS (Block 1).

4.3 Experiments Result

The prediction results of the first blocks of CATS data are shown in Fig. 4. Comparing to the conventional learning method of DBN, i.e., using Hinton's RBM unsupervised learning method [6, 8] and back-propagation (BP), the proposed method which used the reinforcement learning method SGA instead of BP showed its superiority according

Table 1. The comparison of performance between different methods using CATS data

Method	E_1
DBN(SGA) (proposed)	**170**
DBN(BP) + ARIMA [10]	244
DBN[9] (BP)	257
Kalman Smoother (The best of IJCNN '04) [5]	408
DBN[8] (2 RBMs)	1215
MLP[8]	1245
A hierarchical Bayesian Learning Scheme for Autoregressive Neural Networks (The worst of IJCNN '04) [4]	1247

Table 2. Parameters of DBN used for the CATS data (Block 1)

	DBN with SGA (proposed)	DBN with BP [9]
The number of RBMs	3	1
Learning rate of RBM	0.048-0.055-0.026	0.042
Structure of DBN (the number of neurons in each layer)	14-14-18-19-18-2	5-11-2-1
Learning rate of SGA or BP	0.090	0.090
Discount factor γ	0.082	–
Coefficient β	1.320	–

to the measure of the average prediction precision E_1. Additionally, the result by the proposed method achieved the top of rank of all previous studies such as MLP with BP, the best prediction of CATS competition IJCNN'04 [5], the conventional DBNs with BP [8, 9], and hybrid models [10, 11]. The details are shown in Table 1. The optimal parameters obtained by random search method are shown in Table 2.

5 Conclusion

In this paper, we proposed to use a reinforcement learning method "stochastic gradient ascent (SGA)" to realize fine-tuning of a deep belief net (DBN) composed by multiple restricted Boltzmann machines (RBMs) and multi-layer perceptron (MLP). Different from the conventional fine-tuning method using the error backpropagation (BP), the proposed method used a rough judgment of reinforcement learning concept, i.e., "good" output owns positive rewards, and "bad" one with negative reward values, to modify the network. This makes the available of a sparse model building, avoiding to the over-fitting problem which occurs by the lost function with the teacher signal.

Time series prediction was applied to verify the effectiveness of the proposed method, and comparing to the conventional methods, the DBN with SGA showed the highest prediction precision in the case of CATS benchmark data.

The future work of this study is to investigate the effectiveness to the real time series data and other nonlinear systems such as chaotic time series data.

Acknowledgment. This work was supported by JSPS KAKENHI Grant No. 26330254 and No. 25330287.

References

1. Box, G.E.P., Pierce, D.A.: Distribution of residual autocorrelations in autoregressive-integrated moving average time series models. J. Am. Stat. Ass. **65**(332), 1509–1526 (1970)
2. Rumelhart, D.E., Hinton, G.E., Williams, R.J.: Learning representation by back-propagating errors. Nature **232**(9), 533–536 (1986)
3. Casdagli, M.: Nonlinear prediction of chaotic time series. Phys. D **35**, 335–356 (1981)
4. Lendasse, A., Oja, E., Simula, O., Verleysen, M.: Time series prediction competition: the CATS benchmark. In: Proceedings of International Joint Conference on Neural Networks (IJCNN 2004), pp. 1615–1620 (2004)
5. Lendasse, A., Oja, E., Simula, O., Verleysen, M.: Time series prediction competition: the CATS benchmark. Neurocomputing **70**, 2325–2329 (2007)
6. Hinton, G.E., Salakhutdinov, R.R.: Reducing the dimensionality of data with neural networks. Science **313**, 504–507 (2006)
7. Kuremoto, T., Obayashi, M., Kobayashi, M.: Neural forecasting systems. In: Weber, C., Elshaw, M., Mayer, N.M. (eds.) Reinforcement Learning, Theory and Applications, Chap. 1, pp. 1–20. INTECH (2008)
8. Kuremoto, T., Kimura, S., Kobayashi, K., Obayashi, M.: Time series forecasting using a deep belief network with restricted Boltzmann machines. Neurocomputing **137**(5), 47–56 (2014)
9. Kuremoto, T., Hirata, T., Obayashi, M., Mabu, S., Kobayashi, K.: Forecast chaotic time series data by DBNs. In: Proceedings of the 7th International Congress on Image and Signal Processing (CISP 2014), pp. 1304–1309, October 2014
10. Hirata, T., Kuremoto, T., Obayashi, M., Mabu, S.: Time series prediction using DBN and ARIMA. In: International Conference on Computer Application Technologies (CCATS 2015). Matsue, Japan, pp. 24–29, September 2015
11. Zhang, G.P.: Time series forecasting using a hybrid ARIMA and neural network model. Neurocomputing **50**, 159–175 (2003)
12. Bergstra, J., Bengio, Y.: Random search for hyper-parameter optimization. J. Mach. Learn. Res. **13**, 281–305 (2012)
13. Kimura, H., Kobayashi, S.: Reinforcement learning for continuous action using stochastic gradient ascent. In: Proceedings of 5th Intelligent Autonomous Systems, pp. 288–295 (1998)

Neuron-Network Level Problem Decomposition Method for Cooperative Coevolution of Recurrent Networks for Time Series Prediction

Ravneil Nand[✉], Emmenual Reddy, and Mohammed Naseem

School of Computing Information and Mathematical Sciences,
University of South Pacific, Suva, Fiji
ravneiln@yahoo.com

Abstract. The breaking down of a particular problem through problem decomposition has enabled complex problems to be solved efficiently. The two major problem decomposition methods used in cooperative coevolution are synapse and neuron level. The combination of both the problem decomposition as a hybrid problem decomposition has been seen applied in time series prediction. The different problem decomposition methods applied at particular area of a network can share its strengths to solve the problem better, which forms the major motivation. In this paper, we are proposing a problem decomposition method that combines neuron and network level problem decompositions for Elman recurrent neural networks and applied to time series prediction. The results reveal that the proposed method has got better results in few datasets when compared to two popular standalone methods. The results are better in selected cases for proposed method when compared to several other approaches from the literature.

Keywords: Cooperative coevolution · Problem decomposition · Recurrent network · Time series prediction

1 Introduction

In the context of evolutionary computation, cooperative coevolutionary algorithms are gaining increasing attention of most researchers. Cooperative coevolution is an evolutionary method that computes the solution of a large problem by decomposing it into subcomponents and solving them individually [1,2]. An important characteristic of cooperative coevolution is that it offers better diversity using several subcomponents than mainstream evolutionary algorithms [3].

One of the applications of cooperative coevolution have been in the area of neuro-evolution, which include training recurrent neural networks [4,5]. In a recurrent neural network the current network state and input determines the next state and output. Recurrent neural networks trained using cooperative

© Springer International Publishing AG 2016
A. Hirose et al. (Eds.): ICONIP 2016, Part III, LNCS 9949, pp. 38–48, 2016.
DOI: 10.1007/978-3-319-46675-0_5

coevolutionary algorithms are manifested to perform very well for time series prediction problems [3, 6].

Problem decomposition is an important procedure that determines how a neural network is decomposed and encoded as subcomponents [3]. Neuron level and synapse level are the two preeminent problem decomposition methodologies. In neuron level, each neuron in the hidden layer acts as the main reference point for the subcomponents. In synapse level problem decomposition, the network is divided into its lowest level with each weight or synapse in the network forming a subcomponent [2, 7]. Successful application of both decomposition as a hybrid has been seen in [7]. This forms the major motivation for the paper to combine different decomposition methods for successful application to time series prediction.

Time-series analysis is a practicable measure used in various application areas such as the economic models and medical systems. For instance, the prediction of financial markets or the monitoring of physiological signals stemming from a patient during surgery. A crucial objective of time-series analysis is to predict the future behavior of a measurement signal based on its past behavioural observations.

In this paper, we combine neuron and network level problem decomposition to form a new hybrid problem decomposition called Neuron-Network Level (NNL) problem decomposition. NNL is intended for training Elman recurrent neural networks that are chaotic time series problems.

The rest of the paper is organized as follows. In Sect. 2, the proposed problem decomposition is discussed in detail. The Sect. 3 shows the dataset overview and experimental setup. In Sects. 4 and 5, the results and discussion are given respectively. Section 6 concludes the paper with a discussion of future extensions of the paper.

2   Neuron-Network Level Problem Decomposition

In neural networks, Neuron level's emphasis is given on the interacting variables while Network level encoding treats the entire problem as one subset. The problem in cooperative coevolution is to have interacting variables with diversity.

Therefore, we propose a hybrid problem decomposition strategy combining Neuron with Network level decomposition called Neuron-Network level (NNL). NNL breaks down the larger problem into a lower level as in Neuron level problem decomposition. In the proposed method, the subcomponent consists of incoming links associated with neurons in the hidden layer and incoming links to hidden from context layer and one subcomponent for hidden to output layer plus bias as shown in Fig. 1. The total of all the outputs generated at each neuron gives the calculation of the actual output as in all the methods mentioned earlier.

The proposed method seems same as Neuron level but the difference lies in the hidden to output layer. If a problem has more than one output than the proposed method will still have one subcomponent for that area whereas Neuron level will depend on the number of output neurons. In addition, the proposed

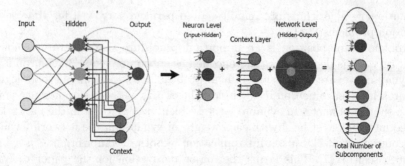

Fig. 1. Neuron-Network Level Problem Decomposition (NNL) breaks down the neural network into specific regions and encodes it into the respective sub-populations.

method includes the total number of bias added to the population size whereas neuron level only includes the number of output bias to the population size.

The total number of sub-components is equal to the total number of hidden layer neurons plus total number on context layer neurons and plus one. The sub-components are implemented as sub-populations.

Algorithm 1 shows the proposed method which is used in training the recurrent network. In the first phase of the algorithm, the problem is broken down in the number of subcomponents according to the proposed NNL method. In the second phase, the problem is encoded based on the number of neurons in the hidden layer and context layer.

In the third phase, random values are used to initialize all the individuals within the population. Each of the individuals fitness is then evaluated based on the selection of arbitrary individuals existing in the rest of the sub-populations.

Algorithm 1. NNL for Training Recurrent Networks

Step 1: Decompose the problem into subcomponents according to NNL. Here problem is decomposed into input-hidden, context, and hidden-output layer subcomponents.
Step 2: Encode each subcomponent in a sub-population according to problem decomposition method used.
Step 3: Initialize and cooperatively evaluate each sub-population.
foreach *Cycle until termination* **do**
    foreach *Sub-population* **do**
        foreach *Depth of n Generations* **do**
            Select and create new offspring using genetic operators
            Cooperative Evaluation the new offspring
            Add new offspring's to the sub-population
        end
    end
end

Hereafter, using genetic operators the evolution for each sub-population also occurs. Each member of the subpopulations fitness evaluation is done cooperatively with the fittest individuals from the other subsets [8].

Evolution of each sub-population is done for a fixed number of generations and one cycle is accomplished when evolution of all sub-populations have been successfully completed. Evolution of each sub-population is done using the generalized generation gap parent centric crossover (G3-PCX) genetic algorithm [9]. This is where the selected parent is having the best fitness while the rest are selected randomly from the subpopulation to generate the new individuals.

In the next phase the fitness of the new individuals are evaluated. At this point, the random selection of two new parents from the main sub-population is carried out for the purpose of comparison. In future, the two new parents can be replaced by new individuals. In the evolution phase, the fitness evaluation of each individual is carried out together with the fittest individuals from the rest of the sub-populations.

The fitness evaluation of any member of the sub-population is done by joining the fittest members from the rest of the sub-populations. The evolution of these sub-populations is carried out for a fixed number of generations. As soon as maximum fitness evaluations of 50,000 are reached for the network, the generalization performance testing begins.

3   Experimental Setup

The following section gives the overview and setup of the dataset.

Reconstructing the dataset before it can be used is allowed by Taken's embedding theorem [10]. The reconstruction of the chaotic time series data into a state space vector is carried out with the two important conditions of *time delay (T)* and *embedding dimension (D)* [10].

Four different datasets are used to test the proposed method. The *Lorenz time series* is the first data set that is used [11]. It is a simulated time series and is chaotic in nature. To train the neural network, the first 500 values are used whereas the remaining 500 values are used for testing.

The *Sunspot time series* is the second dataset that is used [12]. It is a real world problem time series [12]. This dataset contains 2000 points. To train the neural network, the first 1000 values are used whereas the remaining 1000 values are used for training.

The *ACI Worldwide Inc. time series* is the third data set that is used [13]. To obtain the ACI Worldwide Inc. financial time series data set, the NASDAQ stock exchange is used [13]. To train the neural network, the first 400 values are used whereas to test the neural network the remaining 400 values are used.

The Seagate Technology PLC is the last dataset that is used [13]. It is also a financial time series dataset and contains daily closing stock prices from December 2006 to February 2010. To train the neural network, the first 400 values are used whereas to test the neural network the remaining 400 values are used.

The embedding dimension $D = 3$ and $T = 2$ is used to reconstruct the phase space of Lorenz dataset where as $D = 5$ and $T = 2$ is used to reconstruct the phase space of the rest of the dataset.

According to the literature, the scaling of the four time series dataset is done in the range of $[0, 1]$ for ACI Worldwide Inc, Seagate and $[-1, 1]$ for Sunspot and Lorenz in order to provide a fair comparison.

Sigmoid units are employed by the recurrent neural network for the Seagate, and ACI Worldwide Inc. time series while the hyperbolic tangent unit is used for Lorenz and Sunspot time series in the hidden and output layers. Root mean squared error (RMSE) is used in evaluating the performance of the proposed method, as done in [3].

The algorithm is run 50 times. For terminating each run of the algorithm, the maximum number of function evaluations was set at 50,000. According to literature [3], a pool size of 2 parents and 2 offsprings are put in the G3-PCX algorithm. For evolving all the sub-populations in the proposed method, the G3-PCX evolutionary model which uses the *generation gap model* [9] for selection is used since it has shown good results with cooperative neuro-evolution [2].

For dividing the problem into subcomponents, the Neuron level and the Synapse level problem decomposition methods are used for comparison with the proposed method since these two methods are established ones. For each sub-population, the number of generations known as *depth of search* was kept at 1 since this configuration has given optimal results in similar work by [3].

To compute the prediction performance of the proposed method, the Root Mean Squared Error (RMSE) is used, as done in [7]. The Eq. 1 is used.

$$RMSE = \sqrt{\frac{1}{N} \sum_{i=1}^{N} (y_i - \hat{y}_i)^2} \qquad (1)$$

where y_i, is observed data and $\hat{y}_i$ is predicted data. And N is the observed data's length. This performance measure is used to compare the results with the literature.

4   Results

The experimental results based on the performance of proposed method is given in this section.

The Tables 1–4 showcases the results for different number of hidden neurons on the proposed method NNL, being compared to two standalone (NL and SL) methods. The results given in the Tables 1–4 are based on 95 % confidence interval on RMSE and the best results for each algorithm are highlighted in bold. The *Training* shows the train average with train error sum while *Generalisation* is based on test average with test error sum and lastly *Best* shows the best test rmse.

In Table 1, the Lorenz time series problem is been evaluated. It was observed that the NNL has poor performance than the other two methods. It was seen

Table 1. The prediction training and generalization performance (RMSE) of NL, SL and NNL for the Lorenz time series

Method	H	Training	Generalization	Best
RNN-NL	3	0.0177 ± 0.0015	0.0184 ± 0.0016	0.007
	5	**0.0132 ± 0.0014**	**0.0136 ± 0.0015**	**0.003**
	7	0.0143 ± 0.0015	0.0147 ± 0.0016	0.005
RNN-SL	3	0.0209 ± 0.0024	0.0214 ± 0.0025	0.008
	5	0.0168 ± **0.0020**	0.0175 ± **0.0021**	**0.004**
	7	**0.0164** ± 0.0028	**0.0172** ± 0.0030	0.008
RNN-NNL	3	0.0369 ± 0.0041	0.0374 ± 0.0041	0.015
	5	0.0357 ± 0.0100	0.0358 ± 0.0099	0.009
	7	**0.0223 ± 0.0023**	**0.0226 ± 0.0024**	**0.007**

that the generalization performance and training of the NNL and the other two methods increases when the number of the hidden neuron increases. Seven hidden neurons for NNL gave the best result.

Table 2 illustrates the evaluation of the Sunspot time series problem. In this time series, noise is present since it is real-world time series. The proposed method NNL was unable to outperform both of the methods it is compared to but was close in terms of generalization. The best result for NNL was given by seven hidden neurons and even for the other two methods it was seven hidden neurons.

In Table 3, the ACI time series problem results are reported. The time series is real time series where noise is present as in Sunspot time series. For this time series, the NNL didn't perform better than NL method but had similar

Table 2. The prediction training and generalization performance (RMSE) of NL, SL and NNL for the Sunspot time series

Method	H	Training	Generalization	Best
RNN-NL	3	0.0207 ± 0.0022	**0.0512 ± 0.0123**	0.017
	5	0.0179 ± 0.0019	0.0537 ± 0.0128	0.017
	7	**0.0165 ± 0.001**	0.0566 ± 0.0155	**0.015**
RNN-SL	3	0.0207 ± 0.0022	**0.0512 ± 0.0123**	0.017
	5	0.0179 ± 0.0019	0.0537 ± 0.0128	0.017
	7	**0.0165 ± 0.001**	0.0566 ± 0.0155	**0.015**
RNN-NNL	3	0.0332 ± 0.0064	0.0682 ± 0.0132	0.026
	5	0.0246 ± 0.0024	0.0772 ± 0.0184	0.021
	7	**0.0203 ± 0.0022**	**0.0770 ± 0.0238**	**0.019**

Table 3. The prediction training and generalization performance (RMSE) of NL, SL and NNL for the ACI Worldwide Inc. time series

Method	H	Training	Generalization	Best
RNN-NL	3	0.0207 ± 0.0004	0.0212 ± 0.0013	0.019
	5	0.0203 ± 0.0003	0.0204 ± **0.0002**	0.019
	7	**0.0201 ± 0.0002**	**0.0204** ± 0.0004	**0.019**
RNN-SL	3	0.0226 ± 0.0007	0.0219 ± 0.0008	0.019
	5	**0.0220 ± 0.0007**	**0.0211 ± 0.0005**	**0.019**
	7	0.0211 ± **0.0005**	0.0211 ± 0.0006	0.019
RNN-NNL	3	0.0255 ± 0.0009	0.0212 ± 0.0014	0.015
	5	0.0297 ± 0.0027	0.0222 ± 0.0020	**0.014**
	7	**0.0220 ± 0.0008**	**0.0171 ± 0.0009**	0.015

Table 4. The prediction training and generalization performance (RMSE) of NL, SL and NNL for the Seagate time series

Method	H	Training	Generalization	Best
RNN-NL	3	0.0351 ± 0.0549	0.3220 ± 0.0650	0.032
	5	0.0351 ± 0.0448	0.3223 ± 0.0742	0.032
	7	**0.0351 ± 0.0226**	**0.3215 ±0.0551**	**0.032**
RNN-SL	3	0.0351 ± 0.0628	0.3216 ± 0.0653	**0.031**
	5	0.0351 ± 0.0532	**0.3215 ± 0.0562**	0.032
	7	**0.0350 ± 0.0480**	0.3217 ± 0.0654	0.032
RNN-NNL	3	0.0234 ± 0.0015	0.1841 ± 0.0395	0.030
	5	0.0241 ± 0.0038	0.2121 ± 0.0495	0.049
	7	**0.0199 ± 0.0004**	**0.1764 ± 0.0333**	**0.021**

performance to SL method in terms of generalization. Seven hidden neurons have given the best result for the proposed method.

The Seagate time series problem is evaluated in Table 4. For this time series, the NNL method outperformed the other two methods, NL, and SL. For NNL, seven hidden neurons have given best results.

Figure 2 shows the predicted results of RNN-NNL method with original results on Sunspot dataset. The error graph is also given which indicates that better methods will be needed to cater for the chaotic nature of the time series problems at certain time intervals.

The best results from Tables 1–4 with some of the related methods in literature are given in Table 5. The RMSE best run together with NMSE are given for comparison purposes. The proposed NNL method has shown good performance in some of the datasets when compared to other methods in the literature.

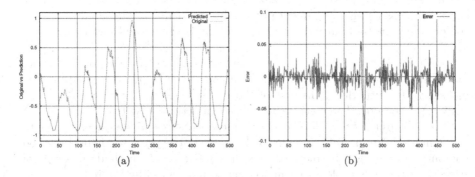

Fig. 2. Typical test dataset prediction by NNL for Sunspot dataset. (a) Performance given by NNL on the testing set. (b) Error on the test dataset given by NNL.

Table 5. A comparison with the results from literature on different time series datasets

Problem	Prediction method	RMSE	NMSE
Lorenz	RBF with orthogonal least squares (2006) [14]		1.41E-09
	Locally linear neuro-fuzzy model (2006) [14]		9.80E-10
	SL-CCRNN [3]	6.36E-03	7.72E-04
	NL-CCRNN [3]	8.20E-03	1.28E-03
	FNN-NSL [7]	2.34E-03	2.87E-05
	Proposed RNN-NNL	7.49E-03	1.09E-03
Sunspot	RBF with orthogonal least squares (2006) [14]		4.60E-02
	Locally linear neuro-fuzzy model (2006) [14]		3.20E-02
	SL-CCRNN [3]	1.66E-02	1.47E-03
	NL-CCRNN [3]	2.60E-02	3.62E-03
	FNN-NSL [7]	1.33E-02	5.38E-04
	Proposed RNN-NNL	1.89E-02	1.90E-03
ACI	FNN-SL [15]	1.92E-02	
Worldwide	FNN-NL [15]	1.91E-02	
	MO-CCFNN-T=2 [16]	1.94E-02	
	MO-CCFNN-T=3 [16]	1.47E-02	
	Neuron-Synapse Level method FNN-NSL [7]	1.51E-02	1.24E-03
	Proposed NNL	1.44E-02	1.98E-03
Seagate	FNN-SL [15]	3.74E-02	
	FNN-NL [15]	2.24E-02	
	Neuron-Synapse Level methodFNN-NSL [7]	2.45E-02	3.56E-03
	Proposed NNL	2.10E-02	4.58E-03

In Table 5 under problem Lorenz, it shows the best result of Lorenz time series problem being compared to works in literature. It has been seen that the proposed method outperformed only NL-CCRNN method and was close to SL-CCRNN method.

In Table 5 under problem Sunspot, the best result of the Sunspot time series problem is compared with results in the literature where the proposed method has shown to outperform the rest of the methods expect for SL-CCRNN and FNN-NSL methods. The method has given competitive results.

The best result of the ACI Worldwide Inc. time series problem is compared to works in literature in Table 5 under problem ACI. The proposed hybrid method has outperformed all the methods expect for FNN-NSL methods in terms of NMSE. Better and stable performance has been achieved by the NNL.

In Table 5, the best result of the Seagate time series problem is compared with results in the literature under problem Seagate. The proposed method has outperformed all the methods expect for FNN-NSL methods in terms of NMSE. The method had similar performance as in ACI Worldwide Inc. dataset.

5   Discussion

The results obtained for the proposed method is competitive when compared to works from literature involving four different data sets. The application of different decomposition method at different stages of network helps in prediction.

The proposed hybrid model has given better performances in some of the datasets used when compared to FNN-NSL. The results for NNL method on financial datsets are better in terms of RMSE.

In some cases, NNL gave better performance than standalone methods based on cooperative coevolution (CCRNN-Synapse Level and CCRNN-Neuron Level).

The limitation in proposed method lies in the area of hidden to the output layer. The context layer can be decomposed by some other decomposition method to enhance the learning. The results of proposed method was unable to beat results of majority datasets when compared to FNN-NL and FNN-NSL. By increasing and decreasing the number of sub-populations associated within the context layer can allow seeing how that region reacts.

One of the advantages of the proposed hybrid method (NNL) is the use of two problem decomposition. The combination of two problem decomposition (NL and NetL) in NNL, allows NL to be used for decision making and NetL for diversity in the search. Therefore, NNL performs better than other methods in some of the cases. The cases where the method was unable to perform is due to either over training or over fitting.

6   Conclusion

This paper has applied a new problem decomposition method (NNL) based on Neuron and Network level problem decomposition for the cooperative coevolution to recurrent neural networks. NNL has been tested on time series prediction.

The investigation began with the testing of the proposed hybrid method with benchmark dataset and later on the financial dataset. As for the hybrid model, it creates fewer subcomponents for the problem when compared to Synapse Level and Neuron level.

The results showed that NNL has been able to achieve the similar solution quality when compared to other methods in terms the generalization performance. In general, NNL has shown better optimization performance in time and success rate than other methods on financial datasets.

In future work, the proposed method can be applied to pattern classification problems and global optimization problems.

References

1. García-Pedrajas, N., Sanz-Tapia, E., Ortiz-Boyer, D., Hervás-Martínez, C.: Introducing multi-objective optimization in cooperative coevolution of neural networks. In: Mira, J., Prieto, A. (eds.) IWANN 2001. LNCS, vol. 2084, pp. 645–652. Springer, Heidelberg (2001). doi:10.1007/3-540-45720-8_77

2. Chandra, R.: Problem decomposition and adaptation in cooperative neuro-evolution (2012)

3. Chandra, R., Zhang, M.: Cooperative coevolution of Elman recurrent neural networks for chaotic time series prediction. Neurocomputing 186, 116–123 (2012)

4. Gomez, F., Mikkulainen, R.: Incremental evolution of complex general behavior. Adapt. Behav. 5(3-4), 317–342 (1997)

5. Chandra, R., Frean, M., Zhang, M.: Adapting modularity during learning in cooperative co-evolutionary recurrent neural networks. Soft Comput. Fusion Found. Methodologies Appl. 16(6), 1009–1020 (2012)

6. Giles, C.L., Lawrence, S., Tsoi, A.C.: Noisy time series prediction using a recurrent neural network and grammatical inference. Mach. Learn. 44, 161–186 (2001)

7. Nand, R., Chandra, R.: Neuron-synapse level problem decomposition method for cooperative neuro-evolution of feedforward networks for time series prediction. In: Arik, S., Huang, T., Lai, W.K., Liu, Q. (eds.) ICONIP 2015. LNCS, vol. 9491, pp. 90–100. Springer, Heidelberg (2015). doi:10.1007/978-3-319-26555-1_11

8. Potter, M.A., Jong, K.A.: A cooperative coevolutionary approach to function optimization. In: Davidor, Y., Schwefel, H.-P., Männer, R. (eds.) PPSN 1994. LNCS, vol. 866, pp. 249–257. Springer, Heidelberg (1994). doi:10.1007/3-540-58484-6_269

9. Deb, K., Anand, A., Joshi, D.: A computationally efficient evolutionary algorithm for real-parameter optimization. Evol. Comput. 10(4), 371–395 (2002)

10. Takens, F.: Detecting strange attractors in turbulence. In: Rand, D., Young, L.-S. (eds.) Dynamical Systems and Turbulence, Warwick 1980. LNM, vol. 898, pp. 366–381. Springer, Heidelberg (1981). doi:10.1007/BFb0091924

11. Lorenz, E.: Deterministic non-periodic flows. J. Atmos. Sci. 20, 267–285 (1963)

12. Sello, S.: Solar cycle forecasting: a nonlinear dynamics approach. Astron. Astrophys. 377, 312–320 (2001)

13. NASDAQ Exchange Daily: 1970–2010 Open, Close, High, Low and Volume. http://www.nasdaq.com/symbol/aciw/stock-chart. Accessed 2 Feb 2015

14. Gholipour, A., Araabi, B.N., Lucas, C.: Predicting chaotic time series using neural and neurofuzzy models: A comparative study. Neural Process. Lett. 24, 217–239 (2006)

15. Chand, S., Chandra, R.: Cooperative coevolution of feed forward neural networks for financial time series problem. In: International Joint Conference on Neural Networks (IJCNN), Beijing, China, pp. 202–209 (2014)
16. Chand, S., Chandra, R.: Multi-objective cooperative coevolution of neural networks for time series prediction. In: 2014 International Joint Conference on Neural Networks (IJCNN), pp. 190–197 (2014)

Data-Driven Approach
for Extracting Latent Features
from Multi-dimensional Data

Yet Another Schatten Norm for Tensor Recovery

Chao Li, Lili Guo, Yu Tao, Jinyu Wang, Lin Qi, and Zheng Dou[(⊠)]

College of Information and Communication Engineering,
Harbin Engineering University, Harbin 150001, Heilongjiang, China
lichao_heu@hotmail.com, douzheng@hrbeu.edu.cn

Abstract. In this paper, we introduce a new class of Schatten norms for tensor recovery. In the new norm, unfoldings of a tensor along not only every single order but also all combinations of orders are taken into account. Additionally, we prove that the proposed tensor norm has similar properties to matrix Schatten norm, and also provides several propositions which is useful in the recovery problem. Furthermore, for reliable recovery of a tensor with Gaussian measurements, we show the necessary size of measurements using the new norm. Compared to using conventional overlapped Schatten norm, the new norm results in less measurements for reliable recovery with high probability. Finally, experimental results demonstrate the efficiency of the new norm in video in-painting.

Keywords: Tensor completion · Video in-painting · Tensor norm · Low-rank decomposition

1 Introduction

The problem that how to reliably recover data from limited measurements has been widely discussed from Compressed Sensing (CS) to Data Completion (DC). Although the basic assumption for CS and DC problems is different, they can be uniformly considered as a linear inverse problem which is generally given by

$$\min_{\mathbf{x}} f(\mathbf{x}), \quad s.t. \, \mathbf{y} = G(\mathbf{x}), \tag{1}$$

where $\mathbf{x} \in \mathcal{R}^N$ denotes target data which we want to recover, and $\mathbf{y} \in \mathcal{R}^M$ denotes the measurement set. $G : \mathcal{R}^N \to \mathcal{R}^M$ is a linear operator which is generally seen as sensing matrix in CS and as sampling operation in DC. $f(\cdot)$ denotes a convex function, which reflects priori structure about the data, such as sparsity, smoothness, low-rankness, and etc. In this paper, we mainly consider a special case that the target data has multi-dimensional low rank structure. It means that $\mathbf{x}$ can be treated as a high order tensor with low rank structure along each order. To depict multi-dimensional rank of a tensor, a well known norm was studied in [9], which is called as overlapped Schatten 1-norm

$$\|\underline{\mathbf{X}}\|_{S_1/1} = \sum_{k=1}^{K} \|\mathbf{X}_{(k)}\|_{S_1} \tag{2}$$

© Springer International Publishing AG 2016
A. Hirose et al. (Eds.): ICONIP 2016, Part III, LNCS 9949, pp. 51–60, 2016.
DOI: 10.1007/978-3-319-46675-0_6

where $\underline{\mathbf{X}} \in \mathcal{R}^{I_1 \times \cdots \times I_K}$ denotes a Kth-order tensor, $\mathbf{X}_{(k)}$ denotes flatting $\underline{\mathbf{X}}$ into a matrix along the kth order, and $\|\cdot\|_{S_1}$ represents Schatten 1-norm over matrix, which is also called as trace norm or nuclear norm in the literature [6].

Many tractable algorithms were developed by using (2) for tensor completion problem, and provide impressive performance in many applications like image/video in-painting [6], denosing [11], and foreground extraction. In [7,11], it has been proved that $O\left(rn^{K-1}\right)$ Gaussian measurements is necessary for reliably recovering the Kth-order tensor $\underline{\mathbf{X}} \in \mathcal{R}^{n \times \cdots \times n}$ with multi-linear rank $(r, \ldots, r)$. However, [7] also shows that a certain (intractable) non-convex formulation needs only $O\left(r^K + nrK\right)$ Gaussian measurements. Therefore, we can believe that it is possible to find a better function (or norm) to reduce the gap of size of measurements for reliable recovery.

In this paper, we introduce a new definition for tensor Schatten norm inspired by (2). In this new frame, the norm obeys important properties such as unitary invariance and norm equivalence as matrix Schatten norm. Additionally, extra multi-linear structures can be exploited by the new norm. Furthermore, it is shown that only $O\left(r^{\llcorner K/2 \lrcorner} n^{\lceil K/2 \rceil}\right)$ Gaussian measurements are necessary for reliable recovery, while $O\left(rn^{K-1}\right)$ is necessary for norm (2) as mentioned above.

As notation, tensors are denoted by boldface underlined capital letters, e.g., $\underline{\mathbf{X}} \in \mathcal{R}^{I_1 \times I_2 \times \cdots \times I_N}$. Matrices are denoted by boldface capital letters, e.g., $\mathbf{X} \in \mathcal{R}^{I \times R}$. Vectors are given by boldface lowercase letters, e.g., $\mathbf{x} \in \mathcal{R}^I$, and overlined lowercase letters denote sets, e.g., $\bar{i}$ and $\bar{j}$. Some types of tensor multiplications are presented by: $\otimes$ for Kronecker, $\odot$ for the Khatri-Rao, $\circ$ for outer product, big symbols $\bigotimes$, $\bigodot$ and $\bigcirc$ denote sequentially Kronecker, Khatri-Rao and outer product, respectively. The process of unfolding flattens a tensor $\underline{\mathbf{X}}$ into a matrix $\mathbf{X}_{(\bar{i})}$ by set-index $\bar{i}$. For example, let $\underline{\mathbf{X}} \in \mathcal{R}^{I_1 \times I_2 \times I_3 \times I_4}$ be a 4th-order tensor and $\bar{i} = \{\{1, 3\}, \{2, 4\}\}$, then the matrix $\mathbf{X}_{(\bar{i})} \in \mathcal{R}^{I \times J}$ represents flattening $1, 3$th mode of $\underline{\mathbf{X}}$ as one dimension and $2, 4$th mode as another dimension of $\mathbf{X}_{(\bar{i})}$ accordingly. Arrangement of entries in unfolding matrix is arbitrary by any convention. Note that, in this paper, such unfolding operation considers all reordered and transposed matrices as an equivalence class. The motivation is that any reordering and transpose operation will not change its Schatten norms. Figure 1 shows diagram representation tensor and unfolding operation. More detail can be found from [3] and references therein.

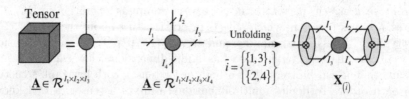

Fig. 1. Basic symbols and unfolding process for tensor diagram

2   Theoretical Results

2.1   New Norm

As a more general formation, overlapped Schatten S_p/q-norm which has been discussed in [11] is written as follows:

$$\|\underline{\mathbf{X}}\|_{S_p/q} = \left(\sum_{k=1}^{K} \|\mathbf{X}_{(k)}\|_{S_p}^q \right)^{1/q}, \tag{3}$$

where $1 \le p, q \le \infty$; here

$$\|\mathbf{X}\|_{S_p} = \left(\sum_{j=1}^{r} \sigma_j^p(\mathbf{X}) \right)^{1/p}, \tag{4}$$

is the Schatten p-norm for matrices, where $\sigma_j(\mathbf{X})$ is the jth largest singular value of $\mathbf{X}$. Note that (3) first maps the tensor into a group of unfolding matrices along each single order, and then calculates the Schatten p-norm of all unfoldings together. It can indeed reflect the structural information from different orders. However, such formulation cannot exploit the dependence among different orders.

To explore the extra dependence structure, (3) is extended by imposing unfoldings along all combinations of orders. Before the definition of the new norm, we first provide a lemma to show how many unfolding matrices can be used by the new norm for a Nth-order tensor.

Lemma 1. *Let $\mathcal{I}$ be the power set of $\{1, , 2, \ldots, N\}$. Construct another set family $\mathcal{J}_N$ in which $\forall \bar{j} \in \mathcal{J}_N$ obeys $\bar{j} = \{\bar{a}, \bar{b}\}$ where $\bar{a} \in \mathcal{I}$, $\bar{b} = \bar{a}^c = \{1, 2, \ldots, N\} \backslash \bar{a}$ and $\bar{a}, \bar{b} \ne \emptyset$. Meanwhile, for $\forall \bar{i} \in \mathcal{I}$ and $\bar{i} \ne \emptyset$, there exists $\bar{j} \in \mathcal{J}_N$ such that $\bar{j} = \{\bar{i}, \bar{i}^c\}$. Then, the order of $\mathcal{J}_N$ is equal to $2^{N-1} - 1$.*

This lemma reveals the fact that we can utilize $2^{N-1} - 1$ unfoldings for any Nth-order tensor. Table 1 shows amount of unfoldings for different orders. For example, there are totally 7 unfoldings for a 4th-order tensor, but (3) can only exploit 4 unfoldings of them. Based on Lemma 1, we give the definition of our new norm as:

Table 1. Number of unfoldings of a tensor along all combination of orders.

Order	1	2	3	4	5	6
Unfoldings	1	2	3	7	15	31
Overlapped Schatten	1	2	3	4	5	6

Definition 1. *Let* $\underline{\mathbf{X}}$ *be a Nth-order tensor, and* $\mathcal{J}_N$ *obeys Lemma 1, then we define the following norm as*

$$\|\underline{\mathbf{X}}\|_{yas_p/q} = \left(\sum_{\bar{j} \in \mathcal{J}_N} \alpha_{\bar{j}} \|\mathbf{X}_{(\bar{j})}\|_{S_p}^q \right)^{1/q}, \tag{5}$$

where $1 \leq p, q \leq \infty$, $\sum_{\bar{j} \in \mathcal{J}_N} \alpha_{\bar{j}} = 1$ *and* $\alpha_{\bar{j}} \geq 0$ *for all* $\bar{j} \in \mathcal{J}_N$,

It is easy to prove that (5) is a norm for the linear space of all Nth-order tensors with specific size:

Proposition 1. *Let* $\underline{\mathbf{X}}, \underline{\mathbf{Y}} \in \mathcal{R}^{I_1 \times I_2 \times \cdots \times I_N}$ *be two Nth-order tensors, and (5) is a function* $\mathcal{R}^{I_1 \times I_2 \times \cdots \times I_N} \to \mathcal{R}$, *then for any* p, q *we have*

1. $\|\underline{\mathbf{X}}\|_{yas_p/q} \geq 0$,
2. $\|\alpha\underline{\mathbf{X}}\|_{yas_p/q} = |\alpha| \|\underline{\mathbf{X}}\|_{yas_p/q}$, *for all* $\alpha \in \mathcal{R}$,
3. $\|\underline{\mathbf{X}} + \underline{\mathbf{Y}}\|_{yas_p/q} \leq \|\underline{\mathbf{X}}\|_{yas_p/q} + \|\underline{\mathbf{Y}}\|_{yas_p/q}$,
4. $\|\underline{\mathbf{X}}\|_{yas_p/q} = 0$ *if and only if all elements of* $\underline{\mathbf{X}}$ *equal zero.*

Note that (5) will degenerate to (2) when the order of tensor is less than or equal to three. But for higher order tensors, (5) imposes more unfolding matrices. Our experiments show that the extra unfolding matrices can provide important information in real applications (e.g. data completion). Figure 2 shows difference between (2) and (5) by diagrams.

One may argue that the proposed norm lead to computational problem when the order of tensor is high, since quantity of unfolding matrices will exponentially increase with the order of tensor. It is worth saying that tuning parameters $\alpha_{\bar{j}}, \forall \bar{j} \in \mathcal{J}_N$ provide an opportunity to control the complexity. Since it is generally unreasonable to assume that all unfoldings are low-rank in practice, we can set several $\alpha_{\bar{j}}$ be zero to "close" the computation of some nuclear norm of unfolding matrices to control computational complexity.

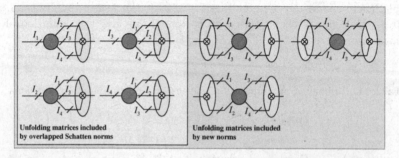

Fig. 2. Comparison between the norm (2) and the proposed one (3). The new norm utilizes unfoldings of all combinations of orders.

2.2 Properties

In matrix theory, Schatten norms have many unique and important properties. For example, all Schatten norms are unitarily invariant, which means that $\|\mathbf{X}\|_{S_p} = \|\mathbf{UXV}\|_{S_p}$ for all matrices $\mathbf{X}$ and all unitary matrices $\mathbf{U}$ and $\mathbf{V}$. In the tensor case, the new norm (3) has the similar properties which is given by

Proposition 2. Let $\underline{\mathbf{X}} \in \mathcal{R}^{I_1 \times \cdots \times I_N}$ be a Nth-order tensor, and $\mathbf{U}_i \in \mathcal{R}^{K_i \times I_i}$, $K_i \geq I_i, i = 1, \ldots, N$ be any semi-orthonormal matrix with orthonormal columns. Assume that $\underline{\mathbf{Y}} = [\![\underline{\mathbf{X}}; \mathbf{U}_1, \ldots, \mathbf{U}_N]\!]$ is tensor-matrix product which is defined in [4]. Then

$$\|\underline{\mathbf{X}}\|_{yas_p/q} = \|\underline{\mathbf{Y}}\|_{yas_p/q} \tag{6}$$

for any $p, q \geq 0$.

For specific p and q, we have the new formulation of tensor spectral norm and nuclear norm. The new spectral norm is defined by

$$\|\underline{\mathbf{X}}\|_{yas_\infty/\infty} = \max_{\bar{j}} \left\{ \frac{1}{\alpha_{\bar{j}}} \|\mathbf{X}_{(\bar{j})}\|_{S_\infty}; \bar{j} \in \mathcal{J}_N \right\}, \tag{7}$$

and the new nuclear norm is defined by

$$\|\underline{\mathbf{X}}\|_\# = \|\underline{\mathbf{X}}\|_{yas_1/1} = \sum_{\bar{j} \in \mathcal{J}_N} \alpha_{\bar{j}} \|\mathbf{X}_{(\bar{j})}\|_{S_1}. \tag{8}$$

Note that (7) is actually the largest singular value of all unfolding matrices, and (8) can be seen as weighted mean of singular values of all unfoldings. Similar to matrix Schatten norm, (7) and (8) provide an upper bound of inner product for any two tensors.

Proposition 3. Let $\underline{\mathbf{X}}, \underline{\mathbf{Y}} \in \mathcal{R}^{I_1 \times I_2 \times \cdots \times I_N}$ be any Nth-order tensor. Then the following inequity always holds

$$\langle \underline{\mathbf{X}}, \underline{\mathbf{Y}} \rangle \leq \|\underline{\mathbf{X}}\|_2 \|\underline{\mathbf{Y}}\|_\#, \tag{9}$$

where $\langle \cdot, \cdot \rangle$ denotes trivial inner product in a linear space. It should be noted that, unlike matrix Schatten norm, the upper bound provided by (9) is not tight enough. In other words, (7) is unfortunately not the dual norm of (8). However, based on the previous work of [11], we can get the dual norm of $\|\underline{\mathbf{X}}\|_\#$ as

Proposition 4. Let $\underline{\mathbf{X}} \in \mathcal{R}^{I_1 \times I_2 \times \cdots \times I_N}$ be a Nth-order tensor. The dual norm of $\|\underline{\mathbf{X}}\|_\#$ is the "latent" spectral norm, which is defined as follows:

$$\|\underline{\mathbf{X}}\|_{\overline{yas_\infty/\infty}} = \inf_{\sum_{\bar{j} \in \mathcal{J}_N} \underline{\mathbf{X}}^{(\bar{j})} = \underline{\mathbf{X}}} \max \left\{ \frac{1}{\alpha_{\bar{j}}} \|\mathbf{X}_{(\bar{j})}^{(\bar{j})}\|_{S_\infty}; \bar{j} \in \mathcal{J}_N \right\}. \tag{10}$$

Here the infimum is taken over the $|\mathcal{J}_N|$-tuple of tensors $\underline{\mathbf{X}}^{(\bar{j})}$ that sums to $\underline{\mathbf{X}}$.

Duality usually plays a key role in optimization problem. For example, in tensor recovery, we usually need to calculate the subgradient of (8). The following proposition gives what conditions the subgradient should obey

Proposition 5. *Let* $\underline{\mathbf{X}}_0 \in \mathcal{R}^{I_1 \times I_2 \times \cdots \times I_N}$ *be a Nth-order tensor, and for each unfolding we have* $\mathbf{X}_{0,(\bar{j})} = \sum_{1 \le k \le r_{\bar{j}}} \sigma_{\bar{j}} \mathbf{u}_{\bar{j},k} \mathbf{v}_{\bar{j},k}^T$. *Then* $\underline{\mathbf{Y}}$ *is a subgradient of* $\|\cdot\|_\#$ *at* $\underline{\mathbf{X}}_0$ *if it is of the form*

$$\underline{\mathbf{Y}} = \sum_{\bar{j} \in \mathcal{J}_N} \alpha_{\bar{j}} ten_{\bar{j}} \left(\sum_{1 \le k \le r_{\bar{j}}} \mathbf{u}_{\bar{j},k} \mathbf{v}_{\bar{j},k}^T + \mathbf{W}_{\bar{j}} \right), \tag{11}$$

where $\mathbf{W}_{\bar{j}}$ *obeys the following two properties:*

- *the column space of* $\mathbf{W}_{\bar{j}}$ *is orthogonal to* $\mathbf{U}_{\bar{j}} \equiv span\left(\mathbf{u}_1, \ldots, \mathbf{u}_{r_{\bar{j}}}\right)$, *and the row space of* $\mathbf{W}_{\bar{j}}$ *is orthogonal to* $\mathbf{V}_{\bar{j}} \equiv span\left(\mathbf{v}_1, \ldots, \mathbf{v}_{r_{\bar{j}}}\right)$ *for all* $\bar{j} \in \mathcal{J}_N$.
- *the spectral norm of* $\mathbf{W}_{\bar{j}}$ *is less than or equal to 1.*

In this proposition, $ten_{\bar{j}}(\cdot)$ denotes tensorlizing the unfolding back to the tensor. There exists relationship between the subgradient $\underline{\mathbf{Y}}$ and the dual norm of $\|\cdot\|_\#$, namely

$$\underline{\mathbf{Y}} \in \left\{ \underline{\mathbf{V}} | \langle \underline{\mathbf{V}}, \underline{\mathbf{X}}_0 \rangle = \|\underline{\mathbf{X}}_0\|_\#, \|\underline{\mathbf{V}}\|_{\overline{yas_\infty/\infty}} \le 1 \right\}. \tag{12}$$

By using (12), we can develop algorithms to minimize the new nuclear norm in tensor recovery problem with low-rankness assumption. since lots of methods have been introduced to efficiently minimize (2), such as [2,6,10], we believe these methods can be straightforwardly extended to the new norm.

2.3 Tensor Recovery

By using the new norm $\|\cdot\|_\#$, we can build a model to recover tensor data with limited measurements. Specifically, let $\mathbf{r} = [r_1, r_2, \ldots, r_N]^T$ denotes multi-linear rank, and let $\mathcal{X}_\mathbf{r}$ denotes an algebraic variety, which is defined by

$$\mathcal{X}_\mathbf{r} = \left\{ \underline{\mathbf{X}} \in \mathcal{R}^{I_1 \times I_2 \times \cdots \times I_N} | rank\left(\mathbf{X}_{(k)}\right) \le r_k, \forall k \right\}. \tag{13}$$

Note that $\mathcal{X}_\mathbf{r}$ denotes the set of all tensors with some low multi-linear rank assumption. Then in order to recover an element $\underline{\mathbf{X}}_0$ of $\mathcal{X}_\mathbf{r}$ from Gaussian measurement G by using the new norm, we need to minimize the following model

$$\min_{\underline{\mathbf{X}}} \|\underline{\mathbf{X}}\|_\#, \quad \mathbf{y} = G\left(\underline{\mathbf{X}}\right), \tag{14}$$

where we set all parameters $\alpha_{\bar{j}} \, \forall \bar{j} \in \mathcal{J}_N$ equals $1/|\mathcal{J}_N|$ for simplicity. It is known in linear inverse problem that to recover $\underline{\mathbf{X}}_0$ from measurements $\mathbf{y}$ is usually ill-posed. But if we achieve sufficient amount of measurements, we can recover $\underline{\mathbf{X}}_0$ uniquely. Based on the results introduced by [1,7], it can be known that how many measurements is necessary for reliable tensor recovery:

Theorem 1. *Let* $\underline{\mathbf{X}}_0 \in \mathcal{X}_{\mathbf{r}}$ *be nonzero. And* $\bar{j} = \{\bar{i}, \bar{i}^c\} \in \mathcal{J}_N$ *is defined as Lemma 1. Assume*

$$\kappa_{\bar{j}} = \frac{\|\mathbf{X}_{0(\bar{j})}\|_{S_1}^2}{\|\mathbf{X}_{0(\bar{j})}\|_{S_2}^2} \max\left\{\prod_{k \in \bar{i}} I_k, \prod_{l \in \bar{i}^c} I_l\right\}, \forall \bar{j} \in \mathcal{J}_N, \tag{15}$$

where

$$\frac{\|\mathbf{X}_{0(\bar{j})}\|_{S_1}^2}{\|\mathbf{X}_{0(\bar{j})}\|_{S_2}^2} \in \left[1, \min\left\{\prod_{k \in \bar{i}} r_k, \prod_{l \in \bar{i}^c} r_l\right\}\right] \tag{16}$$

and set $\kappa = \min_{\bar{j}} \kappa_{\bar{j}}$. *Then if the number of measurement* $m \leq \kappa - 2$, $\underline{\mathbf{X}}_0$ *is not unique solutions to (14), with probability at least* $1 - \exp\left(-\frac{(\kappa-m-2)^2}{16(\kappa-2)}\right)$.

For a special case that $I_1 = I_2 = \ldots = I_N = n$ and $r_1 = r_2 = \ldots = r_N = r$, we have the following corollary

Corollary 1. *There exists* $\underline{\mathbf{X}}_0 \in \mathcal{X}_{\mathbf{r}}$, *for which* $\kappa = r^{\lfloor N/2 \rfloor} n^{\lceil N/2 \rceil}$. *If* $\underline{\mathbf{X}}_0$ *can be decomposed by a rank* r *Canonical Polyadic (CP) model [5], then* $\kappa = rn^{\lceil N/2 \rceil}$.

The corollary shows that only $O\left(r^{\lfloor N/2 \rfloor} n^{\lceil N/2 \rceil}\right)$ Gaussian measurements are necessary for reliable recovery by using the new norm, while $O\left(rn^{N-1}\right)$ is necessary by using conventional norm (2). When $r \ll n$, the new norm lead to less measurements for reliable recovery.

3 Experimental Results

Tensor completion is to estimate missing values from limited observations, which is widely concerned in image/video in-painting, recommender system, relational learning [8], and etc. In the case, we attempt to solve

$$\min_{\underline{\mathbf{X}}} \|\underline{\mathbf{X}}\|_{\#}, \quad s.t. \, P_{\Omega}\left(\underline{\mathbf{X}}\right) = P_{\Omega}\left(\underline{\mathbf{X}}_0\right) \tag{17}$$

Unlike (14), (17) uses sampling operator $P_{\Omega}\left(\cdot\right)$ instead of Gaussian random operator $G\left(\cdot\right)$. However, our simulation results show that the new norm (17) can performs much better performance than using (2).

Since the subgradient of $\|\cdot\|_{\#}$ is given by (12), we can straightforwardly apply Singular Value Threshold (SVT) [2] on the new norm. For comparison, we configure different parameters' value α. For some specific value of α, our norm degenerates to conventional overlapped nuclear norm (2), namely SVT_*, and Square Deal matrix norm (SVT_{SD}). Furthermore, in video in-painting, another two norm-based methods, namely FaLRTC [6] and HardC [10], are also implemented.

In the first experiment, synthetic data is utilized to evaluate completion performance of the new norm. At the beginning, we randomly generate a 5th-order tensor by using rank-$[1,1,1,1,1]$ Tucker model with different data size

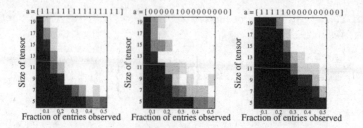

Fig. 3. The colormap indicates the fraction of correct recovery, which increases with brightness from certain failure (black) to certain success (white). The left sub-figure represents that all unfoldings are exploited. The middle sub-figure indicates only the most squared unfolding is utilized (SVT$_{SD}$). The right sub-figure shows that we just consider unfolding along single order (SVT$_*$).

and missing percentage. As Tucker model, random orthonormal factor matrices and Gaussian random core tensor are generated. After that, some entries are randomly removed as missing values according the given percentage. As performance index, we apply Related Square Error ($RSE = \|\underline{\mathbf{X}} - \underline{\mathbf{Y}}\|^2/\|\underline{\mathbf{X}}\|^2$, where $\underline{\mathbf{X}}$ denotes the original tensor and $\underline{\mathbf{Y}}$ is the corresponding estimation.), where smaller value represents better performance. For each simulation, it is assumed that the completion is successful when $RSE \leq 10^{-2}$. Figure 3 gives phase transitions under different conditions where bright blocks indicates higher percentage of successful recovery, where dark blocks indicates opposite results.

In the experiment, we chose three configuration of α. The left sub-figure of Fig. 3 gives the performance using all unfoldings of the data (SVT$_\#$). The middle sub-figure only uses the most squared unfolding (SVT$_{SD}$). The right sub-figure only takes unfoldings along one single order into account which is corresponding conventional overlapped tensor nuclear norm (SVT$_*$). It is shown from Fig. 3 that the method which considers information from all unfoldings outperforms another two conditions where only partial information is utilized.

Next, we consider the influence of recovery performance by different multi-linear rank. Likewise, we randomly generate 4th-order tensor $\underline{\mathbf{X}} \in \mathcal{R}^{20 \times 20 \times 10 \times 10}$ with different multi-linear rank $[x, x, 1, 1]$ where the rank of last two orders is fixed. Figure 4 shows the performance results. It is shown from Fig. 4 that it gives the best performance when $\alpha = [0\,0\,0\,0\,1\,0\,1]$, which is better than SVT$_*$ and SVT$_{SD}$.

In the second experiment, data is selected from three videos. They are talking man, cat, and female with red skirt, respectively. All data is modelled as 4th-order tensors (Height×Width×Channel×Time), and also apply RSE as performance index. In the experiment, a certain percentage of pixels are randomly removed, and set tuning parameters $\alpha_1 = \alpha_2 = \alpha_3 = \alpha_4 = 1$ for SVT$_\#$.

Figure 5 shows the average RSE under 10 independent runs with different missing percentage. It can be seen from Fig. 5 that in some cases SVT$_{SD}$ performs better, but SVT$_\#$ generally outperforms other methods in all three videos under different missing percentage. The bottom right sub-figure of Fig. 5 shows

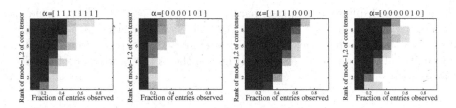

Fig. 4. Phase transition of tensor completion by the new norm with different parameters α. It can be found from the figures that the conventional overlapped Schatten norm (the third sub-figure) cannot generally provide satisfactory performance, and different configuration of α leads to different performance.

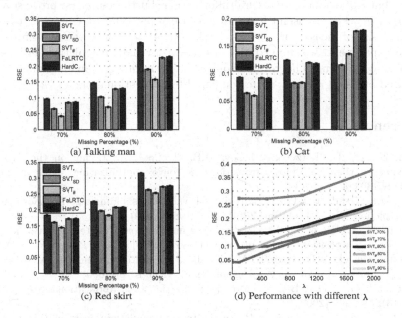

Fig. 5. Comparison between the norm (2) and the proposed one (3). The new norm utilizes unfoldings of all combinations of orders.

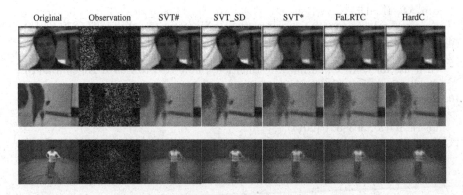

Fig. 6. Performance comparison over three videos.

the influence of different shrinkage parameters λ over $\| \cdot \|_{\#}$ and $\| \cdot \|_*$, which is used in the algorithm. Since $\text{SVT}_{\#}$, SVT_{SD} and SVT_* used the same algorithm, it can be convinced that the proposed Schatten norm can exploit more structural information than conventional norms. Figure 6 shows the estimation of the first frame of three videos.

4   Conclusion

In this paper, a class of new tensor Schatten norms are introduced. The new norm can be considered as an extension of conventional overlapped Schatten norms, but exploit more structural information. Furthermore, we prove several propositions which is useful for tensor recovery problem, and provide a lower bound of size of measurements for reliable recovery by the new Schatten norm. In the experiment, we applied the new norm for data completion and video in-painting problem. The result shows that performance can be significantly improved by using the new norm compared to conventional ones.

References

1. Amelunxen, D., Lotz, M., McCoy, M.B., Tropp, J.A.: Living on the edge: a geometric theory of phase transitions in convex optimization. arXiv preprint arXiv:1303.6672 (2013)
2. Cai, J.F., Cands, E.J., Shen, Z.: A singular value thresholding algorithm for matrix completion. SIAM J. Optim. **20**(4), 1956–1982 (2010)
3. Cichocki, A.: Era of big data processing: a new approach via tensor networks and tensor decompositions. arXiv preprint arXiv:1403.2048 (2014)
4. De Lathauwer, L., De Moor, B., Vandewalle, J.: A multilinear singular value decomposition. SIAM J. Matrix Anal. Appl. **21**(4), 1253–1278 (2000)
5. Kolda, T.G., Bader, B.W.: Tensor decompositions and applications. SIAM Rev. **51**(3), 455–500 (2009)
6. Liu, J., Musialski, P., Wonka, P., Ye, J.: Tensor completion for estimating missing values in visual data. IEEE Trans. Pattern Anal. Mach. Intell. **35**(1), 208–220 (2013)
7. Mu, C., Huang, B., Wright, J., Goldfarb, D.: Square deal: Lower bounds and improved relaxations for tensor recovery. arXiv preprint arXiv:1307.5870 (2013)
8. Nickel, M.: Tensor factorization for relational learning. Ph.D. thesis (2013)
9. Signoretto, M., De Lathauwer, L., Suykens, J.A.: Nuclear norms for tensors and their use for convex multilinear estimation. Submitted to Linear Algebra and Its Applications, vol. 43 (2010)
10. Signoretto, M., Dinh, Q.T., De Lathauwer, L., Suykens, J.A.: Learning with tensors: a framework based on convex optimization and spectral regularization. Mach. Learn. **94**(3), 303–351 (2014)
11. Tomioka, R., Suzuki, T.: Convex tensor decomposition via structured schatten norm regularization. In: Advances in Neural Information Processing Systems, pp. 1331–1339

Memory of Reading Literature in a Hippocampal Network Model Based on Theta Phase Coding

Naoyuki Sato[✉]

Department of Complex and Intelligent Systems,
School of Systems Information Science, Future University Hakodate,
116-2 Kamedanakano-cho, Hakodate-shi, Hokkaido 041-8655, Japan
satonao@fun.ac.jp
http://www.fun.ac.jp/~satonao/

Abstract. Using computer simulations, the authors have demonstrated that temporal compression based on theta phase coding in the hippocampus is essential for the encoding of episodic memory occurring on a behavioral timescale (> a few seconds). In this study, the memory of reading literature was evaluated using a network model based on theta phase coding. Input was derived from an eye movement sequence during reading and each fixated word was encoded by a vector computed from a statistical language model with a large text corpus. The results successfully demonstrated a memory generated by a word sequence during a 6-min reading session and this suggests a general role for theta phase coding in the formation of episodic memory.

Keywords: Neuroscience · Hipppocampus · Neural synchronization · Sequence memory · Episodic memory · Computer simulation

1 Introduction

The hippocampus is known to maintain episodic memory in humans (i.e., part of the declarative memory representing personal events), and its neural mechanisms are a fundamental issue for our understanding of human intelligence. Fortunately, the neural mechanisms operating in the hippocampus are thought to be common to humans and rodents [1], and thus, the computational modeling of episodic memory has been based on evidence from rodents. Yamaguchi [2] proposed a neural coding theory for episodic memory termed theta phase coding, for which the neural dynamics were originally discovered in the rodent hippocampus [3]. According to this theory, cortical input that changes on a behavioral timescale (> a few seconds) is temporally compressed into a sequence represented on a synaptic timescale (~ tens of milliseconds) to produce a stable memory during a behavioral activity (Fig. 1). The authors have successfully demonstrated the importance of theta phase coding in episodic memory formation using a memory sequence that changes on a timescale of tens of seconds [4] and an object–place associative memory encoded by an eye movement sequence [5,6]. However, for the

© Springer International Publishing AG 2016
A. Hirose et al. (Eds.): ICONIP 2016, Part III, LNCS 9949, pp. 61–69, 2016.
DOI: 10.1007/978-3-319-46675-0_7

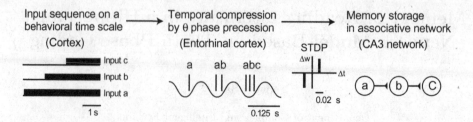

Fig. 1. A theory of memory formation in the hippocampus based on theta phase coding [2].

further understanding of human episodic memory, the model should be developed to include content that is more semantically rich.

A recent brain imaging study by Mitchell et al. [7] revealed that brain activation over distributed cortical regions during the act of seeing words was shown to be correlated with the word features computed using a statistical language model with a large text corpus. From this, we can gain ideas for modeling cortical input, including semantically rich content. Furthermore, the authors recently demonstrated that memory formation during the reading of literature was associated with scalp electroencephalogram (EEG) theta oscillations (4–8 Hz) [8] in accordance with other hippocampus-related memories, such as object–place associative memory [9]. Experience from the reading of literature, which is thought to involve episodic memory, can produce good examples for modeling episodic memory with rich semantic content.

In this study, the ability of theta phase coding to encode a memory was evaluated using the eye movement sequence that occurs during the reading of literature and word encoding computed using the statistical language model with a large text corpus.

2   Computer Simulation

A network model of theta phase coding proposed by the authors [2,4,5] was applied to the eye movement sequence during the reading of a passage of literature, where the parameters of the model were identical to those used previously. The model consists of an input layer, ECII layer (layer II in the entorhinal cortex), and CA3 layer, where each layer consists of 500 word units and 750 sequence units (Fig. 2). The input layer represents a spatio-temporal input pattern with binary values (values 1 and 0 denote activation and resting, respectively). The ECII layer receives one-to-one connections from the input layer and generates a temporally compressed pattern from the input sequence based on theta phase coding. This pattern is transmitted to the CA3 layer and stored into the synaptic strength of recurrent connections among CA3 units based on spike-timing dependent plasticity (STDP).

The computer simulation consisted of encoding and retrieval phases. During encoding, the cortical input sequence determined by eye movements during the

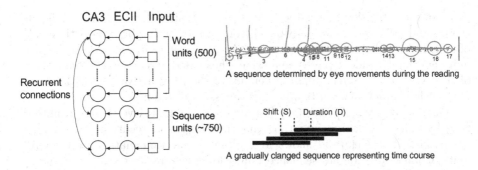

Fig. 2. A network model of the hippocampus based on theta phase coding [4,5] where input for words was included in this study for the first time (see texts for details).

reading of a passage of literature from an experiment (a scientific essay written by T. Terada with the title 'Eagle's eye and olfaction', with 899 words, requiring 360 s of reading, [8]) was given to the network only once. Word units represent the 500-dimensional word features of a fixated word at each time point, and were computed from word-occurrence data within a large text corpus [10] using the algorithm Latent Dirichlet Allocation (LDA). In addition to the word units, sequence units were introduced to represent a gradually changing time course independent of the eye fixations, where one of 750 sequence units randomly selected with an interval of 1 s was activated during 8 s. This interval and duration was given to balance the total activation between the sequence and word units.

During the retrieval phase, an initial activation cue was briefly given to the CA3 network, and the subsequently propagated activation in the CA3 units determined by recurrent connections was evaluated as a retrieval sequence. Its association with the encoded sequence was evaluated using a map of local correlations between the two sequences (defined by the cosine similarity of word unit activations). Additionally, an index for the sequenceness of the retrieval sequence, termed the Sequence Index (S.I.), was calculated as the correlation coefficient for the relationship between the encoding and retrieval times of a set of highly correlated segments in the local correlation map (for which the cosine similarity was >0.5).

3   Results

The input sequence associated with a 360 s reading of literature is shown in Fig. 3a. Sequence units (unit IDs 1–290) were shown to be gradually changed, where the unit ID was ordered according to the activation time for display purposes. Word units (IDs 291–790) were shown to be intermittently activated according to eye fixations on each word. ECII units generated a temporally compressed pattern from the input sequence (Figs. 3b and c). The ECII units showed a periodic pattern of activation while receiving input activation and the phase of the activation relative to a local field potential (LFP) oscillation in the theta band (8 Hz) gradually advanced from late to early phase on a timescale of a second (see activity pattern marked by gray square in Fig. 3c). Each ECII unit shows independent phase precession, thus the difference in the onset time of the input results in the phase difference of the ECII activations within each LFP cycle and actualizes a temporal compression of the input sequence.

The temporally compressed input sequence in the ECII layer was transmitted to CA3 units and stored into the synaptic strength of recurrent connections based on STDP. Figure 3d shows a resultant connection matrix among CA3 units. Connections among sequence units (lower left region) show asymmetry along a diagonal line indicating unidirectional connections associated with the input sequence. Connections from sequence to word units (upper left region) appeared strong in contrast to connections from word to sequence units (lower right region). This kind of asymmetric connection was thought be explained by the difference in their input durations [5]; There were differences in input duration between sequence and word units (8 s for sequence units and typically less than 0.5 s for word units), making the phases of word units earlier than the phases of sequence units on average, and this then results in the formation of asymmetric connections from sequence to word units. Connections among word units (upper right region) also include some asymmetric connections based on a larger-to-shorter duration of input and its detailed structure, to be discussed later.

The memory in the CA3 network was evaluated by retrievals. Figure 4a shows three retrieved patterns in the CA3 network in response to three different initial cue activations. When initial cue activations were applied to both sequence and word units (a1) or only sequence units (a2), the initial activation propagated sequentially following the time series of the encoded sequence. By contrast, initial cue activation applied to only word units (a3) did not evoke any activity propagation. These results indicate that a sequence unit can be sufficient to evoke retrieval sequences, where the unidirectional connections among sequence units are thought to play a dominant role in sequential activation.

The retrieved sequences of word units were compared to the encoded sequence using the cosine similarity of word unit activations. Figure 4b shows a local correlation map between encoded and the retrieval sequences showing a clear correlation between them. Furthermore, the statistical properties of retrievals were tested using a set of different initial cues (Fig. 4c). The results showed that initial cue activation of only sequence units most stably generates clear retrieval sequences in comparison to other initial cue conditions.

The above results showed less dominance by connections among object units in the retrieval phase. However, it is still of interest that asymmetric connections among word units are thought to be associated with their temporal relationships during encoding and those can have a well-organized structure, e.g., a hierarchical spatial structure for object–place associations [5]. Figure 5a shows a network of CA3 word units for which the features were indexed by a set of words, but it was difficult to find any clear semantic structure in it. On the other hand, group

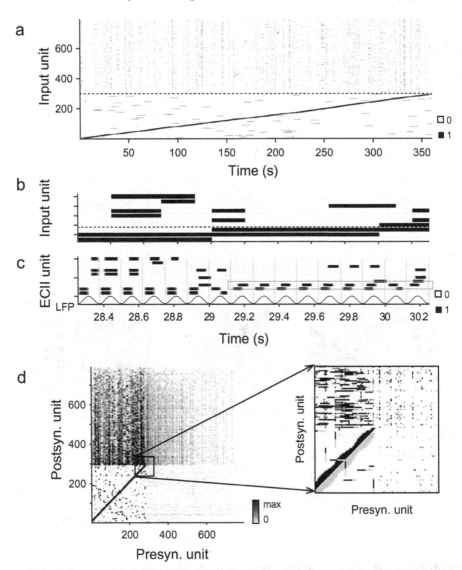

Fig. 3. Results of memory encoding. (a) Input sequence where unit IDs 1–290 denote the sequence units (ordered according to the onset time for their activation) and unit IDs 291–790 denote the word units representing the 500 dimensional features of a fixated word at each time point (the unit ID was ordered according to their total activation). Activations are represented by a gray-scale where white and black indicate resting and activation, respectively. Sequence units without activation during encoding were omitted from the analysis. (b) A part of the input sequence shown in (a). (c) Temporal activation pattern of ECII units for the input shown in (b). Local field potential (LFP) in the theta band (8 Hz) is shown at the bottom, where phases of individual ECII activations to the LFP are gradually advanced from late to early phase (see unit activation marked by gray square). Differences in activation onset times are reflected in the phase difference of the corresponding ECII unit activation that represents a temporally compressed pattern of the input sequence. (c) Synaptic strength of recurrent CA3 connections after one trial of learning the input sequence.

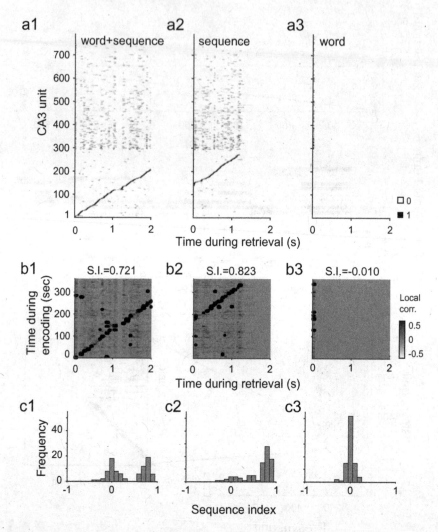

Fig. 4. Results of memory retrieval. (a) Activity propagation in the CA3 network in response to initial cue activation defined by a partial activation to word and sequence units at the beginning of the encoded text (a1), those to only sequence units (a2), and only word units in the middle of the encoded text (a3). Initial cue activation was given briefly during the period 0–0.1 s. (b) Local correlation maps of word unit activations between the encoded and retrieval sequences. Gray and black denote no and high correlations, respectively. The sequence index (S.I.) was calculated from a set of retrieval and encoding times for highly correlated segments (shown in black dots) in the maps. (c) Histogram of S.I. for retrievals initiated by different part of the encoded sequence. The initial cue activation was given to both sequence and word units (c1), only sequence units (c2), or only word units (c3).

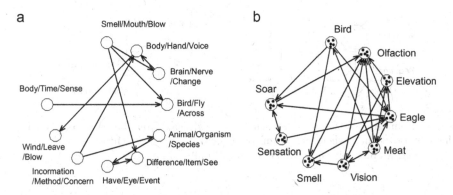

Fig. 5. Network structure of CA3 connections among word units. (a) Network of word units where arrows denote unidirectional connections among them. The word feature of each word unit was represented by a set of words. Nodes for the units were aligned in the order of total activation, clockwise from the top. (b) Network of words with group connections. Nodes for the words were aligned in the order of their frequency, clockwise from the top.

connections for word units were rather shown to be associated with a semantic relationship (Fig. 5b), where unidirectional group connections from 'Bird' to 'Eagle', 'Eagle' to 'Soar' and 'Elevation' were thought to typically represent the order of word appearance during reading. The group connection from the i-th to j-th words (of which word unit activations were given by $\mathbf{v}_i$ and $\mathbf{v}_j$), G_{ij}, was computed by cosine similarity between $\mathbf{v}_j$ and $W\mathbf{v}_i$ where W is the CA3 connection matrix among word units.

4   Discussion

From these results, the model based on theta phase coding was shown to be capable of storing a word sequence defined by the sequence of eye movements during the reading of literature (Figs. 3 and 4) where the gradually changing input (i.e., the input pattern of the sequence units) played a key role in the word sequence memory; the unidirectional connections among sequence units dominantly evoked the sequential activation that further conveyed word unit activations through the unidirectional connections from the sequence to the word units (Fig. 6). This word conveying mechanism showed a new kind of functional contribution by theta phase coding in addition to sequence encoding [2] and the spatial hierarchical encoding of object–place associations [5]. Additionally, the connections among word units were found to have an interesting structure representing their temporal relationship during reading (Fig. 5), but its functional contribution in retrieval was not obvious from the current analysis.

Temporal compression of the retrieval sequence was found to be 180 times that of the encoded sequence. This compression rate was much larger than the physiological evidence obtained from rodent place cells during sleep (20 times

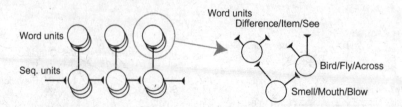

Fig. 6. A mechanism for word sequence retrievals based on the network formed by theta phase coding, where activation of sequence units representing a gradual time course conveys intermittent activation of word units.

for a running sequence in the environment) [11]. In the current model, the speed of retrieval was thought to be a function of the difference in onset time and the duration of activation. These parameters were defined to balance the total activation of word units, but this could be modified to accommodate future experimental evidence. In a future study, however, more sophisticated retrieval mechanisms should be included to simulate human retrieval that can capture the abstract of the literature as a whole, while retaining the temporal structure.

Acknowledgments. This work was supported by JSPS KAKENHI Grant Number 26540069.

References

1. Kahana, M.J., Seelig, D., Madsen, J.R.: Theta returns. Curr. Opin. Neurobiol. **11**(6), 739–744 (2001)
2. Yamaguchi, Y.: A theory of hippocampal memory based on theta phase precession. Biol. Cybern. **89**(1), 1–9 (2003)
3. O'Keefe, J., Recce, M.L.: Phase relationship between hippocampal place units and the EEG theta rhythm. Hippocampus **3**(3), 317–330 (1993)
4. Sato, N., Yamaguchi, Y.: Memory encoding by theta phase precession in the hippocampal network. Neural Comp. **15**, 2379–2397 (2003)
5. Sato, N., Yamaguchi, Y.: Online formation of a hierarchical cognitive map for object-place association by theta phase coding. Hippocampus **15**(7), 963–978 (2005)
6. Sato, N., Yamaguchi, Y.: A computational predictor of human episodic memory based on a theta phase precession network. PloS One **4**, e7536 (2009)
7. Mitchell, T.M., Shinkareva, S.V., Carlson, A., Chang, K.M., Malave, V.L., Mason, R.A., Just, M.A.: Predicting human brain activity associated with the meanings of nouns. Science **320**(5880), 1191–1195 (2008)
8. Sato, N.: Predictability of subsequent retrieval after natural reading of literature: a scalp electroencephalogram study. Program No.171.23. 2015 Neuroscience Meeting Planner. Society for Neuroscience, Chicago (2015)

9. Sato, N., Ozaki, T.J., Someya, Y., Anami, K., Ogawa, S., Mizuhara, H., Yamaguchi, Y.: Subsequent memory-dependent EEG theta correlates to parahippocampal blood oxygenation level-dependent response. Neuroreport **21**(3), 168–172 (2010)

10. Maekawa, K., et al.: Balanced corpus of contemporary written Japanese. Lang. Resour. Eval. **48**, 345–371 (2014)

11. Lee, A.K., Wilson, M.A.: Memory of sequential experience in the hippocampus during slow wave sleep. Neuron **36**(6), 1183–1194 (2002)

Combining Deep Learning and Preference Learning for Object Tracking

Shuchao Pang[1,2], Juan José del Coz[2], Zhezhou Yu[1(✉)], Oscar Luaces[2], and Jorge Díez[2]

[1] College of Computer Science and Technology, Jilin University, Changchun 130012, China
pangshuchao1212@sina.com, yuzz@jlu.edu.cn
[2] Artificial Intelligence Center, University of Oviedo, 33204 Gijón, Spain
{juanjo,oluaces,jdiez}@uniovi.es

Abstract. Object tracking is nowadays a hot topic in computer vision. Generally speaking, its aim is to find a target object in every frame of a video sequence. In order to build a tracking system, this paper proposes to combine two different learning frameworks: deep learning and preference learning. On the one hand, deep learning is used to automatically extract latent features for describing the multi-dimensional raw images. Previous research has shown that deep learning has been successfully applied in different computer vision applications. On the other hand, object tracking can be seen as a ranking problem, in the sense that the regions of an image can be ranked according to their level of overlapping with the target object. Preference learning is used to build the ranking function. The experimental results of our method, called DPL^2 (Deep & Preference Learning), are competitive with respect to the state-of-the-art algorithms.

Keywords: Deep learning · Preference learning · Object tracking

1 Introduction

The goal of object tracking systems is to follow the trajectory of a moving object in a video. This task has attracted substantial research attention because it tackles several interesting applications, including surveillance, vehicle navigation, augmented reality, human-computer interactions and medical imaging, among others [10]. In the past decade, the research of object tracking has made significant progress but it is still a challenging task because a robust tracker must deal with difficult situations in real world environments such as illumination variation or full and partial occlusion of the target object, just to cite a couple of them. In fact, no actual tracker is able to be successful in all possible scenarios and most of them simplify the problem by imposing constraints usually based on prior knowledge about the actual application.

Trackers usually have three main components: object representation including feature selection methods, an object detection mechanism and an update

A. Hirose et al. (Eds.): ICONIP 2016, Part III, LNCS 9949, pp. 70–77, 2016.
DOI: 10.1007/978-3-319-46675-0_8

strategy. Objects can be represented using different characteristics, for instance, their shape, appearance, color, texture, etcetera. These aspects are described by a set of features that can be manually chosen by the user depending on the application domain, automatically using a feature selection algorithm or by the combination of both. The object detection mechanism is responsible for detecting the area in the image occupied by the target object in every frame. Its prediction can be only based on the information of the current frame, but there are also several approaches that take advantage of using the temporal information from previous frames. Due to the unexpected changes in the appearance of the target object, an update strategy is usually required to obtain a robust tracker.

This paper presents a method for object tracking, called DPL^2, that is based on applying two learning frameworks that have been successfully used in several computer vision systems. To the best of our knowledge, they have not been used together in the context of object tracking. Firstly, DPL^2 applies a deep learning architecture to represent the images. Secondly, the model to detect and track the target object is obtained using a supervised ranking algorithm. This selection is motivated by the fact that object tracking is indeed a ranking problem: all the possible bounding boxes of an image could be ranked by their level of overlapping with the target object. Notice that it is not necessary to obtain a perfect total ranking, it is sufficient that the areas very close to the target object rank higher than the rest. Preference learning perfectly fits this goal.

The most related work regarding the use of deep learning in the context of object tracking is [8]. The difference with respect to DPL^2 is that a classifier is used instead of a ranker, thus the representation strategy is the same, but the model follows a completely different approach. In [2] the authors present a tracking system that is based on Laplacian ranking SVM. The tracker incorporates the labeled information of the object in the initial and final frames to deal with drift changes, like occlusion, and the weakly labeled information from the current frame to adapt to substantial changes in appearance. Another difference with our approach is that they use Haar-like features to represent each image patch and we employ a deep learning network. The method described in [3] it is also based on learning to rank, where the authors propose an online learning ranking algorithm in the co-training framework. Two ranking models are built with different types of features and are fused into a semi-supervised learning process. Other references are omitted due to the lack of space.

The rest of the paper is organized as follows. Our proposal based on deep learning and preference learning is presented in Sect. 2. Then, we report a thoroughly experimentation devised to show the benefits of our approach. The paper ends drawing some conclusions.

2   Deep and Preference Learning Tracker

This section is devoted to describe the main components of our proposal. Figure 1 depicts the structure of DPL^2 algorithm that has four main steps. Firstly, according to the position of the target object in the frame, p positive boxes near to the

Fig. 1. The overall framework of our proposed DPL^2 algorithm

target object are selected as positive examples and n negative boxes away from the object as negative examples. Then, DPL^2 uses a deep learning network to extract deep features to describe each example selected in the previous phase. In the third step, DPL^2 applies preference learning to build a ranking model to detect the object. Such model is learned using a set of preferences judgements, formed by all possible pairs between positive and negative examples. Finally, to detect the object in the next frames, DPL^2 uses particle filter to select several particle images around the position of the target object in the last frame. Then, the model outputs the ranking of all particle images; the one that ranks higher is selected as the best position and size of target object in the frame. This last step (number 4 in Fig. 1) is repeated to track the object across each frame of the video sequence. The rest of steps (1–3) are only re-executed when a model update is required as it shall be described in Sect. 2.3.

2.1 Deep Learning

A deep learning method is a representation-learning technique that produces computational models using raw data, obtaining the representation that will be used in the subsequent learning processes. These models, composed of l processing layers ($l > 1$), learn data representations with multiple levels of abstraction; each level transforms the representation at previous level into a more abstract representation [6]. Very complex functions can be learned when l is sufficiently large. The highest layer captures discriminative variations, suppressing the irrelevant ones [11].

In our algorithm we use the SDAE architecture [8] that is composed of two parts, an encoder and a decoder. Each part contains a network, both are usually connected (the last layer of the encoder is the first layer of the decoder). The building block of SDAE is the denoising autoencoder (DAE), a variant of conventional autoencoder. The main property of a DAE network is that it is able to encode the image in a smaller feature space and then recover the original image from the encoded version.

The encoder part is codified by a nonlinear activation function $f_\theta(x)$. The aim of f_θ is to transform the original image vector x_i into a different representation y_i. Here, $\theta = \{W, b\}$ and W denote the weights matrix of the encoder part, whose size depends on the number on layers l and the units of each layer, and b is the bias term. The decoder follows the same formulation and it is described by a function $g_{\theta'}(y_i)$, being $\theta' = \{W', b'\}$, W' the weights matrix of the decoder part and b' its bias term. The goal of the decoder is to map the hidden representation back to a reconstructed input z_i. Obviously, the decoding process is the inverse process of encoding. But the key element is that both networks are trained together trying to minimize the difference between the reconstructed z_i and the original input image x_i. The mathematical formulation is:

$$\min_{W, W', b, b'} \sum_{i=1}^{k} \|x_i - z_i\|_2^2 + \alpha(\|W\|_F^2 + \|W'\|_F^2). \tag{1}$$

Here, the number of original images is k, $y_i = f_\theta(W x_i + b)$, and $z_i = g_{\theta'}(W' y_i + b')$ represents the outputs of both networks (encoder and decoder) for a given example x_i. α is the regularization parameter for trading off between the error and the complexity of both networks, which is measured by means of the Frobenius norm $\|\cdot\|_F$.

Following the training approach described in [8], DPL^2 trains the SDAE model offline. The network architecture used has five layers $l = 5$ with 2560 units in the first hidden layer (overcomplete filters). Then, the number of units is reduced by a half in each layer, so the final layer has just 160 units. A logistic sigmoid activation function is used for f_θ and g_θ.

2.2   Preference Learning

Although there are other approaches to learn preferences, following [1] we will try to induce a real *preference* or *ranking function* $r : \mathbb{R}^d \to \mathbb{R}$, that maps the object from the input space (images represented using SDAE with d=160) to $\mathbb{R}$. r should be learned in such a way that it maximizes the probability of having $r(v) > r(u)$ whenever v is preferable to u $(v \succ u)$. In our case, v is preferable whenever its level of overlapping with the target object is greater than the one of u.

To learn such function r we start from the set of positive and negative examples extracted from the first frame. This set of objects is endowed with a partial order that can be expressed by a collection of preference judgments considering all pairs of positive and negative examples:

$$PJ = \{v_i \succ u_j : i = 1..p, \ j = 1..n\}. \tag{2}$$

In order to induce the ranking function, we look for a function $R : \mathbb{R}^d \times \mathbb{R}^d \to \mathbb{R}$ such that

$$\forall v_i, u_j \in \mathbb{R}^d, R(v_i, u_j) > 0 \Leftrightarrow R(v_i, 0) > R(u_j, 0). \tag{3}$$

Then, the ranking function r can be simply defined by

$$\forall x \in \mathbb{R}^d, r(x) = R(x, 0). \tag{4}$$

Given the set of preference judgments PJ (2), we can specify R by means of the constraints

$$R(v_i, u_j) > 0 \text{ and } R(u_j, v_i) < 0, \quad \forall (v_i, u_j) \in PJ. \tag{5}$$

Therefore, we can apply any binary classifier. DPL^2 uses Support Vector Machines (SVM) [7].

2.3   Model Update

Updating the mechanism to detect the object is crucial for obtaining a robust tracking system [5]. The goal of the update method is to take into account drifting situations that happen along the video sequence, obtaining a final tracking system that is capable of following the object even when there are important variations in the appearance of the object.

DPL^2 updates the model using two different strategies. First, when the description of the predicted region for the current frame is significantly different from the average description of the predicted region of the 10 previous frames. If this situation occurs the preference judgments (2) used to update the model are obtained combining the 10 previous predicted regions (as positive examples) and $n = 30$ negative examples generated randomly. The second rule is even simpler: every t frames the model is updated following the same procedure just described.

3   Experiments

The aim of the experiments reported in this section is to compare the performance of DPL^2 with the best current tracking systems. The selection of such trackers was based on the results of the experimental study recently published in [9]. The best top 14 object tracking algorithms were selected: SCM, Struck, ASLA, L1APG, CT, TLD, IVT, MTT, CSK, MIL, LSK, VTS, VTD and CXT.

In order to fairly compare all the trackers, we used the same datasets, evaluation metrics and the original implementations and results of these trackers reported in [9] but we had to limit the number of experiments due to space restrictions. For the datasets we randomly selected 14 videos of different difficulty degree: hard (Bolt, Coke, Soccer and Woman), middle (David, Deer, Shaking and Trellis) and easy (Car4, CarDark, Fish, Jogging2, Singer1 and Walking). The evaluation method used was OPE (One-Pass Evaluation), that is, the algorithm is initialized with the ground-truth object state in the first frame and the results for the rest of the frames are computed. The scores reported here correspond to two commonly used performance metrics: Success Rate and Precision. Additionally, we also compute Area Under Curve (AUC) scores to measure the overall performance of the trackers.

The parameters used to execute DPL^2 were the following. The number of positive and negative examples were $p = 10$ and $n = 30$ respectively. The examples were generated moving the left top corner of the ground-truth region of the first frame. The positive examples are just moved ± 1 pixels, while in the case of the negatives this distance is always greater than $\pm width/4$ in the X-axis and $\pm height/4$ (Y-axis); $width$ and $heigth$ represent the size of the target object in the first frame. When the model is updated, the same process is applied but instead of using the ground-truth region of each frame, the predicted regions are

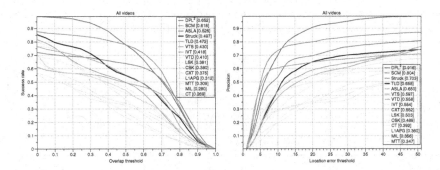

Fig. 2. Average Success Rate (left) and Precision (right) plots of all video frames. Ranks in the legend are computed using AUC of Success Rate and the percentage Precision for 20 pixels

employed. Finally, to detect the object DPL^2 generates randomly 100 particles around the position of the predicted region in previous frame and ranks them; the one that is ranked in first position is returned.

In order to analyze the overall performance of the tracking algorithms considered, we collected together the results of all trackers over the 14 video sequences, a total of 5704 frames, obtaining average values for Success Rate and Precision.

Success Rate measures the overlap rate of each frame. Calculating the percentage of successful frames whose overlap rate is larger than 0.5, we can measure the ability of the trackers to capture most of the area occupied by the target object. However, using just a specific threshold value, it cannot depict the overall performance of each tracker. So we draw the success plot of each tracker by varying the threshold value from 0 to 1. Precision computes the average euclidean distance between the center locations of the predicted regions and the ground-truth positions in all the frames. The Precision percentage is the percentage of frames in which the error of the tracker is less than a given number of pixels. As we can see (Fig. 2), DPL^2 performs quite well for both metrics, its scores are similar to those of SCM, one of the best trackers according to [9]. Although DPL^2 ranks the best for Success Rate when threshold is 0.5, it seems that its performance decays with higher thresholds. Actually, this means that DPL^2 is the best to detect at least part of the object, but it is a little bit worse for capturing a bigger area. This result suggest one of the possible directions to improve DPL^2. Looking at the Precision plots of all videos, we can conclude that when such threshold is smaller than 12 pixels, the Precision scores of SCM and ASLA are slightly better than those of DPL^2. However, when the location error threshold becomes large, DPL^2 clearly outperforms the rest of the trackers. This means that it is a quite robust tracker that rarely losses the track of the object.

In addition, for comprehensively analyzing the overall performance of each tracker from a statistical point of view, Table 1 reports the scores of the Area Under Curve (AUC) of the Success Rate and Precision percentage for all systems

Table 1. Scores and ranking position using two different metrics: AUC of Success Rate (top) and Precision (bottom). The average rank is in the last row of each table

Video	ASLA	LIAPG	CT	TLD	IVT	MTT	CSK	SCM	Struck	MIL	LSK	VTS	VTD	CXT	DPL^2
Bolt	.011(11.5)	.012(10)	.010(14)	.159(5)	.010(14)	.011(11.5)	.019(6)	.016(7.5)	.014(9)	.010(14)	.494(2)	.173(4)	.372(3)	.016(7.5)	.585(1)
Car4	.741(4)	.246(13)	.213(14)	.626(5)	.857(1)	.448(8)	.468(7)	.745(3)	.490(6)	.265(12)	.157(15)	.366(9)	.365(10)	.312(11)	.791(2)
CarDark	.832(3)	.864(2)	.003(15)	.443(13)	.653(10)	.811(7)	.744(8)	.828(4)	.872(1)	.198(14)	.821(5)	.739(9)	.534(12)	.557(11)	.820(6)
Coke	.173(13)	.176(12)	.239(8)	.399(6)	.116(15)	.442(4)	.565(3)	.325(7)	.665(2)	.212(9)	.200(10)	.181(11)	.148(14)	.423(5)	.669(1)
David	.735(1)	.534(10)	.497(11)	.707(3)	.637(5)	.301(14)	.408(13)	.711(2)	.249(15)	.432(12)	.558(8)	.582(7)	.556(9)	.641(4)	.618(6)
Deer	.032(15)	.596(6)	.039(13)	.590(7)	.033(14)	.604(5)	.733(1)	.074(10)	.730(2)	.129(9)	.270(8)	.046(12)	.059(11)	.690(4)	.715(3)
Fish	.833(2)	.349(12)	.705(7)	.796(3)	.758(5)	.181(15)	.222(14)	.736(6)	.840(1)	.451(11)	.313(13)	.684(8.5)	.553(10)	.770(4)	.684(8.5)
Jogging2	.141(8)	.147(6)	.104(15)	.647(3)	.142(7)	.125(13)	.139(9)	.721(2)	.199(5)	.134(10)	.336(4)	.125(13)	.125(13)	.126(11)	.765(1)
Shaking	.465(6)	.081(13)	.109(12)	.394(9)	.037(15)	.049(14)	.572(5)	.680(3)	.356(10)	.430(8)	.459(7)	.698(2)	.700(1)	.126(11)	.652(4)
Singer1	.778(3)	.284(15)	.355(12)	.714(4)	.571(5)	.344(14)	.364(10)	.852(1)	.365(9)	.362(11)	.099(15)	.332(3)	.333(2)	.145(12)	.522(1)
Soccer	.112(14)	.168(8.5)	.128(8.5)	.162(10)	.184(6)	.146(11)	.339(4)	.188(5)	.175(7)	.069(15)	.488(8)	.489(7)	.491(6)	.145(12)	.843(2)
Trellis	.788(1)	.212(15)	.341(11)	.481(8)	.251(12.5)	.220(14)	.479(9)	.665(2.5)	.610(5)	.251(12.5)	.665(2.5)	.494(6)	.451(10)	.649(4)	.488(7)
Walking	.758(1)	.741(3)	.519(12)	.447(14)	.752(2)	.658(6)	.534(11)	.701(4)	.569(6)	.543(10)	.453(13)	.614(7)	.603(8)	.168(15)	.670(5)
Woman	.146(10)	.159(7)	.129(15)	.131(13)	.144(11)	.164(6)	.193(5)	.653(2)	.721(1)	.154(8)	.150(9)	.130(14)	.142(12)	.190(4)	.477(3)
Avg. rank	6.6	9.5	12	7.6	9	9.8	8	4.1	5.7	10.5	8.9	8.1	8.7	7.8	3.6

Video	ASLA	LIAPG	CT	TLD	IVT	MTT	CSK	SCM	Struck	MIL	LSK	VTS	VTD	CXT	DPL^2
Bolt	.017(11)	.017(11)	.011(15)	.306(3.5)	.017(13.5)	.017(11)	.034(6)	.031(7)	.020(9)	.014(13.5)	.977(1)	.089(5)	.306(3.5)	.026(8)	.951(2)
Car4	1.000(2)	.302(13)	.281(14)	.874(6)	1.000(2)	.361(10)	.355(11)	.974(5)	.992(4)	.354(12)	.077(15)	.363(9)	.364(8)	.382(7)	1.000(2)
CarDark	1.000(5)	1.000(5)	.005(15)	.639(13)	.807(10)	1.000(5)	1.000(5)	1.000(5)	1.000(5)	.379(14)	1.000(5)	1.000(5)	.743(11)	.728(12)	1.000(5)
Coke	.165(11)	.265(8)	.113(15)	.684(4)	.131(14)	.660(5)	.873(3)	.430(7)	.948(1)	.151(12)	.258(9)	.189(10)	.148(13)	.653(6)	.928(2)
David	1.000(3)	.805(10)	.815(9)	1.000(3)	1.000(3)	.333(14)	.499(13)	1.000(3)	1.000(3)	.329(15)	.699(11.5)	.962(7)	.943(8)	1.000(3)	.981(6)
Deer	.028(14)	.718(7)	.042(11)	.732(6)	.028(14)	.887(5)	1.000(2)	.028(14)	1.000(2)	.127(9)	.338(8)	.042(11)	.042(11)	1.000(2)	.972(4)
Fish	1.000(3)	.055(13)	.882(7)	1.000(3)	1.000(3)	.042(14.5)	.042(14.5)	.863(8)	1.000(3)	.387(11)	.332(12)	.992(6)	.649(10)	1.000(3)	.811(9)
Jogging2	.182(12)	.186(9)	.166(14)	.857(3)	.199(6)	.173(13)	.186(9)	1.000(1.5)	.254(5)	.186(9)	.544(4)	.186(9)	.186(9)	.163(15)	1.000(1.5)
Shaking	.485(6)	.041(13)	.047(12)	.405(8)	.011(15)	.014(14)	.564(5)	.814(4)	.192(10)	.282(9)	.466(7)	.921(2)	.934(1)	.126(11)	.841(3)
Singer1	1.000(3.5)	.379(14)	.840(9)	1.000(3.5)	.963(9)	.339(15)	.670(10)	1.000(3.5)	.641(11)	.501(12)	.481(13)	1.000(3.5)	1.000(3.5)	.966(7)	1.000(3.5)
Soccer	.122(13)	.207(8)	.219(7)	.115(15)	.173(11)	.184(10)	.135(12)	.268(4)	.253(5)	.191(9)	.481(13)	1.000(3.5)	1.000(3.5)	.232(6)	.798(1)
Trellis	.861(5)	.176(15)	.387(11)	.529(8)	.332(12)	.220(14)	.810(6)	.873(4)	.877(3)	.230(13)	.967(2)	.503(9)	.497(10)	.970(1)	.729(7)
Walking	1.000(6)	1.000(6)	1.000(6)	.964(13)	1.000(6)	1.000(6)	1.000(6)	1.000(6)	1.000(6)	1.000(6)	.658(14)	1.000(6)	1.000(6)	.235(15)	.968(12)
Woman	.203(11.5)	.204(8.5)	.204(8.5)	.191(14)	.201(13)	.204(8.5)	.250(5)	.940(2)	1.000(1)	.206(6)	.204(8.5)	.198(14)	.203(11.5)	.367(4)	.938(5)
Avg. rank	7.6	10	11	7.4	9.3	10.4	7.7	5.3	5.7	10.5	8.9	7	7.8	7.1	4.4

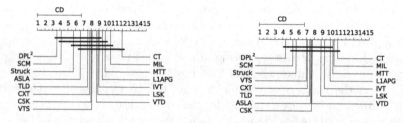

Fig. 3. Comparison using AUC of success rate (left) and precision (right) of all algorithms against each other using the Nemenyi test. Groups of classifiers that are not significantly different at $p = 0.05$ are connected (CD greater than 5.732)

over all videos. To sum up the results, the average rank of each tracker is the last row of each table. Following [4], a two-step statistical test procedure was carried out. The first step consists of a Friedman test of the null hypothesis that states that the trackers perform equally. Such hypothesis is rejected. Then, a Nemenyi test is performed to compare the methods in a pairwise way. Both tests are based on average ranks. The comparison includes 15 trackers over 14 videos, so the critical difference (CD) in the Nemenyi test is 5.732 for significance level of 5 %. The results are in Fig. 3. Notice that DPL^2 ranks higher in both cases. Moreover, only DPL^2 and SCM are significantly better according to a Nemenyi test than the worst trackers (MIL and CT). However, as we can observe in Fig. 3, most of the differences are not statistically significant. This is due in part to the fact that the number of algorithms compared is greater than the number of video sequences, so the critical difference 5.7 is quite difficult to reach. In any case, the overall results discussed in this section are quite promising, supporting the main hypothesis of this work which is that preference learning and deep learning are both well-tailored to tackle object tracking tasks.

4 Conclusions

This paper analyzes the behavior of two very popular techniques, deep learning and preferences learning, working together in the context of object tracking. The performance of the proposed method, DPL^2 is quite competitive with respect to state-of-the-art methods, and sometimes it is even better, both from a quantitative and qualitative point of view. However, one of the most interesting aspects of this study is that it seems that there is still room to improve the accuracy of a tracking system based on the combination of deep learning and preference learning. Moreover, we have introduced a new point of view to approach object tracking tasks: the use of preference learning. The application of this paradigm opens up a new research line that can eventually be explored in the future.

Acknowledgments. This work was funded by Ministerio de Economía y Competitividad de España (grant TIN2015-65069-C2-2-R), Specialized Research Fund for the Doctoral Program of Higher Education of China (grant 20120061110045) and the Project of Science and Technology Development Plan of Jilin Province, China (grant 20150204007GX). The paper was written while Shuchao Pang was visiting the University of Oviedo at Gijón.

References

1. Bahamonde, A., Bayón, G.F., Díez, J., Quevedo, J.R., Luaces, O., del Coz, J.J., Alonso, J., Goyache, F.: Feature subset selection for learning preferences: a case study. In: ACM ICML (2004)
2. Bai, Y., Tang, M.: Robust tracking via weakly supervised ranking svm. In: IEEE CCVPR (2012)
3. Dai, P., Liu, K., Xie, Y., Li, C.: Online co-training ranking svm for visual tracking. In: IEEE ICASSP (2014)
4. Demšar, J.: Statistical comparisons of classifiers over multiple data sets. JMLR **7**, 1–30 (2006)
5. Jepson, A.D., Fleet, D.J., El-Maraghi, T.F.: Robust online appearance models for visual tracking. IEEE PAMI **25**(10), 1296–1311 (2003)
6. LeCun, Y., Bengio, Y., Hinton, G.: Deep learning. Nature **521**(7553), 436–444 (2015)
7. Vapnik, V.: Statistical Learning Theory. Wiley, New York (1998)
8. Wang, N., Yeung, D.Y.: Learning a deep compact image representation for visual tracking. In: NIPS (2013)
9. Wu, Y., Lim, J., Yang, M.: Object tracking benchmark. IEEE PAMI **37**(9), 1834–1848 (2015)
10. Yilmaz, A., Javed, O., Shah, M.: Object tracking: a survey. ACM Comput. Surv. (CSUR) **38**(4), 13 (2006)
11. Zhao, L., Hu, Q., Zhou, Y.: Heterogeneous features integration via semi-supervised multi-modal deep networks. In: Arik, S., Huang, T., Lai, W.K., Liu, Q. (eds.) ICONIP 2015. LNCS, vol. 9492, pp. 11–19. Springer, Heidelberg (2015). doi:10.1007/978-3-319-26561-2_2

A Cost-Sensitive Learning Strategy for Feature Extraction from Imbalanced Data

Ali Braytee[1], Wei Liu[2(⊠)], and Paul Kennedy[1]

[1] Quantum Computation and Intelligent Systems, School of Software,
University of Technology Sydney, Sydney, NSW 2007, Australia
{Ali.Braytee,Paul.Kennedy}@uts.edu.au
[2] Advanced Analytics Institute, University of Technology Sydney,
Sydney, NSW 2007, Australia
Wei.Liu@uts.edu.au

Abstract. In this paper, novel cost-sensitive principal component analysis (CSPCA) and cost-sensitive non-negative matrix factorization (CSNMF) methods are proposed for handling the problem of feature extraction from imbalanced data. The presence of highly imbalanced data misleads existing feature extraction techniques to produce biased features, which results in poor classification performance especially for the minor class problem. To solve this problem, we propose a cost-sensitive learning strategy for feature extraction techniques that uses the imbalance ratio of classes to discount the majority samples. This strategy is adapted to the popular feature extraction methods such as PCA and NMF. The main advantage of the proposed methods is that they are able to lessen the inherent bias of the extracted features to the majority class in existing PCA and NMF algorithms. Experiments on twelve public datasets with different levels of imbalance ratios show that the proposed methods outperformed the state-of-the-art methods on multiple classifiers.

1 Introduction

The class imbalance issue is caused by unequal distributions of the data between class labels [3]. It occurs due to a paucity of cases, for example, patients with a rare disease, or difficulties in collecting samples due to high cost or privacy. The imbalanced class is considered as a crucial issue in machine learning and data mining. Learning from an imbalanced dataset leads to poor classification, because classical data mining algorithms tend to favor classifying examples as belonging to the majority class (negative class).

Moreover, when doing feature extraction from imbalanced data, the extracted features are biased to predict the majority class samples [12]. Principal component analysis (PCA) [6] and non-negative matrix factorization (NMF) [7] are very well-known feature extraction methods. The unsupervised PCA algorithm seeks the orthogonal feature extractors that maximize the total variance. Therefore, the extracted features favors majority class because there number is much

© Springer International Publishing AG 2016
A. Hirose et al. (Eds.): ICONIP 2016, Part III, LNCS 9949, pp. 78–86, 2016.
DOI: 10.1007/978-3-319-46675-0_9

more than the minority class. NMF has recently been shown to be a very effective matrix factorization technique in approximating the high dimensional data [8]. It is a vector space method that uses matrix factorization to find two non-negative reduced-dimension matrices W and H [10]. The factorized matrices W and H will be affected by the imbalance problem and the basis matrix W will be biased to represent the majority class samples, because the magnitude of the residual squared error of negative class instances is much more than the positive ones [11].

To our knowledge, there is no cost-sensitive learning approach to solve the imbalance problem in the supervised feature extraction techniques, especially for NMF and PCA. In this paper, we propose a cost-sensitive approach for classical PCA and NMF feature extractions which is able to solve the imbalanced class problem without modifying the existing base classifiers, or changing the original information of the training datasets. Specifically, **the contributions** of this paper are:

1. We propose an effective cost-sensitive strategy that can improve general feature extraction methods for imbalanced data.
2. We apply the cost-sensitive learning strategy to the popular classical feature extraction methods PCA and NMF to handle the imbalance class problem.
3. We provide comprehensive evaluations of our methods on many real-world imbalanced datasets which show the advantages of our methods.

This rest of this paper is organised as follows. Section 2 describes a motivating example. Section 3 presents the theoretical analysis. Section 4 reports experiments before concluding in Sect. 5.

2   A Motivating Example

A training dataset is comprised of instances $\mathbf{X} \in \mathbb{R}^{n \times m}$ with m features. When the data is imbalanced, the number of majority instances greatly outnumbers

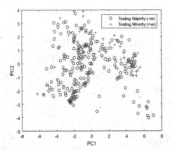

(a) Testing data after projecting the instances using PCA

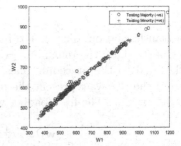

(b) Testing data after projecting the instances using NMF factors

Fig. 1. Applying PCA and NMF on imbalanced Vehicle data leads to overlapping problem

the instances of minority class. In this section, a vehicle dataset is used and we postpone the description of the characteristics of this dataset to Table 1.

Figures 1a and 1b show the distribution of the testing samples on original unbalanced data after a classical PCA and NMF is applied on it respectively. It clearly shows that the classifiers will find it very difficult to distinguish between classes due to the overlap class problem.

3   Theoretical Analysis

In this section, we present the theoretical analysis of our proposed cost-sensitive PCA (CSPCA) and cost-sensitive NMF (CSNMF).

3.1   Imbalance Cost Ratio

Different cost ratios are used in all training examples for the majority and minority classes:

$$C_i = \begin{cases} C_i^- = \frac{(1-\alpha)}{N_-}, \textbf{If } y_i = -1, \text{(Negative class)} \\ C_i^+ = \frac{(1+\alpha)}{N_+}, \textbf{If } y_i = +1, \text{(Positive class)} \end{cases} \tag{1}$$

for $i = 1, ..., N$ samples, N_- and N_+ are the total number of negative and positive samples respectively. $0 \leqslant \alpha < 1$ is a parameter to weight the majority class. If $\alpha = 0$, the majority class is weighted by the ratio of the two class sizes in training data. In the case, where the α value gets close to 1, it gradually represents the learning from the positive examples only.

3.2   Cost-Sensitive Principal Component Analysis (CSPCA)

Principal component analysis (PCA) [6] is one of the most popular feature extraction techniques. It is defined as a statistical procedure that uses an orthogonal transformation to construct a low-dimensional representation of the data known as principal components. The linear transformation aims to maximize the global variance of the data as well as to minimize the least square error of the transformation. The first principal component represents the largest variance of the data; the remaining principal components have smaller variance and orthogonal to the preceding ones. Consider a data matrix $\mathbf{X} \in \mathbb{R}^{n \times m}$, where each of the n rows represent the instances or observations, and m columns are the dimensions. The first loading $\mathbf{w}_1$ is computed by:

$$\mathbf{w}_1 = \arg\max \sum_{ij} (\mathbf{X}_{ij} \cdot \mathbf{w}_{1j})^2 \tag{2}$$

where i and j are the index of rows and columns of $\mathbf{X}$ respectively, $\mathbf{w}_1$ is the first principal component of p dimensions and $p << m$. In the case of the class imbalance issue, the spread of data is dominated by the majority samples, because when the directions of principal axes (components) of both classes are different, the reduced space found by PCA represents the majority space and underrepresents the minority one.

Geometrically, the first step in PCA is to centre the data by subtracting the mean of the data from all points. However, in the case of imbalance class, the global mean may be shifted to the majority samples space. Moreover, PCA computes the covariance matrix of the data which captures the variance of the dataset. But, in the case of highly skewed data, the covariance matrix mostly represents the variance of majority class samples, and the largest variance direction of the data may be captured mostly from the majority space.

Therefore, we propose a cost-sensitive PCA technique (CSPCA) to improve the computations of the principal components with consideration of the imbalanced class issue. In the binary case, assume that the negative and positive samples are discounted by imbalance cost ratio C^- and C^+ respectively. The weighted first principal component becomes:

$$\mathbf{w}_1 = \arg\max \sum_{i:y_i=-1,j} \left(C_i^- \mathbf{X}_{ij} \cdot \mathbf{w}_{1j}\right)^2 + \sum_{i:y_i=+1,j} \left(C_i^+ \mathbf{X}_{ij} \cdot \mathbf{w}_{1j}\right)^2 \tag{3}$$

where C_i^- and C_i^+ from Eq. (1), and j is the dimension index.

3.3   Cost-Sensitive Non-negative Matrix Factorization (CSNMF)

Non-negative Matrix Factorization (NMF) [7] is a matrix factorization technique under the constraint that the values of the input matrix are non-negative. NMF can be described by the following factorization form

$$X_{n \times m} \simeq W_{n \times p} H_{m \times p}^T \tag{4}$$

where n is the number of observations, m is the dimension of the data, p is the desired rank such that $p < \min(m, n)$ and $X \in \mathbb{R}^{+n \times m}$, $W \in \mathbb{R}^{+n \times p}$, $H \in \mathbb{R}^{+p \times m}$.

To find the approximate matrix factorization (4), an optimization function is defined by [7] to *minimize* $\|X - WH\|^2$ with respect to W and H.

In the case of imbalanced data, we propose a cost-sensitive NMF (CSNMF), which injecting the classical unsupervised NMF with labelling information to take into consideration the imbalance class problem.

Our CSNMF function modifies the original matrix to alleviate the effectiveness of the negative samples, a new matrix X' is defined by

$$X' = \begin{bmatrix} \left[C_i^- X_i\right], \textbf{If } y_i = -1 \\ \left[C_i^+ X_i\right], \textbf{If } y_i = +1 \end{bmatrix} \tag{5}$$

where C^- and C^+ is defined in (1), $X' \in \mathbb{R}^{n \times m}$. CSNMF aims to find two non-negative matrices $W = [w_{ip}] \in \mathbb{R}^{n \times p}$ and $H = [h_{jp}] \in \mathbb{R}^{m \times p}$ whose products can estimate the balanced matrix X'. The objective function is the Euclidean distance between two matrices, it can be written as:

$$O = \left\|X' - WH^T\right\|^2 = \left(X' - WH^T\right)\left(X' - WH^T\right)^T \tag{6}$$

The objective function O is convex with respect to W and H separately. Lee and Seung [7] proposed iterative multiplicative update rules to minimise the error of O in Eq. (6):

$$h_{jp} \leftarrow h_{jp} \frac{(X'W)_{jp}}{(HW^TW)_{jp}} w_{ip} \leftarrow w_{ip} \frac{\left(X'^T H\right)_{ip}}{(WH^TH)_{ip}} \tag{7}$$

The convergence of the objective function with X' is the same as the objective with X proof in [7].

3.4 Revisiting the Motivating Example

We visualize the classification improvements after applying our two proposed methods CSPCA and CSNMF on the same dataset that we used in Sect. 2. Figure 2 shows how the classes can be better distinguished compared to Fig. 1.

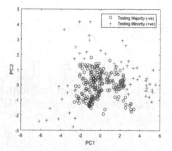

(a) Testing data after projecting the instances using CSPCA

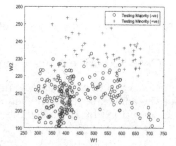

(b) Testing data after projecting the instances using CSNMF factors

Fig. 2. Applying CSPCA and CSNMF on imbalanced Vehicle dataset leads to improve the classification performance.

4 Experiments and Analysis

In this section, we analyse and compare the performance of CSPCA and CSNMF against classical PCA and NMF, existing ADASYN algorithm [5], random undersampling (RU), maximum likelihood cost-sensitive (ML-CST) [4] and CCPDT [9]. We will make use of SVM and kNN as instance based learning approach.

4.1 Experimental Framework

We choose the number of principal components which represent the 90 % of the original data, and the desired rank for $CSNMF$ equal the number of class labels.

Table 1. Characteristics of the data sets. Column #IR is the imbalance ratio (i.e., Neg/Pos)

Dataset	#Instance	#Attributes	#IR	Dataset	#Instances	#Attributes	#IR
Diabetes	768	20	1.86	Blood	748	5	3.2
Glass	214	9	6.38	Colon	62	2000	1.82
Yeast	1484	8	28.1	Survival	306	3	2.77
Spambase	4601	57	1.53	Adeno	86	76	5.34
Breast	198	34	3.21	Ecoli	336	7	8.6
Vehicle	846	18	2.99	Abalone	4174	8	129.43

Also, we set balance factor $\alpha = 0$ for imbalance cost ratio C in (1) to represent equal class proportions in training set.

We evaluated the methods on 12 datasets each with two classes. These datasets are selected from the UCI repository [2] and KEEL-datasets [1]. Details of the data sets are shown in Table 1. The AUC metric is used to measure the accuracy in the experiments, which is widely used metric for evaluations on imbalanced data. Estimates of AUC was averaged over 5-fold cross-validation.

4.2 Analysis and Results

In this section, we analyse the behaviour of classical PCA and NMF with existing algorithms that proposed to handle the imbalanced class issue against our proposed CSPCA and CSNMF methods. Firstly, the state-of-the-art methods are composed of two methods from data sampling techniques: ADASYN [5] and random undersampling, one method from cost-sensitive level: ML-CST [4], and one method from algorithm level: CCPDT [9].

The obtained results in Tables 1 through 5 show that the performance of the classifiers substantially improves for the cost-sensitive version of PCA and NMF. Moreover, the proposed CSPCA and CSNMF with base classifiers outperformed the classical PCA and NMF with the state-of-the-art methods. Therefore, our proposed method can solve the imbalanced class issue at feature extraction level without the need of changing the data distribution or modifying the existing algorithms.

We conduct t-tests between vector results of our methods (Base) against the compared methods for each classifier, under the null hypothesis that the AUC on vectors of the used methods is not significantly different. As shown on the bottom line of Tables 1 through 5, the *p-values* reject the null hypothesis, as most values of our methods (base) are lower than 0.01. This indicates that the proposed methods CSPCA and CSNMF have significantly improved the performance of the classifiers.

We analyse the effects of our CSPCA and CSNMF algorithm on the base classifiers, and we compare it with baseline and existing algorithms that proposed for the imbalanced class problem. Firstly, we apply the data sampling algorithms on the training sets to balance the datasets using ADASYN oversampling method and undersampling method. Then, we apply the PCA and NMF on the training

Table 2. CSPCA classification performance compared to PCA and ADASYN-PCA.

Dataset	PCA	ADASYN - PCA	RU - PCA	CSPCA
Diabetes	.695	.706	.576	**.762**
Spambase	.869	.868	.872	**.904**
Breast	.562	.586	.556	**.620**
Ecoli	.860	.828	.857	**.879**
Abalone	.669	.627	.666	**.683**
Survival	.607	.546	.536	**.611**
Blood	.676	.676	**.677**	.660
Glass	.945	.933	.900	**.951**
Yeast	.850	.831	.839	**.871**
Vehicle	.634	.620	.628	**.681**
Colon	.512	.490	.517	**.531**
Adeno	.579	.525	.497	**.610**
t-test	3.4×10	9.7×10	5.3×10	Base

(a) SVM Performance

Dataset	PCA	ADASYN - PCA	RU - PCA	CSPCA
Diabetes	.632	.633	.536	**.671**
Spambase	.855	.828	.848	**.862**
Breast	.550	.590	.495	**.620**
Ecoli	.580	.823	.844	**.856**
Abalone	.500	.573	.621	**.641**
Survival	.509	.481	.458	**.570**
Blood	.609	.626	.623	**.681**
Glass	.930	.850	.925	**.941**
Yeast	.535	.721	.834	**.880**
Vehicle	.546	.581	**.613**	.600
Colon	.600	.472	.615	**.631**
Adeno	.476	.589	.503	**.591**
t-test	6.7×10	1.1×10	4.2×10	Base

(b) 5-NN Performance

Table 3. CSNMF classification performance compared to NMF, NMF-ADASYN and NMF-RU.

Dataset	NMF	ADASYN - NMF	RU - NMF	CSNMF
Diabetes	.695	.699	.691	**.721**
Spambase	.627	.665	.636	**.825**
Breast	.573	.484	.516	**.597**
Ecoli	.868	.855	.840	**.869**
Abalone	.674	.577	.670	**.693**
Survival	.577	.523	.533	**.621**
Blood	.679	.566	.559	**.756**
Glass	.911	.832	.790	**.925**
Yeast	.844	.833	.817	**.890**
Vehicle	.645	.637	.600	**.740**
Colon	.752	.772	.752	**.798**
Adeno	.537	.620	.613	**.723**
t-test	5.5×10	1.1×10	1.9×10	Base

(a) SVM Performance

Dataset	NMF	ADASYN - NMF	RU - NMF	CSNMF
Diabetes	.631	.643	.658	**.700**
Spambase	.722	.724	.723	**.800**
Breast	.520	.500	.519	**.591**
Ecoli	.733	.831	.850	**.880**
Abalone	.500	.616	.682	**.691**
Survival	.530	.563	.535	**.591**
Blood	.598	.628	.615	**.674**
Glass	.788	.800	.782	.791
Yeast	.497	.789	.795	**.821**
Vehicle	.591	.581	.647	**.691**
Colon	.777	.727	.672	**.791**
Adeno	.460	.570	.586	**.630**
t-test	1.3×10	5.9×10	1.9×10	Base

(b) 5-NN Performance

sets of imbalanced datasets to construct the extracted features. The projected test data is classified using the base classifiers such as SVM and kNN. Secondly, we apply our proposed CSPCA and CSNMF on the training sets of the imbalanced datasets. Then, we use the same base classifiers as in the first case on the projected test data. Tables 2 and 3 show the preference of our CSPCA and CSNMF method over the existing data sampling techniques. The best average values per approach are highlighted in bold.

On the other hand, we compare our CSPCA and CSNMF method with the ML-CST using logistic regression as a base classifier. Then, we conduct another experiment, to compare CSPCA and CSNMF with CCPDT using decision trees as a base classifier, and in both cases our method outperformed the above two methods. Tables 4 and 5 show the quality of using our proposed solution for applying feature extraction on imbalanced datasets, as there is a significant difference between the results of our methods and the standard PCA, NMF and the other compared algorithms. One may also observe the generalization of our new method by improving the performance of classification on a set of the well-known classifiers.

Table 4. CSPCA and CSNMF on logistic regression classifier compared to ML-CST

Logit	PCA-MLCST	CSPCA	NMF-MLCST	CSNMF
Diabetes	.604	**.640**	.625	**.680**
Spambase	.729	**.810**	.716	**.720**
Breast	.523	**.656**	.500	**.580**
Ecoli	.541	**.728**	.500	**.610**
Abalone	.499	**.660**	.500	**.730**
Survival	.543	**.610**	.500	**.690**
Blood	.510	**.610**	.500	**.581**
Glass	**.915**	.660	.500	**.590**
Yeast	.496	**.574**	.500	**.570**
Vehicle	.511	**.634**	.630	**.691**
Colon	.590	**.664**	.602	**.642**
Adeno	.510	**.610**	.500	**.650**
t-test	4.3×10^{-2}	Base	2.9×10^{-4}	Base

Table 5. CSPCA and CSNMF CART classification performance compared to CCPDT

Decision Trees	PCA-CCPDT	CSPCA-DT	NMF-CCPDT	CSNMF-DT
Diabetes	.676	**.689**	.704	**.750**
Spambase	.918	**.836**	.756	**.770**
Breast	.487	**.591**	.487	**.630**
Ecoli	.498	**.852**	.498	**.890**
Abalone	.481	**.615**	.481	**.651**
Survival	.492	**.600**	.492	**.680**
Blood	.721	**.641**	.552	**.710**
Glass	.854	**.891**	.767	**.830**
Yeast	.491	**.812**	.634	**.771**
Vehicle	.652	.625	**.671**	.651
Colon	.472	**.630**	.528	**.751**
Adeno	.433	**.565**	.443	**.620**
t-test	3.3×10^{-2}	Base	9.5×10^{-4}	Base

5 Conclusions and Future Work

This paper proposes a cost-sensitive learning strategy to address the imbalanced class problem that can be applied to well-known feature extraction techniques such as PCA and NMF. We have integrated the cost-sensitive strategy in PCA and NMF and proposed two new methods CSPCA and CSNMF. Our proposed method embeds the labelling information in the classical feature extraction methods to extract the balanced features which improve the accuracy of classification and reduce the overlapping between the classes. Our results show the high-performing quality of our proposed methods on multiple popular classifiers and benchmark datasets. They can deal with different levels of imbalance and sizes of the datasets. In future, we will extend the idea to multi-label classification problems. We also plan to adapt the strategy to other feature extraction techniques.

References

1. Alcalá, J., Fernández, A., et al.: Keel data-mining software tool: data set repository. J. Multiple-Valued Logic Soft Comput. **17**(2–3), 255–287 (2010)
2. Asuncion, A., Newman, D.: Uci machine learning repository (2007)
3. Chawla, N., Japkowicz, N., Kolcz, A.: Special issue on learning from imbalanced datasets, sigkdd explorations. In: ACM SIGKDD (2004)
4. Dmochowski, J.P., Sajda, P., Parra, L.C.: Maximum likelihood in cost-sensitive learning: model specification, approximations, and upper bounds. J. Mach. Learn. Res. **11**, 3313–3332 (2010)
5. He, H., Bai, Y., et al.: Adasyn: adaptive synthetic sampling approach for imbalanced learning. In: IEEE International Joint Conference on Neural Networks, IJCNN 2008, pp. 1322–1328. IEEE (2008)
6. Kirby, M., Sirovich, L.: Application of the karhunen-loeve procedure for the characterization of human faces. IEEE Trans. Pattern Anal. Mach. Intell. **12**(1), 103–108 (1990)
7. Lee, D.D., Seung, H.S.: Algorithms for non-negative matrix factorization. In: Advances in Neural Information Processing Systems, pp. 556–562 (2001)

8. Liu, W., Chan, J., Bailey, J., Leckie, C., Ramamohanarao, K.: Mining labelled tensors by discovering both their common and discriminative subspaces. In: SIAM International Conference on Data Mining (SDM13), pp. 614–622 (2013)

9. Liu, W., Chawla, S., et al.: A robust decision tree algorithm for imbalanced data sets. In: SDM, vol. 10, pp. 766–777. SIAM (2010)

10. Liu, W., Kan, A., Chan, J., Bailey, J., Leckie, C., Pei, J., Kotagiri, R.: On compressing weighted time-evolving graphs. In: Proceedings of the 21st ACM International Conference on Information and Knowledge Management (CIKM 2012), pp. 2319–2322 (2012)

11. Ristanoski, G., Liu, W., Bailey, J.: Discrimination aware classification for imbalanced datasets. In: Proceedings of the 22nd ACM International Conference on Information and Knowledge Management (CIKM 2013), pp. 1529–1532 (2013)

12. Wang, J., You, J., et al.: Extract minimum positive and maximum negative features for imbalanced binary classification. Pattern Recogn. $45(3)$, 1136–1145 (2012)

Nonnegative Tensor Train Decompositions for Multi-domain Feature Extraction and Clustering

Namgil Lee[1](✉), Anh-Huy Phan[1], Fengyu Cong[2,3], and Andrzej Cichocki[1]

[1] Laboratory for Advanced Brain Signal Processing,
RIKEN Brain Science Institute, Wako, Saitama 3510198, Japan
namgil.lee@riken.jp, {phan,cia}@brain.riken.jp
[2] Faculty of Electronic Information and Electrical Engineering,
Department of Biomedical Engineering, Dalian University of Technology,
Dalian 116024, China
cong@dlut.edu.cn
[3] Department of Mathematical Information Technology,
University of Jyväskylä, Jyväskylä, Finland

Abstract. Tensor train (TT) is one of the modern tensor decomposition models for low-rank approximation of high-order tensors. For nonnegative multiway array data analysis, we propose a nonnegative TT (NTT) decomposition algorithm for the NTT model and a hybrid model called the NTT-Tucker model. By employing the hierarchical alternating least squares approach, each fiber vector of core tensors is optimized efficiently at each iteration. We compared the performances of the proposed method with a standard nonnegative Tucker decomposition (NTD) algorithm by using benchmark data sets including event-related potential data and facial image data in multi-domain feature extraction and clustering tasks. It is illustrated that the proposed algorithm extracts physically meaningful features with relatively low storage and computational costs compared to the standard NTD model.

Keywords: EEG · Feature extraction · HALS · Tucker decomposition

1 Introduction

Modern real world data are nowadays often produced in form of multiway arrays, i.e., tensors, requiring to investigate multi-modality and latent low-rank structure in the data. The tensor train (TT) decomposition is a relatively new low-parametric representation of tensors which was introduced for numerical computation with large-scale high-order tensors [9]. The main advantage of the TT is to avoid an exponential rate of growth in computational and storage costs with the order of a tensor. However, the TT has limited applications to data mining yet, especially to signal processing due to lack of interpretability. In this sense, the nonnegative tensor train (NTT) is aimed at both alleviating the curse-of-dimensionality and imposing nonnegativity constraints on its components to enhance interpretability.

© Springer International Publishing AG 2016
A. Hirose et al. (Eds.): ICONIP 2016, Part III, LNCS 9949, pp. 87–95, 2016.
DOI: 10.1007/978-3-319-46675-0_10

On the other hand, traditional tensor decomposition models such as the Candecomp/Parafac (CP) and Tucker models have been widely used in feature extraction for signal processing [2,7]. By incorporating the nonnegativity constraints on their core tensors and factor matrices, the nonnegative CP decomposition and the nonnegative Tucker decomposition (NTD) based data analysis methods have been developed and applied to various real world data in signal processing and data mining; see, e.g., [3,10].

In this paper, we propose a fast NTT decomposition algorithm, which is called the NTT-HALS, for estimating the NTT model and a hybrid model called the NTT-Tucker [5]. The proposed NTT-HALS algorithm is developed based on the alternating linear scheme for TT decomposition [6] and the hierarchical alternating least squares approach for nonnegative matrix/tensor factorization [1,2,10]. In the NTT-HALS, each fiber vectors of core tensors are optimized efficiently without computing any time-consuming matrix inversion.

It is demonstrated that the proposed NTT-Tucker model is a promising alternative to the NTD model by experimental results using benchmark data sets. Through the multi-domain feature extraction task from event-related potential (ERP) data [4] and the clustering task from the ORL database of facial images, it is shown that the proposed NTT-Tucker model using the NTT-HALS algorithm outperforms with respect to the computational and storage costs, and produces physically interpretable features.

2   Nonnegative Tensor Decomposition Models

In this section, we describe nonnegative tensor decomposition models including the NTD, NTT, and NTT-Tucker models.

The notations for tensors and related operations used in this paper are summarized in Table 1. See, e.g., [2,7,8] for further notations. The mode-n multilinear product of a tensor $\underline{\mathbf{A}} \in \mathbb{R}^{I_1 \times \cdots \times I_N}$ with a matrix $\mathbf{B} \in \mathbb{R}^{J_1 \times I_n}$ is defined by $(\underline{\mathbf{A}} \times_n \mathbf{B})_{i_1,\ldots,i_{n-1},j_1,i_{n+1},\ldots,i_N} = \sum_{i_n=1}^{I_n} a_{i_1,\ldots,i_N} b_{j_1,i_n}$. We define the reversed mode-1 multilinear product between two tensors $\underline{\mathbf{A}} \in \mathbb{R}^{I_1 \times \cdots \times I_N}$ and $\underline{\mathbf{B}} \in \mathbb{R}^{J_1 \times \cdots \times J_M}$, $I_N = J_1$, by $(\underline{\mathbf{A}} \times^1 \underline{\mathbf{B}})_{i_1,\ldots,i_{N-1},j_2,\ldots,j_M} = \sum_{i_N=1}^{I_N} a_{i_1,\ldots,i_N} b_{i_N,j_2,\ldots,j_M}$. Note that the reversed 1-mode product is a natural generalization of the matrix-by-matrix multiplication: $\mathbf{A} \times^1 \mathbf{B} \times^1 \mathbf{C} = \mathbf{ABC}$ for matrices $\mathbf{A}, \mathbf{B}$, and $\mathbf{C}$. See [8] for further properties. Figure 1(a) and (b) illustrate graphical representations for the mode-n multilinear product and the reversed mode-1 multilinear product.

2.1   NTD Model

For a given Nth-order tensor $\underline{\mathbf{Y}} \in \mathbb{R}_+^{I_1 \times I_2 \times \cdots \times I_N}$ with nonnegative entries, the NTD model approximately decomposes $\underline{\mathbf{Y}}$ into N nonnegative factor matrices $\mathbf{A}^{(n)} \in \mathbb{R}_+^{I_n \times R_n}, n = 1, \ldots, N$ and a nonnegative core tensor $\underline{\mathbf{G}} \in \mathbb{R}_+^{R_1 \times \cdots \times R_N}$ as

$$\underline{\mathbf{Y}} \approx \underline{\mathbf{G}} \times_1 \mathbf{A}^{(1)} \times_2 \mathbf{A}^{(2)} \times_3 \cdots \times_N \mathbf{A}^{(N)}. \tag{1}$$

The numbers R_n of column vectors of the factor matrices $\mathbf{A}^{(n)}$ are called the (multilinear) rank of the NTD in (1).

Table 1. Notations for tensors and related operations.

Notation	Description
$x, \mathbf{x}, \mathbf{X}$	scalar, vector, matrix
$\underline{\mathbf{X}} \in \mathbb{R}^{I_1 \times I_2 \times \cdots \times I_N}$	Nth-order tensor of size $I_1 \times I_2 \times \cdots \times I_N$
$\mathbf{X}_{(n)} \in \mathbb{R}^{I_n \times I_1 \cdots I_{n-1} I_{n+1} \cdots I_N}$	mode-n matricization of $\underline{\mathbf{X}} \in \mathbb{R}^{I_1 \times \cdots \times I_N}$
$\mathbf{X}_{<n>} \in \mathbb{R}^{I_1 \cdots I_n \times I_{n+1} \cdots I_N}$	mode-$(1, 2, \ldots, n)$ matricization of $\underline{\mathbf{X}} \in \mathbb{R}^{I_1 \times \cdots \times I_N}$
$\underline{\mathbf{A}} \times_n \mathbf{B}$	mode-n multilinear product
$\underline{\mathbf{A}} \times^1 \underline{\mathbf{B}}$	reversed mode-1 multilinear product

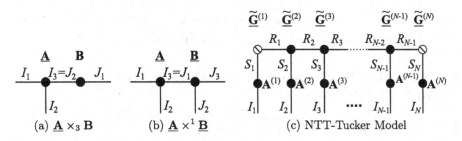

(a) $\underline{\mathbf{A}} \times_3 \mathbf{B}$ (b) $\underline{\mathbf{A}} \times^1 \underline{\mathbf{B}}$ (c) NTT-Tucker Model

Fig. 1. Graphical representations for (a) the mode-3 multilinear product of a tensor $\underline{\mathbf{A}} \in \mathbb{R}^{I_1 \times I_2 \times I_3}$ with a matrix $\mathbf{B} \in \mathbb{R}^{J_1 \times J_2}$, and (b) the reversed mode-1 multilinear product of two 3rd-order tensors $\underline{\mathbf{A}} \in \mathbb{R}^{I_1 \times I_2 \times I_3}$ and $\underline{\mathbf{B}} \in \mathbb{R}^{J_1 \times J_2 \times J_3}$ with $I_3 = J_1$. (c) Graphical representation for the NTT-Tucker decomposition of an Nth-order data tensor $\underline{\mathbf{Y}} \in \mathbb{R}_+^{I_1 \times \cdots \times I_N}$.

2.2 NTT Model

The NTT decomposition approximately represents a high-order tensor as multilinear products of low-order core tensors. Specifically, by the NTT, a nonnegative tensor $\underline{\mathbf{Y}} \in \mathbb{R}_+^{I_1 \times I_2 \times \cdots \times I_N}$ is approximately represented as a multilinear product of low-order core tensors as

$$\underline{\mathbf{Y}} \approx \underline{\mathbf{X}} = \underline{\mathbf{G}}^{(1)} \times^1 \underline{\mathbf{G}}^{(2)} \times^1 \cdots \times^1 \underline{\mathbf{G}}^{(N-1)} \times^1 \underline{\mathbf{G}}^{(N)}, \tag{2}$$

where $\underline{\mathbf{G}}^{(n)} \in \mathbb{R}_+^{R_{n-1} \times I_n \times R_n}, n = 1, \ldots, N$, are 3rd-order nonnegative core tensors called the TT-cores, and the integers $R_1, \ldots, R_{N-1}$ are called the TT-ranks. We define that $R_0 = R_N = 1$. We consider $\underline{\mathbf{G}}^{(1)}$ and $\underline{\mathbf{G}}^{(N)}$ as 3rd-order tensors for notational convenience. Note that the storage complexity for TT is $\mathcal{O}(NIR^2)$, that is linear with the order N, whereas that for the NTD in (1) is $\mathcal{O}(R^N + NIR)$, which is an exponential growth with N, for $I = \max\{I_n\}$ and $R = \max\{R_n\}$.

2.3 NTT-Tucker Model: A Hybrid of the NTD and NTT Models

While the NTT in (2) can achieve smaller storage complexity compared to the NTD in (1) for high-order tensors, the NTT lacks factor matrices which are required for feature extraction and data mining. A hybrid of the NTD and NTT models yields a tensor decomposition model called the NTT-Tucker, and it possesses the favorable properties of the NTD and NTT models; see, Fig. 1(c).

We propose a two-step procedure for the NTT-Tucker decomposition of a data tensor $\underline{\mathbf{Y}} \in \mathbb{R}_+^{I_1 \times \cdots \times I_N}$. First, by the NTT model, $\underline{\mathbf{Y}}$ is represented as a multilinear product of 3rd-order core tensors as in (2). Next, each 3rd-order core tensor is further factored into a 3rd-order core tensor and a factor matrix as

$$\mathbf{G}^{(n)} \approx \widetilde{\mathbf{G}}^{(n)} \times_2 \mathbf{A}^{(n)} \in \mathbb{R}_+^{R_{n-1} \times I_n \times R_n}, \quad n = 1, \ldots, N, \tag{3}$$

with $\widetilde{\mathbf{G}}^{(n)} \in \mathbb{R}_+^{R_{n-1} \times S_n \times R_n}$ and $\mathbf{A}^{(n)} \in \mathbb{R}_+^{I_n \times S_n}$. Since $R_0 = R_N = 1$, we can suppose that $\mathbf{G}^{(1)}$ and $\mathbf{G}^{(N)}$ are reshaped to the matrices $\mathbf{A}^{(1)}$ and $\mathbf{A}^{(N)}$, respectively, and $\widetilde{\mathbf{G}}^{(1)}$ and $\widetilde{\mathbf{G}}^{(N)}$ are equivalent to the identity matrices.

3   NTT-HALS: Proposed Algorithm for NTT and NTT-Tucker

3.1   NTT-HALS for NTT

The NTT-HALS aims at minimizing the sum of squared errors

$$J(\underline{\mathbf{X}}) = \|\underline{\mathbf{Y}} - \underline{\mathbf{X}}\|_F^2, \tag{4}$$

where $\|\cdot\|_F$ is the Frobenius norm, $\underline{\mathbf{Y}} \in \mathbb{R}_+^{I_1 \times \cdots \times I_N}$ is a given nonnegative tensor and $\underline{\mathbf{X}} \in \mathbb{R}_+^{I_1 \times \cdots \times I_N}$ is the NTT tensor in (2). In the NTT-HALS, each of the mode-2 fiber vectors $\mathbf{g}_{r_{n-1},:,r_n}^{(n)} \in \mathbb{R}_+^{I_n}$ of the TT-core $\mathbf{G}^{(n)}$ is updated iteratively for $n = 1, \ldots, N$, $r_{n-1} = 1, \ldots, R_{n-1}$, and $r_n = 1, \ldots, R_n$, assuming that the other components are fixed. For a fixed n, partial multilinear products of the TT-cores are defined as

$$\mathbf{G}^{<n} = \mathbf{G}^{(1)} \times^1 \mathbf{G}^{(2)} \times^1 \cdots \times^1 \mathbf{G}^{(n-1)} \in \mathbb{R}_+^{I_1 \times \cdots \times I_{n-1} \times R_{n-1}},$$

$$\mathbf{G}^{>n} = \mathbf{G}^{(n+1)} \times^1 \mathbf{G}^{(n+2)} \times^1 \cdots \times^1 \mathbf{G}^{(N)} \in \mathbb{R}_+^{R_n \times I_{n+1} \times \cdots \times I_N},$$

and $\mathbf{G}^{<1} = \mathbf{G}^{>N} = 1$. Then, the NTT tensor in (2) can be rewritten as $\underline{\mathbf{X}} = \mathbf{G}^{<n} \times^1 \mathbf{G}^{(n)} \times^1 \mathbf{G}^{>n}$, and its mode-$n$ matricization $\mathbf{X}_{(n)} \in \mathbb{R}_+^{I_n \times I_1 \cdots I_{n-1} I_{n+1} \cdots I_N}$ can be expressed by

$$\mathbf{X}_{(n)} = \mathbf{G}_{(2)}^{(n)} \left(\mathbf{G}_{(1)}^{>n} \otimes \mathbf{G}_{(n)}^{<n} \right). \tag{5}$$

Note that the column vectors of the matrix $\mathbf{G}_{(2)}^{(n)}$ consists of the mode-2 fibers, $\mathbf{g}_{r_{n-1},:,r_n}^{(n)} \in \mathbb{R}_+^{I_n}$, of the nth TT-core $\mathbf{G}^{(n)}$. The cost function (4) can be rewritten by $J = \|\mathbf{Y}_{(n)} - \mathbf{X}_{(n)}\|_F^2$.

For fixed n and (r_{n-1}, r_n), from $\partial J/\partial \mathbf{g}_{r_{n-1},:,r_n}^{(n)} = 0$, we can derive a close form expression for the optimal value of the fiber vector $\mathbf{g}_{r_{n-1},:,r_n}^{(n)}$ as

$$\widehat{\mathbf{g}}_{r_{n-1},:,r_n}^{(n)} = \mathbf{g}_{r_{n-1},:,r_n}^{(n)} + \mathbf{z}_{r_{n-1},:,r_n}^{(n)}, \tag{6}$$

where $\mathbf{z}^{(n)}_{r_{n-1},:,r_n} \in \mathbb{R}^{I_n}$ is the mode-2 fiber vector of the 3rd-order tensor $\underline{\mathbf{Z}}^{(n)} \in \mathbb{R}^{R_{n-1} \times I_n \times R_n}$ defined by

$$\mathbf{Z}^{(n)}_{(2)} = \left(\mathbf{Y}_{(n)} - \mathbf{X}_{(n)}\right) \left(\mathbf{G}^{>n\top}_{(1)} \otimes \mathbf{G}^{<n\top}_{(n)}\right) \left(\mathbf{D}^{>n} \otimes \mathbf{D}^{<n}\right), \qquad (7)$$

with two diagonal matrices $\mathbf{D}^{<n} \in \mathbb{R}^{R_{n-1} \times R_{n-1}}$ and $\mathbf{D}^{>n} \in \mathbb{R}^{R_n \times R_n}$ defined by

$$\mathbf{D}^{<n} = \mathrm{diag}\left(\mathrm{diag}\left(\mathbf{G}^{<n}_{(n)}\mathbf{G}^{<n\top}_{(n)}\right)\right)^{-1}, \mathbf{D}^{>n} = \mathrm{diag}\left(\mathrm{diag}\left(\mathbf{G}^{>n}_{(1)}\mathbf{G}^{>n\top}_{(1)}\right)\right)^{-1}. \quad (8)$$

Then, for the nonnegativity of the solution, we project the estimate onto non-negative values as $\mathbf{g}^{(n)}_{r_{n-1},:,r_n} \leftarrow [\widehat{\mathbf{g}}^{(n)}_{r_{n-1},:,r_n}]_+$. Note that the inversion in (8) does not require any matrix inverse computation.

The error term $\underline{\mathbf{Z}}^{(n)}$ in (7) can be updated efficiently at a low computational cost by a recursive procedure as follows. Suppose that $\mathbf{g}^{(n)}_{r_{n-1},:,r_n}$ is the newly updated fiber vector and $\mathbf{g}^{(n)\mathrm{old}}_{r_{n-1},:,r_n}$ is the previous one. The updated NTT tensor $\underline{\mathbf{X}}$ can also be expressed as, from the matricized form $\mathbf{X}_{(n)}$ in (5), $\mathbf{X}_{(n)} = \mathbf{X}^{\mathrm{old}}_{(n)} + \left(\mathbf{g}^{(n)}_{r_{n-1},:,r_n} - \mathbf{g}^{(n)\mathrm{old}}_{r_{n-1},:,r_n}\right) \left(\mathbf{g}^{>n}_{r_n} \otimes \mathbf{g}^{<n}_{r_{n-1}}\right)^\top$, where $\mathbf{g}^{<n}_{r_{n-1}} = (\mathbf{G}^{<n}_{(n)})_{r_{n-1},:} \in \mathbb{R}^{I_1 \cdots I_{n-1}}_+$ and $\mathbf{g}^{>n}_{r_n} = (\mathbf{G}^{>n}_{(1)})_{r_n,:} \in \mathbb{R}^{I_{n+1} \cdots I_N}_+$ are the r_{n-1}th and r_nth row vectors of $\mathbf{G}^{<n}_{(n)}$ and $\mathbf{G}^{>n}_{(1)}$, respectively. Finally, $\underline{\mathbf{Z}}^{(n)}$ can be updated as

$$\mathbf{Z}^{(n)}_{(2)} = \mathbf{Z}^{(n)\mathrm{old}}_{(2)} - \left(\mathbf{g}^{(n)}_{r_{n-1},:,r_n} - \mathbf{g}^{(n)\mathrm{old}}_{r_{n-1},:,r_n}\right) \left(\mathbf{D}^{>n}\mathbf{G}^{>n}_{(1)}\mathbf{g}^{>n}_{r_n} \otimes \mathbf{D}^{<n}\mathbf{G}^{<n}_{(n)}\mathbf{g}^{<n}_{r_{n-1}}\right)^\top.$$
$$(9)$$

Note that $\mathbf{G}^{>n}_{(1)}\mathbf{g}^{>n}_{r_n} = (\mathbf{G}^{>n}_{(1)}\mathbf{G}^{>n\top}_{(1)})_{r_n,:}$ and $\mathbf{G}^{<n}_{(n)}\mathbf{g}^{<n}_{r_{n-1}} = (\mathbf{G}^{<n}_{(n)}\mathbf{G}^{<n\top}_{(n)})_{r_{n-1},:}$ are row vectors of the symmetric matrices which can be efficiently calculated during the iteration process in a recursive manner.

Algorithm 1 describes the NTT-HALS algorithm for the NTT decomposition.

3.2   NTT-HALS for NTT-Tucker

For the NTT-Tucker decomposition, we propose to perform the two-step decomposition process, as described in Sect. 2.3, by applying the NTT-HALS algorithm. On the other hand, an "one-step" HALS approach could also be developed for the NTT-Tucker model, instead of the proposed two-step process, however, in that case convergence speed would slow down due to optimization of each "entry" of core tensors having no "fiber vectors".

4   Experiments

We run and compared the standard NTD algorithm [12] and the proposed NTT-Tucker decomposition based on the NTT-HALS algorithm. The relative approximation error is calculated by: $\|\underline{\mathbf{Y}} - \underline{\mathbf{X}}\|_\mathrm{F}/\|\underline{\mathbf{Y}}\|_\mathrm{F} \times 100$ (%). We repeatedly run the algorithms 100 times independently and averaged the results. Experiments were performed on a desktop computer with an Intel CPU at 3.60 GHz using 32.0 GB of RAM running Windows 7 Professional and Matlab R2010b.

Algorithm 1. NTT-HALS for the NTT Decomposition

Input: $\underline{\mathbf{Y}} \in \mathbb{R}_+^{I_1 \times \cdots \times I_N}$, TT-ranks $R_n, n = 1, \ldots, N-1$.

Output: NTT tensor $\underline{\mathbf{X}}$ in (2) with nonnegative TT-cores $\underline{\mathbf{G}}^{(n)}$.

1 Initialize $\underline{\mathbf{X}}$ by ALS [6]. Set $g_{r_{n-1},i_n,r_n}^{(n)} \leftarrow |g_{r_{n-1},i_n,r_n}^{(n)}|$ and $\mathbf{G}^{<1} = \mathbf{G}^{>N} = 1$.

2 **repeat**

3   **for** $n = N, N-1, \ldots, 2$ **do**

4   Normalization: for each $r_{n-1} = 1, \ldots, R_{n-1}$, update

$$\mathbf{G}_{r_{n-1},:,:}^{(n)} \leftarrow \mathbf{G}_{r_{n-1},:,:}^{(n)}/\|\mathbf{G}_{r_{n-1},:,:}^{(n)}\|_{\mathrm{F}}, \quad \mathbf{G}_{:,:,r_{n-1}}^{(n-1)} \leftarrow \mathbf{G}_{:,:,r_{n-1}}^{(n-1)} \cdot \|\mathbf{G}_{r_{n-1},:,:}^{(n)}\|_{\mathrm{F}}.$$

5   Compute $\mathbf{Q}^{>n-1} \equiv \mathbf{G}_{(1)}^{>n-1} \mathbf{G}_{(1)}^{>n-1\top}$ recursively.

6   **end**

7   **for** $n = 1, 2, \ldots, N$ **do**

8   Compute $\mathbf{Q}^{<n} \equiv \mathbf{G}_{(n)}^{<n} \mathbf{G}_{(n)}^{<n\top}$ recursively.

9   Let $\mathbf{D}^{<n} = \mathrm{diag}(\mathrm{diag}(\mathbf{Q}^{<n}))^{-1}$, $\mathbf{D}^{>n} = \mathrm{diag}(\mathrm{diag}(\mathbf{Q}^{>n}))^{-1}$.

10   Compute $\underline{\mathbf{W}}_1^{(n)}, \underline{\mathbf{W}}_2^{(n)} \in \mathbb{R}_+^{R_{n-1} \times I_n \times R_n}$ by

$$\mathbf{W}_{1,(2)}^{(n)} = \mathbf{Y}_{(n)}\left(\mathbf{G}_{(1)}^{>n\top} \otimes \mathbf{G}_{(n)}^{<n\top}\right) \text{ and } \underline{\mathbf{W}}_2^{(n)} = \underline{\mathbf{G}}^{(n)} \times_1 \mathbf{Q}^{<n} \times_3 \mathbf{Q}^{>n}.$$

11   Calculate $\underline{\mathbf{Z}}^{(n)} = \left(\underline{\mathbf{W}}_1^{(n)} - \underline{\mathbf{W}}_2^{(n)}\right) \times_1 \mathbf{D}^{<n} \times_3 \mathbf{D}^{>n}$ as in (7).

12   **for** $r_{n-1} = 1, \ldots, R_{n-1}, r_n = 1, \ldots, R_n$ **do**

13             $g_{r_{n-1},:,r_n}^{(n)} \leftarrow [g_{r_{n-1},:,r_n}^{(n)} + z_{r_{n-1},:,r_n}^{(n)}]_+$

14   Update $\underline{\mathbf{Z}}^{(n)}$ by (9).

15   **end**

16   Normalization: If $n < N$, for each $r_n = 1, \ldots, R_n$, update

$$\mathbf{G}_{:,:,r_n}^{(n)} \leftarrow \mathbf{G}_{:,:,r_n}^{(n)}/\|\mathbf{G}_{:,:,r_n}^{(n)}\|_{\mathrm{F}}, \quad \mathbf{G}_{r_n,:,:}^{(n+1)} \leftarrow \mathbf{G}_{r_n,:,:}^{(n+1)} \cdot \|\mathbf{G}_{:,:,r_n}^{(n)}\|_{\mathrm{F}}.$$

17   **end**

18 **until** *a stopping criterion is met;*

4.1 Multi-domain Feature Extraction from ERP Data

The original data is a power spectrum obtained from the event-related potential (ERP) data [4]. It is given in the form of a fourth order tensor of size $71 \times 60 \times 42 \times 9$. Each dimension corresponds to the spectral (Hz), temporal (ms), subject, and spatial modes. There is no missing entry. The goal is to find three factor matrices and a Tucker core tensor so that the data tensor is approximated by

$$\underline{\mathbf{Y}} \approx \underline{\mathbf{G}} \times_1 \mathbf{A}^{(1)} \times_2 \mathbf{A}^{(2)} \times_4 \mathbf{A}^{(4)}, \quad \mathbf{A}^{(1)} \in \mathbb{R}_+^{71 \times 4}, \; \mathbf{A}^{(2)} \in \mathbb{R}_+^{60 \times 8}, \; \mathbf{A}^{(4)} \in \mathbb{R}_+^{9 \times 6}.$$

The sizes of the three factor matrices were determined in [4]. The rest factor matrix, e.g., $\mathbf{A}^{(3)}$, is not estimated in this case. Instead, the experimenter can compare a difference between two groups (normal and abnormal) of subjects based on the core tensor, i.e., $\underline{\mathbf{G}}(:,:,i,:)$, $i = 1, \ldots, 42$, where the index i corresponds to each subject. Via the NTT-Tucker (3), the Tucker core can be computed as $\underline{\mathbf{G}} = \widetilde{\underline{\mathbf{G}}}^{(1)} \times^1 \cdots \times^1 \widetilde{\underline{\mathbf{G}}}^{(4)}$. The sizes of $\mathbf{A}^{(1)}$ and $\mathbf{A}^{(4)}$ determine the two of the three TT-ranks, i.e., $R_1 = 4$ and $R_3 = 6$. The TT-rank R_2 was selected by $R_2 = 10$ from the result of the truncated SVD of the unfolded data tensor, $\mathbf{Y}_{<2>} \in \mathbb{R}^{I_1 I_2 \times I_3 I_4}$ with truncation parameter $\delta = \frac{0.3}{\sqrt{N-1}}$, $N = 4$ [9].

Table 2. Comparison of performances and experimental results of the NTD and NTT-Tucker for multi-domain feature extraction from the ERP data.

	NTD	NTT-Tucker
Size of Core Tensor	$\{4 \times 8 \times 42 \times 6\}$	$\{4 \times 8 \times 10, 10 \times 42 \times 6\}$
(number of entries)	(8064)	(2840)
Computation Time (s, $\pm$std)	10.96 ± 0.10	4.40 ± 0.90
Relative Error (%, $\pm$std)	13.46 ± 0.67	13.80 ± 0.25
p-Value	0.0314 ± 0.0001	0.0300 ± 0.0110
Significant Component#: (see, [4, Sect. 4.3] and Fig. 2)		
spectral# (peak at)	#2 (2 Hz)	#1 (3 Hz)
temporal# (peak at)	#8 (150 ms)	#1 (150 ms)
spatial# (peat at)	#4 (C3)	#6 (C3)
p-value	0.0312	0.0121

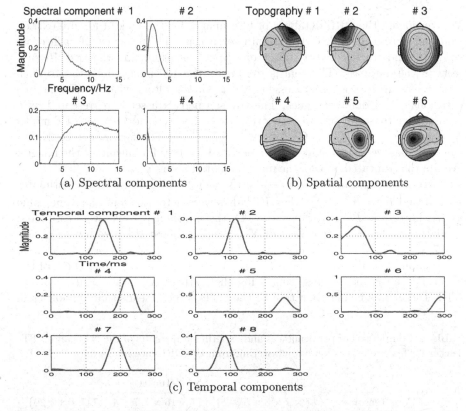

(a) Spectral components

(b) Spatial components

(c) Temporal components

Fig. 2. Illustration of the components extracted by the NTT-Tucker: (a) spectral, (b) spatial, and (c) temporal components.

The results are summarized in Table 2. The computational and storage costs of the NTT-Tucker were much lower than those of the NTD. Both the NTD and NTT-Tucker extracted similar factor matrices (i.e., components) with a slight difference in the spectral and spatial components; see Fig. 2 for the estimated components. See also, [4, Sect. 4.3] for the components extracted by the NTD. Note that the NTT-Tucker could extract the spatial components with its peak at the left or right eye separately, which is in tune with the results obtained by the NTF [4, Sect. 4.2]. For a statistical test of difference between the two groups of subjects, the most significant component was selected in each of the factor matrices; see, Table 2. With a slight difference in the selected spectral components, the NTT-Tucker model obtained smaller p-values than those obtained by the NTD. This suggests that the core tensor of the NTD has a low-rank structure which can be exploited better by the NTT-Tucker model.

4.2 Feature Extraction and Clustering from ORL Database of Face Images

We considered the ORL database of face images for the task of feature extraction and clustering [11]. Original data set consists of 10 images of each of 40 subjects, and we selected a subset of images corresponding the first 10 subjects in the database. The goal is to classify the 100 images into the 10 clusters. Each image was preprocessed by the Gabor filters with 8 different orientations and 4 different scales. The transformed data set is represented by a nonnegative tensor of size $16 \times 16 \times 4 \times 8 \times 100$. We considered the NTT model $\underline{\mathbf{Y}} \approx \underline{\mathbf{G}}^{(1)} \times^1 \cdots \times^1 \underline{\mathbf{G}}^{(5)}$, $\underline{\mathbf{G}}^{(5)} \equiv \mathbf{A}^{(5)} \in \mathbb{R}_+^{100 \times S_5}$. The factor matrix $\mathbf{A}^{(5)}$ is considered as the extracted lower dimensional features for clustering the images. We run the truncated SVD of the unfolded data tensor, $\mathbf{Y}_{<n>} \in \mathbb{R}^{I_1 \cdots I_n \times I_{n+1} \cdots I_5}$ with truncation parameter $\delta = \frac{0.3}{\sqrt{N-1}}$, $N = 5$ [9], and obtained the estimates $R_1 = 7$ and $R_4 = S_5 = 20$. Other TT-ranks were determined as the same value as R_1, i.e., $R_2 = R_3 = 7$. For the NTD, the multilinear ranks were estimated similarly by the truncated SVD of $\mathbf{Y}_{(n)}$, and obtained the core tensor of size $7 \times 7 \times 3 \times 5 \times 20$. The results are summarized in Table 3. Both the storage and computational costs of the NTT-Tucker were much lower than those of the NTD. By the K-means clustering using the features $\mathbf{A}^{(5)}$, the accuracy of the NTT-Tucker was higher than that of the NTD, which means that the low-rank

Table 3. Comparison of performances and experimental results of the NTD and NTT-Tucker for feature extraction and clustering from the ORL face image data.

	NTD	NTT-Tucker
Size of Core Tensor	$\{7 \times 7 \times 3 \times 5 \times 20\}$	$\{7 \times 16 \times 7, 7 \times 4 \times 7, 7 \times 8 \times 20\}$
(number of entries)	(14700)	(2100)
Computation Time (s)	154.41 ± 8.84	2.27 ± 0.70
Clustering Accuracy (%)	90.62 ± 5.39	96.06 ± 3.29

constraints on the TT-core tensors helped improve the clustering accuracy and extract meaningful features from images.

5 Discussion

We proposed the NTT and NTT-Tucker decomposition models for feature extraction and clustering from multiway nonnegative data. The NTT/NTT-Tucker can extract low-dimensional features from high-order nonnegative tensor data, which are useful for multi-dimensional data analysis and data mining.

Compared to a standard NTD method, the proposed NTT-HALS method achieved high accuracy with much lower storage costs for real world data analysis tasks. It implies that the low-rank assumption on the core tensors is helpful for extracting meaningful features from high-order data tensors. To develop a method for adaptively determining TT-ranks is left as a future work.

References

1. Cichocki, A., Zdunek, R., Amari, S.: Hierarchical ALS algorithms for nonnegative matrix and 3D tensor factorization. In: Davies, M.E., James, C.J., Abdallah, S.A., Plumbley, M.D. (eds.) ICA 2007. LNCS, vol. 4666, pp. 169–176. Springer, Heidelberg (2007)
2. Cichocki, A., Zdunek, R., Phan, A.H., Amari, S.: Nonnegative Matrix and Tensor Factorizations: Applications to Exploratory Multi-way Data Analysis and Blind Source Separation. Wiley, Chichester (2009)
3. Cong, F., Phan, A.-H., Astikainen, P., Zhao, Q., Wu, Q., Hietanen, J.K., Ristaniemi, T., Cichocki, A.: Multi-domain feature extraction for small event-related potentials through nonnegative multi-way array decomposition from low dense array EEG. Int. J. Neural Syst. **23**, 1350006 (2013)
4. Cong, F., Lin, Q.-H., Kuang, L.-D., Gong, X.-F., Astikainen, P., Ristaniemi, T.: Tensor decomposition of EEG signals: a brief review. J. Neurosci. Methods **248**, 59–69 (2015)
5. Dolgov, S., Khoromskij, B.: Two-level QTT-Tucker format for optimized tensor calculus. SIAM J. Matrix Anal. Appl. **34**, 593–623 (2013)
6. Holtz, S., Rohwedder, T., Schneider, R.: The alternating linear scheme for tensor optimization in the tensor train format. SIAM J. Sci. Comput. **34**, A683–A713 (2012)
7. Kolda, T.G., Bader, B.W.: Tensor decompositions and applications. SIAM Rev. **51**, 455–500 (2009)
8. Lee, N., Cichocki, A.: Fundamental tensor operations for large-scale data analysis in tensor train formats. ArXiv preprint arXiv:1405.7786 (2014)
9. Oseledets, I.V.: Tensor-train decomposition. SIAM J. Sci. Comput. **33**, 2295–2317 (2011)
10. Phan, A.H., Cichocki, A.: Extended HALS algorithm for nonnegative Tucker decomposition and its applications for multiway analysis and classification. Neurocomputing **74**, 1956–1969 (2011)
11. Samaria, F.S., Harter, A.C.: Parameterisation of a stochastic model for human face identification. In: Proceedings of the Second IEEE Workshop on Applications of Computer Vision, pp. 138–142. IEEE Computer Society Press (1994)
12. Zhou, G., Cichocki, A., Xie, S.: Fast nonnegative matrix/tensor factorization based on low-rank approximation. IEEE T. Signal Process. **60**, 2928–2940 (2012)

Hyper-parameter Optimization of Sticky HDP-HMM Through an Enhanced Particle Swarm Optimization

Jiaxi Li[✉], Junfu Yin, Yuk Ying Chung, and Feng Sha

School of Information Technologies, University of Sydney,
Sydney, NSW 2006, Australia
jili2506@uni.sydney.edu.au, {junfu.yin,vera.chung,feng.sha}@sydney.edu.au

Abstract. Faced with the problem of uncertainties in object trajectory and pattern recognition in terms of the non-parametric Bayesian approach, we have derived that 2 major methods of optimizing hierarchical Dirichlet process hidden Markov model (HDP-HMM) for the task. HDP-HMM suffers from poor performance not only on moderate dimensional data, but also sensitivity to its parameter settings. For the purpose of optimizing HDP-HMM on dimensional data, test for optimized results will be carried on the Tum Kitchen dataset [7], which was provided for the purpose of research the motion and activity recognitions. The optimization techniques capture the best hyper-parameters which then produce optimal solution to the task given in a certain search space.

Keywords: Non-parametric Bayes · HDP-HMM · Pattern recognition · Model selection · Optimization · Hyper-parameters

1 Introduction

Hierarchical Dirichlet Process is a core component of the infinite Hidden Markov Model, and it defines a prior distribution on transaction probabilities over infinite state space. As a result, the Hierarchical Dirichlet Process Hidden Markov Model (HDP-HMM) solves the puzzle of unbounded number, and provides promising results in variety of applications, including visual recognition [6] and genetic recombination modeling [8]. Unfortunately, one of the most serious limitations of HDP-HMM provides an inadequate model of transitioning states. In other, the resulting state sequence from HDP-HMM is not persistent.

Under this context of problems, Fox et al. [5] has proposed the sticky HDP-HMM which can promote its self-transitioning or stickiness by using appropriate priors in Bayesian non-parametric approach. The most significant drawbacks of sticky HDP-HMM, which is also the nature of non-parametric Bayesian, are sensitive to the existence of its many hyperparameters and extremely large searching space with its hyperparameter tuning.

In this paper, we present three existent tuning methodologies for hyperparameter and also propose another approach on improving particle swarm optimization (PSO) in Sect. 4, called ring-based particle swarm optimization (RPSO),

© Springer International Publishing AG 2016
A. Hirose et al. (Eds.): ICONIP 2016, Part III, LNCS 9949, pp. 96–103, 2016.
DOI: 10.1007/978-3-319-46675-0_11

which provides an enhanced performance on hyperparameter tuning. To explain the purpose of tuning hyperparameters, the related work of sticky HDP-HMM is covered in Sect. 2, and the problem statement is proposed in Sect. 3. In Sect. 4, potential solutions are introduced, following by experimental results in Sect. 5. At last, a conclusion and an indication for future work is given in Sect. 6.

2   Related Work

Hierarchical Dirichlet Process Hidden Markov Model (HDP-HMM) is a resulting combination of HDP and HMM, which can be used to compute and sample from posterior distributions over the number of model states. In addition, HDP-HMM allows the number of state to be unknown in prior and learns from training data. For HDP-HMM, the formal definition is given by Fox et al. in doctoral dissertation [2]. Consider the following relations from Eqs. 1 to 5:

$$\beta|\gamma \sim GEM(\gamma) \tag{1}$$

$$\pi_j|\beta,\alpha \sim DP(\alpha,\beta) \quad j = 1,2,... \tag{2}$$

$$\theta_j|H,\lambda \sim H(\lambda) \quad j = 1,2,... \tag{3}$$

$$z_t|\{\pi_j\}_{j=1}^{\infty}, z_{t-1} \sim \pi_{z_{t-1}} \quad t = 1,...,T \tag{4}$$

$$y_t|\{\theta_j\}_{j=1}^{\infty}, z_{t-1} \sim F(\theta_{z_t}) \quad t = 1,...,T \tag{5}$$

$GEM(\cdot)$, shown in the Eq. 1, is the stick-breaking construction of a Dirichlet process, it denotes a special case of Dirichlet process with a concentration parameter γ. When we defined $\pi_j \sim DP(\alpha,\beta)$ (Eq. 2), the HDP prior makes all states to have a similar transition probability. Then, Let z_t represents the state of the Markov model at time t, where $t = 1,2,...,T$, and π_j denotes the specific distribution of state transition for state j. The observations are represented as y_t, which are conditional independent given the state sequence $\{\theta_j\}$ (Eq. 5).

For a sticky HDP-HMM (Fig. 1), a sticky parameter κ is introduced when sampling group-specific transition distribution π_j. It can be addressed following relations:

$$\pi_j|\alpha,\kappa,\beta \sim DP\left(\alpha+\kappa, \frac{\alpha\beta+\kappa\delta_j}{\alpha+\kappa}\right) \tag{6}$$

In Eq. 6, $(\alpha\beta + +\kappa\delta_j)$ shows that there is additional amount is added to the j_{th} component of $\alpha\beta$ (Eq. 2) by κ ($\kappa > 0$), where δ_j denotes a unit-mass measure concentrated at j. In other, the expected probability of self-transition is increased by a number proportional to κ in terms of the unit-mass δ_j. More formally [2], over a finite subset $Z_1, Z_2, ..., Z_K$ of positive integers, the theorem of Dirichlet process implies that the prior of distribution π_j is added additional amount κ to a small subset containing j, by giving it self-transition. That is,

$$\pi_j(Z_1),...,\pi_j(Z_K)|\alpha,\beta \sim Dir(\alpha\beta(Z_1)+\kappa\delta_j(Z_1),...,\alpha\beta(Z_K)+\kappa\delta_j(Z_K)) \tag{7}$$

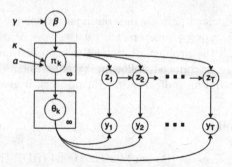

Fig. 1. *Sticky HDP-HMM* represented as graphic model. The state transaction can be illustrated as $z_{z+1}|\{\pi_k\}_{k=1}^{\infty}, z_t \sim \pi_{zt}$, where $\pi_k|\alpha, \kappa, \beta \sim DP(\alpha+\kappa, (\alpha\beta+\kappa\delta_k)/(\alpha+\kappa)$, $\beta|\gamma \sim GEM(\gamma)$. The observations are generated as $y_t|\theta_k, z_t \sim F(\theta_{z_t})$. Note that, the original HDP-HMM has $\kappa = 0$.

3   Problem Statement

As mentioned in [4], HMM and HDP-HMM have applied into many fields of study in temporal pattern recognition. However there are limitations of HMM and HDP-HMM. HMM is not compatible with unbounded number of states in this representation. On the another hand, HDP-HMM, which develops HMM with an infinite state space, has the property of unrealistic fast dynamics allowing state sequences to have much wider posterior probability [2]. That is, HDP-HMM does not differentiate its self-transitions from transitions between different states. In this case, the identifications on observation of a model with redundant states can be hindered.

With the limitations of HMM and HDP-HMM, the sticky HDP-HMM appears to be an elegant Bayesian model. Recall the Stick HDP-HMM in Sect. 2, an extra parameter κ is introduced to prevent the rapidly transition among the redundant states. Referring to Eq. 6, the transition probabilities between states is dependent on the parameters α, κ, and the distribution β, where β is a stick-breaking process corresponding to parameter γ. According to Eqs. 1 and 6, a more intuitive relation can be introduced:

$$\pi_j|\alpha, \kappa, \gamma \sim DP\left(\alpha + \kappa, \frac{\alpha GEM(\gamma) + \kappa\delta_j}{\alpha + \kappa}\right) \qquad (8)$$

The distribution described in Eq. 8 indicates every selection of state θ_k follows a Dirichlet process, which has a concentration parameter with value of $\alpha + \kappa$ and a modified stick-breaking construction as its base distribution. That is, the 3 real positive, α, κ, γ, extremely impact on the state transition. As the consequence, the resulting observations differ in terms of selections on hyperparameters.

As Emily et al. mentioned in [3], the way of generating the hyperparameters are random draws from a Gamma distribution. However, this random Gamma measure can be inefficient and takes extended time to discover optimal solution due to the arbitrary nature of the stochastic distribution family.

4   Method: Hyperparameter Optimization

In order to discover the optimal setting of hyperparameter for fitting a data, several optimization techniques have already been developed. Such as grid search, random search and particle swarm optimization (PSO). In addition, an enhanced tweak on PSO is introduced at the end of this section.

4.1   Random Search

Random search is a variation on grid search. Instead of searching over the entire grid step by step, random search only evaluates randomized sampling of points in the grid, where the number of randomized samples is fixed. Besides, random search provides more efficient approach than grid search in high dimensional space. Bergastra et al. concludes in [1] as "granting random search the same budget of computation, random search finds better model by effectively searching a much larger, but also less promising grid" [1].

4.2   Particle Swarm Optimization

Particle Swarm Optimization (PSO) is a computational approach to optimize a problem by improving a possible solution repeatedly to some quality measurements. Similar to the basic PSO algorithm described in [9], a vector $\lambda = \{\gamma, \alpha, \kappa\}$ can be used to represent a sticky HDP-HMM model, where γ, α, κ are the hyperparameters. This vector λ acts like a particle, and the value of γ, α and κ represents the position where this particle is in the space. For each particle p_i with a position vector $\lambda_i = \{\gamma, \alpha, \kappa\}$, it is associated with a velocity vector $V_i = \{\Delta_\gamma, \Delta_\alpha, \Delta_\kappa\}$, which implies the capacity of the particle to move from a position λ_i^t to another position λ_i^{t+1}. For each position vector λ_i, there will be a evaluation score s_i produced by evaluation function $(F(\cdot))$.

For each particle in the swarm, the best position (with the highest s_i) of this particle ever reached in the space is stored in $pbest$, where

$$pbest_{p_i}^k = argmax(F(\lambda_i^t)) \quad t = 1, 2, \ldots \tag{9}$$

In addition, the best position among all the particle is stored in the variable $gbest$, where

$$gbest^k = argmax(F(\lambda_i^t)) \quad i = 1, 2, \ldots \tag{10}$$

After finding the 2 best positions in the space, particles updates their velocity with the following Eq. 11.

$$v_i^{t+1} = v_i^t + c_1 \cdot r_1 \cdot (pbest_i^k - \lambda_i) + c_2 \cdot r_2 \cdot (gbest^k - \lambda_i) \tag{11}$$

r_1 and r_2 are 2 uniformly distributed random positive between 0 and 1. c_1 and c_2 are constants named cognition and social acceleration [9].

The particle positions are then updated in Eq. 12 (Fig. 2).

$$\lambda_i^{t+1} = \lambda_i^t + v_i^{t+1} \tag{12}$$

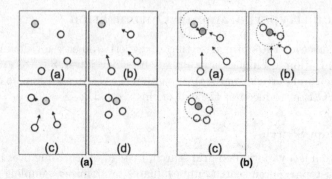

Fig. 2. (a) *PSO* represented in graphical model. (b) Ring-based PSO enhancement represented in graphical model.

4.3   Ring-Based Particle Swarm Optimization

A ring-based Particle Swarm Optimization (RPSO) enhancement, true to its name, places an additional searching region around the *gbest* (Sect. 4.2) among the swarm particles. Given the fact that, the distribution of a sticky HDP-HMM may have multiple optimal solutions in the space. That is, whenever a best particle *gbest* found, there is a possibility that there exists a global optimal solution close to this particle.

Recall the velocity updates in Eq. 11, when the particle that has the best position *gbest* in the space, it also holds the local best position *pbest* which it has ever reached. As a consequence, the updating velocity for this particular particle is 0 which then resulting the particle stays at the same position and waits other particles to move towards it.

Before the velocity updates, other particles are committed according to the Eq. 11. It is an additional process added to PSO to enhance its performance. After the *gbest* particle position is found by Eq. 10, searching vector $v_n'^k = \{\Delta_{\lambda_n'}\}$ will be introduced to the methodology, $n = 1, 2, ..., N$ where N represents the number of searching points around *gbest*. The searching request will then be processed as following:

$$gbest'^k = argmax(F(\lambda_{gbest^k}), F(\lambda_{gbest^k} + v_n'^k)) \qquad n = 1, 2, ..., N \qquad (13)$$

Therefore, the velocity update equation of each particle can be reproduced:

$$v_i^{t+1} = v_i^t + c_1 \cdot r_1 \cdot (pbest_i^k - \lambda_i) + c_2 \cdot r_2 \cdot (gbest'^k - \lambda_i) \qquad (14)$$

5   Experiment

Two characterizing tasks are carried on both synthetic and real-world multivariate data using the optimization methodology. For comparison, the results of applying the original Gamma prior sampling from [5] will be presented as well as the other optimization approaches in Sect. 4.

5.1   Dataset

The first dataset is the synthetic dataset which was generated using a HMM model. It only has three transition states with equal start probabilities in a two . dimensional space. The transition matrix can be viewed as follow:

$$A = \begin{pmatrix} 0.90 \ 0.05 \ 0.05 \\ 0.05 \ 0.90 \ 0.05 \\ 0.00 \ 0.10 \ 0.90 \end{pmatrix}$$

Note that, there is no transitions between stage 1 and 3. By using this model, we generated ten-thousand points in the space for the later experiment. The data example can be viewed in Fig. 3.

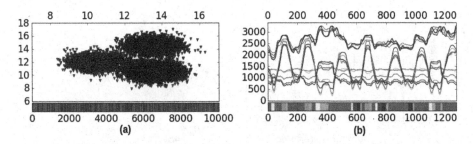

Fig. 3. (*a*) Synthetic data visualized in 2-dimension. (*b*) Tum kitchen data visualized in time series, where each line denotes a dimension.

The second dataset is the *Tum Kitchen* dataset [7], which is a comprehensive collection of activity sequences recorded in a kitchen environment with a person wearing multiple complementary sensors. The recorded data consists of multiple tasks that are naturally encountered in everyday activity of human life. The motion data is captured through the full-body MeMoMan trackers with a sample rate of 2 Hz as in the Fig. 3. The entire data is collected in eighty-four signals including the motion of arms, legs and trunk. To simplify the experiment, only the motion of the human left hand, which consists of fifteen signals recorded in a three-dimension format of (X, Y, Z), is being evaluated.

5.2   Synthetic Data Results

The results from synthetic dataset, which is generated from a three-state HMM, are shown in the left column of Table 1.

5.3   Tum Kitchen Dataset Results

Results from the real-world Tum Kitchen dataset are shown in the right of Table 1.

Table 1. Results from synthetic dataset (left) and Tum Kitchen dataset (right)

Method	γ	α	κ	Acc.	Perf.
Grid	7.55	12.32	112.89	98.92%	N/A
Random	60.82	49.60	205.22	98.96%	10m32.08s
Gamma	39.54	87.43	294.26	98.97%	5m32.17s
PSO	144.52	43.99	118.97	99.09%	22m59.05s
RPSO	125.26	109.46	441.73	99.08%	8m52.65s

Acc. = Accuracy in percentage (%)

Perf. = Time consumed to find the optimal solution

N/A = Extremely high computation

Method	γ	α	κ	Acc.	Perf.
Grid	21.59	23.90	100.00	83.99%	N/A
Random	292.97	1500.00	0.10	88.88%	193m44.08s
Gamma	73.57	93.61	276.83	92.26%	41m19.68s
PSO	306.17	727.66	49.55	94.42%	55m17.02s
RPSO	85.31	468.82	821.29	95.43%	11m29.91s

Acc. = Accuracy in percentage (%)

Perf. = Time consumed to find the optimal solution

N/A = Extremely high computation

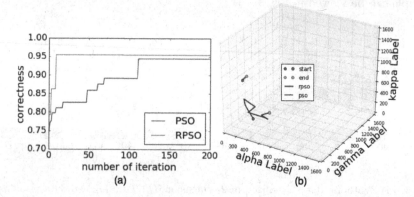

Fig. 4. Demonstration of the global best particle score in the swarm for *PSO* and *RPSO*. (a) X-axis represents the number of updating iterations, whereas Y-axis indicates the best particle score in current position. (b) A trace of the global best particle (*gbest*) in the searching space.

There are Gamma resampling, PSO and ring-based PSO achieved fairly high accuracy in the kitchen data. In PSO and Gamma sampling, PSO produced slightly higher accurates result than Gamma sampling, while Gamma sampling provides a slightly faster performance. Among all, ring-based PSO achieved acceptable results with 95.43 % accurate in shortest time (approximately 11 min).

In terms of PSO and RPSO, the ring-based enhancement takes less route to discover the optimal solution than PSO (shown in Fig. 4b). In addition, ring-based PSO converges faster than PSO meanwhile keeping the similar accuracy (shown in Fig. 4a). That is, RPSO takes less updating iterations to discover the optimal solution in an efficient amount of time.

6   Conclusion

In this paper, we have presented an enhancement of particle swarm optimization (PSO) for hyperparameter tuning for the sticky HDP-HMM. Ring-based particle swarm optimization (RPSO) can achieve global optimal solution for tuning hyperparameters. Moreover, RPSO reveals faster approach to reach convergence and produces an optimal solution in a relatively less amount of computation when dealing with high dimensional data. That is, the ring-based PSO enhancement are carried out in an optimal and efficient manner.

Comparing to the Gamma prior, which does not guarantee an optimal solution, the ring-based PSO approach is an efficient and effective way to discover the optimal setting for sticky HDP-HMM hyper parameters.

References

1. Bergstra, J., Bengio, Y.: Random search for hyper-parameter optimization. J. Mach. Learn. Res. **13**, 281–305 (2012)
2. Fox, E.B.: Bayesian nonparametric learning of complex dynamical phenomena. PhD thesis, Massachusetts Institute of Technology, September 2009
3. Fox, E.B., Sudderth, E.B., Jordan, M.I., Willsky, A.S.: The sticky HDP-HMM: Bayesian nonparametric hidden Markov models with persistent states. Arxiv preprint (2007)
4. Fox, E.B., Sudderth, E.B., Jordan, M.I., Willsky, A.S.: An HDP-HMM for systems with state persistence. In: Proceedings of the 25th International Conference on Machine Learning, pp. 312–319. ACM (2008)
5. Fox, E.B., Sudderth, E.B., Jordan, M.I., Willsky, A.S.: A sticky HDP-HMM with application to speaker diarization. ArXiv e-prints, May 2011
6. Kivinen, J.J., Sudderth, E.B., Jordan, M.I.: Learning multiscale representations of natural scenes using Dirichlet processes. In: IEEE 11th International Conference on Computer Vision, 2007. ICCV 2007, pp. 1–8. IEEE (2007)
7. Tenorth, M., Bandouch, J., Beetz, M.: The TUM kitchen data set of everyday manipulation activities for motion tracking and action recognition. In: IEEE International Workshop on Tracking Humans for the Evaluation of their Motion in Image Sequences (THEMIS), in conjunction with ICCV 2009 (2009)
8. Xing, E.P., Sohn, K.-A., et al.: Hidden Markov Dirichlet process: modeling genetic inference in open ancestral space. Bayesian Anal. **2**(3), 501–527 (2007)
9. Xue, L., Yin, J., Ji, Z., Jiang, L.: A particle swarm optimization for hidden Markov model training. In: 8th International Conference on Signal Processing, vol. 1. IEEE (2006)

Approximate Inference Method for Dynamic Interactions in Larger Neural Populations

Christian Donner[1,2] and Hideaki Shimazaki[3(⊠)]

[1] Bernstein Center for Computational Neuroscience, Berlin, Germany
christian.donner@bccn-berlin.de
[2] Neural Information Processing Group,
Technische Universität Berlin, Berlin, Germany
[3] RIKEN Brain Science Institute, Saitama, Japan
shimazaki@brain.riken.jp

Abstract. The maximum entropy method has been successfully employed to explain stationary spiking activity of a neural population by using fewer features than the number of possible activity patterns. Modeling network activity *in vivo*, however, has been challenging because features such as spike-rates and interactions can change according to sensory stimulation, behavior, or brain state. To capture the time-dependent activity, Shimazaki *et al.* (PLOS Comp Biol, 2012) previously introduced a state-space framework for the latent dynamics of neural interactions. However, the exact method suffers from computational cost; therefore its application was limited to only ~15 neurons. Here we introduce the pseudolikelihood method combined with the TAP or Bethe approximation to the state-space model, and make it possible to estimate dynamic pairwise interactions of up to 30 neurons. These analytic approximations allow analyses of time-varying activity of larger networks in relation to stimuli or behavior.

1   Introduction

As a consequence of recent advances in extracellular recording techniques, the parallel recording of activity from many neurons challenges standard statistical techniques to analyze population activity. Maximum entropy (ME) models - such as the Ising model or the Boltzmann machine - have been shown to provide useful descriptions of neuronal ensembles to reduce dimension of the system [7,11]. For a large system, however, exact inference of these models becomes computationally infeasible. Thus researchers have employed approximation or sampling methods. Another fundamental problem for the conventional ME models is the stationary assumption: the models assume temporarily constant features in neural activity patterns such as firing rates of individual neurons. This assumption is not valid, e.g., when *in vivo* activity is recorded while an animal performs a behavioral task. Ignoring such dynamics might result in erroneous model estimates and misleading interpretations [2,6].

The time-dependence of neural activity may be explained by including stimulus signals in the model, e.g., for analyses of early sensory cells [4]. However,

© Springer International Publishing AG 2016
A. Hirose et al. (Eds.): ICONIP 2016, Part III, LNCS 9949, pp. 104–110, 2016.
DOI: 10.1007/978-3-319-46675-0_12

the approach may become infeasible when analyzing neurons in higher brain areas in which receptive fields of neurons are not well defined. Alternatively, the model proposed in [8–10] suggested the sequential Bayesian estimation of dynamic neural interactions using a general smoothness prior. However, as previously mentioned, this dynamic interaction model is restricted by the problem of computational cost. Therefore it can only be utilized to analyze small populations ($N \leq 15$). Here, by combining this model with approximation methods, we provide a method for estimating interactions of neuronal populations consisting of up to 30 neurons. In addition to the point estimates of interaction parameters, the model provides credible intervals of those estimates. The analysis methods for a larger number of neurons allow us to better understand macroscopic quantities of a neural population, such as entropy and free energy, all in a time-resolved manner and with confidence intervals. This provides a new way to investigate effects of stimuli and behavior on neuronal populations in local circuitries.

2   Methods

2.1   The State-Space Model of Neural Interactions

We analyze simultaneous spiking activity of N neurons. To this end, we discretize time into T bins of length Δt, and represent the spiking activity of individual neurons as binary data X_i^t for $t = 1, \ldots, T$, where $X_i^t = 1$ if the neuron i spiked in time bin t and $X_i^t = 0$ if it was silent. For each time bin t, this yields an N-dimensional vector $\mathbf{X}^t = (X_1^t, \ldots, X_N^t)'$ describing the joint spike pattern of the neural population. Here $\mathbf{v}'$ indicates the transpose of vector $\mathbf{v}$.

Let $\mathbf{x} = (x_1, \ldots, x_N)'$ be an N-tuple binary variable. The probability of observing a pattern $\mathbf{x}$ is dictated by the joint probability mass function $p(\mathbf{x})$. The ME method constructs the most unstructured probability function given specified constraints. First, let us construct the model that is consistent with the observed spike-rates of the neurons $y_i = \frac{1}{T} \sum_{t=1}^{T} X_i^t$ ($i = 1, \ldots, N$) and joint spike-rates of two neurons $y_{ij} = \frac{1}{T} \sum_{t=1}^{T} X_i^t X_j^t$ ($i > j$), assuming that these rates are constant across the time-bins. In the following 'spike rates' denotes both, single and joint rates, unless stated otherwise. Let $\mathbf{y}$ be a vector of the constraints, $\mathbf{y} = (y_1, \ldots, y_N, y_{12}, \ldots, y_{N-1,N})'$. The ME distribution is then obtained by maximizing the following Lagrangian function:

$$\mathcal{L}[p] = -\sum_{\mathbf{x}} p(\mathbf{x}) \log p(\mathbf{x}) - \boldsymbol{\theta}' \left\{ \sum_{\mathbf{x}} \mathbf{F}(\mathbf{x}) p(\mathbf{x}) - \mathbf{y} \right\} - \alpha \left\{ 1 - \sum_{\mathbf{x}} p(\mathbf{x}) \right\}, \quad (1)$$

where $\mathbf{F}(\mathbf{x}) = (x_1, \ldots, x_N, x_1 x_2, \ldots, x_{N-1} x_N)'$, and $\boldsymbol{\theta}$ is a vector of the Lagrangian multipliers which we defined as $\boldsymbol{\theta} = (\theta_1, \ldots, \theta_N, \theta_{12}, \ldots, \theta_{N-1,N})'$. The last term in Eq. 1 ensures that $p(\mathbf{x})$ is normalized. The $p(\mathbf{x})$ maximizing Eq. 1 is an exponential family distribution,

$$p(\mathbf{x}|\boldsymbol{\theta}) = \exp\left[\boldsymbol{\theta}' \mathbf{F}(\mathbf{x}) - \psi(\boldsymbol{\theta})\right]. \quad (2)$$

The parameters $\boldsymbol{\theta}$ are adjusted to satisfy the constraints, $\sum_{\mathbf{x}} \mathbf{F}(\mathbf{x})p(\mathbf{x}|\boldsymbol{\theta}) = \mathbf{y}$. The same constraints are obtained by maximizing the likelihood of the recorded data for the model Eq. 2, and can be solved by conventional gradient ascent algorithms. The function $\psi(\boldsymbol{\theta})$ was substituted for $1 - \alpha$ and is called the *log-partition* as well as the free energy of the system.

The stationary assumption, however, is seldom appropriate in *in vivo* recordings in which an animal performs behavioral tasks. In a typical setup the animal repeats a task multiple times under the same conditions. Each repeated task, or *trial*, has a duration of time $T\Delta t$. If we again bin the data of the rth trial $(r = 1, \ldots, R)$ into time intervals of Δt, we obtain the binary data, $\mathbf{X}^{r,t} = (X_1^{r,t}, \ldots, X_N^{r,t})'$ for $t = 1, \ldots, T$. Assuming that the statistics of the neurons' activity at time t are the same for all trials (*stationary across-trials*), we obtain time-dependent spike-rates averaged across trials as $\mathbf{y}_t = \frac{1}{R}\sum_{r=1}^{R}\mathbf{F}(\mathbf{X}^{r,t})$. These data typically do not have enough observations R to reliably estimate the corresponding parameter $\boldsymbol{\theta}_t$ that is a solution under the maximum entropy (or maximum likelihood) principle independently applied at each time step.

To reliably estimate the time-varying model parameters in Eq. 2, a state-space framework was proposed by assuming that $\boldsymbol{\theta}_t$ follows smooth dynamics [9,10]. In the state-space model, the dynamics of $\boldsymbol{\theta}_t$ is modeled as a Markov chain dictated by a Gaussian prior $p(\boldsymbol{\theta}_t|\boldsymbol{\theta}_{t-1}) = \mathcal{N}(\boldsymbol{\theta}_{t-1}, \mathbf{Q})$ for $t = 2, \ldots, T$, where the covariance matrix $\mathbf{Q}$ is a diagonal matrix regulating the temporal smoothness. The prior for the first time-bin is $p(\boldsymbol{\theta}_1) = \mathcal{N}(\boldsymbol{\mu}, \boldsymbol{\Sigma})$. Given $\boldsymbol{\theta}_t$, the population activity at time t is assumed to be sampled from the pairwise interaction model $p(\mathbf{x}|\boldsymbol{\theta}_t)$ as given by Eq. 2. The approximate forward filtering and backward smoothing algorithms were developed for the non-Gaussian observation model, that were combined with optimization of the hyperparameters, $\mathbf{w} = \{\mathbf{Q}, \boldsymbol{\mu}, \boldsymbol{\Sigma}\}$, via an expectation-maximization (EM) algorithm [9,10]. In the filtering step, the filter density of $\boldsymbol{\theta}_t$ given the data up to the time-bin t, $\mathbf{X}^{1:t} \equiv \{\mathbf{X}^{r,1}, \ldots, \mathbf{X}^{r,T}\}_{r=1}^{R}$, is computed as

$$p(\boldsymbol{\theta}_t|\mathbf{X}^{1:t}, \mathbf{w}) = \frac{p(\mathbf{X}^t|\boldsymbol{\theta}_t)p(\boldsymbol{\theta}_t|\mathbf{X}^{1:t-1}, \mathbf{w})}{p(\mathbf{X}^t|\mathbf{X}^{1:t-1}, \mathbf{w})}$$

$$\propto \prod_{r=1}^{R} p(\mathbf{X}^{r,t}|\boldsymbol{\theta}_t)p(\boldsymbol{\theta}_t|\mathbf{X}^{1:t-1}, \mathbf{w}), \tag{3}$$

where $p(\boldsymbol{\theta}_t|\mathbf{X}^{1:t-1}, \mathbf{w})$ is called the one-step prediction density. Equation 3 can be recursively computed for $t = 2, \ldots, T$ because the one-step prediction density is computed using the prior and the filter density at the previous time-bin via the Chapman-Kolmogorov equation,

$$p(\boldsymbol{\theta}_t|\mathbf{X}^{1:t-1}, \mathbf{w}) = \int p(\boldsymbol{\theta}_t|\boldsymbol{\theta}_{t-1}, \mathbf{w})p(\boldsymbol{\theta}_{t-1}|\mathbf{X}^{1:t-1}, \mathbf{w})d\boldsymbol{\theta}_{t-1}. \tag{4}$$

The nonlinear recursive formulae were developed by utilizing the Laplace method that approximates the posterior density (Eq. 3) as a Gaussian distribution. Once the approximate filter density was constructed, the backward smoothing algorithm is applied to obtain the posterior density, $p(\boldsymbol{\theta}_t|\mathbf{X}^{1:T}, \mathbf{w})$ [3]. This recursive

Bayesian estimation is combined with the EM algorithm to optimize the hyperparameters $\mathbf{w}$ that maximizes the marginal likelihood,

$$l(\mathbf{X}^{1:T}|\mathbf{w}) = p(\mathbf{X}^1|\boldsymbol{\mu}, \boldsymbol{\Sigma}) \prod_{t=2}^{T} p(\mathbf{X}^t|\mathbf{X}^{1:t-1}, \mathbf{w}). \tag{5}$$

Construction of the posterior density and optimization of the hyperparameters were alternately performed until the marginal likelihood (Eq. 5) converges.

2.2 Approximation Methods for a Large-Scale Analysis

In the method described in the previous section, the maximum a posteriori (MAP) estimate for the parameters $\boldsymbol{\theta}_t$ was obtained by solving $\partial_{\boldsymbol{\theta}_t} \log p(\boldsymbol{\theta}_t|\mathbf{X}^{1:t}, \mathbf{w}) = \mathbf{0}$. In this calculation, the derivative of the logarithm of the first term in Eq. 3 includes $\partial_{\boldsymbol{\theta}_t} \psi(\boldsymbol{\theta}_t)$, which results in expectation of $\mathbf{F}(\mathbf{x})$ by $p(\mathbf{x}|\boldsymbol{\theta}_t)$, namely

$$\frac{\partial \psi(\boldsymbol{\theta}_t)}{\partial \boldsymbol{\theta}_t} = \sum_{\mathbf{x}} \mathbf{F}(\mathbf{x}) \exp\left[\boldsymbol{\theta}'_t \mathbf{F}(\mathbf{x}) - \psi(\boldsymbol{\theta}_t)\right] \equiv \boldsymbol{\eta}_t. \tag{6}$$

Furthermore, when we approximate the filter density by a Gauss distribution (the Laplace approximation), its precision matrix is identified with the negative Hessian of $\log p(\boldsymbol{\theta}_t|\mathbf{X}^{1:t}, \mathbf{w})$. This again includes the second derivative of $\psi(\boldsymbol{\theta}_t)$,

$$\frac{\partial^2 \psi(\boldsymbol{\theta}_t)}{\partial \boldsymbol{\theta}_t \partial \boldsymbol{\theta}'_t} = \sum_{\mathbf{x}} \mathbf{F}(\mathbf{x})\mathbf{F}(\mathbf{x})' \exp\left[\boldsymbol{\theta}'_t \mathbf{F}(\mathbf{x}) - \psi(\boldsymbol{\theta}_t)\right] - \boldsymbol{\eta}_t \boldsymbol{\eta}'_t, \tag{7}$$

which is known as the Fisher information matrix. The calculation of Eq. 7 involves not only the second order correlations such as the expectation of $x_i x_j$ but up to the fourth order.

Both Eqs. 6 and 7 involve summation of all 2^N possible patterns $\mathbf{x}$, which makes inference for larger networks ($N > 15$) infeasible. To obtain the MAP estimate of $\boldsymbol{\theta}_t$ without explicit computation of Eq. 6, we replaced $p(\mathbf{X}^t|\boldsymbol{\theta}_t)$ in Eq. 3 with the pseudolikelihood that has proven to well approximate the maximum likelihood estimate of $\boldsymbol{\theta}_t$ [1,5]. The pseudolikelihood, however, does not provide Eq. 7. To obtain it, we made further approximations. First, to limit the required statistics to at most the second-order, we approximate Eq. 7 by a diagonal matrix. We then tried two different approximations to obtain the diagonal of Eq. 7. The first is the Thouless-Anderson-Palmer (TAP) mean-field approximation [12,13]. This analytic approximation method is based on an asymptotic expansion of the model near the independent model, therefore works well if the second-order interactions are small and fails if they are large. The second approach is the Bethe approximation, which ignores loops in the model to ensure tractability. To find a solution of the Bethe approximation, we implemented three different procedures. The first is the belief propagation (BP) [16], which is computationally less expensive but does not garantee the convergence to the solution. The second algorithm is the concave-convex procedure (CCCP) [17] that is guaranteed to converge. To get the merits of the both procedures, we also tested a *hybrid* procedure which uses the BP to find the solution of the Bethe approximation and falls back to the CCCP whenever the BP fails to converge.

3   Results

We tested the approximation methods with a dynamic network consisting of 30 neurons simulated for 500 time bins. Underlying model parameters θ_t of Eq. 2 were generated as Gaussian processes using a smooth Gauss kernel function. The means of the processes for θ_i and θ_{ij} are normally distributed as $\mathcal{N}(-3, 0.1)$ and $\mathcal{N}(0, 1)$, respectively. 200 trials of binary data are sampled from the time-dependent model (data not shown). Figure 1 displays the result using the Bethe hybrid method that was combined with the pseudolikelihood to find the MAP estimate. In Fig. 1**A** the estimates of the model parameters (credible intervals as shaded area) track the underlying values (lines). Note that only 3 exemplary θ_{ij}'s are shown while there are many more (snapshots of the network activity are shown at Top). The approximation method allows us to smoothly estimate dynamics of the population rate and the probability that the network is silent (i.e., $p(\mathbf{x} = \mathbf{0}) = \exp(-\psi)$) (Fig. 1**B** Top) as well as the entropy (Fig. 1**B** Bottom).

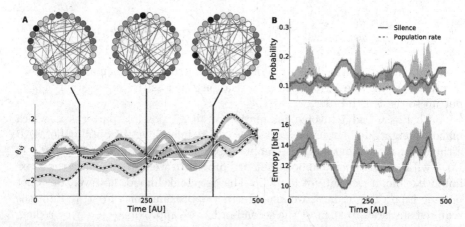

Fig. 1. Network dynamics inferred from spiking events of 30 neurons that are repeatedly measured 200 times. **A** Top: Network states at $t = 100, 250, 400$. Nodes represent neurons with color coding θ_i. Color of a link between nodes indicates strength of their interaction (dark blue '−2', dark red '+2'). Bottom: Dynamics of 3 exemplary interaction parameters, θ_{ij}. The lines denote the ground truth from which the binary data were sampled. The shaded areas are a 99 % credible interval. **B** Top: Estimated population rate and probability of network silence (solid lines) with 5 % and 95 % quantiles (shaded areas) obtained by resampling θ from the fitted distribution. Bottom: Entropy of the neural population. (Color figure online)

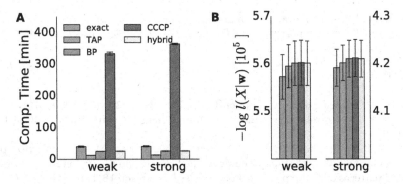

Fig. 2. Performance comparison of the different approximation methods applied to a network of 15 neurons. The pseudolikelihood method was used to obtain the MAP estimate, except for the exact method. **A**: Computation time required for the methods (from left to right: exact, TAP, BP, CCCP, and hybrid) to fit the model to data containing weak or strong interactions. **B**: The negative log-marginal likelihoods. In both plots error bars show standard deviation of 10 fitted networks.

To compare performance of the different approximation methods, the procedure described above is repeated for networks of 15 neurons, for which we can compute the exact marginal likelihood. We tested the methods to data with weak (means of θ_{ij} distributed according to $\mathcal{N}(0, 0.3)$) and strong ($\mathcal{N}(0, 1)$) interactions. The results show that the TAP and BP methods are much faster than the exact or CCCP method (Fig. 2**A**) while these approximation methods perform with similar goodness-of-fit (Fig. 2**B**). However, since the TAP and BP occasionally fail to provide a solution when fit to a population of larger than 15 neurons (data not shown), we conclude that the Bethe hybrid method provides robust and fast estimates for pairwise interactions of a large neural population.

4   Conclusion

Applying ME models to larger neural ensembles allowed to approximate network behavior in the limit of large N [14,15]. However, so far this was not possible for time-dependent spiking data, since this may lead to erroneous inference of functional network structure [2,6]. By combining the Bethe approximation methods realized by belief propagation and CCCP with the state-space framework, we are now allowed to investigate larger networks operating *in vivo* in a time-resolved manner. The larger scale analysis based on the exponential family distribution makes it possible to characterize macroscopic network properties of observed neuronal circuits, such as the free energy, entropy and criticality (i.e., maximized heat capacity). Characterizing these network properties in response to e.g., sensory stimuli and motor action is expected to bring new insights into processing of the neural systems.

Acknowledgments. CD acknowledges T. Toyoizumi for hosting his stay in RIKEN Brain Science Institute and K. Obermayer for valuable ideas and discussions. The custom-made Python programs were developed based on the code originally written by T. Sharp. CD was supported by the Deutsche Forschungsgemeinschaft GRK1589/2.

References

1. Besag, J.: Statistical analysis of non-lattice data. Stat. **24**, 179–195 (1975)
2. Brody, C.D.: Correlations without synchrony. Neural Comput. **11**(7), 1537–1551 (1999)
3. Brown, E.N., Frank, L.M., Tang, D., Quirk, M.C., Wilson, M.A.: A statistical paradigm for neural spike train decoding applied to position prediction from ensemble firing patterns of rat hippocampal place cells. J. Neurosci. **18**(18), 7411–7425 (1998)
4. Granot-Atedgi, E., Tkacik, G., Segev, R., Schneidman, E.: Stimulus-dependent maximum entropy models of neural population codes. PLoS Comput. Biol. **9**(3), e1002922 (2013)
5. Höfling, H., Tibshirani, R.: Estimation of sparse binary pairwise Markov networks using pseudo-likelihoods. J. Mach. Learn. Res. **10**, 883–906 (2009)
6. Renart, A., de la Rocha, J., Bartho, P., Hollender, L., Parga, N., Reyes, A., Harris, K.D.: The asynchronous state in cortical circuits. Science **327**(5965), 587–590 (2010)
7. Schneidman, E., Berry, M.J., Segev, R., Bialek, W.: Weak pairwise correlations imply strongly correlated network states in a neural population. Nature **440**(7087), 1007–1012 (2006)
8. Shimazaki, H.: Single-trial estimation of stimulus and spike-history effects on time-varying ensemble spiking activity of multiple neurons: a simulation study. J. Phys. Conf. Ser. **473**(1), 012009 (2013)
9. Shimazaki, H., Amari, S.I., Brown, E.N., Grün, S.: State-space analysis on time-varying correlations in parallel spike sequences. In: IEEE International Conference on Acoustics, Speech and Signal Processing (ICASSP) 2009, pp. 3501–3504. IEEE (2009)
10. Shimazaki, H., Amari, S.I., Brown, E.N., Grün, S.: State-space analysis of time-varying higher-order spike correlation for multiple neural spike train data. PLoS Comput. Biol. **8**(3), e1002385 (2012)
11. Shlens, J., Field, G.D., Gauthier, J.L., Grivich, M.I., Petrusca, D., Sher, A., Litke, A.M., Chichilnisky, E.: The structure of multi-neuron firing patterns in primate retina. J. Neurosci. **26**(32), 8254–8266 (2006)
12. Tanaka, T.: Mean-field theory of Boltzmann machine learning. Phys. Rev. E **58**(2), 2302 (1998)
13. Thouless, D.J., Anderson, P.W., Palmer, R.G.: Solution of 'solvable model of a spin glass'. Philos. Mag. **35**(3), 593–601 (1977)
14. Tkačik, G., Marre, O., Amodei, D., Schneidman, E., Bialek, W., Berry II, M.J.: Searching for collective behavior in a large network of sensory neurons. PLoS Comput. Biol. **10**(1), e1003408 (2014)
15. Tkačik, G., Mora, T., Marre, O., Amodei, D., Palmer, S.E., Berry, M.J., Bialek, W.: Thermodynamics and signatures of criticality in a network of neurons. Proc. Natl. Acad. Sci. **112**(37), 11508–11513 (2015)
16. Yedidia, J.S., Freeman, W.T., Weiss, Y.: Understanding belief propagation and its generalizations. Explor. Artif. Intell. New Millenn. **8**, 236–239 (2003)
17. Yuille, A.L.: CCCP algorithms to minimize the Bethe and Kikuchi free energies: convergent alternatives to belief propagation. Neural Comput. **14**(7), 1691–1722 (2002)

Features Learning and Transformation Based on Deep Autoencoders

Eric Janvier[1], Thierry Couronne[1], and Nistor Grozavu[2(✉)]

[1] Mindlytix, 33 Avenue Robert Andr Vivien, 94160 Saint-Mand, France
{e.janvier,t.couronne}@mindlytix.com
[2] LIPN CNRS UMR 7030, CNRS - Université Paris 13,
99, av. J-B Clement, 93430 Villetaneuse, France
Nistor.Grozavu@lipn.univ-paris13.fr

Abstract. Tag recommendation has become one of the most important ways of an organization to index online resources like articles, movies, and music in order to recommend it to potential users. Since recommendation information is usually very sparse, effective learning of the content representation for these resources is crucial to accurate the recommendation.

One of the issue of this problem is features transformation or features learning. In one hand, the projection methods allows to find new representations of the data, but it is not adapted for non-linear data or very sparse datasets. In another hand, unsupervised feature learning with deep networks has been widely studied in the recent years. Despite the progress, most existing models would be fragile to non-Gaussian noises, outliers or high dimensional sparse data. In this paper, we propose a study on the use of deep denoising autoencoders and other dimensional reduction techniques to learn relevant representations of the data in order to increase the quality of the clustering model.

In this paper, we propose an hybrid framework with a deep learning model called stacked denoising autoencoder (SDAE), the SVD and Diffusion Maps to learn more effective content representation. The proposed framework is tested on real tag recommendation dataset which was validated by using internal clustering indexes and by experts.

1 Introduction

Data mining, or knowledge discovery in databases (KDD), an evolving area in information technology, has received much interest in recent studies. The aim of data mining is to extract knowledge from data. The data size can be measured in two dimensions, the size of features and the size of observations. Both dimensions can take very high values, which can cause problems during the exploration and analysis of the dataset [12]. Models and tools are therefore required to process data for an improved understanding. Indeed, datasets with a large dimension (size of features) display small differences between the most similar and the least similar data. In such cases it is thus very difficult for a learning algorithm to detect the similarity of variables that define the clusters [9].

© Springer International Publishing AG 2016
A. Hirose et al. (Eds.): ICONIP 2016, Part III, LNCS 9949, pp. 111–118, 2016.
DOI: 10.1007/978-3-319-46675-0_13

In hybrid methods, learning of item representations (also called item latent factors in some models) is crucial for the recommendation accuracy especially when the tag-item matrix is extremely sparse [15].

The main purpose of unsupervised learning methods is to extract generally useful features from unlabelled data, to detect and remove input redundancies, and to preserve only essential aspects of the data in robust and discriminative representations. Unsupervised methods have been routinely used in many scientific and industrial applications. In the context of neural network architectures, unsupervised layers can be stacked on top of each other to build deep hierarchies [7].

Unsupervised feature learning algorithms aim to find good representations for data, which can be used for different tasks i.e. classification, clustering, reconstruction, visualization,... Recently, deep networks such as stacked autoencoders (SAE) and diffusion maps (DM) have shown high feature learning performance [8].

Despite the progress, robust feature learning is still faced with challenges due to noise and outliers which are commonly appeared in the real-world data. In order to improve the antinoise ability of the deep networks, a new method was proposed by modifying the traditional stacked autoencoder to learn useful features from corrupted data and developed the stacked denoising autoencoder (SDAE) [8,14]. By corrupting the input data and using denoising criterion, the SDAE could learn robust representations and achieve good performance under different types of noises and to learn only the relevant features structure.

In this study, we focus on reducing the dimensions of the feature space as part of the unsupervised learning through different methods: Singular Value Decomposition (SVD), Diffusion Maps (DM) and Stacked Denoising Autoencoder (SDAE).

After transforming the features space, the new dataset will be clustered in order to detect relevant groups of tags which will be used furtherer for the recommendation. In this work a two-level topological clustering linked with the hierarchical clustering is used to visualize the results and to improve the computational time of the clustering model.

The rest of this paper is organized as follows: we present the proposed feature learning framework in Sect. 3 after introducing the feature transformation problem in Sect. 2. Section 3 introduces the use of the two-level topological clustering: the Self-Organizing Maps (SOM) with the Hierarchical Clustering which further is used for the tag recommendation. In the Sect. 4 we show the first experimental results on a real dataset. Finally we drew some conclusions and the possibilities of further research in this area.

2   Unsupervised Transformation of the Feature Space

Predictive models capable to classify new objects generally require learning by using labeled data. Unfortunately, only a small amount of labeled learning data may be available because of the cost of manual annotation of the data. Recent research has been focused on the use of large amounts of available unlabeled data, including: the transformation, the reduction of dimensionality, hierarchical representations of the variables ("deep learning"), kernel based learning, etc.

The unsupervised learning is often used for clustering data and rarely as a data preprocessing method. However, there are many methods that produce new data representations from unlabeled data. These unsupervised methods are sometimes used as a preprocessing tool for supervised or unsupervised learning models [4].

Given a data matrix represented as vectors of variables (p observations and n features), the goal of the unsupervised transformation of feature space is to produce another data matrix of dimension (p, n') (the transformed representation of n' new latent variables) or a similarity matrix between the data of size (p, p). Applying a model on the transformed matrix should provide better results compared to the original dataset.

The transformation of the feature space is done in two steps. First, we decompose the sparse data matrix using a normalization method and the SVD (Singular Value Decomposition). Then the matrix of latent variables obtained after this decomposition is used to learn the feature representation space using the Diffusion Maps and the SDAE method.

2.1 Matrix Decomposition and Normalization

The approximate factorization and tensor factorization (or decomposition) of a matrix have a main contribution in the improvement of data and the extraction of latent components. A common point for noisy detection, reduction of the model, the reconstruction of feasibility is to replace original data by an approximate representation of reduced dimensions obtained via a matrix factorization or decomposition. The concept of matrix factorization is used in a wide range of important applications and each matrix factorization is a different assumption about the components (factors) of matrices and their underlying structures, and this choice is an essential process in each application domain [4].

Very often, the datasets to be analyzed are nonnegative (or partially positive), and sometimes they also have a sparse representation. For these datasets, it is better to take into account these constraints in the analysis and to extract factors with physical meaning or a reasonable interpretation, and thus to avoid absurd or unpredictable results.

The singular value decomposition (SVD) treats the rows and columns in a symmetrical manner, and thus provides more information on the data matrix. This method also allows us to sort the information in the matrix so that, in general, the relevant part becomes visible. This property makes the SVD so useful in data mining and many other areas.

The bidiagonalisation GK (Golub-Kahan) method was originally formulated [3] for computing the SVD. This method can be also used to calculate a partial bidiagonalisation:

$$AQ_k = P_{k+1}B_{k+1}$$

where A is the data matrix, B_{k+1} are bidiagonal, and the clones Q_k and P_{k+1} are orthonormal.

With this decomposition, the approximations of singular values and singular vectors can be calculated similarly by tridiagonalisation. Indeed, it can be shown that the procedure of the GK bidiagonalisation is equivalent to applying the Lanczos tridiagonalisation on a symmetric matrix with a particular initial vector.

In our method we use this technique for the sparse data and Principal Component Analysis (PCA) is used for non-sparse datasets.

2.2 Diffusion Maps

The diffusion maps (DM) are based on defining a Markov random walk on the graph of the data. By performing the random walk for a number of timesteps, a measure for the proximity of the datapoints is obtained. The DM distance is defined using this measure. The key idea behind the diffusion distance is that it is based on integrating over all paths through the graph. This makes the diffusion distance more robust than, e.g., the geodesic distance employed in Isomap [7].

2.3 Deep Autoencoders

Denoising autoencoder (DAE) was proposed to overcome the limitations of autoencoders by reconstructing denoised inputs x from corrupted, noisy inputs $\tilde{x}$. DAEs avoids overfitting and learns better, non-trivial features by introducing stochastic noises to training samples. To generate corrupted inputs $\tilde{x}$ from their original value x it can be done with several different stochastic corruption criteria $q_D(\tilde{x}|x)$, including adding Gaussian random noise, randomly masking dimensions to zero, etc.

The objective function of DAEs remains the same as typical autoencoders. Note that the objective function minimizes the discrepancy between reconstructions and original, uncorrupted inputs x, not the corrupted inputs $\tilde{x}$ [6].

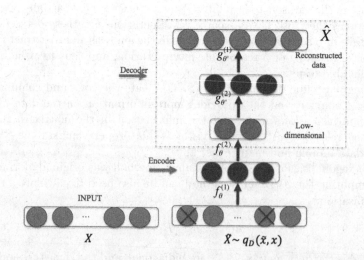

Fig. 1. Denoising autonecoder

Algorithme 1 . Transformation of the feature space and data coding

---

Inputs:
Learning (Training) data
Output:
New representation of the dataset)
Begin
1. Apply the diagonalization and factorization of the initial matrix (training data);
2. Apply the Diffusion Maps on the dataset;
3. Train the DSAE on the dataset;
4. Concatenate the obtained factors;
2-levels Clustering:
5. Construct the prototypes matrix using the SOM algorithm
6. Apply the hierarchical clustering on the prototypes map.
End

Stacking DAES (Fig. 1) on top of each other allows the model to learn more complex mapping from input to hidden representations. Just as other deep models including deep belief networks, training stacked DAEs is also done in twophase: layerwise, greedy pre-training and fine-tuning.

Unlike typical deep models that are extended by adding layers from bottom to top in pre-training, stacked DAEs are extended by adding layers in the middle of them. More specifically, the pre-training of stacked DAEs is done by the following steps. First, train bottom layer DAE with encoding function $y^{(1)} = f^{(1)}(x, \theta_f^{(1)})$ and decoding function $z^{(1)} = g^{(1)}(y^{(1)}, \theta_g^{(1)})$.

Train more DAEs in a similar way until the desired number of layers is achieved. After pre-training, the weights and biases of stacked DAE are fine-tuned by back-propagation as ordinary neural networks [6,7].

For a training dataset A, the first step of the proposed method is presented as following:

1. Normalization: $\widehat{A} = A * diag(std(A))^{\frac{1}{2}}$
2. Dimensionality reduction of the dataset $\widehat{A}$ by matrix factorisation: $svd(\widehat{A}) = [U_{\widehat{A}} S_{\widehat{A}} V_{\widehat{A}}]$
   For each column of $U_{\widehat{A}}$, $U_k = \frac{U_k}{\|U_k\|}$, where k is the number of retained eigenvectors

In the following (Algorithm 1) we present the proposed unsupervised learning algorithm for feature space transformation.

3   Topological Clustering

Topological learning is a recent direction in Machine Learning which aims to develop methods grounded on statistics to recover the topological invariants from the observed data points. Most of the existed topological learning approaches are based on graph theory or graph-based clustering methods. The topological learning is one of the most known technique which allow clustering and visualization simultaneously. At the end of the topographic learning, the "similar" data will be collect in clusters, which correspond to the sets of similar observations. These clusters can be represented by more concise information than the brutal listing of their patterns, such as their gravity center or different statistical moments. As expected, this information is easier to manipulate than the original data points. The neural networks based techniques are the most adapted to topological learning as these approaches represent already a network (graph) [5]. The models that interest us in this paper are those that could make at the same time the dimensionality reduction and clustering using Self-Organizing Maps (SOM) [10] in order to characterize clusters. SOM models are often used for visualization and unsupervised topological clustering. Its allow projection in small spaces that are generally two dimensional. Some extensions and reformulations of the SOM model have been described in the literature [1,5,11].

For map clustering we use traditional hierarchical clustering combined with Davides-bouldin index to choose optimal partition [13].

4   Experimental Results

This experiment is conducted with 3 datasets containing the description of 54000 web domains made with 800 topics. The 800 topics are extracted via a semantic analysis of words crawled on each domain, then the count of specific words is used to profile the domains by topic. The matrix is quite empty with a vast majority of domains qualified by very few topics. Topics are also classified into 50 groups according to level of similarity/dissimilarity.

Data set 1 contains all the domains/lines (54000) and all the topics/columns (800). Data set 2 (short) contains only the domains where largest number of topics/lines is informed (4910) and all the topics/columns (800). Data set 2 (short) contains only the domains where largest number of topics/lines is informed (4910) and topics/columns with highest weight (280).

Since clustering is an unsupervised process and most of theses algorithms are very sensitive to their initial assumptions, some evaluation is required to describe/analyze the clustering results [2]. Cluster validity represents the goodness measure of a clustering result relative to others created by other clustering algorithms, or by the same algorithm using different parameter values.

In general, there are three fundamental criteria to investigate the cluster validity: external criteria, internal criteria, and relative criteria. In the following we show main clustering validity indices.

Table 1 shows the quality of the clustering results in terms of the Davies-Bouldin index. It is easy to see that the proposed method outperforms the classical clustering for different numbers of clusters. The expert validated the results by indicated that the number of clusters should be 50, that means that the method proposed here outperforms a lot the clustering results for 50 clusters.

The same analysis can be made for the quality of the obtained topological map, where the quantization and topographic error decrease by using the deep learning features transformation (Table 2).

Table 1. Davies-Bouldin index obtained on the datasets

Method	Proposed model				Classical clustering			
nb cl.	5	15	30	50	5	15	30	50
dataset1	0.9259	0.7911	0.7301	0.7445	0.9736	0.8107	0.8005	0.7742
dataset2	0.5775	0.5870	0.7316	0.6867	0.5840	0.6776	0.7806	21.8992
dataset3	0.9177	0.8524	0.7511	6.5137	0.9682	0.9404	0.7911	27.8449

Table 2. Topological and quantization errors of the maps

Method	Proposed model		Classical clustering	
nb cl.	Quantization error	Topographic error	Quantization error	Topographic error
dataset1	0.56	0.41	3.245	0.73
dataset2	0.059	0.044	2.974	0.047
dataset3	0.124	0.142	3.436	0.071

5 Conclusion

In this work, we proposed k with a deep learning model called stacked denoising autoencoder (SDAE), the SVD and Diffusion Maps to learn more effective content representation. The transformed data was clustered using a two-level clustering model: SOM and hierarchical clustering to cluster the users web behaviour. The results on a real tag recommender dataset show that this approach aouperforms the classical clustering method and was also validated by the experts. As future works, we plan to test this method on different synthetic datasets and to compare it with other approaches. Also, some current work is made on the evaluation of the recommender system which use this approach.

References

1. Bishop, C.M., Svensén, M., Williams, C.K.I.: GTM: The generative topographic mapping. Neural Comput. **10**(1), 215–234 (1998)
2. Saporta, G.: Probabilits, analyse des donnes et statistiques. Editions Technip (2006)
3. Golub, G.H., Kahan, W.: Calculating the singular values and pseudo-inverse of a matrix. SIAM J. Numer. Anal. **2**, 205–224 (1965)
4. Grozavu, N., Bennani, Y., Labiod, L.: Feature space transformation for transfer learning. In: The 2012 International Joint Conference on Neural Networks (IJCNN), Brisbane, 10–15 June 2012, pp. 1–6 (2012)
5. Grozavu, N., Bennani, Y., Lebbah, M.: From variable weighting to cluster characterization in topographic unsupervised learning. In: Proceedings of International Joint Conference on Neural Network. IJCNN (2009)
6. Kang, L., Lee, K.T., Eun, J., Park, S.E., Choi, S.: Stacked denoising autoencoders for face pose normalization. In: Lee, M., Hirose, A., Hou, Z.-G., Kil, R.M. (eds.) Neural Information Processing. Theoretical Computer Science and General Issues, vol. 8227, pp. 241–248. Springer, Heidelberg (2013)

7. Van der Maaten, L., Postma, E., Van den Herik, H.: Dimensionality reduction: a comparative review. Technical report TiCC TR 2009–005 (2009)
8. Qi, Y., Wang, Y., Zheng, X., Wu, Z.: Robust feature learning by stacked autoencoder with maximum correntropy criterion. In: IEEE International Conference on Acoustics, Speech and Signal Processing, ICASSP 2014, Florence, 4–9 May 2014, pp. 6716–6720 (2014). doi:10.1109/ICASSp.2014.6854900
9. Roth, V., Lange, T.: Feature selection in clustering problems. In: Thrun, S., Saul, L., Schölkopf, B. (eds.) Advances in Neural Information Processing Systems, vol. 16. MIT Press, Cambridge (2003)
10. Kohonen, T.: Self-organizing Maps. Springer, Heidelberg (2001)
11. Verbeek, J., Vlassis, N., Krose, B.: Self-organizing mixture models. Neurocomputing **63**, 99–123 (2005)
12. Verleysen, M., Francois, D., Simon, G., Wertz, V.: On the effects of dimensionality on data analysis with neural networks. In: Mira, J., Álvarez, J.R. (eds.) IWANN 2003. LNCS, vol. 2687, pp. 105–112. Springer, Heidelberg (2003). doi:10.1007/3-540-44869-1_14
13. Vesanto, J., Alhoniemi, E.: Clustering of the self-organizing map. IEEE Trans. Neural Netw. **11**(3), 586–600 (2000)
14. Vincent, P., Larochelle, H., Bengio, Y., Manzagol, P.A.: Extracting and composing robust features with denoising autoencoders. In: Proceedings of the 25th International Conference on Machine Learning. ICML 2008, pp. 1096–1103. ACM, New York (2008)
15. Wang, H., Shi, X., Yeung, D.Y.: Relational stacked denoising autoencoder for tag recommendation. In: Proceedings of the Twenty-Ninth AAAI Conference on Artificial Intelligence. AAAI 2015, pp. 3052–3058. AAAI Press (2015). http://dl.acm.org/citation.cfm?id=2888116.2888141

t-Distributed Stochastic Neighbor Embedding with Inhomogeneous Degrees of Freedom

Jun Kitazono[1]([envelope]), Nistor Grozavu[2], Nicoleta Rogovschi[3], Toshiaki Omori[1], and Seiichi Ozawa[1]

[1] Graduate School of Engineering, Kobe University,
1-1, Rokkodai-cho, Nada, Kobe, Hyogo, Japan
{kitazono,omori}@eedept.kobe-u.ac.jp, ozawasei@kobe-u.ac.jp
[2] LIPN UMR CNRS 7030, Sorbonne Paris Cité, Université Paris 13,
99 av. J-B Clément, 93430 Villetaneuse, France
nistor.grozavu@lipn.univ-paris13.fr
[3] LIPADE, Université Paris Descartes, 45 Rue des Saints-Peres, 75006 Paris, France
nicoleta.rogovschi@parisdescartes.fr

Abstract. One of the dimension reduction (DR) methods for data-visualization, t-distributed stochastic neighbor embedding (t-SNE), has drawn increasing attention. t-SNE gives us better visualization than conventional DR methods, by relieving so-called crowding problem. The crowding problem is one of the curses of dimensionality, which is caused by discrepancy between high and low dimensional spaces. However, in t-SNE, it is assumed that the strength of the discrepancy is the same for all samples in all datasets regardless of ununiformity of distributions or the difference in dimensions, and this assumption sometimes ruins visualization. Here we propose a new DR method inhomogeneous t-SNE, in which the strength is estimated for each point and dataset. Experimental results show that such pointwise estimation is important for reasonable visualization and that the proposed method achieves better visualization than the original t-SNE.

Keywords: SNE · t-SNE · Dimensionality reduction · Degrees of freedom

1 Introduction

In the last decades, a number of dimension reduction methods for data-visualization have been proposed such as Isomap [1], Locally linear embedding [2], Laplacian eigenmaps [3], Stochastic Neighbor Embedding (SNE) [4], and Diffusion maps [5]. Recently, a DR method, t-Distributed SNE (t-SNE) [6], has received increasing attention. t-SNE give us better visualization than conventional DR methods, especially when dataset contains high-dimensional data manifold. For instance, t-SNE can reveal natural clusters of the MNIST handwritten digit dataset without supervised class information, whereas other conventional methods fail [6]. The reason why t-SNE is better than other methods

© Springer International Publishing AG 2016
A. Hirose et al. (Eds.): ICONIP 2016, Part III, LNCS 9949, pp. 119–128, 2016.
DOI: 10.1007/978-3-319-46675-0_14

for high-dimensional data is that it alleviate so-called crowding problem [6]. The crowding problem is one of the curses of dimensionality that low-dimensional space has narrower area for accommodating points than high-dimensional space. Due to this problem, in the conventional DR methods, positional relationship among points is disordered and clusters are merged. To compensate for this discrepancy between high and low dimensions, t-SNE puts emphasis on repelling points far apart in the low-dimensional space that are at moderate distance in the original high-dimensional space, and secure the area for modeling the positional relationship. However, in t-SNE, the strength of the repelling power is constant independent of dataset. This is not reasonable because the level of the discrepancy changes depending on the dimension of intrinsic data manifold.

In [7], the strength is optimized by estimating a parameter of t-distribution, the number of degree of freedom. However, as shown in [7], this optimization does not necessarily improve visualization compared to the case when the value is predefined. This can be because although it is assumed that the strength is the same for all samples, it is not true when the intrinsic dimension of the data manifold is inhomogeneous; for example, let us consider a dataset that consists of two clusters: the intrinsic dimension of one cluster is high and that of the other is low. In this case, the power should be stronger for samples in the first cluster, and weaker for those in the second cluster. Moreover, in other cases, the intrinsic dimension may differ sample by sample.

In this paper, we propose a new DR method, inhomogeneous t-SNE, in which the strength is optimized for each point, by estimating pointwise number of degree of freedom. SNE and t-SNE are considered as special cases of the proposed method.

The remainder of this paper is organized as follows. In Sects. 2 and 3, we describe SNE and t-SNE, which are the bases of the proposed method, inhomogeneous t-SNE. In Sect. 4, we present the proposed method. Experimental results are shown in Sect. 5. Section 6 concludes the paper.

2   Stochastic Neighbor Embedding

Given a dataset $X = \{x_1, \ldots, x_N\}$ that consists of N samples in a high-dimensional space, the common goal among dimension reduction methods for data-visualization is to map X to $Y = \{y_1, \ldots, y_N\}$ in a low (two or three)-dimensional space while preserving some aspects of positional relationship among X. Stochastic Neighbor Embedding (SNE) begins by converting a distance $d(x_i, x_j)$ (e.g. the Euclidean distance $\|x_i - x_j\|$) to probability $p_{j|i}$ that represents similarity of x_j to x_i, using normalized Gaussian kernel as follows:

$$p_{j|i} = \frac{\exp(-d(x_i, x_j)^2/2\sigma_i^2)}{\sum_{k \neq i} \exp(-d(x_i, x_k)^2/2\sigma_i^2)}, \ p_{i|i} = 0, \tag{1}$$

where σ_i^2 is the variance of the Gaussian kernel. The value of σ_i is determined so that the perplexity $2^{-\sum_j p_{j|i} \log_2 p_{j|i}}$ equals a user-defined value u, by a binary search [4] or robust root-finding method [8]. Then, also in the low-dimensional space, probability $q_{j|i}$ is computed from the Euclidean distance $||\boldsymbol{y}_i - \boldsymbol{y}_j||$ using normalized Gaussian kernel:

$$q_{j|i} = \frac{\exp(-||\boldsymbol{y}_i - \boldsymbol{y}_j||^2)}{\sum_{k \neq i} \exp(-||\boldsymbol{y}_i - \boldsymbol{y}_k||^2)}, \ q_{i|i} = 0. \tag{2}$$

The coordinates of $\boldsymbol{y}_i$ are determined by minimizing KL divergence between the distribution P_i and Q_i:

$$C_{\text{SNE}}(Y) = \sum_i KL(P_i||Q_i) = \sum_i \sum_{j \neq i} p_{j|i} \log \frac{p_{j|i}}{q_{j|i}}. \tag{3}$$

Typically, the minimization of Eq. (3) is performed using the following gradient:

$$\frac{\partial C_{\text{SNE}}(Y)}{\partial \boldsymbol{y}_i} = 2 \sum_{j \neq i} (p_{j|i} - q_{j|i} + p_{i|j} - q_{i|j})(\boldsymbol{y}_i - \boldsymbol{y}_j). \tag{4}$$

3   t-Distributed Stochastic Neighbor Embedding

t-SNE differs from SNE in two respects. First, t-SNE utilizes symmetrized probability $p_{ij} = \frac{p_{j|i} + p_{i|j}}{2N}$. Although this symmetrization makes the gradient of the cost function somewhat simpler, not the first but the second one is the key to alleviate the crowding problem: in t-SNE, probability q_{ij} in low-dimensional space that measure the similarity is defined not by the Gaussian but by t-distribution kernel:

$$q_{ij} = \frac{(1 + ||\boldsymbol{y}_i - \boldsymbol{y}_j||^2)^{-1}}{\sum_{k,l(k \neq l)} (1 + ||\boldsymbol{y}_k - \boldsymbol{y}_l||^2)^{-1}}, \ q_{ii} = 0. \tag{5}$$

The cost function and its gradient are given by

$$C_{\text{tSNE}}(Y) = KL(P||Q) = \sum_{i,j(i \neq j)} p_{ij} \log \frac{p_{ij}}{q_{ij}}, \tag{6}$$

$$\frac{\partial C_{\text{tSNE}}(Y)}{\partial \boldsymbol{y}_i} = 4 \sum_{j \neq i} (p_{ij} - q_{ij})(1 + ||\boldsymbol{y}_i - \boldsymbol{y}_j||^2)^{-1}(\boldsymbol{y}_i - \boldsymbol{y}_j). \tag{7}$$

t-Distribution has much thicker tale than Gaussian distribution. Therefore, to model small value of p_{ij}, $\boldsymbol{y}_i$ and $\boldsymbol{y}_j$ must be put far away each other. This is known as the mechanism in t-SNE to compensate for the crowding problem that low-dimensional space has narrower space than high-dimensional one for accommodating points.

4   Inhomogeneous t-SNE

Although t-SNE can relieve the crowding problem by utilizing t-distribution, the strength of putting away $\boldsymbol{y}_i$ and $\boldsymbol{y}_j$ with small p_{ij} is the same for all the samples

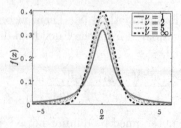

Fig. 1. t-Distributions for different values of ν.

and for any dataset. Therefore, it can be considered that controlling the strength by adapting the heaviness of the tail of q_{ij} can improve visualization by t-SNE. Here, we propose a method that adopts more general formula of t-distribution than in the original t-SNE. In the formula, there is a parameter called the number of degree of freedom, which control the heaviness of the tail. By optimizing the parameter for each point, the proposed method is expected to tune the strength depending on data-dimension and to capture intrinsic inhomogeneity in datasets. In the following subsections, we first describe the general formula of t-distribution. Then, we define the cost function of the proposed method and explain the way to optimize it.

4.1 Degrees of Freedom

Generally, the probability density function of t-distribution is given by

$$f(x) = \frac{\Gamma\left(\frac{\nu+1}{2}\right)}{\sqrt{\nu\pi}\Gamma\left(\frac{\nu}{2}\right)}\left(1 + \frac{x^2}{\nu}\right)^{-\frac{\nu+1}{2}}, \tag{8}$$

where $\Gamma(\cdot)$ is the gamma function and ν is the parameter called the number of degrees of freedom. The density functions for several νs are shown in Fig. 1. As ν decreases, the tail of the distribution becomes heavier. When $\nu = 1$, the density function is $f(x) \propto (1 + x^2)^{-1}$ and is used in the original t-SNE. When $\nu = \infty$, $f(x)$ becomes the Gaussian distribution: $\frac{1}{\sqrt{2\pi}}\exp(-\frac{x^2}{2})$.

4.2 Cost Function and Its Gradient

In the proposed method, we define the probability in the low-dimensional space as

$$q_{j|i} = \frac{\left(1 + \frac{\|\boldsymbol{y}_i - \boldsymbol{y}_j\|^2}{\nu_i}\right)^{-\frac{\nu_i+1}{2}}}{\sum_{k\neq i}\left(1 + \frac{\|\boldsymbol{y}_i - \boldsymbol{y}_k\|^2}{\nu_i}\right)^{-\frac{\nu_i+1}{2}}}, \quad q_{i|i} = 0. \tag{9}$$

The cost function is given by the same form as that of SNE (Eq. (3)): $C(Y) = \sum_i \sum_{j\neq i} p_{j|i} \log \frac{p_{j|i}}{q_{j|i}}$. In the above definition of $q_{j|i}$, pointwise parameter ν_i is introduced. By optimizing this parameter, the proposed method is expected to tune the thickness of the tail for each dataset and for each point.

After some algebra, we get the gradient of the cost function:

$$\frac{\partial C(Y)}{\partial \boldsymbol{y}_i} = \sum_{j \neq i} \left\{ \frac{\nu_i + 1}{\nu_i}(p_{j|i} - q_{j|i})\left(1 + \frac{d_{ij}^2}{\nu_i}\right)^{-1} \right.$$
$$\left. + \frac{\nu_j + 1}{\nu_j}(p_{i|j} - q_{i|j})\left(1 + \frac{d_{ij}^2}{\nu_j}\right)^{-1} \right\}(\boldsymbol{y}_i - \boldsymbol{y}_j), \tag{10}$$

$$\frac{\partial C(Y)}{\partial \nu_i} = \sum_{j \neq i} \left\{ \frac{1}{2}\log\left(1 + \frac{d_{ij}^2}{\nu_i}\right) - \frac{\nu_i + 1}{2\nu_i^2}d_{ij}^2\left(1 + \frac{d_{ij}^2}{\nu_i}\right)^{-1} \right\}(p_{j|i} - q_{j|i}), \tag{11}$$

where d_{ij} is $\|\boldsymbol{y}_i - \boldsymbol{y}_j\|$. When we fix $\nu_i = \infty$ for all i, the proposed method coincides with SNE. When $\nu_i = 1$ for all i, the proposed method becomes an asymmetric version of t-SNE.

4.3 Optimization

In this paper, we optimize the cost function using one of quasi-Newton methods, Limited-memory Broyden–Fletcher–Goldfarb–Shanno (L-BFGS) method as in Vladymyrov and Carreira-Perpinán [9]. Although simply applying L-BFGS can achieve good visualization results, we found in preliminary experiments that the following three-step procedure further improve results. In the first step, all of the probability $p_{j|i}$ is multiplied by some constant (e.g. 4). This multiplication is known as "early exaggeration", which was introduced in the original t-SNE paper [6]. Additionally in the first step, all of the ν_i are fixed to a relatively higher value (e.g. 10) and only $\boldsymbol{y}_i$ $(i = 1\ldots, N)$ are optimized. In the second step, $p_{j|i}$ is turned back to the original value, and $\boldsymbol{y}_i$ are optimized. Finally in the third step, both $\boldsymbol{y}_i$ and ν_i are optimized. In all the experiments in this paper, we multiplied $p_{j|i}$ by 4 and set the initial value of ν_i to 10. Additionally, to keep ν positive, we represent ν as $\epsilon + \xi^2$, where ϵ is an infinitesimal positive constant, and optimize ξ. In this paper, we set $\epsilon = 10^{-3}$.

5 Experiments

To show the validity of the concept of the proposed method, we first investigate how the value of ν changes depending on the dimension of data, using normally-distributed samples. Then, we apply the proposed method to four datasets and compare it to SNE and t-SNE. In all the experiments in this paper, if the dimension of the dataset was higher than 50, we preprocessed the dataset by using PCA to reduce the dimensionality to 50 before applying the methods. We set the perplexity $u = 30$. The initial value of y is sampled from a normal distribution with zero mean and variance 10^{-8}. We used the toolbox of L-BFGS method provided by Mark Schmidt, publicly available on http://www.cs.ubc.ca/~schmidtm. For SNE and t-SNE, we utilize the software provided by Laurence van der Maaten, publicly available on https://lvdmaaten.github.io/drtoolbox/.

5.1 Experiment 1

We drew $N = 1,000$ samples from 3, 10, 20, and 30 dimensional normal distributions, and applied the proposed method to the samples to check how the value of ν changes with increase in the dimension of the data. Figure 2 shows the histograms of ν. From this figure, we can see that ν becomes smaller as the dimension gets higher. This results indicate that ν reflect the dimension of the data and that ν must be optimized in concert with data distributions, to correctly visualize the relationship of samples. Moreover, for each of 3, 10, 20, and 30 dimensional samples, the variance of ν is substantial. This is because the density and points' distribution around each point is not uniform. Thus, even in simple datasets such as normally-distributed samples, the inhomogeneity appears.

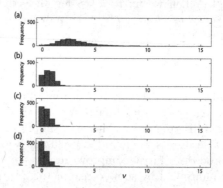

Fig. 2. Results of Experiment 1. Histograms of ν for normally distributed data. (a) 3-dimensional, (b) 10-dimensional, (c) 20-dimensional, and (d) 30-dimensional cases.

5.2 Experiment 2

We applied SNE, t-SNE, and the proposed method to four datasets. We briefly explain the datasets below.

Swiss Roll. We generated 3,000 samples in three-dimensional space via the equation $x_i = (t_i \cos(t_i), t_i \sin(t_i), 30w_i)$, where $t_i = 3\pi(1 + 2v_i)/2$, and v_i and w_i are random numbers drawn from a uniform distribution with the interval $(0, 1)$. Data lie on two-dimensional manifold.

COIL-20. The COIL-20 dataset contains images of 20 objects. Each object was placed on a turntable and rotated, and images were taken at every 5 degrees (72 images for each object). In total, it contains 1,440 grayscale images. The size of each image is 128×128 pixels.

2D sheet and 5D Gaussian. This is a five-dimensional dataset composed of two clusters. The cluster 1 is first generated from a uniform distribution in two-dimensional space $((0, 1)^2)$, and zeros are concatenated to make the cluster five-dimensional. The cluster 2 is generated from five-dimensional Gaussian distribution with mean $(0.5, \ldots, 0.5)$ and covariance I (identity matrix). 500 samples are generated for each cluster.

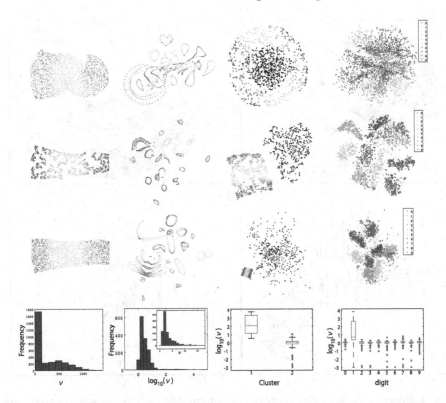

Fig. 3. Results of Experiment 2. The columns 1–4 correspond to the results for the Swiss-roll, COIL-20, 2D sheet and 5D Gaussian, and MNIST datasets, respectively. The rows 1–3 represents results of SNE, t-SNE, and the proposed method, respectively. The first and second panels of the last row shows the histograms of ν for the Swiss-roll and COI-20 datasets. The inset in the second panel is a zoom-in around small ν on a linear scale. The third panel of the last row shows boxplot of ν for each cluster. The fourth panel of the last row shows boxplot of ν for each digit. The rainbow-like color in the first and third columns indicate the value of the random number t_i and the value of the first coordinate of the cluster 1, respectively.

MNIST. The MNIST dataset consists of grayscale images of handwritten digits. The size of each image is $28 \times 28 = 784$ pixels. Each image corresponds to one of ten classes (0–9). We selected 300 images from each of the class.

The first column of Fig. 3 shows the results for the Swiss-roll dataset. It can be seen that in SNE and the proposed method, the roll is successfully unfolded and the sheet-like structure (i.e., positional relationship among samples on the two-dimensional manifold) is preserved. In contrast in t-SNE, the sheet is teared, and false clusters that do not exist in the original data emerge. This is because t-SNE overly puts points far away that are at moderate distance in the original high-dimensional space.

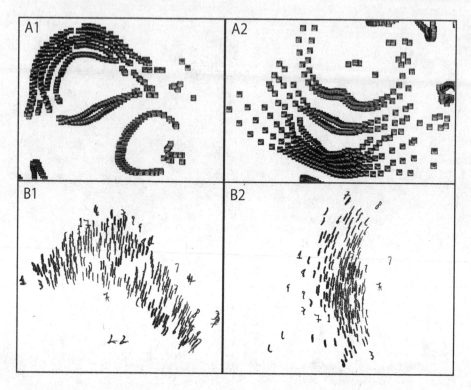

Fig. 4. Zoom-in views of the COIL-20 dataset and the MNIST dataset. The left and right columns show results given by t-SNE and the proposed method. In the top row, the areas where the cars and boxes are mapped are displayed. In the bottom row, the areas around the cluster of digit 1 are shown.

The second column of Fig. 3 shows the results for the COIL-20 dataset. In SNE, lines and circles that correspond to respective objects overlap one another, whereas in t-SNE and the proposed method, they are well separated. Then we focus on the area around car and box images and compare results by t-SNE and the proposed method. The zoom-in views are presented in Fig. 4 A1 and A2, in which the original images are shown. The cars and boxes are arranged close to each other, because the images of car and box are similar to one another. In the proposed method, the lines of car and box images are aligned parallel. However, in t-SNE, although the lines of car images are aligned parallel to one another, the lines of box images are not parallel to those of car images. This can be because t-SNE pushes the cars and boxes far apart a bit too strongly. Thus, optimizing the strength of repelling power is considered to be important.

The third column of Fig. 3 shows the results for the 2D sheet and 5D Gaussian dataset. In SNE, two clusters are merged: the 2D sheet is cleaved and has a gaping hole, and the other cluster is at the center of the hole. In t-SNE, 2D sheet is teared and false clusters that do not exist in the original data emerge, as is in the Swiss-roll dataset. In contrast, the proposed method achieves a reasonable result. The fourth column of Fig. 3 shows the results for the MNIST dataset. We can see that in t-SNE and the proposed method, clusters of respective digits are revealed and there are clear gaps among them. In contrast in SNE, clusters are merged and gaps disappear. This is a typical example of the crowding problem.

Next, we take a look at the values of ν. The first panel in the last row in Fig. 3 shows the histogram of estimated values of ν for the Swiss-roll dataset. We can see that νs take larger values compared to those in Fig. 2, reflecting the fact that the dimension of the data manifold is 2. Similarly, in the COIL-20 dataset, the values of ν are relatively larger.

Then, we focus on the values of ν for the 2D sheet and 5D Gaussian, and MNIST datasets, and investigate the difference in ν among digits. The third and fourth panel in the last row in Fig. 3 shows the boxplots of ν for respective clusters and digits. In 2D sheet and 5D Gaussian dataset, ν for the cluster 1 is higher than that of the cluster 2, because the cluster 1 lies on two-dimensional manifold and the cluster 2 generated from five-dimensional Gaussian distribution. In a similar way, ν for the digit 1 takes the larger value than those for other digits. This can be because samples of digit 1 almost lie on two-dimensional manifold, that is, the difference among images of digit 1 are mainly explained by 2 components: orientation and thickness of segments. The zoom-in views are shown in Fig. 4 B1 and B2. We can see from these figures that in the proposed method two-dimensional (i.e. orientation and thickness) manifold structure is revealed: orientation changes gradually along vertical axis and segment becomes thinner from left to right. In contrast, in t-SNE, the structure is teared as in the Swiss-roll and 2D sheet and 5D Gaussian datasets.

These results can be summarized as follows: SNE is good when the dimension of the data manifold is low, and is suffered from the crowding problem when the dimension is high; t-SNE is good for high-dimensional dataset, but is not for low-dimensional one; the proposed method performs well for both low and high-dimensional datasets, by adapting the values of ν. Additionally, the superiority of the proposed method becomes pronounced when datasets are inhomogeneous.

6   Summary and Discussion

In this paper, we have proposed a dimension reduction method, inhomogeneous t-SNE. In the proposed method, the pointwise parameter ν_i of t-distribution called the number of degrees of freedom is introduced and is optimized to capture the inhomogeneity in dataset. Experimental results have shown that the proposed method overcomes the weakness of SNE and t-SNE and performs well for both datasets with low and high-dimensional manifold.

One of the remaining tasks is speed-up of the method. Although the computational complexity of the proposed method is $\mathcal{O}(N^2)$ and is the same as those of SNE and t-SNE, this prevent it from being applied to large datasets. For t-SNE, several approximate optimization methods are proposed that require only $\mathcal{O}(N \log N)$ computation and $\mathcal{O}(N)$ memory, using algorithms for N-body simulation [9–12]. In a similar manner, it is expected that the proposed method can be accelerated.

Acknowledgements. This work was partially supported by the Grant for Enhancement of International Research from Kobe University, and Grants-in-Aid for Young Scientists (B) [No. 15K16064 (J.K.)] from the MEXT of Japan.

References

1. Tenenbaum, J.B., De Silva, V., Langford, J.C.: A global geometric framework for nonlinear dimensionality reduction. Science **290**(5500), 2319–2323 (2000)
2. Roweis, S.T., Saul, L.K.: Nonlinear dimensionality reduction by locally linear embedding. Science **290**(5500), 2323–2326 (2000)
3. Belkin, M., Niyogi, P.: Laplacian eigenmaps for dimensionality reduction and data representation. Neural Comput. **15**(6), 1373–1396 (2003)
4. Hinton, G.E., Roweis, S.T.: Stochastic neighbor embedding. In: Advances in Neural Information Processing Systems, vol. 15, pp. 833–840. MIT Press, Cambridge (2002)
5. Lafon, S., Lee, A.B.: Diffusion maps and coarse-graining: a unified framework for dimensionality reduction, graph partitioning, and data set parameterization. IEEE Trans. Pattern Anal. Mach. Intell. **28**(9), 1393–1403 (2006)
6. van der Maaten, L., Hinton, G.: Visualizing data using t-SNE. J. Mach. Learn. Res. **9**, 2579–2605 (2008)
7. van der Maaten, L.: Learning a parametric embedding by preserving local structure. In: International Conference on Artificial Intelligence and Statistics, JMLR W&CP, vol. 5 (2009)
8. Vladymyrov, M., Carreira-Perpinán, M.: Entropic affinities: properties and efficient numerical computation. In: Proceedings of the 30th International Conference on Machine Learning, pp. 477–485 (2013)
9. Vladymyrov, M., Carreira-Perpinán, M.A.: Linear-time training of nonlinear low-dimensional embeddings. In: Proceedings of AISTATS 2014, International Conference on Artificial Intelligence and Statistics, JMLR W&CP, vol. 33, pp. 968–977 (2014)
10. van der Maaten, L.: Barnes-hut-sne. arXiv preprint arXiv:1301.3342 (2013)
11. van der Maaten, L.: Accelerating t-sne using tree-based algorithms. J. Mach. Learn. Res. **15**(1), 3221–3245 (2014)
12. Parviainen, E.: A Graph-based n-body Approximation with Application to Stochastic Neighbor Embedding. Neural Netw. **75**, 1–11 (2016)

Topological and Graph Based Clustering Methods

Parcellating Whole Brain for Individuals by Simple Linear Iterative Clustering

Jing Wang[1], Zilan Hu[2], and Haixian Wang[1(✉)]

[1] Key Lab of Child Development and Learning Science
of Ministry of Education, Research Center for Learning Science,
Southeast University, Nanjing 210096, Jiangsu, China
{wangjing0,hxwang}@seu.edu.cn
[2] School of Mathematics and Physics, Anhui University of Technology,
Maanshan 243002, Anhui, China
hu_107@ahut.edu.cn

Abstract. This paper utilizes a supervoxel method called simple linear iterative clustering (SLIC) to parcellate whole brain into functional subunits using resting-state fMRI data. The parcellation algorithm is directly applied on the resting-state fMRI time series without feature extraction, and the parcellation is conducted on the individual subject level. In order to obtain parcellations with multiple granularities, we vary the cluster number in a wide range. To demonstrate the reasonability of the proposed approach, we compare it with a state-of-the-art whole brain parcellation approach, i.e., the normalized cuts (Ncut) approach. The experimental results show that the proposed approach achieves satisfying performances in terms of spatial contiguity, functional homogeneity and reproducibility. The proposed approach could be used to generate individualized brain atlases for applications such as personalized medicine.

Keywords: Whole brain parcellation · Supervoxel · Resting-state fMRI · Functional connectivity · Individualized brain atlas

1 Introduction

Since the manifestation of brain functional connectivity [1], studies have shown that the brain could be characterized as a network. To construct the brain network, an atlas should be defined in prior. It is usually chosen from the standardized atlases such as the automated anatomical labeling (AAL) atlas [2] and the Harvard-Oxford (HO) atlas. However, these atlases are generated based on structural criteria, and cannot guarantee the functional homogeneity of the fMRI time series in each node. Parcellating the brain based on resting-state functional connectivity (RSFC) could avoid the problem, and has attracted exploding attentions in recent years.

The majority of studies concerning RSFC-based parcellation are focusing on a region of interest (ROI) rather than the whole brain. Only a few studies [3–7] generate whole brain atlases. Among them, the normalized cuts (Ncut) [8, 9] is one of the most successful approaches and being widely applied.

© Springer International Publishing AG 2016
A. Hirose et al. (Eds.): ICONIP 2016, Part III, LNCS 9949, pp. 131–139, 2016.
DOI: 10.1007/978-3-319-46675-0_15

This paper employs a supervoxel method called simple linear iterative clustering (SLIC) [10, 11] to parcellate whole brain for individuals. By varying the initialized cluster number, we generate brain atlases with multiple granularities. The Ncut approach is also applied to parcellate the same dataset. Finally, we make a comparison between the two kinds of parcellation approaches under different evaluation metrics.

2   Materials and Methods

2.1   Subjects

In the study, we use data from the 1000 Functional Connectomes Project (http://www.nitrc.org/projects/fcon_1000/) [12] that is publicly available online. Specifically, we use the structural and resting-state fMRI data acquired from 18 subjects in the Beijing_Zang dataset. The demographics of the subjects could be found online. The dataset is preprocessed by the Data Processing Assistant for Resting-State fMRI (DPARSF) [13]. The preprocessing steps include: discarding the first ten volumes, slice timing correction, motion correction, coregistration, segmenting the structural images, normalizing the functional images to the Montreal Neurological Institute (MNI) space at $4 \times 4 \times 4$ mm^3 resolution; smoothing with a 6 mm FWHM Gaussian kernel; linear detrending; bandpass filtering with passband 0.01–0.08 Hz; regressing out nuisance covariates. No subject is excluded due to excessive head motion under the excluding criteria 2.0 mm and 2.0°. The global signal regression (GSR) is not included since its effect is still controversial.

2.2   Simple Linear Iterative Clustering (SLIC)

SLIC could be used as a superpixel method [10] or a supervoxel method [11], which is determined by whether the target image is 2D or 3D. The common idea is to separate an image into perceptually meaningful patches. SLIC is actually an adaptation of K-means. Two important differences between SLIC and K-means are that SLIC limits the search space to the neighborhood of a cluster center and creates a unified distance by integrating the intensity distance and the spatial distance. SLIC has become very popular in the field of computer vision in recent years due to its simplicity, effectiveness and good clustering performance. In this study, we apply it on resting-state fMRI data to carry out whole brain parcellation.

The algorithm procedure of SLIC is stated as follows. To parcellate the brain into K clusters, we first initialize K cluster centers periodically in the 3D space, as shown in Fig. 1A. Assume that the number of voxels in the gray matter is N. Then the average length of a supervoxel is $S = \sqrt[3]{N/K}$. For each voxel in the $3S \times 3S \times 3S$ region around a cluster center, a distance between the voxel and the cluster center is calculated. This distance is assigned to the voxel as a measure to judge which cluster it should belong to. If the distance decreases comparing to the result in the previous iteration, then associate the voxel to the current cluster center. This procedure is repeated for all cluster centers. Once completed, each cluster center is updated to be a vector formed by averaging the fMRI time series and coordinates of voxels in that cluster. The above assignment and update steps are repeated until the change of the cluster centers is lower than a certain threshold. The resultant clusters or supervoxels make up the final brain atlas. The algorithm procedure is summarized in Table 1. An illustration of the initializing and searching steps is shown in Fig. 1.

Table 1. The algorithm procedure of the SLIC approach

Input: the resting-state fMRI time series and the initialized cluster number.
Output: the cluster labels.

Initialize the cluster centers.
Initialize label $l(i) = -1$ for each pixel i.
Initialize distance $d(i) = \infty$ for each pixel i.
while not converged **do**
    for each cluster center C_k **do**
        for each voxel i in the $3S \times 3S \times 3S$ region around C_k **do**
            Compute the unified distance D between C_k and i.
            if $D < d(i)$ **then**
                Set $l(i) = k$.
                Set $d(i) = D$.
            end if
        end for
    end for
    Compute new cluster centers.
end while

How to define the unified distance is very important in the clustering procedure. For the ith voxel, assume that its fMRI time series is v_i and its coordinates in the MNI space is u_i, $i = 1, 2, \ldots, N$, then the unified distance between two voxels could be defined as

$$d_{ij} = \sqrt{\frac{\|v_i - v_j\|_2^2}{m^2} + \frac{\|u_i - u_j\|_2^2}{s^2}}, \tag{1}$$

where m and S are two tuning parameters which normalize the functional distance and the spatial distance respectively. The parameter m could be chosen around the median of all functional distances, and we fix it to be 40 empirically. The parameter S is fixed to be the average length of the supervoxels. Though the algorithm in [11] and in this study are targeting at parcellating the 3D space, a major difference exists between them. That is, the functional distance is calculated between the image intensity of two voxels in [11] while it is calculated between the fMRI time series of two voxels in this study. Since the functional distance is incorporated in the clustering procedure, the proposed approach could be regarded as a RSFC-based parcellation approach.

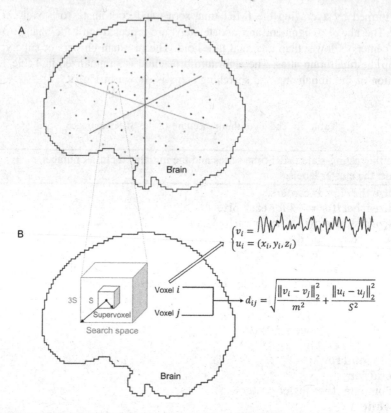

Fig. 1. Illustration of the SLIC approach on whole brain parcellation. (A) Initializing the cluster centers periodically in the 3D space. The three straight lines denote the xyz-axis system. (B) For each cluster, SLIC searches in the $3S \times 3S \times 3S$ region around its center to update the labels of all voxels in the search space. A unified distance is calculated between each voxel and the cluster center to judge whether the voxel should belong the cluster. The unified distance is composed of the functional distance and the spatial distance, wherein the functional distance is calculated between the fMRI time series of two voxels. Note that the supervoxel is unnecessary to be a cube. It is displayed as a cube for simplicity.

We choose Ncut as the competing approach because it has achieved great success in whole brain parcellation. For that approach, the definition of the individual subject level weight matrix and the implementation of the multiclass spectral clustering (MSC) algorithm [14] are kept the same as in [4] in order to make a fair comparison. Only the individual subject level parcellations of the two approaches are generated and compared in this study. A comparison of the pipeline of the Ncut approach and the SLIC approach is shown in Fig. 2. Without confusion, we use MSC to denote the clustering algorithm that operates after extracting features by Ncut. Since SLIC is directly applied on the fMRI time series, it only needs a single step to generate parcellations.

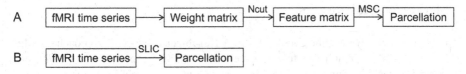

Fig. 2. The pipeline of the two parcellation approaches. (A) Ncut. (B) SLIC.

2.3 Evaluation Metrics

The clusters in a brain parcellation result should be spatially contiguous, functionally homogeneous and reproducible [4, 5]. For spatial contiguity, we treat the spatially discrete regions that belong to the same cluster as separate clusters and count the increased cluster number. The increased cluster number is referred to as the spatial discontiguity index. For functional homogeneity, we first average similarities across all pairs of voxels within a cluster and then average the obtained results across clusters. Assume that the voxel number in the kth cluster C_k is n_k, $k = 1, 2, \ldots, K$. The similarity between voxels i and j is s_{ij}, $i, j = 1, 2, \ldots, N$. The average similarity within the kth cluster is

$$a(k) = \frac{1}{n_k(n_k - 1)} \sum_{i,j \in C_k, i \neq j} s_{ij}. \tag{2}$$

Then the functional homogeneity of the brain atlas is

$$\frac{1}{K} \sum_{k=1}^{K} a(k). \tag{3}$$

To avoid circular analysis, we train an atlas on one subject and calculate the functional homogeneity based on this atlas and the resting-state fMRI data of other subjects. For reproducibility, we calculate the Dice coefficient between different brain atlases that are generated from different subjects. As a prior step, we should calculate an adjacency matrix for each brain atlas. An adjacency matrix is A a $N \times N$ symmetric matrix that is calculated by setting its elements a_{ij} is set to be one if voxels i and j belong to the same cluster in the brain atlas, and zero otherwise. For two adjacency matrices A and B derived from two atlases, the Dice coefficient between them is

$$\frac{2|A \cap B|}{|A| + |B|}, \tag{4}$$

where $|\cdot|$ denotes the number of ones in an adjacency matrix, $A \cap B$ denotes the union of the two adjacency matrices.

3   Experimental Results

In the experiment, we use the fMRI data from 18 subjects. The Ncut approach and the SLIC approach are employed to do parcellation. Then the parcellation results are compared under difference evaluation metrics. The initialized cluster number is set to be [50:50:1000] in order to generate parcellations with multiple granularities. For each subject, each parcellation approach and each cluster number, one atlas is obtained. Figure 3 shows the atlases when the first subject is parcellated into 100, 300 and 800 clusters by Ncut and SLIC.

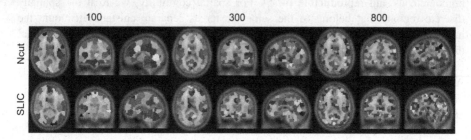

Fig. 3. Illustration of the atlases generated by Ncut and SLIC. Each atlas is represented by its three orthogonal cross sections. The initialized cluster numbers are 100, 300 and 800 from left to right. The colormap for each atlas is randomly generated, and each color represents a cluster. (Color figure online)

For a brain parcellation approach, the actual cluster number should be close to the initialized cluster number in order to obtain the granularity we have expected. By subtracting the initialized cluster number from the average actual cluster number for each parcellation approach, we could obtain their differences, as shown in Fig. 4A. The results show that SLIC outperforms Ncut in approximating the initialized cluster number.

To evaluate spatial contiguity, we calculate the spatial discontiguity index for each brain atlas and then average the results across subjects, as shown in Fig. 4B. A smaller result indicates that the brain atlases are more spatially contiguous. Ncut outperforms SLIC in spatial contiguity. The reason is that Ncut incorporates spatial constraint in the parcellation procedure that could guarantee to obtain spatially contiguous clusters [4]. The spatial constraint is a strong spatial structure, which weakens the influences of data structure and renders the generated atlases to have comparable shapes and sizes, as displayed in Fig. 3. This brings quite a lot of doubts to the Ncut approach [3, 5]. For SLIC, the spatial discontiguity index generally decreases with increasing cluster number. When the actual cluster number is larger than 200, there are only few spatially discontiguous regions in each atlas, which is also satisfactory.

To evaluate functional homogeneity, we train a brain atlas on one subject and calculate homogeneity based on this atlas and the resting-state fMRI data of the remaining subjects. The homogeneity results of each parcellation approach and each cluster number are averaged, as shown in Fig. 4C. The curves correspond to the two

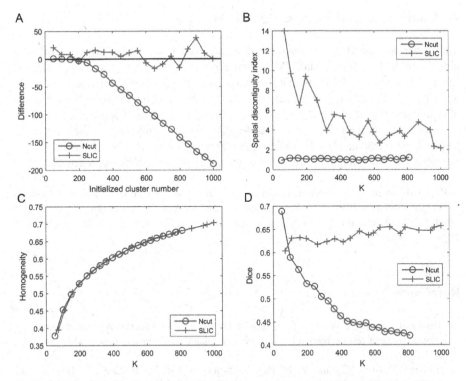

Fig. 4. The results of different evaluation metrics for Ncut and SLIC. (A) The difference between the initialized cluster number and the actual cluster number. (B) Spatial discontiguity index. (C) Functional homogeneity. (D) Dice coefficient. The first metric is plotted against the initialized cluster number while the other three metrics are plotted against the actual cluster number that is denoted by K.

approaches are very close, which indicates that the two approaches obtain similar homogeneity results. Homogeneity increases with increasing cluster number. This is consistent with [4, 5, 15].

To evaluate reproducibility, we randomly choose two from the eighteen subjects and calculate Dice coefficient between their corresponding atlases when the parcellation approach and the initialized cluster number are fixed. This procedure is repeated for twenty times. The twenty results are averaged to yield a single Dice coefficient for each parcellation approach and each cluster number. The averaged results are shown in Fig. 4D. The Dice coefficients of SLIC are higher than the Dice coefficients of Ncut except when the initialized cluster number is 50. The result demonstrates that the atlases generated by SLIC have higher reproducibility across individuals than the atlases generated by Ncut. The Dice coefficient of Ncut decreases with increasing cluster number, which is consistent with [3–5]

For reproducibility, the source codes of this study have been made publicly available at https://github.com/yuzhounh/SLIC_individual.

4   Conclusion and Future Directions

This paper applies SLIC to generated individualized brain atlases. The algorithm is directly applied on resting-state fMRI time series without feature extraction in prior. The experimental results show that the proposed approach obtains satisfying results in terms of spatial contiguity and functional homogeneity, and outperforms Ncut in terms of reproducibility. It demonstrates the rationality of the proposed approach. For future directions, the individualized brain atlas might find its application in fields such as personalized medicine [16]. In addition, the proposed approach has the potential to be extended from individual subject level to group level as the Ncut approach [4].

Acknowledgements. This work was supported in part by the National Basic Research Program of China under Grant 2015CB351704, the National Natural Science Foundation of China under Grant 61375118, and the Research Foundation for Young Teachers in Anhui University of Technology under Grant QZ201516.

References

1. Biswal, B., Zerrin Yetkin, F., Haughton, V.M., Hyde, J.S.: Functional connectivity in the motor cortex of resting human brain using echo-planar MRI. Magn. Reson. Med. **34**, 537–541 (1995)
2. Tzourio-Mazoyer, N., Landeau, B., Papathanassiou, D., Crivello, F., Etard, O., Delcroix, N., Mazoyer, B., Joliot, M.: Automated anatomical labeling of activations in SPM using a macroscopic anatomical parcellation of the MNI MRI single-subject brain. Neuroimage **15**, 273–289 (2002)
3. Blumensath, T., Jbabdi, S., Glasser, M.F., Van Essen, D.C., Ugurbil, K., Behrens, T.E.J., Smith, S.M.: Spatially constrained hierarchical parcellation of the brain with resting-state fMRI. Neuroimage **76**, 313–324 (2013)
4. Craddock, R.C., James, G.A., Holtzheimer, P.E., Hu, X.P.P., Mayberg, H.S.: A whole brain fMRI atlas generated via spatially constrained spectral clustering. Hum. Brain Mapp. **33**, 1914–1928 (2012)
5. Shen, X., Tokoglu, F., Papademetris, X., Constable, R.T.: Groupwise whole-brain parcellation from resting-state fMRI data for network node identification. Neuroimage **82**, 403–415 (2013)
6. van den Heuvel, M., Mandl, R., Pol, H.H.: Normalized cut group clustering of resting-state fMRI data. PLoS ONE **3**, e2001 (2008)
7. Moreno-Dominguez, D., Anwander, A., Knosche, T.R.: A hierarchical method for whole-brain connectivity-based parcellation. Hum. Brain Mapp. **35**, 5000–5025 (2014)
8. Shi, J.B., Malik, J.: Normalized cuts and image segmentation. IEEE Trans. Pattern Anal. **22**, 888–905 (2000)
9. von Luxburg, U.: A tutorial on spectral clustering. Stat. Comput. **17**, 395–416 (2007)
10. Achanta, R., Shaji, A., Smith, K., Lucchi, A., Fua, P., Susstrunk, S.: SLIC superpixels compared to state-of-the-art superpixel methods. IEEE Trans. Pattern Anal. **34**, 2274–2282 (2012)
11. Lucchi, A., Smith, K., Achanta, R., Knott, G., Fua, P.: Supervoxel-based segmentation of mitochondria in EM image stacks with learned shape features. IEEE Trans. Med. Imaging **31**, 474–486 (2012)

12. Biswal, B.B., Mennes, M., Zuo, X.N., et al.: Toward discovery science of human brain function. Proc. Natl. Acad. Sci. USA **107**, 4734–4739 (2010)

13. Yan, C.G., Zang, Y.F.: DPARSF: a MATLAB toolbox for "pipeline" data analysis of resting-state fMRI. Front. Syst. Neurosci. **4**, 1–7 (2010)

14. Yu, S.X., Shi, J.B.: Multiclass spectral clustering. In: Proceedings of IEEE International Conference on Computer Vision, pp. 313–319 (2003)

15. Gordon, E.M., Laumann, T.O., Adeyemo, B., Huckins, J.F., Kelley, W.M., Petersen, S.E.: Generation and evaluation of a cortical area parcellation from resting-state correlations. Cereb. Cortex **26**(1), 288–303 (2016)

16. Wang, D.H., Buckner, R.L., Fox, M.D., et al.: Parcellating cortical functional networks in individuals. Nat. Neurosci. **18**, 1853–1860 (2015)

Overlapping Community Structure Detection of Brain Functional Network Using Non-negative Matrix Factorization

Xuan Li[1], Zilan Hu[2], and Haixian Wang[1(✉)]

[1] Key Lab of Child Development and Learning Science of Ministry of Education, Research Center for Learning Science, Southeast University, Nanjing 210096, Jiangsu, People's Republic of China
{xuanli,hxwang}@seu.edu.cn
[2] School of Mathematics and Physics, Anhui University of Technology, Maanshan 243002, Anhui, People's Republic of China

Abstract. Community structure, as a main feature of a complex network, has been investigated recently under the assumption that the identified communities are non-overlapping. However, few studies have revealed the overlapping community structure of the brain functional network, despite the fact that communities of most real networks overlap. In this paper, we propose a novel framework to identify the overlapping community structure of the brain functional network by using the symmetric non-negative matrix factorization (SNMF), in which we develop a non-negative adaptive sparse representation (NASR) to produce an association matrix. Experimental results on fMRI data sets show that, compared with modularity optimization, normalized cuts and affinity propagation, SNMF identifies the community structure more accurately and can shed new light on the understanding of brain functional systems.

Keywords: Overlapping community · Non-negative matrix factorization (nmf) · Brain functional network · fMRI

1 Introduction

Community structure is a signature of a complex network [1]. It reflects the underlying organization of the network, where nodes are highly connected within one community and sparsely connected between communities. Recent studies have revealed non-overlapping community structures of the complex brain functional and structural networks, using popular spectral clustering methods [2,3]. However, these methods are usually incapable of detecting the overlapping community structure, which is an important characteristic of most real networks. In this paper, we apply a variant of the standard non-negative matrix factorization (NMF) [4], specifically the symmetric NMF (SNMF) [5], to identify the overlapping community structure of the brain functional network. SNMF learns a part-based representation of the original data, grouping coherent parts together

© Springer International Publishing AG 2016
A. Hirose et al. (Eds.): ICONIP 2016, Part III, LNCS 9949, pp. 140–147, 2016.
DOI: 10.1007/978-3-319-46675-0_16

and only allows for additive combination due to its non-negativity. Importantly, it allows overlapping between different parts, which can be regarded as a soft clustering method. Another important issue influencing the community detection is the construction of the brain functionally connected network, which is denoted by an association matrix. In our previous work [6], we used the adaptive sparse representation (ASR) [7] to depict the brain functional network, which achieved better performance than the methods of Pearson correlation and partial correlation. In this paper, we convert ASR into a non-negative ASR (NASR) by adding a non-negative constraint to construct an association matrix with only non-negative entries. This feature of NASR ensures that only the data that contributes positively are selected for the representation, which is in accordance with the additive property of neural process. Furthermore, the resulting non-negative association matrix can be directly factorized by SNMF.

In short, the contribution of this paper is two-fold. Firstly, we propose the NASR to construct a non-negative association matrix, which aims to achieve a physically meaningful depiction of the brain functional network. Secondly, we apply SNMF to identify the overlapping community structure of the brain functional network.

2   Methods and Material

2.1   Association Matrix Construction with NASR

Given a sample $y \in \mathbb{R}^d$ and a dictionary matrix $X \in \mathbb{R}^{d \times n}$, ASR aims to obtain a sparse solution $w \in \mathbb{R}^n$ such that it represents y with Xw. Specifically, ASR uses the trace LASSO norm $\|X\mathrm{Diag}(w)\|_*$ [8] as a regularizer in optimization, which adaptively mediates between the great sparsity of ℓ_1-norm and the grouping effect of ℓ_2-norm. Here, $\mathrm{Diag}(w)$ means a diagonal matrix with w as its diagonal elements, and the trace norm $\|X\mathrm{Diag}(w)\|_*$ sums up the singular values of $X\mathrm{Diag}(w)$. We propose the NASR by adding a non-negative constraint on w, and the problem is formulated as

$$\min_{w} \|X\mathrm{Diag}(w)\|_*, \ s.t. \ \|y - Xw\|_2 \le \varepsilon, w \ge 0, \tag{1}$$

where ε denotes a given tolerance. Such non-negative constraint ensures that all terms in the dictionary are additive for the representation of one sample, making it easier to interpret the results. This optimization problem can be reformulated as

$$\min_{w \ge 0} \frac{1}{2}\|y - Xw\|_2^2 + \lambda\|X\mathrm{Diag}(w)\|_*, \tag{2}$$

where $\lambda > 0$ is a regularization parameter. A globally optimal solution of Eq. (2) can be achieved by alternating direction method (ADM), the same method as used in solving the ASR problem [6,7]. With NASR, we thus construct the association matrix. Specifically, given a normalized fMRI data matrix $X = (x_1, \ldots, x_n) \in \mathbb{R}^{d \times n}$ with n nodes and d time points, the corresponding dictionary of sample x_i (i.e., the time series of the ith node) is the remaining part of X excluding x_i itself, i.e., $X_i = (x_1, \ldots, x_{i-1}, x_{i+1}, \ldots, x_n) \in \mathbb{R}^{d \times (n-1)}$. The resulting non-negative sparse solution $w_i \in \mathbb{R}^{n-1}$ characterizes the degree of other samples' contribution to the representation of x_i. The vector w_i is then padded with a zero in its i th position, denoted as $\tilde{w}_i \in \mathbb{R}^n$, indicating that the association between x_i and itself is set to zero. All $\tilde{w}_i$ $(i = 1, \ldots, n)$ are stacked as columns of a matrix $\tilde{W} \in \mathbb{R}^{n \times n}$. Then a symmetric association matrix G is constructed by $G = (\tilde{W} + \tilde{W}^T)/2$.

2.2   Community Detection with SNMF

SNMF can symmetrically factorize any non-negative symmetric matrix, which is especially useful to detect the community structure of a network. The network can be characterized by an undirected graph, which is then expressed by an association matrix. Suppose the undirected graph $\mathcal{G}$ consists of k communities. Then its association matrix G can be approximately represented as

$$G \approx s_1 h_1 h_1^T + s_2 h_2 h_2^T + \ldots + s_k h_k h_k^T, \tag{3}$$

where the binary vector h_i indicates the membership of community i and s_i denotes the weight of community i (assuming nodes in the same community share the same weight). Let $H = (h_1, \ldots, h_k)$, and S be the diagonal matrix: $S = \text{diag}(s_1, \ldots, s_k)$. Then Eq. (3) can be seen as a factorization of G: $G \approx HSH^T$. Since S is symmetric, this factorization can be converted into the following SNMF problem [5]

$$\min_{\hat{H} \geq 0} \|G - \hat{H}\hat{H}^T\|_F^2, \tag{4}$$

where $\hat{H} = HS^{\frac{1}{2}}$. That is, the SNMF method is directly applied to the non-negative symmetric association matrix constructed by NASR. Once the membership matrix $\hat{H}$ is obtained, we normalize its row vectors. The resulting $\hat{h}_{ij}$ can be considered as a posterior probability. Finally, a threshold τ of these probabilities is used to determine the membership of each community. In this sense, SNMF allows for assigning one node into more than one community. Consequently, it discovers the overlapping community structure of the brain functional network.

2.3   Data Preparation

In order to comprehensively test the performance of the proposed scheme, i.e. NASR+SNMF, two simulated fMRI data sets ($S1$ and $S2$) and a real resting-state fMRI data set are used.

Simulated data $S1$ of 40 subjects and $S2$ of 50 subjects are available from [9] and [10] respectively. They both have a network size of 50 nodes and are generated at a TR value of 3 s by using dynamic causal modeling (DCM) [10]. The main difference between the two data sets is whether the community structure is overlapping or not. The overall network structure of $S1$ consists of 9 overlapping communities of different sizes. Besides, the number of communities included in the data of each subject varies from 6 to 9. Details of this data set can be found in Fig. 2 of [9]. Note that a slight adjustment is made on the ground-truth of the original data set.[1] By contrast, the underlying network structure of $S2$ consists of 10 non-overlapping clusters with the same size of 5 nodes.

The real resting-state fMRI data set, named Beijing_Zang, can be downloaded from the 1000 Functional Connectomes Project online database [11]. A subset of 20 subjects out of all the 198 healthy adults are used in this experiment. The TR value is set as 2 s and the length of each time-series is 225 time points. The real fMRI data is preprocessed in the same way as [6], using the Data Processing Assistant for Resting-State fMRI (DPARSF) [12] toolbox with Statistical Parametric Mapping package (SPM8). Then the widely used AAL template of 90 regions of interest (ROIs) are defined as nodes of the network, and time series of each node are extracted.

2.4   Experiment

We use NASR to construct the association matrix, in which the parameter λ is empirically set to 0.1 for all three data sets. Based on the association matrix, SNMF is compared with three other commonly used clustering algorithms, i.e., modularity optimization [3], normalized cut (N-cut) [13] and affinity propagation (AP) [14]. Another parameter to be pre-determined for SNMF, N-cut and AP is the number of communities, k. For the simulated data, k is set individually for each subject according to their ground truth; while for the real fMRI data, k is set to 8 according to [15] that around 8 functionally-linked resting-state networks have been consistently reported by a number of studies. For the simulated data, performance of all algorithms are evaluated quantitatively due to the availability of the ground truth. Three metrics are used to compare the correspondence between the identified communities and the ground truth, i.e., the averaged F1-score [16], normalized mutual information (NMI) designed for overlapping communities [17], and Omega index [18].

3   Results and Discussion

3.1   Simulated Data Set

Clustering results as well as the ground truth on $S1$ from two randomly selected subjects are shown in Fig. 1. As can be seen, SNMF naturally obtains a sparse solution and groups the most coherent nodes into one community. By appropriately setting a threshold to the original results, SNMF can identify the overlapping nodes. By contrast, the other three methods only assign one node into one community, and thus are incapable of identifying the overlapping structure. In

---

[1] Since the relationship between different communities is not the concern of this paper, we take the cluster of the last five nodes as an independent community, thus changing the number of communities in the overall ground truth from 8 to 9.

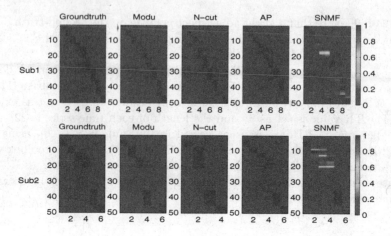

Fig. 1. Examples of identified community structure on the overlapping simulated data set. It shows the results from two randomly selected subjects (*upper and lower panels*). The *x- and y-axis* denotes the community index and node index respectively.

some occasions, they may scatter the inner structure of the community (e.g., AP of Sub2) or result in a large community by combining overlapping communities (e.g., N-Cut of Sub2).

The quantitative measures averaged over all subjects for each method on both $S1$ and $S2$ are summarized in Table 1. Here, we select the best threshold τ of SNMF individually for each subject according to their average F1-scores. The NMI and Omega indices are then calculated based on the resulting community structure after applying the best τ. It shows that SNMF achieves the highest scores in terms of all measures on the data set $S1$. Interestingly, even on the non-overlapping data set $S2$, SNMF still performs substantially better than the others. More precisely, 16 out of 18 two-sample t-tests comparing SNMF with the other methods indicate that SNMF performs significantly better than others ($p < .05$ marked with $*$) in detecting the overlapping and non-overlapping community structure.

Table 1. Measures of F1-score, NMI, and Omega averaged over all subjects on the two simulated data sets.

	$S1$ (Overlapping)			$S2$ (Non-overlapping)		
	F1-score	NMI	Omega	F1-score	NMI	Omega
Modu	0.932*	0.844	0.884	0.817*	0.654*	0.659*
NC	0.900*	0.758*	0.824*	0.597*	0.353*	0.362*
AP	0.928*	0.823*	0.852*	0.651*	0.364*	0.333*
SNMF	0.940	0.855	0.885	0.861	0.730	0.733

3.2   Real Resting-State fMRI Data Set

The communities identified on the group-averaged association matrix over 20 subjects are shown in Fig. 2. Here, τ is empirically set to 0.22 in SNMF. Similar communities are revealed including the basal ganglia, limbic, visual and the bilateral frontal-parietal networks (referring to Comm. 3, 4, 5 and 6 respectively) for all methods. However, great differences between SNMF and the other methods are displayed on the rest communities. Specifically, only SNMF separates the auditory network (Comm. 7) from the sensory-motor network (Comm. 1) [19] and identifies the main parts of the default mode network (Comm. 2) including posterior cingulate cortex (PCC), precuneus and angular gyrus [19] more completely. Furthermore, SNMF uniquely identifies a community including prefrontal cortex and anterior cingulate cortex (ACC) (Comm. 8), which is thought to be relevant with emotion processing [20].

The overlapping community structure is further analyzed by identifying the overlapping nodes. The top 30 % most reported overlapping nodes obtained by SNMF is firstly computed under each of the 5 different thresholds ($\tau = 0.2, 0.22, 0.25, 0.28, 0.3$) and combined together under all the thresholds. Then

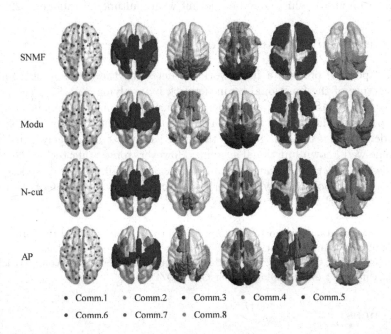

Fig. 2. Community structure based on the mean association matrix over 20 subjects on the real fMRI data set. Each row displays the results produced by one method. The overall community structures are drawn *in the left column* and individual communities are shown in the remaining columns. The pictures on the right are viewed in ROIs from the dorsal axial except for the last column that is from the ventral axial view. For each method, members of one community are drawn in the same color and overlapping nodes of SNMF are marked in *bigger red nodes*. (Color figure online)

Table 2. Overlapping nodes occurring in both the individual and group-averaged community structures on the real fMRI data set.

Lobe	Frontal					Limbic		Occipital
Region	SFGdor.R	ORBinf.L	ORBinf.R	ORBsupmed.L	ORBsupmed.R	DCG.L	DCG.R	FFG.L
Community	6, 8	6, 8	6, 8	4, 8	4, 8	2, 8	2, 8	4, 5
Lobe	Parietal				Sub-cortical	Temporal		
Region	SPG.L	SMG.R	ANG.L	ANG.R	CAU.R	TPOsup.L	TPOsup.R	TPOmid.R
Community	1, 2, 6	6, 7	2, 6	2, 6	3, 8	4, 7	4, 7	4, 8

among the resulting nodes, the top 30 % nodes (27) are selected and compared with the overlapping nodes (25) identified in the group-averaged results. Finally, 16 overlapping nodes occurring in both individual and group community structures are identified and listed in Table 2. In general, most overlapping nodes are located in the frontal and parietal lobes, which play important roles in multiple communities. For example, the overlapping nodes of the frontal lobe in Comm. 8 that are relevant with emotion processing are also in charge of the central executive function of Comm. 6, especially the node of SFGdor.R [21]. In addition, the bilateral angular gyrus of DMN (Comm. 2) are parts of the fronto-parietal network (Comm. 6), which may be relevant with multimodal thoughts [22].

4   Conclusion and Limitation

In this paper, we propose a framework for detecting the overlapping community structure of the functional brain network by applying SNMF to the graph constructed by NASR. Compared with modularity optimization, N-cut and AP, SNMF performs the best on two simulated data sets. The overlapping communities identified on the real resting-state fMRI data set are largely consistent with some well-known functional systems. However, some other techniques may be needed to select the parameters automatically. In all, our proposed framework is a useful tool to study the overlapping community structure of the brain functional network.

Acknowledgments. This work was supported in part by the National Basic Research Program of China under Grant 2015CB351704, the National Natural Science Foundation of China under Grant 61375118, and the Research Foundation for Young Teachers in Anhui University of Technology under Grant QZ201516.

References

1. Bullmore, E., Sporns, O.: Complex brain networks: graph theoretical analysis of structural and functional systems. Nat. Rev. Neurosci. **10**(3), 186–198 (2009)
2. Chen, Z.J., He, Y., Rosa-Neto, P., Germann, J., Evans, A.C.: Revealing modular architecture of human brain structural networks by using cortical thickness from MRI. Cereb. Cortex **18**(10), 2374–2381 (2008)
3. Newman, M.E.: Modularity and community structure in networks. Proc. Natl. Acad. Sci. **103**(23), 8577–8582 (2006)

4. Lee, D.D., Seung, H.S.: Learning the parts of objects by non-negative matrix factorization. Nature **401**(6755), 788–791 (1999)
5. Wang, F., Li, T., Wang, X., Zhu, S., Ding, C.: Community discovery using nonnegative matrix factorization. Data Min. Knowl. Discov. **22**(3), 493–521 (2011)
6. Li, X., Wang, H.: Identification of functional networks in resting state fMRI data using adaptive sparse representation and affinity propagation clustering. Front. Neurosci. **9**, 383 (2015)
7. Lu, C., Feng, J., Lin, Z., Yan, S.: Correlation adaptive subspace segmentation by trace LASSO. In: 2013 IEEE International Conference on Computer Vision, pp. 1345–1352 (2013)
8. Grave, E., Obozinski, G.R., Bach, F.R.: Trace lasso: a trace norm regularization for correlated designs. In: Advances in Neural Information Processing Systems, pp. 2187–2195 (2011)
9. Eavani, H., Satterthwaite, T.D., Filipovych, R., Gur, R.E., Gur, R.C., Davatzikos, C.: Identifying sparse connectivity patterns in the brain using resting-state fMRI. Neuroimage **105**, 286–299 (2015)
10. Smith, S.M., Miller, K.L., Salimi-Khorshidi, G., Webster, M., Beckmann, C.F., Nichols, T.E., Ramsey, J.D., Woolrich, M.W.: Network modelling methods for fMRI. Neuroimage **54**(2), 875–891 (2011)
11. Biswal, B.B., Mennes, M., Zuo, X.N., et al.: Toward discovery science of human brain function. Proc. Natl. Acad. Sci. **107**, 4734–4739 (2010)
12. Yan, C., Zang, Y.: DPARSF: a MATLAB toolbox for pipeline data analysis of resting-state fMRI. Front. Syst. Neurosci. **4**, 13 (2010)
13. Van Den Heuvel, M., Mandl, R., Pol, H.H.: Normalized cut group clustering of resting-state fMRI data. PLoS ONE **3**(4), e2001 (2008)
14. Frey, B.J., Dueck, D.: Clustering by passing messages between data points. Science **315**, 972–976 (2007)
15. Van den Heuvel, M.P., Pol, H.E.H.: Exploring the brain network: a review on resting-state fMRI functional connectivity. Eur. Neuropsychopharmacol. **20**, 519–534 (2010)
16. Yang, J., Leskovec, J.: Overlapping community detection at scale: a nonnegative matrix factorization approach. In: Proceedings of the 6th ACM International Conference on Web Search and Data Mining, pp. 587–596. ACM (2013)
17. McDaid, A.F., Greene, D., Hurley, N.: Normalized mutual information to evaluate overlapping community finding algorithms (2011)
18. Gregory, S.: Fuzzy overlapping communities in networks. J. Stat. Mech. Theory and Exp. **2011**, 02017 (2011)
19. Mantini, D., Perrucci, M.G., Del Gratta, C., Romani, G.L., Corbetta, M.: Electrophysiological signatures of resting state networks in the human brain. Proc. Natl. Acad. Sci. **104**(32), 13170–13175 (2007)
20. Etkin, A., Egner, T., Kalisch, R.: Emotional processing in anterior cingulate and medial prefrontal cortex. Trends Cogn. Sci. **15**(2), 85–93 (2011)
21. Sridharan, D., Levitin, D.J., Menon, V.: A critical role for the right rronto-insular cortex in switching between central-executive and default-mode networks. Proc. Natl. Acad. Sci. **105**(34), 12569–12574 (2008)
22. Mazoyer, B., Zago, L., Mellet, E., Bricogne, S., Etard, O., Houde, O., Crivello, F., Joliot, M., Petit, L., Tzourio-Mazoyer, N.: Cortical Networks for Working Memory and Executive Functions Sustain the Conscious Resting State in Man. Brain Res. Bull. **54**(3), 287–298 (2001)

Collaborative-Based Multi-scale Clustering in Very High Resolution Satellite Images

Jérémie Sublime[1,2,3]($\boxtimes$), Antoine Cornuéjols[1], and Younès Bennani[2]

[1] UMR MIA-Paris, AgroParisTech, INRA, Université Paris-Saclay,
75005 Paris, France
{jeremie.sublime,antoine.cornuejols}@agroparistech.fr
[2] Université Paris 13 - Sorbonne Paris Cité, Laboratoire d'Informatique de
Paris-Nord - CNRS (UMR 7030), 99 Avenue Jean Baptiste Clément,
93430 Villetaneuse, France
Younes.bennani@lipn.univ-paris13.fr
[3] LISITE Laboratory - RDI Team, ISEP, Paris, France
jeremie.sublime@isep.fr

Abstract. In this article, we show an application of collaborative clustering applied to real data from very high resolution images. Our proposed method makes it possible to have several algorithms working at different scales of details while exchanging their information on the clusters.

Our method that aims at strengthening the hierarchical links between the clusters extracted at different level of detail has shown good results in terms of clustering quality based on common unsupervised learning indexes, but also when using external indexes: We compared our results with other algorithms and analyzed them based on an expert ground truth.

Keywords: Multi-scale clustering · Cluster analysis · Image segmentation

1 Introduction

With the booming number of available satellite images and data, the automatic interpretation of remotely sensed images has become an increasingly active domain. With sensors now capable of getting images with a very high resolution (VHR) on a large spectral resolution, it is more and more difficult to design algorithms and methods capable of efficiently processing such data in a reasonable amount of time.

Such process usually contains two steps: (1) A segmentation step that consists in grouping together connected groups of pixels with the goal of finding homogeneous segments the borders of which will be a good estimation of the objects present in the image [3,5]. The segments created using this process are supposed to be relevant and match the real objects that can be found in the picture. (2) A clustering step the aim of which is to analyze the segmented objects and create groups of elements that are similar.

A. Hirose et al. (Eds.): ICONIP 2016, Part III, LNCS 9949, pp. 148–155, 2016.
DOI: 10.1007/978-3-319-46675-0_17

In this article, we will focus on the clustering of already pre-processed images. The clustering of this type of image data set is challenging for several reasons:

- The quality of the initial segmentation can have a huge impact on the quality of the data set, and therefore of the final clusters. While an *over-segmentation* (splitting elements of interest into several segments) can be fixed during the clustering process, in case of *under-segmentation* (several elements of interest into a single segments) several objects of interest and their clusters may be lost for good.
- This type of data has spacial dependencies between the data that don't exist in regular data sets. Another important aspect is that when the segments and regions are created, this adds a large number of new attributes that may be taken into considerations: surface of the region, perimeter and elongation; extrema, variance and average values of the attributes in a given region, contrast with the neighboring regions, etc. Quite often, these attributes are redundant.

However, this type of data sets also have several advantages: The results can easily be evaluated simply by projecting them. Another common issue which consists in finding the right number of clusters is less of a problem when it comes to clustering of satellite images. With the recent progress in satellite imaging, there are several possible level of interest available on a very high resolution satellite image: At the first level, we can usually distinguish three main types of objects, namely water bodies, vegetation areas and urban areas. At a second level we can separate different types of urban blocs, different types of vegetation areas, and start to distinguish elements such as roads. When zooming even more, very high resolutions images enable detecting small urban elements such as individual houses, cars, trees, or swimming pools.

As one can see, there is an obvious hierarchical relationship between the different objects of interests that can be detected when searching for different numbers of clusters. However, the huge size of these data sets usually makes them ineligible for hierarchical clustering algorithms because of their high computational complexity.

Our idea in this article is to propose a framework in which different clustering algorithms suited for image data sets and looking for different numbers of clusters can exchange their information in order to cross-validate the structures they respectively find. Our proposed method is inspired from previous works done in collaborative clustering [4,6,7] that allow different clustering algorithms to exchange with a goal of mutual improvement.

The difference between our proposed method and existing methods in the literature can be summed up in two points:

- Our proposed method is generic and can work with any clustering algorithm. In particular, it is suited for clustering algorithms used in image data sets.
- Unlike previously proposed collaborative frameworks, we don't have the limitation that all collaborators should be looking for the same number of clusters. In our case, our framework is specifically designed for multi-scale analysis and therefore each collaborator will be looking for a different number of clusters.

The remainder of this article is organized as follows: In Sect. 2, we introduce our proposed collaborative method for multi-scale analysis in satellite images. In Sect. 3, we introduce our data set and the experimental results of our method when compared with other algorithms. Finally, this article ends with a conclusion and perspectives on future works.

2   Multi-scale Communication Between Different Algorithms

Let us consider $X = \{x_1, ..., x_N\}, x_n \in \mathbb{R}^d$ a data set containing N segments of a pre-processed image described by d numerical feature attributes. We consider that we have J clustering algorithms $\{\mathcal{A}^1, ..., \mathcal{A}^J\}$ working on this data set. Each algorithm $\mathcal{A}^i$ has access to the same features but is looking for K_i clusters, K_i being a different number for each algorithm. Likewise, each algorithm $\mathcal{A}^i$ will output its own clustering solution $S^i = \{s_1, ..., s_N\}, s_n \in [C_1^i..C_{K_i}^i]$.

Since we want these algorithms to communicate in order to exchange information on there clusters with a goal of mutual improvement, following an idea from earlier works we define $\Omega^{i,j}$ the Probabilistic Confusion Matrix (PCM) mapping the clusters of any algorithms $\mathcal{A}^i$ and $\mathcal{A}^j$ as described in Eq. (1) where $\omega_{a,b}^{i,j}$ represents the percentage of data from the cluster a of algorithm $\mathcal{A}^i$ that are in the cluster b of algorithm $\mathcal{A}^j$.

$$\Omega^{i,j} = \begin{pmatrix} \omega_{1,1}^{i,j} & \cdots & \omega_{1,K_j}^{i,j} \\ \vdots & \ddots & \vdots \\ \omega_{K_i,1}^{i,j} & \cdots & \omega_{K_i,K_j}^{i,j} \end{pmatrix} \text{ where } \omega_{a,b}^{i,j} = \frac{|C_a^i \cap C_b^j|}{|C_a^i|} \tag{1}$$

The PCM $\Omega^{i,j}$ makes it possible to know whether or not the objects of two results have similar clusters, or if the two clustering results are dissimilar. Since our goal in this article is to have the different algorithms working on a same satellite image while exchanging their information, and since we know that the clusters to be found at different scales have a hierarchical structure, our aim will be to influence each local algorithm into tweaking its clusters in order to enforce this hierarchical structure.

To do so, our proposed method consists in an iterative framework where all algorithms are optimized in parallel following a two-step process similar to the process in the Expectation-Maximization algorithm [2]. In our case, we use the Iterated Conditional Modes [1] with a local Gaussian Mixture model and a neighborhood function adapted to very high resolution satellite images [9]. Under these conditions, each local clustering algorithm $\mathcal{A}^i$ has a set of local parameters $\Theta^i = \{\theta_1^i, ..., \theta_{K_i}^i\}, \theta_c^i = \{\pi_c^i, \mu_c^i, \Sigma_c^i\}$ where μ_c^i is the mean vector of cluster c for the algorithm $\mathcal{A}^i$, Σ_c^i its covariance matrix and π_c^i the mixing proportion for this cluster.

Algorithm 1. Collaborative multi-scale analysis: General algorithm

---
Local step:
forall the *clustering algorithms* $\mathcal{A}^i$ **do**
    Initialize each algorithm $\mathcal{A}^i$ on the data X.
    $\rightarrow$ Learn the local parameters Θ_0^i, initialize S_0^i
end
Collaborative step:
while *the system's global entropy* $\mathcal{H}$ *decreases* **do**
    Meta E-Step:
    forall the *clustering algorithms* $\mathcal{A}^i$ **do**
        For each data x_n, assess $s_n^i|_{t+1}$ using Eq. (2)
    end
    Meta M-Step:
    forall the *clustering algorithms* $\mathcal{A}^i$ **do**
        Update parameters $\Theta^i|_{t+1}$ using Eq. (4)
    end
    $t++$
end
Return S

Our proposed algorithm then runs the following two steps in an iterative way:

- *Expectation step:* Equation (2) is optimized in parallel for each algorithm and each data. In this equation $s_{n,c}^i(\Theta^i|_t)$ is the local algorithm responsibility function given in Eq. (3), and the remainder of this equation is a collaborative term the aim of which is to strengthen the hierarchy between the different clustering scales. $Z(i,n) = \sum_{c=1}^{K_i} \sum_{j \neq i}^{J} \omega_{q,c}^{j,i}$ is a normalization constant.

- *Maximization step:* For each algorithms, the parameters Θ^i are updated depending on the new solution vectors S^i. This is done using a simple Maximum a posteriori approximation as shown in Eq. (4).

$$s_n^i|_{t+1} = \underset{c}{\text{Argmax}} \left(s_{n,c}^i(\Theta^i|_t) + \frac{1}{Z(i,n)} \sum_{j \neq i}^{J} \omega_{q,c}^{j,i} \right), \quad q = s_n^j|_t \qquad (2)$$

$$s_{n,c}^i(\Theta^i) = \frac{\pi_c^i}{Z} \mathcal{N}(\mu_c^i, \Sigma_c^i, x_n) \cdot \exp^{f(V_x,c)} \qquad (3)$$

In Eq. (3), V_x is the neighborhood of the data segment x_n, and $f(V_x, c)$ is an energy function used to weight the likelihood of the cluster c being chosen given the neighbor segments [9]. In Fig. 1(b), an example of such neighborhood is shown with a segment and its neighborhood being highlighted in yellow and red respectively.

$$\Theta^i|_{t+1} = \underset{\Theta}{\mathrm{Argmax}}\, p(X|\Theta) : \forall c \begin{cases} N_c = \sum_{n=1}^{N} s_n^i(c) \\ \mu_c^i = \frac{1}{N_c} \sum_{n=1}^{N} s_n^i(c) \cdot x_n \\ \Sigma_c^i = \frac{1}{N_c} \sum_{n=1}^{N} s_n^i(c) \cdot (x_n - \mu_k^i)(x_n - \mu_k^i)^T \\ \pi_c^i = \frac{N_c}{N} \end{cases}$$

(4)

In Algorithm 1, we show the general framework of our algorithm. The collaborative step is iterated until the global entropy $\mathcal{H}$ described in Eq. (5) stabilizes. Using this stopping criterion, we ensure that the hierarchy between the clusters of the different algorithms will be maximized when the algorithm stops.

$$\mathcal{H} = \sum_{i=1}^{J} \sum_{j \neq i}^{J} \frac{-1}{K_i \times \ln(K_j)} \sum_{l=1}^{K_i} \sum_{m=1}^{K_j} \omega_{l,m}^{i,j} \ln(\omega_{l,m}^{i,j})$$

(5)

3 Experimental Results

3.1 Description of the Data

The VHR Strasbourg data set is a set made from a very high resolution image of the French city of Strasbourg, $1\,px = (50\,\mathrm{cm})^2$, an extract of which is shown in Fig. 1(a).

 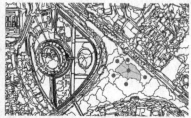

(a) Extract of the original image   (b) Example of a segment neighborhood

Fig. 1. Original image and segmentation example in Central Strasbourg (Color figure online)

This image has been pre-processed into a data set made of 187058 segments, each of them described by 27 attributes either geometric or radiometric [8]. These attributes include the geographic position of the segment, the surface of the area covered by the segment, the mean RGB values, the contrast compared to neighbor pixels and segments, the brightness, and the standard deviations, among others.

In addition to this information, this data set provides the neighborhood dependencies between the segments: number of neighbors, id number of each neighbor segment, and relative percentage of shared border.

The segments in this data set have highly irregular shapes (cf. Figure 1(b)), and consequently the neighborhoods themselves are also irregular. Unlike in the pixel model where 1st order neighborhoods usually include 4 or 8 neighbors, in this data set each segment can have 1 to 15 neighbors depending on its shape and position.

In order to validate our results, we had to find a ground truth. It was however too tedious a task to manually label the 187058 segments. Therefore we decided to rely on maps of the area made by expert geographers. The ground-truth that we use to assess our results was produced using a hybrid methodology, mixing data from topographic databases for roads and buildings, a supervised classification for different types of water and vegetation, as well as further manual refinement in order to reduce classification errors.

The expert geographers provided 15 different classes, this number was reduced to 9 classes by regrouping those that are very similar or technically impossible to distinguish for an unsupervised algorithm (such as different types of crops, or similar vegetation classes with different heights).

3.2   Results

In our experiments, we used three algorithms: the EM algorithm for the Gaussian Mixture Model [2], the regular SR-ICM algorithm [9] and our proposed collaborative Co C-ICM algorithm. The algorithms were run searching for 3, 6 and 9 clusters, separately in the case of the EM and the SR-ICM algorithm, and collaboratively for our method.

The results were assessed using the Davies-Bouldin index as a an internal criterion. This index assesses the compactness of the clusters and how well they are separated. The Davies-Bouldin index is not normalized and a lower value indicates better clustering results. It is worth mentioning that the Davies-Bouldin index usually gives better results with less clusters. As for the external index, we used the Rand Index to compare our results with the expert ground truth. We did not used the Adjusted version of the Rand Index because it was giving really poor results for the clusterings with 3 and 6 clusters which is much lower than in the ground truth. The Rand Index expresses in percentage how much the result matches with the expert ground-truth.

The results of this experiments over a dozen simulations for each algorithm are shown in Table 1, where the best result for each number of cluster is highlighted in bold.

As one can see, our proposed method performs better on average in 5 cases out of 6. We can also mention that the improvement on the Davies-Bouldin Index are more remarkable than these on the Rand Index. These results were to be expected since our proposed method doesn't change the unsupervised nature of the base clustering algorithms that we use. It is therefore logical that we have better results with unsupervised indexes. We believe that the same explanation applies to justify that the SR-ICM algorithm achieves somewhat identical Rand Index results with these of our proposed method when dealing with only 3 clusters.

Table 1. Experimental results

Algorithm	Davies-Bouldin index	Rand index
EM 3	2.36928	0.67454
SR-ICM 3	2.32855	**0.67606**
Co C-ICM 3	**2.32674**	0.67435
EM 6	2.88014	0.75867
SR-ICM 6	2.67816	0.76935
Co C-ICM 6	**2.49726**	**0.77068**
EM 9	2.62786	0.78225
SR-ICM 9	2.94065	0.79063
Co C-ICM 9	**2.58836**	**0.792187**

In Figs. 2(a) and (b), we show extracts of the results for the clusterings with 6 and 9 clusters. As can be seen on a color large scale version of the images, we achieve a decent segmentation at both scales with most elements being recognizable. On interesting remark is that while we did get a hierarchical structure between the clusters at different scales, the 3 initial clusters were not those that we would have had, had we used a supervised method. The 3 clusters found at the first scale by all algorithms (collaborative or not) were: vegetation areas, industrial buildings, and a cluster containing both urban elements and water bodies. Actually, water bodies only became a distinct cluster at the scale with 9 clusters.

Nevertheless, the unsupervised methods that we used found clusters that overall made sense. In the case of our proposed method, the results were not only better when using internal and external criteria, we also had a better hierarchical structure between the clusters from different scales.

(a) Collaborative C-ICM result for 6 clusters   (b) Collaborative C-ICM result for 9 clusters

Fig. 2. Clustering extracts (Color figure online)

4   Conclusion

In this article we have propose a simple yet powerful method that makes it possible for several clustering algorithm to work together with the goal of achieving a multi-scale clustering of the same data set. Our proposed method based on collaborative techniques lifts up previous limitations where collaboration was only possible when the different algorithms were looking for the same number of clusters. By doing so, our proposed framework can find the hierarchical structures between the clusters found at different scales.

We have tested our method on a very high resolution image data set for multi-scale clustering purposes, and we have compared our results with other specialized and more generalist methods available in the literature.

While this article was focused on an application for clustering algorithms used with very high resolution satellite images -data sets for which our method seemed well suited-, it is our strong belief that our proposed method is generic and can easily be extended for any other types of clustering algorithms. Our future research will therefore most likely focus on trying this method with other clustering algorithms and application contexts.

Acknowledgements. This work has been supported by the ANR Project COCLICO, ANR-12-MONU-0001.

References

1. Besag, J.: On the statistical analysis of dirty pictures. J. R. Stat. Soc. Ser. B **48**(3), 259–302 (1986)
2. Dempster, A.P., Laird, N.M., Rubin, D.B.: Maximum likelihood from incomplete data via the EM algorithm. J. R. Stat. Soc. Ser. B **39**(1), 1–38 (1977)
3. Gonzalez, R.C., Woods, R.E.: Digital Image Processing, 3rd edn. Prentice-Hall Inc., Upper Saddle River (2006)
4. Grozavu, N., Bennani, Y.: Topological collaborative clustering. Aust. J. Intell. Inf. Proces. Syst. **12**(3) (2010). https://cs.anu.edu.au/ojs/index.php/ajiips/issue/view/156
5. Pal, N.R., Pal, S.K.: A review on image segmentation techniques. Pattern Recognit. **26**(9), 1277–1294 (1993)
6. Pedrycz, W.: Collaborative fuzzy clustering. Pattern Recognit. Lett. **23**(14), 1675–1686 (2002)
7. Pedrycz, W.: Interpretation of clusters in the framework of shadowed sets. Pattern Recogn. Lett. **26**(15), 2439–2449 (2005)
8. Rougier, S., Puissant, A.: Improvements of urban vegetation segmentation and classification using multi-temporal pleiades images. In: 5th International Conference on Geographic Object-Based Image Analysis, p. 6 (2014)
9. Sublime, J., Troya-Galvis, A., Bennani, Y., Gancarski, P., Cornuéjols, A.: Semantic rich ICM algorithm for VHR satellite image segmentation. In: IAPR International Conference on Machine Vision Applications, Tokyo (2015)

Towards Ontology Reasoning for Topological Cluster Labeling

Hatim Chahdi[1,2]([✉]), Nistor Grozavu[2], Isabelle Mougenot[1], Younès Bennani[2], and Laure Berti-Equille[1,3]

[1] Espace-Dev UMR 228, IRD - Université de Montpellier, 500 Rue J.F. Breton, 34090 Montpellier, France
{hatim.chahdi,isabelle.mougenot,laure.berti-Equille}@ird.fr
[2] LIPN CNRS UMR 7030, CNRS - Université Paris 13, 99, av. J-B Clement, 93430 Villetaneuse, France
{hatim.chahdi,nistor.grozavu,younes.bennani}@lipn.univ-paris13.fr
[3] Qatar Computing Research Institute, Hamad Bin Khalifa University, Doha, Qatar

Abstract. In this paper, we present a new approach combining topological unsupervised learning with ontology based reasoning to achieve both: (i) automatic interpretation of clustering, and (ii) scaling ontology reasoning over large datasets. The interest of such approach holds on the use of expert knowledge to automate cluster labeling and gives them high level semantics that meets the user interest. The proposed approach is based on two steps. The first step performs a topographic unsupervised learning based on the SOM (Self-Organizing Maps) algorithm. The second step integrates expert knowledge in the map using ontology reasoning over the prototypes and provides an automatic interpretation of the clusters. We apply our approach to the real problem of satellite image classification. The experiments highlight the capacity of our approach to obtain a semantically labeled topographic map and the obtained results show very promising performances.

1 Introduction and Motivations

Clustering is a very important step in the process of knowledge extraction and discovery. When no labeled instances are available, unsupervised learning aims to discover new structures and group instances according to similarity, density, and proximity. Clustering has multiple applications [10]. We are interested in this work in topological clustering which allows clustering and visualization simultaneously i.e. the Self-Organizing Map (SOM) [11] algorithm. SOM is an unsupervised artificial neural network that produces a low-dimensional (generally two-dimensional) map from an unlabeled dataset. The nodes of the map represents a summarized version of the original dataset via a set of reference vectors (prototypes) spatially organized. This makes SOM suitable for lot of practical applications, including data visualization, data summarization and compression. However, as an unsupervised learning algorithm, the labeling and interpretation of the resulting map have to be performed manually, which can be difficult and

© Springer International Publishing AG 2016
A. Hirose et al. (Eds.): ICONIP 2016, Part III, LNCS 9949, pp. 156–164, 2016.
DOI: 10.1007/978-3-319-46675-0_18

time consuming. For example, in the field of remote sensing images analysis, where domain expertise of the user have to be very high, the interpretation of the results is not an easy task and only experts can handle it. In addition, the fact that images are encoded using low level features (numerical data) makes the interpretation even harder. This issue is known in the literature as the semantic gap [7].

The field of knowledge representation has been subject to lot of researches last years. These researches, supported by the emergence of the semantic Web [18], have lead to the development of the OWL 2 [6] as a standard language of ontology modelling. Based on description logics [1], OWL offers a standardized way to represent rich and complex knowledge. It comes with standard elements with precise meaning and formal semantics. This formal basis gives reasoners [5,20] the possibility to automatically process the ontologies and propose a set of inferences services that deduce new knowledge by calculating the logical consequences of the present facts, like the instances classes. However, when it comes to use ontology to classify a large number of instances, the reasoners fail to scale [8]. This issue can be very problematic when dealing with real world problems.

One of the motivations of our work is to use ontology reasoning to automate the labeling of a topological map. In our approach, the clusters are labeled using the concepts defined in the ontology. The use of an ontology as a support of the expert knowledge introduces also a modularity to our approach. The reasoning is a procedural process and the results change automatically when the concepts of the ontology change. This allow our approach to give interpretations that automatically meet the interest of the user following the expressed concepts in the ontology. Another motivation is the possibility to scale reasoning over large datasets. The SOM algorithm represents the input data using a finite set of prototypes. By reasoning over the prototypes and not all the input data, the reasoning can be performed over large datasets and in a shorter time.

The next sections are organized as following, we will first present related work in Sect. 2, followed by some preliminaries about ontology and reasoning in Sect. 3. The description of the different steps of our approach are given in Sect. 4. Section 5 highlights the experiments we made on the UCI wine dataset, it also shows the application of our approach on the real problem of remote sensing images classification. Conclusion and future work conclude the paper in Sect. 6.

2   Related Work

The problem of cluster labeling has been subject to different interesting researches in the literature [4, 12, 15–17]. These researches have explored different techniques to achieve cluster labeling. A version of SOM dedicated to textual data, called LabelSOM [17] was proposed by Rauber and Merkl. The authors labeled the trained map with a set of features of the data input. In another research, Treeratpituk et al. [21] proposed a method to label hierarchical text clustering. The presented algorithm assigns few labels to the clusters based on the cluster analysis information, the parent cluster and statistics about the corpus.

Recently, Li et al. [12] proposed an hybrid approach, combining linguistic and statistical techniques to achieve an automated labeling of the clusters. Although these approaches are very interesting, they are all dedicated to textual data and cannot work on quantitative data.

When it comes to methods that deal with numerical data, most of the approaches use *a priori* knowledge on the data to propose candidate labels of the clusters [3,4]. The work presented in this paper is focused in proposing an hybrid approach producing a semantic interpretation of the SOM's map based on the ontology reasoning. This makes our approach different from the other approaches present in the literature. Only few approaches are capable to propose a high level labels on quantitative data, and these methods are usually not adapted for topological clustering.

3   Preliminaries About Ontology and Reasoning

Before we present the proposed approach, we introduce in this section some core concepts related to ontology and description logics reasoning. We adopt in our work the Web Ontology Language (OWL 2) [6] as a standard language for ontology formalization. OWL was introduced and is now maintained by the World Wide Web Consortium. The aim of OWL is to give users a simple way to represent rich and complex knowledge. OWL introduces standardized elements with precise meaning and formal semantics. The formal part of OWL is mainly based on description logics [1], which is a family of knowledge representation.

In the following, we define an ontology $\mathcal{O}$ as a set of axioms (facts) describing a particular situation in the world from a specific domain point-of-view[1]. Formally, an ontology consists of three sets: the set of classes (concepts) denoted $\mathcal{N}_C$, the set of properties (roles) denoted $\mathcal{N}_P$, and the set of instances (individuals) denoted $\mathcal{N}_I$. Conceptually, it is often divided into two parts: TBox $\mathcal{T}$ and ABox $\mathcal{A}$, where the TBox contains axioms about classes (Domain knowledge) and ABox contains axioms about instances (data), such as:

$$\mathcal{O} = <\mathcal{T}, \mathcal{A}> = <\mathcal{N}_C, \mathcal{N}_P, \mathcal{N}_I> \qquad (1)$$

The formalization of the knowledge using formal semantics allows automatic interpretation. This is done by computing the logical consequences of the explicitly stated axioms in $\mathcal{O}$ to infer new knowledge [9]. An interpretation $\mathcal{I}$ of an ontology $\mathcal{O}$ consists of $(\Delta^I, \cdot^I)$, where Δ^I is the domain of I, and $\cdot^I$ the interpretation function of I that maps every class to a subset of Δ^I, every property to a subset of $\Delta^I \times \Delta^I$, and every instance a to an element $a^I \in \Delta^I$.

The interpretation of the ontology is computed using DL reasoners, which provides a set of inference services. Each inference service represents a specific reasoning task. This capability makes OWL very powerful for both knowledge modeling and knowledge processing. One of the inference services proposed by the reasoner is *instance checking*. This task can be performed to check if an

---

[1] In DL literature, an ontology is considered to be equivalent to a Knowledge Base.

instance a_i belongs to a concept $C \in \mathcal{T}$ based on the definition of the later. By empowering the OWL 2 modelling capacities, the concepts of the ontology can exploit qualified number restrictions over data properties [13] and logical operators to bridge the semantic gap and permit an efficient instance labeling.

4   Hybrid Approach: SOM Ontology Based Labeling

In this paper, we present a new hybrid approach using the available expert knowledge to semantically label the generated SOM map. Given an unlabeled dataset X and a TBox $\mathcal{T}$ of the ontology O, our goal is, to build a labeled map that reduces data dimension and at the same time gives them a semantic labeling. This can bring an understandable view of the results based on the users point of interest. To achieve this automatic labeling, we propose a two steps approach. The first step performs an unsupervised learning based on the SOM algorithm and generate a spatially organized map that summarizes the data in terms of a set of prototypes. In the second step, we use a dedicated process that transforms the prototypes of the map to OWL instances, inject them in a reasoner with the TBox (formalized expert knowledge) and performs a deductive reasoning that produces a semantic labeling of the prototypes based on the concepts present in the ontology. The rest of this section will detail the two steps of our approach.

4.1   Topological Unsupervised Learning Step

The first step of our approach performs an unsupervised learning over the input dataset using the Self-organizing maps. We used the basic model proposed by Kohonen. It consists of a discrete set C of cells called "map". This map has a discrete topology defined by an undirected graph, which usually is a regular grid in two dimensions. For each pair of cells (j, k) on the map, the distance $\delta(j, k)$ is defined as the length of the shortest chain linking cells j and k on the grid. For each cell j this distance defines a neighbor cell; in order to control the neighborhood area, we introduce a kernel positive function $\mathcal{K}$ ($\mathcal{K} \geq 0$ and $\lim_{|y| \to \infty} \mathcal{K}(y) = 0$). We define the mutual influence of two cells j and k by $\mathcal{K}_{j,k}$. In practice, as for traditional topological maps we use a smooth function to control the size of the neighborhood as $\mathcal{K}_{j,k} = \exp(\frac{-\delta(j,k)}{T})$. Using this kernel function, T becomes a parameter of the model. As in the Kohonen algorithm, we decrease T from an initial value T_{max} to a final value T_{min}.

Let $\mathbb{R}^d$ be the euclidean data space and $X = \{\mathbf{x}_i; i = 1, \ldots, N\}$ a set of observations, where each observation $\mathbf{x}_i = (x_i^1, x_i^2, \ldots, x_i^d)$ is a vector in $\mathbb{R}^d$. For each cell j of the grid (map), we associate a referent vector (prototype) $\mathbf{w}_i = (w_i^1, w_i^2, \ldots, w_i^d)$ which characterizes one cluster associated to cell i. We denote by $\mathcal{W} = \{\mathbf{w}_j, \mathbf{w}_j \in \mathbb{R}^d\}_{j=1}^{|\mathcal{W}|}$ the set of the referent vectors. The set of parameter $\mathcal{W}$ has to be estimated iteratively by minimizing the classical objective function defined as follows:

$$R(\chi, \mathcal{W}) = \sum_{i=1}^{N} \sum_{j=1}^{|\mathcal{W}|} \mathcal{K}_{j,\chi(\mathbf{x}_i)} \|\mathbf{x}_i - \mathbf{w}_j\|^2 \tag{2}$$

where χ assigns each observation $\mathbf{x}_i$ to a single cell in the map $\mathcal{C}$. This cost function can be minimized using both stochastic and batch techniques [11].

4.2 Ontology Based Map Labeling

Once we obtain the SOM's map, we extract the set of referent vectors $\mathcal{W}$ (prototypes), and transform them to OWL instances. In fact, before injecting the prototypes in the reasoner. They have to be transformed to OWL axioms, where each prototype is presented as an OWL instance and described using the properties present in the TBox of the Ontology O. We have designed and implemented a semi-automatic process (Algorithm 1) that performs this projection. As shown in the algorithm, our process takes as inputs the TBox of the ontology (formalized expert knowledge), and the prototypes obtained by SOM $\mathcal{W}$. Based on the properties $\mathcal{N}_\mathcal{P}$ of the TBox and the set of variables (features) V describing the data, the process suggests a mapping between the inputs (Algorithm 1, line: 2). Once the mapping is established, our process generates OWL axioms that represent the data. Each prototype $\mathbf{w}_j$ is represented as an OWL instance a_i (Algorithm 1), where a_i is described by the properties available in TBox (Algorithm 1, line: 9–11), and where these properties get their values from the prototypes vector (Algorithm 1, line: 12). At this point, all the required components to perform reasoning are available. Once we transform the prototypes

---

Algorithme 1. Semi-Automatic Projection of the prototypes in the ontology

---

Inputs:
    Set of prototypes $\mathcal{W} = \{\mathbf{w}_i, \mathbf{w}_i \in \mathbb{R}^d\}_{i=1}^{|\mathcal{W}|}$ described by $V = \{v_j\}_{j=1}^d$
    Domain Knowledge : $\mathcal{T} = <\mathcal{N}_\mathcal{C}, \mathcal{N}_\mathcal{P}>$
Output:
    ABox : $\mathcal{A} = \{a_i\}_{i=1}^n$
Method:
1: **for all** p_k in $\mathcal{N}_\mathcal{P}$ **and** v_j in V **do**
2:    **Boolean** Query = Does p_k correspond to v_j
3:    **if** Query.isTrue() **then**
4:        $map(\mathcal{N}_\mathcal{P}, V)$.add($p_k, v_j$)
5:    **end if**
6: **end for**
7: **for all** $\mathbf{w}_i$ in $\mathcal{W}$ **do**
8:    $a_i := createOWLInstance()$;
9:    **for all** p_k in $map(\mathcal{N}_\mathcal{P}, V)$ **do**
10:      $a_i.addProperty(p_k)$
11:      $a_i.setPropertyType(p_k, \mathcal{T}.getPropertyType(p_k))$
12:      $a_i.setPropertyValue(p_k, \mathbf{w}_i.getValueOf(v_k))$
13:    **end for**
14:    **return** a_i : OWL representation of $\mathbf{w}_i$
        $\mathcal{A}.add(a_i)$
15: **end for**

---

using our algorithm, we obtain an ABox containing all the OWL instances $a_i \in$ ABox representing the prototypes $w_i \in \mathcal{W}$ of the SOM's map. We inject the ABox with the TBox in a DL reasoner to obtain our knowledge base. We use the Pellet [20] reasoner in our approach because it effectively implements the instance checking task and supports OWL 2 specifications. As mentioned above, instance checking consists in finding the most specific concept which a given instance belongs to. Performing this reasoning task over the constructed knowledge base will label the SOM's map with the concepts formalized in the TBox of the ontology.

5 Experiments

In this section, we present the conducted experimentations and the results we obtained. The purpose of our evaluation is to highlight the effectiveness of our approach to automatically label the SOM's map based on ontology reasoning. We apply our method on two datasets.

The first one is the UCI wine dataset[2], which consists of 178 instances. Each instance is described with 13 variables that represent the quantities of 13 constituents (e.g. alcohol, Mg...) found in each of the wines. The inputs used in our method are the unlabeled dataset, and an ontology about three concepts, those concepts have been constructed following a similar approach to the one proposed by Sheeren et al. [19]. Each concept is defined in OWL 2 using qualified number restrictions over the properties. We apply the different steps described in our approach. First, we applied the SOM algorithm to obtain the map. We fixed the size to 6×11. Then we extract the prototypes, transform them using the Algorithm 1 and perform an *instance checking* with the Hermit [5] reasoner to label them based on the three concepts. We evaluate the results using purity and the labeling percentage. The labeling percentage is important as it shows the efficiency of our ontology to give automatic interpretation of the results. The purity of our map is 96,62 % and 61 (92,42 %) prototypes were correctly labeled.

5.1 Satellite Images Classification

We also apply our approach to a real-world problem of satellite image classification. The image we used is an extract of a Landsat 5 TM image. The Landsat program is a joint NASA/USGS program[3] that freely provides satellite images covering all the earth surface. The image can be downloaded from the USGS Earth Explorer[4]. The Landsat 5 TM have a spatial resolution of 30 meters and seven spectral bands. The size of our image is of 760×680 pixels. The images concern the region of the river Rio Tapajos in the Amazon, Brazil. The input TBox of our ontology contains 2 thematic concepts: Water and Vegetation. To build the corresponding TBox, several spectral bands and indices were used.

---

[2] Wine dataset: http://archive.ics.uci.edu/ml/datasets/Wine.

[3] Landsat Science: http://landsat.gsfc.nasa.gov/.

[4] USGS Earth Explorer: http://earthexplorer.usgs.gov/.

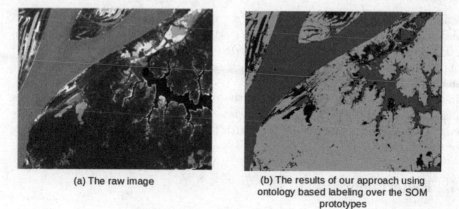

(a) The raw image

(b) The results of our approach using ontology based labeling over the SOM prototypes

■ #Water ■ #Vegetation   ■ Unlabeled

Fig. 1. Application of our approach for the classification of a Landsat satellite image

The concepts were defined using the seven bands: TM1,...,TM7 and the spectral indices NDVI (Normalized Difference Vegetation Index) [2] and NDWI (Normalized Difference Water Index) [14]. For example, the water concept is defined as follows:

$$Water\_Pixel \equiv Pixel \wedge ((\exists TM4. < 0.05 \wedge \exists ndvi. < 0.01) \vee (\exists TM4. < 0.11 \wedge \exists ndvi. < 0.001))$$

We applied our approach as described before with a fixed map of 20×20. The Fig. 1 shows visually the obtained results projected using the satellite image. We evaluated the purity of the results using a reference classification made by a domain expert. We obtained 98,45 % for the purity index, with 76 % of the prototypes correctly labeled using ontology reasoning. This experiment illustrates how using SOM helped the reasoning to scale as it was performed only on the 400 prototypes instead of the 460.000 pixels of the image. If we had to reason over all the dataset, the reasoner would not have been able to scale [8].

6   Conclusion and Future Work

We have presented in this paper a new hybrid approach combining the unsupervised topographic learning with ontology reasoning in order to semantically label clustering results. Combining both deductive and inductive reasoning, our method can automate the interpretation of clustering based on the ontology and in the same time scales and speed up the reasoning process by exploiting the proposed prototype label propagation. We have applied our approach to multiple datasets and evaluate the results. We have also shown how our approach can be used in the real-world problem of satellite images classification. As future work, we plan to extend our approach by using a constraints based on the ontology reasoning results to modify the obtained maps and improve its semantic coherence.

Acknowledgment. This work was supported by the French Agence Nationale de la Recherche under Grant ANR-12-MONU-0001.

References

1. Baader, F.: The Description Logic Handbook: Theory, Implementation, and Applications. Cambridge University Press, New York (2003)
2. DeFries, R., Townshend, J.: NDVI-derived land cover classifications at a global scale. Int. J. Remote Sens. **15**(17), 3567–3586 (1994)
3. Durand, N., Derivaux, S., Forestier, G., Wemmert, C., Gançarski, P., Boussaid, O., Puissant, A.: Ontology-based object recognition for remote sensing image interpretation. In: 19th IEEE International Conference on Tools with Artificial Intelligence, 2007. ICTAI 2007, vol. 1, pp. 472–479. IEEE (2007)
4. Forestier, G., Puissant, A., Wemmert, C., Gançarski, P.: Knowledge-based region labeling for remote sensing image interpretation. Comput. Environ. Urban Syst. **36**(5), 470–480 (2012)
5. Glimm, B., Horrocks, I., Motik, B., Stoilos, G., Wang, Z.: HermiT: an OWL 2 reasoner. J. Autom. Reason. **53**(3), 245–269 (2014)
6. Group, W.O.W., et al.: OWL 2 web ontology language document overview (2009)
7. Hare, J.S., Lewis, P.H., Enser, P.G., Sandom, C.J.: Mind the gap: another look at the problem of the semantic gap in image retrieval. In: Electronic Imaging 2006, p. 607309. International Society for Optics and Photonics (2006)
8. Horrocks, I., Li, L., Turi, D., Bechhofer, S.: The instance store: Dl reasoning with large numbers of individuals. In: Proceedings of the 2004 Description Logic Workshop (DL 2004), pp. 31–40 (2004)
9. Horrocks, I., Sattler, U.: Ontology reasoning in the SHOQ (D) description logic. IJCAI **1**, 199–204 (2001)
10. Jain, A.K., Murty, M.N., Flynn, P.J.: Data clustering: a review. ACM Comput. Surv. (CSUR) **31**(3), 264–323 (1999)
11. Kohonen, T.: The self-organizing map. Proc. IEEE **78**(9), 1464–1480 (1990)
12. Li, Z., Li, J., Liao, Y., Wen, S., Tang, J.: Labeling clusters from both linguistic and statistical perspectives: a hybrid approach. Knowl. Based Syst. **76**, 219–227 (2015)
13. Lutz, C.: Description logics with concrete domains-a survey (2003)
14. McFeeters, S.: The use of the normalized difference water index (NDWI) in the delineation of open water features. Int. J. Remote Sens. **17**(7), 1425–1432 (1996)
15. Mei, Q., Shen, X., Zhai, C.: Automatic labeling of multinomial topic models. In: Proceedings of the 13th ACM SIGKDD International Conference on Knowledge Discovery and Data Mining, pp. 490–499. ACM (2007)
16. Popescul, A., Ungar, L.H.: Automatic labeling of document clusters. Unpublished manuscript (2000). http://citeseer.nj.nec.com/popescul00automatic.html
17. Rauber, A., Merkl, D.: Automatic labeling of self-organizing maps: making a treasure-map reveal its secrets. In: Zhong, N., Zhou, L. (eds.) PAKDD 1999. LNCS (LNAI), vol. 1574, pp. 228–237. Springer, Heidelberg (1999)
18. Shadbolt, N., Berners-Lee, T., Hall, W.: The semantic web revisited. IEEE Intell. Syst. **21**(3), 96–101 (2006). doi:10.1109/MIS.2006.62

19. Sheeren, D., Quirin, A., Puissant, A., Gançarski, P., Weber, C.: Discovering rules with genetic algorithms to classify urban remotely sensed data. In: Proceedings of IEEE International Geoscience and Remote Sensing Symposium (IGARSS 2006), pp. 3919–3922 (2006)
20. Sirin, E., Parsia, B., Grau, B.C., Kalyanpur, A., Katz, Y.: Pellet: a practical OWL-DL reasoner. Web Semant. Sci. Serv. Agents. World Wide Web 5(2), 51–53 (2007)
21. Treeratpituk, P., Callan, J.: Automatically labeling hierarchical clusters. In: Proceedings of the 2006 International Conference on Digital Government Research, pp. 167–176. Digital Government Society of North America (2006)

Overlapping Community Detection Using Core Label Propagation and Belonging Function

Jean-Philippe Attal[1]([envelope]), Maria Malek[1], and Marc Zolghadri[2]

[1] Quartz Laboratory, EISTI, 95000 Cergy, France
jal@eisti.eu
[2] Quartz Laboratory, SUPMECA, 93407 Saint-Ouen, France

Abstract. Label propagation is one of the fastest methods for community detection, with a near linear time complexity. It acts locally. Each node interacts with neighbours to change its own label by a majority vote. But this method has three major drawbacks: (i) it can lead to huge communities without sense called also monster communities, (ii) it is unstable, and (iii) it is unable to detect overlapping communities.

In this paper, we suggest new techniques that improve considerably the basic technique by using an existing core detection label propagation technique. It is then possible to detect overlapping communities through a belonging function which qualifies the belonging degree of nodes to several communities.

Nodes are assigned and replicated by the function a number of times to communities which are found automatically. User may also interact with the technique by imposing and freezing the number of communities a node may belong to. A comparative analysis will be done.

1 Introduction

Networks are powerful tools to model complex systems in many fields such as biology (protein-protein interaction), anthropology, sport, etc. Most of the networks representing complex systems show specific characteristics, with dense groups of nodes. The nodes of a group have lots of interconnections while they are loosely linked to the rest of the graph. These highly connected groups of nodes are called *communities*. However, some nodes can also belong to several communities, this means that they have overlaps and are known as *overlapping community detection problem*.

In this paper, we expose new overlapping community detection algorithms, based on a core label propagation. We discuss label propagation highlighting some of the open issues in Sect. 2. Section 3 shows our proposed algorithms. In Sect. 4, we show the results of our experiments on different graphs and provide a benchmark comparing with some of the existing algorithms of the literature. Finally, Sect. 5 provides some conclusions and perspectives.

© Springer International Publishing AG 2016
A. Hirose et al. (Eds.): ICONIP 2016, Part III, LNCS 9949, pp. 165–174, 2016.
DOI: 10.1007/978-3-319-46675-0_19

2   Label Propagation Algorithm

2.1   Standard Label Propagation

Label propagation algorithm (LPA) is an iterative algorithm based on local information of neighbouring nodes [1]. Let us consider an undirected graph $G = (V, E)$, with V the set of vertices and E the set of edges. The neighbors of the node x are gathered in $V(x) = \{x_1, ..., x_k\}$. Through an iterative approach, at each step, every node makes an updating of its label according to the label of its immediate neighbours, by voting mechanism. The label of x is then changed to the label of the majority of the labels of its neighbors. More formally, if c_x stands for x's label and $N^l(x)$ for the set of x's neighbours having the label l, then the new label of x is obtained by $c_x = \arg\max_l |N^l(x)|$. At the end of the process, nodes with the same label represent a community. This method can be applied in a *synchronous way* (new label assignment or label propagation is performed in parallel on all nodes) or in an *asynchronous way* (label propagation step is performed sequentially). In general, asynchronous propagation has a better stability.

2.2   Label Propagation with Dams and Core Detection

In [2], authors proposed a new label propagation technique named LPA with dams (LPWD). The goal was to find edges that separate the densest groups of nodes by putting first artificial dams and running the label propagation algorithm afterwards. This method avoids bad label propagation. We have chosen the edge betweenness centrality to put dams since edges having a strong score connect dense community structures and are the main cause to bad label propagations. This method has been applied in a "core way" to stabilize the LPA. The key idea to stabilize a non-deterministic method is to consider nodes that appear most of the time in the same community after the application of the algorithm several times. These nodes are called *core* nodes. To find out these nodes we study the frequency of pairs of nodes in the same community after several trials of a non-deterministic algorithm. We used the technique of frequency of co-occurrence communities defined by Seifi et al. [3]. The method consists of launching $\mathcal{N}$ times the non-deterministic algorithm and creating a matrix $P_{ij}^{\mathcal{N}} = [p_{ij}]_{n \times n}^{\mathcal{N}}$ such that p_{ij} represents the frequency nodes that i and j are in the same communities. A new graph $G' = (V, E')$ is then created using a threshold $\alpha \in [0, 1]$. Pairs of nodes of the matrix $P_{ij}^{\mathcal{N}}$ having a weight smaller than α are not considered to construct the edges of G'. The study led by Seifi et al. shows that weak values of α give a small number of connected components with a more important density than strong values of α giving a more important number of connected components. Connected components of G' correspond to communities.

3 Proposed Methods for Detection of Overlapping Communities

The methodology we suggest here to detect overlapping community is based on the graph G'. G' is first projected on the original graph G. This means that those edges present in G' but not in G are removed to exclude irrelevant edges (i.e. preserving the topological structure of the original graph). We use the information stored in $P_{ij}^{\mathcal{N}}$ to weight the graph G. This allows to see the nodes with a higher probability to be together in communities. We obtain the *edge-between communities*, noted EBC (the set of the edges linking different communities), with nodes connected to different communities. These nodes, connected to several communities by their outgoing edges, are the possible candidates for the overlapping communities. To know if these nodes can be overlapped, we propose several belonging functions, all based on the topology of the communities.

Considering a candidate node x, the idea is to measure the possibility of belonging of x regarding the surrounding communities and their topological properties. We note $\mathcal{C}_x = \{\mathcal{C}_1^x, ..., \mathcal{C}_K^x\}$, the K different communities x belongs to. As x is linked to K different communities, we note $|\mathcal{C}_x| = K$.

3.1 Function 1: Membership Function Based on the Density

We consider the edge density of the communities by using the weight of the edges which link the node x to them. Considering the K different communities in the neighbourhood of x, we propose the *belonging density measure* $f_d : x \times \{C_1^x, ..., C_K^x\} \longmapsto \mathbb{R}_+$ such that:

$$f_d(x, \{C_1^x, ..., C_K^x\}) = \max_{c \in \binom{\mathcal{C}_K^x}{j}, j \in \{1, ..., |C_K^x|\}} \left(\frac{1}{|c|} \times \sum_{i \in c} \omega_{x,i} d^S(i) \right)$$

where $\omega_{x,i}$ is the weight of the edge linking node x to i, c is the list of combination grouping communities, and $d^S(i)$ is the density of the subgraph S where node i is. The binomial coefficient term $\binom{\mathcal{C}_K^x}{j}$ allows to compute the j-combinations in a set of $\mathcal{C}_K^x$ communities to which x may belong. The configuration with the highest score will duplicate the node x to the associated communities, creating overlapping communities.

A node with a strong weight linked to high density communities has more chance to overlap that a node linked to weak density communities with a weak weight. Overlapping may also be refused. This could be the case if the edges linking node x to the other communities have a weak weight or if f_d is weak, with a small density. Based on this fact, the overlapping will be done if and only if $f_d(x, \{C_1^x, ..., C_K^x\}) \geq \frac{1}{|c|} \sum_{S \in c} d^S$ where d^S is the density of the subgraph S (here the community). It is also possible to force overlappings. To do so, the domain of variations of j which was greater than 1 (i.e. $j \in \{1, ..., |C_K^x|\}$) is modified to L, i.e. $j \in \{L, ..., |C_K^x|\}$. This will force the node x to belong simultaneously to L communities, respecting the density constraint.

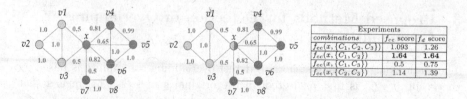

Fig. 1. After computing the function f_d and f_{cc} on the node x, x belongs to 2 communities

Let us give an example to illustrate these steps. Suppose that we have the following graph Fig. 1, with the partition of communities $P = \{C_1, C_2, C_3, C_4\}$ such that $C_1 = \{v_1, v_2, v_3\}$, $C_2 = \{v_4, v_5, v_6\}$, $C_3 = \{v_7, v_8\}$ and $C_4 = \{x\}$, resulting of the core label propagation with the frequency matrix. The question is to know whether x can belong to several communities. In Fig. 1, the best configuration to obtain an overlapping community detection on x is given with C_1 and C_2. The node x will be replicated where the density is the highest and the relation between x and the other communities, given by the weight $\omega_{x,i}$ (i being a community linked to x), is strong.

3.2 Function 2 Membership Function Based on the Local Clustering Coefficient

The clustering coefficient (CC) [4] is a social network measure dealing with the clustering of nodes in a network. It computes the probability that two individuals linking to an other person are also linked together. We define the *belonging coefficient clustering measure* : $f_{cc} : x \times \{C_1^x, ..., C_K^x\} \longmapsto \mathbb{R}_+$ such that:

$$f_{cc}(x, \{C_1^x, ..., C_K^x\}) = \max_{c \in \binom{C_K^x}{j}, j \in \{1, ..., |C_K^x|\}} \left(\frac{1}{|c|} \times \sum_{i \in c} \omega_{x,i} CC^S(i) \right)$$

where $CC^S(i)$ is the average clustering coefficient of the subgraph S of G, $\omega_{x,i}$ is the weight of the edge linking node x to i and c the combination grouping communities. Again, overlapping may be refused. This can be the case if the edges linking a node x to the other communities have a weak weight or clustering coefficient. The overlapping is then done if and only if $f_{cc}(x, \{C_1^x, ..., C_K^x\}) \geq \frac{1}{|c|} \sum_{S \in c} CC^S$, where CC^S is the average clustering coefficient of the subgraph S and c. In the example of the Fig. 1, the best configuration to obtain an overlapping community detection on x is given by joining x to C_1 and C_2. This is due to the null value of the clustering coefficient of C_3. The idea behind f_{CC} is to assign nodes to communities with important nodes, linked to several triangles, well connected, which is a characteristic of scale−free networks (which degree distribution follows a power law).

3.3 Proposed Community Detection Algorithms

The label propagation with core detection for the overlapping (CDLPOV) is explained in Algorithm 1. $f_{belonging}$ refers to the function chosen by the user, namely f_d of f_{CC}.

Algorithm 1. The CDLP with belonging measure function (CDLPOV)

Input: A graph $G = (V, E)$, the threshold α, $\mathcal{N}$ the number of runs
Output: Overlapping communities of G

1: Allocate an empty co-occurrence frequency matrix
2: Run $\mathcal{N}$ times the asynchronous label propagation
3: Fill the co-occurrence frequency matrix with the results of the asynchronous label propagation
4: Create a new graph $G' = (V, E')$ from $P_{ij}^{\mathcal{N}}$ with edges having a weight superior or equal to α
5: Project G' on G with the weight (but by removing edges in G' and not in G)
6: Create the partitions $P = \{P_1, ..., P_C\}$ by considering the $\mathcal{C}$ connected components as cores
7: Compute the set of edge between communities (EBC) using the partition P
8: $Cand \leftarrow []$ $\{Cand$ is a list of candidates$\}$
9: **for** every node x having an edge in EBC **do**
10:     $Cand$.append(x)
11: **end for**
12: $P^{Ov} \leftarrow P$
13: **for** every node x in $Cand$ **do**
14:     $Cand$.append(x)
15: **end for**
16: **for** every node x in $Cand$ **do**
17:     compute $f_{belonging}(x, \{C_1^x, ..., C_K^x\})$
18:     **if** $f_{belonging}(x, \{C_1^x, ..., C_K^x\}) \geq \sum_{S \in \{C_1^x, ..., C_K^x\}} SNM(S)$ **then**
19:         Duplicate the node x in the corresponding communities of P^{Ov}
20:     **end if**
21: **end for**
22: **return** The partition $P^{Ov} = \{P_1^{Ov}, ..., P_C^{Ov}\}$.

$SNM(S)$ stands for the value of a social network measure regarding the topological aspect of the community S, namely, the density or the clustering coefficient. By varying α in an interval with a specific step, we can get an overlapping dendrogram.

4 Evaluation Measures of Community Detection Algorithm, Benchmarks, Experiments and Discussion

To test the validity of our algorithms, we use some measures exclusively for overlapping community detection problem. It exists two kinds of evaluation measures. For the **unsupervised measures**, we use an overlapping version of the *modularity* [5]. To compute **supervised measures** (knowing the ground-truth communities), we use the *normalized mutual information* (NMI) [6] with its extended overlapping version by Lancichinetti et al. the *omega-index*, an overlapping version of the adjusted rand index [7] and the F_1 score. We give also the *edge between communities* (EBC) in percentage and the relative number of communities #. We are interested in the characteristics that a node needs regarding its neighborhood communities to be replicated and the difference in term of results between the density and clustering coefficient. For each of our experiments, we use $\mathcal{N} = 100$ to compute the CDLP. We do variate the threshold

Table 1. Characteristics of some networks with the Average Transitivity (AT) and the number of communities (#)

| Characteristics of some networks | | | | | | | | | |
Networks	$|V|$ and $|E|$	Density	Diameter	AT	Networks	$|V|$ and $|E|$	Density	Diameter	AT
Zachary	34\78	0.139	5.0	0.256	Pol	105\441	0.081	7.0	0.348
Foot	115\615	0.094	4.0	0.407	NS	1589\2742	0.002	17.0	0.693
Dol	62\159	0.084	8.0	0.309	Jazz	198\2742	0.140	6	0.52

α (see Sect. 2.2), allowing to create a new weighted graph, on an interval allowing to have more than one single community. The networks on which we run our algorithms, Table 1, are the Zachary Karate Club network [8] (Zac), the football club network [9] (Foot), the political book network [10] (Pol), the dolphins network [11] (Dol), a coauthorship network of scientists [12] (NS) a musician network [13] (jazz).

4.1   Experiments

Zachary Karate Club: We obtain the following result with the Zachary Karate Club.
  In Fig. 2, candidates for the overlapping using f_d and f_{cc} are the same. For $\alpha \leq 0.6$, there is just the node 10 which is overlapped once (f_{cc} (C_2)) by two communities with f_d (C_3 and C_4). C_2 has a bigger number of triangles than C_4. For $\alpha \geq 0.8$, the node 3 which is known to be in overlapping communities in the literature, belongs to two communities and is replicated in two communities with f_d in C_5 and C_4, but just in one community using f_{cc} (C_4). The node 1 is replicated in one community in C_2 for each method. In Table 2, the higher is α, the more will be the number of candidates for the overlapping. Even if the number of candidates is the same until $\alpha \geq 0.9$, the quality of results is better using f_d than f_{cc}. The highest score of the modularity is for $\alpha \geq 0.7$ (0.62 for each of the methods) with the highest NMI and the highest Ω index.

Dolphins network: The graph is composed of two communities. On Table 3, for $\alpha \geq 0.5$, the system well finds the two communities, without overlapping community with an NMI, an Omega index and a F_1 score of 1.0. By increasing the value of α, the size of the communities decreases while increasing the number of possible candidates for the overlapping. We see that the percentage of replicated nodes is the same from $\alpha \geq 0.6$ to $\alpha \geq 0.8$. Nevertheless, the two methods do not replicate the candidate

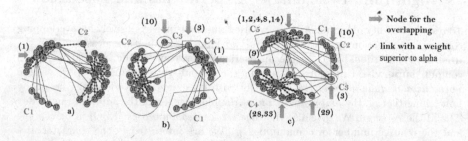

Fig. 2. Graph with different values of α using f_d and f_{cc}, (a) $\alpha \geq 0.6$, (b) $\alpha \geq 0.7$ (c) $\alpha \geq 0.8$

Table 2. Cand: Possible candidates, CandOv: Percentage of overlapping nodes

Results with f_d and f_{cc} on Zachary Karate Club

f_d	Cand	CandOv	EBC	Q_{Ov}^{Nic}	Ω	F_1	NMI	#
$\alpha \geq 0.6$	47.058 %	2.94 % (1)	17.95 %	0.3986	0.0645	0.65	0.2365	2
$\alpha \geq 0.7$	41.17 %	8.8235 % (3)	16.0 %	0.6210	0.7110	0.857	0.5178	4
$\alpha \geq 0.8$	55.88 %	32.352 % (11)	26.92 %	0.4202	0.4923	0.7499	0.3488	5
$\alpha \geq 0.9$	55.88 %	32.352 % (11)	26.92 %	0.4202	0.4923	0.7499	0.3488	5
f_{cc}	Cand	CandOv	EBC	Q_{Ov}^{Nic}	Ω	F_1	NMI	#
$\alpha \geq 0.6$	47.058 %	2.941 % (1)	17.948 %	0.3986	0.064	0.65	0.2365	2
$\alpha \geq 0.7$	41.176 %	8.823 % (3)	16.0 %	0.6210	0.7110	0.8571	0.5178	3
$\alpha \geq 0.8$	55.882%	32.353 % (11)	26.923 %	0.4202	0.4923	0.7499	0.3488	5
$\alpha \geq 0.9$	55.882%	32.353% (11)	26.923 %	0.4202	0.4923	0.7499	0.3488	5

Table 3. Cand: Possible candidates, CandOv: Percentage of overlapping nodes

Results with f_d and f_{cc} on Dolphins network

f_d	Cand	CandOv	EBC	Q_{Ov}^{Nic}	Ω	F_1	NMI	#
$\alpha \geq 0.5$	51.61 % %	0.0 %	20.38 %	0.7959	1.0	1.0	1.0	2
$\alpha \geq 0.6$	54.838 %	6.451 % (4)	24.050 %	0.7502	0.6165	.8571	0.5936	4
$\alpha \geq 0.7$	64.51 %	8.0645 % (5)	30.57 %	0.7144	0.4777	0.7499	0.457	5
$\alpha \geq 0.8$	61.29 %	19.3548 % (12)	29.30 %	0.6052	0.4777	0.6184	0.4421	8
$\alpha \geq 0.9$	77.41 %	25.81 % (16)	43.94 %	0.5415	0.3549	0.5333	0.2456	12
f_{cc}	Cand	CandOv	EBC	Q_{Ov}^{Nic}	Ω	F_1	NMI	#
$\alpha \geq 0.5$	51.613%	0.0 %	20.382 %	0.7959	1.0	1.0	1.0	2
$\alpha \geq 0.6$	54.838 %	6.451 % (4)	24.051 %	0.7502	0.6125	.8571	0.5936	4
$\alpha \geq 0.7$	64.516 %	8.0645 % (5)	30.57 %	0.7144	0.4294	0.7499	0.457	5
$\alpha \geq 0.8$	61.29 %	19.3548 % (12)	29.299 %	0.6062	0.4777	0.6184	0.4421	8
$\alpha \geq 0.9$	77.419 %	35.483 % (22)	43.949 %	0.4412	0.5882	0.5489	0.2772	12

in the same way. f_{cc} function produces the same quality in term of communities but replicates more nodes than f_d for a high value of α.

Netscience: The Netscience graph is characterised by a very weak density of 0.0021. The size of the detected communities is very small. Regarding Table 4, it is for $\alpha \geq 0.4$ that the first overlapping nodes appear for both methods. The EBC is relatively weak from 2.39 % of edges with $\alpha \geq 0.2$ to 18.27 % with $\alpha \geq 0.8$, implying a small percentage of candidates. Just only 0,61 % of nodes are candidates with $\alpha \geq 0.4$ whilst it is 6.36 % for f_d. For both methods, the number of overlapping nodes is very similar and relatively small. Observing the modularity values, results are both similar. We can conclude that for a weak density graph, results are very similar using belonging functions based on density or clustering coefficient.

Table 4. Cand: Possible candidates, CandOv: Percentage of overlapping nodes

Results with f_d and f_{cc} on Netscience network

	f_d					f_{cc}				
	Cand	CandOv	EBC	Q_{Ov}^{Nic}	#	Cand	CandOv	EBC	Q_{Ov}^{Nic}	#
$\alpha \geq 0.2$	3.2854 %	0.0 %	2.3956 %	0.9768	293	3.2854 %	0.0 %	2.3956 %	0.9768	293
$\alpha \geq 0.3$	5.5441 %	0.0 %	4.1752 %	0.9724	297	5.5441 %	0.0 %	4.1752 %	0.9724	297
$\alpha \geq 0.4$	7.3921 %	0.6160 % (9)	5.794 %	0.9476	308	7.3921 %	0.5475 % (8)	3.1460 %	0.9492	308
$\alpha \geq 0.5$	9.2402 %	0.6844 % (10)	7.7344 %	0.9401	315	9.2402 %	0.6160 % (9)	4.2322 %	0.9417	315
$\alpha \geq 0.6$	14.099 %	2.3956 % (35)	12.7310 %	0.8862	342	14.0999 %	2.1640 % (36)	6.9662 %	0.8859	342
$\alpha \geq 0.7$	16.0164 %	4.5174 % (72)	15.3319 %	0.8450	360	16.0164 %	5.3388 % (85)	8.3895 %	0.8550	360
$\alpha \geq 0.8$	18.0698 %	6.3655 % (101)	18.2751 %	0.8173	371	18.0698 %	7.0499 % (112)	10.0 %	0.8130	371

4.2 Comparative Analysis

We compare our proposed methods with some of overlapping community detection algorithms coming from the state of the art, namely: CFinder [14], COPRA ($\nu = 2$ and $\nu = 3$ [15], ν being the number of communities to which nodes could belong), and OSLOM [16], SLPA [17], and CONGA [18].

Table 5. (*) algorithms based on the label propagation

Networks	F_1	Ω	NMI	Q_{Ov}^{Nic}	#	%	Networks	F_1	Ω	NMI	Q_{Ov}^{Nic}	#	%
Zac #2							**Dol #2**						
CFinder	0.48	0.35	0.18	0.52	3	5.88%	CFinder	0.57	0.35	0.26	0.66	4	3.72%
OSLOM	**0.86**	**0.84**	**0.80**	**0.748**	2	2.94%	OSLOM	**1.0**	**0.914**	**0.852**	0.742	2	1.61%
CONGA	0.65	0.113	0.274	0.441	2	2.94%	CONGA	0.85	0.892	0.821	0.746	2	3.22%
$COPRA_2$*	0.281	0.266	0.228	0.414	11.3	5.58%	$COPRA_2$*	0.933	0.788	0.751	0.693	10.8	0.52%
$COPRA_3$*	0.684	0.359	0.347	0.452	6.4	12.64%	$COPRA_3$*	0.893	0.767	0.701	0.677	3.7	7.73%
SLPA*	0.86	0.633	0.564	**0.608**	2.12	2.20%	SLPA*	0.56	0.754	0.632	0.742	3.44	2.00%
CDLPOV f_d*	0.852	0.711	**0.518**	0.621	4	8.82%	CDLPOV f_d*	**1.0**	**1.0**	**1.0**	**0.796**	2	0.0%
CDLPOV f_{cc}*	0.852	0.711	**0.518**	0.621	4	8.82%	CDLPOV f_{cc}*	**1.0**	**1.0**	**1.0**	**0.796**	2	0.0%
Foot #12							**Pol #4**						
CFinder	0.701	0.64	0.55	0.51	13	6.9%	CFinder	**0.855**	**0.740**	**0.79**	0.884	4	(9)
OSLOM	**0.954**	**0.802**	**0.759**	0.696	12	0.0%	OSLOM	**0.814**	**0.704**	**0.55**	**0.847**	2	1.90%
CONGA	0.823	0.321	0.423	0.451	11	60.0%	CONGA	0.688	0.651	0.49	0.779	4	4.16%
$COPRA_2$*	0.933	0.788	0.705	0.693	10.8	0.52%	$COPRA_2$*	0.687	0.637	0.385	0.825	3	1.05%
$COPRA_3$*	0.944	0.747	0.712	0.668	11.2	2.52%	$COPRA_3$*	0.702	0.649	0.416	0.827	2.8	6.47%
SLPA*	0.748	0.684	0.612	**0.715**	10.30	1.69%	SLPA*	0.755	0.648	0.497	0.83	3.40	12.5%
CDLPOV f_d*	0.854	0.865	**0.751**	0.699	11	0.0%	CDLPOV f_d*	0.784	0.654	0.495	0.844	3	1.90%
CDLPOV f_{cc}*	0.854	0.865	**0.751**	0.699	11	0.0%	CDLPOV f_{cc}*	**0.788**	**0.667**	**0.503**	**0.834**	2	0.0%
NS							**Jazz**						
CDLPOV f_d*				0.977	293	0.0%	CDLPOV f_d*				0.64	2	0.50%
CDLPOV f_{cc}*				0.977	293	0.0%	CDLPOV f_{cc}*				0.64	2	0.50%

We show the results obtained by our methods having the highest unsupervised scores in Table 5. Our proposed algorithms give relatively good results in term of quality. We have a better quality than COPRA, and a better stabilization. Even if label propagation based algorithms produce more communities, the CDLP with f_d and f_{cc} produces less communities than other label propagation approaches. We explain this fact by the frequency matrix which stabilizes the label propagations.

5 Perspectives and Conclusion

In this article, we have proposed two methods applied to an existing core label propagation with frequency matrix to find overlapping communities. Each of the methods uses the weight of the frequency matrix and the social characteristics of the communities to know if some possible nodes could be assigned to several communities. Users have the choice between letting the algorithm assign some nodes to several communities found automatically or giving a number of communities for which the nodes could belong. The belonging functions are based on the density (f_d) and the clustering coefficient (f_{cc}) of the communities found by the CDLP. Results of the two different methods are similar, with the same number of candidates and the quality is almost the same. Nevertheless, there is more replication using f_d rather than f_{cc}. Regarding the runtime, the higher will be the density of the graph, with a high number of possible candidates, the higher will be the runtime. We have observed that computing the frequency matrix with 100 LPA was fast enough (1 s for Zachary, and 6 s for NS). Nevertheless, the time to compute f_{cc} is bigger than the one to compute f_d, due to the time of computing the triangles for the clustering coefficient. The time to compute the functions increases with α, due to the increasing number of candidates and the EBC. For $\alpha \geq 0.5$ we need near 30 s to compute f_d and 60 s for f_{cc} on NS network. For $\alpha \geq 0.8$, we need more than one hundred seconds to compute f_d and f_{cc}. For Zacahry, Dolphins or political books, with α between 0.1 and 0.8, we need around 10 s for f_d and near 20 s with f_{cc}.

We work now on a method to reduce the number of combinations that the functions have to test. The time being important for large graphs, we have developped a Hadoop and a Spark version of the CDLP. We are currently developping the functions on these two distributed models. We plan to work on multiplex graphs, in the biology field, where nodes are of the same types, but edges can have several meanings.

References

1. Raghavan, U.N., Albert, R., Kumara, S.: Near linear time algorithm to detect community structures in large-scale networks. Phys. Rev. E **76**(3), 036106 (2007)
2. Attal, J.-P., Malek, M.: A new label propagation with dams. In: Proceedings of the 2015 IEEE/ACM International Conference on Advances in Social Networks Analysis and Mining, pp. 1292–1299. ACM (2015)
3. Seifi, M., Junier, I., Rouquier, J.-B., Iskrov, S., Guillaume, J.-L.: Stable community cores in complex networks. In: Menezes, R., Evsukoff, A., González, M.C. (eds.) Complex Networks. SCI, vol. 424, pp. 87–98. Springer, Heidelberg (2013)
4. Watts, D., Strogatz, S.: Collective dynamics of small-world networks. Nature **393**, 440–442 (1998)
5. Nicosia, V., Mangioni, G., Carchiolo, V., Malgeri, M.: Extending the definition of modularity to directed graphs with overlapping communities. J. Stat. Mech. Theory Exp. **2009**(03), P03024 (2009)
6. Ana, L., Jain, A.K.: Robust data clustering. In: Proceedings of 2003 IEEE Computer Society Conference on Computer Vision and Pattern Recognition, vol. 2, p. II-128. IEEE (2003)
7. Hubert, L., Arabie, P.: Comparing partitions. J. Classif. **2**(1), 193–218 (1985)
8. Zachary, W.: An information flow model for conflict and fission in small groups. J. Anthropol. Res. **33**, 452–473 (1977)
9. Girvan, M., Newman, M.E.J.: Community structure in social and biological networks. Proc. Natl. Acad. Sci. **99**(12), 7821–7826 (2002)
10. Krebs, V.: Books about US politics (2004). http://www.orgnet.com/
11. Lusseau, D., Schneider, K., Boisseau, O.J., Haase, P., Slooten, E., Dawson, S.M.: The bottlenose dolphin community of doubtful sound features a large proportion of long-lasting associations. Behav. Ecol. Sociobiol. **54**(4), 396–405 (2003)

12. Newman, M.E.J.: Finding community structure in networks using the eigenvectors of matrices. Phys. Rev. E **74**(3), 22 pages, 8 figures, minor corrections in this version. http://arxiv.org/abs/physics/0605087

13. Gleiser, P., Danon, L.: Community structure in jazz. In: Advances in Complex Systems, vol. 6, p. 565 (2003). doi:10.1142/S0219525903001067

14. Palla, G., Derenyi, I., Farkas, I., Vicsek, T.: Uncovering the overlapping community structure of complex networks in nature, society. Nature, **435**(7043), 814–818. http://dx.doi.org/10.1038/nature03607

15. Gregory, S.: Finding overlapping communities in networks by label propagation. New J. Phys. **12**(10), 103018 (2010)

16. Lancichinetti, A., Radicchi, F., Ramasco, J.J., Fortunato, S.: Finding statistically significant communities in networks. PLoS ONE **6**(4), e18961 (2011)

17. Xie, J., Szymanski, B.K., Liu, X.: SLPA: uncovering overlapping communities in social networks via a speaker-listener interaction dynamic process. In: 2011 IEEE 11th International Conference on Data Mining Workshops (ICDMW), pp. 344–349. IEEE (2011)

18. Gregory, S.: An algorithm to find overlapping community structure in networks. In: Kok, J.N., Koronacki, J., Lopez de Mantaras, R., Matwin, S., Mladenič, D., Skowron, A. (eds.) PKDD 2007. LNCS (LNAI), vol. 4702, pp. 91–102. Springer, Heidelberg (2007)

A New Clustering Algorithm for Dynamic Data

Parisa Rastin[✉], Tong Zhang, and Guénaël Cabanes

LIPN UMR CNRS 7030, Université Paris 13 Sorbonne Paris Cité,
99, avenue Jean-Baptiste Clément, 93430 Villetaneuse, France
`parisa.rastin@lipn.uni-paris13.fr`

Abstract. In this paper, we propose an algorithm for the discovery and the monitoring of clusters in dynamic datasets. The proposed method is based on a Growing Neural Gas and learns simultaneously the prototypes and their segmentation using and estimation of the local density of data to detect the boundaries between clusters. The quality of our algorithm is evaluated on a set of artificial datasets presenting a set of static and dynamic cluster structures.

Keywords: Growing neural gas · Clustering · Density · Dynamic data

1 Introduction

Clustering task is a challenging problem. It aims at regrouping data into clusters based on their similarity, without any external or a priori knowledge. Several difficulties are known for this task, including the selection of a suitable number of clusters to represent the data structure and the capability to detect non-convex clusters [4]. In addition, in many cases, databases are characterized by a dynamic structure over time, new data constantly arriving. Sometimes, the evolution and the mass of data are so important that it is not possible to store it in a database and we need "on the fly" analyses. These processes have been the subject of numerous studies in recent years because of the important potential applications in many fields [3,7,10]. However, it is a difficult problem because of the computation and memory cost associated with the volume of data involved.

To deal with that type of data, it is essential to construct a condensed description of the dynamic properties of the data [8,12], and we need a method to detect the variations in the data structure [1,5]. As this type of data can be seen as an infinite process constantly changing over time, the segmentation of these data (i.e. the model of their structure) must also constantly evolve. An adapted clustering algorithm must therefore be able to compute the segmentation of the dataset over different periods of time, and be able to compare the data structures over these different periods.

The objective of this work is to develop a method of clustering based on the Growing Neural Gas algorithm (GNG) [6,9] to train prototypes capable of following the dynamic of the data. The idea is to learn dynamically an estimation of the local density of data for each prototype, then to use this density to

© Springer International Publishing AG 2016
A. Hirose et al. (Eds.): ICONIP 2016, Part III, LNCS 9949, pp. 175–182, 2016.
DOI: 10.1007/978-3-319-46675-0_20

detect clusters boundaries. The number of clusters is detected automatically and clusters of any type of structure can be detected.

The remaining of the paper is organized as follows: in Sect. 2 we describe the GNG algorithm, in Sect. 3 we present the proposed method, in Sect. 4 the experimental results of the proposed method are discussed, finally in Sect. 5 a conclusion and some perspectives are given.

2   Growing Neural Gas

Growing Neural Gas (GNG) is an algorithm proposed by [9], which is able to compute dynamically a set of prototypes to represent the data in a condensed form. GNG is able to adapt the number of prototypes to the representation need and compute a neighbourhood network between prototypes by linking nodes representing similar data. In addition, each connected component of this graph can be seen as a representation of a cluster. Its greatest weakness, however, is its inability to adapt to rapidly changing distributions.

The algorithm GNG + Utility [6] on the other hand, is designed to handle this type of data by removing periodically the least useful nodes and adding relevant new nodes when needed. Basically, it removes nodes that contribute little to reducing the error, and inserts new nodes in the graph when they contribute significantly to the error reduction. The difference between the GNG and GNG-U algorithm is not large but the impact on performance is significant [6].

The GNG-U algorithm can be described as follow:

1. Initialization: create two nodes positioned randomly with an $error$ value and an utility value U set to 0, connected with an edge with $age = 0$.
2. For a data point $\bar{x}$: locate the two nodes s and t the closest to $\bar{x}$, that is, the two nodes with reference vectors $\bar{w}_s$ and $\bar{w}_t$ such that $\|\bar{w}_s - \bar{x}\|^2$ is the smallest value and $\|\bar{w}_t - \bar{x}\|^2$ is the second smallest for all k nodes.
3. Update the $error$ and U values of s: $error_s \leftarrow error_s + \|\bar{w}_s - \bar{x}\|^2$ and $U_s \leftarrow U_s + error_t - error_s$
4. Update prototype w_s and its topological neighbours (i.e. all nodes connected to s by an edge) using leaning steps e_w and e_n, with $e_w, e_n \in [0,1]$: $\bar{w}_s \leftarrow \bar{w}_s + e_w(\bar{x} - \bar{w}_s)$ and $\bar{w}_n \leftarrow \bar{w}_n + e_n(\bar{x} - \bar{w}_n), \forall n \in Neighbour(s)$
5. Increment by 1 the age of all edges between s and its topological neighbours. If s and t are connected by an edge, then set the age of this edge to 0. If they are not connected, create an edge between them and set it's age to 0.
6. If there are edges with age over the age_{max} threshold, remove them. If, after this, there are nodes without edges, remove these nodes. Then remove node i with the smallest value U_i if: $\frac{error_j}{U_i} > k$, where j is the node with the biggest error, and k is a constant parameter (we remove node i if the utility falls below a certain fraction of the error).
7. If the current iteration is a multiple of a parameter λ, then insert a new node r as follows:
    (a) Find the node u with the highest $error$.
    (b) Among the neighbours of u, find the node v with the highest $error$.

(c) Insert the new node r between u and v as follows: $\bar{w}_r \leftarrow \frac{\bar{w}_u + \bar{w}_v}{2}$.

(d) Create edges between u and r and v and r, and remove the edge between u and v.

(e) Reduce errors in variables u and v and define the error of r : $error_u \leftarrow \alpha \times error_u$, $error_v \leftarrow \alpha \times error_v$ and $error_r \leftarrow error_u$

(f) Initialise the utility value of the new node r: $U_r \leftarrow \frac{U_u + U_v}{2}$

8. Reduce the $errors$ and U of all nodes by a factor β: $error_k \leftarrow error_k - \beta \times error_k$ and $U_k \leftarrow U_k - \beta \times U_k$

9. If the stop criterion is not met then repeat from 2.

3   A New Two-Level Clustering Algorithm for GNG

Here we propose a new algorithm that simultaneously learns the prototypes of GNG and their segmentation using information on the data densities. The principle is inspired from DS2L-SOM [4], a clustering algorithm adapted to static datasets. The new algorithm estimates the local density of data for each node of the GNG, in order to detect the density fluctuations that characterize the boundaries between data groups.

The first part of the new algorithm is based on the same assumptions as GNG + U with the addition of:

- For every node n, we include a local variable, the density D_n of this node.
- In GNG + U, Step 3, after the local error and utility of node s have been updated we now add a rule to update the density D_j for each node j.

$$D_j(t) = D_j(t-1) + e^{-\frac{\|\bar{x} - \bar{w}_j\|^2}{2\sigma^2}} \tag{1}$$

with σ a width parameter.

- In Step 7, when a node is created, initialise the density of the new node r as follow:

$$D_r \leftarrow \frac{D_u + D_v}{2} \tag{2}$$

- In Step 9, the density for all nodes is updated in the same manner as $error$ and U, with the same decay constant.

$$D_k \leftarrow D_k - \beta \times D_k \tag{3}$$

The second part of the new algorithm is the off-line computation of the segmentation (i.e. the clusters), using the density information to detect the boundaries between clusters. The idea is to detect low density zones within the L connected components C of the GNG+U, in order to characterize the subgroups defined by density. We use, for each pair of adjacent subgroups, a "density-dependent" index is computed to determine whether a low density area is a reliable indicator of the data structure, or whether it should be regarded as a random fluctuation in density.

Inputs: $P = C_{i=1..L}$ and $D_{j=1..M}$.
Output: Segmentation (clusters).

1. **for each component $C_k \in P$ do:**
   - Determine the set $M(Ck)$ of local maximal density:

$$M(C_k) = \{i \in C_k | D_i \geqslant D_j, \forall j \text{ neighbour of } i\} \tag{4}$$

   - Compute the threshold matrix:

$$S = [S(i,j)]_{i,j=1...|M(C_k)|} \text{ with } S(i,j) = \left(\frac{1}{D_i} + \frac{1}{D_j}\right)^{-1} \tag{5}$$

   - For each node $i \in C_k$, label i with an element $label(i)$ of $M(C_k)$, according to an ascending gradient of density along topological connections.
   - For each pair of neighbour node (i, j), if $label(i) \neq label(j)$ and if $D_i > S(i,j)$ and $D_j > S(i,j)$ then merge the two groups (the small variation of density is considered as noise).
2. **Return the segmentation** (i.e. the clusters).

4   Experimental Results

We compared the proposed algorithm (named "Den") with other two-level methods in which traditional clustering algorithms [2] are applied on the nodes of GNG+U: K-means ("KM"), Hierarchical Ascendant Clustering with Ward ("Ward") or Average Link ("Avg") distances, affinity propagation ("Aff"), the density-based DBSCAN algorithm ("DBScan") and GNG-U alone ("GNG") with cluster defined by connected component of the graph. The parameters used for these algorithms are $e_w = 0.2$, $e_n = 0.006$, $age_{max} = 50$, $\lambda = 100$, $\alpha = 0.5$ and $k = 30$. The methods were compared on six artificial datasets: "Static_Noconv_2", "Static_Gauss_2", "Static_Gauss_10" and "Dyn_Noconv_2", "Dyn_Gauss_2", "Dyn_Gauss_10". Three of them are static, the three others are dynamic, their distribution changing over time. The clusters can have a Gaussian or a non-convex distribution. The dimensions vary from 2 to 10, the total number of data point is 10000. The clusters' density is variable (see Figs. 1 and 2).

Table 1 presents the value of the Adjusted Rand Index (ARI) [11] for the 7 different algorithms for each data set in each of the 10 periods (a time period is represented by 1000 data points). Results based on the Normalized Mutual Information and the Jaccard index are highly similar and are not presented here. We used the true number of clusters for the three algorithms that need this number as a parameter or to extract the clusters from the dendrogram ("KM", "Ward" and "Avg"). The others parameters were chosen to give the best results. Figure 3 represents the mean quality of the algorithms over the 6 datasets, for each time period.

The results show that our algorithm provides a generally better solution than the other algorithms. For the three algorithms "Km", "Ward" and "Avg",

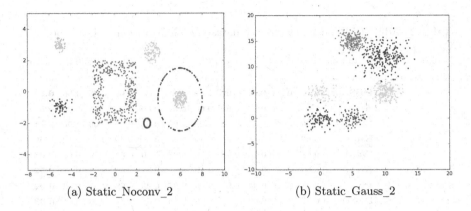

(a) Static_Noconv_2   (b) Static_Gauss_2

Fig. 1. Two-dimensional static datasets.

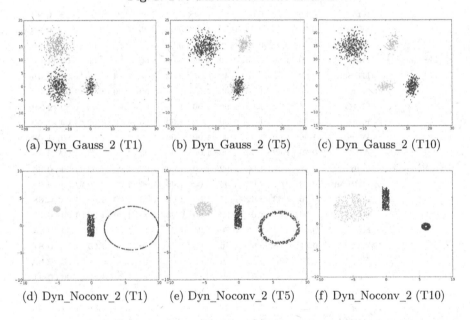

(a) Dyn_Gauss_2 (T1)   (b) Dyn_Gauss_2 (T5)   (c) Dyn_Gauss_2 (T10)

(d) Dyn_Noconv_2 (T1)   (e) Dyn_Noconv_2 (T5)   (f) Dyn_Noconv_2 (T10)

Fig. 2. Two-dimensional dynamic datasets. Clusters can change in shape and position, as well as appearing or disappearing.

satisfactory quality is observed in general, especially for static Gaussian datasets, but the number of cluster must be known a priori, which is rarely the case in reality. Amongst algorithms that don't need this value, our algorithm is the only one that perform well for all the dataset, as shown in Fig. 4. On the contrary, we can observe that the algorithm "GNG" can not separate the clusters in contact because of the connections between prototypes; "DBSCAN" struggles with clusters of different density; as for "Aff" algorithm, it does not work well on non-convex data sets.

Table 1. Adjusted Rand Index for each dataset, each algorithm and each period

Dataset	Algorithm	T1	T2	T3	T4	T5	T6	T7	T8	T9	T10	Mean
Static_Noconv_2	Den	0.62	0.66	0.72	0.73	0.98	1.00	0.95	0.79	1.00	0.98	**0.84**
	DBScan	0.17	0.23	0.95	1.00	1.00	0.98	0.98	1.00	0.98	1.00	**0.83**
	Avg	0.65	0.51	0.53	0.67	0.65	0.62	0.66	0.53	0.54	0.66	**0.60**
	GNG	0.00	0.08	0.51	0.76	1.00	1.00	0.98	0.79	1.00	1.00	**0.71**
	Km	0.62	0.49	0.46	0.48	0.50	0.50	0.48	0.55	0.45	0.52	**0.50**
	Ward	0.68	0.66	0.55	0.65	0.66	0.63	0.66	0.66	0.50	0.65	**0.63**
	Aff	0.80	0.48	0.47	0.48	0.49	0.46	0.54	0.53	0.48	0.50	**0.52**
Static_Gauss_2	Den	0.81	0.89	0.94	0.94	0.90	0.97	0.93	0.98	0.95	0.93	**0.92**
	DBScan	0.67	0.68	0.69	0.31	0.68	0.65	0.68	0.64	0.66	0.64	**0.63**
	Avg	0.93	0.95	0.96	0.95	0.92	0.95	0.95	0.98	0.97	0.94	**0.95**
	GNG	0.00	0.00	0.00	0.28	0.68	0.26	0.60	0.59	0.29	0.64	**0.33**
	Km	0.89	0.92	0.94	0.95	0.88	0.85	0.94	0.95	0.95	0.93	**0.92**
	Ward	0.93	0.95	0.96	0.95	0.92	0.95	0.95	0.98	0.97	0.95	**0.95**
	Aff	0.85	0.92	0.87	0.88	0.88	0.86	0.94	0.88	0.93	0.85	**0.89**
Static_Gauss_10	Den	1.00	1.00	1.00	1.00	1.00	1.00	1.00	1.00	1.00	1.00	**1.00**
	DBScan	0.02	1.00	0.76	0.49	0.25	0.53	0.36	0.44	0.36	0.30	**0.45**
	Avg	1.00	1.00	1.00	1.00	1.00	1.00	1.00	1.00	1.00	1.00	**1.00**
	GNG	0.00	1.00	1.00	1.00	1.00	1.00	1.00	1.00	1.00	0.89	**0.89**
	Km	1.00	1.00	1.00	1.00	1.00	1.00	1.00	1.00	1.00	1.00	**1.00**
	Ward	1.00	1.00	1.00	1.00	1.00	1.00	1.00	1.00	1.00	1.00	**1.00**
	Aff	0.79	1.00	1.00	1.00	1.00	1.00	1.00	1.00	1.00	1.00	**0.98**
Dyn_Nocov_2	Den	0.58	1.00	1.00	0.92	1.00	1.00	1.00	1.00	1.00	1.00	**0.95**
	DBScan	0.83	0.61	1.00	1.00	1.00	1.00	1.00	1.00	1.00	1.00	**0.94**
	Avg	0.76	0.63	0.62	0.77	1.00	1.00	1.00	1.00	1.00	1.00	**0.88**
	GNG	0.58	1.00	1.00	0.92	1.00	1.00	1.00	1.00	1.00	1.00	**0.95**
	Km	0.66	0.72	0.70	0.70	1.00	1.00	1.00	1.00	1.00	1.00	**0.88**
	Ward	0.61	0.78	0.68	0.66	1.00	1.00	1.00	1.00	1.00	1.00	**0.87**
	Aff	0.71	0.73	0.29	0.80	0.66	0.69	0.46	0.58	0.68	0.65	**0.63**
Dyn_Gauss_2	Den	0.91	0.91	0.94	0.69	0.84	0.82	0.96	1.00	1.00	1.00	**0.91**
	DBScan	0.91	0.91	0.51	0.69	0.84	0.82	0.82	0.82	1.00	1.00	**0.83**
	Avg	0.88	0.74	0.79	0.46	0.67	0.62	0.81	0.80	1.00	1.00	**0.78**
	GNG	0.00	0.00	0.29	0.69	0.84	0.82	0.82	0.82	1.00	1.00	**0.63**
	Km	0.76	0.73	0.78	0.42	0.59	0.59	0.59	1.00	1.00	1.00	**0.75**
	Ward	0.80	0.76	0.79	0.41	0.70	0.59	0.60	1.00	1.00	1.00	**0.77**
	Aff	0.91	0.73	0.61	0.42	0.48	0.50	0.60	0.76	0.76	0.76	**0.65**
Dyn_Gauss_10	Den	1.00	1.00	1.00	0.51	1.00	1.00	0.32	1.00	1.00	1.00	**0.88**
	DBScan	0.72	1.00	1.00	0.29	1.00	1.00	0.00	1.00	1.00	1.00	**0.80**
	Avg	1.00	1.00	1.00	0.61	1.00	1.00	0.31	1.00	1.00	1.00	**0.89**
	GNG	0.37	1.00	1.00	0.08	1.00	1.00	0.00	1.00	1.00	1.00	**0.74**
	Km	1.00	1.00	1.00	0.61	1.00	1.00	0.38	1.00	1.00	1.00	**0.90**
	Ward	1.00	1.00	1.00	0.61	1.00	1.00	0.38	1.00	1.00	1.00	**0.90**
	Aff	0.85	1.00	1.00	0.52	1.00	1.00	0.33	1.00	1.00	1.00	**0.87**

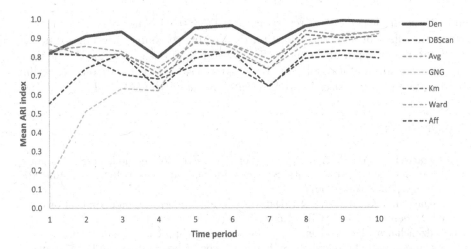

Fig. 3. Mean value of the Adjusted Rand Index over the 6 datasets, for each time period and each algorithm.

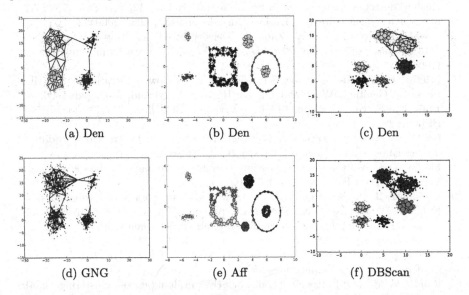

Fig. 4. Example of clustering results according to the algorithms used. Each color represent a cluster. The graph of GNG+U is shown with node size proportional to the estimated density used in the proposed algorithm.

5 Conclusions and Perspectives

We proposed in this paper a new two-level algorithm based on GNG-U, able to represent the data structure in real-time and automatically discover the number of clusters at each moment. We shown that our algorithm works well on a set of static and dynamic datasets and produces better quality results than classical algorithms applied on the prototypes of a GNG-U. In the future, we plan to study the effect of different algorithm parameters to analyse the importance of each parameter and find the most effective combinations depending on the data size and the speed of its evolution. It is also necessary to compare our algorithm on real data of larger sizes in order to validate the method for real applications.

References

1. Aggarwal, C., Yu, P.: A survey of synopsis construction methods in data streams. In: Aggarwal, C. (ed.) Data Streams: Models and Algorithms, pp. 169–207. Springer, New York (2007)
2. Aggarwal, C.C., Reddy, C.K.: Data Clustering: Algorithms and Applications, 1st edn. Chapman & Hall/CRC, Boca Raton (2013)
3. Balzanella, A., Lechevallier, Y., Verde, R.: A new approach for clustering multiple streams of data. In: Ingrassia, S., Rocci, R. (eds.) Classification and Data Analysis, pp. 417–420 (2009)
4. Cabanes, G., Bennani, Y., Fresneau, D.: Enriched topological learning for cluster detection and visualization. Neural Netw. **32**(1), 186–195 (2012). http://www.sciencedirect.com/science/article/pii/S0893608012000482
5. Cao, F., Ester, M., Qian, W., Zhou, A.: Density-based clustering over an evolving data stream with noise. In: 2006 SIAM Conference on Data Mining, pp. 328–339 (2006)
6. Fritzke, B.: A self-organizing network that can follow non-stationary distributions. In: Gerstner, W., Germond, A., Hasler, M., Nicoud, J.-D. (eds.) ICANN 1997. LNCS, vol. 1327, pp. 613–618. Springer, Heidelberg (1997). doi:10.1007/BFb0020222
7. Guha, S., Harb, B.: Approximation algorithms for wavelet transform coding of data streams. IEEE Trans. Inf. Theory **54**(2), 811–830 (2008)
8. Manku, G.S., Motwani, R.: Approximate frequency counts over data streams. In: Very Large Data Base, pp. 346–357 (2002)
9. Martinetz, T.M., Schulten, K.J.: A "neural-gas" network learns topologies. In: Kohonen, T., Mäkisara, K., Simula, O., Kangas, J. (eds.) Artificial Neural Networks, pp. 397–402. Elsevier Science Publishers, Amsterdam (1991)
10. O'Callaghan, L., Mishra, N., Meyerson, A., Guha, S., Motwani, R.: Streaming-data algorithms for high-quality clustering. In: Proceedings of IEEE International Conference on Data Engineering (2002)
11. Rand, W.M.: Objective criteria for the evaluation of clustering methods. J. Am. Stat. Assoc. **66**(336), 846–850 (1971). http://links.jstor.org/sici?sici=0162-1459%28197112%2966%3A336%3C846%3AOCFTEO%3E2.0.CO%3B2-T
12. Verde, R., de Carvalho, F., Lechevallier, Y.: A dynamical clustering algorithm for multi-nominal data. In: Kiers, H., et al. (eds.) Data Analysis, Classification, and Related Methods, pp. 387–393. Springer, Heidelberg (2000)

Reinforcement Learning

Decentralized Stabilization for Nonlinear Systems with Unknown Mismatched Interconnections

Bo Zhao[1]([✉]), Ding Wang[1], Guang Shi[1], Derong Liu[2], and Yuanchun Li[3]

[1] The State Key Laboratory of Management and Control for Complex Systems,
Institute of Automation, Chinese Academy of Sciences, Beijing 100190, China
{zhaobo,ding.wang,shiguang2012}@ia.ac.cn
[2] School of Automation and Electrical Engineering,
University of Science and Technology Beijing, Beijing 100083, China
derong@ustb.edu.cn
[3] Department of Control Science and Engineering,
Changchun University of Technology, Changchun 130012, China
liyc@ccut.edu.cn

Abstract. This paper establishes a neural network and policy iteration based decentralized control scheme to stabilize large-scale nonlinear systems with unknown mismatched interconnections. For relaxing the common assumption of upper boundedness on interconnections when designing the decentralized optimal control, interconnections are approximated by neural networks with local signals of isolated subsystem and replaced reference signals of coupled subsystems. By using the adaptive estimation term, the performance index function is constructed to reflect the replacement error. Hereafter, it is proven that the developed decentralized optimal control policies can guarantee the closed-loop large-scale nonlinear system to be uniformly ultimately bounded. The effectiveness of the developed scheme is verified by a simulation example.

Keywords: Adaptive Dynamic Programming · Decentralized control · Unknown mismatched interconnections · Policy iteration

1 Introduction

The modern systems are becoming large-scale and complex. To overcome the difficulties on analyzing and designing of centralized controllers, decentralized control strategy has received considered attention.

Recently, adaptive dynamic programming (ADP) algorithms were further developed to solve the optimal control problem of continuous-time systems [1], discrete-time systems [2], external disturbances and uncertainties [3], trajectory tracking [4], constraints [5], etc. Recently, some local controllers were designed in decentralized control strategy for large-scale nonlinear systems with unknown parameters and dynamic uncertainties [6], unmatched uncertainties [7], unknown

© Springer International Publishing AG 2016
A. Hirose et al. (Eds.): ICONIP 2016, Part III, LNCS 9949, pp. 185–192, 2016.
DOI: 10.1007/978-3-319-46675-0_21

dynamic model [8]. These schemes were widely applied to coordinated control systems [9], power systems [10] and Markov decision process [11], but they focused on the controlled system in linear or satisfying assumed matching conditions. Actually, interconnections are always unknown and mismatched.

This paper solves the decentralized control problem for large-scale nonlinear systems with unknown mismatched interconnections. The main contribution is that the developed scheme avoids the common assumptions of matching condition and upper boundedness on interconnections of large-scale nonlinear systems.

2   Problem Statement

A large-scale nonlinear system is composed of N interconnected subsystems as

$$\dot{x}_i(t) = f_i(x_i(t)) + g_i(x_i(t))u_i(x_i(t)) + h_i(x(t)) \tag{1}$$

where $i = 1, 2, \ldots, N$, $x_i(t) \in \mathrm{R}^{n_i}$ and $u_i(x_i(t)) \in \mathrm{R}^{m_i}$ are the state and input vectors of the ith subsystem, respectively. $x = [x_1, x_2, \ldots, x_N]^\mathsf{T} \in \mathrm{R}^n$ with $n = \sum_{i=1}^{N} n_i$ denotes the overall system state, and $u_1(x_1), u_2(x_2), \ldots, u_N(x_N)$ are local control inputs. For the ith subsystem, $f_i(\cdot)$ and $g_i(\cdot)$ are locally Lipschitz in their augments with $f_i(0) = 0$, the unknown mismatched interconnection $h_i(x(t))$ is approximated by radial basis function NN (RBFNN) as

$$h_i(x(t)) = W_{ih}^\mathsf{T} \sigma_{ih}(x(t)) + \varepsilon_i(x(t)) \tag{2}$$

where $\sigma_{ih}(x(t))$ is the radial basis function, $W_{ih} = [w_{i1}, w_{i2}, \ldots, w_{ik}]^\mathsf{T}$ is optimal weight vector and $\varepsilon_i(x)$ is the NN approximation error.

Assumption 1. *The NN approximation error ε_i is bounded as $|\varepsilon_i(x)| \leq \phi_{i1}$, where $\phi_{i1} > 0$ is an unknown constant.*

To remove the assumption of upper boundedness on interconnections, we approximate the interconnection $h_i(x(t))$ via RBFNN using the signals of isolated subsystem and the reference signals of the coupled subsystems, i.e., $h_i(x) = h_{id}(x_{iD}) + \Delta_i(x_{iD}) + \varepsilon_i(x_i)$, where $x_{iD} = [x_{1d}, x_{2d}, \ldots, x_i, \ldots, x_{Nd}]^\mathsf{T}$, $h_{id}(x_{iD}) = W_{ih}^\mathsf{T} \sigma_{ih}(x_{iD})$, $\Delta_i(x_{iD}) = W_{ih}^\mathsf{T} \sigma_{ih}(x) - W_{ih}^\mathsf{T} \sigma_{ih}(x_{iD})$ is the replacement error arises from the replacement of NN inputs. According to [12], the Gaussian function $\sigma_{ih}(x_i)$ satisfies the global Lipschitz condition, which thus implies $\|\Delta_i\| \leq \sum_{j=1, j\neq i}^{N} d_{ij} E_j$, where $E_j = \|x_j - x_{jd}\|$, and $d_{ij} > 0$ is a global Lipschitz constant.

Consider the isolated subsystem

$$\dot{x}_i(t) = f_i(x_i(t)) + g_i(x_i(t))u_i(x_i(t)) + h_{id}(x_{iD}) \tag{3}$$

where $f_i(\cdot)$ and $g_i(\cdot)$ are locally Lipschitz continuous. Our objective is to find the optimal control policy $u_i^*(x_i)$ of the ith subsystem by finding the feedback control policy $u_i(x_i)$ to minimize the infinite horizon performance index function

$$J_i(x_i) = \int_0^\infty \left(\hat{\delta}_i \left\| \nabla J_i^\mathsf{T}(x_i) \right\| E_i + U_i(x_i, u_i) \right) \mathrm{d}\tau \tag{4}$$

where $U_i(x_i, u_i) = x_i^\mathsf{T} Q_i x_i + u_i^\mathsf{T} R_i u_i$ is the utility function, $U_i(0,0) = 0$, and $U_i(x_i, u_i) \geq 0$ for all x_i and u_i, in which $Q_i \in \mathrm{R}^{n_i \times n_i}$ and $R_i \in \mathrm{R}^{m_i \times m_i}$ are positive definite matrices. $\hat{\delta}_i$ is the estimation of δ_i, which is a positive function and will be defined later. And $\nabla J_i(x_i) = \frac{\partial J_i(x_i)}{\partial x_i}$.

3 Decentralized Controller Design and Stability Analysis

3.1 Optimal Control

Consider the ith isolated subsystem (3), if the performance index function

$$V_i(x_i) = \int_0^\infty \left(\hat{\delta}_i \left\| \nabla V_i^\mathsf{T}(x_i) \right\| E_i + U_i(x_i, u_i) \right) \mathrm{d}\tau \tag{5}$$

is continuously differentiable, then the infinitesimal version of (5) is the so-called nonlinear Lyapunov equation

$$0 = \hat{\delta}_i \left\| \nabla V_i^\mathsf{T} \right\| E_i + U_i(x_i, u_i) + \nabla V_i^\mathsf{T} \left(f_i(x_i) + g_i(x_i)u_i(x_i) + h_{id}(x_{iD}) \right) \tag{6}$$

with $V_i(0) = 0$. Define the Hamiltonian function as

$$H_i\left(x_i, u_i, \nabla V_i(x_i)\right) = \hat{\delta}_i \left\| \nabla V_i^\mathsf{T}(x_i) \right\| E_i + U_i + \nabla V_i^\mathsf{T}(x_i) \left(f_i + g_i u_i + h_{id}(x_{iD}) \right)$$

and the optimal performance index function as

$$V_i^*(x_i) = \min_{u_i \in \Psi_i(\Omega_i)} \int_0^\infty \left(\hat{\delta}_i \left\| \nabla V_i^\mathsf{T}(x_i) \right\| E_i + U_i(x_i, u_i) \right) \mathrm{d}\tau \tag{7}$$

and let $V_i^*(x_i)$ be the local optimal performance index function, then

$$0 = \min_{u_i \in \Psi_i(\Omega_i)} H_i\left(x_i, u_i, \nabla V_i^*(x_i)\right) \tag{8}$$

where $\nabla V_i^*(x_i) = \frac{\partial V_i^*(x_i)}{\partial x_i}$. If the solution $V_i^*(x_i)$ exists and is continuously differentiable, the local optimal control can be expressed as

$$u_i^*(x_i) = -\frac{1}{2} R_i^{-1} g_i^\mathsf{T}(x_i) \nabla V_i^*(x_i) \tag{9}$$

Theorem 1. *For the ith interconnected subsystem (1) with the performance index function (5), $V_i^*(x_i)$ is the optimal solution of the HJB Eq. (8).*

Proof. The theorem can be proved by showing that $V_i^*(x_i)$ is a Lyapunov function. From (7), $V_i^*(x_i) \geq 0$ for any x_i, so $V_i^*(x_i)$ is a positive definite function. Considering (6) and Assumption 1, the state derivative of $V_i^*(x_i)$ is

$$\dot{V}_i^*(x_i) = \left(\nabla V_i^*(x_i)\right)^\mathsf{T} \left(f_i(x_i) + g_i(x_i)u_i(x_i) + h_{iD}(x_{iD}) + \Delta_i + \varepsilon_i \right)$$

$$\leq -\hat{\delta}_i \left\| \nabla V_i^{*\mathsf{T}} \right\| E_i - U_i + \left\| \nabla V_i^{*\mathsf{T}} \right\| \phi_{i1} + \max_{ij}\{d_{ij}\} \left\| \nabla V_i^{*\mathsf{T}} \right\| \sum_{j=1, j \neq i}^N E_j$$

Since $\dot{V}^* = \sum_{i=1}^{N} V_i^*$ and $E_i = \|x_i - x_{id}\| \geq 0$, by using the Chebyshev inequality $\sum_{i=1}^{N} \|\nabla V_i^{*\mathsf{T}}\| \sum_{j=1}^{N} E_j \leq N \sum_{i=1}^{N} \|\nabla V_i^{*\mathsf{T}}\| E_i$, we have

$$\dot{V}^* \leq \sum_{i=1}^{N} \left(-\hat{\delta}_i \|\nabla V_i^{*\mathsf{T}}\| E_i - U_i + \|\nabla V_i^{*\mathsf{T}}\| \phi_i \right) + \max_{ij} \{d_{ij}\} \sum_{i=1}^{N} \|\nabla V_i^{*\mathsf{T}}\| \sum_{j=1}^{N} E_j$$

$$\leq \sum_{i=1}^{N} \left(\tilde{\delta}_i \|\nabla V_i^{*\mathsf{T}}\| E_i - U_i(x_i, u_i) + \|\nabla V_i^{*\mathsf{T}}\| \phi_{i1} \right) \tag{10}$$

where $\delta_i = N \cdot \max_{ij}\{d_{ij}\}$, $\tilde{\delta}_i = \delta_i - \hat{\delta}_i$. Assuming $\left\| \tilde{\delta}_i \|\nabla V_i^{*\mathsf{T}}\| E_i + \|\nabla V_i^{*\mathsf{T}}\| \phi_{i1} \right\| \leq \Phi_i$, where Φ_i is a small positive constant. So, $\dot{V}^* \leq \sum_{i=1}^{N} \left(\Phi_i - \lambda_{\min}(Q_i) \|x_i\|^2 \right)$, where $\lambda_i(\cdot)$ denotes the minimum eigenvalue of the matrix. Hence, $\dot{V}^* \leq 0$ as long as $\|x_i\| \geq \sqrt{\frac{\Phi_i}{\lambda_{\min}(Q_i)}}$ holds, i.e., $V^*(x)$ is a Lyapunov function.

3.2   Neural Network Implementation

Approximation of the Interconnection. In this part, a state observer is employed to observe the state of interconnected subsystem (1) as

$$\dot{\hat{x}}_i(t) = f_i + g_i u_i + \hat{h}_i + l_i e_i \tag{11}$$

where $e_i = x_i - \hat{x}_i$ is the observation error, and $l_i = \mathrm{diag}[l_{i1}, l_{i2}, \ldots, l_{in}] \in \mathbb{R}^{n_i * n_i}$ denotes the observer gain matrix. Noticing that $\hat{h}_i$ is expressed as

$$\hat{h}_i = \hat{W}_{ih}^{\mathsf{T}} \sigma_{ih}(x_{iD}) \tag{12}$$

where the weight vector can be updated by

$$\dot{\hat{W}}_{ih} = \Gamma_{ih} e_i^{\mathsf{T}} \sigma_{ih}(x_{iD}) \tag{13}$$

with $\Gamma_{ih} > 0$ is a positive constant.

Theorem 2. *For the interconnected subsystem (1), with the help of the approximation of the unknown mismatched interconnection (12) with the updated law as (13), the observation error e_i which is obtained by combining (1) with the state observer (11) is guaranteed to be uniformly ultimately bounded (UUB).*

Proof. Select the Lyapunov candidate function as $L_{i1} = \frac{1}{2} e_i^{\mathsf{T}} e_i + \frac{1}{2} \tilde{W}_{ih}^{\mathsf{T}} \Gamma_{ih}^{-1} \tilde{W}_{ih}$, where $\tilde{W}_{ih} = W_{ih} - \hat{W}_{ih}$. Its time derivative is

$$\dot{L}_{i1} = e_i^{\mathsf{T}} \left(\tilde{W}_{ih}^{\mathsf{T}} \sigma_{ih}(x_{iD}) + \Delta_i(x_{iD}) + \varepsilon_i - l_i e_i \right) - \tilde{W}_{ih}^{\mathsf{T}} \Gamma_{ih}^{-1} \dot{\hat{W}}_{ih}$$

Suppose that $\|\Delta_i(x_{iD}) + \varepsilon_i\| \leq \phi_{i2}$ with $\phi_{i2} > 0$, and considering (12), we have $\dot{L}_{i1} \leq \|e_i^{\mathsf{T}}\| \left(-l_i \|e_i^{\mathsf{T}}\| + \phi_{i2} \right)$, so $\dot{L}_{i1} \leq 0$ as long as $\|e_i\| \geq \frac{\phi_{i2}}{l_i}$, the observation error is UUB.

The Critic Neural Network. The unknown part $\nabla V_i(x_i)$ is the gradient of the critic NN, it can be obtained by approximating $V_i(x_i)$ with a NN as

$$V_i(x_i) = W_{ic}^\mathsf{T} \sigma_{ic}(x_i) + \varepsilon_{ic}(x_i) \tag{14}$$

where $W_{ic} \in \mathrm{R}^{l_i}$ is the ideal weight vector, $\sigma_i(x_i) \in \mathrm{R}^{l_i}$ is the activation function, l_i is the number of neurons in the hidden-layer, and $\varepsilon_{ic}(x_i)$ is the approximation error. Then its gradient along with corresponding x_i is

$$\nabla V_i(x_i) = (\nabla \sigma_{ic}(x_i))^\mathsf{T} W_{ic} + \nabla \varepsilon_{ic}^\mathsf{T}(x_i) \tag{15}$$

where $\nabla \sigma_{ic}(x_i)$ and $\nabla \varepsilon_{ic}(x_i)$ are the gradients of the activation function and the approximation error, respectively. The critic NN can be approximated by $\hat{V}_i(x_i) = \hat{W}_{ic}^\mathsf{T} \sigma_{ic}(x_i)$, then the gradient of $\hat{V}_i(x_i)$ along the corresponding state is $\nabla \hat{V}_i(x_i) = (\nabla \sigma_{ic}(x_i))^\mathsf{T} \hat{W}_{ic}$. The approximate Hamiltonian function is

$$H_i\left(x_i, u_i, \hat{W}_{ic}\right) = \hat{\delta}_i \left\| \left(\nabla \hat{V}_i(x_i)\right)^\mathsf{T} \right\| E_i + U_i(x_i, u_i) + \hat{W}_{ic}^\mathsf{T} \nabla \sigma_{ic}(x_i) \dot{x}_i = e_{ic}$$

For the isolated subsystem (3), substituting (15) into (6), we have

$$0 = \hat{\delta}_i \left\| \nabla V_i^\mathsf{T} \right\| E_i + U_i(x_i, u_i) + (W_{ic} \nabla \sigma_{ic} + \nabla \varepsilon_{ic})^\mathsf{T} (f_i + g_i u_i + h_{id}(x_{iD}))$$

Let $v_i = \left\| (\nabla V_i(x_i))^\mathsf{T} \right\| - \left\| \left(\nabla \hat{V}_i(x_i)\right)^\mathsf{T} \right\|$, for the interconnected subsystem (1), the Hamiltonian function can be expressed as

$$H_i(x_i, u_i, W_{ic}) = \hat{\delta}_i \left\| \left(\nabla \hat{V}_i(x_i)\right)^\mathsf{T} \right\| E_i + U_i(x_i, u_i) + W_{ic}^\mathsf{T} \nabla \sigma_{ic}(x_i) \dot{x}_i$$

$$= -\hat{\delta}_i v_i E_i - \nabla \varepsilon_{ic}^\mathsf{T}(x_i) \dot{x}_i = e_{icH} \tag{16}$$

where e_{icH} is the approximation error of the critic NN.

Let $\theta_i = \nabla \sigma_{ic}(x_i) \dot{x}_i$, we minimize the objective function $E_{ic} = \frac{1}{2} e_{ic}^\mathsf{T} e_{ic}$ by the steepest descent algorithm to tune the critic NN weight vector $\hat{W}_{ic}$ by

$$\dot{\hat{W}}_{ic} = -l_{ic} e_{ic} \theta_i \tag{17}$$

where $l_{ic} > 0$ is the learning rate of the critic NN.

Define $\tilde{W}_{ic} = W_{ic} - \hat{W}_{ic}$, according to (16) and (17), $e_{ic} = e_{icH} - \tilde{W}_{ic}^\mathsf{T} \theta_i$. The weight approximation error can be updated by

$$\dot{\tilde{W}}_{ic} = -\dot{\hat{W}}_{ic} = l_{ic}(e_{icH} - \tilde{W}_{ic}^\mathsf{T} \theta_i) \theta_i \tag{18}$$

Hence, according to (9) and (14), the ideal control policy can be described as

$$u_i(x_i) = -\frac{1}{2} R_i^{-1} g_i^\mathsf{T}(x_i) \left((\nabla \sigma_{ic}(x_i))^\mathsf{T} W_{ic} + \nabla \varepsilon_{ic}^\mathsf{T} \right)$$

And it can be approximated as

$$\hat{u}_i(x_i) = -\frac{1}{2} R_i^{-1} g_i^\mathsf{T}(x_i) (\nabla \sigma_{ic}(x_i))^\mathsf{T} \hat{W}_{ic} \tag{19}$$

The UUB stability proof of the weight approximation error vector is similar to that of Theorem 3 in [6], so it is omitted here.

3.3 Stability Analysis

Theorem 3. *Consider the interconnected subsystem (1) with the performance index function (5), where $\hat{\delta}_i$ is updated by*

$$\dot{\hat{\delta}}_i = \Gamma_{i\delta} \left\| (\nabla V_i^*(x_i))^\mathsf{T} \right\| E_i \tag{20}$$

with $\Gamma_{i\delta} > 0$ is a constant, the N approximated decentralized control policies developed by (19) guarantee that the closed-loop large-scale nonlinear system is asymptotically stable. That is to say, the control pair $u_1(x_1), u_2(x_2), \ldots, u_N(x_N)$ is the decentralized control law of the large-scale nonlinear system.

Proof. Select the Lyapunov candidate function as $L_3 = \sum\limits_{i=1}^{N} \left(V_i^* + \frac{1}{2} \tilde{\delta}_i^\mathsf{T} \Gamma_{i\delta}^{-1} \tilde{\delta}_i \right)$,

and its time derivative is $\dot{L}_3 = \sum\limits_{i=1}^{N} \left(\nabla V_i^{*\mathsf{T}} \dot{x}_i - \tilde{\delta}_i^\mathsf{T} \Gamma_{i\delta}^{-1} \dot{\hat{\delta}}_i \right)$. Considering (6), (20)

and Algorithm 1, we can obtain $\dot{L}_3 \leq \sum\limits_{i=1}^{N} \left(-\lambda_{\min}(Q_i) \|x_i\|^2 + \left\| \nabla V_i^{*\mathsf{T}} \right\| \phi_i \right)$,

Thus, $\dot{L}_3 \leq 0$ as long as $\|x_i\| \geq \sqrt{\frac{\|\nabla V_i^{*\mathsf{T}}\| \phi_i}{\lambda_{\min}(Q_i)}}$. According to Lyapunov stability theorem, the closed-loop large-scale nonlinear system is UUB under the action of the control pair $u_1(x_1), u_2(x_2), \ldots, u_N(x_N)$.

4   Simulation Study

Consider the following large-scale nonlinear system:

$$\dot{x}_1 = \begin{bmatrix} x_{12} - x_{11} \\ -0.5(x_{11} + x_{12}) - 0.5x_{12}\left(\cos(2x_{11}) + 2\right)^2 \end{bmatrix} + \begin{bmatrix} 0 \\ \cos(2x_1) + 2 \end{bmatrix} u_1(x_1)$$
$$+ \begin{bmatrix} 0 \\ 4(x_{11} + x_{22})\sin(x_{12}^3)\cos(0.5x_{21}) \end{bmatrix}$$

$$\dot{x}_2 = \begin{bmatrix} x_{22} \\ -x_{21} - 0.5x_{22} + 0.5x_{21}^2 x_{22} \end{bmatrix} + \begin{bmatrix} 0 \\ x_{21} \end{bmatrix} u_2(x_2) + \begin{bmatrix} 0 \\ 0.5(x_{12} + x_{22})\cos(e^{x_{21}^2}) \end{bmatrix}$$

where $x_i = [x_{i1}, x_{i2}]^\mathsf{T} \in \mathrm{R}^2$ and $u_i(x_i) \in \mathrm{R}$ are the state and control input of ith subsystem, respectively. The interconnections are unknown and mismatched.

The performance index function (4) is approximated by a critic NN, whose activation function is chosen as $\sigma_{ic}(x_i) = [x_{i1}^3, x_{i1}^2 x_{i2}, x_{i1}x_{i2}^2, x_{i1}^2, x_{i1}x_{i2}, x_{i1}, x_{i2}, x_{i2}^2, x_{i2}^3]$. Let $Q_i = 40I_2$, $R_i = 10I$, the weight learning rates of the approximated interconnection and the critic NN be $\Gamma_{ih} = 10$, $l_{ic} = 0.06$ and $\Gamma_i = 10$, the updated rate of $\hat{\delta}_i$ be $\Gamma_{i\delta} = 0.0001$, the state observer gain matrix be $l_i = 10I_2$, where I_n denotes the identify matrix with n dimensions, respectively.

The simulation results are described as Figs. 1 and 2. From Fig. 1, we can see that the unknown interconnection can be approximated online. By using the improved performance index function (4) and the proposed PI algorithm, the system states can converge to zero as shown in Fig. 2. Therefore, the simulation results demonstrate the effectiveness of the proposed scheme.

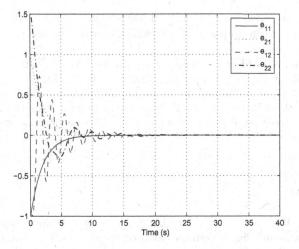

Fig. 1. State observation errors

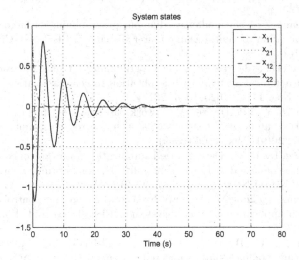

Fig. 2. System states

5   Conclusion

This paper proposes a decentralized control scheme via NN and PI algorithm for nonlinear systems with unknown mismatched interconnections. To avoid the assumption of upper boundedness on the interconnection, a RBFNN with the help of the local signals of isolated subsystem and the replaced reference signals of coupled subsystems is employed to approximate the interconnection term. Then, an improved performance index function is constructed to reflect the NN replacement error. At last, the closed-loop large-scale nonlinear system is guaranteed to be UUB via the Lyapunov stability theorem. Simulation results demonstrate the effectiveness of the proposed decentralized control scheme.

Acknowledgments. This work was supported in part by the National Natural Science Foundation of China under Grants 61233001, 61273140, 61304086, 61374105, 61374051, 61533017, 61603387 and U1501251, in part by the Scientific and Technological Development Plan Project in Jilin Province of China under Grants 20150520112JH and 20160414033GH, and in part by Beijing Natural Science Foundation under Grant 4162065.

References

1. Vrabie, D., Pastravanu, O., Abu-Khalaf, M., et al.: Adaptive optimal control for continuous-time linear systems based on policy iteration. Automatica **45**(2), 477–484 (2009)
2. Wang, D., Liu, D., Wei, Q., et al.: Optimal control of unknown nonaffine nonlinear discrete-time systems based on adaptive dynamic programming. Automatica **48**(8), 1825–1832 (2012)
3. Jiang, Y., Jiang, Z.: Robust adaptive dynamic programming for large-scale systems with an application to multimachine power systems. IEEE Trans. Circ. II **59**(10), 693–697 (2012)
4. Wei, Q., Liu, D.: Adaptive dynamic programming for optimal tracking control of unknown nonlinear systems with application to coal gasification. IEEE Trans. Autom. Sci. Eng. **11**(4), 1020–1036 (2014)
5. Zhang, H., Luo, Y., Liu, D.: Neural-network-based near-optimal control for a class of discrete-time affine nonlinear systems with control constraints. IEEE Trans. Neural Netw. **20**(9), 1490–1503 (2009)
6. Liu, D., Wang, D., Li, H.: Decentralized stabilization for a class of continuous-time nonlinear interconnected systems using online learning optimal control approach. IEEE Trans. Neural Netw. Learn. Syst. **25**(2), 418–428 (2014)
7. Bian, T., Jiang, Y., Jiang, Z.: Decentralized adaptive optimal control of large-scale systems with application to power systems. IEEE Trans. Ind. Electron. **62**(4), 2439–2447 (2015)
8. Liu, D., Li, C., Li, H., et al.: Neural-network-based decentralized control of continuous-time nonlinear interconnected systems with unknown dynamics. Neurocomputing **165**, 90–98 (2015)
9. Lu, C., Si, J., Xie, X.: Direct heuristic dynamic programming for damping oscillations in a large power system. IEEE Trans. Syst. Man Cybern. Part B Cybern. **38**(4), 1008–1013 (2008)
10. Molina, D., Venayagamoorthy, G., Liang, J., et al.: Intelligent local area signals based damping of power system oscillations using virtual generators and approximate dynamic programming. IEEE Trans. Smart Grid. **4**(1), 498–508 (2013)
11. Bernstein, D., Amato, C., Hansen, E., et al.: Policy iteration for decentralized control of Markov decision processes. J. Artif. Intell. Res. **34**(1), 89–132 (2009)
12. Chen, W., Li, J.: Decentralized output-feedback neural control for systems with unknown interconnections. IEEE Trans. Syst. Man Cybern. Part B Cybern. **38**(1), 258–266 (2008)

Optimal Constrained Neuro-Dynamic Programming Based Self-learning Battery Management in Microgrids

Qinglai Wei[1(✉)] and Derong Liu[2]

[1] The State Key Laboratory of Management and Control for Complex Systems, Institute of Automation, Chinese Academy of Sciences, Beijing 100190, China
qinglai.wei@ia.ac.cn
[2] School of Automation and Electrical Engineering, University of Science and Technology, Beijing 100083, China

Abstract. In this paper, a novel optimal self-learning battery sequential control scheme is investigated for smart home energy systems. Using the iterative adaptive dynamic programming (ADP) technique, the optimal battery control can be obtained iteratively. Considering the power constraints of the battery, a new non-quadratic form performance index function is established, which guarantees the value of the iterative control law not to exceed the maximum charging/discharging power of the battery to extend the service life of the battery. Simulation results are given to illustrate the performance of the presented method.

Keywords: Adaptive dynamic programming · Approximate dynamic programming · Energy management system · Smart home · Optimal control

1 Introduction

Adaptive dynamic programming (ADP), proposed by Werbos [25,26], overcomes the curse of dimensionality problem in dynamic programming by approximating the performance index function forward-in-time and becomes an important brain-like intelligent method of approximate optimal control for nonlinear systems. Iterative methods are widely used in ADP to obtain the solution of HJB equation indirectly [5,8,12,13,16–19,21–24,28,29]. Policy and value iterations are two primary iterative ADP algorithms [6]. Policy iteration algorithms for optimal control of continuous-time (CT) systems with continuous states and action spaces were first given in [1]. In [9], a complex-valued ADP algorithm was discussed, where for the first time the optimal control problem of complex-valued nonlinear systems was successfully solved by ADP. In [10], according to

This work was supported in part by the National Natural Science Foundation of China under Grants 61233001, 61273140, 61304086, 61374105, 61503377, 61503379, 61304079, 61533017, and U1501251.

neurocognitive psychology, a novel controller based on multiple actor-critic structures was developed for unknown systems and the proposed controller traded off fast actions based on stored behavior patterns with real-time exploration using current input-output data. In [11], an effective off-policy learning based integral reinforcement learning (IRL) algorithm was presented, which successfully solved the optimal control problem for completely unknown continuous-time systems with unknown disturbances. In [4], a Q-learning based ADP algorithm was proposed to obtain the optimal control law of the battery, which solved the optimal energy management for the microgrids of smart homes. In [2], considering renewable electricity, such as electricity generation of wind and solar energies, the optimal control for the battery was solved by ADP method. In [3], a particle swarm optimization (PSO) method was proposed to pre-train the weights the action and critic neural networks, which facilitates the implementation of the ADP method for the optimal control for the battery. In [20], an effective dual Q-learning based iterative ADP algorithm was developed to obtain the optimal battery management for the microgrids of smart homes, where the convergence and optimality of the dual Q-learning based ADP algorithm was proven to guarantee the optimal battery control. However, in [20], the charging/discharging constraints of the battery was not considered. Actually, for all the energy management systems of the microgrids of smart homes, the charging and discharging power of the battery cannot reach the infinity. Hence, the optimal control of the battery with power constraints of the battery is a key technique for real-world the smart home energy management system.

In this paper, in spirted by [20], a new iterative ADP algorithm is developed to solve the optimal battery control for the smart home energy management system, where the charging/discharging constraint of the battery is considered. First, the models of the smart home energy systems and the battery are established, where the efficiency of the battery is considered. Second, inspired by [1], a new non-quadratic performance index function is constructed, where the charging/discharging power of the battery is defined in the performance index function. Then, the iterative ADP algorithm is derived for the optimal control law for the battery. Via the system transformation and the definition of the performance index function, the expression of the iterative control law for the battery can be obtained. The convergence of the algorithm is presented, which guarantees that the iterative value function will converge to the optimal performance index function, as the iteration index increases to infinity.

2   Problem Formulation

In this paper, the optimal battery control problem is treated as a discrete time problem with the time step of 1 h and it is assumed that the home load varies hourly. The battery will make decisions to meet the demand of the home load and according to the real-time electricity rate. There are three operational modes for the battery of the home energy system, including charging mode, idle mode, and discharging mode. The battery model used in this work is based on [4,27] where battery efficiency is considered to extend the battery's lifetime as far as possible. Let E_{bt} be the battery energy at time t and let $\eta(\cdot)$ be the charging/discharging efficiency of the battery. Then, the battery model can be expressed as

$$E_{b(t+1)} = E_{bt} - P_{bt} \times \eta(P_{bt}), \tag{1}$$

where P_{bt} is the battery power output at time t. Let $P_{bt} > 0$ denote battery discharging. Let $P_{bt} < 0$ denote battery charging and let $P_{bt} = 0$ denote that the battery is idle. Let the efficiency of battery charging/discharging be derived as

$$\eta(P_{bt}) = 0.898 - 0.173|P_{bt}|/P_{\text{rate}}, \tag{2}$$

where $P_{\text{rate}} > 0$ is the rated power output of the battery.

3 Iterative ADP Algorithm for Battery Management System

In this section, inspired by [20], considering the power efficiency and the power constraint of the battery, a new system function will be constructed and a new performance index function will be established. First, we define the system states. Let $x_{1t} = P_{gt}$ and $P_{bt} = u_t$. Letting $x_t = [x_{1t}, x_{2t}]^{\mathsf{T}}$, the equation of the home energy system can be written as

$$x_{t+1} = F(x_t, u_t, t) = \begin{pmatrix} P_{Lt} - u_t\eta(u_t) \\ x_{2t} - u_t\eta(u_t) \end{pmatrix}. \tag{3}$$

From (3), we can see that the power constraint of the battery appears in both of the system function. In [20], the performance index function was defined while the power constraint of the battery was not considered. In this paper, inspired by [1], a non-quadratic performance index function will be defined for the battery management system, which is expressed as

$$\sum_{t=0}^{\infty} \left(m_1(C_t P_{gt})^2 + m_2(E_{bt} - E_b^o)^2 + m_3 \int_0^{P_{bt}\eta(P_{bt})} (\Phi^{-1}(s))^{\mathsf{T}} ds \right). \tag{4}$$

where $\Phi(\cdot)$ is a monotonic odd function with its first derivative bounded by a constant $\mathcal{M}$. An example is the hyperbolic tangent $\Phi(\cdot) = \tanh(\cdot)$. R is positive definite matrix.

Let $\underline{u}_t = (u_t, u_{t+1}, \dots)$ denote the control sequence from t to ∞. Let $M_t = \begin{bmatrix} m_1 C_t^2 & 0 \\ 0 & m_2 \end{bmatrix}$. Let x_0 be the initial states. Then, the performance index function (4) can be written as $J(x_0, \underline{u}_0, 0) = \sum_{t=0}^{\infty} U(x_t, u_t, t)$, where the utility function is expressed as

$$U(x_t, u_t, t) = x_t^T M_t x_t + m_3 \int_0^{u_t\eta(u_t)} (\Phi^{-1}(s))^{\mathsf{T}} ds. \tag{5}$$

3.1   Derivations of the Iterative ADP Algorithm

From (3) we know that the battery management system is time-varying. It means that the control law is also time-varying. This makes the controller design difficult. To overcome this difficulty, in [15, 20], defining a new sequence of control for a period, the time-varying optimal control is transformed into a time-invariant one, which makes the computation burden much relax. In this paper, inspired by [15, 20], we will define the sequence of control for a period, where the constraints of the battery is considered. For any $k = 0, 1, \ldots$, we define $\mathcal{U}_k$ as the control sequence from k to $k + \lambda - 1$, i.e., $\mathcal{U}_k = (u_k, u_{k+1}, \ldots, u_{k+\lambda-1})$. We can define a new utility function as $\Lambda(x_k, \mathcal{U}_k) = \sum_{\theta=0}^{\lambda-1} U(x_{k+\theta}, u_{k+\theta}, \theta)$, Then, for any $k = 0, 1, \ldots$, the optimal performance index function is obtained as

$$J^*(x_k) = \min_{\mathcal{U}_k} \left\{ \Lambda(x_k, \mathcal{U}_k) + \bar{\gamma} J^*(x_{k+\lambda}) \right\}, \tag{6}$$

where $\bar{\gamma} = \gamma^\lambda$.

Define a iteration index $i = 0, 1, \ldots$. The iterative value function is defined as

$$V_{i+1}(x_k) = \min_{\mathcal{U}_k} \{ \Lambda(x_k, \mathcal{U}_k) + \bar{\gamma} V_i(x_{k+\lambda}) \}, \tag{7}$$

where $V_0(x_k) = \Psi(x_k)$ and $\Psi(x_k)$ is a positive semi-definite function. The iterative control law sequence $\mathcal{U}_i$ can be computed as follows

$$\mathcal{U}_i(x_k) = \arg \min_{\mathcal{U}_k} \{ \Lambda(x_k, \mathcal{U}_k) + \bar{\gamma} V_i(x_{k+\lambda}) \}. \tag{8}$$

For any $i = 0, 1, \ldots$, we define a new iteration index $j = 0, 1, \ldots, 23$. We can get

$$\begin{aligned} V_i^{j+1}(x_k) &= \min_{u_k} \{ U(x_k, u_k, j) + V_i^j(x_{k+1}) \} \\ &= U(x_k, u_i^j(x_k), j) + V_i^j(x_{k+1}) \end{aligned} \tag{9}$$

and

$$u_i^j(x_k) = \arg \min_{u_k} \{ U(x_k, u_k, j) + V_i^j(x_{k+1}) \}, \tag{10}$$

where we let $V_i^0(x_k) = V_i(x_k)$ and $V_{i+1}(x_k) = V_i^{23}(x_k)$. The system function is defined as

$$x_{k+1} = f(x_k, j) + g v_k, \tag{11}$$

where $f(x_k, j) = [P_{L(\lambda-1-j)}, x_{2k}]^\mathsf{T}$, $g = [-1, -1]^\mathsf{T}$, and $v_k = u_k \eta(x_k)$. According to the principle of optimality, for any $i = 0, 1, \ldots$ and $j = 0, 1, \ldots, 23$, the iterative control law $v_i^j(x_k)$ satisfies $\frac{\partial V_i^{j+1}(x_k)}{\partial v_i^j(x_k)} = 0$. Then, we can obtain that

$$v_i^j(x_k) = -\Phi \left(\frac{1}{2} m_3^{-1} g^\mathsf{T} \frac{dV_i^j(x_{k+1})}{dx_{k+1}} \right), \tag{12}$$

where $v_i^j(x_k) = u_i^j(x_k)\eta(u_i^j(x_k))$. Next, the convergence properties of the developed iterative ADP algorithm will be developed.

Theorem 1. *For $i = 0, 1, \ldots$ and $j = 0, 1, \ldots, 23$, let the iterative value function $V_i^j(x_k)$ and the iterative control law $v_i^j(x_k)$ be obtained by (9)–(10). Then, the iterative control law sequence can be expressed by*

$$\mathcal{U}_i(x_k) = \{u_i^{23}(x_k), u_i^{22}(x_{k+1}), \ldots, u_i^0(x_{k+23})\}. \tag{13}$$

Theorem 2. *For $i = 0, 1, \ldots$, let $V_{i+1}(x_k)$ and $\mathcal{U}_i(x_k)$ be obtained by (7)–(8). Then, the iterative value function $V_i(x_k)$ converges to the optimal performance index function, i.e.,*

$$\lim_{i \to \infty} V_i(x_k) = J^*(x_k). \tag{14}$$

Proof. According to the sequence of the control law $\mathcal{U}_k = (u_k, u_{k+1}, \ldots, u_{k+\lambda-1})$, we can obtain $x_{k+\lambda} = \mathcal{F}(x_k, \mathcal{U}_k)$. Inspired by [7], let ψ_1, ψ_2, and λ be constants, such that $0 \leq \psi_1 \leq 1 \leq \psi_2 < \infty$ and $0 < \chi < \infty$, which satisfies $J^*(\mathcal{F}(x_k, \mathcal{U}_k)) \leq \chi U(x_k, \mathcal{U}_k)$, and $\psi_1 J^*(x_k) \leq V_0(x_k) \leq \psi_2 J^*(x_k)$. For $i = 0, 1, \ldots$, the iterative value function $V_i(x_k)$ satisfies

$$\left(1 + \frac{\psi_1 - 1}{(1 + \chi^{-1})^i}\right) J^*(x_k) \leq V_i(x_k) \leq \left(1 + \frac{\psi_2 - 1}{(1 + \chi^{-1})^i}\right) J^*(x_k). \tag{15}$$

Letting $i \to \infty$, we can obtain the conclusion.

4   Simulation Analysis

In this section, the performance of the iterative ADP algorithm with constraints will be examined by numerical experiments. The simulation results for the developed iterative ADP algorithm with constraints will be compared with the dual Q-learning algorithm in [20]. The profiles of the home load demand (kW) and the real-time electricity rate (cents) are taken from [4]. Neural networks are introduced to implement the iterative ADP algorithm. There are two neural networks, which are critic and action networks, respectively, to implement the developed algorithm. Both neural networks are chosen as three-layer back-propagation (BP) network. The structures of the critic and action networks are chosen as 2–8–1 and 2–8–1, respectively. According to these data, we implement the iterative ADP algorithm (7)–(12) for 15 iterations to guarantee the computation precision $\varepsilon = 10^{-2}$. The plots of the iterative value function are shown in Fig. 1. From Fig. 1, we can see that under the power constraint of the battery, the iterative value function converges to the optimum under 15 iterations. The optimal

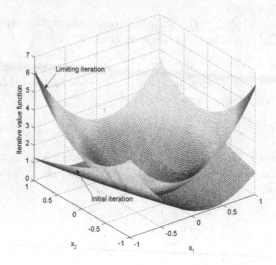

Fig. 1. The trajectories of the iterative Q function.

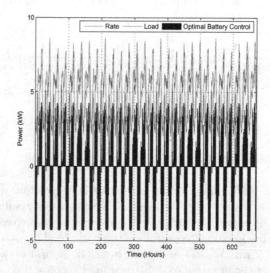

Fig. 2. Optimal management of battery with constraints in one week

battery charging/discharging management is shown in Fig. 2. From Fig. 2, we can see that the charging and discharging power of the battery cannot exceed the boundary 4.5 kW, which shows the effectiveness of the developed algorithm. The plot of battery energy is shown in Fig. 3. From Fig. 3, we can see that the energy of the battery does not exceed the maximum energy of the battery.

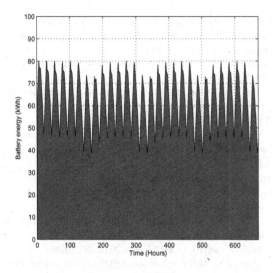

Fig. 3. Optimal battery energy with constraints in four weeks

5   Conclusion

In this paper, an iterative constraint ADP algorithm is developed to solve the optimal battery control problem for the smart home energy systems. To solve the constraint of the power, a new non-quadratic form performance index function is established, which guarantees the iterative control law not to exceed the upper bound of the battery. In the iterative ADP algorithm, the expressions of the iterative control law can be obtained. Finally, simulation results are given to illustrate the performance of the presented method.

References

1. Abu-Khalaf, M., Lewis, F.L.: Nearly optimal control laws for nonlinear systems with saturating actuators using a neural network HJB approach. Automatica **41**, 779–791 (2005)
2. Boaro, M., Fuselli, D., Angelis, F.D., Liu, D., Wei, Q., Piazza, F.: Adaptive dynamic programming algorithm for renewable energy scheduling and battery management. Cogn. Comput. **5**, 264–277 (2013)
3. Fuselli, D., Angelis, F.D., Boaro, M., Liu, D., Wei, Q., Squartini, S., Piazza, F.: Action dependent heuristic dynamic programming for home energy resource scheduling. Int. J. Electr. Power Energy Syst. **48**, 148–160 (2013)
4. Huang, T., Liu, D.: A self-learning scheme for residential energy system control and management. Neural Comput. Appl. **22**, 259–269 (2013)
5. Jiang, Y., Jiang, Z.P.: Robust adaptive dynamic programming and feedback stabilization of nonlinear systems. IEEE Trans. Neural Netw. Learn. Syst. **25**, 882–893 (2014)

6. Lewis, F.L., Vrabie, D., Vamvoudakis, K.G.: Reinforcement learning and feedback control: using natural decision methods to design optimal adaptive controllers. IEEE Control Syst. **32**, 76–105 (2012)
7. Lincoln, B., Rantzer, A.: Relaxing dynamic programming. IEEE Trans. Autom. Control **51**, 1249–1260 (2006)
8. Modares, H., Lewis, F.L.: Optimal tracking control of nonlinear partially-unknown constrained-input systems using integral reinforcement learning. Automatica **50**, 1780–1792 (2014)
9. Song, R., Xiao, W., Zhang, H., Sun, C.: Adaptive dynamic programming for a class of complex-valued nonlinear systems. IEEE Trans. Neural Netw. Learn. Syst. **25**, 1733–1739 (2014)
10. Song, R., Lewis, F.L., Wei, Q., Zhang, H., Jiang, Z.P., Levine, D.: Multiple actor-critic structures for continuous-time optimal control using input-output data. IEEE Trans. Neural Netw. Learn. Syst. **26**, 851–865 (2015)
11. Song, R., Lewis, F.L., Wei, Q., Zhang, H.: Off-policy actor-critic structure for optimal control of unknown systems with disturbances. IEEE Transactions on Cybernetics (2015, in press). doi:10.1109/TCYB.2015.2421338
12. Wei, Q., Liu, D., Yang, X.: Infinite horizon self-learning optimal control of nonaffine discrete-time nonlinear systems. IEEE Trans. Neural Netw. Learn. Syst. **26**, 866–879 (2015)
13. Wei, Q., Wang, F., Liu, D., Yang, X.: Finite-approximation-error based discrete-time iterative adaptive dynamic programming. IEEE Trans. Cybern. **44**, 2820–2833 (2014)
14. Wei, Q., Liu, D.: Data-driven neuro-optimal temperature control of water gas shift reaction using stable iterative adaptive dynamic programming. IEEE Trans. Ind. Electron. **61**, 6399–6408 (2014)
15. Wei, Q., Liu, D., Shi, G., Liu, Y.: Optimal multi-battery coordination control for home energy management systems via distributed iterative adaptive dynamic programming. IEEE Trans. Ind. Electron. **42**, 4203–4214 (2015)
16. Wei, Q., Song, R., Yan, P.: Data-driven zero-sum neuro-optimal control for a class of continuous-time unknown nonlinear systems with disturbance using ADP. IEEE Trans. Neural Netw. Learn. Syst. **27**, 444–458 (2016)
17. Wei, Q., Liu, D., Lin, H.: Value iteration adaptive dynamic programming for optimal control of discrete-time nonlinear systems. IEEE Trans. Cybern. **46**, 840–853 (2016)
18. Wei, Q., Liu, D.: A novel iterative θ-adaptive dynamic programming for discrete-time nonlinear systems. IEEE Trans. Autom. Sci. Eng. **11**, 1176–1190 (2014)
19. Wei, Q., Liu, D.: Adaptive dynamic programming for optimal tracking control of unknown nonlinear systems with application to coal gasification. IEEE Trans. Autom. Sci. Eng. **11**, 1020–1036 (2014)
20. Wei, Q., Liu, D., Shi, G.: A novel dual iterative Q-learning method for optimal battery management in smart residential environments. IEEE Trans. Ind. Electron. **62**, 2509–2518 (2015)
21. Wei, Q., Liu, D., Lewis, F.L.: Optimal distributed synchronization control for continuous-time heterogeneous multi-agent differential graphical games. Inf. Sci. **317**, 96–113 (2015)
22. Wei, Q., Liu, D.: A novel policy iteration based deterministic Q-learning for discrete-time nonlinear systems. Sci. China Inf. Sci. **58**, 1–15 (2015)
23. Wei, Q., Lewis, F.L., Sun, Q., Yan, P., Song, R.: Discrete-time deterministic Q-learning: a novel convergence analysis. IEEE Transactions on Cybernetics (2016, in press)

24. Wei, Q., Liu, D., Lin, Q., Song, R.: Discrete-time optimal control via local policy iteration adaptive dynamic programming. IEEE Transactions on Cybernetics (2016, in press)
25. Werbos, P.J.: Advanced forecasting methods for global crisis warning and models of intelligence. General Syst. Yearbook **22**, 25–38 (1977)
26. Werbos, P.J.: A menu of designs for reinforcement learning over time. In: Miller, W.T., Sutton, R.S., Werbos, P.J. (eds.) Neural Networks for Control, pp. 67–95. MIT Press, Cambridge (1991)
27. Yau, T., Walker, L.N., Graham, H.L., Raithel, R.: Effects of battery storage devices on power system dispatch. IEEE Trans. Power Apparatus Syst. **100**, 375–383 (1981)
28. Zhang, H., Qing, C., Luo, Y.: Neural-network-based constrained optimal control scheme for discrete-time switched nonlinear system using dual heuristic programming. IEEE Trans. Autom. Sci. Eng. **11**, 839–849 (2014)
29. Zhao, Q., Xu, H., Jagannathan, S.: Near optimal output feedback control of nonlinear discrete-time systems based on reinforcement neural network learning. IEEE/CAA J. Automatica Sin. **1**, 372–384 (2014)

Risk Sensitive Reinforcement Learning Scheme Is Suitable for Learning on a Budget

Kazuyoshi Kato[✉] and Koichiro Yamauchi

Department of Computer Science, Chubu University,
Matsumoto 1200, Kasugai, Aichi, Japan
tp15010-6353@sti.chubu.ac.jp, yamauchi@cs.chubu.ac.jp
http://sakura.cs.chubu.ac.jp/

Abstract. Risk-sensitive reinforcement learning (Risk-sensitiveRL) has been studied by many researchers. The methods are based on a prospect method, which imitates the value function of a human. Although they are mainly intended at imitating human behaviors, there are fewer discussions about the engineering meaning of it. In this paper, we show that Risk-sensitiveRL is useful for using online-learning machines whose resources are limited. In such a learning method, a part of the learned memories should be removed to create space for recording a new important instance. The experimental results show that risk-sensitive RL is superior to normal RL. This might mean that the human brain is also constructed by a limited number of neurons, so that humans hire the risk-sensitive value function for the learning.

1 Introduction

Learning methods on a budget are effective tools to construct intelligent embedded systems. Although the computational power of embedded systems is improving, their storage capacities are still limited. Under such restricted environments, the learning machines have to work on a limited number of kernels without compromising on their flexibility.

Moreover, such learning methods on a budget are also useful for imitation learning applications for a large number of crowd-workers [1,2]. To increase the number of workers, the computational power for the learning machine should be minimal to execute them on a server. However, if we apply such a learning method for reinforcement learning (RL) applications, we usually face a problem; the learner usually forgets important negative experiences.

RL is known as the general framework to explain biological learning systems. The past reinforcement learning algorithms can roughly be divided into two groups: risk-averse and risk-neutral groups. Risk-neutral groups are designed to realize RL for maximizing expected rewards [3], whereas risk-averse models are designed to realize an economically rational learning system [3]. The prospect theory [4] describes the human risk-averse model well, and it is well matched with human behavior data [3]. Almost all the risk-averse models try to construct an economically rational suitable model. However, the reason that the humans

A. Hirose et al. (Eds.): ICONIP 2016, Part III, LNCS 9949, pp. 202–210, 2016.
DOI: 10.1007/978-3-319-46675-0_23

employ the risk-averse policy from the problem solving viewpoint is unclear. Frank et al. [5] showed that a risk seeking reinforcement learning agent outperforms the risk adverse learning agents for investment tasks in a noisy market. Although they did not use the risk-sensitive RL method, the results suggested that risk-related behavior considerably affects the performances of agents. However, their model did not focus on the restriction in the learning method itself.

In this paper, an actor-critic based RL, which is an effective and realistic reinforcement learning method, is used. The actor and critic are constructed by the learners that realize continuous function approximation. To construct the actor-critic based RL in an embedded system, the actor and critic are to be executed in limited resources. If the learner has a limited amount of resources, the learner usually faces the risk of overflow. Essentially, learning is usually achieved through a resource allocation process that tends to lack its flexibility when no resources remain. To maintain flexibility, the learner has to forget a part of memory to create space for learning new instances.

However, if such a learner realizes the critic functions, the learner usually fails to record important negative experiences properly into the resultant value function. This is because when the negative experiences are recorded by the critic, the agent avoids revisiting to the negative situation, thus resulting in fewer number of experiences for the negative situation for positive ones. If the learning on a budget algorithm is employed the least recently used negative memories are pruned.

This paper demonstrates that the learning machine using an important weight depending on the risk-sensitive utility function enables avoiding such failures.

2   Incremental Learning on a Budget

The learning machine whose learning resource is bounded, is usually called "learning on a budget." Traditionally, the multi-layered perceptron consists of a fixed number of neuron-models, and their parameters are adjusted using the steepest descent rule. Although there are sophisticated learning methods for the perceptron such as those proposed in [6], they cannot support one-pass learning, which is usually needed for practical applications.

In this paper, a kernel regression model is considered. This model supports one-pass learning by simply adding a new kernel to record the current instance. The kernel perceptron learning on a budget continues the learning with a bounded number of kernels [7–9]. Similarly, kernel regression model also supports learning on a budget [10,11].

The kernel regression model behavior is similar to that of the nearest neighbors so that it realizes stable learning. According to the survey paper [12], a number of prototype selection methods of nearest neighbors have been developed for classifications.

The learning methods listed above reduce redundant kernels or prototypes. However, the learning machine, which has no more redundant kernel or prototypes, is usually under pressure to adapt to a new approaching environment.

Under such a situation, the learning machine has to dispose a part of kernels or prototypes to create space to record new instances. In reinforcement learning environments, similar situations frequently occur, for which the learning machine has to dispose a part of the memory.

Therefore, in this paper, we focus on the performances of the learning machine, which realizes an adaptive forgetting. For simplicity, the learning machines used in the experiments do not evaluate the redundancy of each kernel or prototype for forgetting. Instead, they simply estimate a modified version of the least recently used or least frequently used policies.

2.1 Kernel Regression Based Learning on a Budget

The kernel regression model is based on a kernel method. The kernel regression model is based on a kernel method. Using the kernel method, the learning machine can be described using a function vector on Hilbert space as follows:

$$f_{\theta}[x] \equiv \langle f_t, K(x, \cdot) \rangle, \tag{1}$$

where $\langle \cdot, \cdot \rangle$ denotes the dot-product of two function vectors, and f_t denotes the function vector after the t-th learning. Now, let us assume that f_t is described by the weighted sum of several reproduction kernels:

$$f_t \equiv \sum_{i=1}^{B} w_i K(x_i, \cdot), \tag{2}$$

where $K(x_i, \cdot)$ denotes the i-th kernel function. According to the theorem of reproductive kernel, the dot-product of the two kernel functions can be calculated and described as follows.

$$\langle K(x_i, \cdot), K(x, \cdot) \rangle = \exp \left(-\frac{\|x_i - x\|^2}{2\sigma^2} \right), \tag{3}$$

where σ denotes the standard deviation of the kernel.

Actually, the above equation is a linear regression model. The output of the kernel regression is described using two function vectors as follows:

$$f_{\theta}[x] \equiv \frac{\langle f_t, K(x, \cdot) \rangle}{\langle g_t, K(x, \cdot) \rangle}, \tag{4}$$

where f_t is equivalent to Eq. (2) and g_t is $g_t = \sum_i K(x_i, \cdot)$. The output function of kernel regression is a soft-max function. This means that the kernel-regression behaviors are similar to that of k-nearest neighbors. Algorithm 1 is the pseudo algorithm.

2.2 Kernel Replacement Algorithm

As mentioned before, we focus on the performances of the learning machines that realize an adaptive forgetting. The learning machines simply estimate a modified

---

Algorithm 1. Algorithm of a kernel regression on a budget

Require: new learning sample (x_t, y_t), The size of the support set: B, The previous learning result f_t and g_t. Importance weight for this instance I_w, error threshold ϵ_e, the threshold for replacement ϵ_R.

$e_t = \|y_t - f_\theta[x]\|^2$ (see Eq. (4)).

if $e_t > \epsilon_e$ **then**

   if $|s_{t-1}| < B$ **then**

      $f_t = f_{t-1} + y_t k(x_t, \cdot)$

      $g_t = g_{t-1} + k(x_t, \cdot)$

      $s_t = s_{t-1} \cup \{t\}$

   else

      {replace the most ineffective hidden unit with a new one}

      $i = $ most ineffective kernel. (see Sect. 2.2)

      if $e(t) \cdot I_w > \epsilon_R$ **then**

         $f_t = f_{t-1} - w_i k(x_i, \cdot) + y_t k(x_t, \cdot)$

         $g_t = g_{t-1} - k(x_i, \cdot) + k(x_t, \cdot)$

         $s_t = (s_{t-1} \setminus \{i\}) \cup \{t\}$

      else

         $f_t = f_{t-1}, g_t = g_{t-1}.$

      end if

   end if

end if

RETURN f_t, g_t

---

version of the least recently used or least frequently used policies. These two types of learning machines employed the same pruning algorithm: least recently and frequently used (LRFU) method [13]. In operating systems, an LRFU policy [13] is a combination of the least recently and least frequently used policies for page replacement in virtual memory system. Similar approaches are used by some authors in [14,15].

In operating systems, an LRFU policy [13] is a combination of the least recently and least frequently used policies for the page replacement algorithm. Hence, once the main-memory has reached its capacity, the operating system moves the most useless memory page into the swap space on the hard-disk drive. LRFU is a useful policy for selecting the most useless memory page.

However, a modified version of LRFU proposed in this paper introduces an important weight determined by the risk-sensitive utility function.

In this study, an LRFU policy is used for choosing the most ineffective kernel in the network. The LRFU policy is useful for detecting the least activated and oldest kernel. Such kernels may probably not activate again for an extended time, indicating that the cumulative error for the learner remains low even if the kernel is pruned.

LRFU estimates the recency of each kernel's activation time by calculating the difference between the time of activation, when the kernel center is the nearest to the input, and the current time. Let t_c be the current time and t_j be the j-th The kernel's effectiveness is measured by the following weighting function $f_j[.]$.

$$C_j(t_c) \equiv \sum_{k=1}^{w} f[t_c - t_{jk}], \quad where \quad f[t_c - t_j] \equiv \left(\frac{1}{2}\right)^{\lambda(t_c - t_j)} \tag{5}$$

Note that $0 < f[n_j] \leq 1$ and $0 \leq \lambda \leq 1$. From this equation, the following recurrence formula can be obtained.

$$C[j] := \begin{cases} |U_t| + \left(\frac{1}{2}\right)^{\lambda} C[j] & \text{if j-th kernel center is the nearest to the input} \\ \left(\frac{1}{2}\right)^{\lambda} C[j] & \text{otherwise} \end{cases} \tag{6}$$

where U_t denotes the utility function output at time t. Note that U_t is enlarged when the learning machine feels some loss as described in the next Sect. 3.

Thus, we can detect the most ineffective kernel in terms of its recency as $i = \arg\min_j C[j]$. λ was set 0.0005 in the experiments.

The pseudo algorithm for calculation of outputs and estimation of each kernel is shown in Algorithm 2, where the importance weight I_w is to be set to $|U_t|$.

Algorithm 2. Algorithm for calculation of $f_\theta[x]$

Require: new learning sample x_t, Current learning result f_t and g_t. Importance
   weight for this instance I_w
   $minDistance = MAX$.
   for each kernel center x_i **do**
      diff= $\|x_t - x_i\|^2$
      if diff< $minDistane$ **then**
         $ActiveKernel = i$;
         $minDistance = diff$;
      end if
   end for
   for each Kernel index j **do**
      if $j == ActiveKernel$ **then**
         $C[j] = I_w + (1/2)^{\lambda} C[j]$
      else
         $C[j] = (1/2)^{\lambda} C[j]$
      end if
   end for
   $f_\theta[x] = \frac{\langle f_t, K(x, \cdot)\rangle}{\langle g_t, K(x, \cdot)\rangle}$
   return $f_\theta[x]$.

---

3   Risk Sensitive Reinforcement Learning

The risk-sensitive RL such as proposed in [3] uses human specific utility function based on the prospect theory [4]. The specialized utility function gives an extremely large importance weight for risk related experience. Therefore,

$$u(x) = \begin{cases} k_+ x^{l+} & x \geq 0 \\ -k_- x^{l-} & x < 0 \end{cases} \tag{7}$$

In this study, an actor-critic model is used for the experiments so that the utility function estimates the temporal difference error.

$$\delta = r + \gamma V(s^{'}) - V(s),\tag{8}$$

where $V(s)$ denotes the value function of current state-s, $s^{'}$ denotes the previous state, and γ denotes the discount factor. The critic describes $V(s)$ and the predicted $V(s)$ is used for the learning of actor and critic. Thus, the error signal for the critic is $u(\delta)$, so that the desired output for the critic is

$$\hat{y}_t^{critic}(s) = V(s) + u(\delta)\tag{9}$$

Note that $V(s) = f_\theta^{critic}[s]$, which is the output of the kernel regression for critic.
    For the actor, the desired output is the previous action.

$$\hat{y}_t^{actor} = \left[\frac{\gamma y_{t-1} + (1-\gamma)f_{t-1}^{actor}(S_{t-1})}{\sigma}\right]\tag{10}$$

Note that the actor output is derived by

$$y_t(x) = f_\theta^{actor}[x_t] + n,\tag{11}$$

where $n \sim \mathcal{N}(0, \Sigma)$ and $\Sigma = Diag(\sigma)$. $f_\theta^{actor}[x_t]$ denotes the output of the kernel regression. σ is also the additional output of kernel regression: $\sigma = f^{actor\ 2}[x_t]$. The actor learns the desired vector $[y_t(x_t)\ \eta\sigma]^T$, the importance weight for the actor and critic:I_a is set as $I_w = |U_t|$, where $U_t = u(\delta)$

4   Experimental Results

The performances of the reinforcement learning with and without the risk-sensitive utility function were examined. Benchmark test for kernel-regression LRFU have been conducted on Puddle-World-Tasks. The details of the puddle-world-tasks were explained in the literature [16,17].
    In the Puddle-Work-Tasks, the agent searches for the goal by avoiding the puddle. The puddle is located on a two-dimensional field $[0,0]$ to $[1,1]$. agent's movement is restricted to this region. In this area, the goal field is set to $[0.9, 0.9] - [1, 1]$. If the agent falls in the puddle, the agent gives negative rewards (Fig. 1).
    The agent receives a negative reward, -1, if it moves in one of the four directions: up, down, right, and left. However, if the agent get in the puddle, the agent gives negative reward -400 times the distance into the puddle (distance to the nearest edge). If the agent reaches the goal or the number of steps is larger than 10000, the learning process is terminated. The puddle-world-task were repeated 100 times by randomly changing the start point. This learning process was repeated 10 times and the cumulative rewards for the 10 trials were averaged. The parameters used for the reinforcement learning were $\alpha = 0.5$ and $\gamma = 0.5$.

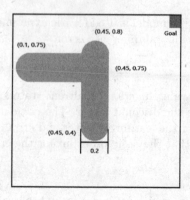

Fig. 1. Puddle-world-task

4.1 Results

Figures 2 and 3 show the comparison of the systems whose maximum number of kernels is 10 and 20, with and without the risk-sensitive utility function.

The resultant value functions were also examined. Figures 4 and 5 are resultant value functions in the case of risk-neutral and risk-sensitive RL, respectively. We can see that the value function caused by risk-neutral RL cannot represent

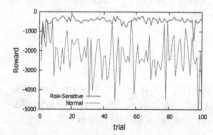

Fig. 2. Comparison of cumulative rewards in the case of $B = 10$

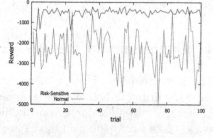

Fig. 3. Comparison of cumulative rewards in the case of $B = 20$

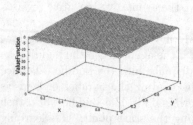

Fig. 4. Resultant value function of risk-neutral RL $B = 20$

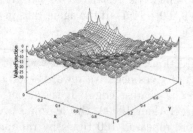

Fig. 5. Resultant value function of risk-sensitive RL $B = 20$

the negative reward of getting into the puddle well, while that of risk-sensitive RL represents it clearly.

5   Conclusion

In this paper, we pointed out that reinforcement learning by using a learning method on a budget needs risk-sensitive utility function. This is because the learning machine on a budget usually disposes a part of the memory to create space to record a new instance. As a result, the learning machine usually forgets rare experiences, such as negative experiences. This inhibits the learning to take an action to avoid a dangerous situation. The preliminary results shown in this paper suggest that the kernel regression with a fixed budget works well on the puddle-world-task. This scheme may be applied to various learning algorithms on a budget to investigate their performances as a future work.

Acknowledgement. This research has been supported by Grant-in-Aid for Scientific Research(c) 12008012.

References

1. Igushi, K., Ogiso, T., Yamauchi, K.: Acceleration of reinforcement learning via game-based renewal energy management system. In: SCISISIS 2014, pp. 415–420. The Institute of Electrical and Electronics Engineers, Inc., New York, December 2014
2. Ogiso, T., Yamauchi, K., Ishii, N., Suzuki, Y.: Co-learning system for humans and machines using a weighted majority-based method. Int. J. Hybrid Intell. Syst. **13**, 63–76 (2016)
3. Shen, Y., Tobia, M.J., Sommer, T., Obermayer, K.: Risk-sensitive reinforcement learning. Neural Comput. **26**, 1298–1328 (2014)
4. Kahneman, D., Tversky, A.: Prospect theory: an analysis of decision under risk. Econometrica **47**(2), 263–291 (1979)
5. Walter, F.E., Schweitzer, F.: Risk-seeking versus risk-avoiding investments in noisy periodic environments. Int. J. Mod. Phys. C **19**(6), 971–994 (2008)
6. Amari, S., Park, H., Fukumizu, K.: Adaptive method of realizing natural gradient learning for multilayer perceptrons. Neural Comput. **12**, 1399–1409 (2000)
7. Dekel, O., Shalev-Shwartz, S., Singer, Y.: The forgetron: a kernel-based perceptron on a budget. SIAM J. Comput. (SICOMP) **37**(5), 1342–1372 (2008)
8. Orabona, F., Keshet, J., Caputo, B.: The projectron: a bounded kernel-based perceptron. In: ICML 2008, pp. 720–727 (2008)
9. He, W., Si, W.: A kernel-based perceptron with dynamic memory. Neural Netw. **25**, 105–113 (2011)
10. Yamauchi, K.: Pruning with replacement and automatic distance metric detection in limited general regression neural networks. In: IJCNN 2011, pp. 899–906. The Institute of Electrical and Electronics Engineers, Inc., New York, July 2011
11. Yamauchi, K.: Incremental learning on a budget and its application to quick maximum power point tracking of photovoltaic systems. J. Adv. Comput. Intell. Intell. Inform. **18**(4), 682–696 (2014)

12. Garcìa, S., Derrac, J., Cano, J.R., Herrera, F.: Prototype selection for nearest neighbor classification: taxonomy and empirical study. IEEE Trans. Pattern Anal. Mach. Intell. **34**(3), 417–435 (2012)

13. Lee, D., Noh, S.H., Min, S.L., Choi, J., Kim, J.H., Cho, Y., Sang, K.C.: LRFU: a spectrum of policies that subsumes the least recently used and least frequently used policies. IEEE Trans. Comput. **50**(12), 1352–1361 (2001)

14. Kondo, Y., Yamauchi, K.: A dynamic pruning strategy for incremental learning on a budget. In: Loo, C.K., Yap, K.S., Wong, K.W., Teoh, A., Huang, K. (eds.) ICONIP 2014, Part I. LNCS, vol. 8834, pp. 295–303. Springer, Heidelberg (2014)

15. Kato, H., Yamauchi, K.: Quick MPPT microconverter using a limited general regression neural network with adaptive forgetting. In: 2015 International Conference on Sustainable Energy Engineering and Application (ICSEEA), pp. 42–48. The Institute of Electrical and Electronics Engineers, Inc., New York, February 2016

16. Sutton, R.S.: Generalization in reinforcement learning: successful examples using sparse coarse coding. Adv. Neural Inf. Process. Syst. **8**, 1038–1044 (1995)

17. Ryuichi UEDA: Comparison of data amount for representing decision making policy. In: Burgard, W., Dillmann, R., Plagemann, C., Vahrenkamp, N. (eds.) Intelligent Autonomous Systems 10 (IAS 2010), vol. 10, pp. 26–35. IOS Press (2008)

A Kernel-Based Sarsa(λ) Algorithm with Clustering-Based Sample Sparsification

Haijun Zhu[1], Fei Zhu[1,2], Yuchen Fu[1,3($\boxtimes$)], Quan Liu[1],
Jianwei Zhai[1], Cijia Sun[1], and Peng Zhang[1]

[1] School of Computer Science and Technology,
Soochow University, Suzhou 215000, China
[2] Provincial Key Laboratory for Computer Information Processing Technology,
Soochow University, Suzhou, China
[3] School of Computer Science and Engineering,
Changshu Institute of Technology, Changshu 215500, China
yuchenfu@suda.edu.cn

Abstract. In the past several decades, as a significant class of solutions to the large scale or continuous space control problems, kernel-based reinforcement learning (KBRL) methods have been a research hotspot. While the existing sample sparsification methods of KBRL exist the problems of low time efficiency and poor effect. For this problem, we propose a new sample sparsification method, clustering-based novelty criterion (CNC), which combines a clustering algorithm with a distance-based novelty criterion. Besides, we propose a clustering-based selective kernel Sarsa(λ) (CSKS(λ)) on the basis of CNC, which applies Sarsa(λ) to learning parameters of the selective kernel-based value function based on local validity. Finally, we illustrate that our CSKS(λ) surpasses other state-of-the-art algorithms by Acrobot experiment.

Keywords: Reinforcement learning · Kernel method · Sample sparsification · Clustering · Sarsa(λ)

1 Introduction

Machine learning, an important domain of computer science, can be typically classified into three broad categories: supervised learning, unsupervised learning and reinforcement learning (RL). Supervised learning algorithms who learns from the labeled samples provided by a knowledgable external supervisor, have been proven to be effective. However, the labeled data are scarcity and supervised learning algorithms are not adequate for learning from interaction. Unsupervised learning algorithms can find the structure hidden in unlabeled data, while it cannot use this structure to deal with the objective tasks. In order to succeed to the advantages of supervised learning and unsupervised learning, and overcome their shortcomings, RL came into being.

As a solution to the sequential decision problems, reinforcement learning (RL) offers a trial-and-error learning framework to chase the maximal cumulative

© Springer International Publishing AG 2016
A. Hirose et al. (Eds.): ICONIP 2016, Part III, LNCS 9949, pp. 211–220, 2016.
DOI: 10.1007/978-3-319-46675-0_24

payoffs [10] by continually interacting with the environment. In the past several years, there have been some respectable successful applications, such as AlphaGo [9] and upper limb rehabilitation robot [5]. However, how to solve the large scale or continuous state space problem is still an open problem, which is the so-called dimensionality curse. The common solutions include linear function approximation [10], kernel methods [6], regression tree methods [4], and neural networks [9]. In this paper, we focus on the kernel methods.

As an important domain of RL, kernel-based RL (KBRL) has been intensively studied. The sparsification process plays a significant role in KBRL, which directly influences the time of calculation and the accuracy of prediction. The common methods of sparsification include approximate linear dependence (ALD) [3] and novelty criterion (NC) [2]. ALD method can keep the feature vectors in the dictionary approximately linear independent with each other. The famous algorithms using ALD include kernel recursive least-squares (KRLS) [3] and kernel-based least squares policy iteration (KLSPI) [14]. However, the complexity of ALD is $O(n^2)$, which is computational exhausted for the real-time applications. NC method considers both the minimum distance and prediction error. Its complexity is $O(n)$, which is competent for online sparsification. However, the accuracy of prediction using NC is not ideal. Online selective kernel-based temporal difference (OSKTD) [2] is a famous algorithm using NC.

In view of the above problems in NC and ALD, we propose a new sparsification method, which is named clustering-based NC (CNC). This method combines K-means with a distance-based NC. K-means is the most popular clustering algorithm, which can find the hidden structure behind the sample set. This structure can improve the accuracy of estimate of algorithms. And the distance-based NC enables our method to do online sparsification in the learning stage. On the basis of CNC, we propose a clustering-based selective kernel Sarsa(λ) (CSKS(λ)) algorithm for the large scale or continuous state space problems. Besides of this sparsification process, our algorithm also has the parameters learning process. In the parameters learning phase, we apply a simple but efficient Sarsa(λ) to update the parameters of the selective kernel-based value function (SKVF). This kind of value function, which is based on local validity, will have better generalization accuracy than the traditional kernel-based value function when the data dictionary is constructed by our sparsification method.

This paper has three main contributions. First, we propose a new sample sparsification method named CNC. Second, on basis of the sample sparsification method, we propose an efficient kernel-based RL algorithm (CSKS(λ)). Third, we illustrate that our CNC method and CSKS(λ) algorithm have outstanding performance by Acrobot experiment.

2   RL and Kernel Method

RL problems can be modeled by a Markov Decision Process (MDP) [7], where the MDP is a tuple $M = < X, A, P, R >$. X denotes the state space, A denotes the action space, P denotes the transition model describing the dynamics of MDP, and R denotes the reward model assigning rewards to state-action pairs. In RL, the agent interacts with the environment continuously to learn the optimal or near-optimal policy. The policy of the MDPs is a mapping $\pi : X \rightarrow A$. And the optimal policy π^* is a fixed point of the Bellman equation [1] that satisfies

$$J_{\pi^*} = \max J_\pi = \max E^\pi \left[\sum_{t=0}^\infty \gamma^t r_t \right], \tag{1}$$

where γ is the discount factor $(0 \leq \gamma \leq 1)$, r_t is the immediate reward at time t, J_π is the expected reward over the total trajectories, and $E^\pi[\cdot]$ is the expectation with respect to the policy π and the transition model P.

In KBRL, the kernel-based value function (KVF) is defined as

$$\widetilde{Q}(s) = \boldsymbol{\Theta}^\top \boldsymbol{k}(s) = \sum_{i=1}^M \theta_i k(s, s_i), \tag{2}$$

where $s = (x, a)$ is a state-action pair, $\boldsymbol{\Theta}$ is a weight vector, M denotes the size of the data dictionary, $k(\cdot, \cdot)$ is a kernel function, and

$$\boldsymbol{k}(\cdot) = [k(\cdot, s_1), k(\cdot, s_2), ..., k(\cdot, s_M)]^\top. \tag{3}$$

After getting the value function, it is easily to obtain the policy

$$\pi(x) = \arg \max_a Q(s). \tag{4}$$

3 Clustering-Based Selective Kernel Sarsa(λ)

In this section, we state the CSKS(λ) algorithm, which uses the new sparsification method CNC to simplify a sample set and construct a data dictionary. With this dictionary, CSKS(λ) can apply Sarsa(λ) to learn parameters more efficiently. In the following, we first introduce how our CNC method works. Then CSKS(λ) is proposed for the large or continuous state problems.

3.1 Clustering-Based Novelty Criterion

In KBRL, How to filtrate samples efficiently and hold the accuracy of prediction is still an open problem. Sample sparsification can reduce the computation time of learning. However, as an extra component, the sparsification process itself also needs some computing time. So it is important to accomplish the sparsification process as quickly as possible. Besides, the terrible selections of samples will result in the low accuracy of prediction. Hence, the sparsification needs to consider the tradeoff between the accuracy and the computing time.

Aiming at the issue above, we propose a CNC method to keep the samples spare. In CNC, the K-means method is used to initialize the data dictionary D. Then we perfect this dictionary D by a distance-based NC method.

For samples $S = \{s_1, s_2, \ldots s_n\}$, the K-means method allocates each sample s_i to one of the K clusters to minimize the cluster variance

$$G = \sum_{k=1}^{K} \sum_{s_i \in A_k} g(s_i, c_k), \tag{5}$$

where A_k is the kth subset of S, $g(\cdot, \cdot)$ is a similarity metric between a sample point and a centroid and the c_k is the centroid of A_k, which can be defined as

$$c_k = \frac{1}{n_k} \sum_{s_i \in A_k} s_i, \tag{6}$$

where n_k is the size of set A_k. In this paper, we use the Euclidean distance $\| s_i - c_k \|^2$ as the similarity metric.

K-means method can find the hidden structure of the data. Yet, the improper selection of initial centers may result in K-means getting a local optimal solution. To obtain a better result, we select some samples as the initial centers, which are far from each other. Besides, the output of K-means are not the sample points but the centroids or weighted centroids of clusters. Therefore, we needs/to choose the sample points from the original sample set, closest to the corresponding centroids of clusters, to initializes the data dictionary D.

The initial data dictionary represents the hidden structure of state-action space. This structure can improve the whole performance of our algorithm. However, only using this structure can not guarantee the accuracy of estimation. To improve this accuracy, we use a distance-based NC method to optimize this data dictionary.

In KBRL, the KVF can be rewritten as

$$\tilde{Q}(s) = \sum_{i=1}^{M} w_i \tilde{Q}_i(s), \tag{7}$$

where w_i is a weight ($0 \leq w_i \leq 1$ and $\sum_{i=1}^{M} w_i = 1$), and $\tilde{Q}_i(s) = \frac{1}{w_i} \theta_i k(s, s_i)$.

Lemma 1. *According to selective ensemble learning [12], the approximation error of value function $\tilde{Q}$ is*

$$E = \frac{1}{M^2} \sum_{i=1}^{M} \sum_{j=1}^{M} C_{ij}, \tag{8}$$

where C_{ij} is the covariance between estimator $\tilde{Q}_i$ and $\tilde{Q}_j$.

Lemma 1 shows that the less correlation between estimators, the less approximation error is.

Lemma 2. *The covariance has the property $2C_{ij} \leq E_i + E_j$, where E_i is the generalization error of ith estimator $\tilde{Q}_i$*

Lemma 2 indicates that the closer the performances of the two estimators are, the higher the covariance they get. If two samples are close to each other and have similar performance, which brings perishing learning result. Thus, enlarging the distance between the samples in the data dictionary can decrease approximation error. So the distance-based NC method can help improve the accuracy of prediction of learning algorithm.

Supposes $D = \{s_1, s_2, ..., s_M\}$ have been obtained, where M is the size of D. For a new sample s, if $\delta^l(s) = \min_{s_i \in D} \| \phi(s) - \phi(s_i) \|^2 \geq \mu$, then add it to the data dictionary, where $\phi(\cdot)$ is a mapping from original space to a Hilbert space. Using the kernel trick $k(s_i, s_j) = \phi(s_i)^\top \phi(s_j)$, we can obtain

$$\delta^l(s) = \min_{s_i \in D} (2 - 2k(s_i, s)). \tag{9}$$

The update rule of the data dictionary D is

$$D \leftarrow \begin{cases} D & \text{if } \delta^l(s) < \mu \\ D \bigcup \{s\} & \text{otherwise} \end{cases}, \tag{10}$$

where μ is a distance threshold value.

This method enables the sample sparsification online, and its computational complexity is just $O(n)$ in the learning stage. However, the method can not guarantee that the feature vectors in the data dictionary are linear independent with each other, which is a challenge for the convergence of RL algorithm. To alleviate this problem and improve the generalization ability of samples, we use the SKVF instead of traditional KVF, which is described in the next section.

3.2 Framework of CSKS(λ)

Before we proceed to CSKS(λ), we introduce the SKVF. SKVF is a kind of kernel-based approximate value function. When approximate the value function of the target sample s, this method eliminates some samples in the dictionary, which are less relevant to the sample s. Yet, as described in Eq. 3, the KVF considers the whole dictionary. According to SKVF, we change the KVF into

$$\boldsymbol{k}(s) = [K(s, s_1), K(s, s_2), ..., K(s, s_M)]^\top, \tag{11}$$

where $K(s, s_i)$ is a variant of kernel function, which is defined as

$$K(s, s_i) = \beta(s, s_i)k(s, s_i), \tag{12}$$

where $k(s, s_i)$ is a conventional kernel function, and $\beta(\cdot, \cdot) \in \{0, 1\}$ is an indicator function. $\beta(s, s_i) = 1$ shows that the sample s_i can contribute to the accuracy of $\widetilde{Q}(s)$. While $\beta(s, s_i) = 0$ indicates that the sample s_i is rarely helpful for this accuracy or even hurt it. $\beta(\cdot, \cdot)$ can be defined by the prior knowledge, local validity and so on. In this paper, we only consider the local validity. For a sample s_i in the dictionary, if it is too far from the target sample s in the reproducing kernel Hilbert space, then s_i has less generalization ability for the sample s and should be ignored. The judgment rule of $\beta(\cdot, \cdot)$ is

$$\beta(s, s_i) \leftarrow \begin{cases} 1 & \text{if } (1 - k(s_i, s)) < \mu_1 \\ 0 & \text{otherwise} \end{cases}, \tag{13}$$

where μ_1 is a threshold value.

Sarsa(λ) is a simple but useful TD algorithm, and its keys are TD error and eligibility traces. The TD error is

$$\delta_t \leftarrow r_t + \gamma \tilde{Q}(s_{t+1}) - \tilde{Q}(s_t). \tag{14}$$

The eligibility traces record the occurrence of an event, such as visiting a state or taking an action, which can improve the learning efficiency. There are three kinds of eligibility traces: accumulating traces, replacing traces and decaying eligibility traces [8]. In this paper, we only consider accumulating traces, which is defined as

$$e_t \leftarrow \gamma \lambda e_{t-1} + k(s_t). \tag{15}$$

Finally, the update rule is

$$\boldsymbol{\Theta}_{t+1} \leftarrow \boldsymbol{\Theta}_t + \alpha_t \delta_t e_t, \tag{16}$$

where α_t is a learning factor. We use this simple but useful method to learn the parameters of SKVF.

As shown in Algorithm 1, CSKS(λ) is an on-policy TD control method for the large or continuous state space problems. On-policy means that the samples used to learn are the same as the samples taken by the agent. If we make $a^* = \arg\max_b \tilde{Q}(x', b)$, our algorithm will be an off-policy Q(λ) algorithm. However, the data dictionary D obtained by the CNC method can not satisfy the convergence assumption[1] of the TD method [11]. So the convergence of our algorithm needs further research.

4   Experiment and Results

In this section, the Acrobot [14] is given to demonstrate the effectivenesses of our algorithm. We compare our method with others in the following four aspects: different algorithms, different sparsification methods, using SKVF or KVF and different numbers of clusters.

4.1   Settings of Acrobot

Acrobot is a classical learning control problem shown in Fig. 1a. Agent forces at the elbow B to swing up the pendulum from the straight-down equilibrium

---

[1] The matrix $N \geq M$ and $\mathcal{K} = [k(s_1), k(s_2), ..., k(s_N)]$ has full rank, where N is the size of the state-action space.

Algorithm 1. Clustering-Based Selective Kernel Sarsa(λ)

Initialize:
   kernel function $k(\cdot, \cdot)$ and its parameters, samples $S = \{s_t\}_{t=1}^n$
Process:
   using K-means method to initialize the data dictionary D
   repeat
      initialize x, a
      repeat
         take a, observe next state x' and reward r
         $a' \leftarrow \arg\max_b \widetilde{Q}(x', b) + greedy(\epsilon)$
         $a^* \leftarrow a'$
         $\delta \leftarrow r + \gamma \widetilde{Q}(x', a^*) - \widetilde{Q}(x, a)$
         for each element $s_i \in D$
            $e_i \leftarrow \gamma \lambda e_i + \beta(s, s_i) k(s, s_i)$
            $\theta_i \leftarrow \theta_i + \alpha \delta e_i$
         end for
         $\delta^l(s) = \min_{s_i \in D} (2 - 2k(s, s_i))$
         update D according to Eq. 10
         $x \leftarrow x', a \leftarrow a'$
      until x is terminal
   until the terminal criterion is satisfied

---

point to the neighborhood of the straight-up equilibrium point. The state of this problem can be denoted as $[\theta_1, \theta_2, \dot{\theta}_1, \dot{\theta}_2]$, where θ_1 and θ_2 are shown in the figure, and $\dot{\theta}_1$ and $\dot{\theta}_2$ are the angle velocities of θ_1 and θ_2, respectively. The action space of this problem is $\{-1N, 0N, 1N\}$. The goal of the agent is to reach the target line, and it will receive a reward 0 when achieve the goal, otherwise the reward of each step is -1.

The dynamic model of the acrobot system is described as

$$\ddot{\theta}_1 = -\left(d_2\ddot{\theta}_2 + \phi_1\right)/d_1, \tag{17}$$

$$\ddot{\theta}_2 = \tau + d_2\phi_1/d_1 - \phi_2, \tag{18}$$

where

$$d_1 = m_1 l_{c1}^2 + m_2 \left(l_1^2 + l_{c2}^2 + 2l_1 l_{c2} \cos\theta_2\right) + I_1 + I_2,$$
$$d_2 = m_2 \left(l_{c2}^2 + l_1 l_{c2} \cos\theta_2\right) + I_2,$$
$$\phi_1 = -m_2 l_1 l_{c2}\dot{\theta}_2^2 \sin\theta_2 - 2m_2 l_1 l_{c2}\dot{\theta}_1\dot{\theta}_2 \sin\theta_2$$
$$+ (m_1 l_{c1} + m_2 l_1) g \cos(\theta_1 - \pi/2) + \phi_2,$$
$$\phi_2 = m_2 l_{c2} g \cos(\theta_1 + \theta_2 - \pi/2),$$

where the parameters $m_1 = m_2 = 1\,\text{kg}$, $I_1 = I_2 = 1\,\text{kg} \cdot m^2$, $l_{c_1} = l_{c_2} = 0.5\,\text{m}$, $l_1 = l_2 = 1\,\text{m}$, and $\tau \in \{-1N, 0N, 1N\}$ is torque.

4.2 Results

Before presenting the results, we state the settings in the implementation of the CSKS(λ) algorithm, which are summarized as follows: 1) Gaussian kernel function $k(x_i, x_j) = e^{-\|x_i - x_j\|^2/2\sigma^2}$, where $\sigma = 10$; 2) The parameter $\lambda = 0.4$, $\epsilon = 0.1$, $\mu = 2.5$, $\mu_1 = 6$ and $\gamma = 0.9$. Moreover, the x-axis and y-axis of Fig. 1(b)-(d) denote the training episodes and the steps of each episode, respectively.

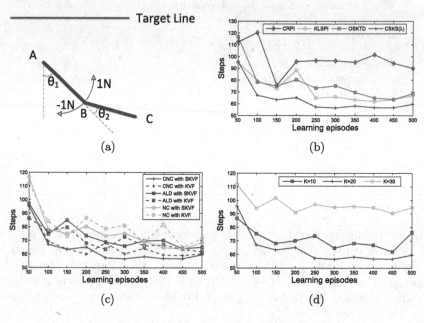

Fig. 1. (a) An underactuated double-link pendulum. (b)-(d) Steps of different algorithms in Acrobot

We compare the performances of our method CSKS(λ) with other state-of-the-art algorithms OSKTD, KLSPI and CRPI [13]. Among them, our algorithm, OSKTD and KLSPI are kernel-based methods, while CRPI is not. As shown in Fig. 1b, the kernel methods perform better than the non-kernel method CRPI. In addition, we can find that CSKS(λ) performs better than the other two kernel-based algorithms. That is mainly because of the use of CNC and SKVF.

In order to analyse the effects of different sparsification methods (ALD, NC and CNC) and two kinds of value function (SKVF and KVF), we combine Sarsa(λ) with this three sparsification methods and two value functions respectively. As shown in Fig. 1c, no matter using KVF or SKVF, CNC has the better performances than NC. That is mainly because CNC makes the best of the hidden structure of the samples found by K-means. The early performance of CNC is inferior to that of ALD, while it defeats ALD soon. In addition, we also find SKVF improves the performances of Sarsa(λ) using the distance-based sparsification methods (CNC and NC) rather than ALD method. Because ALD concerns less about the distribution of samples in dictionary. When using SKVF to select samples, there may be less or even no samples to be chosen. While the distance-based sparsification methods can make the samples in the dictionary distribute uniformly. To sum up, our CNC method and SKVF can improve the performance of the learning algorithms.

Finally, we compare the influence of different $K = 10, 20 or 30$ on our method. As shown in Fig. 1d, different K has different results. So we hope to find a heuristic rule that can determine K and guarantee the remarkable performance for some problems, which needs further research.

5   Conclusion

In this paper, we propose a clustering-based selective kernel Sarsa(λ) algorithm, which combines a clustering-based sparsification method with the kernel version of Sarsa(λ), for the large or continuous state space problems. We use the clustering-based NC method for the sample sparsification. In the procedure of parameters learning, Sarsa(λ) is used to update the parameters of value functions based on the local validity. And we evaluate this algorithm about a continuous state space control problem and the results show that it performs better than other state-of-the-art algorithms.

However, there are some questions, which can be the future work. First, we need to find a heuristic rule about the number of cluster K. Second, the convergence of CSKS(λ) needs to be analyzed theoretically. Third, we hope to combine our algorithm with the Actor-Critic framework to solve the continuous action space problems.

Acknowledgement. This work was funded by National Science Foundation of China (61303108, 61373094, 61472262), Natural Science Foundation of Jiangsu (BK2012616), High School Natural Foundation of Jiangsu (13KJB520020), Key Laboratory of Symbolic Computation and Knowledge Engineering of Ministry of Education, Jilin University (93K172014K04), Suzhou Industrial application of basic research program part (SYG201422), Provincial Key Laboratory for Computer Information Processing Technology, Soochow University (KJS1524).

References

1. Bellman, R.: Dynamic Programming. Princeton University Press, Princeton (1957)
2. Chen, X., Gao, Y., Wang, R.: Online selective kernel-based temporal difference learning. IEEE Trans. Neural Netw. Learn. Syst. **24**(12), 1944–1956 (2013)
3. Engel, Y., Mannor, S., Meir, R.: The kernel recursive least-squares algorithm. IEEE Trans. Signal Process. **52**(8), 2275–2285 (2004)
4. Ernst, D., Geurts, P., Wehenkel, L.: Tree-based batch mode reinforcement learning. J. Mach. Learn. Res. **6**(2), 503–556 (2005)
5. Fan-Cheng, M., Ya-Ping, D.: Reinforcement learning adaptive control for upper limb rehabilitation robot based on fuzzy neural network. In: Control Conference (CCC), pp. 5157–5161 (2012)

6. Ormoneit, D., Sen, Ś.: Kernel-based reinforcement learning. Mach. Learn. **49**(2–3), 161–178 (2002)
7. Puterman, M.L.: Markov Decision Processes: Discrete Stochastic Dynamic Programming. Wiley, New York (2014)
8. van Seijen, H., Sutton, R.S.: True online TD(λ). In: Proceedings of the 31st International Conference on Machine Learning (ICML), pp. 692–700 (2014)
9. Silver, D., Huang, A., Maddison, C.J., Guez, A., Sifre, L., Driessche, G.V.D., Schrittwieser, J., Antonoglou, I., Panneershelvam, V., Lanctot, M.: Mastering the game of go with deep neural networks and tree search. Nature **529**(7587), 484–489 (2016)
10. Sutton, R.S., Barto, A.G.: Reinforcement Learning: An Introduction. MIT Press, Cambridge (1998)
11. Tsitsiklis, J.N., Van Roy, B.: An analysis of temporal-difference learning with function approximation. IEEE Trans. Autom. Control **42**(5), 674–690 (1996)
12. Xiaoyang, T., Songcan, C., Zhi-Hua, Z., Fuyan, Z.: Recognizing partially occluded, expression variant faces from single training image per person with som and soft k-NN ensemble. IEEE Trans. Neural Netw. **16**(4), 875–886 (2005)
13. Xu, X., Huang, Z., Graves, D., Pedrycz, W.: A clustering-based graph Laplacian framework for value function approximation in reinforcement learning. IEEE Trans. Cybern. **44**(12), 2613–2625 (2014)
14. Xu, X., Hu, D., Lu, X.: Kernel-based least squares policy iteration for reinforcement learning. IEEE Trans. Neural Netw. **18**(4), 973–992 (2007)

Sparse Kernel-Based Least Squares Temporal Difference with Prioritized Sweeping

Cijia Sun, Xinghong Ling[✉], Yuchen Fu, Quan Liu, Haijun Zhu, Jianwei Zhai, and Peng Zhang

School of Computer Science and Technology,
Soochow University, Suzhou 215000, China
lingxinghong@suda.edu.cn

Abstract. How to improve the efficiency of the algorithms to solve the large scale or continuous space reinforcement learning (RL) problems has been a hot research. Kernel-based least squares temporal difference(KLSTD) algorithm can solve continuous space RL problems. But it has the problem of high computational complexity because of kernel-based and complex matrix computation. For the problem, this paper proposes an algorithm named sparse kernel-based least squares temporal difference with prioritized sweeping (PS-SKLSTD). PS-SKLSTD consists of two parts: learning and planning. In the learning process, we exploit the ALD-based sparse kernel function to represent value function and update the parameter vectors based on the Sherman-Morrison equation. In the planning process, we use prioritized sweeping method to select the current updated state-action pair. The experimental results demonstrate that PS-SKLSTD has better performance on convergence and calculation efficiency than KLSTD.

Keywords: Reinforcement learning · Prioritized sweeping · Sparse kernel · Least squares temporal difference

1 Introduction

Reinforcement learning (RL) is an important content of machine learning which learns how to map situations to actions with the goal of maximizing the cumulative rewards [1]. In RL, agent can exploit the learning algorithms in the interaction with the environment to strengthen their learned experience and maximize the cumulative rewards [2]. Temporal difference (TD) algorithm is one of the classical RL algorithms. It has the advantages of Monte Carlo (MC) and Dynamic Programming (DP). TD with eligibility traces is an efficient and incremental algorithm which is useful to solve the problem of time credit assignment [3].

The real-world RL applications usually have large or continuous state space [4]. Both the look-up table and function approximation have been widely used in TD algorithm to solve the real-world RL applications. However using the look-up table will face great difficulties in calculation and storage which will

© Springer International Publishing AG 2016
A. Hirose et al. (Eds.): ICONIP 2016, Part III, LNCS 9949, pp. 221–230, 2016.
DOI: 10.1007/978-3-319-46675-0_25

result in the dimension disaster problem. Kernel function approximation is a classical method of nonparametric function approximation. It can be applied to approximately represent the state-action value function. The parameters of kernel function are produced and adjusted by the samples entirely. Therefore kernel function approximation is more flexible and freer, but its computational complexity increases rapidly with the increase of the samples.

The kernel-based TD algorithm can be used to solve the real-world RL applications. In 2005, Xu et al. proposed a kernel-based least squares temporal difference (KLSTD) algorithm. This algorithm has a better performance on the evaluation of nonlinear value function and policy than the linear TD algorithm [5]. But its computational complexity increases quickly with the increase of the number of samples. In 2007, Xu et al. proposed the kernel-based least squares policy iteration (KLSPI) algorithm based on the KLSTD algorithm. An approximate linear dependency (ALD) kernel sparse method is performed in the algorithm to reduce the time and space complexity of the algorithm [6]. However, during update process, matrix inversion needs a large number of computations which impacts the speed of calculation and convergence.

Prioritized sweeping method is implemented on the basis of the Dyna architecture which combines the learning process with the planning process [7,8]. This method determines the order of updating the state-action value function by comparing the priorities of the state-action pairs.

In this paper, we propose an algorithm named sparse kernel-based least squares temporal difference with prioritized sweeping (PS-SKLSTD). The goal of the algorithm is accelerating the efficiency of calculation and convergence. This algorithm has two processes: the learning process and the planning process. In the learning process, we exploit the ALD-based sparse kernel function to represent value function approximately and update the parameter vector of the value function based on the Sherman-Morrison equation. In the planning process, we utilize prioritized sweeping method to select the current updated state-action pair. It has a good performance on speeding up the convergence of the algorithm. Experimentally, we compare PS-SKLSTD with KLSTD in terms of convergence efficiency on Puddle World and Cart-Pole problems to demonstrate that PS-SKLSTD has a good performance of convergence. We compare the convergence efficiency of PS-SKLSTD with varying prioritized sweeping parameter to demonstrate that the parameters of prioritized sweeping effect the convergence rate of PS-SKLSTD sensitively.

The rest of this paper is organized as follows. First, we describe the background in Sect. 2. Second, we describe the PS-SKLSTD algorithm in detail in Sect. 3. Third, we describe and analyze the experimental results of PS-SKLSTD and KLSTD on the Puddle World and Cart-Pole problems in Sect. 4. Finally, we make a conclusion in Sect. 5.

2   Background

In the field of the reinforcement learning, learning task can be modeled as a Markov Decision Process(MDP). A MDP is denoted as a tuple $< X, U, \rho, f >$, where X denotes the state space, U denotes the action space, ρ denotes the reward function and f denotes the state transition function. The policy of the MDP is defined as h. In MDP, agent has the goal of maximizing the cumulative rewards, the cumulative discount reward can be expressed as:

$$R_t = r_{t+1} + \gamma r_{t+2} + \gamma^2 r_{t+3} + \ldots = \sum_{k=0}^{\infty} \gamma^k r_{t+k+1} = \sum_{k=0}^{\infty} \gamma^k \rho(x_{t+k}, h(x_{t+k})), \quad (1)$$

where $\gamma (0 \leq \gamma \leq 1)$ is the discount factor and r_t is the immediate reward at time-step t.

The state-action value function of a MDP under policy h can be defined as:

$$Q^h(x, u) = E^h \left[\sum_{t=0}^{\infty} \gamma^t r_t | x_0 = x, u_0 = u \right]$$

$$= E^h \left[\rho(x_t, u_t) + \gamma \sum_{u_{t+1} \in U} f^h(x_{t+1}, u_{t+1}) Q^h(x_{t+1}, u_{t+1}) \right]. \quad (2)$$

The optimal state-action value function is:

$$Q^*(x, u) = \max_h Q^h(x, u). \quad (3)$$

Based on the nature of the mercer kernel function, kernel function can be defined as [6]:

$$k((x, u), (x', u')) = \phi(z), \phi(z') >= \phi(z)\phi(z'). \quad (4)$$

If the state space extends to the continuous state space, the kernel-based representation of state-action value function can be defined as:

$$\hat{Q}(z) = \sum_{l=1}^{n} k_l((z), (z_l))\theta_l = k^{\top}(z)\theta, \quad (5)$$

where $k(z_i) = (k(z_1, z_i), k(z_2, z_i), ..., k(z_T, z_i))^{\top}$, $z_i = (x_i, u_i)$ denotes the state-action pair and $\theta = [\theta_1, \theta_2, ..., \theta_n]^{\top}$ denotes the parameter vector.

3   Sparse Kernel-Based Least Squares Temporal Difference with Prioritized Sweeping (PS-SKLSTD)

This section describes the PS-SKLSTD algorithm in detail. All the methods taken to improve the calculation and convergence efficiency of the algorithms will be introduced in the following sections.

3.1 Sparse Kernel-Based Least Squares Temporal Difference

The regression equation of the least squares temporal difference learning is:

$$E_0\{\phi(z_i)[Q(z_i) - \gamma Q(z_{i+1})]\} = E_0[\phi(z_i)r_i], \tag{6}$$

where $Q(z_t) = \phi^\top(z_t)W$, $W = \sum_{i=1}^{T} \phi(z_i)\theta_i$ is the weight vector, θ_i is the weighted coefficient and $\phi(z_i)$ is the feature of the observed state-action pairs. then the regression equation (6) can be converted to:

$$E_0\{\phi(z_i)[\phi^\top(z_i) - \gamma \phi^\top(z_{i+1})]\}W = E_0[\phi(z_i)r_i]. \tag{7}$$

If we use the weighted average approximation to calculate the mathematical expectation, then the regression equation (7) can be converted to:

$$\sum_{i=1}^{T}\{\phi(z_i)[\phi^\top(z_i) - \gamma\phi^\top(z_{i+1})]\}\sum_{j=1}^{T}\phi(z_j)\theta_j = \sum_{i=1}^{T}\phi(z_i)r_i, \tag{8}$$

set $\Phi_T = (\phi^\top(z_i), \phi^\top(z_2), ..., \phi^\top(z_T))^\top$. If both sides of the Eq. (8) are multiplied by Φ_T, then the equation can be converted to:

$$\sum_{i=1}^{T} k(z_i)[k^\top(z_i) - \gamma k^\top(z_{i+1})]\theta = \sum_{i=1}^{T} k(z_i)r_i. \tag{9}$$

Let

$$A_t = \sum_{i=1}^{t} k(z_i)[k^\top(z_i) - \gamma k^\top(z_{i+1})] = A_{t-1} + k(z_t)[k^\top(z_t) - \gamma k^\top(z_{t+1})], \tag{10}$$

$$b_t = \sum_{i=1}^{t} k(z_i)r_i = b_{t-1} + k(z_t)r_t, \tag{11}$$

then the coefficient vector θ can be calculated as: $\theta = A_t^{-1}b_t$.

Calculating θ needs to complete the complex matrix inverse calculation. Therefore, we use the Sherman-Morrison equation to incrementally calculate θ [9,10]. The specific form of the Sherman-Morrison equation is:

$$(C + uv^\top)^{-1} = C^{-1} - \frac{C^{-1}uv^\top C^{-1}}{1 + v^\top C^{-1}u}. \tag{12}$$

The Eq. (10) is in accordance with the Sherman-Morrison equation, then A_t^{-1} can be solved incrementally as:

$$A_t^{-1} = A_{t-1}^{-1} - \frac{A_{t-1}^{-1}k(z_t)[k(z_t) - \gamma k(z_{t+1})]^\top A_{t-1}^{-1}}{1 + [k(z_t) - \gamma k(z_{t+1})]^\top A_{t-1}^{-1}k(z_t)}, \tag{13}$$

where $\{1 + [k(z_t) - \gamma k(z_{t+1})]^\top A_{t-1}^{-1}k(z_t)\} \neq 0$.

In order to reduce the time and space complexity of the kernel calculation further. We use the approximate linear dependency (ALD) analysis to realize the kernel sparsification [11]. An approximately linearly independent data dictionary in lower dimension can be produced. The space of the basis functions $\{\phi(z_i)\}_{i=1}^{t}$ can be represented by linearly independent feature vectors. We construct a smaller ALD-based sample space D and expand D incrementally. For a new data mapping, we use the least square method to calculate the approximate linear correlation. The specific calculation process can be shown as:

$$
\begin{aligned}
\xi_t &= \min_a || \sum_j a_j\phi(x_j) - \phi(x_t)||^2 \\
&= \min_a (\sum_{i,j} a_i a_j <\phi(x_i), \phi(x_j) > -2\sum_i a_i <\phi(x_i), \phi(x_t) > \\
&\quad + <\phi(x_t), \phi(x_t) >) \\
&= \min_a (a^\top K_{t-1}a - 2a^\top k_{t-1}(x_t) + k_{tt}),
\end{aligned}
\tag{14}
$$

where $a = (a_1, a_2, ..., a_d)^\top$, $x_i(i = 1, 2, ..., d)$ is the element of the data dictionary, d is the number of elements, $[K_{t-1}]_{i,j} = k(x_i, x_j)$, $k_{tt} = k(x_t, x_t)$ and $k_{t-1}(x_t) = (k(x_1, x_t), k(x_2, x_t), ..., k(x_d, x_t))^\top$. Supposing we take the derivation of the Eq. (14), the optimal solution is:

$$
a_t = K_{t-1}^{-1}k_{t-1}(x_t), \tag{15}
$$

$$
\xi_t = k_{tt} - k_{t-1}^{-1}(x_t)c_t, \tag{16}
$$

The threshold of the sparsification is μ. If $\xi_t < \mu$, the dictionary is unchanged, else insert the new sample z_t into the dictionary. Then incrementally recursive equation of KLSTD after the sparsification is:

$$
A_t^d = A_{t-1}^d + k^d(z_t)[k^d(z_t) - \gamma k^d(z_{t+1})]^\top, \tag{17}
$$

$$
b_t^d = b_{t-1}^d + k^d(z_t)r_t, \tag{18}
$$

$$
(A_t^d)^{-1} = (A_{t-1}^d)^{-1} - \frac{(A_{t-1}^d)^{-1}k(z_t)[k(z_t) - \gamma k(z_{t+1})]^\top (A_{t-1}^d)^{-1}}{1 + [k(z_t) - \gamma k(z_{t+1})]^\top (A_{t-1}^d)^{-1}k(z_t)}, \tag{19}
$$

then the coefficient vector θ can be calculated as:

$$
\theta = (A_t^d)^{-1}b_t^d. \tag{20}
$$

3.2 Kernel-Based Prioritized Sweeping

The prioritized sweeping method determines the order of updating state-action value function by comparing the priorities of the state-action pairs. This method can speed up the convergence efficiency of the algorithm. The priorities of the state-action pairs can be calculated as:

$$
p \leftarrow |r + \gamma \max_u Q(x', u) - Q(x, u)|. \tag{21}
$$

Supposing we represent the Q-value function by kernel function approximately, likely the priorities of the state-action pairs in continuous space can be calculated as:

$$p \leftarrow |r + \gamma \max_u k^\top(z')\theta - k^\top(z)\theta|. \tag{22}$$

3.3   PS-SKLSTD Algorithm

Algorithm 1 gives the specific PS-SKLSTD algorithm. In PS-SKLSTD, the value functions of the state-action pairs are represented by ALD-based kernel function approximately. The automatic feature selection using ALD-based kernel sparsification reduces the complexity of the samples. We establish a priority queue to store the important samples and update the priorities of the samples at every time step. In addition, we start with the first sample of the priority queue to update the parameter vector of the value function without the matrix inversion.

Algorithm 1. PS-SKLSTD

1: Initialize kernel function k, sample set data $\{(x_i, u_i, r_i, x'_i, u'_i)\}(i = 1, 2, ..., n)$, matrix A, b, the threshold parameter of prioritized sweeping v, the threshold parameter of kernel sparsification μ, learning policy h with $M(M > 0)$ episodes and $PQueue$ to empty, set the parameter vector θ to 0.
2: Sparse Process:
3: Initial $i = 0$ and $D = \{\}$
4: **repeat**(for the whole data of M episodes)
5:     For all (x_i, u_i), use Eqs. (15) and (16) to compute a_i and ξ_i
6:     If $\xi_i < \mu$, $D = D$, else $D = D \cup z(x_i, u_i)$
7:     For (x'_i, u'_i), perform the same computation and update the dictionary
8: **until** complete the kernel sparsification
9: **repeat**
10:     $x \leftarrow$ current state, $u \leftarrow h(x)$
11:     Execute u, achieve r, x' and u'
12:     Calculate p of (x, u, r, x', u') according to the Eq. (22)
13:     If $p > v$,insert (x, u, r, x', u') into $PQueue$ with p
14:     **repeat**(while $PQueue$ is not empty)
15:         $(x, u, r, x', u') \leftarrow first(PQueue)$
16:         Update A, b, A^{-1}, θ according to Eqs. (17), (18), (19), (20)
17:         **repeat**(for all samples predicted to lead to current samples)
18:             Calculate p according to the Eq. (22)
19:             If $p > v$, then insert the sample into $PQueue$ with p
20:         **until** all samples has calculated their priorities
21:     **until** the number of repeats satisfies the terminal condition
22:     $x = x', u = u'$
23: **until** θ satisfies the terminal condition

---

4   Experimental Results

This section presents some experimental results on two continuous problems, Puddle World [12] and Cart-Pole [10] to confirm the effectiveness of PS-SKLSTD. Firstly we compare the convergence efficiency of PS-SKLSTD with KLSTD. Secondly we study the influence of varying prioritized sweeping parameters on the convergence efficiency of PS-SKLSTD.

4.1   Puddle World

Figure 1a shows a Puddle World. It has a goal in the upper-right corner and the goal region satisfies $x + y \geq 0.95 + 0.95$. There are two puddles which extend with radius 0.1 from two line segments, one from (0.1, 0.75) to (0.45, 0.75) and the other from (0.45, 0.4) to (0.45, 0.8). The two state variables are the x and y coordinates. The four actions are up, down, left and right which correspond to four cardinal directions. Each action moves the agent 0.05 in the indicated direction. The agent receives a -1 reward for each action outside of the two puddles. Being in a puddle agent receives a negative reward equal to 400 times the distance inside the puddle.

In this section, we firstly compare the convergence efficiency of PS-SKLSTD with KLSTD. Further We study the influence of varying ν on PS-SKLSTD. Before that all parameters set can be summarized as follows: Gaussian kernel function $k(x_i, x_j) = e^{-||x_i - x_j||^2/2\sigma^2}$, where $\sigma = 0.01$. The parameters $\gamma = 0.95$, $\varepsilon = 0.01$, $\mu = 0.001$, $\nu = 0.01$. The probability of exploration will attenuate at the 0.99 times speed in order to reduce explorations and improve exploits.

Figure 1b shows the convergence results of PS-SKLSTD and KLSTD. From Fig. 1b we can see that PS-SKLSTD converges after 131 episodes while KLSTD converges after 264 episodes. PS-SKLSTD can converge to an optimal policy more quickly than KLSTD on the Puddle World problem.

Figure 1c shows the convergence results of PS-SKLSTD when ν is 0.01, 0.05 and 0.09. From Fig. 1c we can see that PS-SKLSTD with different ν can converge, but PS-SKLSTD can converge to a optimal policy more quickly when

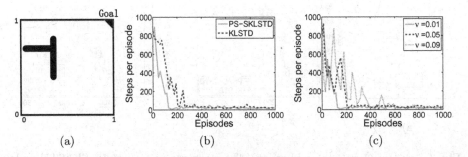

(a)   (b)   (c)

Fig. 1. (a) The Puddle World problem. (b) The steps per episode of PS-SKLSTD and KLSTD. (c) The steps per episode of PS-SKLSTD with different ν.

the ν is 0.01. The influence of varying prioritized sweeping parameters on the convergence efficiency of PS-SKLSTD is sensitive.

4.2   Cart-Pole

The Fig. 2a shows a Cart-Pole problem. From Fig. 2a we can see that a cart is placed in the horizontal orbit and a pole is fixed on the cart. There is an angle between the pole and the vertical direction. The purpose is to make the angle between the pole and the vertical direction in the range $[-\pi/4, \pi/4]$. In order to keep the balance of the pole, in every 0.1 s we need to exert a force F in the horizontal direction on the cart. There are three different actions $\{-50N, 0N, 50N\}$. The positive direction of the force is the right direction. At the same time, the horizontal direction will be disturbed by the random noise in the range $[-10N, 10N]$. The state space of the Cart-Pole is continuous. The state of the Cart-Pole problem can be represented by the χ and $\dot{\chi}$. χ denotes the angle between the pole and the vertical direction. $\dot{\chi}$ denotes the current angular velocity. The calculation equation of the angular acceleration $\ddot{\chi}$ of the pole can be shown as:

$$\ddot{\chi} = \frac{g\sin(\chi) - \eta al(\dot{\chi})^2/2 - \eta\cos(\chi)F}{4l/3 - \eta al\cos^2(\chi)}, \tag{23}$$

where g denotes gravitational acceleration ($g = 9.8\,\mathrm{m/s^2}$), a denotes the quality of the pole ($a = 2.0\,\mathrm{kg}$), b denotes the quality of the cart ($b = 8.0\,\mathrm{kg}$), l denotes the length of the pole($l = 0.5\,\mathrm{m}$) and η denotes constant ($\eta = 1/(a+b)$), $\dot{\chi} = \dot{\chi} + \ddot{\chi}\Delta t$, $\chi = \chi + \dot{\chi}\Delta t$. Episode starts with $\chi = 0$, $\dot{\chi} = 0$. When the angle of the next state is greater than $\pi/4$, the episode will end and agent will receive a -1 reward. When agent complete 3000 steps in a episode, the episode will end and agent will receive a 1 reward.

In this section, we firstly compare the convergence efficiency of PS-SKLSTD with KLSTD. Further we study the effect of varying ν on PS-SKLSTD. Before that all parameters set can be summarized as follows: Gaussian kernel function $k(x_i, x_j) = e^{-||x_i - x_j||^2/2\sigma^2}$, where $\sigma = 10$. The parameters $\gamma = 0.9$, $\varepsilon = 0.1$,

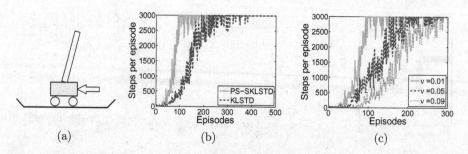

Fig. 2. (a) The Cart-Pole problem. (b) The steps per episode of PS-SKLSTD and KLSTD. (c) The steps per episode of PS-SKLSTD with different ν.

$\mu = 0.001$, $\nu = 0.01$. The probability of exploration will attenuate at the 0.99 times speed in order to reduce explorations and improve exploits.

Figure 2b shows the convergence results of PS-SKLSTD and KLSTD. From Fig. 2b we can see that PS-SKLSTD firstly completes 3000 steps in 92th episode and converges after 258 episodes. However, KLSTD firstly completes 3000 steps in 199th episode and converges after 384 episodes. PS-SKLSTD can converge to an optimal policy more quickly than KLSTD on the Cart-Pole problem.

Figure 2c shows the convergence results of PS-SKLSTD when ν is 0.01, 0.05 and 0.09. From Fig. 2c we can see that when ν is 0.09, PS-SKLSTD firstly completes 3000 steps in 224th episode, but can not converge in the first 300 episodes. When ν is 0.05, PS-SKLSTD firstly completes 3000 steps in 153th episode and converges after 283 episodes. When ν is 0.01, PS-SKLSTD firstly completes 3000 steps in 92th episode and converges after 258 episodes. In conclusion, the influence of varying prioritized sweeping parameters on the convergence efficiency of PS-SKLSTD is sensitive.

5   Conclusions

This paper proposes the PS-SKLSTD algorithm which combines the learning process with the planning process to solve the continuous RL problems efficiently. In the learning process, the ALD-based kernel sparsification reduces the complexity of the samples. When update the parameter vector, the Sherman-Morrison equation used to replace the matrix inverse calculation reduces the computational complexity. In the planning process, prioritized sweeping method helps the algorithm more efficiently selecting the current updated samples. The experimental results on the Puddle World and Cart-Pole demonstrate that PS-SKLSTD has better performance than KLSTD in terms of convergence and calculation. The threshold parameter of prioritized sweeping effects the convergence efficiency of PS-SKLSTD sensitively.

Acknowledgments. This paper was funded by National Natural Science Foundation (61103045, 61272005, 61272244, 61303108, 61373094, 61472262). Natural Science Foundation of Jiangsu (BK2012616), High School Natural Foundation of Jiangsu (13KJB520020), Key Laboratory of Symbolic Computation and Knowledge Engineering of Ministry of Education, Jilin University (93K172014K04), Suzhou Industrial application of basic research program part (SYG201422).

References

1. Sutton, R.S., Barto, A.G.: Reinforcement Learning: An Introduction. MIT Press, Cambridge (1998)
2. Busoniu, L., Babuska, R., Schutter, B.D., Ernst, D.: Reinforcement Learning and Dynamic Programming Using Function Approximators. CRC Press, Boca Raton (2010)
3. Wiering, M., van Otterlo, M.: Reinforcement learning: state-of-the-art. Phillip Journal Fr Restaurative Zahnmedizin (2012)

4. van Hasselt, H.: Reinforcement learning in continuous state and action spaces. In: Wiering, M., van Otterlo, M. (eds.) Reinforcement Learning. ALO, vol. 12, pp. 205–248. Springer, Heidelberg (2012)
5. Xu, X., Xie, X., Hu, D.: Kernel least-squares temporal difference learning. Int. J. Inf. Technol. **11**(9), 54–63 (2005)
6. Xu, X., Hu, D.: Kernel-based least squares policy iteration for reinforcement learning. IEEE Trans. Neural Netw. **18**(4), 973–992 (2007)
7. Moore, A.W., Atkeson, C.G.: Prioritized sweeping: reinforcement learning with less data and less time. Mach. Learn. **13**(1), 103–130 (1993)
8. Sutton, R.S., Szepesvari, C., Geramifard, A., Bowling, M.P.: Dyna-style planning with linear function approximation and prioritized sweeping. In: Conference on Uncertainty in Artificial Intelligence (2008)
9. Lagoudakis, M.G., Parr, R.: Least-squares policy iteration. J. Mach. Learn. Res. **4**(6), 1107–1149 (2010)
10. Liu, Q., Zhou, X., Zhu, F., Fu, Q., Fu, Y.: Experience replay for least-squares policy iteration. IEEE/CAA J. Autom. Sin. **1**(3), 274–281 (2014). IEEE
11. Xu, X.: A sparse kernel-based least-squares temporal difference algorithm for reinforcement learning. In: Jiao, L., Wang, L., Gao, X., Liu, J., Wu, F. (eds.) ICNC 2006. LNCS, vol. 4221, pp. 47–56. Springer, Heidelberg (2006). doi:10.1007/11881070_8
12. Jong, N., Stone, P.: Kernel-based models for reinforcement learning. In: ICML Workshop on Kernel Machines and Reinforcement Learning (2006)
13. Engel, Y., Mannor, S., Meir, R.: Bayes meets Bellman: the Gaussian process approach to temporal difference learning. In: 20th International Conference on Machine Learning, pp. 154–161. American Association for Artificial Intelligence (2003)
14. Lazaric, A., Ghavamzadeh, M., Munos, R.: Finite-sample analysis of least-squares policy iteration. J. Mach. Learn. Res. **13**(1), 3041–3074 (2012). Microtome Publishing

Computational Intelligence

Vietnamese POS Tagging for Social Media Text

Ngo Xuan Bach[1,2(✉)], Nguyen Dieu Linh[1], and Tu Minh Phuong[1,2]

[1] Department of Computer Science, Posts and Telecommunications Institute
of Technology, Hanoi, Vietnam
{bachnx,ndlinh,phuongtm}@ptit.edu.vn
[2] FPT Software Research Lab, Hanoi, Vietnam

Abstract. This paper presents an empirical study on Vietnamese part-of-speech (POS) tagging for social media text, which shows several challenges compared with tagging for general text. Social media text does not always conform to formal grammars and correct spelling. It also uses abbreviations, foreign words, and icons frequently. A POS tagger developed for conventional, edited text would perform poorly on such noisy data. We address this problem by proposing a tagging model based on Conditional random fields with various kinds of features for Vietnamese social media text. We introduce a corpus for POS tagging, which consists of more than four thousands sentences from Facebook, the most popular social network in Vietnam. Using this corpus, we performed a series of experiments to evaluate the proposed model. Our model achieved 88.26 % tagging accuracy, which is 11.27 % improvement over a state-of-the-art Vietnamese POS tagger developed for general, conventional text.

Keywords: Part-of-Speech tagging · Social media text · Conditional Random Fields

1 Introduction

Part-of-speech (POS) tagging is a fundamental task in natural language processing (NLP), which provides useful information not only to other NLP problems such as syntactic parsing, semantic role labeling, and semantic parsing but also to NLP applications, including information extraction, question answering, and machine translation. The goal of this task is to assign a POS tag (such as "noun", "verb", etc.) to each word in a sentence depending on the word's context.

POS tagging for general text has been studied intensively for several decades, not only for English [4,18,20,21] but also other languages, including Japanese [11,12], Chinese [9,25], Arabic [2], and Vietnamese [3,16,22,23]. Most state-of-the-art POS taggers are statistical or machine learning based models trained on annotated corpora of conventional text, which usually consist of news articles, such as the Wall Street Journal corpus of the Penn Treebank [10] for English, the Kyoto corpus [6] for Japanese, and the Viet Treebank corpus [15] for Vietnamese.

Web 2.0 platforms such as blogs, forums, wikis, and social networks have facilitated the generation of a huge volume of user-generated text. These data have

© Springer International Publishing AG 2016
A. Hirose et al. (Eds.): ICONIP 2016, Part III, LNCS 9949, pp. 233–242, 2016.
DOI: 10.1007/978-3-319-46675-0_26

#	Facebook sentences	Expected sentences	Translated sentences	Problems
1	em đọc đc ấy mà a	em đọc được ấy mà anh	I can read it	abbreviation
2	nó good vậyyy	nó giỏi vậy	He is so good	foreign word, typo
3	Ng đàn ông mặc áo đen cơ. :))	Người đàn ông mặc áo đen cơ. :))	Must be the guy with a black shirt	abbreviation, emoticon
4	toi thich cái màu trắng	tôi thích cái màu trắng	I like the white one	word without tone mark
5	Tks chị nhìu	Cảm ơn chị nhiều	Thank you very much	abbreviation, foreign word, typo

Fig. 1. Examples of sentences in Vietnamese Facebook

become an important source for both data mining and NLP communities, and at the same time require appropriate tools for text analysis. Although available POS taggers can achieve high accuracy on conventional data, the performance usually degrades on noisy, unconventional text generated by social users. Social media text in Vietnamese poses even more challenges. It is not always written conforming to the formal grammar and correct spelling. As shown in the examples in Fig. 1, it also uses abbreviations, wrong capital letters, foreign words (usually English), words without tone marks, typos, and icons frequently.

In this paper, we consider the task of Vietnamese POS tagging for social media text. We describe a tagging model using Conditional random fields (CRFs), and explore various kinds of features specially designed for this type of text. To train the model and compare it again existing taggers, we construct a corpus with more than four thousand sentences from Facebook. A series of experiments were conducted to explore the challenges that social media text poses to conventional taggers, and verify the effectiveness of our method. To the best of our knowledge, this is the first attempt to deal with Vietnamese POS tagging for social media text.

Our contributions are as follows:

- We have developed a tagging model with various types of linguistic features and a new POS tagset, and empirically shown the effectiveness of the method on data from Vietnamese Facebook.
- We have constructed an annotated corpus for Vietnamese POS tagging with 4150 sentences collected from Facebook. Both annotated corpus and trained POS tagger are made available to the research community (from the first author).

The rest of this paper is structured as follows. Section 2 describes related work. Section 3 introduces our dataset for Vietnamese POS tagging for Facebook. Section 4 presents our tagging method. Section 5 describes experiments and error analysis. Finally, Sect. 6 concludes the paper.

2 Related Work

Several studies have been conducted for POS tagging for social media text. Gimpel et al. [5] present a study on POS tagging for Twitter in English. Their method uses CRFs and achieves 89.37 % tagging accuracy on a corpus consisting of 1,827 tweets. Owoputi et al. [17] extract extra word features from word clusters to improve POS tagging for online conversational text. For English Twitter POS tagging, their system achieves an accuracy improvement of more than 3 % over a baseline by learning word clusters from 56 million tweets.

Among the work on POS tagging for social media text for languages other than English, Albogamy and Ramsay [1] conduct an empirical study on POS Tagging for Arabic Tweets. They report 79 % accuracy on a corpus consisting of 390 Arabic tweets. Neunerdt et al. [13] investigate the task for German social media text. They evaluate four state-of-the-art German taggers, i.e. TreeTagger, TnT, Stanford, and SVMTool [13] on multiple types of data sources, including chat messages, blog comments, Merkur comments, and YouTube comments. Vyas and Gella [24] consider the task of POS tagging of English-Hindi code-mixed social media content. They build an annotated corpus of Hindi-English code-mixed text collated from Facebook forums, and explore language identification, back-transliteration, normalization and POS tagging of this data.

In Vietnamese, most studies on POS tagging focus on general text. They usually consider the task as a sequence labeling problem, and various kinds of learning models have been investigated, including Maximum entropy models [8,22,23], Support vector machines [14], Conditional random fields [16,22], Guided online learning [16], and Dual decomposition [3]. These methods achieved relatively good results on the Vietnamese POS tagging task. Unlike previous work, the aim of our work is to build a Vietnamese POS tagger for social media text.

3 Dataset

To build the corpus, we extracted textual content of posts and comments from Vietnamese Facebook. By "Vietnamese Facebook" we mean Facebook posts and comments, whose majority language is Vietnamese. The data were collected from friends of several users with different professions and ages to provide better coverage and diversity. The statistics of the data are shown in Table 1.

The first analysis of the extracted text shows that it contains a number of deviations from conventional text, including abbreviations, wrong capital letters, foreign language words, words without tone marks, typos, and icons. A traditional POS tagset for general text does not contain specific tags for such kinds of words. To build an annotated corpus from such data, we first extended the conventional POS tagset to cover those variations. Figure 2 lists our POS tagset, most of them are inherited from the tagset of vnTagger [8], the most popular and a state-of-the-art Vietnamese POS tagger trained on Viet TreeBank [15]. We added eight tags for specific purposes, including AB for abbreviation words, FL for foreign words, five tags for emoticons i.e. AR (angry), CF (confuse), HP (happy), IL (involved), SD (sad), and PUN for punctuation.

Old tags from vnTagger				New tags	
Tag	Description	Tag	Description	Tag	Description
Np	Proper noun	L	Determiner	AB	Abbreviation
Nc	Classifier	M	Numeral	FL	Foreign language
Nu	Unit noun	E	Preposition	AR	Angry (emoticon)
N	Common noun	C	Subordinating conjunction	CF	Confused (emoticon)
V	Verb	CC	Coordinating conjunction	HP	Happy (emoticon)
A	Adjective	I	Interjection	IL	Involved (emoticon)
P	Pronoun	T	Auxiliary, modal words	SD	Sad (emoticon)
R	Adverb	X	Unknown	PUN	Punctuation

Fig. 2. A POS tagset for Vietnamese social media text

Table 1. Corpus statistical information

#of sentences	#of unique words	#of tokens	#of tokens/sentence
4150	6416	38498	9.3

Next, to annotate the data, we performed the following steps:

- **Step 0: Preprocessing**. The following preprocessing steps were performed: divide status and comments into sentences; remove very short sentences such as the ones consisting of only icons; remove sentences consisting of all typos or grammar errors; remove sentences with wrong font format; and replace icons with @ character such as @@ or @_@ by special symbol (?).
- **Step 1: Automatic Tagging**. Preprocessed sentences were then tokenized and tagged by using vnTagger. This is done to speed up the manual tagging in the next step.
- **Step 2: Manual Tagging**. In this step, sentences with tags from step 2 were manually checked and corrected by two annotators separately. Sentences with disagreement between two annotators were examined and corrected by the third annotator.

To measure the inter-annotator agreement, we used the Cohen's kappa coefficient as follows:

$$K = \frac{Pr(a) - Pr(e)}{1 - Pr(e)}$$

where $Pr(a)$ is the relative observed agreement between two annotators, and $Pr(e)$ is the hypothetical probability of chance agreement. The Cohen's kappa coefficient of our corpus was 0.84, which can be interpreted as almost perfect agreement. The statistics of annotated corpus are shown in Table 2.

Table 2. Statistical information about the tagset

Tag	Appearances	Percentage	Tag	Appearances	Percentage
V	7390	19.20	HP	910	2.36
N	7105	18.46	X	674	1.75
PUN	3409	8.86	I	595	1.55
R	3097	8.04	FL	505	1.31
A	2572	6.68	Nc	444	1.15
P	2266	5.89	L	357	0.93
AB	2014	5.23	CF	303	0.79
E	1461	3.80	CC	190	0.49
Np	1361	3.54	Nu	125	0.32
T	1355	3.52	SD	114	0.30
C	1222	3.17	IL	82	0.21
M	926	2.41	AR	21	0.05

4   Tagging Method

Our tagging model employs Conditional random fields (CRFs) [7], a powerful and efficient framework for sequence labeling tasks, such as POS tagging, named entity recognition, information retrieval, and information extraction [7,19]. It also has been shown to be an effective model for building a POS tagger for social media text [5,24].

Our model encodes various types of features to use with CRFs as follows:

– **Basic features:** Basic features consist of unigrams, bigrams, and trigrams of words extracted in the window of size 5 centered at the current word.
– **Enhanced features:** Enhanced features include: a feature that checks whether the word is a special character (hyphen, punctuation, mathematical operation, dash, and so on); a feature that checks whether the word is an icon or emoticon; a feature that detects whether the word contains digits; features looking at capitalization patterns in the word; and features that detect hashtags and URLs.
– **METAPH feature:** Similar to the work of Gimpel et al. [5], we used the Metaphone algorithm[1] to create a coarse phonetic normalization of words to simpler keys. Metaphone consists of 19 rules that rewrite consonants and delete vowels.
– **GENTAG features:** We also use the output (the predicted POS tags) of vnTagger[2], a Vietnamese POS tagger trained on general text, as extra features. Similar to the word features, GENTAG features include unigrams, bigrams, and trigrams of POS tags extracted in windows of size 5.

---

[1] http://commons.apache.org/codec/.
[2] http://mim.hus.vnu.edu.vn/phuonglh/softwares/vnTagger.

5   Experiments

5.1   Experimental Setting

We evaluated the effectiveness of the proposed model using 10-fold cross-validation. The performance of tagging models were measured using accuracy, precision, recall, and the F_1 score. Accuracy was computed over all kinds of POS tags as follows:

$$Accuracy = \frac{\#\text{of words correctly tagged}}{\#\text{of words}}.$$

Precision, recall, and the F_1 score were computed for each kind of POS tag. Let's consider an example of tag N (noun). Suppose that A and B are the set of words that the system predicted as N and the set of words labeled as N by human respectively. The precision, recall, and the F_1 score for tag N can be computed as follows (similarly for other tags):

$$Precision = \frac{|A \cap B|}{|A|}, Recall = \frac{|A \cap B|}{|B|}, \text{and } F_1 = \frac{2 * Precision * Recall}{Precision + Recall}.$$

5.2   Models to Compare

We conducted experiments to compare the performances of the following models:

- **Baseline1:** We used the output of vnTagger as the first baseline. The purpose of this experiment was to evaluate the performance of a state-of-the-art POS tagger trained on general text for the task.
- **Baseline2:** Because vnTagger uses a POS tagset different from ours, it cannot assign correctly the new tags in our tagset. Our second baseline was an improvement of vnTagger, in which we used a list of icons[3] to automatically correct the output of vnTagger.
- **CRF1:** This model used CRFs with basic features.
- **CRF2:** This model was similar to CRF1, but adding the enhanced feature set.
- **CRF3:** This model was similar to CRF2, but adding the METAPH feature.
- **CRF4:** This model was similar to CRF3, but adding the output of Baseline1 as features.
- **CRF5:** This model was similar to CRF3, but adding the output of Baseline2 as features.

---

[3] We collected the list from two links:
http://kenh76.vn/ki-tu-ky-hieu-bieu-tuong-tren-facebook-chat-cap-nhat-2014.html, and https://en.wikipedia.org/wiki/List_of_emoticons.

5.3   Results

Table 3 shows experimental results of the tagging models. Two baselines achieved 76.99 % and 80.69 % accuracies, respectively, which are much lower than the reported accuracy of vnTagger on general text (which is higher than 90 %) [8]. These results show the need for the tagging models that take into account the noisy nature of social media text. The superiority of baseline2 also shows the importance of processing icons separately. It is not surprising, given the popularity of icons in user-generated text.

Our models using CRFs outperformed both baselines, in some cases by large margins: our best model achieved 88.26 % accuracy. Even the CRFs model with basic features trained on social media text (CRF1) performed better than a state-of-the-art POS tagger trained on general text. Adding enhanced features further improved the accuracy of the tagger. Finally, the output of vnTagger, although not as accurate as for general text, provides important information to boost the accuracy when used with other features within the CRF models.

Table 3. Accuracies of different tagging models

Model	Baseline1	Baseline2	CRF1	CRF2	CRF3	CRF4	CRF5
Accuracy(%)	76.99	80.69	81.39	83.53	83.86	88.16	88.26

Tagging performance of CRF5 model for different types of tags is shown in Table 4. The system achieved F_1 score of 100 % for all icons' tags (HP, SD, AR, CF, and IL). This is reasonable because we have a list of icons and have a feature to check if a word is in that list or not. Other tags with a high F_1 score are closed-set words, including punctuation (PUN, 99.71 %), coordinating conjunctions (CC, 98.41 %), and prepositions (E, 96.14 %). On the other side, most of tags with a low F_1 score are open set words, including proper nouns (Np, 76.86 %), foreign language words (FL, 52.45 %), and unknown words (X, 46.93 %). By analyzing the dataset, we found that many proper nouns were written without capital letters and most unknown words were typos. It was difficult to predict correct labels for foreign language words. They can be seen as typos or proper nouns in some cases.

Error Analysis. In POS tagging, it is important to analyze the nature of wrong tags: which tags are most often confused with each other. For example, we want to know which tag was the most frequently (wrongly) assigned to proper nouns. In Table 5, we summarize the most frequent confused tag pairs for each kind of words. The most frequent errors were assigning V and N to non-verbs and non-nouns respectively. Notice that V and N are two most popular tags in the corpus, with 19.20 % and 18.46 %, respectively. Another typical case is that proper nouns not written with capitalization, were confused with nouns.

Table 4. Tagging performance for individual types of POS tags

Tag	Precision(%)	Recall(%)	F_1(%)	Tag	Precision(%)	Recall(%)	F_1(%)
Np	78.18	75.58	76.86	CC	98.86	97.97	98.41
Nc	90.13	79.86	84.68	I	83.78	61.37	70.85
Nu	84.89	80.62	82.70	T	82.97	73.02	77.68
N	80.70	93.19	86.50	X	82.18	32.84	46.93
V	84.22	94.27	88.96	PUN	99.48	99.94	99.71
A	86.14	79.25	82.56	AB	95.47	79.58	86.80
P	93.90	90.87	92.36	FL	88.31	37.30	52.45
R	93.97	88.31	91.05	AR	100.0	100.0	100.0
L	98.15	88.57	93.12	CF	100.0	100.0	100.0
M	95.52	96.40	95.96	HP	100.0	100.0	100.0
E	97.57	94.75	96.14	IL	100.0	100.0	100.0
C	90.34	89.96	90.15	SD	100.0	100.0	100.0

Table 5. Statistical information on confused tags

Tag	Recall(%)	Confused	Tag	Recall(%)	Confused
Np	75.58	N	E	94.75	N
Nc	79.86	N	C	89.96	T
Nu	80.62	N	CC	97.97	V
N	93.19	V	I	61.37	N
V	94.27	N	T	73.02	P
A	79.25	V	X	32.84	N
P	90.87	N	PUN	99.94	M
R	88.31	V	AB	79.58	N
L	88.57	N	FL	37.30	N
M	96.40	N			

6 Conclusion

We have developed a Vietnamese part-of-speech tagging system for social media text. We showed that our tagger, which incorporates various types of linguistic features in a CRF framework, outperformed a state-of-the-art Vietnamese POS tagger trained on general text by a large margin. We believe that our tagger as well as the annotated data can be useful for further research not only on POS tagging but also other NLP tasks for Vietnamese social media text.

Acknowledgements. This work was partially supported by "2016 PTIT Research Grant", Posts and Telecommunications Institute of Technology, Vietnam. We also would like to thank FPT for financial support which made this work possible.

References

1. Albogamy, F., Ramsay, A.: POS tagging for Arabic tweets. In: Proceedings of RANLP, pp. 1–8 (2015)
2. Aldarmaki, H., Diab, M.: Robust part-of-speech tagging of Arabic text. In: Proceedings of the 2nd Workshop on Arabic NLP, pp. 173–182 (2015)
3. Bach, N.X., Hiraishi, K., Minh, N.L., Shimazu, A.: Dual decomposition for Vietnamese part-of-speech tagging. In: Proceedings of KES, pp. 123–131 (2013)
4. Brill, E.: Transformation-based error-driven learning and natural language processing: a case study in part of speech tagging. Comput. Linguist. **21**(4), 543–565 (1995)
5. Gimpel, K., Schneider, N., O'Connor, B., Das, D., Mills, D., Eisenstein, J., Heilman, M., Yogatama, D., Flanigan, J., Smith, N.A.: Part-of-speech tagging for twitter: annotation, features, and experiments. In: Proceedings of ACL, pp. 42–47 (2011)
6. Kawahara, D., Kurohashi, S., Hasida, K.: Construction of a Japanese relevance-tagged corpus. In: Proceedings of LREC, pp. 2008–2013 (2002)
7. Lafferty, J., McCallum, A., Pereira, F.: Conditional random fields: probabilistic models for segmenting and labeling sequence data. In: Proceedings of ICML, pp. 282–289 (2001)
8. Le, H.P., Roussanaly, A., Nguyen, T.M.H., Rossignol, M.: An empirical study of maximum entropy approach for part-of-speech tagging of Vietnamese texts. In: Proceedings of TALN (2010)
9. Li, Z., Zhang, M., Che, W., Liu, T., Chen, W., Li, H.: Joint models for Chinese POS tagging and dependency parsing. In: Proceedings of EMNLP, pp. 1180–1191 (2011)
10. Marcus, M.P., Marcinkiewicz, M.A., Santorini, B.: Building a large annotated corpus of English: the Penn Treebank. Comput. Linguist. **19**(2), 313–330 (1993)
11. Nakagawa, T., Kudo, T., Matsumoto, Y.: Revision learning and its application to part-of-speech tagging. In: Proceedings of ACL, pp. 497–450 (2002)
12. Nakagawa, T., Uchimoto, K.: A hybrid approach to word segmentation and POS tagging. In: Proceedings of ACL, pp. 217–220 (2007)
13. Neunerdt, M., Trevisan, B., Reyer, M., Mathar, R.: Part-of-speech tagging for social media texts. In: Gurevych, I., Biemann, C., Zesch, T. (eds.) GSCL. LNCS, vol. 8105, pp. 139–150. Springer, Heidelberg (2013)
14. Nghiem, M., Dinh, D., Nguyen, M.: Improving Vietnamese POS tagging by integrating a rich feature set and support vector machines. In: Proceedings of RIVF, pp. 128–133 (2008)
15. Nguyen, P.T., Vu, X.L., Nguyen, T.M.H., Nguyen, V.H., Le, H.P.: Building a large syntactically-annotated corpus of Vietnamese. In: Proceedings of the Third Linguistic Annotation Workshop, ACL-IJCNLP, pp. 182–185 (2009)
16. Nguyen, L.M., Xuan, B.N., Viet, C.N., Nhat, M.P.Q., Shimazu, A.: A semi-supervised learning method for Vietnamese part-of-speech tagging. In: Proceedings of KSE, pp. 141–146 (2010)

17. Owoputi, O., O'Connor, B., Dyer, C., Gimpel, K., Schneider, N., Smith, N.A.: Improved part-of-speech tagging for online conversational text with word clusters. In: Proceedings of NAACL, pp. 380–390 (2013)
18. Ratnaparkhi, A.: A maximum entropy model for part-of-speech tagging. In: Proceedings of EMNLP, pp. 133–142 (1996)
19. Sha, F.P.: Shallow parsing with conditional random fields. In: Proceedings of NAACL, pp. 213–220 (2003)
20. Toutanova, K., Manning, C.: Enriching the knowledge sources used in a maximum entropy part-of-speech tagger. In: Proceedings of EMNLP, pp. 63–70 (2000)
21. Toutanova, K., Klein, D., Manning, C., Singer, Y.: Feature-rich part-of-speech tagging with a cyclic dependency network. In: Proceedings of NAACL, pp. 252–259 (2003)
22. Tran, T.O., Le, A.C., Ha, Q.T., Le, H.Q.: An experimental study on Vietnamese POS tagging. In: Proceedings of IALP, pp. 23–27 (2009)
23. Tran, T.O., Le, A.C., Ha, Q.T.: Improving Vietnamese word segmentation and POS tagging using MEM with various kinds of resources. J. Nat. Lang. Process. **17**(3), 41–60 (2010)
24. Vyas, Y., Gella, S.: POS tagging of English-Hindi code-mixed social media content. In: Proceedings of EMNLP, pp. 974–979 (2014)
25. Zheng, X., Chen, H., Xu, T.: Deep learning for Chinese word segmentation and POS tagging. In: Proceedings of EMNLP, pp. 647–657 (2013)

Scaled Conjugate Gradient Learning for Quaternion-Valued Neural Networks

Călin-Adrian Popa[✉]

Department of Computer and Software Engineering,
Polytechnic University Timişoara,
Blvd. V. Pârvan, No. 2, 300223 Timişoara, Romania
calin.popa@cs.upt.ro

Abstract. This paper presents the deduction of the scaled conjugate gradient method for training quaternion-valued feedforward neural networks, using the framework of the HR calculus. The performances of the scaled conjugate algorithm in the real- and complex-valued cases lead to the idea of extending it to the quaternion domain, also. Experiments done using the proposed training method on time series prediction applications showed a significant performance improvement over the quaternion gradient descent and quaternion conjugate gradient algorithms.

Keywords: Quaternion-valued neural networks · Scaled conjugate gradient algorithm · Time series prediction

1 Introduction

The domain of quaternion-valued neural networks has received an increasing interest over the last few years. Some popular applications of these networks include chaotic time-series prediction [2], color image compression [8], color night vision [11], polarized signal classification [4], and 3D wind forecasting [9,20,21].

Some signals in the 3D and 4D domains can more naturally be expressed in quaternion-valued form. Thus, these networks appear as a natural choice for solving problems such as time series prediction. Several methods have been proposed to increase the efficiency of learning in quaternion-valued neural networks. These methods include different network architectures and different learning algorithms, some of which are specially designed for this type of networks, while others are extended from the real-valued case.

The scaled conjugate gradient learning method has been proven to be very efficient in the real-valued and complex-valued [15] cases. First proposed by [13], the scaled conjugate gradient method represents a popular algorithm for training feedforward neural networks. Taking these facts into account, it seems natural to extend this learning algorithm to quaternion-valued neural networks, also.

In this paper, we deduce the quaternion-valued scaled conjugate gradient algorithm starting from the real-valued case, by using the framework of the HR calculus. We test the proposed scaled conjugate gradient algorithm on linear and chaotic time series prediction problems.

© Springer International Publishing AG 2016
A. Hirose et al. (Eds.): ICONIP 2016, Part III, LNCS 9949, pp. 243–252, 2016.
DOI: 10.1007/978-3-319-46675-0_27

The remainder of this paper is organized as follows: Sect. 2 introduces the $\mathbb{HR}$ calculus, which is a type of calculus used for extending real-valued algorithms to the quaternion-valued domain. Section 3 gives the derivation of the quaternion-valued conjugate gradient algorithms. Then, Sect. 4 presents the scaled conjugate algorithm for quaternion-valued feedforward neural networks. The experimental results of the three applications of the proposed algorithms are discussed in Sect. 5, together with a description of the nature of each problem. Section 6 is dedicated to presenting the conclusions of the study.

2 The $\mathbb{HR}$ Calculus

First, we will present the basic ideas of the $\mathbb{HR}$ calculus [23], which will be later used to deduce the conjugate gradient and scaled conjugate gradient algorithms for optimizing a quaternion domain error function.

Let $\mathbb{H} = \{q_a + iq_b + jq_c + kq_d | q_a, q_b, q_c, q_d \in \mathbb{R}\}$ be the algebra of quaternions, where i, j, k are the imaginary units which satisfy $i^2 = j^2 = k^2 = ijk = -1$. We define the operation $q^\mu := \mu q \mu^{-1}$, for any $\mu \in \mathbb{H}$. Using this operation, for any $q = q_a + iq_b + jq_c + kq_d \in \mathbb{H}$, we have $q^i = iqi^{-1} = q_a + iq_b - jq_c - kq_d$, $q^j = jqj^{-1} = q_a - iq_b + jq_c - kq_d$, $q^k = kqk^{-1} = q_a - iq_b - jq_c + kq_d$. For a function $f : \mathbb{H} \to \mathbb{H}$, we can define the $\mathbb{HR}$ derivatives of f by

$$
\begin{pmatrix} \frac{\partial f}{\partial q} \\ \frac{\partial f}{\partial q^i} \\ \frac{\partial f}{\partial q^j} \\ \frac{\partial f}{\partial q^k} \end{pmatrix} := \frac{1}{4} \begin{pmatrix} 1 & -i & -j & -k \\ 1 & -i & j & k \\ 1 & i & -j & k \\ 1 & i & j & -k \end{pmatrix} \begin{pmatrix} \frac{\partial f}{\partial q_a} \\ \frac{\partial f}{\partial q_b} \\ \frac{\partial f}{\partial q_c} \\ \frac{\partial f}{\partial q_d} \end{pmatrix}.
$$

Consider now a quaternion vector $\mathbf{q} = (q_1, q_2, \ldots, q_N)^T \in \mathbb{H}^N$, which can be written as $\mathbf{q} = \mathbf{q}_a + i\mathbf{q}_b + j\mathbf{q}_c + k\mathbf{q}_d \in \mathbb{H}^N$, where $\mathbf{q}_a, \mathbf{q}_b, \mathbf{q}_c, \mathbf{q}_d \in \mathbb{R}^N$. We have $\mathbf{q}^i = i\mathbf{q}i^{-1} = \mathbf{q}_a + i\mathbf{q}_b - j\mathbf{q}_c - k\mathbf{q}_d$, $\mathbf{q}^j = j\mathbf{q}j^{-1} = \mathbf{q}_a - i\mathbf{q}_b + j\mathbf{q}_c - k\mathbf{q}_d$, $\mathbf{q}^k = k\mathbf{q}k^{-1} = \mathbf{q}_a - i\mathbf{q}_b - j\mathbf{q}_c + k\mathbf{q}_d \in \mathbb{H}^N$, or, equivalently,

$$
\begin{pmatrix} \mathbf{q} \\ \mathbf{q}^i \\ \mathbf{q}^j \\ \mathbf{q}^k \end{pmatrix} = \begin{pmatrix} \mathbf{I}_N & i\mathbf{I}_N & j\mathbf{I}_N & k\mathbf{I}_N \\ \mathbf{I}_N & i\mathbf{I}_N & -j\mathbf{I}_N & -k\mathbf{I}_N \\ \mathbf{I}_N & -i\mathbf{I}_N & j\mathbf{I}_N & -k\mathbf{I}_N \\ \mathbf{I}_N & -i\mathbf{I}_N & -j\mathbf{I}_N & k\mathbf{I}_N \end{pmatrix} \begin{pmatrix} \mathbf{q}_a \\ \mathbf{q}_b \\ \mathbf{q}_c \\ \mathbf{q}_d \end{pmatrix},
$$

where $\mathbf{I}_N$ is the $N \times N$ identity matrix. By denoting

$$
\overset{\mathcal{H}}{\mathbf{q}} := \begin{pmatrix} \mathbf{q} \\ \mathbf{q}^i \\ \mathbf{q}^j \\ \mathbf{q}^k \end{pmatrix} \in \mathbb{H}^{4N}, \quad \overset{\mathcal{R}}{\mathbf{q}} := \begin{pmatrix} \mathbf{q}_a \\ \mathbf{q}_b \\ \mathbf{q}_c \\ \mathbf{q}_d \end{pmatrix} \in \mathbb{R}^{4N}, \quad \mathbf{J} := \begin{pmatrix} \mathbf{I}_N & i\mathbf{I}_N & j\mathbf{I}_N & k\mathbf{I}_N \\ \mathbf{I}_N & i\mathbf{I}_N & -j\mathbf{I}_N & -k\mathbf{I}_N \\ \mathbf{I}_N & -i\mathbf{I}_N & j\mathbf{I}_N & -k\mathbf{I}_N \\ \mathbf{I}_N & -i\mathbf{I}_N & -j\mathbf{I}_N & k\mathbf{I}_N \end{pmatrix},
$$

the above relation becomes

$$
\overset{\mathcal{H}}{\mathbf{q}} = \mathbf{J} \overset{\mathcal{R}}{\mathbf{q}}.
$$

It can be checked that $\mathbf{J}^H\mathbf{J} = \mathbf{J}\mathbf{J}^H = 4\mathbf{I}_{4N}$, and so we deduce that

$$\overset{\mathcal{R}}{\mathbf{q}} = \frac{1}{4}\mathbf{J}^H\overset{\mathcal{H}}{\mathbf{q}}. \tag{1}$$

A function $f : \mathbb{H}^N \to \mathbb{R}$ can now be seen in three equivalent forms

$$f(\mathbf{q}) \Leftrightarrow f(\overset{\mathcal{H}}{\mathbf{q}}) := f(\mathbf{q}, \mathbf{q}^i, \mathbf{q}^j, \mathbf{q}^k) \Leftrightarrow f(\overset{\mathcal{R}}{\mathbf{q}}) := f(\mathbf{q}_a, \mathbf{q}_b, \mathbf{q}_c, \mathbf{q}_d).$$

If we define

$$\frac{\partial f}{\partial \mathbf{q}} := \left(\frac{\partial f}{\partial q_1}, \dots, \frac{\partial f}{\partial q_N} \right),$$

$$\frac{\partial f}{\partial \overset{\mathcal{H}}{\mathbf{q}}} := \left(\frac{\partial f}{\partial \mathbf{q}}, \frac{\partial f}{\partial \mathbf{q}^i}, \frac{\partial f}{\partial \mathbf{q}^j}, \frac{\partial f}{\partial \mathbf{q}^k} \right),$$

$$\frac{\partial f}{\partial \overset{\mathcal{R}}{\mathbf{q}}} := \left(\frac{\partial f}{\partial \mathbf{q}_a}, \frac{\partial f}{\partial \mathbf{q}_b}, \frac{\partial f}{\partial \mathbf{q}_c}, \frac{\partial f}{\partial \mathbf{q}_d} \right),$$

we have, from the chain rule, that

$$\frac{\partial f}{\partial \overset{\mathcal{H}}{\mathbf{q}}} = \frac{1}{4}\frac{\partial f}{\partial \overset{\mathcal{R}}{\mathbf{q}}}\mathbf{J}^H \Leftrightarrow \frac{\partial f}{\partial \overset{\mathcal{R}}{\mathbf{q}}} = \frac{\partial f}{\partial \overset{\mathcal{H}}{\mathbf{q}}}\mathbf{J}.$$

If we now define $\nabla_{\mathbf{q}} f := \left(\frac{\partial f}{\partial \mathbf{q}} \right)^H$, $\nabla_{\overset{\mathcal{H}}{\mathbf{q}}} f := \left(\frac{\partial f}{\partial \overset{\mathcal{H}}{\mathbf{q}}} \right)^H$, $\nabla_{\overset{\mathcal{R}}{\mathbf{q}}} f := \left(\frac{\partial f}{\partial \overset{\mathcal{R}}{\mathbf{q}}} \right)^T$, where $(\cdot)^T$ and $(\cdot)^H$ represent the transpose and the Hermitian transpose, respectively, the above relations can be written as

$$\nabla_{\overset{\mathcal{H}}{\mathbf{q}}} f = \frac{1}{4}\mathbf{J}\nabla_{\overset{\mathcal{R}}{\mathbf{q}}} f \Leftrightarrow \nabla_{\overset{\mathcal{R}}{\mathbf{q}}} f = \mathbf{J}^H\nabla_{\overset{\mathcal{H}}{\mathbf{q}}} f. \tag{2}$$

3   Conjugate Gradient Algorithms

Conjugate gradient methods belong to the larger class of line search algorithms, which replace the negative gradient of the gradient descent method with some particular search direction, and then determine the minimum of the error function in that direction, thus yielding a real number that tells us how far we should move in the search direction, which replaces the learning rate. These methods generally perform better than the classical gradient descent. For the full deduction of conjugate gradient algorithms in the real-valued case, see [3,10].

Let's assume that we have a quaternion-valued neural network with an error function denoted by $E : \mathbb{H}^N \to \mathbb{R}$, and an N-dimensional weight vector denoted by $\mathbf{w} \in \mathbb{H}^N$. We start with the conjugate gradient algorithm for the real-valued case, in which the function $E(\mathbf{w})$ can be viewed as $E(\overset{\mathcal{R}}{\mathbf{w}})$. The iteration for calculating the minimum of the function $E(\overset{\mathcal{R}}{\mathbf{w}})$ is

$$\overset{\mathcal{R}}{\mathbf{w}}_{k+1} = \overset{\mathcal{R}}{\mathbf{w}}_k + \alpha_k \overset{\mathcal{R}}{\mathbf{p}}_k, \tag{3}$$

where $\overset{\mathcal{R}}{\mathbf{p}}_k \in \mathbb{R}^{4N}$ represents the search direction. The value of $\alpha_k \in \mathbb{R}$ is determined using the formula

$$\alpha_k = -\frac{\overset{\mathcal{R}}{\mathbf{p}}_k^T \overset{\mathcal{R}}{\mathbf{g}}_k}{\overset{\mathcal{R}}{\mathbf{p}}_k^T \overset{\mathcal{R}}{\mathbf{H}}_k \overset{\mathcal{R}}{\mathbf{p}}_k}, \tag{4}$$

where $\overset{\mathcal{R}}{\mathbf{g}}_k := \nabla_{\overset{\mathcal{R}}{\mathbf{w}}_k} E$ and $\overset{\mathcal{R}}{\mathbf{H}}_k := \nabla^2_{\overset{\mathcal{R}}{\mathbf{w}}_k} E$.

Now, the iteration for the next search direction is given by

$$\overset{\mathcal{R}}{\mathbf{p}}_{k+1} = -\overset{\mathcal{R}}{\mathbf{g}}_{k+1} + \beta_k \overset{\mathcal{R}}{\mathbf{p}}_k, \tag{5}$$

where $\beta_k \in \mathbb{R}$ has different expressions, depending on the type of the conjugate gradient algorithm. For example, the *Hestenes-Stiefel* update expression (see [7]) for β_k is:

$$\beta_k = \frac{\overset{\mathcal{R}}{\mathbf{g}}_{k+1}^T (\overset{\mathcal{R}}{\mathbf{g}}_{k+1} - \overset{\mathcal{R}}{\mathbf{g}}_k)}{\overset{\mathcal{R}}{\mathbf{p}}_k^T (\overset{\mathcal{R}}{\mathbf{g}}_{k+1} - \overset{\mathcal{R}}{\mathbf{g}}_k)}. \tag{6}$$

From (1) and (2), we observe that (5) can be written as

$$\frac{1}{4} \mathbf{J}^H \overset{\mathcal{H}}{\mathbf{p}}_{k+1} = -\mathbf{J}^H \overset{\mathcal{H}}{\mathbf{g}}_{k+1} + \beta_k \frac{1}{4} \mathbf{J}^H \overset{\mathcal{H}}{\mathbf{p}}_k,$$

or, equivalently, as

$$\overset{\mathcal{H}}{\mathbf{p}}_{k+1} = -\overset{\mathcal{H}}{\mathbf{g}}_k + \beta_k \overset{\mathcal{H}}{\mathbf{p}}_k, \tag{7}$$

because the $\frac{1}{4}$ factor can be absorbed into $\overset{\mathcal{H}}{\mathbf{p}}_k$.

By observing that

$$\overset{\mathcal{R}}{\mathbf{q}}_1^T \overset{\mathcal{R}}{\mathbf{q}}_2 = \overset{\mathcal{R}}{\mathbf{q}}_1^H \overset{\mathcal{R}}{\mathbf{q}}_2 = \left(\frac{1}{4} \mathbf{J}^H \overset{\mathcal{H}}{\mathbf{q}}_1\right)^H \frac{1}{4} \mathbf{J}^H \overset{\mathcal{H}}{\mathbf{q}}_2 = \frac{1}{16} \overset{\mathcal{H}}{\mathbf{q}}_1^H \mathbf{J}\mathbf{J}^H \overset{\mathcal{H}}{\mathbf{q}}_2 = \frac{1}{4} \overset{\mathcal{H}}{\mathbf{q}}_1^H \overset{\mathcal{H}}{\mathbf{q}}_2,$$

the update expression (6) can be written as

$$\beta_k = \frac{\overset{\mathcal{H}}{\mathbf{g}}_{k+1}^H (\overset{\mathcal{H}}{\mathbf{g}}_{k+1} - \overset{\mathcal{H}}{\mathbf{g}}_k)}{\overset{\mathcal{H}}{\mathbf{p}}_k^H (\overset{\mathcal{H}}{\mathbf{g}}_{k+1} - \overset{\mathcal{H}}{\mathbf{g}}_k)}. \tag{8}$$

Until now we have worked with vectors from $\mathbb{H}^{4N}$, but we would like to work with vectors directly in $\mathbb{H}^N$. From the definition of $\overset{\mathcal{H}}{\mathbf{q}}$ for $\mathbf{q} \in \mathbb{H}^N$, we see that this is done by taking the first N elements of the vector $\overset{\mathcal{H}}{\mathbf{q}}$. In this setting, Eqs. (3) and (7) become, respectively,

$$\mathbf{w}_{k+1} = \mathbf{w}_k + \alpha_k \mathbf{p}_k, \tag{9}$$

$$\mathbf{p}_{k+1} = -\mathbf{g}_{k+1} + \beta_k \mathbf{p}_k, \tag{10}$$

where $\mathbf{w}_k$, $\mathbf{p}_k$, $\mathbf{g}_k := \nabla_{\mathbf{w}_k} E \in \mathbb{H}^N$. We can also see that

$$\frac{1}{4}\mathbf{q}_1^{\mathcal{H}H\mathcal{H}}\mathbf{q}_2 = \mathrm{Re}\left(\mathbf{q}_1^H \mathbf{q}_2\right),$$

where $\mathrm{Re}(q)$ represents the real part of the quaternion q, i.e. $\mathrm{Re}(q) = q_a$, if $q = q_a + iq_b + jq_c + kq_d \in \mathbb{H}$, and so (8) can be written as.

$$\beta_k = \frac{\mathrm{Re}\left(\mathbf{g}_{k+1}^H(\mathbf{g}_{k+1} - \mathbf{g}_k)\right)}{\mathrm{Re}\left(\mathbf{p}_k^H(\mathbf{g}_{k+1} - \mathbf{g}_k)\right)}. \tag{11}$$

Relations (9), (10), and (11) now define the quaternion-valued *Hestenes-Stiefel* algorithm.

Similar calculations give the quaternion-valued *Polak-Ribiere* update expression (see [14]):

$$\beta_k = \frac{\mathrm{Re}\left(\mathbf{g}_{k+1}^H(\mathbf{g}_{k+1} - \mathbf{g}_k)\right)}{\mathbf{g}_k^H \mathbf{g}_k},$$

and the quaternion-valued *Fletcher-Reeves* update formula (see [16]):

$$\beta_k = \frac{\mathbf{g}_{k+1}^H \mathbf{g}_{k+1}}{\mathbf{g}_k^H \mathbf{g}_k}.$$

4   Scaled Conjugate Gradient Algorithm

Møller proposed in [13] the scaled conjugate algorithm, which brings two improvements to the conjugate gradient algorithm. The first one uses the model trust region method known from the Levenberg-Marquardt algorithm to ensure the positive definiteness of the Hessian matrix by adding to it a sufficiently large positive constant λ_k multiplied by the identity matrix. Thus, the formula for the step length given in (4), becomes in this case

$$\alpha_k = -\frac{\mathbf{P}_k^{\mathcal{R}^T\mathcal{R}} \mathbf{g}_k}{\delta_k}, \tag{12}$$

where we denoted by $\delta_k := \mathbf{P}_k^{\mathcal{R}^T\mathcal{R}}\mathbf{H}_k\mathbf{P}_k^{\mathcal{R}} + \lambda_k\mathbf{P}_k^{\mathcal{R}^T\mathcal{R}}\mathbf{P}_k$. If the Hessian matrix is positive definite, we have that $\delta_k > 0$. But if $\delta_k \leq 0$, then we should increase the value of δ_k in order to make it positive. If we denote by $\overline{\delta_k}$ the new value of δ_k, and by $\overline{\lambda_k}$ the new value of λ_k, then $\overline{\delta_k}$ is given by

$$\overline{\delta_k} = \delta_k + (\overline{\lambda_k} - \lambda_k)\mathbf{P}_k^{\mathcal{R}^T\mathcal{R}}\mathbf{P}_k, \tag{13}$$

where we take $\overline{\lambda_k} = 2\left(\lambda_k - \frac{\delta_k}{\mathbf{P}_k^{\mathcal{R}^T\mathcal{R}}\mathbf{P}_k}\right)$, which insures that $\overline{\delta_k} > 0$, and will be used in (12) to calculate the value of α_k.

The second improvement uses a comparison parameter to evaluate how good the quadratic approximation for the error function E really is, in the conjugate gradient algorithm. This parameter is defined by

$$\Delta_k = \frac{2(E(\overset{\mathcal{R}}{\mathbf{w}}_k) - E(\overset{\mathcal{R}}{\mathbf{w}}_k + \alpha_k \overset{\mathcal{R}}{\mathbf{p}}_k))}{\alpha_k \overset{\mathcal{R}}{\mathbf{p}}_k^T \overset{\mathcal{R}}{\mathbf{g}}_k}. \tag{14}$$

Based on the value of Δ_k, the parameter λ_k is then updated in the following way:

$$\lambda_{k+1} = \begin{cases} \lambda_k/2, & \text{if } \Delta_k > 0.75 \\ 4\lambda_k, & \text{if } \Delta_k < 0.25 \,, \\ \lambda_k, & \text{else} \end{cases}$$

in order to ensure a better quadratic approximation.

Thus, there are two stages for updating λ_k: one to ensure that $\delta_k > 0$ and one according to the validity of the local quadratic approximation. The two stages are applied successively after each weight update.

Using the same ideas as above, relations (12), (13), and (14), become, respectively,

$$\alpha_k = -\frac{\mathrm{Re}(\mathbf{p}_k^H \mathbf{g}_k)}{\delta_k},$$

$$\delta_k = \mathrm{Re}(\mathbf{p}_k^H \mathbf{H}_k \mathbf{p}_k) + \lambda_k \mathbf{p}_k^H \mathbf{p}_k,$$

$$\overline{\delta_k} = \delta_k + (\overline{\lambda_k} - \lambda_k)\mathbf{p}_k^H \mathbf{p}_k,$$

$$\overline{\lambda_k} = 2\left(\lambda_k - \frac{\delta_k}{\mathbf{p}_k^H \mathbf{p}_k}\right),$$

$$\Delta_k = \frac{2(E(\mathbf{w}_k) - E(\mathbf{w}_k + \alpha_k \mathbf{p}_k))}{\alpha_k \mathrm{Re}(\mathbf{p}_k^H \mathbf{g}_k)},$$

which, together, give the quaternion-valued scaled conjugate gradient algorithm.

5   Experimental Results

5.1   Linear Autoregressive Process with Circular Noise

An important application of quaternion-valued neural networks is at signal prediction. A known benchmark proposed in [12], and used for testing quaternion-valued neural networks in [6,19,22], is the prediction of the quaternion-valued circular white noise $n(k) = n_a(k) + in_b(k) + jn_c(k) + kn_d(k)$, where $n_a(k)$, $n_b(k)$, $n_c(k)$, $n_d(k) \sim \mathcal{N}(0,1)$, passed through the stable autoregressive filter given by $y(k) = 1.79y(k-1) - 1.85y(k-2) + 1.27y(k-3) - 0.41y(k-4) + n(k)$.

In our experiments, we trained quaternion-valued feedforward neural networks using the classical gradient descent algorithm (abbreviated GD), the conjugate gradient algorithm with Hestenes-Stiefel updates (CGHS), the conjugate gradient algorithm with Polak-Ribiere updates (CGPR), the conjugate gradient algorithm with Fletcher-Reeves updates (CGFR), and the scaled conjugate gradient algorithm (SCG).

The tap input of the filter was 4, so the networks had 4 inputs, 4 hidden neurons on a single hidden layer, and one output. The activation function for the hidden layer was the fully quaternion hyperbolic tangent function, given by $G^2(q) = \tanh q = \frac{e^q - e^{-q}}{e^q + e^{-q}}$, and the activation function for the output layer was the identity $G^3(q) = q$. Training was done for 5000 epochs with 5000 training samples.

To evaluate the effectiveness of the algorithms, we used a measure of performance called *prediction gain*, defined by $R_p = 10 \log_{10} \frac{\sigma_x^2}{\sigma_e^2}$, where σ_x^2 represents the variance of the input signal and σ_e^2 represents the variance of the prediction error. The prediction gain is given in dB and it is obvious that, because of the way it is defined, a bigger prediction gain means better performance. After running each algorithm 50 times, the prediction gains are given in Table 1.

The best algorithm was clearly SCG, followed by the conjugate gradient algorithms, and lastly by the gradient descent algorithm.

Table 1. Experimental results for linear autoregressive process with circular noise

Algorithm	Prediction gain
GD	4.51
CGHS	5.17
CGPR	5.19
CGFR	6.91
SCG	**7.36**

5.2 3D Lorenz System

The 3D Lorenz system is given by the ordinary differential equations

$$\frac{dx}{dt} = \alpha(y - x)$$

$$\frac{dy}{dt} = -xz + \rho x - y$$

$$\frac{dz}{dt} = xy - \beta z,$$

where $\alpha = 10$, $\rho = 28$ and $\beta = 2/3$. This represents a chaotic time series prediction problem, and was used to test quaternion-valued neural networks in [2,17,22].

The tap input of the filter was 4, and so the networks had 4 inputs, 4 hidden neurons, and one output neuron. The networks were trained for 5000 epochs with 1337 training samples.

Table 2. Experimental results for the 3D Lorenz system

Algorithm	Prediction gain
GD	7.56
CGHS	10.04
CGPR	11.31
CGFR	10.69
SCG	**12.58**

The results after 50 runs of each algorithm are given in Table 2. The measure of performance was the prediction gain, like in the above experiment.

In this experiment also, SCG had better results than the conjugate gradient and gradient descent algorithms.

5.3   4D Saito Chaotic Circuit

Lastly, we experimented on the 4D Saito chaotic circuit given by

$$\begin{bmatrix} \frac{dx_1}{dt} \\ \frac{dy_1}{dt} \end{bmatrix} = \begin{bmatrix} -1 & 1 \\ -\alpha_1 & \alpha_1\beta_1 \end{bmatrix} \begin{bmatrix} x_1 - \eta\rho_1 h(z) \\ y_1 - \eta\frac{\rho_1}{\beta_1}h(z) \end{bmatrix}$$

$$\begin{bmatrix} \frac{dx_2}{dt} \\ \frac{dy_2}{dt} \end{bmatrix} = \begin{bmatrix} -1 & 1 \\ -\alpha_2 & \alpha_2\beta_2 \end{bmatrix} \begin{bmatrix} x_2 - \eta\rho_2 h(z) \\ y_2 - \eta\frac{\rho_2}{\beta_2}h(z) \end{bmatrix},$$

where $h(z) = \begin{cases} 1, & z \geq 1 \\ -1, & z \leq 1 \end{cases}$, is the normalized hysteresis value, and $z = x_1 + x_2$, $\rho_1 = \frac{\beta_1}{1-\beta_1}$, $\rho_2 = \frac{\beta_2}{1-\beta_2}$. The parameters are given by $(\alpha_1, \beta_1, \alpha_2, \beta_2, \eta) = (7.5, 0.16, 15, 0.097, 1.3)$. This is also a chaotic time series prediction problem, and was used as a benchmark for quaternion-valued neural networks in [1, 2, 5, 6, 18, 19].

The networks had the same architectures as the ones described earlier, and were trained for 5000 epochs with 5249 training samples.

The prediction gains after 50 runs of each algorithm are given in Table 3.

Table 3. Experimental results for the 4D Saito chaotic circuit

Algorithm	Prediction gain
GD	5.76
CGHS	11.59
CGPR	13.64
CGFR	12.08
SCG	**15.32**

The performances of the algorithms were similar to the ones in the previous experiments.

6   Conclusions

The deductions of the scaled conjugate gradient algorithm and of the most known variants of the conjugate gradient algorithm for training quaternion-valued feed-forward neural networks were presented, starting from the real-valued case and using the framework of $\mathbb{HR}$ calculus for the extension to the quaternion-valued case. The three variants of the conjugate gradient algorithm with different update rules and the scaled conjugate gradient algorithm for optimizing the error function were applied for training networks used to solve three time series prediction problems.

Experimental results showed that the scaled conjugate gradient method performed better on the proposed problems than the classical gradient descent, in some cases as much as 10 dB better in terms of prediction gain. The scaled conjugate gradient algorithm was generally better than the classical variants of the conjugate gradient algorithm, also.

We can conclude that the scaled conjugate gradient algorithm represents an effective method for training feedforward quaternion-valued neural networks, as it was shown by its performance in solving different time series prediction problems.

References

1. Arena, P., Fortuna, L., Muscato, G., Xibilia, M.: Multilayer perceptrons to approximate quaternion valued functions. Neural Netw. **10**(2), 335–342 (1997)
2. Arena, P., Fortuna, L., Muscato, G., Xibilia, M.: Neural Networks in Multidimensional Domains Fundamentals and New Trends in Modelling and Control. Lecture Notes in Control and Information Sciences, vol. 234. Springer, London (1998)
3. Bishop, C.: Neural Networks for Pattern Recognition. Oxford University Press Inc., New York (1995)
4. Buchholz, S., Le Bihan, N.: Polarized signal classification by complex and quaternionic multi-layer perceptrons. Int. J. Neural Syst. **18**(2), 75–85 (2008)
5. Che Ujang, B., Took, C., Mandic, D.: Split quaternion nonlinear adaptive filtering. Neural Netw. **23**(3), 426–434 (2010)
6. Che Ujang, B., Took, C.: Quaternion-valued nonlinear adaptive filtering. IEEE Trans. Neural Netw. **22**(8), 1193–1206 (2011)
7. Hestenes, M., Stiefel, E.: Methods of conjugate gradients for solving linear systems. J. Res. Nat. Bur. Stan. **49**(6), 409–436 (1952)
8. Isokawa, T., Kusakabe, T., Matsui, N., Peper, F.: Quaternion neural network and its application. In: Palade, V., Howlett, R.J., Jain, L. (eds.) KES 2003. LNCS (LNAI), vol. 2774, pp. 318–324. Springer, Heidelberg (2003). doi:10.1007/978-3-540-45226-3_44
9. Jahanchahi, C., Took, C., Mandic, D.: On hr calculus, quaternion valued stochastic gradient, and adaptive three dimensional wind forecasting. In: International Joint Conference on Neural Networks (IJCNN), pp. 1–5. IEEE, July 2010

10. Johansson, E., Dowla, F., Goodman, D.: Backpropagation learning for multilayer feed-forward neural networks using the conjugate gradient method. Int. J. Neural Syst. **2**(4), 291–301 (1991)
11. Kusamichi, H., Isokawa, T., Matsui, N., Ogawa, Y., Maeda, K.: A new scheme for color night vision by quaternion neural network. In: International Conference on Autonomous Robots and Agents, pp. 101–106, December 2004
12. Mandic, D., Chambers, J.: Recurrent Neural Networks for Prediction: Learning Algorithms, Architectures and Stability. Wiley, New York (2001)
13. Møller, M.: A scaled conjugate gradient algorithm for fast supervised learning. Neural Netw. **6**(4), 525–533 (1993)
14. Polak, E., Ribiere, G.: Note sur la convergence de méthodes de directions conjuguées. Revue Française d'Informatique et de Recherche Opérationnelle **3**(16), 35–43 (1969)
15. Popa, C.-A.: Scaled conjugate gradient learning for complex-valued neural networks. In: Matoušek, R. (ed.) Mendel 2015: Recent Advances in Soft Computing. AISC, vol. 378, pp. 221–233. Springer, Heidelberg (2015). doi:10.1007/978-3-319-19824-8_18
16. Reeves, C., Fletcher, R.: Function minimization by conjugate gradients. Comput. J. **7**(2), 149–154 (1964)
17. Took, C., Mandic, D.: The quaternion LMS algorithm for adaptive filtering of hypercomplex processes. IEEE Trans. Sig. Process. **57**(4), 1316–1327 (2009)
18. Took, C., Mandic, D.: Quaternion-valued stochastic gradient-based adaptive IIR filtering. IEEE Tran. Sig. Process. **58**(7), 3895–3901 (2010)
19. Took, C., Mandic, D.: A quaternion widely linear adaptive filter. IEEE Trans. Sig. Process. **58**(8), 4427–4431 (2010)
20. Took, C., Mandic, D., Aihara, K.: Quaternion-valued short term forecasting of wind profile. In: International Joint Conference on Neural Networks (IJCNN), pp. 1–6. IEEE, July 2010
21. Took, C., Strbac, G., Aihara, K., Mandic, D.: Quaternion-valued short-term joint forecasting of three-dimensional wind and atmospheric parameters. Renew. Energy **36**(6), 1754–1760 (2011)
22. Xia, Y., Jahanchahi, C., Mandic, D.: Quaternion-valued echo state networks. IEEE Trans. Neural Netw. Learn. Syst. **26**(4), 663–673 (2015)
23. Xu, D., Xia, Y., Mandic, D.: Optimization in quaternion dynamic systems: gradient, Hessian, and learning algorithms. IEEE Trans. Neural Netw. Learn. Syst. **27**(2), 249–261 (2016)

Performance of Qubit Neural Network in Chaotic Time Series Forecasting

Taisei Ueguchi[✉], Nobuyuki Matsui, and Teijiro Isokawa

Graduate School of Engineering, University of Hyogo,
2167 Shosha, Himeji, Hyogo 671-2280, Japan
ei15v004@steng.u-hyogo.ac.jp, {matsui,isokawa}@eng.u-hyogo.ac.jp

Abstract. In recent years, quantum inspired neural networks have been applied to various practical problems since their proposal. Here we investigate whether our qubit neural network(QNN) leads to an advantage over the conventional (real-valued) neural network(NN) in the forecasting of chaotic time series. QNN is constructed from a set of qubit neuron, of which internal state is a coherent superposition of qubit states. In this paper, we evaluate the performance of QNN through a prediction of well-known Lorentz attractor, which produces chaotic time series by three dynamical systems. The experimental results show that QNN can forecast time series more precisely, compared with the conventional NN. In addition, we found that QNN outperforms the conventional NN by reconstructing the trajectories of Lorentz attractor.

Keywords: Quantum information processing · Qubit · Neural network · Chaotic time series forecasting

1 Introduction

Much attention has been given recently to the quantum computational intelligence (QCI) that is one of the promising methods to improve the information processing capacity of the conventional artificial intelligence [1]. In various QCI studies, the study that became those starting point is quantum-neuro computing [2,3]. In recent years, some researchers have developed the quantum-neuro computing in which the algorithm of quantum computation is used to improve the efficiency of neural computing system. The quantum state and the operator of quantum computation are both important to realize parallelisms and plasticity respectively in information processing systems. Complex valued representation of these quantum concepts allow neural computation system to advance in learning abilities and to enlarge its possibility of further practical applications. In a series of papers [4–9], we have proposed Qubit neural network (QNN) model and investigated its characteristic features, such as the effects of quantum superposition and probabilistic interpretation in the way of applying quantum computing to neural network.

One of the applications for neural networks is to forecast time-series from dynamical systems [10,11], especially from chaotic systems by their various

© Springer International Publishing AG 2016
A. Hirose et al. (Eds.): ICONIP 2016, Part III, LNCS 9949, pp. 253–260, 2016.
DOI: 10.1007/978-3-319-46675-0_28

applications. Several researches have been available concerning the prediction of chaotic time-series by neural networks [12,13], but there are few attempts for prediction by Qubit-based neural networks.

In this paper, we investigate the performances of QNN through a problem of prediction of chaotic system, called Lorentz attractor. The experiments are conducted by comparing with the conventional (real-valued) neural network (NN) in terms of learning iterations, learning success rates, and prediction errors for the test data. In addition, we reconstruct the trajectories of Lorentz attractor based on Takens' theorem [14] in order to clarify the forecasting accuracy.

The rest of the paper is organized as follows. Section 2 gives the model of qubit neuron and multilayered perceptron-type network. The experimental results are explored in Sect. 3. We finish with the conclusion in Sect. 4.

2   Qubit Neural Network

This section recapitulates the qubit neural network (QNN) model used in this paper. A qubit neuron is a basic component of QNN [6]. The internal state of a qubit neuron is described as a quantum phase in the complex-valued representation. A firing neuron state is defined as a qubit state $|1\rangle$, a non-firing neuron state is defined as a qubit state $|0\rangle$. In quantum computing, the concept of 'qubit' has been introduced as the counterpart of the classical concept of 'bit' in conventional computers [15,16]. The arbitrary qubit state $|\phi\rangle$ maintains a coherent superposition of states $|0\rangle$ and $|1\rangle$:

$$|\phi\rangle = \cos\theta\,|0\rangle + e^{i\psi}\sin\theta\,|1\rangle, \tag{1}$$

which is called Bloch-sphere representation. In this work, we adopt the complex-valued representation of the quantum state. That is, we use the following representation instead of Eq. (1) [9]:

$$f(\theta) = e^{i\theta} = \cos\theta + i\sin\theta, \tag{2}$$

where i is an imaginary unit.

2.1   Qubit Neuron Model

In the qubit neuron model with complex-valued representation, the output state of the neuron k, denoted as x_k, is given as

$$x_k = f(y_k) = \cos y_k + i\sin y_k = e^{iy_k}. \tag{3}$$

The internal state u_k and the phase input y_k for the neuron k are calculated

$$u_k = \sum_{l=1}^{L} e^{i\theta_{l,k}} \cdot x_l - e^{i\lambda_k}, \tag{4}$$

$$y_k = \frac{\pi}{2} \cdot g(\delta_k) + \{1 - 2g(\delta_k)\} \cdot \arg(u_k), \tag{5}$$

respectively. In these equations, $\theta_{l,k}$ and λ_k are phase parameters that corresponds to the connection weight and threshold in the conventional neural network, respectively. δ_k controls the degree of phase inversion. $g(\cdot)$ is a real-valued sigmoidal function.

2.2 Multilayered Network of QNN

QNN is a network organized with Qubit neurons. In this paper, we adopt a structure of the multilayered perceptron. The network consists of an input layer with L qubit neurons denoted by $\{I_l\}$ $(l = 1, \cdots, L)$, a hidden layer with M qubit neurons denoted by $\{H_m\}$ $(m = 1, \cdots, M)$, and an output layer with N qubit neurons denoted by $\{O_n\}$ $(n = 1, \cdots, N)$. The structure of the network is described by L-M-$N(D)$ where D is the number of parameters that can be adjusted by learning.

Input values (denoted by $input_l$) in the range $[0, 1]$ in the input layer are converted into quantum states with phase values in the range $[0, \pi/2]$. The phase input of the l-th neuron in the input layer, y_l^I, thus

$$y_l^I = \frac{\pi}{2} \cdot input_l. \tag{6}$$

We obtain the output of the network, denoted by $output_n$, with the probability of the state in which $|1\rangle$ is observed from the n-th neuron in the output layer,

$$output_n = |\text{Im}(x_n^O)|^2, \tag{7}$$

where x_n^O is the output from the neuron n in the output layer.

In this network, we adopt a quantum version of the well-known Back Propagation algorithm as its learning rule [6,7]. This rule is expressed by the following equations:

$$w_k^{new} = w_k^{old} - \eta \frac{\partial E_{total}}{\partial w_k^{old}}, \quad w \in \{\theta, \lambda, \delta\}, \tag{8}$$

where η is a leaning rate. E_{total} is the squared error function defined as

$$E_{total} = \frac{1}{2} \sum_p^K \sum_n^N (t_n^p - output_{n,p})^2, \tag{9}$$

which is to be minimized as part of the learning process. Here, t_n^p and $output_{n,p}$ is the target signal for the n-th neuron and output signal respectively when it learns the p-th pattern.

3 Experimental Results

The performances of the presented QNN are investigated in this section, by comparing with conventional (real-valued) neural network. The tasks for these networks are the prediction of three-dimensional chaotic signals called the Lorenz system. In addition, we reconstruct the trajectories of Lorentz attractor based on Takens' theorem in order to clarify the forecasting accuracy.

3.1   Setup

The Lorenz system is a system with three ordinary differential equations [17]:

$$\frac{dx}{dt} = \sigma(y - x), \tag{10}$$

$$\frac{dy}{dt} = x(\rho - z) - y, \tag{11}$$

$$\frac{dz}{dt} = xy - \beta z, \tag{12}$$

where x, y, and z are the states of the system with time t, and σ, ρ, and β are the parameters of the system. This system shows chaotic behaviors for particular sets of parameters, e.g. for a set of parameters $\sigma = 10$, $\rho = 28$, and $\beta = 8/3$, the system has two fixed attractors, called the Lorenz attractors.

To predict the states of the system, a three-layered QNN and real-valued neural network are prepared. The structure of these networks are 9-45-3(636) for the QNN and 9-49-3(640) for the NN. These networks are chosen so that the numbers of trainable parameters for these networks become almost same.

These networks are trained so that the prediction state at time $(t + \Delta t)$ is come up from the system states at time t, $(t - 1)$, $(t - 2)$ presented on the input of the network. The parameter Δt takes a integer value from 1 to 10. A set of 3000 training and test data is prepared from a Lorenz system evolved by Euler method with the time step 0.01 and with the initial configuration of $x = 1.0$, $y = 1.0$, and $z = 1.0$. First 2000 data is used for training the networks, and the rest is used for the test. Each component for the input data is normalized in the range $[0, 1]$.

The trainings for networks are conducted until the squared error from the network (E_{total}) is less than $ER = 0.1$ or the learning epoch exceeds to $LM = 30,000$. Where, ER is a truncation error and LM is a maximum number of learning iteration.

3.2   Results

Typical evolutions of the squared errors (denoted as error rate) from QNN and real-valued neural network (NN) are shown in Figs. 1(a), (b), (c), and (d), for $\Delta t = 1$, $\Delta t = 4$, $\Delta t = 7$, and $\Delta t = 10$, respectively. In Fig. 1(a), the error rate from QNN quickly falls to ER, on the other hand the one from NN converges quite slowly. The best numbers of training epochs for QNN and NN are about 400 and 10,500, respectively, showing that QNN can acquire the input-output relations from the learning data very quickly than NN. In addition, these results show that NN tends to fail the learning when Δt increases whereas QNN can perform learning correctly.

Δt dependences of the learning iteration and learning success rates are shown in Fig. 2. In this experiment, the learning coefficient $\eta = 0.035$ for both networks is adopted. For each Δt, 50 trials are conducted. The results show that QNN can acquire the input-output relations for wide range of Δt with smaller iterations

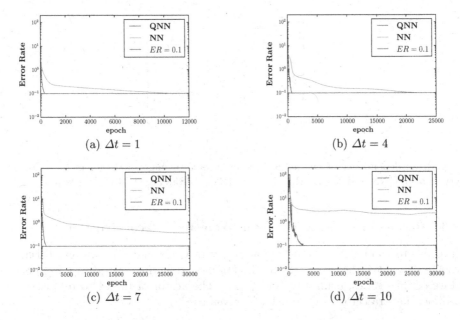

Fig. 1. Evolutions of error rates for QNN and NN, with different Δt conditions.

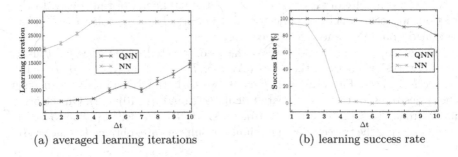

Fig. 2. Δt-dependence of the averaged learning iterations and success rates for QNN and NN.

for training. For each Δt, the prediction errors for the test data are calculated for QNN and NN. The maximum, minimum, and average mean squared errors are shown in Fig. 3. These MSEs show that variances as well as the averaged errors by QNN are much smaller than NN. Finally, the averaged MSEs with their maximum and minimum values with respect to Δt are shown in Fig. 3. MSEs are displayed in log scale, as in Fig. 3. The predictions by the real-valued NN degrade by increasing the durations of inputs for time series, but the QNN can maintain the prediction accuracies. Thus, QNN can predict the time series with fewer input information.

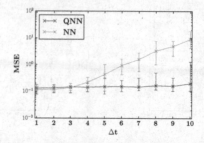

Fig. 3. Δt-dependence of maximum, minimum, and average values for mean squared errors from QNN and NN for the test data (these MSEs are shown in log scale).

3.3 Results of Lorentz Attractor Trajectory Reconstruction

In order to clarify the forecasting accuracy of QNN and NN, we use Takens' theorem. Takens' theorem is the method of reconstructing a trajectory from one state of time series, using time delay [14]. The vector of time t to reconstruct d-dimension trajectory, denoted V_t, is given as

$$V_t = (v(t), v(t - T), v(t - 2T), \ldots, v(t - (d - 1)T)), \tag{13}$$

where $v(\cdot)$ is a state of time series and T is the parameter of time delay. In this section, we reconstruct the trajectories of Lorentz attractor using the V_t data obtained from QNN and NN, respectively.

For example, Figs. 4 and 5 show the reconstructed trajectories for y using the networks' V_t data in the cases $(\Delta t = 1, T = 10)$ and $(\Delta t = 10, T = 10)$, respectively. From Fig. 4 $(\Delta t = 1, T = 10)$, we see both QNN and NN reconstruct the trajectories for the test V_t data calculated from Eqs. (10)–(12). On the other hand, from Fig. 5 $(\Delta t = 10, T = 10)$, we see QNN is much superior to NN in reconstructing the trajectories. The similar results are also obtained in the case of

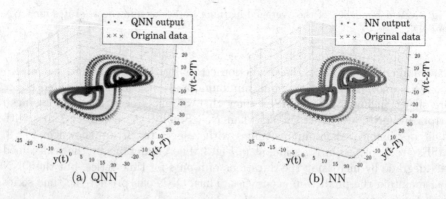

(a) QNN   (b) NN

Fig. 4. A reconstructed trajectory by Takens' theorem for y, produced by QNN output, NN output and test data $(\Delta t = 1, T = 10)$.

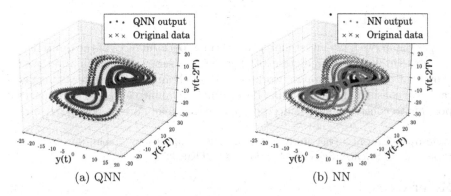

Fig. 5. A reconstructed trajectory by Takens' theorem for y, produced by QNN output, NN output and test data ($\Delta t = 10$, $T = 10$).

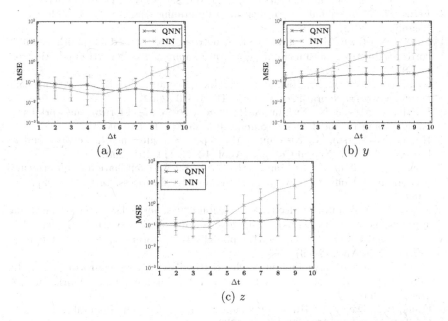

Fig. 6. Δt-dependence of maximum, minimum, and average values for mean squared errors from QNN and NN, separately x, y and z, for the test data (these MSEs are shown in log scale).

z data. The averaged MSEs with their maximum and minimum values, separately x, y and z, with respect to Δt are shown in Fig. 6. MSEs are displayed in log scale, as in Fig. 6. In all values, the predictions by the real-valued NN degrade by increasing the durations of inputs for time series, however, the QNN can maintain the prediction accuracies.

4   Conclusion

In this paper, we have investigated the learning performances of the qubit neural network through the prediction of chaotic time series. Our simulation results show that our qubit neural network can effectively improve the forecasting accuracy. We consider that the further development of qubit neural network makes it possible to construct more enforced applications than current neural networks.

Acknowledgment. This study was financially supported by Japan Society for the Promotion of Science (Scientific Research (C) 16K00337).

References

1. Manju, A., Nigam, M.J.: Applications of quantum inspired computational intelligence: a survey. Artif. Intell. Rev. **42**(1), 79–156 (2014)
2. Kak, S.C.: On quantum neural computing. Inf. Sci. **83**(3), 143–160 (1995)
3. Perus, M.: Neuro-quantum parallelism in brain-mind and computers. Informatica (Ljubljana) **20**(2), 173–184 (1996)
4. Matsui, N., Takai, M., Nishimura, H.: A network model based on qubitlike neuron corresponding to quantum circuit. Electron. Commun. Jpn Part III Fund. Electron. Sci. **83**(10), 67–73 (2000)
5. Kouda, N., Matsui, N., Nishimura, H.: Image compression by layered quantum neural networks. Neural Process. Lett. **16**, 67–80 (2002)
6. Kouda, N., Matsui, N., Nishimura, H.: A multilayered feed-forward network based on qubit neuron model. Syst. Comput. Jpn **35**(13), 43–51 (2004)
7. Kouda, N., Matsui, N., Nishimura, H., Peper, F.: Qubit neural network and its learning efficiency. Neural Comput. Appl. **14**(2), 114–121 (2005)
8. Kouda, N., Matsui, N., Nishimura, H., Peper, F.: An examination of qubit neural network in controlling an inverted pendulum. Neural Process. Lett. **22**(3), 277–290 (2005)
9. Matsui, N., Nishimura, H., Isokawa, T.: Qubit neural networks: its performance and applications. In: Nitta, T. (ed.) Complex-Valued Neural Networks: Utilizing High-Dimensional Parameters, chap. XIII, pp. 325–351. Information Science Reference, Hershey, New York (2009)
10. Kido, K.: Short-term prediction on chaotic time-series using neurocomputing. In: Proceedings of the 1st Western Pacific and 3rd Australia-Japan Workshop, pp. 278–284 (1999)
11. Frank, R.J., Davey, N., Hunt, S.P.: Time series prediction and neural networks. J. Intell. Robot. Syst. **31**(1–3), 91–103 (2001)
12. Han, M., Xi, J., Xu, S., Yin, F.L.: Prediction of chaotic time series based on the recurrent predictor neural network. IEEE Trans. Sig. Process. **52**(12), 3409–3416 (2004)
13. Karunasinghea, D.S.K., Liongb, S.Y.: Chaotic time series prediction with a global model: artificial neural network. J. Hydrol. **323**(1–4), 92–105 (2006)
14. Takens, F.: Detecting Strange Attractors in Turbulence. Springer, Heidelberg (1981)
15. Bennett, C.H.: Quantum information and computation. Phys. Today **48**(10), 24–31 (1995)
16. Berman, G.P., Doolen, G.D., Mainieri, R., Tsifrinovich, V.I.: Introduction to Quantum Computers. World Scientific, River Edge (1998)
17. Lorenz, E.N.: Deterministic nonperiodic flow. J. Atmos. Sci. **20**(2), 130–141 (1963)

The Evolutionary Process of Image Transition in Conjunction with Box and Strip Mutation

Aneta Neumann[(✉)], Bradley Alexander, and Frank Neumann

Optimisation and Logistics, School of Computer Science,
The University of Adelaide, Adelaide, Australia
`aneta.neumann@adelaide.edu.au`

Abstract. Evolutionary algorithms have been used in many ways to generate digital art. We study how evolutionary processes are used for evolutionary art and present a new approach to the transition of images. Our main idea is to define evolutionary processes for digital image transition, combining different variants of mutation and evolutionary mechanisms. We introduce box and strip mutation operators which are specifically designed for image transition. Our experimental results show that the process of an evolutionary algorithm in combination with these mutation operators can be used as a valuable way to produce unique generative art.

1 Introduction

The field of evolutionary algorithms (EAs) has been successfully applied to the areas of modern art [1,2]. Evolutionary computation is an interesting approach to the generation of novel images and is beginning to have a broader impact on artistic fields. More generally, evolutionary methods applied to problem solving in creative fields is an exciting and fast developing topic in computer science. In earlier years research has been conducted using evolutionary algorithms for interactive generation of art. Dawkins [3] and Smith [4] demonstrated the potential of Darwinian variation to evolve biomorphs graphic objects. Following this steps Sims [5], Latham and Todd [6] have combined evolutionary techniques and computer graphics to create artistic images and to reproduce computer creatures of great complexity. More recently Unemi [7], Hart [8] created a collection of images using refined means of combining individual images and their genotypes. These works have not only influenced the field of evolutionary computation, but were also displayed in galleries and museums around the world. Furthermore, research in computational art has employed deep neural networks to create artistic images [9,10].

In our work we introduce a novel approach to define evolutionary processes for evolutionary image transition[1]. The main idea comprises of using well-known evolutionary processes and adapting these in an artistic way to create innovative images.

---

[1] Images and videos are available at https://evolutionary-art.blogspot.com.

© Springer International Publishing AG 2016
A. Hirose et al. (Eds.): ICONIP 2016, Part III, LNCS 9949, pp. 261–268, 2016.
DOI: 10.1007/978-3-319-46675-0_29

The evolutionary transition process starts from a given starting image S and, through a stochastic process, evolves it towards a target image T. Our goal is to maximise the fitness function where we count the number of the pixels matching those of the target image T. This problem mirrors the classical OneMax problem, which has been widely studied in academic literature [11,12].

We start by introducing a variant of the asymmetric mutation operator studied in the context of minimum spanning trees [13] and pseudo-Boolean optimization [12]. This operator is based on flipping pixels with a given probability in each step and avoids the coupon collectors effect [14] by using standard bit mutation applied to OneMax. Afterwards, we introduce specific mutation operators for evolutionary image transition, namely strip and box mutations, which pick parts of the image consisting of strips and boxes and set them to the target image. Our experimental results show the artistic effects which can be produced by applying evolutionary process to an image via the operators described below.

The rest of the paper is structured as follows. In Sect. 2, we introduce evolutionary transition process and examine the behaviour of $(1+1)$ EA for image transition in Sect. 3. Section 4 gives results of how *strip mutation, combined strip mutation* and *box mutation* can be used for evolutionary image transition process. In Sect. 5 we investigate the use of *asymmetric mutation* as part of mutation operators and study their combinations with pixel-based mutations during the evolutionary process. Finally, we conclude with our overview about the insights from our investigations.

2 Evolutionary Image Transition

Artificial intelligence has the potential to change our perception of the process of creating art. Using evolutionary algorithms can lead to significant novelty in the interaction between the computer and the artist in order to produce an artwork. The application of evolutionary algorithms in an innovative way can change the apprehension of human-machine creativity. Artificial intelligence as a medium for creating art has a long and well-established tradition. The growing influence and demand of artificial intelligence in our global industrial life has been discussed throughout a wide spectrum of our society. So comes a meaningful question: how can computational creativity perform as a valuable addition to digital art to produce unique generative art?

We consider an evolutionary transition process that transforms a given image X of size $m \times n$ into a given target image T of size $m \times n$. Our goal is to study different ways of carrying out this evolutionary transformation based on random processes from an artistic perspective. We start our process with a starting image $S = (S_{ij})$. Our algorithms evolve S towards T and have at each point in time an image X where $X_{ij} \in \{S_{ij}, T_{ij}\}$. We say that pixel X_{ij} is in state s if $X_{ij} = S_{ij}$, and X_{ij} is in state t if $X_{ij} = T_{ij}$. For our process we assume that $S_{ij} \neq T_{ij}$ as pixels with $S_{ij} = T_{ij}$ can not change values and therefore do not have to be considered in the evolutionary process. To illustrate the effect of the different methods presented in this paper, we consider two images of famous Kyoto Temples (see Fig. 1).

Fig. 1. Starting image S (Kinkaku-Ji Temple Kyoto) and target image T (Daigo-ji Temple Kyoto)

Algorithm 1. Evolutionary algorithm for image transition

 - Let S be the starting image and T be the target image.
 - Set X:=S.
 - Evaluate $f(X,T)$.
 - while (not termination condition)
   - Obtain image Y from X by mutation.
   - Evaluate $f(Y,T)$
   - If $f(Y,T) \geq f(X,T)$, set $X := Y$.

The fitness function of an evolutionary algorithm guides its search process and determines how to move between images. Therefore, the fitness function itself has a strong influence on the artistic behaviour of the evolutionary image transition process. An important property for evolutionary image transition, in this work, is that images close to the target image get a higher fitness score. We measure the fitness of an image X as the number of pixels where X and T agree. This fitness function is isomorphic to that of the OneMax problem when interpreting the pixels of S as 0's and the pixels of T as 1's. Formally, we define the fitness of X with respect to T as

$$f(X,T) = |\{X_{ij} \in X \mid X_{ij} = T_{ij}\}|.$$

We consider simple variants of the classical $(1+1)$ EA in the context of image transition. The algorithm is using mutation only and accepts an offspring if it is at least as good as its parent according to the fitness function. The approach is given in Algorithm 1. Using this algorithm has the advantage that parents and offspring do not differ too much in terms of pixel which ensures a smooth process for transitioning the starting image towards the target.

All experimental results in this paper are shown for the process of moving from the starting image to the target image given in Fig. 1 where the images are of size 200×200 pixels. The algorithms have been implemented in Matlab. In order to visualize the process, we show the images obtained when the evolutionary process reaches 12.5%, 37.5%, 62.5% and 87.5% of pixels evolved towards the target image.

3   Asymmetric Mutation

We consider a simple evolutionary algorithm that has been well studied in the area of runtime analysis, namely variants of the classical $(1+1)$ EA [11]. As already mentioned, our setting for the image transition process is equivalent to the optimization process for the classical benchmark function OneMax. The standard variant of the $(1+1)$ EA flips each pixel with probability $1/|X|$ where $|X|$ is the total number of pixels in the given image. Using this mutation operator, the algorithm encounters the well-known coupon collector's effect which means that the additive drift towards the target image when having k missing target pixels is $\Theta(k/n)$ [15].

In order to avoid the coupon collector's effect but obtaining images where all pixel are treated symmetrically, we use the *asymmetric mutation* operator introduced in [13]. Jansen and Sudholt [12] have shown that the $(1+1)$ EA using *asymmetric mutation* optimizes OneMax in time $\Theta(n)$ which improves upon the usual bound of $\Theta(n \log n)$ when using standard bit mutations. We denote by $|X|_T$ the number of pixels where X and T agree. Similarly, we denote by $|X|_S$ the number of pixels where X and S agree. Each pixel is starting state s is flipped with probability $c_s/(2|X|_S)$ and each pixel in target state t is flipped with probability $c_t/(2|X|_T)$. The mutation operator is shown in Algorithm 2. The mutation operator differs from the one given in [12] by the two constants c_s and c_t which allow one to determine the expected number of new pixels from the starting image and the target image, respectively. The choice of c_s and c_t determines the expected number of pixel in the starting state and target state to be flipped. To be precise, the expected number of pixels currently in the starting state s to be flipped is $c_s/2$ and the number of pixels in the target state t to be flipped is $c_t/2$ as long as the current solution X contains at least that many pixel of the corresponding type. In [12] the case $c_s = c_t = 1$ has been investigated which ensures that at each point in time has an additive drift of $\Theta(1)$ and therefore avoiding the coupon collector effect at the end of the process. Using different values for c_s and c_t allows us to change the speed of transition as well as the relation of the number of pixels switching from the starting image to the target and vice versa while still ensuring that there is constant progress towards the target. For all our experiments we use $c_s = 100$ and $c_t = 50$.

Algorithm 2. Asymmetric mutation

– Obtain Y from X by flipping each pixel X_{ij} of X independently of the others with probability $c_s/(2|X|_S)$ if $X_{ij} = S_{ij}$, and flip X_{ij} with probability $c_t/(2|X|_T)$ if $X_{ij} = T_{ij}$, where $c_s \geq 1$ and $c_t \geq 1$ are constants.

---

4   Strip and Box Mutation

We now introduce specific mutation operators for image transition. The design of our three mutation operators, namely *strip mutation, combined strip mutation* and *block mutation*, is strongly oriented towards production of interesting artistic images.

All of these mutation operators transition a region of the current image X to an area of the target image T, starting at randomly chosen pixel position X_{ij}. *Strip mutation* mutates a vertical strip of pixels from X_{ij} for a length of 180 pixels or to the boundary of the image whichever comes first. *The strip mutation* imitates well-know technique in generative art called a glitch. This effect intentionally corrupts data of an image by encoding the JPEG process. All experiments shows that in each new generation the initial image and the target image overlay through a series of randomly placed strips. In Fig. 2 we show the experimental results of *strip mutation*. The effect of *strip mutation* creates generative art with highly interesting components. Furthermore, we have conceptualised a *combined strip mutation* operator where both horizontal and vertical strips gradually overlay the original image in random locations. The parameter settings for *the combined strip mutation* are 200 pixels in width and 40 pixels in height for the horizontal strip and one pixel in width and 200 pixels in height for the vertical orientation. Figure 3 shows a very interesting transition from the point of view of an image processing which also has artistic value.

Our third mutation operator is *block mutation*. This operator randomly selects a position in the space of the matrix $S(i,j)$ and flips a block of 3×3

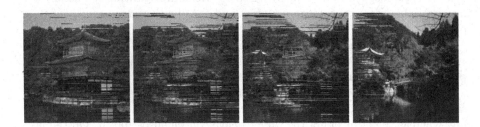

Fig. 2. Image transition for strip mutation

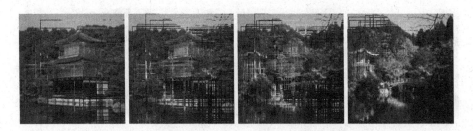

Fig. 3. Image transition for combined strip mutation

Fig. 4. Image transition for box mutation

pixels. Because the position chosen for mutation can be any point X_{ij} we can observe that mutated blocks can overlap.

Figure 4 shows the experimental results of *block mutation*. We have executed a smooth transition so as to interest the viewer. Additionally, we have made the changes in the image clearly visible.

5   Combined Approaches

The asymmetric, strip, combined strip and *box mutation* operators have quite distinct behaviours when applied to image transition. We now study the effect of combining the approaches for evolutionary image transition in order to obtain a more artistic evolutionary process. We explore the combination of *the asymmetric mutation* operator and *strip, combined strip* and *box mutation*. Here we run *the asymmetric mutation* operator as described in Algorithm 2. The process alternates between *asymmetric mutation* and one of either *strip, combined strip* or *box mutation*.

Figure 5 shows the results of the experiments for *combined asymmetric* and *strip mutation*. The transitional images are significantly different to the images produced in *strip mutation* or *box mutation*.

Interesting blurred elements randomly emerge over the image during the transition stage. Similar elements are shown as a characteristically steadily changing, whereas the elements of the target image continuously appear. On the last image we can almost completely see the details of the target image. The blurred

Fig. 5. Image transition for combined asymmetric and strip mutation

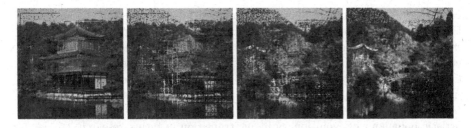

Fig. 6. Image transition for combined asymmetric and combined strip mutation

Fig. 7. Image transition for combined asymmetric and box mutation

elements seem to be created through the bright autumn colour in the target image in this stage of the sequence.

Figure 6 shows the results of the experiments for *combined asymmetric* and *combined strip mutation*. The lower sequence of the picture is similar to the top in terms of the behaviour of the patches. More differences can be seen as the transition progresses, but the picture continues to be partially unrecognisable during the transition. New elements can be discovered on the affected image which slowly appear and create an element of surprise.

We see less of the dotted image pattern, but now observe combined blurred elements randomly occurring in the upper sequence of the images. The image provides an interesting comparison to the old linen pictures of the Rembrandt epoch. In the advanced stages, the patches are well integrated into the images using *asymmetric mutation*. Finally, the last image we see is well integrated and shows more detail in all areas of the image.

Figure 7 shows the results of the experiments for *combined asymmetric* and *box mutation*. In the advanced stages, the patches are well integrated into the images using *asymmetric mutation* to produce a smooth transition process. In summary, the image transition of *combined asymmetric* and *box mutation* differs significantly from *the box mutation* transition introduced in Sect. 4.

6   Conclusions and Future Work

In this paper we have investigated how evolutionary algorithms and related random processes can be used for image transition. We have shown how *the asymmetric mutation* operator, introduced originally for the optimization of OneMax, can be applied to our problem. We examined how the varying mutation settings influences our results through *box mutation, strip mutation, combined strip mutation* and combined approaches with *asymmetric mutation*. By investigating combinations of the different approaches, we have shown a variety of interesting evolutionary image transition processes. All our investigations are based on a fitness function that is equivalent to the well-known OneMax problem. It would be interesting to study more complex fitness functions and their impact on the artistic behaviour of evolutionary image transition in future research.

References

1. Bentley, P., Corne, D.: Creative Evolutionary Systems. Evolutionary Computation Series. Morgan Kaufmann, San Francisco (2002)
2. McCormack, J., d'Inverno, M. (eds.): Computers and Creativity. Springer, Heidelberg (2012)
3. Dawkins, R.: The Blind Watchmaker: Why the Evidence of Evolution Reveals a Universe Without Design. National Bestseller. Science. Norton, New York (1986)
4. Smith, J.: Evolution and the Theory of Games. Cambridge University Press, Cambridge (1982)
5. Sims, K.: Artificial evolution for computer graphics. In: Thomas, J.J. (eds.) Proceedings of the 18th Annual Conference on Computer Graphics and Interactive Techniques, SIGGRAPH 1991, pp. 319–328. ACM (1991)
6. Todd, S., Latham, W.: Evolutionary Art and Computers. Academic Press Inc., Orlando (1994)
7. Unemi, T.: Sbart4 for an automatic evolutionary art. In: Proceedings of the IEEECongress on Evolutionary Computation, CEC 2012, Brisbane, Australia, June 10-15, 2012, pp. 1–8. IEEE (2012)
8. Hart, D.: Toward greater artistic control for interactive evolution of images and animation. In: Finnegan, J.W., Pfister, H. (eds.) 33 International Conference on Computer Graphics and Interactive Techniques, SIGGRAPH 2006. ACM 2(2006)
9. Champandard, A.J.: Semantic style transfer and turning two-bit doodles into fine artworks. CoRR abs/1603.01768 (2016)
10. Li, C., Wand, M.: Combining Markov random fields and convolutional neural networks for image synthesis. CoRR abs/1601.04589 (2016)
11. Droste, S., Jansen, T., Wegener, I.: On the analysis of the $(1 + 1)$ evolutionary algorithm. Theor. Comput. Sci. **276**(1–2), 51–81 (2002)
12. Jansen, T., Sudholt, D.: Analysis of an asymmetric mutation operator. Evol. Comput. **18**(1), 1–26 (2010)
13. Neumann, F., Wegener, I.: Randomized local search, evolutionary algorithms, and the minimum spanning tree problem. Theor. Comput. Sci. **378**(1), 32–40 (2007)
14. Mitzenmacher, M., Upfal, E.: Probability and Computing: Randomized Algorithms and Probabilistic Analysis. Cambridge University Press, New York (2005)
15. He, J., Yao, X.: Drift analysis and average time complexity of evolutionary algorithms. Artif. Intell. **127**(1), 57–85 (2001)

A Preliminary Model for Understanding How Life Experiences Generate Human Emotions and Behavioural Responses

D.A. Irosh P. Fernando[1(⊠)] and Björn Rüffer[2]

[1] School of Electrical Engineering and Computer Science, School of Medicine,
Public Health University of Newcastle, Callaghan, NSW, Australia
irosh.fernando@uon.edu.au
[2] School of Mathematical and Physical Sciences, Faculty of Science and
Information Technology, University of Newcastle, Callaghan, NSW, Australia
bjorn.ruffer@newcastle.edu.au

Abstract. Whilst human emotional and behaviour responses are generated via a complex mechanism, understanding this process is important for a broader range of applications that span over clinical disciplines including psychiatry and psychology, and computer science. Even though there is a large body of literature and established findings in clinical disciplines, these are under-utilised in developing more realistic computational models. This paper presents a preliminary model based on the integration of a number of established theories in clinical psychology and psychiatry through an interdisciplinary research effort.

Keywords: Modelling human behavior and emotions · Emotional computing · Affective computing · Computational psychiatry

1 Introduction

In humans, life experiences generate various emotional and behavioural responses via a complex process, which involves interactions between memory of past life experiences, beliefs, needs, cognitive control, and emotional state. Modelling this process can yield a broader range of applications not only in the field of computer science but also in clinical disciplines including psychiatry and psychology. Whilst it will provides a theoretical foundation for developing systems that mimic human like behaviour in computer applications, it will also assist in understanding the pathogenesis of psychiatric illnesses that result in abnormal emotional and behavioural responses.

There have been a number of computational models for simulating emotions and human behaviour including: Affective Reasoner [1]; FLAME [2]; An Architecture for Action, Emotion and Social Behaviour [3]; Cathexis model [4]; PsychSim [5]; GEmA [6]. Whilst most of these models have used the Cognitive Structure of Emotions [7] as the theoretical basis, others have used Emotion Regulation Theory [8] and the Theory of Mind [9]. Also, an information processing model, which is based on four main cognitive functions consisting of perception, interpretation, planning, and execution, have also been proposed from systems and control perspective in order to explain

© Springer International Publishing AG 2016
A. Hirose et al. (Eds.): ICONIP 2016, Part III, LNCS 9949, pp. 269–278, 2016.
DOI: 10.1007/978-3-319-46675-0_30

human behaviour [10]. Even though these models have been able to deliver useful application in areas such as gaming [11] and tutoring [12], they do not have the adequate theoretical foundation for simulating broader human emotional and behaviour responses or studying emotional disorders.

It is important to note that there has been a large body of literature on various aspects of human emotions and behaviour in clinical disciplines, which have been under-utilised in developing computational models. This situation is understandable due to practical difficulties in transferring this clinical knowledge to researchers in computer science who may not have the necessary clinical background. However, an interdisciplinary endeavour to develop computational models based on established clinical knowledge is important since it will benefit both disciplines.

Using a number of theories from clinical psychology including Cognitive Schema Theory(CST) [13, 14] as the conceptual foundation, this paper presents a preliminary model, which is expected to be further developed via ongoing research.

2 Conceptual Model

Human emotions and behaviour can be understood as responses to various stimuli that could be either external or internal [15]. We refer to life experiences that originate outside of human body as external events, which results in during interactions with the environment including social interactions. On the other hand, internal events originate within human body mainly due to physiological changes resulting in emotional and behavioural changes. For example, a person can experience sadness as a result of an underactive thyroid gland [16]. Internal events also include recollection of past external events that are stored in human memory without any direct external stimuli. In order to maintain simplicity in this preliminary model, we only consider external events for which we use the term life events in this paper.

Beliefs can be considered as an essential core component that mediates behavioural and emotional responses according to the Attribution and Motivation theory [17] (we use the term belief rather than attribution, which is used in Attribution theory). Whilst beliefs can be of broader categories, for the purpose of this preliminary model we have restricted them to those defined in CST in relation to cognitive schema. The CST defines 18 different cognitive schemas, which are complex structures consisting of beliefs, emotions, and memories (see the cited publications on the CST for details of 18 different belief categories/schemas). It is important to note that beliefs are in dynamic states [14], which can be shifted by life events. For example, a belief state that one can change his/her adverse circumstance can evolve in to a state of helplessness as a result of repeated failed attempts to change the situation [18].

One of the key question is 'what entails a life event?' For the purpose of this pre-liminary model we have abstracted life events as a force that impact on each belief differently. For example, consider two beliefs: (1) 'I am not good enough'; (2) 'people are not trustworthy'. A life event resulting in a failure of an exam is likely to affect the first belief by making it stronger more than the second belief. As the states of beliefs undergo changes in response to life events, it results in various emotion and behaviour responses.

In the above example, the individual is likely to experience sadness as the result, and might choose a behavioural response from a range of options including reattempting and quitting. Whilst there are six categories of emotions, which include joy, sadness, love, anger, fear, and surprise [19], CST also defines 3 types of behaviour responses consisting of surrender, avoidance, and overcompensation [14].

The process by which life events generates emotional and behavioural responses can be understood using the concept of cognitive appraisal, which is described in the Stress and Coping Theory [20]. The cognitive appraisal of life events involves a complex process consisting of primary and secondary appraisal by which the perception of a life event is processed and its impact on the individual is determined [20]. The primary cognitive appraisal is mediated by beliefs, memories of past life events, and the cognitive control (also known as executive control), which is exerted by the individual in regulating the impact of life events [21, 22]. Individuals, which can exert an increased level of cognitive control can change their beliefs and emotions, and minimise the negative impact from life events. The process of primary appraisal is described in Fig. 1.

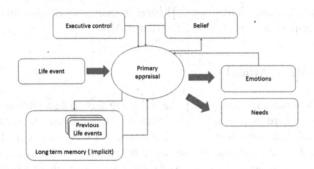

Fig. 1. The primary appraisal of life events

Secondary cognitive appraisal involves evaluating what actions can be taken in response to the life event [20]. As described in Fig. 2, the secondary appraisal can be conceptualised as a process mediated by a number of factors including cognitive control, beliefs, emotions, and needs. Whilst there can be a large number of needs they can be grouped in to several categories including physical and emotional needs. For example, CST recognises five types of core emotional needs [14], which include: secure attachment to others; autonomy, competency and sense of identity; freedom to express; spontaneity and play; realistic limits and self-control.

It is reasonable to assume that the most appropriate action in response to a life event is primarily determined by needs, which are altered according to the changes in beliefs in the primary appraisal. Whilst a need grows initially as the corresponding belief becomes stronger, it will start to diminish after a certain point. For example, as was demonstrated in experiments related to the concept of learned helplessness in which an aversive stimulus (e.g. electric shock) is given to the experimental subjects confined to

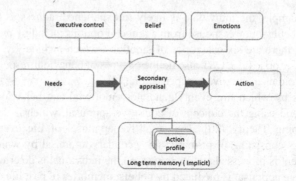

Fig. 2. The secondary appraisal of life events

the situation, the need for an escape increases as the belief of being confined increases until a certain point when the subjects start to give up as they find themselves helpless [23, 24].

Therefore, when the degree of need reaches a certain threshold, an action is taken. According to the Cognitive Dissonance Theory, an individual chooses an action which is in most agreement with his/her belief system [25].

3   Mathematical Model

In developing the mathematical model, the following section first introduces notations and definitions that are used in subsequent sections in describing emotional responses, changes in beliefs, and behaviours (i.e. actions) in response to life events. Based on the above described conceptual model, we have defined needs, beliefs, emotions, life events, cognitive control, and actions as follows. For maintaining simplicity in this preliminary model, the cognitive control has been kept a constant even though it can change due to various biological and psychological factors. Since life events can be considered as the main independent variable or input to the primary appraisal process (see Fig. 1), the emotional responses and changes in beliefs are modelled as a system of difference equations that changes from the nth state to the (n + 1)th state after the nth life event. Beliefs can be considered as the core variables that links life events to emotions and actions. Therefore, beliefs, emotions and life events are represented as vectors in R^{18} in relation to 18 different schemas in the conceptual model from a clinical perspective. A life event is represented as a vector of real values representing the rates at which the corresponding beliefs get changed after experiencing the life event. It can be considered that an individual has an access to a repertoire of various actions, which can be categorised in relation to the previously stated three coping mechanisms that were described in the CST.

N_i: i th need, $i = 1, 2, 3, \ldots$
$q(N_i) \in [0, 1]$: degree of i th need.
B_j: j th belief category, $j = 1, 2, 3, \ldots 18$.

$q(B_j) \in [-1, 1]$: amount of belief in j th belief category (i.e. degree of conviction). It is important to note that beliefs can have opposite states corresponding to positive and negative values. For example, the belief 'I am a failure' can be considered as the opposite of 'I am invincible'. Also, a belief may take different forms ranging from 'just a thought' (e.g. when $q(B_j)$ is closer to zero) to a fact or delusion (e.g. when $q(B_j) = 1$) depending on $q(B_j)$.

$\overrightarrow{q(B_n)} = <q(B_{n1}), \ldots, q(B_{n18})>$: state of the belief system after n th life event.

$\overrightarrow{L_n} = <l_{n1} \ldots, l_{n18}>$: n th life event where $l_{nj} \in [-1, 1]$ is the rate at which jth belief gets changed, and can have either positive or negative value depending on whether the life event is going to increase or decrease the belief state. For example, getting a work promotion can decrease the belief 'I am a failure' whereas a relationship breakup may increase it.

E_k: kth emotion, $k = 1, 2, \ldots, 6$. $q(E_k) \in [0, 1]$: degree of kth emotion.

$\overrightarrow{q(E_n)} = <q(E_{n1}), \ldots, q(E_{n18})>$: emotional state after nth life event.

Ψ: amount of executive control

$A = [\overrightarrow{A_1} \overrightarrow{A_2} \overrightarrow{A_3} \ldots]$: all possible actions where $\overrightarrow{A_i} = <a_k(B_1), \ldots, a_k(B_{18})>$, $i = 1, 2, 3 \ldots$ and $a_i(B_j) \in [-1, 1]$ is the level of agreement between belief B_j and the action $\overrightarrow{A_i}$. Whilst $a_i(B_j) = 1$ represents the full agreement, $a_i(B_j) = -1$ represents a complete disagreement; $a_i(j) = 0$ represents neutral state (i.e. neither agrees or disagrees).

Impact of a life event in changing a belief is determined by a cumulative effect of similar past experiences, emotional state and the cognitive control. It can be conceptualised that the influence of past life experiences in changing a belief is mediated via the corresponding beliefs that progressively undergo changes or evolve as life events occur. For example, the belief 'I am not good enough' can progressively evolve as a result of life events that can be perceived as a failure (e.g. relationship breakup, failing an exam). Based on clinical observations, when a belief is stronger, life events tend to make it stronger compared to weaker belief. Similarly, beliefs tend to change proportionately to relevant emotional states. For example, if an individual is feeling sad, his/her belief 'I am not good enough' tends to grow stronger after an adverse life event compared to state of feeling happy. On the other hand, the greater the cognitive control the lesser is the change in belief. Based on these proportionality relations, a difference equation for change in belief state $\Delta q(B_j) = q(B_{(n+1)j}) - q(B_{nj})$ can be modelled as a follows.

$$\Delta q(B_j) \propto q(B_j); \Delta q(B_j) \propto q(E_k); \Delta q(B_j) \propto -\Psi$$

$$q(B_{(n+1)j}) - q(B_{nj}) = l_{nj}q(B_{nj}) + \sum_{k=1}^{k=6} e_{kj}q(E_{nk}) - \Psi$$

where e_{kj} is the rate at which the jth belief is changed by the kth emotion. Since we restrict degree of belief to $[0, 1]$,

$$q\left(B_{(n+1)j}\right) = \begin{cases} 1, \text{ if } q(B_{nj}) > 1 \\ (1+l_{nj})q(B_{nj}) + \sum_{k=1}^{k=6} e_{kj}q(E_{nk}) - \Psi, \text{ if } 1 \geq q(B_{nj}) \geq 0 \\ -1, \text{ if } q(B_{nj}) < -1 \end{cases}$$

Based on clinical observations, belief and emotions feed each other. Whilst beliefs tend to change proportionately to relevant emotional states, emotions also tend to change in proportion to belief state. For example, the stronger the belief 'I am not good enough', the higher the experience of sadness compared to a weaker belief. Therefore, this proportionality relation can be used to derive a difference equation for change in emotions $\Delta q(E_k) = q\left(E_{(n+1)k}\right) - q(E_{nk})$ after the nth life event as follows.

$$\Delta q(E_k) \propto q(B_j)$$

$$q\left(E_{(n+1)k}\right) - q(E_{nk}) = \sum_{j=1}^{k=18} b_{jk}q(B_{nj})$$

where b_{jk} is the rate at which the kth emotion is changed by the jth belief. Since we restrict emotional states to $[0, 1]$,

$$q\left(E_{(n+1)k}\right) = \begin{cases} 1, \text{ if } q(E_{nk}) > 1 \\ q(E_{nk}) + \sum_{j=1}^{j=18} b_{jk}q(B_{nj}), \text{ if } 1 \geq q(E_{nk}) \geq 0 \\ 0, \text{ if } q(E_{nk}) < 0 \end{cases}$$

In order to elaborate the above described changes in beliefs and emotional states, consider the following simple example consisting of only 2 emotions and 2 beliefs as follows.

B_1: defectiveness, with the initial value: $q(B_{01}) = 0.5$ (e.g. "I am not good enough"). B_2: mistrust, with the initial value $q(B_{02}) = 0.5$ (e.g. "People are not trustworthy"). E_1: Sadness, with the initial value $q(E_{01}) = 0.5$. E_2: Anger, with the initial value $q(E_{02}) = 0.5$

Based on the above described models, the difference equations for change in the two beliefs can be derived as follows.

$$q\left(B_{(n+1)1}\right) = (1+l_{n1})q(B_{n1}) + e_{11}q(E_{n1}) + e_{21}q(E_{n2}) - \Psi$$
$$q\left(B_{(n+1)2}\right) = (1+l_{n2})q(B_{n2}) + e_{12}q(E_{n1}) + e_{22}q(E_{n2}) - \Psi$$

Similarly, the difference equations for change in the two emotional states can be derived as follows.

$$q\left(E_{(n+1)1}\right) = q(E_{n1}) + b_{11}q(B_{n1}) + b_{21}q(B_{n2})$$

$$q\left(E_{(n+1)2}\right) = q(E_{n2}) + b_{12}q(B_{n1}) + b_{22}q(B_{n2})$$

Based on clinical observations, sadness tends to increase the belief 'I am not good enough' more than anger, and therefore $e_{11} > e_{21}$. Similarly, anger directed at other people tends to increase the belief 'people are not trustworthy' more than sadness, which means $e_{22} > e_{12}$. In relation to changes in emotions, sadness is increased more by the belief 'I am not good enough' compared to the belief 'people are not trustworthy' meaning $b_{11} > b_{21}$. Similarly, anger is increased more by the belief 'people are not trustworthy' compared to the belief 'I am not good enough' meaning $b_{22} > b_{12}$.

A simulation result of the above model using 20 random life events and the following arbitrary values are shown in Fig. 3. It is important to note that even though the following values are arbitrary, they were chosen in a meaningful way using clinical intuition and expertise.

$$e_{jk} = \begin{bmatrix} 0.05 & 0.02 \\ 0.03 & 0.05 \end{bmatrix}, b_{jk} = \begin{bmatrix} 0.06 & 0.02 \\ 0.03 & 0.05 \end{bmatrix}, \Psi = 0.08.$$

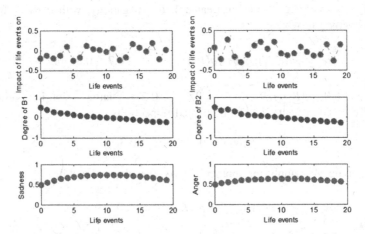

Fig. 3. A pattern of slowly decreasing belief states resulting in an overall stable emotional states as life events occur

In determining what actions to be taken, beliefs and needs play an important role. It is reasonable to consider that an individual will decide to act once the need grows surpassing a threshold value t as modelled by the following function.

$$\Phi(q(N_i)) = \begin{cases} Choose\ an\ action, q(N_i) > t \\ nil\ action, q(N_i) \le t \end{cases}$$

Given that the belief system and actions are represented as vectors, the agreement between the belief system $\overrightarrow{q(B_n)}$ and a given action $\overrightarrow{A_i}$ can be determined via the vector projection method [26], which can be used to derive the similarity between the two vectors as follows.

$$L\left(\overrightarrow{q(B_n)}, \overrightarrow{A_i}\right) = \frac{\left|\overrightarrow{A_i}\right| \cdot \left|\overrightarrow{q(B_n)}\right|}{\left|\overrightarrow{q(B_n)}\right|^2}$$

where $\left|\overrightarrow{A_i}\right|$ and $\left|q(B_n)\right|$ represents the lengths of these vectors, and $L\left(\overrightarrow{q(B_n)}, \overrightarrow{A_i}\right)$ is the agreement between the belief system $\overrightarrow{q(B_n)}$ and the action $\overrightarrow{A_i}$. $L\left(\overrightarrow{q(B_n)}, \overrightarrow{A_i}\right)$ is calculated for each $\overrightarrow{A_i}, i = 1, 2, 3, \ldots$ and the action that is associated with the highest value is selected as the most appropriate action.

In order to elaborate on the process described above, consider the following simple example consisting of a set of only 3 beliefs categories from the CST [14], 3 actions, and one life event. The secondary appraisal involves choosing the action, which is in most agreement with the system of 3 beliefs.

B_1 : defectiveness, with $q(B_1) = -0.7$ (e.g. "I am not good enough"). B_2 : mistrust, with $q(B_2) = -0.85$(e.g. "I get mistreated"). B_3 : negativity, with $q(B_3) = 0.9$ (e.g. "nothing is going to work").

$$\overrightarrow{q(B)} = <q(B_1), q(B_2), q(B_3) > = < -0.7, -0.85, 0.9 >$$

$\overrightarrow{L(t)} = <l(B_1), l(B_2), l(B_3) > = <0.6, 0.2, 0.2 > $: failing a job interview
N_1 : To be competent. N_2 : To be treated respectfully.

$$q(N_1) = 0.5; q(N_2) = 0.3$$

A_1 : give up. A_2 : apply for another job. A_3 : develop more skills

$$\overrightarrow{A_1} = <a_1(B_1), a_1(B_2), a_1(B_3) > = <0.8, 0.85, 0.9 >$$

$$\overrightarrow{A_2} = <a_2(B_1), a_2(B_2), a_2(B_3) > = <0.5, 0.6, -0.7 >$$

$$\overrightarrow{A_3} = <a_3(B_1), a_3(B_2), a_3(B_3) > = <0.9, 0.5, -0.7 >$$

$$L\left(\overrightarrow{q(B)}, \overrightarrow{A_1}\right) = 1.0364; L\left(\overrightarrow{q(B)}, \overrightarrow{A_2}\right) = 0.7375; L\left(\overrightarrow{q(B)}, \overrightarrow{A_3}\right) = 0.8754$$

Therefore, A_1 which has the highest agreement with the belief set will be chosen as the preferred action.

4 Conclusion

Whilst the above described model is able to explain behavioural and emotional changes that are caused by life events as a result of changes in beliefs and emotional states, we acknowledge its limitations and simplicity. Firstly, we have deliberately omitted internal events that can also cause behavioural and internal changes, in order to maintain the simplicity. The model assumes that the variables related to these internal events remains constant, and the initial states of the model variables embody the effect of individual genetic makeup, which is also known to influence behavioural and emotional responses [27, 28]. Also, cognitive/executive control is included as a constant for simplicity whilst it can vary over time.

The model can be used to study emotional disorders in psychiatry by simulating various clinical scenarios as demonstrated in Fig. 3. Even though the model has been developed mainly from a clinical perspective, it can be adopted to develop software systems such as autonomous agents that mimic human behaviour. In adapting or extending the model, it is possible to abstract the main concepts of the model including life events and cognitive schema in different ways depending on the purpose that the model is going to serve. Also, whilst the beliefs have been restricted to only 18 different categories as per the CST from a clinical perspective, these can be expanded for the purpose of non-clinical applications.

Finally, the model coefficients and constant parameters (e.g. rates at which a life event can change a belief) were chosen arbitrarily and approximated using clinical expertise. It will be possible to determine some of the parameters more accurately by model fitting as this preliminary model will be further enhanced and tested with real data in future. It is expected that the model will serve as the foundation for further work in clinical research and in developing more realistic computational models for specific applications.

References

1. Elliott, C.: The Affective Reasoner: A Process Model of Emotions in a Multi-Agent System. (1992) http://condor.depaul.edu/elliott/ar/papers/dis/elliott-phd.html
2. El-Nasr, M.S., Yen, J., Ioerger, T.R.: FLAME—fuzzy logic adaptive model of emotions. Auton. Agent. Multi-Agent Syst. **3**, 219–257 (2000)
3. Bates, J., Loyall, A., Reilly, W.: An architecture for action, emotion, and social behavior. In: 4th European Workshop on Modelling Autonomous Agents in a Multi-Agent World, Artificial Social Systems, pp. 55–68 (1994)
4. Velásquez, J.D.: Modeling emotions and other motivations in synthetic agents. In: Proceedings of the Fourteenth National Conference on Artificial Intelligence and Ninth Conference on Innovative Applications of Artificial Intelligence, Providence, Rhode Island (1997)
5. Pynadath, D.V., Marsella, S.C.: PsychSim: modeling theory of mind with decision-theoretic agents. In: Proceedings of the 19th International Joint Conference on Artificial Intelligence, Edinburgh, Scotland (2005)
6. Kazemifard, M., Ghasem-Aghaee, N., Ören, T.I.: Design and implementation of GEmA: A generic emotional agent. Expert Syst. Appl. **38**, 2640–2652 (2011)
7. Ortony, A., Clore, G.L., Collins, A.: The Cognitive Structure of Emotions. Cambridge University Press, Cambridge (1990)
8. Gross, J.J.: The emerging field of emotion regulation: An integrative review. Rev. General Psychol. **2**, 271–299 (1998)

9. Whiten, A.: Natural Theories of Mind: Evolution, Development, and Simulation of Everyday Mindreading: B. Blackwell, Oxford (1991)
10. Cacciabue, P.C.: Modelling and Simulation of Human Behaviour in System Control. Springer, Heidelberg (1998)
11. Klein, J., Moon, Y., Picard, R.W.: This computer responds to user frustration: Theory, design, and results. Interact. Comput. **14**, 119–140 (2002)
12. Maldonado, H., Lee, J.-E.R., Brave, S., Nass, C., Nakajima, H., Yamada, R., et al.: We learn better together: enhancing eLearning with emotional characters. In: Proceedings of th 2005 Conference on Computer Support for Collaborative Learning: Learning 2005: the next 10 years!, Taipei, Taiwan (2005)
13. Young, J.E.: Cognitive Therapy for Personality Disorders: A Schema-Focused Approach: Professional Resource Exchange (1999)
14. Young, J.E., Klosko, J.S., Weishaar, M.E.: Schema Therapy: A Practitioner's Guide. Guilford Press (2003)
15. Greger, R., Windhorst, U.: Comprehensive Human Physiology: From Cellular Mechanisms to Integration. Springer, Heidelberg (2013)
16. Hage, M.P., Azar, S.T.: The link between thyroid function and depression. J. Thyroid Res. **2012**, 590–648 (2012)
17. Weiner, B.: An attributional theory of achievement motivation and emotion. Psychol. Rev. **92**, 548 (1985)
18. Garber, J., Seligman, M.E.: Human Helplessness: Theory and Applications. Cambridge Univ Press, Cambridge (1980)
19. Shaver, P., Schwartz, J., Kirson, D., O'Connor, C.: Emotion knowledge: further exploration of a prototype approach. J. Pers. Soc. Psychol. **52**, 1061–1086 (1987)
20. Folkman, S., Lazarus, R.S., Dunkel-Schetter, C., DeLongis, A., Gruen, R.J.: Dynamics of a stressful encounter: cognitive appraisal, coping, and encounter outcomes. J. Pers. Soc. Psychol. **50**, 992 (1986)
21. Botvinick, M.M., Braver, T.S., Barch, D.M., Carter, C.S., Cohen, J.D.: Conflict monitoring and cognitive control. Psychol. Rev. **108**, 624–652 (2001)
22. Casey, B.J., Tottenham, N., Fossella, J.: Clinical, imaging, lesion, and genetic approaches toward a model of cognitive control. Dev. Psychobiol. **40**, 237–254 (2002)
23. Seligman, M.E., Beagley, G.: Learned helplessness in the rat. J. Comp. Physiol. Psychol. **88**, 534 (1975)
24. Hiroto, D.S., Seligman, M.E.: Generality of learned helplessness in man. J. Pers. Soc. Psychol. **31**, 311 (1975)
25. Festinger, L.: A Theory of Cognitive Dissonance, vol. 2. Stanford University Press, Stanford (1962)
26. Fernando, D.A.I., Henskens, F.A.: A modified case-based reasoning approach for triaging psychiatric patients using a similarity measure derived from orthogonal vector projection. In: Chalup, S.K., Blair, A.D., Randall, M. (eds.) ACALCI 2015. LNCS, vol. 8955, pp. 360–372. Springer, Heidelberg (2015)
27. Minnis, H., Reekie, J., Young, D., O'connor, T., Ronald, A., Gray, A., et al.: Genetic, environmental and gender influences on attachment disorder behaviours. Br. J. Psychiatry **190**, 490–495 (2007). 2007-06-01 00:00:00
28. Breed, M., Sanchez, L.: Both environment and genetic makeup influence behavior. Nat. Educ. Knowl. **3**, 68 (2010)

Artificial Bee Colony Algorithm Based on Neighboring Information Learning

Laizhong Cui[1(✉)], Genghui Li[1], Qiuzhen Lin[1], Jianyong Chen[1], Nan Lu[1], and Guanjing Zhang[2]

[1] College of Computer Science and Software Engineering, Shenzhen University, Shenzhen 518060, Guangdong, China
{cuilz,qiuzhlin,jychen,lunan}@szu.edu.cn, ligenghuigm@gmail.com
[2] E-Techco Information Technologies Co., Ltd., Shenzhen 518060, China
john.zhang@e-techco.com

Abstract. Artificial bee colony (ABC) algorithm is one of the most effective and efficient swarm intelligence algorithms for global numerical optimization, which is inspired by the intelligent foraging behavior of honey bees and has shown good performance in most case. However, due to its solution search equation is good at exploration but poor at exploitation, ABC often suffers from a slow convergence speed. In order to solve this concerning issue, in this paper, we propose a novel artificial bee colony algorithm based on neighboring information learning (called NILABC), in which the employed bees and onlooker bees search candidate food source by learning the valuable information from the best food source among their neighbors. Furthermore, the size of the neighbors is linearly increased with the evolutionary process, which is used to ensure the employed bees and onlooker bees obtain the guidance from the best solution in local area at the early stage and the best solution in the global area at the late stage. Through the comparison of NILABC with the basic ABC and some other variants of ABC on 22 benchmark functions, the experimental results demonstrate that NILABC is better than the compared algorithms on most cases in terms of solution quality, robustness and convergence speed.

Keywords: Evolutionary algorithm · Artificial bee colony algorithm · Neighboring information learning · Ranking-based probability selection · Global numerical optimization

1 Introduction

Global optimization is a basic research field with extensive applications involved almost all areas of science and engineering, and therefore optimization techniques have been continuously playing a very important role. In general, some global optimizations problems can be characterized as non-convexity, discontinuity, non-differentiability and multimodality, but it has been becoming a very difficult task that they are solved by traditional derivative-based techniques.

© Springer International Publishing AG 2016
A. Hirose et al. (Eds.): ICONIP 2016, Part III, LNCS 9949, pp. 279–289, 2016.
DOI: 10.1007/978-3-319-46675-0_31

Enlightened by nature selection and survival of fittest, many derivative-free methods have been proposed, such as genetic algorithm (GA) [1], differential evolution (DE) [2], particle swarm optimization (PSO) [3] and artificial bee colony (ABC) [4] algorithm, which have shown enormous potentiality to address complex optimization problems.

In this paper, we focus on the study of ABC, which is presented by Karaboga [4] and inspired by the intelligent foraging behavior of honey bees. The performance of ABC has been validated by the comparison of ABC and others Evolutionary algorithms (EAs) [4,5]. Although ABC has shown outstanding performance, it is also facing the challenge of slow convergence likes other EAs. It is mainly caused by its solution search equation, which does well in exploration but badly in exploitation [7]. To address this concerning issue, many ABC variants have been developed in the last decade, which can be mainly divided into two categories, namely invention of the new solution search equations and combination with other techniques. For instance, Zhu et al. [6] presented a gbest-guided ABC by designing a new search equation, which exploits the information of the current best solution to enhance the exploitation ability of ABC. Xiang et al. [7] invented a novel search equation, which employs the information of the best solution and other good solutions. Banharnsakun et al. [8] put forward an ABC variant, which includes a best-so-far selection method, an adjustable search radius and an objective-value-based comparison method to enhance exploitation and exploration of ABC. Recently, Karaboga et al. [10] proposed a quick ABC, in which a new version of search equation exploiting the information of the best solution among the neighbors was introduced into the onlooker bee phase. In addition, since different search equations have distinct advantages and perform differently on different problems or at different stage on the same problems, some methods using multiple equations were proposed to enhance the comprehensive performance of ABC, such as MEABC [11] and ABCVSS [12]. On the other hand, Kang et al. [13] proposed a Rosenbrock ABC, in which the exploration phase was realized by ABC and the exploitation phase was completed by the rotational direction method. Kang et al. [14] proposed a memetic algorithm, which combined Hooke-Jeeves pattern search with ABC. Gao et al. proposed the orthogonal learning method in [9], and the chaotic map and opposition-based learning method in [15] to improve the performance of ABC. In this paper, we concentrate on the first category, and propose an improved ABC called NILABC with a new solution search equation, which learns from the beneficial information from the best solution among neighbors, which can effectively enhance the performance of ABC.

The remainder of this paper is organized as follow. Section 2 introduces the original ABC algorithm briefly. The details of our proposed algorithms are described in Sect. 3. The experimental evaluations of our proposed algorithm are presented in Sect. 4. Finally, Sect. 5 concludes this paper.

2   The Basic ABC Algorithm

ABC is a swarm-based stochastic optimization method, which is inspired by the intelligent foraging behavior of honey bees [5]. In ABC, the position of a food source represents a possible solution of the optimization problem, and the nectar amount of each food source corresponds to the quality (fitness) of the associated solution. The colony includes three groups of bees. The first half of the colony consists of employed bees and the second half of the colony is composed of onlooker bees. If the quality of a food source is not improved by a preset number of times (*limit*), this food source is abandoned by its employed bee, and then this employed bee becomes a scout bee that begins to seek a new random food source. The original ABC algorithm includes four phases, i.e., initialization phase, employed bee phase, onlooker bee phase and scout bee phase. After the initialization phase, ABC turns into a circulation of employed bee phase, onlooker bee phase and scout bee phase until the termination condition is satisfied. The detailed process of each phase is described as follows.

Initialization phase: At the beginning of ABC, the initial food sources are generated randomly according to Eq. 1

$$X_{i,j}^0 = X_j^L + r_j \cdot (X_j^U - X_j^L), \quad i = 1, ..., SN, \quad j = 1, ..., D \qquad (1)$$

where SN is the number of food sources, and is equal to the number of employed bees and onlooker bees; D is the dimension (variables) of the search space; X_j^L and X_j^U are the lower and upper bounds of the jth variable respectively; r_j is a random real number in the range of [0,1]. The fitness values of the food sources are obtained by Eq. 2

$$fit_i = \begin{cases} \dfrac{1}{1 + f_i} & \text{if } (f_i \geq 0) \\ 1 + |f_i| & \text{otherwise} \end{cases} \qquad (2)$$

where fit_i is the fitness value of the ith food source X_i, and f_i is the objective function value of food source X_i.

Employed bee phase: Each employed bee flies to a distinct food source (solution) and finds a candidate food source (new solution) by using Eq. 3

$$V_{i,j}^G = X_{i,j}^G + \phi_{i,j} \cdot (X_{i,j}^G - X_{k,j}^G) \qquad (3)$$

where G is the generation count; k is picked up from $\{1, 2, ..., SN\}$ randomly and $k \neq i$; j is randomly selected from $\{1, 2, ..., D\}$; $\phi_{i,j}$ is a random real number in the range of [−1,1]. If the candidate food source V_i is better than the old food source X_i, V_i will replace X_i. Otherwise, the old food source is kept.

Onlooker bee phase: when all employed bees have completed their searches, they will share the quality information of the food sources with onlooker bees. Each onlooker bee will apply the roulette method to select a candidate food source with a probability P_i. The probability P_i associated with ith food source is calculated as Eq. 4. Similarly, if the candidate food source obtained by the onlooker bee is better than the old food source, the old food source will be replaced by the new one. Otherwise, the old food source is kept.

$$P_i = fit_i / \sum_{i=1}^{SN} fit_i \tag{4}$$

Scout bee phase: If any better food source cannot be found through a predetermined number of search times in the neighborhood of a certain source position, the corresponding food source will be discarded by its employed bee, and this employed bee will become a scout bee to search a new food source randomly according to Eq. 1. Note that in each generation, there is only one scout bee at most.

Note that if the jth variable of the ith candidate food source $V_{i,j}^G$ violates the boundary constraints on employed bee phase and onlooker bee phase, $V_{i,j}^G$ will be reset according to Eq. 1.

3   The Proposed Algorithm (NILABC)

In this section, in order to accelerate the convergence speed without losing diversity of ABC and improve the performance of ABC, we present a new ABC variant, called NILABC. In NILABC, we design a new solution search equation by introducing a novel neighboring information learning model, which exploits the valuable information of the best solution among some local areas.

In original ABC, the employed bees in employed bee phase only conduct independent search, and the valuable information is not fully shared between them, which also exists in onlooker bee phase. Therefore, this kind of independent searching model makes ABC do well in exploration but badly in exploitation. Actually, in real honey bee colonies, employed bees (or onlooker bees) could exchange useful information with each other. Based on this fact, when an employed bee (or onlooker bee) finds a candidate food source, it should exploits the useful information from other employed bees (or onlooker bees). Considering this phenomenon, we propose a novel neighboring information learning model as follows.

When searching a candidate food source, an employed bee (or onlooker bee) will be guided by the existing best solution in its visual scope. The novel neighboring information learning model is defined as Eq. 5

$$V_{i,j}^G = X_{nbest,j}^G + \phi_{i,j} \cdot (X_{i,j}^G - X_{k,j}^G) \tag{5}$$

In this model, X_{nbest}^G represents the best solution among the neighbors of X_i^G; k is randomly selected from $\{1, 2, ..., SN\}$ randomly and $k \neq i$; j is randomly selected from $\{1, 2, ..., D\}$; $\phi_{i,j}$ is a random real number in the range of $[-1,1]$. For the sake of simplicity, we assume that each bee has the same visual scope and each honey bee selects ns closest bees based on the Euclidean distance as its neighbors. The size of neighbors ns is defined as Eq. 6

$$ns = floor((0.1 + 0.9 \cdot FES/maxFES) \cdot SN)\tag{6}$$

Algorithm 1. The procedure of the NILABC algorithm

1: **Initialization**: Generate SN solutions that contain D variables according to Eq. 1
2: **while** $FES < maxFES$ **do**
3:     Calculate ns using Eq. 6
4:     **for** $i = 1$ to SN **do**
5:         Select the best solution X_{best}^G among the neighbours of X_i^G
6:         Generate a new solution V_i^G using Eq. 5
7:         **if** $f(V_i^G) \le f(X_i^G)$ **then**
8:             Replace X_i^G by V_i^G
9:             $counter(i) = 0$
10:         **else**
11:             $counter(i) = counter(i) + 1$
12:         **end if**
13:     **end for**// end employed bee phase
14:     Calculate probability P using Eq. 7
15:     **for** $i = 1$ to SN **do**
16:         Select a food source X_s^G for the ith onlooker bee with the probability P
17:         Select the best solution X_{best}^G among the neighbours of X_s^G
18:         Generate a new solution V_s^G using Eq. 5
19:         **if** $f(V_s^G) \le f(X_s^G)$ **then**
20:             Replace X_s^G by V_s^G
21:             $counter(s) = 0$
22:         **else**
23:             $counter(s) = counter(s) + 1$
24:         **end if**
25:     **end for**// end onlooker phase
26:     $FES = FES + 2SN$
27:     Select the solution X_{max}^G with the max $counter$ value
28:     **if** $counter(max) > limit$ **then**
29:         Replace X_{max}^G by a new solution generated by Eq. 1
30:         $FES = FES + 1, counter(max) = 0$
31:     **end if**//end scout phase
32: **end while**

---

where FES represents the current number function evaluation, and $maxFES$ denotes the maximal number of function evaluation. $floor(x)$ is a function returning the largest integer that is smaller than its parameter x. By this way, ns is increased with the number of function evaluation FES in $[0.1 \cdot SN, SN]$, which controls the honey bee to obtain the guidance from the current local best solution among its local area in the early stage and the current global best solution in the late stage.

Note that in onlooker bee phase of original ABC, a food source will be searched with the probability P shown in Eq. 4, which is proportional to the fitness value or inversely proportional to the objective function value. In theory, the food source with higher quality will attract more onlooker bee to search. However, if there is no significant difference between the fitness values, all food sources will obtain nearly identical and very small probability P. In this case, a food source is selected by the wheel roulette method with probability P, which is a time-consuming process. In order to make up for the defect, we introduces a probability calculating method based on the ranking of the food sources, which is defined as Eq. 7

$$P_i = \frac{R_i}{SN} \tag{7}$$

while SN is the number of food sources, and R_i is the ranking of the food source i in ascending order among all food sources. For example, the ranking of the current worst food source and best food source are 1 and SN respectively, and therefore the probability of the worst food source and best food source are $1/SN$ and 1 respectively.

Our novel neighboring information learning model serves as the solution search equation and the ranking-based probability calculating method are integrated with the framework of original ABC to form NILABC algorithm. The pseudo-code of the complete algorithm NILABC is demonstrated in Algorithm 1.

4   Experiments and Results

In this section, to evaluate the performance of NILABC, we use a set of 22 scalable benchmark functions with dimensions $D = 30$ [9], which are summarized in Table 1. The search range, the global optimal value, and the "acceptable value" of each function are shown in column 3, 4 and 5 of Table 1 respectively. When the objective function value of the best solution obtained by an algorithm in a run is less than the acceptable value, the run is regarded as a *successful run*. In our performance evaluation, three metrics are considered, described as follows.

(1) The mean and standard deviation of the best objective function value are obtained by each algorithm, which are used to evaluate the quality or accuracy of the solutions. The smaller the value of this metric is, the higher quality/accuracy the solution has.
(2) The average FES (AVEN) is required to reach the acceptable value, which is employed to evaluate the convergence speed. Note that AVEN will be only calculated for the successful runs. If an algorithm cannot find any solution whose objective function value is smaller than the acceptable value in all runs, AVEN will be denoted by "NA".
(3) The successful rate (SR%) of the 25 independent runs is utilized to evaluate the robustness or reliability of different algorithms.

Table 1. Benchmark functions in experiments

Name	Function	Range	Min	Accept
Sphere	$f_1(x) = \sum_{i=1}^{D} x_i^2$	$[-100, 100]^D$	0	1×10^{-8}
Elliptic	$f_2(x) = \sum_{i=1}^{D} (10^6)^{\frac{i-1}{D-1}} x_i^2$	$[-100, 100]^D$	0	1×10^{-8}
SumSquare	$f_3(x) = \sum_{i=1}^{D} i x_i^2$	$[-10, 10]^D$	0	1×10^{-8}
sumPower	$f_4(x) = \sum_{i=1}^{D} |x_i|^{(i+1)}$	$[-1, 1]^D$	0	1×10^{-8}
Schwefel 2.22	$f_5(x) = \sum_{i=1}^{D} |x_i| + \prod_{i=1}^{D} |x_i|$	$[-10, 10]^D$	0	1×10^{-8}
Schwefel 2.21	$f_6(x) = max\{|x_i|, 1 \leq i \leq n\}$	$[-100, 100]^D$	0	1×10^{-0}
Step	$f_7(x) = \sum_{i=1}^{D} (\lfloor x_i + 0.5 \rfloor)^2$	$[-100, 100]^D$	0	1×10^{-8}
Exponential	$f_8(x) = exp(0.5 * \sum_{i=1}^{D} x_i)$	$[-10, 10]^D$	0	1×10^{-8}
Quartic	$f_9(x) = \sum_{i=1}^{D} i x_i^4 + random[0, 1]$	$[-1.28, 1.28]^D$	0	1×10^{-1}
Rosenbrock	$f_{10}(x) = \sum_{i=1}^{D-1} [100(x_{i+1} - x_i^2) + (x_i - 1)^2]$	$[-5, 10]^D$	0	1×10^{-1}
Rastrigin	$f_{11}(x) = \sum_{i=1}^{D} [x_i^2 - 10\cos(2\pi x_i) + 10]$ $f_{12}(x) = \sum_{i=1}^{D} [y_i^2 - 10\cos(2\pi y_i) + 10]$	$[-5.12, 5.12]^D$	0	1×10^{-8}
NCRastrigin	$y_i = \begin{cases} x_i, & |x_i| < \frac{1}{2} \\ round(2x_i)/2, & |x_i| \geq \frac{1}{2} \end{cases}$	$[-5.12, 5.12]^D$	0	1×10^{-8}
Griewank	$f_{13}(x) = 1/4000 \sum_{i=1}^{D} x_i^2 - \prod_{i=1}^{D} \cos(\frac{x_i}{\sqrt{i}}) + 1$	$[-600, 600]^D$	0	1×10^{-8}
Schwefel 2.26	$f_{14}(x) = 418.98288727243380 * D - \sum_{i=1}^{D} x_i \sin(\sqrt{|x_i|})$	$[-500, 500]^D$	0	1×10^{-8}
Ackley	$f_{15}(x) = 20 + e - 20 exp(-0.2\sqrt{\frac{1}{D}\sum_{i=1}^{D} x_i^2}) - exp(\frac{1}{D}\sum_{i=1}^{D} \cos(2\pi x_i))$	$[-50, 50]^D$	0	1×10^{-8}
Penalized1	$f_{16}(x) = \frac{\pi}{D}\{10\sin^2(\pi y_1) + \sum_{i=1}^{D-1}(y_i - 1)^2[1 + \sin^2(\pi y_{i+1})] + (y_D - 1)^2\} + \sum_{i=1}^{D} u(x_i, 10, 100, 4)$ $y_i = 1 + 1/4(x_i + 1),$ $u_{x_i, a, k, m} = \begin{cases} k(x_i - a)^m, & x_i > a \\ 0, & -a \leq x_i \leq a \\ k(-x_i - a)^m, & x_i < a \end{cases}$	$[-100, 100]^D$	0	1×10^{-8}
Penalized2	$f_{17}(x) = \frac{1}{10}\{\sin^2(\pi x_1) + \sum_{i=1}^{D-1}(x_i - 1)^2[1 + \sin^2(3\pi x_{i+1})] + (x_n - 1)^2[1 + \sin^2(2\pi x_{i+1})]\} + \sum_{i=1}^{D} u(x_i, 5, 100, 4)$	$[-100, 100]^D$	0	1×10^{-8}
Alpine	$f_{18}(x) = \sum_{i=1}^{D-1} |x_i \cdot \sin(x_i) + 0.1 \cdot x_i|$	$[-10, 10]^D$	0	1×10^{-8}
Levy	$f_{19}(x) = \sum_{i=1}^{D-1}(x_i - 1)^2[1 + sin^2(3\pi x_{i+1})] + sin^2(3\pi_1) + |x_D - 1|[1 + sin^2(3\pi x_D)]$	$[-10, 10]^D$	0	1×10^{-8}
Weierstrass	$f_{20}(x) = \sum_{i=1}^{D}(\sum_{k=0}^{kmax}[a^k \cos(2\pi b^k(x_i + 0.5))] - D\sum_{k=0}^{kmax}[a^k \cos(2\pi b^k 0.5)], a = 0.5, b = 3, kmax = 20$	$[-1, 1]^D$	0	1×10^{-8}
Himmelblau	$f_{21}(x) = 1/D \sum_{i=1}^{D}(x_i^4 - 16x_i^2 + 5x_i)$	$[-5, 5]^D$	-78.33236	-78
Michalewicz	$f_{22}(x) = -\sum_{i=1}^{n} \sin(x_i) \sin^{20}(\frac{i \times x_i^2}{\pi})$	$[0, \pi]^D$	-30,-50	-29

NILABC is compared with four ABC variants, i.e., original ABC, GABC [6], best-so-far ABC [8], qABC [10]. For all the compared algorithms, the initial population is generated randomly according to Eq. 1, and the *maxFES* is used as the termination condition, which is set to 5000*D*. The parameter settings of all the compared algorithms are set as used in the original papers, except that *SN* and *limit* are set to 50 and $SN \cdot D$ respectively. In order to show the differences between NILABC and other algorithms, the Wilcoxon's rank sum test is conducted on the experimental results at a 5 % significant level. The experimental results are presented in Table 2 and the best results are marked with boldface. It is noted that the symbols "-", "+", "=" denote that the performance of the corresponding algorithm is worse than, better than and similar to that of NILABC respectively.

Table 2. Comparisons of NILABC and other ABC variants on 22 test functions with 30D

Alg	ABC mean(std) SR/AVEN	GABC mean(std) SR/AVEN	Best-so-far ABC mean(std) SR/AVEN	qABC mean(std) SR/AVEN	NILABC mean(std) SR/AVEN
f_1	1.04e-17(1.20e-17)- 100/83702	1.07e-30(6.09e-31)- 100/53134	6.03e-17(7.53e-17)- 100/95594	3.38e-15(5.42e-15)- 100/72922	**7.89e-37(1.25e-36)** 100/47282
f_2	4.38e-10(4.72e-10)- 100/136290	2.64e-24(2.48e-24)- 100/81406	9.30e-11(2.52e-10)- 100/128730	1.31e-10(2.21e-10)- 100/124310	**1.38e-33(8.60e-34)** 100/60286
f_3	1.14e-19(9.89e-20)- 100/75402	6.14e-32(4.74e-32)- 100/48446	5.82e-19(6.63e-19)- 100/87154	2.47e-16(2.43e-16)- 100/63730	**1.00e-37(8.94e-38)** 100/44894
f_4	2.02e-31(5.30e-31)- 100/23578	6.78e-50(2.91e-49)- 100/14890	6.33e-30(2.06e-29)- 100/22014	2.99e-21(5.31e-21)- 100/15710	**6.56e-62(2.71e-61)** 100/12702
f_5	7.69e-11(3.04e-11)- 100/124870	8.30e-17(3.10e-17)- 100/81006	5.08e-10(1.76e-10)- 100/135350	1.17e-08(3.31e-09)- 36/148340	**1.08e-19(4.32e-20)** 100/72066
f_6	4.39e+00(1.07e+00)- 0/NA	2.47e-01(5.89e-02)= 100/110270	1.51e+01(1.62e+00)- 0/NA	**9.87e-02(2.30e-02)+** 100/37742	2.14e-01(1.03e-01) 100/105190
f_7	0(0)= 100/10994	0(0)= 100/10606	0(0)= 100/63458	0(0)= 100/7298	0(0) 100/19102
f_8	7.18e-66(5.21e-73)= 100/150	7.18e-66(1.07e-75)= 100/150	7.18e-66(5.50e-71)= 100/150	7.18e-66(3.50e-72)= 100/150	**7.18e-66(2.36e-78)** 100/150
f_9	6.02e-02(1.09e-02)= 100/91786	3.08e-02(6.44e-03)- 100/48050	7.41e-02(1.44e-02)- 96/113930	2.73e-02(6.83e-03)- 100/12562	**2.57e-02(5.02e-03)** 100/42346
f_{10}	**5.45e-02(5.86e-02)+** 88/11014	4.78e+00(1.54e+01)- 60/82303	3.39e-01(2.09e-01)- 12/149450	5.47e-01(5.00e-01)= 24/82817	3.51e-01(7.29e-01) 52/127170
f_{11}	3.50e-14(1.35e-13)- 100/99134	0(0)= 100/74522	2.11e+00(6.70e-01)- 0/NA	1.33e-10(2.15e-10)- 100/109150	0(0) 100/62090
f_{12}	1.70e-12(4.36e-12)- 100/112080	8.38e-15(4.19e-14)- 100/83602	3.33e+00(1.04e+00)- 0/NA	5.28e-10(5.37e-10)- 100/122610	0(0) 100/68510
f_{13}	2.36e-14(5.62e-14)- 100/94862	1.48e-11(7.39e-11)- 100/71778	6.00e-04(2.97e-03)- 84/118800	5.47e-12(2.39e-11)- 100/95710	**4.89e-17(2.44e-16)** 100/59078
f_{14}	4.58e-12(1.59e-12)- 100/82946	3.42e-12(6.03e-13)- 100/65126	5.12e+01(6.56e+01)- 0/NA	4.16e-10(1.05e-09)- 100/116770	0(0) 100/61414
f_{15}	4.31e-09(1.85e-09)- 100/145410	3.71e-14(5.78e-15)- 100/95798	4.44e-05(3.24e-05)- 0/NA	1.67e-06(8.22e-07)- 0/NA	**6.22e-15(0.00e+00)** 100/80242
f_{16}	1.03e-18(6.90e-19)- 100/77346	7.53e-32(3.06e-32)- 100/48178	5.18e-19(8.51e-19)- 100/85538	2.31e-14(9.81e-14)- 100/61070	**1.57e-32(5.59e-48)** 100/42394
f_{17}	4.88e-18(5.03e-18)- 100/86542	4.83e-31(2.50e-31)- 100/52566	2.62e-16(3.60e-16)- 100/101900	1.72e-15(1.86e-15)- 100/75426	**1.50e-33(0.00e+00)** 100/48294
f_{18}	2.35e-06(1.66e-06)- 0/NA	5.22e-07(1.12e-06)- 20/144750	3.16e-04(3.55e-04)- 0/NA	8.87e-06(2.10e-05)- 0/NA	**1.56e-14(7.70e-14)** 100/87166
f_{19}	4.46e-14(5.39e-14)- 100/90558	1.40e-30(1.63e-30)- 100/52834	3.41e-15(4.40e-15)- 100/98006	2.40e-09(3.34e-09)- 96/125240	**1.35e-31(2.23e-47)** 100/46242
f_{20}	2.06e-02(2.35e-02)- 0/NA	3.62e-02(0.00e+00)- 0/NA	1.90e-01(6.55e-02)- 0/NA	1.10e-02(1.03e-02)- 0/NA	**9.77e-09(1.92e-08)** 76/136770
f_{21}	-78.332(0.00e+00)= 100/26594	-78.332(5.02e-15)- 100/16442	-78.332(4.14e-10)= 100/49602	-78.332(4.10e-15)- 100/6738	-78.332(6.49e-15) 100/13310
f_{22}	-29.999(6.36e-04)- 100/25458	-29.997(1.72e-03)- 100/22410	-29.997(1.17e-03)- 100/34526	**-30.000(1.03e-05)=** 100/2070	-30.000(6.92e-07) 100/17666
+/=/-	1/4/17	0/6/16	0/4/18	1/6/15	

"-", "+" , and "=" respectively denote that the performance of the corresponding algorithm is worse than, better than, and similar to that of NILABC according to the Wilcoxon's rank SUM test at a 0.05 significance level.

It can be clearly observed from Table 2 that NILABC is significantly better than all the other algorithms in terms of solution accuracy, convergence rate and robustness on most cases. To be specific, NILABC outperforms all compared algorithms on all unimodal functions ($f_1 - f_6$ and f_8) except f_6 and f_8; on f_6 and f_{10}, NILABC only beaten by qABC and basic ABC respectively, and on f_7 and f_8, all algorithms obtain similar results. Concerning all multimodal functions ($f_{11} - f_{22}$), NILABC is also superior to all competitors on all cases, excluding f_{21} and f_{22}, on which all algorithms perform similarly. Overall, NILABC is better than ABC, GABC, best-so-far ABC and qABC on 17, 18, 16 and 15 case out of 22 functions respectively, while NILABC is only beaten by qABC and the basic ABC on f_6 and f_{10} respectively. Regarding the remainder functions, all algorithms show similar performance. To clear show the advantages of ILABC, the convergence curves of mean objective function value for some typical functions are plotted in Fig. 1. Evidently, NILABC has better solution accuracy, convergence rate and robustness than all the competitors on majority of cases.

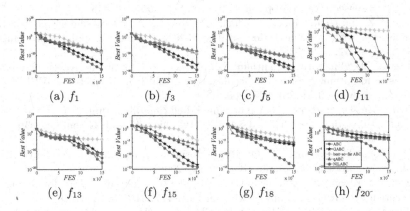

(a) f_1 (b) f_3 (c) f_5 (d) f_{11}

(e) f_{13} (f) f_{15} (g) f_{18} (h) f_{20}

Fig. 1. Convergence performance of different ABCs on some typical test functions

5 Conclusion and Future Works

In this paper, we propose a novel neighboring information learning model, in which the honey bees get the guidance from the best food source among its neighbors to search new food source. Moreover, the size of neighbors is linearly increased with the evolutionary process, which can ensure that all honey bees learn beneficial information from different local best solutions in the early stage and all honey bees exploit the valuable information of the current global best solution in the late stage. Besides, we introduce a ranking-based probability calculating method in onlooker phase. The performance of our proposed algorithm (NILABC) is validated by the comparisons with the basic ABC and other three state-of-the-art ABC variants on 22 test function. Our experimental results show that the performance of NILABC is much better than that of original ABC and some ABC variants. In the future, we will study how to choose the neighbors and design effective solution search equations based on neighboring information.

Acknowledgement. This work is supported by the National Natural Science Foundation of China under Grants 61402291, 61402294, and 61170283, National High-Technology Research and Development Program (863 Program) of China under Grant 2013AA01A212, Ministry of Education in the New Century Excellent Talents Support Program under Grant NCET-12-0649, Foundation for Distinguished Young Talents in Higher Education of Guangdong, China under Grant 2013LYM_0076 and 2014KQNCX129, Major Fundamental Research Project in the Science and Technology Plan of Shenzhen under Grants JCYJ20140828163633977 and JCYJ20140418181958501.

References

1. Chuang, Y.C., Chen, C.T., Hwang, C.: A real-code genetic algorithm with a direction-based crossover operator. Inform. Sci. **305**, 320–348 (2015)
2. Cui, L.Z., Li, G.H., Lin, Q.Z., Chen, J.Y., Lu, N.: Adaptive differential evolution algorithm with novel mutation strategies in multiple sub-populations. Comput. Oper. Res. **67**, 155–173 (2016)
3. Liang, J.J., Qin, A.K., Suganthan, P.N., Baskar, S.: Comprehensive learning particle swarm optimizer for global optimization of multimodal functions. IEEE Trans. Evol. Comput. **10**(3), 281–295 (2006)
4. Karaboga, D., Basturk, B.: A powerful and efficient algorithm for numerical function optimization: artificial bee colony (ABC) algorithm. J. Global Opt. **39**, 459–471 (2007)
5. Karaboga, D., Basturk, B.: On the performance of artificial bee colony (ABC) algorithm. Appl. Soft Comput. **8**, 687–697 (2008)
6. Zhu, G., Kwong, S.: Gbest-guided artificial bee colony algorithm for numerical function optimization. Appl. Math. Comput. **217**, 3166–3173 (2010)
7. Xiang, Y., Peng, Y.M., Zhong, Y.B., Chen, Z.Y., Lu, X.W., Zhong, X.J.: A particle swarm inspired multi-elite artificial bee colony algorithm for real-parameter optimization. Comput. Optim. Appl. **57**, 493–516 (2014)
8. Banharnsakun, A., Achalakul, T., Sirinaovakul, B.: The best-so-far selection in artificial bee colony algorithm. Appl. Soft. Comput. **11**, 2888–2901 (2010)
9. Gao, W.F., Liu, S.Y., Huang, L.L.: A novel artificial bee colony algorithm based on modified search equation and orthogonal learning. IEEE Trans. Cybern. **43**, 1011–1024 (2013)
10. Karaboga, D., Gorkemli, B.: A quick artificial bee colony (qABC) algorithm and its performance on optimization problems. Appl. Soft Comput. **23**, 227–238 (2014)
11. Wang, H., Wu, Z.J., Rahnamayan, S., Sun, H., Liu, Y., Pan, J.S.: Multi-strategy ensemble artificial bee colony algorithm. Inform. Sci. **279**, 587–603 (2014)
12. Kiran, M.S., Hakli, H., Guanduz, M., Uguz, H.: Artificial bee colony algorithm with variable search strategy for continuous optimization. Inform. Sci. **300**, 140–157 (2015)
13. Kang, F., Li, J.J., Ma, Z.Y.: Rosenbrock artificial bee colony algorithm for accurate global optimization of numerical functions. Inform. Sci. **12**, 3508–3531 (2011)

14. Kang, F., Li, J.J., Li, H.J.: Artificial bee colony algorithm and pattern search hybridized for global optimization. Appl. Soft Comput. **13**, 1781–1791 (2013)
15. Gao, W.F., Liu, S.Y.: A modified artificial bee colony algorithm. Comput. Oper. Res. **39**, 687–697 (2012)

Data-Driven Design of Type-2 Fuzzy Logic System by Merging Type-1 Fuzzy Logic Systems

Chengdong Li$^{(\boxtimes)}$, Li Wang, Zixiang Ding, and Guiqing Zhang

School of Information and Electrical Engineering,
Shandong Jianzhu University, Jinan, China
chengdong.li@foxmail.com

Abstract. Type-2 fuzzy logic systems (T2 FLSs) have shown their superiorities in many real-world applications. With the exponential growth of data, it is a time consuming task to directly design a satisfactory T2 FLS through data-driven methods. This paper presents an ensembling approach based data-driven method to construct T2 FLS through merging type-1 fuzzy logic systems (T1 FLSs) which are generated using the popular ANFIS method. Firstly, T1FLSs are constructed using the ANFIS method based on the sub-data sets. Then, an ensembling approach is proposed to merge the constructed T1 FLSs in order to generate a T2 FLS. Finally, the constructed T2 FLS is applied to a wind speed prediction problem. Simulation and comparison results show that, compared with the well-known BPNN and ANFIS, the proposed method have similar performance but greatly reduced training time.

Keywords: Data-driven method · Fuzzy logic system · ANFIS · Wind speed prediction

1 Introduction

Fuzzy sets and fuzzy logic [1, 2] have powerful abilities to describe and deal with the uncertain information and have found lots of applications. Specifically, in the modeling and control research domains, fuzzy logic systems (FLSs) have been widely used to effectively tackle the prior knowledge and expert knowledge to achieve satisfied performance [3, 4]. At present, FLSs utilized as models and controllers mainly include the type-1 fuzzy logic system (T1 FLS) and its extensions. As a popular and widely-used extension, type-2 fuzzy logic systems (T2 FLSs) adopt type-2 fuzzy sets (T2FSs) to replace the type-1 fuzzy sets (T1FSs) in T1 FLSs. Compared to T1FSs, T2FSs have more parameters and freedom to deal with the high levels of uncertainties [5–9]. As a result, better performance can be accomplished by the T2 FLSs.

Currently, data-driven methods are usually used to construct T2 FLSs as models and controllers [10–12]. However, due to the large number of parameters and their complex input-output mappings, it is a heaven burden to obtain the optimized T2 FLSs. Particularly, with the exponential growth of data, it is a time consuming task to directly design an excellent T2 FLS through data-driven methods.

© Springer International Publishing AG 2016
A. Hirose et al. (Eds.): ICONIP 2016, Part III, LNCS 9949, pp. 290–298, 2016.
DOI: 10.1007/978-3-319-46675-0_32

To solve such issues, this study proposes an ensembling approach based data-driven method for T2 FLS through merging the optimized T1 FLSs. In our approach, we firstly partition the whole data set to some sub-data sets. Then, for each sub-dataset, we construct one T1 FLS by the popular ANFIS method [13–15]. Finally, we merge all the constructed T1 FLSs to generate the T2 FLS. One thing to be mentioned is that the design processes of the T1 FLSs can be tackled and run in parallel. Hence, the training time can be expected to be reduced greatly. To verify the proposed method, we apply it to the wind prediction problem. Also, comparisons with other popular methods, such as the backward propagation neural network (BPNN) [16, 17] and ANFIS, are made.

The rest of this paper is organized as follows. In Sect. 2, the FLSs including T1 FLS and T2 FLS will be introduced briefly. In Sect. 3 the proposed ensembling approach based data-driven method will be presented. In Sect. 4, simulation and comparisons will be made. At last, conclusions will be drawn in Sect. 5.

2 Fuzzy Logic Systems

In this section, we will briefly review T1 FLS and T2 FLS. For simplicity, only multi-input-single-output FISs are taken into account.

2.1 T1 FLS

Suppose that there are n input variables $x = (x_1, x_2, \ldots, x_n)$. The ith fuzzy rule of the T1 FLS can be written as follows [2, 4, 7, 8]

$$\text{Rule } i: \ \text{if } x_1 \text{ is } A_1^i, x_2 \text{ is } A_2^i, \cdots x_n \text{ is } A_n^i, \text{ then } y_0(x) \text{ is } \omega^i$$

where, $i = 1, 2, \ldots, M$, $A_j^i s$ are the antecedent T1FSs for the input variables, $\omega^i s$ are the crisp consequent weights.

Using the singleton fuzzifier and by the product operation, we can obtain the firing strength of Rule i as [2, 4, 7, 8]

$$f^i(x) = \prod_{j=1}^{n} \mu_{A_j^i}(x_j) \tag{1}$$

Finally, the output of T1 FLS can be calculated as [2, 4, 7, 8]

$$y_0(x) = \frac{\sum_{i=1}^{M} f^i(x)\omega^i}{\sum_{i=1}^{M} f^i(x)} = \frac{\sum_{i=1}^{M} \omega^i \prod_{j=1}^{n} \mu_{A_j^i}(x_j)}{\sum_{i=1}^{M} \prod_{j=1}^{n} \mu_{A_j^i}(x_j)} \tag{2}$$

2.2 T2 FLS

Compared with T1 FLS, T2 FLS has a type-reducer in its output processing block to transform T2FSs to T1FSs. The fuzzy rules of T2 FLS can be obtained by blurring the T1FSs and crisp weights in the rule base of T1 FLS, and can be expressed as follows [5–8]

$$\text{Rule } i: \quad \textit{if } x_1 \textit{ is } \tilde{A}_1^i, x_2 \textit{ is } \tilde{A}_2^i, \cdots x_n \textit{ is } \tilde{A}_n^i, \textit{ then } y_0(x) \textit{ is } \left[\underline{\omega}^i, \bar{\omega}^i\right]$$

where $i = 1, 2, \ldots, M$, $\tilde{A}_j^i s$ are the antecedent T2FSs for the input variables, $[\underline{\omega}^i, \bar{\omega}^i]s$ are the interval consequent weights which can be seen as the centroid of the consequent T2 FSs [5–8].

Once a crisp value $x = (x_1, x_2, \ldots, x_n)$ is input to T2 FLS, through the singleton fuzzifier and by the product operation, the firing strength of Rule i can be expressed as

$$F^i(x) = \left[\underline{f}^i(x), \bar{f}^i(x)\right] = \left[\prod_{j=1}^{n} \underline{\mu}_{\tilde{A}_j^i}(x_j), \prod_{j=1}^{n} \bar{\mu}_{\tilde{A}_j^i}(x_j)\right] \tag{3}$$

where $\underline{\mu}_{\tilde{A}_j^i}(x_j)$ and $\bar{\mu}_{\tilde{A}_j^i}(x_j)$ are the lower membership function (LMF) and upper membership function (UMF) of the T2FS $\tilde{A}_j^i$.

Using the most widely-used Karnik-Mendel type-reduction method and the Center-Of-Sets(COS) defuzzification method, the final output of T2 FLS can be calculated as [5–8]

$$y_0(x) = \frac{1}{2}[y_l(x) + y_r(x)] \tag{4}$$

where

$$y_l(x) = \frac{\sum_{i=1}^{L} \bar{f}^i(x)\underline{\omega}^i + \sum_{i=L+1}^{M} \underline{f}^i(x)\underline{\omega}^i}{\sum_{i=1}^{L} \bar{f}^i(x) + \sum_{i=L+1}^{M} \underline{f}^i(x)} \tag{5}$$

$$y_r(x) = \frac{\sum_{i=1}^{R} \underline{f}^i(x)\bar{\omega}^i + \sum_{i=R+1}^{M} \bar{f}^i(x)\bar{\omega}^i}{\sum_{i=1}^{R} \underline{f}^i(x) + \sum_{i=R+1}^{M} \bar{f}^i(x)} \tag{6}$$

in which L and R are switch points which can be computed using the iterative Karnik-Mendel algorithm [5–7].

Obviously, the input-output mapping of T2 FLS is much more complex than that of T1 FLS. And, to obtain the crisp output from T2 FLS, iterative algorithm is needed to compute the switch points, so T2 FLS is more time consuming than T1 FLS.

3 Data-Driven Design of T2 FLS

In recent years, data has increased to a large scale in various fields. How to effectively deal with such large scaled data is becoming more and more important in many applications. In the fuzzy research fields, FLSs are usually designed by data driven methods. However, it will cost too much time if the whole large scaled dataset is used for FLSs' learning. So, we need to accelerate the learning speed of the training algorithms. There are two ways to realize this objective. One is to develop novel rapid learning algorithms. And, the other one is to use the parallel learning strategies.

In this study, we adopt the second approach. We will firstly partition the dataset to several sub-data sets, and then for each sub-dataset, we will generate one T1 FLS using the ANFIS method. The T1FLSs for different sub-data sets are learned in parallel. At last, such trained T1 FLSs will be merged to generate the T2 FLS. The whole flowchart of this data-driven strategy is depicted in Fig. 1.

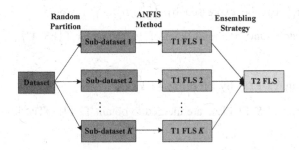

Fig. 1. The proposed data-driven strategy

3.1 Design of T1 FLS for Each Sub-Dataset Using ANFIS

In order to improve efficiency, the random partition will be applied to the whole dataset to obtain the distributed sub-data sets which are named as

$$\text{sub-dataset } 1, \text{sub-dataset } 2, \cdots, \text{sub-dataset } K$$

where K is the number of the sub-datasets.

For each sub-data set, we generate one T1 FLS using the ANFIS method [13, 14]. ANFIS is one of the most popular fuzzy methods used in modeling and control applications. ANFIS combines T1 FLS and neural network to make use of their respective advantages. In the ANFIS, TSK type FLSs are generated through data-driven methods. TSK FLSs include the T1 FLS discussed in Subsect. 2.1, which can be seen as the zero order TSK FLS. The T1 FLSs for all the K sub-data sets utilize the same initial fuzzy rule base.

In the ANFIS, the parameters of the antecedent and consequent parts are optimized by a hybrid learning algorithm which has well-performed learning ability and can generate well-performed T1 FLSs. The hybrid learning algorithm utilizes both the steepest descent method and the least square method to train the parameters of T1 FLSs. In the ANFIS, the antecedent parameters of T1 FLS are learned using the steepest descent method, while the consequent parameters are optimized through both the steepest descent method and the least square method. Applications of ANFIS demonstrated that it can achieve excellent approximation and generalization performances [15].

Suppose that, through the ANFIS method, the fuzzy rule base of the kth T1 FLS for sub-data set k is obtained as

$$x_1 = A_{1,k}^{j_{1,k}}, x_2 = A_{2,k}^{j_{2,k}}, \cdots x_n \text{ is } A_{n,k}^{j_{n,k}} \rightarrow y = y_k^{j_{1,k}\cdots j_{n,k}}$$

where $j_{s,k} \in \{1, 2, \ldots m_{s,k}\}$, $m_{s,k}$ is the number of fuzzy sets for the input variable x_s. For simplicity, we assume that $m_{s,1} = m_{s,2} = \ldots = m_{s,K} = m_s$. Hence, in all the T1 FLSs, there are $M = \prod_{s=1}^{n} m_s$ fuzzy rules. In our study, triangular fuzzy sets are adopted, so the T1FS $A_{s,k}^{j_{s,k}}$ can be expressed by $\left(d_{s,k}^{j_{s,k}}, b_{s,k}^{j_{s,k}}, c_{s,k}^{j_{s,k}}\right)$ where $d_{s,k}^{j_{s,k}}, b_{s,k}^{j_{s,k}}, c_{s,k}^{j_{s,k}}$ are the left-end point, center and right-end point of the triangular T1FS $A_{s,k}^{j_{s,k}}$.

3.2   Generating T2 FLS by Merging T1 FLSs

Then, the constructed K T1 FLSs are merged to obtain T2 FLS. The fuzzy rules of the obtained T2 FLS are

$$x_1 = \tilde{A}_1^{j_1}, x_2 = \tilde{A}_2^{j_2}, \cdots x_n \text{ is } \tilde{A}_n^{j_n} \rightarrow y = \tilde{y}^{j_1\cdots j_n} \tag{7}$$

where the antecedent T2FS $\tilde{A}_s^{j_s}$ is constructed by merging the T1FSs $A_{s,1}^{j_{s,1}}, \ldots A_{s,K}^{j_{s,K}}$ in the constructed K T1 FLSs. And, the consequent interval weight $\tilde{y}^{j_1\cdots j_n}$ of the corresponding fuzzy rule is obtained by ensembling the crisp weights $y_1^{j_{1,1}\cdots j_{n,1}}, \ldots, y_K^{j_{1,K}\cdots j_{n,K}}$ as

$$\tilde{y}^{j_1\cdots j_n} = \left[\min_{k=1}^{K} y_k^{j_{1,k}\cdots j_{n,k}}, \max_{k=1}^{K} y_k^{j_{1,k}\cdots j_{n,k}}\right] \tag{8}$$

In this study, we use the trapezoidal T2FS as shown in Fig. 2 to merge the triangular T1FSs. The detailed merging strategy is also depicted in this figure. The trapezoidal T2FS $\tilde{A}_s^{j_s}$ can be depicted by eight parameters $\left(z_s^{j_s^1}, z_s^{j_s^2}, z_s^{j_s^3}, t_s^{j_s^1}, t_s^{j_s^2}, t_s^{j_s^3}, t_s^{j_s^4}, h_s^{j_s}\right)$ where $t_s^{j_s^1}, t_s^{j_s^2}, t_s^{j_s^3}, t_s^{j_s^4}$ are the parameters of the UMF of T2FS $\tilde{A}_s^{j_s}$, and $z_s^{j_s^1}, z_s^{j_s^2}, z_s^{j_s^3}, h_s^{j_s}$ are the parameters of the LMF of T2FS $\tilde{A}_s^{j_s}$. Such parameters can be determined as shown Fig. 2.

Through such strategies, different T1 FLSs can be ensembled into one T2 FLS model. As demonstrated above, the T2 FLS have T2FSs with uncertain MFs as antecedents and interval weights as consequents. Hence, the generated T2 FLS can model the uncertainties from these different T1 FLSs.

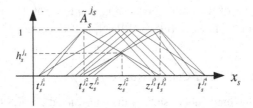

Fig. 2. Merging strategy for the antecedent T2FSs

4 Application to Wind Speed Prediction

Accurate prediction of wind speed is very helpful to the maintenance scheduling and the dispatch planning [18, 19]. However, due to the unstable and random natures of the wind, the accurate wind speed prediction is a challenging task. In this section, we will apply the proposed data-driven method to the wind speed prediction. Also, comparisons with ANFIS and BPNN are made to show the effectiveness and fast learning speed of the proposed method.

4.1 Problem Description

In our experiment, the wind speed data were downloaded from the website: "http://climate.weather.gc.ca/". The data set comprises wind speeds collected every hour between July 1, 2014 and July 31, 2015 in Regina, Saskatchewan, which is a region of Canada. Totally, there are 9504 samples. And, the data during July 1, 2014 and June 30, 2015 will be selected as the training data, while those during July 1, 2015 and July 31, 2015 will be chosen as the testing data. In other words, the first 8760 samples are utilized for training while the left 744 samples are for testing.

In our experiments, we use the wind speeds of the previous four hours to predict the speed of the next hour, i.e. the models for this application have four inputs and one output. For the ANFIS, the whole training data set are utilized. And, for the T2 FLS, the training dataset is randomly partitioned to form four sub-data sets. In the ANFIS and the four T1 FLSs for the four sub-datasets, we assign 3 triangular fuzzy sets to partition each input domain, and their maximum learning epochs are all set to be 5. At last, 81 fuzzy rules will be used in both ANFIS and T2 FLS. For the BPNN, we choose one hidden layer with 80 nodes. And, it is trained by the BP algorithm.

The three methods are run at the same computer platform. We will compare their performances based on three aspects: learning speed, approximation ability and generalization ability. The approximation and generalization abilities are tested based on the following root mean square error (RMSE) index

$$RMSE = \sqrt{\frac{1}{N}\sum_{k=1}^{N}|y(X^k) - y^k|^2} \tag{17}$$

where N is the number of the training or testing input-output data pairs. $y(X^k)$ is the predicted output from T2 FLS, ANFIS or BPNN, and y^k is the real wind speed.

4.2 Simulation Results and Comparisons

After being trained, the prediction results of the constructed T2 FLS for the testing data are depicted in Fig. 3. From this figure, we can observe that satisfactory performance can be achieved by the proposed method.

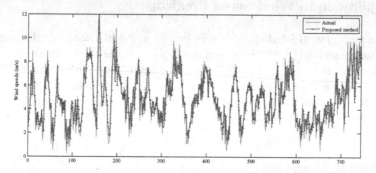

Fig. 3. Testing data and the predicted output of the proposed method

Table 1. Comparisons of the three models

Indices	Training Time		Training Accuracy		Testing Accuracy	
	Average	STD	Average	STD	Average	STD
T2FLS	**0.1435**	0.0637	1.4126	0.0682	1.0707	0.0037
ANFIS	5.9936	0.1143	1.3017	0.0000	1.0620	0.0000
BPNN	3.6723	0.4250	1.4005	0.0290	1.1314	0.0152

For comparison, the learning process has been run for 5 times. The comparison results of the three models are listed in Table 1 in detail. From this table, we can observe that, compared with the ANFIS and the BPNN, the data-driven T2 FLS can give similar performance. However, the learning speed of the proposed method is greatly improved. The proposed method has about 40 times faster learning speed than the ANFIS and about 25 times faster learning speed than the BPNN. The reason for this is that, in the proposed method, the design processes of the corresponding T1 FLSs can be tackled with smaller sub-data sets and run in parallel.

5 Conclusion

This study proposed a data-driven method to design IT2 FLSs for modeling, prediction and identification applications. The proposed method can greatly reduce the learning time of the design process. To enhance the learning accuracy, the ANFIS method was adopted to construct T1 FLSs for all the sub-data sets. And, a ensembling strategy was proposed to merge the differences in the learned T1 FLSs. The proposed method has also been verified by an real-world application. Comparisons with ANFIS and BPNN showed that the proposed scheme can give similar performance but have much less learning time. It is suitable for big data and distributed data, but its accuracy needs to be improved further. This will be done in our future studies.

Acknowledgments. This work is supported by National Natural Science Foundation of China (61473176, and 61573225), and the Natural Science Fund of Shandong Province for Outstanding Young Talents in Provincial Universities (ZR2015JL021).

References

1. Zadeh, L.A.: Fuzzy sets. Inf. Control **8**(3), 338–353 (1965)
2. Zadeh, L.A.: Fuzzy logic–a personal perspective. Fuzzy Sets Syst. **281**, 4–20 (2015)
3. Pedrycz, W., Izakian, H.: Cluster-centric fuzzy modeling. IEEE Trans. Fuzzy Syst. **22**(6), 1585–1597 (2014)
4. Yager, R.R., Zadeh, L.A.: An introduction to fuzzy logic applications in intelligent systems, p. 165. Springer Science & Business Media, Heidelberg (2012)
5. Liang, Q., Mendel, J.M.: Interval type-2 fuzzy logic systems: Theory and design. IEEE Trans. Fuzzy Syst. **8**(5), 535–550 (2000)
6. Mendel, J.M., John, R., Liu, F.: Interval type-2 fuzzy logic systems made simple. IEEE Trans. Fuzzy Syst. **14**(6), 808–821 (2006)
7. Mendel, J.M.: Uncertain rule-based fuzzy logic system: introduction and new directions (2001)
8. Li, C., Yi, J., Zhang, G.: On the monotonicity of interval type-2 fuzzy logic systems. IEEE Trans. Fuzzy Syst. **22**(5), 1197–1212 (2014)
9. Li, C., Zhang, G., Wang, M., Yi, J.: Data-driven modeling and optimization of thermal comfort and energy consumption using type-2 fuzzy method. Soft. Comput. **17**(11), 2075–2088 (2013)
10. Lin, Y.Y., Chang, J.Y., Lin, C.T.: A TSK-type-based self-evolving compensatory interval type-2 fuzzy neural network (TSCIT2FNN) and its applications. Industrial Electronics. IEEE Trans. Fuzzy Syst. **61**(1), 447–459 (2014)
11. Gaxiola, F., et al.: Interval type-2 fuzzy weight adjustment for backpropagation neural networks with application in time series prediction. Inf. Sci. **260**, 1–14 (2014)
12. Wu, G.D., Huang, P.H.: A vectorization-optimization-method-based type-2 fuzzy neural network for noisy data classification. IEEE Trans. Fuzzy Syst. **21**(1), 1–15 (2013)
13. Jang, J.S.R.: ANFIS adaptive-network-based fuzzy inference system. IEEE Trans. Fuzzy Syst. Man Cybern. **23**(3), 665–685 (1993)
14. Melin, P., et al.: A new approach for time series prediction using ensembles of ANFIS models. Expert Syst. Appl. **39**(3), 3494–3506 (2012)

15. Svalina, I., et al.: An adaptive network-based fuzzy inference system (ANFIS) for the forecasting: The case of close price indices. Expert Syst. Appl. **40**(15), 6055–6063 (2013)
16. Goh, A.T.C.: Back-propagation neural networks for modeling complex systems. Artif. Intell. Eng. **9**(3), 143–151 (1995)
17. Wang, L., Zeng, Y., Chen, T.: Back propagation neural network with adaptive differential evolution algorithm for time series forecasting. Expert Syst. Appl. **42**(2), 855–863 (2015)
18. Soder, L., Hofmann, L., Orths, A., Holttinen, H., Wan, Y.H., Tuohy, A.: Experience from wind integration in some high penetration areas. IEEE Trans. Energy Convers. **22**(1), 4–12 (2007)
19. Tascikaraoglu, A., Uzunoglu, M.: A review of combined approaches for prediction of short-term wind speed and power. Renew. Sustain. Energy Rev. **34**, 243–254 (2014)

Memetic Cooperative Neuro-Evolution for Chaotic Time Series Prediction

Gary Wong[1], Rohitash Chandra[2(✉)], and Anuraganand Sharma[1]

[1] School of Computing Information and Mathematical Sciences,
University of the South Pacific, Suva, Fiji
s11085179@student.usp.ac.fj, anuraganand.sharma@usp.ac.fj
[2] Artificial Intelligence and Cybernetics Research Group,
Software Foundation, Nausori, Fiji
c.rohitash@gmail.com

Abstract. Cooperative neuro-evolution has shown to be promising for chaotic time series problem as it provides global search features using evolutionary algorithms. Back-propagation features gradient descent as a local search method that has the ability to give competing results. A synergy between the methods is needed in order to exploit their features and achieve better performance. Memetic algorithms incorporate local search methods for enhancing the balance between diversification and intensification. We present a memetic cooperative neuro-evolution method that features gradient descent for chaotic time series prediction. The results show that the proposed method utilizes lower computational costs while achieving higher prediction accuracy when compared to related methods. In comparison to related methods from the literature, the proposed method has favorable results for highly noisy and chaotic time series problems.

Keywords: Memetic algorithms · Cooperative neuro-evolution · Backpropagation · Gradient descent · Feedforward networks

1 Introduction

Evolutionary Algorithms (EAs) have achieved significant success as search and optimization techniques across various domains [8,11]. In particular, they are best suited for non-linear and noisy systems [4], and have gained success for training neural networks which are also known as neuro-evolution [25]. A major drawback of EAs is convergence due to computational costs as they are essentially black-box optimization methods. Gradient-based training methods provide much faster convergence which has been successful for convex optimization, howsoever, premature convergence and over-training have been a recurring challenge [2].

The balance between diversification and intensification for neuro-evolution is imperative in order to provide more emphasis on promising solutions in the search space [14,17,20]. Memetic algorithms (MAs) [17] take advantage of the

© Springer International Publishing AG 2016
A. Hirose et al. (Eds.): ICONIP 2016, Part III, LNCS 9949, pp. 299–308, 2016.
DOI: 10.1007/978-3-319-46675-0_33

strengths of evolutionary algorithms and local search methods while eliminating their weaknesses [2,4,9,13,20]. Successful implementations of MAs span over many fields including plane wing design [20], image classification [1], continuous function optimization [2], and pattern classification [9].

Cooperative coevolution decomposes a problem into subcomponents [21] which are typically implemented as sub-populations that evolve in isolation and cooperation takes place for fitness evaluation. Cooperative neuro-evolution (CNE) refers to the application of cooperative coevolution for training neural networks. CNE has been applied successfully for time series prediction [8]. A memetic CNE method was presented with crossover-based local search and was able to reduce the computational time while achieving high-quality solutions for pattern classification problems [9]. Moreover, it has been shown that gradient descent algorithms converge faster with high multi-modal problems that complement in developing memetic algorithms [14]. There has not been much work done that features local search in improving CNE for chaotic time series problems.

In this paper, we present memetic cooperative neuro-evolution method that features gradient descent for chaotic time series problems. We evaluate the performance of the approach on simulated and real-world benchmark chaotic time series and also compare the results with related methods from the literature.

The rest of the paper is organized as follows. Section 2 presents the proposed method and Sect. 3 focuses on the experiment design and results. Consequently, Sect. 4 presents conclusion and discussion for future work.

2   Memetic Cooperative Neuro-Evolution with Gradient Local Search

As discussed earlier, Chandra et al. presented a framework for memetic cooperative coevolution for pattern classification [9] where it was highlighted that the challenge lies in terms of implementation due to multiple sub-populations in cooperative neuro-evolution. We note that conventional memetic algorithms feature a single population and hence local search incorporation is relatively easier when compared to several decomposed sub-populations in cooperative neuro-evolution.

We employ a similar strategy for memetic cooperative neuro-evolution presented by Chandra et al. [9] where the best individuals from all the respective sub-populations are concatenated after a given number of cycles for local refinement. Note that a *cycle* refers to the completion of evolution of the respective sub-populations in a round-robin fashion. We employ problem decomposition in cooperative neuro-evolution that is based on neuron level decomposition (CNE-NL) [10]. Each neuron acts as a reference for a subcomponent that contains all the weight connections from the previous layer as shown in Fig. 1. The subcomponents are created from reference to all the neurons in the hidden and output layers. These subcomponents are implemented as *sub-populations*.

We present the implementation details of memetic cooperative neuro-evolution for feedforward networks (MCNE) shown in Algorithm 1 with

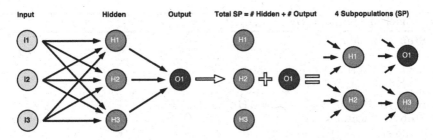

Fig. 1. Neuron level decomposition: a feedforward neural network with 3 input neurons, 3 hidden and 1 output neuron. Neurons in the hidden and output layers act as reference points for new sub-populations, hence the creation of several sub-populations after decomposition.

additional details from Table 1 and Fig. 1. In Step 1, the algorithm begins by configuring the number of subcomponents needed which is determined by the number of h and o neurons used. The function $\mathtt{rand}(L, s, \mu, \overline{w}, \underline{w})$ then initializes the subcomponents in L with a population size of μ containing randomized values between $\underline{w}$ and $\overline{w}$, respectively. Then follows the evaluation phase in $\mathtt{eval}(L, s)$ where the fitness values of the sub-population individuals are computed. This initialization operation cost also contributes to the total number of function evaluations, Γ_{CC}.

The loop on lines 7–16 holds the entire part of evolution that encompasses cooperative neuro-evolution with local refinement by gradient descent which is implemented through back-propagation. The algorithm halts if the specified minimum error ε_t, or if the maximum function evaluations Γ_{max} has been reached. Global search is employed in Step 3 where the sub-populations are evolved in a round-robin fashion using evolutionary operators given by the designated evolutionary algorithm. Hence, the subcomponent D_j is evaluated through function, $\mathtt{evolve}(b, D_j, \gamma, \alpha)$ which returns a training error given by the *root-mean-squared-error (RMSE)*. The total number of function evaluations Γ_{CC} for is then updated. This step is repeated according to the local search frequency lsf.

After completion of the evolution *cycle*, the meme δ^* that contains the fittest individuals from each sub-population in L is retrieved and concatenated from a call to function $\mathtt{getbestsolution}(L, s)$. This calls the local refinement function $\mathtt{bp}(\delta^*, \lambda)$ shown in Step 4. This is executed for lsi number of iterations where the refined meme δ^* is then broken up in $\mathtt{replaceworst}(L, s, \delta^*, b)$ as shown in line 16 and returned in order to be encoded into the respective sub-populations L, replacing only the individuals with the lowest fitness. Finally, the total number of function evaluations Γ_{CC} is updated in consideration of the lsi. Although gradient descent has been used for local refinement, the proposed algorithm is flexible, and hence other local refinement procedures can also be used.

Algorithm 1: mcnefnn(i, h, o, α, μ, Γ_{max}, λ, ε, lsf, lsi, δ^*)

---

1 **Step 1:** Initialization
2 $s = \text{h} + \text{o}$;
3 **for** $k \in \{1,..,h\}$ & $k \in \{1,..,o\}$ **do**
4 $\quad \lfloor\; L_k \mathrel{+}= new\ C$;
5 rand(μ, $\overline{w}$, $\underline{w}$, L, s); $b = $ eval(L, s); $\Gamma_{CC} = \mu$;
6 **Step 2:** Evolution
7 **while** $\varepsilon_t > \varepsilon_{min}$ & $\Gamma_{CC} < \Gamma_{max}$ **do**
8 $\quad$ **Step 3:** Global Search
9 $\quad$ **for** $k \in \{1,..,lsi\}$ *(Cycles)* **do**
10 $\quad\quad$ **for** $j \in \{1,..,s\}$ *(Subcomponents)* **do**
11 $\quad\quad\quad$ $b = $ evolve(b, L_j, γ, α);
12 $\quad\quad\quad$ $\Gamma_{CC} \mathrel{+}= \mu \times (\gamma + 1)$;
13 $\quad$ $\delta^* = $ getbestsolution(L, s);
14 $\quad$ **Step 4:** Local Refinement
15 $\quad$ **for** $k \in \{1,..,lsi\}$ **do**
16 $\quad\quad$ $\varepsilon_t = $ bp(δ^*, λ);
17 $\quad$ replaceworst(L, s, δ^*, b); $\Gamma_{CC} \mathrel{+}= lsi$;

Table 1. Variables

Variable	Description	Variable	Description
α	species mutation rate	ε_t	test error
μ	the population size	lsf	local search frequency
Γ_{max}	max function evaluations	lsi	local search intensity
Γ_{CC}	total algorithm evaluations	δ^*	best meme
λ	backpropagation learning rate	L	decomposition components set
γ	subcomponent optimization time	C	decomposition component
i	number of input neurons	s	L size
h	number of hidden neurons	$\overline{w}$	upper weight boundary
o	number of output neurons	$\underline{w}$	lower weight boundary
ε_{min}	minimum error needed		

3   Simulation and Analysis

This section presents an experimental study on the performance of MCNE for chaotic time series problems. We first provide details of the parameters used in the implementation of the algorithms and then present experimental design.

The sub-populations in cooperative neuro-evolution employ a pool of 200 individuals (μ) that feature the G3-PCX evolutionary algorithm (generalized generation gap with parent-centric crossover) [11]. It employs a mating pool size of 2 offspring and 2 parents. The respective sub-populations are initialized within $\{\underline{w}:-5, \overline{w}:5\}$. We use fixed local search intensity $(lsi = 200)$ and local search frequency $(lsf = 10)$ that gave good performance in trial experiments. Additionally, back-propagation (gradient descent) employs a learning rate of $\lambda = 0.1$.

$$RMSE = \sqrt{\frac{1}{N}\sum_{i=1}^{N}(y_i - \hat{y}_i)^2)} \quad (1) \qquad NMSE = \left(\frac{\sum_{i=1}^{N}(y_i - \hat{y}_i)^2}{\sum_{i=1}^{N}(y_i - \bar{y}_i)^2}\right) \quad (2)$$

The embedding dimension D determines the number of input neurons while the number of hidden neurons are varied (from 3 to 9) in order to test for robustness and scalability. The number of function evaluations (Γ_{max}) was set at 5000. The $(RMSE)$ and normalized mean squared error $(NMSE)$ given in Eqs. 1 and 2, provide the main performance measures. Note that y_i refers to observed data, $\hat{y}_i$ the predicted data, $\bar{y}_i$ the average of observed data, and N the size of the data.

The datasets employed consist of simulated (Mackey-Glass) and real-world (Sunspot, Laser, and ACI finance) chaotic time series problems. The ACI finance dataset contains closing stock prices of ACI Worldwide Inc. from December 2006 to February 2010 (800 data points [19]). The Laser dataset [24] contains 1000 data points along with the Mackey-Glass [16] while the Sunspot problem consists of 2000 data points [5].

Each problem is divided into a 60-20-20 structure to provide training, validation and test set, respectively. The Taken's embedding theorem [23] is employed to reconstruct original time series data into a usable phase space for training the neural network. The time-lag T determines the interval at which the data points are to be picked while the embedding dimension D, specifies the size of the sliding window. The reconstruction settings for this study are as follows; Mackey-Glass and Laser $\{D:3, T:2\}$, Sunspot $\{D:5, T:3\}$, and ACI $\{D:5, T:2\}$.

3.1 Results

The results are presented in Fig. 2 that highlights the training time (function evaluations) and generalization performance (RMSE) of the given methods. The respective histograms show mean and error bars from with 95 % confidence interval from 50 independent experimental runs. In Fig. 2(a), MCNE shows significant convergence time improvement over CNE, with the exception of BP. The confidence interval of MCNE is the best for the different number of hidden neurons when we consider the RMSE which shows that it has been the most robust method for this problem. Moreover, MCNE has achieved the highest accuracy in prediction (with lowest values of RMSE).

MCNE has shown to get the lowest training time for the ACI-finance problem shown in Fig. 2(b). It also achieves best prediction accuracy in terms of lowest RMSE for Sunspot, Mackey-Glass, and Laser problems. MCNE achieves a competing performance with BP for prediction performance for ACI-finance problem where BP takes most training time when compared to the other methods. This

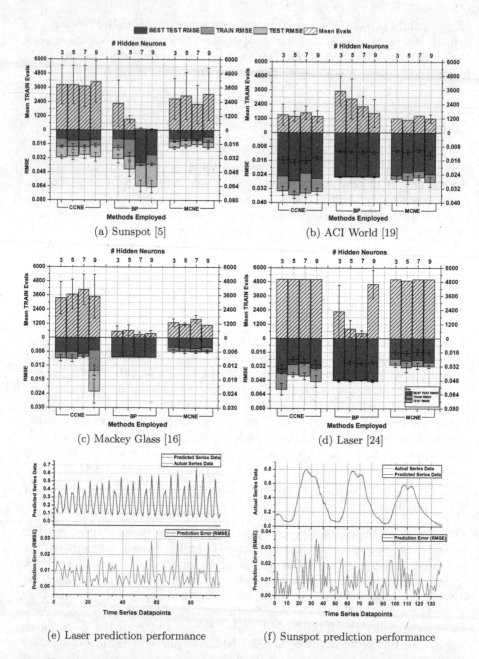

Fig. 2. The training time (function evaluations) and generalization performance (RMSE) of the given methods for the Sunspot, ACI finance, Mackey-Glass and Laser time series. Note that the values below the x-axis are not negative. Additionally, (e) and (f) shows the prediction accuracy of MCNE for a single test run for the Sunspot and Laser problems.

Table 2. Comparison with related methods from the literature

Problem	Prediction method	RMSE	NMSE
Sunspot [5]	RBF-OLS (2006) [12]		46.0E–03
	LLNF-LoLiMot (2006) [12]		32.0E–03
	SL-CCRNN (2012) [8]	1.66E–02	1.47E–03
	NL-CCRNN (2012) [8]	2.60E–02	3.62E–03
	MO-CCFNN-T = 2 (2014) [6]	1.84E–02	1.02E–03
	MO-CCFNN-T = 3 (2014) [6]	1.81E–02	0.998E–03
	CICC-RNN (2015) [7]	1.57E–02	1.31E–03
	CCFNN-CSFR (2016) [18]	1.58E–02	**0.756E–03**
	MCNE	**1.01E–02**	1.24E–03
ACI [19]	CICC-RNN (2009) [7]	1.92E–02	
	MO-CCFNN-T = 2 (2014) [6]	1.94E–02	
	MO-CCFNN-T = 3 (2014) [6]	1.47E–02	
	CCFNN-CSFR (2016) [18]	**1.34E–02**	0.995E–03
	MCNE	2.21E–02	1.139E–03
Mackey [16]	RBF-OLS (2006) [12]	**1.02E–03**	460E–04
	LLNF-LoLiMot (2006) [12]	0.961E–03	320E–04
	PS0 (2009) [15]	21.0E–03	
	Neural fuzzy network and DE (2009) [15]	16.2E–03	
	Neural fuzzy network and GA (2009) [15]	16.3E–03	
	SL-CCRNN (2012) [8]	6.33E–03	2.79E–04
	NL-CCRNN (2012) [8]	8.28E–03	4.77E–04
	MO-CCFNN-T = 2 (2014) [6]	3.84E–03	0.28E–04
	MO-CCFNN-T = 3 (2014) [6]	3.77E–03	0.27E–04
	CICC-RNN (2015) [7]	3.99E–03	1.11E–04
	CCFNN-CSFR (2016) [18]	2.90E–03	**0.016E–04**
	MCNE	4.57E–03	1.23E–04
Laser [24]	RNN-BPTT (2002) [3]		1.54E–02
	RNN-CBPTT (2002) [3]		2.24E–02
	Echo State Network (SN) with IS 0.5 (2011) [22]		9.83E–02
	Echo State Network (SN) with IS 1 (2011) [22]		10.58E–02
	MCNE	**2.32E–02**	**0.589E–02**

case shows the inefficiency of BP convergence although it is perceived as a faster method when compared to evolutionary techniques. In terms of robustness, 5 hidden neurons provide the best convergence and prediction performance for all the problems.

Table 2 presents a comparison with related methods from the literature. We note that MCNE shows the best performance for Sunspot and Laser problems however the algorithm did not perform very well for the ACI finance and Mackey problems. One of the problems with direct comparison with other methods is the difference in training time and related parameters of the experimental design.

We employed maximum of 5000 function evaluations for training while compared to Nand and Chandra [18] (CSFR in Table 2) used 50,000 function evaluations. Hence, more training time could be attributed to better performance. Additionally, recurrent neural networks seem to be better suited for time series prediction [8].

3.2   Discussion

The results show that the proposed method performs very well on the real-world problems (Sunspot and Laser) when compared to standalone methods. The improvement can be credited to solution transfer between global and local search, especially through the use of gradient information. We note that extra emphasis on global search increases computational costs that produce lower accuracy since neuro-evolution is essentially a black-box training method. Real world problems are often non-linear and noisy thus the diversity of solutions through exchange between local and global search has been beneficial. A major advantage of local refinement has been the decrease in the number of function evaluations in reaching a promising solution. The results show that local refinement in evolution speeds convergence and produces better accuracy for chaotic time series prediction.

Although the method has shown to perform better than its counterparts, the results face some challenges when compared to some of the methods from the literature. We note that it is not fair for a straight-forward comparison with results from the literature as some of the methods rely on different approaches that include hybrid methods, ensembles, and feature extraction. However, an extensive study on the balance between the local search frequency and intensity in our proposed method is needed for further improvement.

4   Conclusion and Future Work

We presented a memetic cooperative neuro-evolution method where backpropagation was used for local refinement. The method was applied to selected problems in chaotic time series prediction and the results showed promising when compared to related methods in the literature. This was mainly for real-world application problems that feature noise and are chaotic in nature. Cooperative neuro-evolution provided diversification while backpropagation refined the promising solutions that lead to rapid convergence. We highlight that the proposed method has been feasible even with limited training time.

The proposed method is flexible and can be easily adapted to incorporate other local search methods. A group of local search methods can also be used that can exploit different regions of search space at different stages of the learning process.

A limitation of this study has been the lack of extensive study on the balance between diversification and intensification. Hence, future work can focus on this in the context of time series prediction. Furthermore, the results motivate real-world applications to domains that range from finance to climate change. Further improvements in the results could be achieved with different local refinement procedures. Moreover, the method can be extended to multi-step-ahead time series prediction.

References

1. Alsmadi, M., Omar, K., Noah, S., Almarashdeh, I.: A hybrid memetic algorithm with back-propagation classifier for fish classification based on robust features extraction from PLGF and shape measurements. Inf. Technol. J. **10**(5), 944–954 (2011). ISSN 1812–5638
2. Arab, A., Alfi, A.: An adaptive gradient descent-based local search in memetic algorithm applied to optimal controller design. Inf. Sci. **299**, 117–142 (2015). ISSN 0020–0255
3. Boné, R., Crucianu, M., de Beauville, J.-P.A.: Learning long-term dependencies by the selective addition of time-delayed connections to recurrent neural networks. Neurocomputing **48**(1), 251–266 (2002)
4. Castillo, P., Arenas, M., Castellano, J., Merelo, J., Prieto, A., Rivas, V., Romero, G.: Lamarckian evolution, the baldwin effect in evolutionary neural networks. arXiv:cs/0603004 (2006)
5. SILSO, World Data Center: Sunspot Number and Long-term Solar Observations, Royal Observatory of Belgium, on-line Sunspot Number catalogue. http://www.sidc.be/silso
6. Chand, S., Chandra, R.: Multi-objective cooperative coevolution of neural networks for time series prediction. In 2014 International Joint Conference on Neural Networks (IJCNN), pp. 190–197. IEEE (2014)
7. Chandra, R.: Competition and collaboration in cooperative coevolution of elman recurrent neural networks for time-series prediction. IEEE Trans. Neural Netw. Learn. Syst. **26**(12), 3123–3136 (2015)
8. Chandra, R., Zhang, M.: Cooperative coevolution of elman recurrent neural networks for chaotic time series prediction. Neurocomputing **86**, 116–123 (2012). ISSN 0925–2312
9. Chandra, R., Frean, M., Zhang, M.: Crossover-based local search in cooperative co-evolutionary feedforward neural networks. Appl. Soft Comput. **12**(9), 2924–2932 (2012a). ISSN 1568–4946
10. Chandra, R., Frean, M., Zhang, M.: On the issue of separability for problem decomposition in cooperative neuro-evolution. Neurocomputing **87**, 33–40 (2012b)
11. Deb, K., Anand, A., Joshi, D.: A computationally efficient evolutionary algorithm for real-parameter optimization. Evol. Comput. **10**(4), 371–395 (2002). ISSN 1063–6560
12. Gholipour, A., Araabi, B.N., Lucas, C.: Predicting chaotic time series using neural and neurofuzzy models: a comparative study. Neural Process. Lett. **24**(3), 217–239 (2006)
13. Kazarlis, S.A., Papadakis, S.E., Theocharis, J., Petridis, V.: Microgenetic algorithms as generalized hill-climbing operators for GA optimization. IEEE Trans. Evol. Comput. **5**(3), 204–217 (2001). ISSN 1089–778X
14. Li, B., Ong, Y.-S., Le, M.N., Goh, C.K.: Memetic gradient search. In: IEEE Congress on Evolutionary Computation, 2008. CEC 2008. (IEEE World Congress on Computational Intelligence), pp. 2894–2901. IEEE (2008). ISBN: 1424418224

15. Lin, C.-J., Chen, C.-H., Lin, C.-T.: A hybrid of cooperative particle swarm optimization and cultural algorithm for neural fuzzy networks and its prediction applications. IEEE Trans. Syst. Man. Cybern. Part C Appl. Rev. **39**(1), 55–68 (2009)
16. Mackey, M.C., Glass, L.: Oscillation and chaos in physiological control systems. Science **197**(4300), 287–289 (1977)
17. Moscato, P.: On evolution, search, optimization, genetic algorithms, martial arts: Towards memetic algorithms. Technical report 826, Caltech Concurrent Computation Program (1989)
18. Nand, R., Chandra, R.: Coevolutionary feature selection and reconstruction in neuro-evolution for time series prediction. In: Ray, T., Sarker, R., Li, X. (eds.) ACALCI 2016. LNCS (LNAI), vol. 9592, pp. 285–297. Springer, Heidelberg (2016). doi:10.1007/978-3-319-28270-1_24
19. NASDAQ. http://www.nasdaq.com/symbol/aciw/stock-chart
20. Ong, Y.S., Keane, A.J.: Meta-Lamarckian learning in memetic algorithms. IEEE Trans. Evol. Comput. **8**(2), 99–110 (2004). ISSN 1089–778X
21. Potter, M.A., De Jong, K.A.: Cooperative coevolution: an architecture for evolving coadapted subcomponents. Evol. Comput. **8**(1), 1–29 (2000)
22. Rodan, A., Tiño, P.: Minimum complexity echo state network. IEEE Trans. Neural Netw. **22**(1), 131–144 (2011)
23. Takens, F.: Detecting strange attractors in turbulence. In: Rand, D., Young, L.-S. (eds.) Dynamical Systems and Turbulence, Warwick 1980. Lecture Notes in Mathematics, vol. 898. Springer, Heidelberg (1981). http://www-psych.stanford.edu/~andreas/Time-Series/SantaFe.html
24. Weigend, A.S., Gershenfeld, N.A.: Laser problem dataset: the Santa Fe time series competition data (1994). http://www-psych.stanford.edu/~andreas/Time-Series/SantaFe.html
25. Yao, X.: Evolving artificial neural networks. Proc. IEEE **87**(9), 1423–1447 (1999)

SLA Management Framework to Avoid Violation in Cloud

Walayat Hussain[1(✉)], Farookh Khadeer Hussain[1],
and Omar Khadeer Hussain[2]

[1] Decision Support and e-Service Intelligence Lab, School of Software,
Centre for Quantum Computation and Intelligent Systems,
University of Technology Sydney, Sydney, NSW 2007, Australia
{walayat.hussain,farookh.hussain}@uts.edu.au
[2] School of Business, University of New South Wales, Canberra, Australia
o.hussain@adfa.edu.au

Abstract. Cloud computing is an emerging technology that have a broad scope to offers a wide range of services to revolutionize the existing IT infrastructure. This internet based technology offers a services like – on demand service, shared resources, multitenant architecture, scalability, portability, elasticity and giving an illusion of having an infinite resource by a consumer through virtualization. Because of the elastic nature of a cloud it is very critical of a service provider specially for a small/medium cloud provider to form a viable SLA with a consumer to avoid any service violation. SLA is a key agreement that need to be intelligently form and monitor, and if there is a chance of service violation then a provider should be informed to take necessary remedial action to avoid violation. In this paper we propose our viable SLA management framework that comprise of two time phases – pre-interaction time phase and post-interaction time phase. Our viable SLA framework help a service provider in making a decision of a consumer request, offer the amount of resources to consumer, predict QoS parameters, monitor run time QoS parameters and take an appropriate action to mitigate risks when there is a variation between a predicted and an agreed QoS parameters.

Keywords: Cloud computing · SLA management framework · SLA monitoring · Risk management in cloud

1 Introduction

Cloud computing is a recent technology trend that is attracting the attention of a wide range of businesses and enterprise. A cloud computing is combination of different technologies – grid computing, parallel computing, distributed computing, virtualization and multitenant architecture [1]. Due to its wide range of services small and large enterprises are shifting their businesses on cloud. According to a recent press release by Gartner, Inc [2] that state that the expected growth for a public cloud service market is 16.5 % with a total amount of $204 billion increased from $175 billion in 2015. Just in IaaS the market growth is expected from 31.9 % in 38.4 % in 2016 and a global cloud service market is predicted to grow by 13.6 % in 2016 that will reach to $90.3 billion.

© Springer International Publishing AG 2016
A. Hirose et al. (Eds.): ICONIP 2016, Part III, LNCS 9949, pp. 309–316, 2016.
DOI: 10.1007/978-3-319-46675-0_34

This increase in cloud market raises new challenges for cloud providers. One of the main issue is the formation of a viable SLA and predicting likely violation to alarm a service provider for an early remedial actions.

Service level agreement (SLA) is a key business agreement that bond a consumer and a provider for their commitment and promises for a specified period of a time. To avoid from violation penalties a service provider always need a system which intelligently predict a risk of a likely service violation and generate recommendations to manage those risks. There are number of approaches to monitor SLA violation [3] but the problem with most of the existing approaches is that it start monitoring when a provider and a consumer execute their SLA, however for an optimal SLA management a system need to assure the SLA violation avoidance from a pre-interaction phase.

In our previous work [4] we proposed a viable SLA management framework that comprise of two time phases – pre-interaction time phase and a post-interaction time phase. A pre-interaction phase consists of all steps before SLA execution and when both parties agreed and signed the agreement then a post-interaction phase start. A proposed SLA management framework is presented in Fig. 1. The pre-interaction phase comprises of two modules Identity manager module (IMM) and a viable SLA module (VSLAM). We described the pre-interaction section in our previous work [5, 6]. The post-interaction section is comprised of four modules – threshold formation module, runtime QoS monitoring module, QoS prediction module and risk management module. Modules in pre-interaction phase are responsible to authenticate requesting consumers and by considering their previous profile take a decision on consumer request for marginal resources and the amount of resources offer to them. In post-interaction phase, a threshold formation module (TFM) form a threshold and by observing a runtime behaviour of a consumer the Qos prediction module (PQoSM) QoS parameters for future intervals. If a system finds a difference between a values of PQoSM and the agreed QoS parameters then the risk management module (RMM) is activated, which consider reliability of a consumer, risk attitude of a provider and the predicted trajectory to decide an appropriate action.

The rest of the paper is organized as follows. Section 2 discuss related literature. Section 3 describe different components of our framework. Section 4 discuss implemention of a framework and Sect. 5 conclude a paper.

2 Related Work

Authors in [7] proposed risk based model to ensure the fulfilment of a SLA by a service provider and to maximize financial competence by considering a risk in a decision making process. A system identifies a risk and categorizes it into one of three levels – average risk, less risk and very less risk. Authors proposed three policies to minimize cost and risk both at node level and at graph level. To calculate the probability of a failure authors used statistical information from history data, however there is no comparisons for the optimality of a method with other methods like machine learning or non-linear regression methods. Authors did not use systematic estimation methods to estimate business values. Authors in [8] proposed a lightweight cloud platform for quickly access of changing resource information like CPU, memory etc. and to identify a specific need of resources. The platform helps for efficient monitoring of SLA violation. Their framework is comprised of five modules and the SLA management module is responsible to monitor SLA violations in the application layer based on existing approach CASViD. Morin et al. in [9] identified the problems and challenges linked with the SLA violation in cloud. The exhaustive use of the Internet of Services raises serious issues about data security and privacy. They proposed that due to frequent changes of the status of services the existing information security risk management methods are insufficient which need to be improved for better performance. Consumer's profile history plays a key role to identify service violation. In our previous work [10], we categorized requesting consumer's based on their previous transactions. A consumer who has previous already communicated with a provider has a transaction record. We consider its track record to calculate its transaction trend and for all new consumers we find its nearest neighbors and calculate the transaction trend of a new consumer based on its nearest neighbor's transaction history. From evaluation results we found that a profile history has a significant impact in prediction of SLA violation. To avoid SLA violation a provider, need an optimal prediction method which can intelligently predict likely violation and alarm service provider to mitigate it. There are a number of prediction methods that have different prediction accuracy depending on a type of a dataset. In our previous work [11, 12] we considered a cloud dataset from Amazon and applied neural network, stochastic and other time series prediction methods to evaluate their prediction accuracy. From the evaluation results we found that ARIMA method has the most optimal results.

3 SLA Management Framework

In this section we discuss about our SLA management framework. As presented in Fig. 1, the framework is comprised of two time phases- pre-interaction and post-interaction time phase. Modules in each time phase is explained below:

3.1 Pre-interaction Time Phase

There are two modules in pre-interaction time phase, identity manager module and a viable SLA module. These are explained below:

- Identity manager module (IMM): This is the first module in our framework which is responsible for authentication and validation of a consumer. The transaction record of each consumer is stored in a profile repository. When a module receives a request of a consumer after validation it passes the request to viable SLA module along with its previous history.
- Viable SLA module (VSLAM): The module receives a consumer request from IMM along with the transaction history if it is an existing consumer. For a new consumer the module selects a transaction history from its top-K nearest neighbors. The module use FIS at two levels. First it finds the suitability of a consumer by considering a reliability of a consumer and contract duration and then it combines

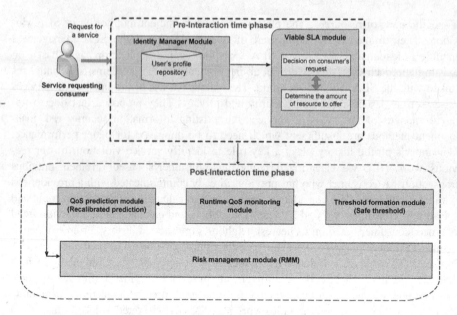

Fig. 1. Viable SLA management framework

the output with the risk attitude of a provider to decide the amount of resources offer to them. The detail of pre-interaction is explained in our previous work [5, 6].

3.2 Post-interaction Time Phase

The post-interaction phase is comprised of four modules – threshold formation module (TFM), runtime QoS monitoring module (RQoSM), QoS prediction module (QoSPM) and risk management module (RMM). The modules are explained below:

- Threshold formation module (TFM): Once both parties signed and execute a SLA, and then based on all agreed QoS parameters a provider defined its violation threshold. We propose two thresholds one is agreed threshold (Ta) which is defined in SLA and the second is safe threshold (Ts) which is a provider's threat threshold. Ts is more strict than a Ta, and when the runtime behavior reaches or exceed this threshold then it alert a service provider for managing a risk of service violation.
- Runtime QoS monitoring module (RQoSM): The module is responsible for monitoring runtime QoS parameters and send it to QoS prediction module (QoSPM) for recalibrated results.
- QoS prediction module (QoSPM): The module takes the input from a RQoSM and predict expected QoS parameters for future intervals. The value of QoSPM is compared with the agreed QoS parameters. If it finds that the value of QoSPM is reached or exceed the Ts value then the risk management module (RMM) is activated to manage risk.

- Risk management module (RMM): The module is started when a system find that predicted QoS parameters has reached or exceed the threat threshold. The module use FIS and take inputs – risk attitude of a provider, reliability of a consumer and predicted trajectory to determine an appropriate action to mitigate a risk. The action is either immediate action, delayed action or no action.

4 Implementation and Evaluation

In this section we evaluate our framework. We use two datasets from different sources for two phases of our framework. One from an existing dataset [13] which comprised of 142 users using 4,532 web services. We consider a throughput and a response time for 10 web services. For a second dataset we consider Amazon EC2 IaaS cloud services – EC2 US East, collected from cloudclimate [14] through the PRTG monitoring service [15]. We consider a CPU performance with 5 min of intervals for a duration of 4 days starting from 21 April 2015 to 25th April 2015 with 1007 observations. For evaluation we used Microsoft Visual Studio 2010 with Microsoft SQL Server Management Studio 2008 for the databases and MATLAB to design the FIS application. Figure 2 presents CPU, memory and I/O for the mentioned period.

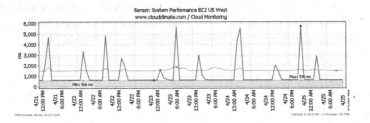

Fig. 2. Data from EC2 US West [15]

We consider a scenario in which a provider receives two request from a consumer A with a consumer ID 806 and from a consumer B with consumer ID 809. We divide the working of our framework into different steps which are explained below:

Step1: When a viable SLA management framework receives a request it is forwarded to the IMM module where it is checked from its stored repository for the identity and its previous history. In this case the IMM module found that both consumers have a profile history. The IMM forward the request to the VSLAM to decide about the request and the amount of resources offer to accepted requests.

Step 2: The VSLAM is a key module in our framework that decide about the request of consumer and the amount of resources offer to them. VSLAM receives previous history of consumers and based on their previous profile history it calculates the Ttrend of both consumers. Which is then compared with the threshold value. A provider has fixed a threshold value of 50 %. The T_{trend} calculated from the previous record of consumer 806 and 809 are 59.22 % and 44.88 % respectively.

In a case of first consumer the T_{trend} value is greater than the threshold however for second consumer the T_{trend} value is less than the threshold value hence a request is rejected. After a decision the VSLAM use the FIS by taking contract duration, reliability of consumer to calculate suitability of a consumer and then by considering the risk attitude of a provider then calculate the amount of resources offer to consumer A. The output of the VSLAM for consumer A is 49.01 % that means a system offers 49.01 % of the requested resource for the marginal resources [6]. The output by VSLAM is presented in Table 1.

Table 1. Request determination and resource allocation of requesting consumers [5]

ID	Transaction trend	Reliability	Suitability of consumer	Current suitability value			Risk attitude of provider	Risk propensity		Required suitability value	Resource allocation	Decision of consumer request
806	59.22 %	42	45.17	Med = 0.90	Low = 0.10		0.6	RN = 0.80	RT = 0.20	Medium = 0.7	49.01 %	Accept
809	44.88 %	13.6	–	–	–		–	–	–	–	–	Reject

Step 3: In a third step a consumer and a provider both are agreed on each service level objectives and each QoS parameters, mentioned in SLA. At this stage the process of post-interaction start. The first module in post-interaction phase is a threshold formation module. We considered EC2 cloud dataset and consider QoS parameter - CPU. The agreed threshold T_a value between consumer and a provider is 290 ms and a provider set its T_s value as 250 ms. For this phase we considered 10 time intervals that starts from 5:00 AM and end at 7:30 AM. The T_s and T_a is presented in Fig. 3.

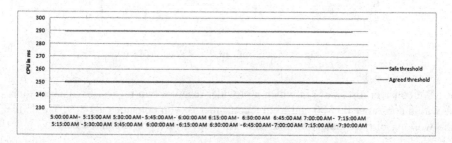

Fig. 3. Agreed and safe threshold

Step 4: When the transection start the runtime behaviour of consumer is recored and the value of RQoSM is forwarded to QoSPM, that use an intelligent prediction method to predict for future intervals. For this experiment we considered ARIMA method because it has an optimal result [11]. The predicted result for 10 intervals are presented in Fig. 4.

Step 5: From a Fig. 4 we see that till 6:00:00 AM the predicted result is below the T_s value so framework let a system to execute, but we see at time interval 6:15:00 AM a predicted result has touch T_s value and but at the next interval 6:20:00 AM come back below the T_s value. At time interval 7:15:00 AM we see that predicted result exceed T_s value and moving towards T_a. At this instance the risk management module is activated to manage a risk.

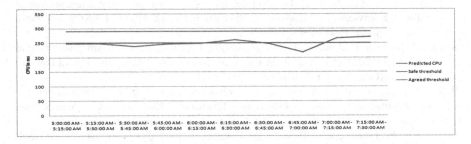

Fig. 4. Prediction result for 10 intervals

Step 6: The RMM take reliability of consumer, risk attitude of a provider and predicted trajectory and use FIS to decide either to take immediate action, delayed action or no action.

5 Conclusion

SLA is a crucial contract between a consumer and a provider that let them for executing their business. To enhance its trust value and to avoid from penalties a service provider need a framework that help in decision making for SLA formation, its monitoring and a mechanism that should inform a service provider to take immediate action when there is a risk of a service violation. Our viable SLA management framework assist a service provider to achieve all mentioned objectives. From the evaluation result we observed that our framework not only enable a service provider to monitor SLA in both phases of SLA life cycle but it also helps a service provider to manage a risk of a service violation.

References

1. Bahtovski, A., Gusev, M.: Analysis of cloud portability. In: The 10th Conference for Informatics and Information Technology, pp. 280–284 (2013)
2. Gartner, I.: Forecast: public cloud services. Worldwide, 2013-2019 (2016). (Gartner)
3. Hussain, W., Hussain, F.K., Hussain, O.K.: Maintaining trust in cloud computing through SLA monitoring. In: Loo, C.K., Yap, K.S., Wong, K.W., Beng Jin, A.T., Huang, K. (eds.) ICONIP 2014, Part III. LNCS, vol. 8836, pp. 690–697. Springer, Heidelberg (2014)

4. Hussain, W., et al.: Profile-based viable service level agreement (SLA) violation prediction model in the cloud. In: 2015 10th International Conference on P2P, Parallel, Grid, Cloud and Internet Computing (3PGCIC), Krakow, Poland, pp. 268–272. IEEE (2015)

5. Hussain, W., Hussain, F.K., Hussain, O.: Allocating optimized resources in the cloud by a viable SLA model. In: 2016 IEEE International Conference on Fuzzy Systems (FUZZ-IEEE). IEEE, Vancouver, Canada (2016)

6. Hussain, W., et al.: Provider-based optimized personalized viable SLA (OPV-SLA) framework to prevent SLA violation. Comput. J., 24 (2016)

7. Macias, M., Guitart, J.: A risk-based model for service level agreement differentiation in cloud market providers. In: Magoutis, K., Pietzuch, P. (eds.) DAIS 2014. LNCS, vol. 8460, pp. 1–15. Springer, Heidelberg (2014)

8. Liu, D., Kanabar, U., Lung, C.-H.: A light weight SLA management infrastructure for cloud computing. In: 2013 26th Annual IEEE Canadian Conference on Electrical and Computer Engineering (CCECE). IEEE (2013)

9. Morin, J.-H., Aubert, J., Gateau, B.: Towards cloud computing SLA risk management: issues and challenges. In: 2012 45th Hawaii International Conference on System Science (HICSS). IEEE (2012)

10. Hussain, W., Hussain, F.K., Hussain, O.: Comparative analysis of consumer profile-based methods to predict SLA violation. In: FUZZ-IEEE. IEEE, Istanbul Turkey (2015)

11. Hussain, W., Hussain, F.K., Hussain, O.: QoS prediction methods to avoid SLA violation in post-interaction time phase. In: 11th IEEE Conference on Industrial Electronics and Applications (ICIEA 2016). IEEE, Hefei (2016)

12. Hussain, W., Hussain, F.K., Hussain, O.K.: Towards soft computing approaches for formulating viable service level agreements in cloud. In: Arik, S., Huang, T., Lai, W.K., Liu, Q. (eds.) ICONIP 2015. LNCS, vol. 9492, pp. 639–646. Springer, Heidelberg (2015). doi:10.1007/978-3-319-26561-2_75

13. Zhang, Y., Zheng, Z., Lyu, M.R.: WSPred: a time-aware personalized QoS prediction framework for Web services. In: 2011 IEEE 22nd International Symposium on Software Reliability Engineering (ISSRE). IEEE (2011)

14. CloudClimate: Watching the Cloud. http://www.cloudclimate.com/

15. Monitor, P.N.: PRTG Network Monitor. https://prtg.paessler.com/

Pattern Retrieval by Quaternionic Associative Memory with Dual Connections

Toshifumi Minemoto[1]([✉]), Teijiro Isokawa[1], Masaki Kobayashi[2],
Haruhiko Nishimura[3], and Nobuyuki Matsui[1]

[1] Graduate School of Engineering, University of Hyogo, Himeji, Hyogo, Japan
eu14n001@steng.u-hyogo.ac.jp
[2] Yamanashi University, Kofu, Yamanashi, Japan
[3] Graduate School of Applied Informatics, University of Hyogo, Kobe, Hyogo, Japan

Abstract. An associative memory based on Hopfield-type neural network, called Quaternionic Hopfield Associative Memory with Dual Connection (QHAMDC), is presented and analyzed in this paper. The state of a neuron, input, output, and connection weights are encoded by quaternion, a class of hypercomplex number systems with non-commutativity for its multiplications. In QHAMDC, calculation for an internal state of a neuron is conducted by two types of multiplications for neuron's output and connection weight. This makes robustness of the proposed associative memory for retrieval of patterns. The experimental results show that the performances of retrieving patterns by QHAMDC are superior to those by the previous QHAM.

1 Introduction

Extensions in complex and hypercomplex number systems on Hopfield-type associative memories (HAMs) have been extensively investigated [1]. Representations for states in the networks are enriched due to the degrees of freedom in these systems. Complex-valued extension of Hopfield Associative memories, called CHAMs, were proposed [2–5] and their learning schemes for embedding the patterns in the networks were also investigated [6–8]. Many CHAMs adopt the representation for the state of a neuron as a discrete point in the complex plane, so multilevel values, such as the intensity of a pixel in the gray-scaled image, can be naturally represented in these networks. A quaternionic extension of HAMs is called QHAMs where the state of a neuron is represented by three kinds of phases in distinct complex planes [9,10]. QHAMs are expected to process three-dimensional multilevel values such as intensities of RGB color components or body coordinates in three dimensional space.

One difficulty in CHAMs is the existence of rotated patterns in the network by degenerated states of embedded patterns. When a pattern is to be embedded to a CHAM network where a neuron takes K of quantization levels, $K - 1$ rotated patterns are also to be embedded in the network. These rotated patterns makes their mixture patterns, and they become spurious patterns in the network, leading to reduce the noise robustness in retrieving patterns

© Springer International Publishing AG 2016
A. Hirose et al. (Eds.): ICONIP 2016, Part III, LNCS 9949, pp. 317–325, 2016.
DOI: 10.1007/978-3-319-46675-0_35

from the network. QHAMs adopt a similar principle in representing the patterns in the network, thus they also suffer from the rotated and mixture patterns. One possible solution to cope with rotated patterns is to compose a heterogeneous network where real-valued neurons are incorporated in the complex or quaternionic networks. A pattern in the real-valued HAM has only one rotated pattern in which each of elements takes the inverted value. A combination of real-valued and complex-valued/quaternionic patterns makes fewer rotated patterns. Such networks are presented and investigated as Complex-valued Bipartite Auto-Associative Memory (CBAAM) [11] and Quaternionic Bipartite Auto-Associative Memory (QBAAM) [12]. It is shown that the retrieval performances for CBAAM and QBAAM are superior to those in CHAM and QHAM.

This paper presents a QHAM with another type of heterogeneousness, called QHAM with dual connections (QHAMDC). This comes from the nature of quaternion, i.e., non-commutative of multiplications between quaternions. The internal state of a neuron in the network is calculated according to dual connection of neurons, i.e., both by the inputs being multiplied by the connection weights and by the connection weights multiplied by inputs, both of which are identical in real and complex numbers but not in quaternions.

It is shown that this network has more robustness than QHAM networks. In [13], neuronal states in the proposed network are binary type, and the learning scheme for embedding patterns is Hebbian rule. So, the effects for multivalued-state neurons and for an advanced learning scheme have not been investigated. The experimental results performed in this paper show that the proposed QHAMDC has better performances than QHAM in retrieving patterns, where multivalued patterns are embedded to these networks by using projection rule.

2   Preliminaries

2.1   Quaternion Algebra and Its Representation

Quaternions form a class of hypercomplex numbers that consist of a real number and three kinds of imaginary number, i, j, k. Formally, a quaternion number is defined as a vector x in a 4-dimensional vector space, $x = x^{(e)} + x^{(i)}i + x^{(j)}j + x^{(k)}k$, where $x^{(e)}$, $x^{(i)}$, $x^{(j)}$, and $x^{(k)}$ are real numbers. $\mathbb{H}$, the division ring of quaternions, thus constitutes the four-dimensional vector space over the real numbers with the bases $1, i, j, k$. These bases satisfy the following identities: $i^2 = j^2 = k^2 = ijk = -1, ij = -ji = k, jk = -kj = i, ki = -ik = j$. From these rules it follows immediately that multiplication of quaternions is not commutative.

Quaternion is also written using 4-tuple notation $x = (x^{(e)}, x^{(i)}, x^{(j)}, x^{(k)})$ or 2-tuple notation $(x^{(e)}, \vec{x})$, where $\vec{x} = \{x^{(i)}, x^{(j)}, x^{(k)}\}$. In this representation $x^{(e)}$ is the scalar part of x and $\vec{x}$ forms the vector part. The quaternion conjugate is defined as $x^* = (x^{(e)}, -\vec{x}) = x^{(e)} - x^{(i)}i - x^{(j)}j - x^{(k)}k$.

The operations between quaternions $p = (p^{(e)}, \vec{p}) = (p^{(e)}, p^{(i)}, p^{(j)}, p^{(k)})$ and $q = (q^{(e)}, \vec{q}) = (q^{(e)}, q^{(i)}, q^{(j)}, q^{(k)})$ are defined as follows. The addition and subtraction of quaternions are defined, in the same manner as those of complex-valued numbers or vectors, by $p \pm q = (p^{(e)} \pm q^{(e)}, \vec{p} \pm \vec{q})$. With regard to the multiplication, the product between p and q is determined as $pq = (p^{(e)}q^{(e)} - \vec{p} \cdot \vec{q}, p^{(e)}\vec{q} + q^{(e)}\vec{p} + \vec{p} \times \vec{q})$ where $\vec{p} \cdot \vec{q}$ and $\vec{p} \times \vec{q}$ denote the dot and cross products respectively between three dimensional vectors $\vec{p}$ and $\vec{q}$. The conjugate of product holds the relation of $(pq)^* = q^*p^*$. The quaternion norm of x, notation $|x|$, is defined by $|x| = \sqrt{xx^*}$.

A complex value $c = c^{(e)} + ic^{(i)}$ in Cartesian form can also be represented as one in polar form: $c = r \cdot e^{i\theta}$, where $r = \sqrt{c^{(e)^2} + c^{(i)^2}}$ and $\theta = \tan^{-1} c^{(i)}/c^{(e)}$. In a similar way, a quaternion x in Cartesian form can be transformed into one in polar form. We adopt the representation defined in [14, 15], and a quaternion x in a polar form can be represented as $x = |x|e^{i\varphi}e^{k\psi}e^{j\theta}$, where $e^{i\varphi} = \cos\varphi + i\sin\varphi$, $e^{j\theta} = \cos\theta + j\sin\theta$, and $e^{k\psi} = \cos\psi + k\sin\psi$. Three kinds of phase angles φ, θ, ψ are defined within the interval $-\pi \leq \varphi < \pi$, $-\pi/2 \leq \theta < \pi/2$, and $-\pi/4 \leq \psi \leq \pi/4$, respectively.

2.2 Quaternionic Hopfield Associative Memory

We assume a quaternionic Hopfield associative memory (QHAM) with N quaternionic multistate neurons. The state of a p-th neuron in QHAM is represented in the polar form,

$$u_p = e^{i\varphi_p}e^{k\psi_p}e^{j\theta_p}, \tag{1}$$

where $|u_p| = 1$. The internal state of the p-th neuron at a discrete time t is defined as

$$h_p(t) = \sum_{q=1}^{N} w_{pq}u_q(t) = \sum_{q=1}^{N} w_{pq}e^{i\varphi_q(t)}e^{k\psi_q(t)}e^{j\theta_q(t)}, \tag{2}$$

where $w_{pq} \in \mathbb{H}$ denotes the connection weight from the p-th neuron to the q-th neuron.

We use qsign function which consists of three types of complex-valued multistate signum function as an activation function for quaternionic multistate neurons. Thus, the state of a neuron p at time t is updated as

$$u_p(t+1) = \text{qsign}\left(h_p(t)\right), \tag{3}$$

where

$$\text{qsign}\left(h_p\right) = \text{qsign}\left(e^{i\varphi_p}\,e^{k\psi_p}\,e^{j\theta_p}\right) \tag{4}$$

$$= csign_A\left(e^{i\varphi_p}\right)\,csign_B\left(e^{k\psi_p}\right)\,csign_C\left(e^{j\theta_p}\right). \tag{5}$$

The function $csign_A$ is used for updating φ, and it is defined as

$$\text{csign}_A\left(e^{i\varphi}\right) = \begin{cases} q_0^{(\varphi)} & \text{for } -\pi \leq \arg\left(e^{i\varphi}\right) < -\pi + \varphi_0 \\ q_1^{(\varphi)} & \text{for } -\pi + \varphi_0 \leq \arg\left(e^{i\varphi}\right) < -\pi + 2\varphi_0 \\ \quad\vdots \\ q_{A-1}^{(\varphi)} & \text{for } -\pi + (A-1)\varphi_0 \leq \arg\left(e^{i\varphi}\right) < -\pi + A\varphi_0 \end{cases}, \tag{6}$$

where A is the quantization level for φ, and φ_0 denotes a quantization unit which is defined by $\varphi_0 = 2\pi/A$. $q_a^{(\varphi)}$ is a distinct point on a unit circle which is defined as $q_a^{(\varphi)} = \exp(i(-\pi + a\varphi_0 + \frac{\varphi_0}{2}))$. Therefore the function csign_A outputs the closest quaternion in $\{q_0^{(\varphi)}, \ldots, q_{A-1}^{(\varphi)}\}$ corresponding to the input. Similarly, the function csign_B for updating ψ and the function csign_C for updating θ are defined as follows:

$$
\mathrm{csign}_B\left(e^{k\psi}\right) =
\begin{cases}
q_0^{(\psi)} & \text{for } -\frac{\pi}{4} \leq \arg\left(e^{k\psi}\right) < -\frac{\pi}{4} + \psi_0 \\
q_1^{(\psi)} & \text{for } -\frac{\pi}{4} + \psi_0 \leq \arg\left(e^{k\psi}\right) < -\frac{\pi}{4} + 2\psi_0 \\
\quad\vdots \\
q_{B-1}^{(\psi)} & \text{for } -\frac{\pi}{4} + (B-1)\psi_0 \leq \arg\left(e^{k\psi}\right) < -\frac{\pi}{4} + B\psi_0
\end{cases}
, \quad (7)
$$

$$
\mathrm{csign}_C\left(e^{j\theta}\right) =
\begin{cases}
q_0^{(\theta)} & \text{for } -\frac{\pi}{2} \leq \arg\left(e^{j\theta}\right) < -\frac{\pi}{2} + \theta_0 \\
q_1^{(\theta)} & \text{for } -\frac{\pi}{2} + \theta_0 \leq \arg\left(e^{j\theta}\right) < -\frac{\pi}{2} + 2\theta_0 \\
\quad\vdots \\
q_{C-1}^{(\theta)} & \text{for } -\frac{\pi}{2} + (C-1)\theta_0 \leq \arg\left(e^{j\theta}\right) < -\frac{\pi}{2} + C\theta_0
\end{cases}
, \quad (8)
$$

where B and C are the quantization levels for φ and θ, respectively. The quantization units ψ_0 and θ_0 are defined by $\psi_0 = \pi/2B$ and $\theta_0 = \pi/C$, respectively. $q_b^{(\psi)}$ and $q_c^{(\theta)}$ are also defined as follows: $q_b^{(\psi)} = \exp(j(-\frac{\pi}{4} + b\psi_0 + \frac{\psi_0}{2}))$, $q_c^{(\theta)} = \exp(k(-\frac{\pi}{2} + c\theta_0 + \frac{\theta_0}{2}))$. Hence, a quaternionic neuron takes a total of $A \times B \times C$ states. An example of the quantized output points of the quaternionic neuron is shown in Fig. 1, where $A = 4$, $B = 2$, and $C = 3$. The update of neurons is conducted in an asynchronous manner which at most one neuron is updated at a time.

The energy function of QHAM takes real value when the connection weights w_{pq} satisfy the following conditions:

$$
w_{pq} = w_{qp}^*, \quad w_{pp} = w_{pp}^* = (w^{(e)}, \vec{0}), \quad w^{(e)} \geq 0. \quad (9)
$$

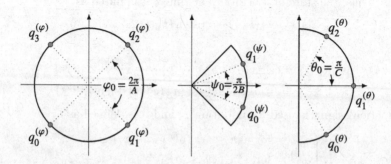

Fig. 1. An example of quantized output points in the quaternionic multistate neuron $(A = 4, B = 2, C = 3)$

The function monotonically decreases under the condition $|\Delta\varphi| < \varphi_0, |\Delta\varphi| < \psi_0, |\Delta\psi| < \varphi_0$. $\Delta\varphi$, $\Delta\psi$, $\Delta\theta$ are a phase difference between the state at time $t+1$ and the internal state at time t for the neuron undergoing its update [9].

2.3   Projection Rule for Quaternionic Hopfield Associative Memory

We describe the projection rule for QHAM [16], which is a learning scheme that can embed non-orthogonal (correlated) memory patterns in a network. A key idea of the projection rule is that non-orthogonal patterns are first projected onto orthogonal ones, and then the Hebbian rule is applied to these projected patterns.

Let $\boldsymbol{\xi}^p = (\xi_1^p, \dots, \xi_N^p)^{\mathrm{T}}$ be a p-th memory pattern ($p = 1, \dots, P$), where $\xi_i^p \in \{q_a^{(\varphi)} q_b^{(\psi)} q_c^{(\theta)} | a = 1, \dots, A, \ b = 1, \dots, B, \ c = 1, \dots, C\}$. Then the training pattern matrix $\boldsymbol{X}$ is represented as

$$\boldsymbol{X} = (\boldsymbol{\xi}^1, \boldsymbol{\xi}^2, \dots, \boldsymbol{\xi}^p). \tag{10}$$

The connection weight matrix $\boldsymbol{W}$ whose element at (p, q) is denoted as w_{pq} is given as:

$$\boldsymbol{W} = \boldsymbol{A} - \mathrm{diag}(\boldsymbol{A}), \quad \boldsymbol{A} = \boldsymbol{X}(\boldsymbol{X}^*\boldsymbol{X})^{-1}\boldsymbol{X}^*, \tag{11}$$

where $\boldsymbol{X}^*$ denotes the conjugate transpose of $\boldsymbol{X}$, and $\mathrm{diag}(\boldsymbol{A})$ is the diagonal matrix whose diagonal elements equal to the diagonal of $\boldsymbol{A}$. The number of patterns that can be stored in the network equals to the number of neurons in the network by using the projection rule.

3   Quaternionic Hopfield Associative Memory with Dual Connections

This section presents quaternionic Hopfield associative memory with dual connections (QHAMDC). We can consider two types of multiplication order of operations to calculate weighted input because of non-commutative nature of quaternion algebra. In QHAMDC, two types of connection weights are used for updating the internal states of neurons:

$$h_p(t) = \sum_{q=1}^{N} \left(w_{pq}^{\mathrm{L}} u_q(t) + u_q(t) w_{pq}^{\mathrm{R}} \right), \tag{12}$$

where w_{pq}^{L} is a connection weight from p-th neuron to q-th neuron which is multiplied from the left and w_{pq}^{R} is one which is multiplied from the right. The connection weight matrix $\boldsymbol{W}^{\mathrm{L}}$ whose element is w_{pq}^{L} is given by Eq. (11). The connection weight matrix $\boldsymbol{W}^{\mathrm{R}}$ whose element is w_{pq}^{R} is given as follows:

$$\boldsymbol{W}^{\mathrm{R}} = \boldsymbol{B} - \mathrm{diag}(\boldsymbol{B}), \quad \boldsymbol{B} = \boldsymbol{Y}^*(\boldsymbol{Y}\boldsymbol{Y}^*)^{-1}\boldsymbol{Y}, \quad \boldsymbol{Y} = \boldsymbol{X}^{\mathrm{T}}, \tag{13}$$

where X is the training pattern matrix. Thus, W^R denotes the connection weights obtained by using the projection rule with multiplications in the reverse order. The state of a neuron in QHAMDC is updated by Eq. (3) in the same manner as in QHAM.

We define the energy function of the network with dual connections as

$$E(t) = -\frac{1}{2} \sum_{p \neq q} \left(u_p^*(t) w_{pq}^L u_q(t) + u_q(t) w_{pq}^R u_p(t)^* \right), \tag{14}$$

where E takes real value ($E = E^*$) when w_{pq}^L and w_{pq}^R satisfy the condition Eq. (9). Since $\text{Re}(xy) = \text{Re}(yx)$ holds for $x, y \in \mathbb{H}$, the energy function can be represented as

$$E(t) = -\frac{1}{2} \text{Re} \sum_{p \neq q} u_p^*(t) \left(w_{pq}^L u_q(t) + u_q(t) w_{pq}^R \right). \tag{15}$$

Suppose that the state of the r-th neuron is updated at time $t+1$ according to the Eqs. (12) and (3). The energy gap ΔE between the $E(t+1)$ and $E(t)$ finally becomes as follows:

$$\Delta E = - \text{Re} \sum_{r \neq q} u_r^*(t+1) \left(w_{rq}^L u_q(t) + u_q(t) w_{rq}^R \right)$$
$$+ \text{Re} \sum_{r \neq q} u_r^*(t) \left(w_{rq}^L u_q(t) + u_q(t) w_{rq}^R \right). \tag{16}$$

ΔE becomes non-positive because $u_r^*(t)$ maximize $\text{Re}(u_r^*(t) h_r(t))$ according to the activation function. Therefore, the energy decreases monotonically.

4   Experimental Results

In this section, we demonstrate the performances of QHAMDC in terms of the noise robustness of retrieving embedded patterns under several conditions. The performances of QHAM are also shown for the comparisons with QHAMDC.

The experiments for QHAMs and QHAMDCs are conducted with the following parameters:

1. The number of neurons in the network (N) is set to 100.
2. The number of patterns to be embedded to the network (P) is set to 10, 30, and 50.
3. The quantization levels for quaternionic neurons (A, B, C) are set to $(8, 2, 4)$, $(16, 4, 8)$, and $(32, 8, 16)$.
4. The noise level (r) changes to 0.00, 0.05, $\cdots$, and 0.80.
5. The update of neurons in the network is repeated for 1000 iterations.
6. For each parameter, 1000 experiments are conducted with randomly generated patterns being embedded.

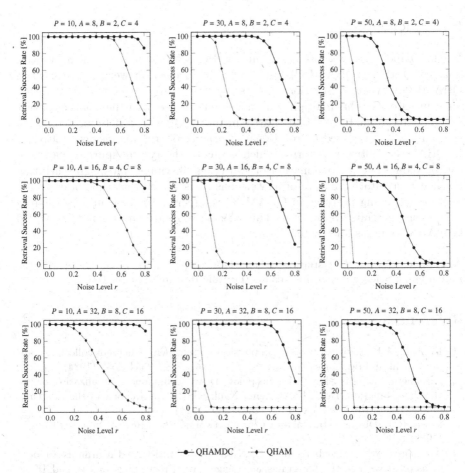

Fig. 2. Noise robustness for retrieving patterns for QHAM and QHAMDC in the case of $N = 100$.

The input pattern to the network is chosen as one of the embedded patterns with some noises being imposed with the level r. The values for $\lfloor rN \rfloor$ elements in the chosen pattern are randomly changed, and the resultant pattern is used for the input pattern to the network.

The retrieval performances against the noise level of QHAM and QHAMDC are shown in Fig. 2. It is shown that the retrieval success rates for QHAMs get worse when the number of embedded patterns increases or the quantization levels increases. Also, the quantization levels are particularly sensitive parameters for QHAMs. In contrast, the retrieval success rates for QHAMDCs are robust against the quantization levels. The quantization levels have little effect on the retrieval performances of QHAMDC as compared to the performances of QHAM. From these results, we can say that the effects of rotated patterns emerged from training patterns in QHAM are reduced by adopting dual connections.

5   Conclusion

In this paper, we propose quaternionic Hopfield associative memory with dual connections (QHAMDC). Two types of connection weights are used in QHAMDC: one are connection weights multiplied from the left and the other are connection weights multiplied from the right. Noise robustness of the conventional quaternionic Hopfield associative memory (QHAM) is deteriorated by the spurious patterns caused by rotational invariance of training memory patterns. In QHAMDC, rotated patterns, which is one of the typical spurious patterns, can be reduced by a combination of two types of connection weights utilizing the non-commutative nature of quaternions. It is shown that the noise robustness for retrieving patterns in QHAMDC is superior to those in QHAM from experimental results. More detailed analysis on spurious patterns in QHAM and QHAMDC remains for our future work.

Acknowledgment. This study was financially supported by Japan Society for the Promotion of Science (Grant-in-Aids for Scientific Research (C) 16K00337).

References

1. Hopfield, J.J.: Neural networks and physical systems with emergent collective computational abilities. Proc. Natl. Acad. Sci. USA **79**(8), 2554–2558 (1984)
2. Aizenberg, N.N., Ivaskiv, Y.L., Pospelov, D.A.: About one generalization of the threshold function. Doklady Akademii Nauk SSSR (The Reports of the Academy of Sciences of the USSR) **196**(6), 1287–1290 (1971)
3. Noest, A.J.: Discrete-state phasor neural networks. Phys. Rev. A **38**(4), 2196–2199 (1988)
4. Aizenberg, N.N., Aizenberg, I.N.: CNN based on multi-valued neuron as a model of associative memory for gray-scale images. In: Proceedings of the 2nd IEEE International Workshop on Cellular Neural Networks and their Applications, pp, 36–41 (1992)
5. Jankowski, S., Lozowski, A., Zurada, J.M.: Complex-valued multistate neural associative memory. IEEE Trans. Neural Netw. **7**(6), 1491–1496 (1996)
6. Aoki, H., Kosugi, Y.: An image storage system using complex-valued associative memories. Proc. Int. Conf. Pattern Recognit. **2**, 626–629 (2000)
7. Müezzinoğlu, M.K., Güzeliş, C., Zurada, J.M.: A new design method for the complex-valued multistate Hopfield associative memory. IEEE Trans. Neural Netw. **14**(4), 891–899 (2003)
8. Lee, D.L.: Improvements of complex-valued Hopfield associative memory by using generalized projection rules. IEEE Trans. Neural Netw. **17**(5), 1341–1347 (2006)
9. Isokawa, T., Nishimura, H., Saitoh, A., Kamiura, N., Matsui, N.: On the scheme of quaternionic multistate Hopfield neural network. In: Proceedings of of Joint 4th International Conference on Soft Computing and Intelligent Systems and 9th International Symposium on advanced Intelligent Systems, pp. 809–813 (2008)
10. Minemoto, T., Isokawa, T., Nishimura, H., Matsui, N.: Quaternionic multistate Hopfield neural network with extended projection rule. Artif. Life Robot. **21**(1), 106–111 (2016)

11. Suzuki, Y., Kobayashi, M.: Complex-valued bipartite auto-associative memory. IEICE Trans. Fund. Electron. Commun. Comput. Sci. **97**(8), 1680–1687 (2014)
12. Minemoto, T., Isokawa, T., Matsui, N., Kobayashi, M., Nishimura, H.: On the performance of quaternionic bidirectional auto-associative memory. In: Proceedings of International Joint Conference on Neural Networks, #15594, 6 pages (2015)
13. Kobayashi, M.: Hybrid quaternionic Hopfield neural network. IEICE Trans. Fund. Electron. Commun. Comput. Sci. **98**(7), 1512–1518 (2015)
14. Bülow, T.: Hypercomplex spectral signal representations for the processing and analysis of images. Ph.D. thesis, Christian-Albrechts-Universität zu Kiel (1999)
15. Bülow, T., Sommer, G.: Hypercomplex signals–a novel extension of the analytic signal to the multidimensional case. IEEE Trans. Signal Process. **49**(11), 2844–2852 (2001)
16. Isokawa, T., Nishimura, H., Matsui, N.: Quaternionic neural networks for associative memories. In: Hirose, A. (ed.) Complex-Valued Neural Networks: Advances and Applications, pp. 103–131. Wiley-IEEE Press (2013)

A GPU Implementation of a Bat Algorithm Trained Neural Network

Amit Roy Choudhury$^{(\boxtimes)}$, Rishabh Jain, and Kapil Sharma

Department of Computer Engineering, Delhi Technological University,
New Delhi, India
amitrc17@gmail.com

Abstract. In recent times, there has been an exponential growth in the viability of Neural Networks (NN) as a Machine Learning tool. Most standard training algorithms for NNs, like gradient descent and its variants fall prey to local optima. Metaheuristics have been found to be a viable alternative to traditional training methods. Among these metaheuristics the Bat Algorithm (BA), has been shown to be superior. Even though BA promises better results, yet being a population based metaheuristic, it forces us to involve many Neural Networks and evaluate them on nearly every iteration. This makes the already computationally expensive task of training a NN even more so. To overcome this problem, we exploit the inherent concurrent characteristics of both NNs as well as BA to design a framework which utilizes the massively parallel architecture of Graphics Processing Units (GPUs). Our framework is able to offer speed-ups of upto 47× depending on the architecture of the NN.

Keywords: Neural Networks · Bat Algorithm · GPU

1 Introduction

Neural Networks have received immense attention from the international research community owing to their universality and adaptability. There has been a growth in the use of Artificial Neural Networks (ANN) for real world applications [1] in recent years, mostly because of availability of powerful hardware. ANNs find their use in both classification and regression tasks [2] but in this paper we focus on the classification application of feed-forward Neural Networks. These NNs consist of an input layer, an output layer and one or more hidden layers in between them composed of nodes called neurons. The number of nodes in the input layer is dictated by the number of dimensions of input. Similarly, the number of nodes in the output layer is equal to the number of output classes. The number of nodes in any hidden layer is a parameter of the system and is set empirically. Each node is connected to all nodes of the next layer. Each connection is associated with a weight. The value associated to any neuron, called it's activation value, can be calculated as follows

$$V = g(\sum_{i=0}^{n} I_i.W_i) \tag{1}$$

© Springer International Publishing AG 2016
A. Hirose et al. (Eds.): ICONIP 2016, Part III, LNCS 9949, pp. 326–334, 2016.
DOI: 10.1007/978-3-319-46675-0_36

where W_i is the weight associated with a connection from input i to current node, I_i is input value i and $g(x)$ is called the activation function. The activation function is generally taken to be a non-linear function such as logistic function, hyperbolic tangent or rectified linear units.

Training of neural networks has been a widely researched topic [3,4]. The most popular methods of training are gradient based algorithms like Gradient Descent or Conjugate Gradient. Though these are widely used, they are known to fall prey to local minima [5]. Other Metaheuristics have shown comparable or better results in comparison to these methods [4,6]. Among them, the Bat Algorithm (BA) has been shown to be superior [7].

BA is a population based optimization technique. Such techniques are used for optimization problems i.e. problems in which the goal is to find the minimum (or maximum) of an objective function. BA is inspired by the hunting behaviour of bats [8]. It uses a process similar to the echolocation used by bats to identify and track down its prey. We consider a population of N bats. The ith of which is flying with a velocity of V_i^t and is at a position X_i^t at time t. The bats vary their frequency f (and thus wavelength) and loudness A to locate prey. They alter their frequencies and velocities as follows

$$f = f_{min} + (f_{max} - f_{min}) * \delta \tag{2}$$
$$V_i^t = V_i^{t-1} + (X_i^t - X_{gBest}^t) * f_i \tag{3}$$

where X_{gBest} (global best) is the position of the bat which is closest to the prey, f_{max} and f_{min} are maximum and minimum frequencies respectively which are parameters of the system that are decided empirically and $\delta \in [0,1]$ is a randomly generated number. The best bat is identified by the value of the fitness function at its position. This fitness function is the function which is being optimized. The position of a bat varies with time given by

$$X_i^t = X_i^{t-1} + V_i^t \tag{4}$$

Any bat may perform a local search instead of moving along its velocity according to its pulse rate. For the local search procedure each bat takes a random walk creating a new position (solution) for itself, this is given by

$$X_{new} = X_{old} + \alpha.A_{avg} \tag{5}$$

where A_{avg} is the average loudness of the population and $\alpha \in [-1,1]$ is a random number. The loudness of a bat decreases and the pulse rate increases as the bat gets closer to its prey.

$$A^{t+1} = \alpha.A^t \tag{6}$$
$$r^{t+1} = r_0.(1 - e^{-\gamma.t}) \tag{7}$$

where alpha and gamma are constants which are decided empirically and r_0 is the final pulse rate generally equal to 1.

BA (or any other population based metaheuristic) requires many evaluations of the fitness function. When applying BA for the training of NNs we must perform evaluations of NNs multiple times in an iteration. This is a very computationally expensive task and might not be feasible for large networks. Thus we propose a parallel approach to this training mechanism, by performing the entire process on a Graphical Processing Unit (GPU). We use the Compute Unified Device Architecture (CUDA) [9] which is NVIDIA's proprietary framework for general purpose GPU computing. It is based on the Single Instruction Multiple Thread (SIMT) model of parallelism. The same instruction is run simultaneously on all threads of execution. We use the CUDA C/C++ for programming which is entirely like C/C++ except for the fact that we can distinguish which functions to run on the GPU(device) and which to run on the CPU(host). Functions that run on the GPU are called kernels. We limit our discussion on CUDA here, further details are available in [9].

In this paper we provide a mapping of BA to the training procedure of an NN and we go on to propose an approach that can be used to parallelize the process. We discuss this approach in detail and also carry out extensive testing on various standard as well as non - standard datasets. We are able to achieve a maximum speedup of 47× on dense networks trained using large bat populations.

The rest of the paper is organized as follows. In Sect. 2 we discuss the implementation details of our method. In Sect. 3 we analyze our framework and test it on various datasets which are themselves described as well. Then, in Sect. 4 we conclude and mention some future work that may emerge.

2 Implementation

Each bat is represented as a complete Neural Net. To minimize memory transfers between the CPU and GPU, the entire population of bats (the neural networks) is kept on the GPU Global memory. The fitness function for BA used was the accuracy of the neural net i.e. the number of correctly classified instances. The position of a bat is considered to be defined by each weight of each weight matrix of the neural net. Hence the number of dimensions of the position (and velocity) of a bat is given by

$$\sum_{i=0}^{n-1} L_i.L_{i+1} \tag{8}$$

where L_i is the number of neurons in the ith layer of the NN, with L_0 being the size of input layer, L_n being size of output layer and n the number of layers. We now discuss our approach to designing a parallel framework for the training of NNs.

In our approach we parallelize both BA and the NN so that both are able to utilize the GPU to its maximum potential. This can be done using the fact that, each bat is independent of the others. This means each neural network can be evaluated in parallel. We exploit this fact by giving the responsibility of a neural network to a block of GPU threads, so that each block utilizes its threads to forward propagate the input.

The algorithm is broken down into several kernel calls. First we use a kernel to update all the bats' positions according to their velocities and their velocities as

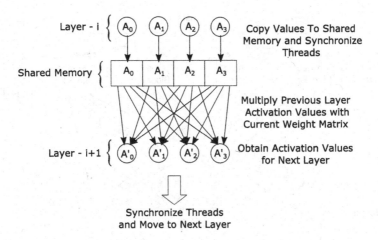

Fig. 1. Procedure for forward-propagate kernel. Synchronization is done twice per layer, once after copying data to shared memory and again after computing activation values.

well using Eqs. (2)–(4). Only a single block is launched for this kernel containing as many threads as there are bats.

Next we need to calculate the fitness values for the bats at their new positions. For this, we use 2 kernels for each input. The first one aligns the current input with all of the NNs as shown in Fig. 2. The ith block copies the ith dimension of input from its shared memory to the input layer of the NNs. Within each block the jth thread copies to the jth NN. So this kernel will launch as many blocks as there are dimensions of input and the number of threads in each block will be equal to N (population size).

The next kernel is used to forward-propagate the input through the NNs. As each NN is independent of any other so each block of this kernel is assigned to each NN. Within each block work done is as shown in Fig. 1. A single thread performs all computations required to calculate the activation value of a single neuron of the next layer according to Eq. (1). All threads within a block are synchronized before continuing on to propagate to the next layer. This ensures that previous values have been assigned before they are used. The activation values of the previous layer are kept in shared memory as they will need to be repeatedly accessed and shared memory has much faster access times in comparison to global memory [9].

Now we check the condition required for random walk of each bat using another kernel. As we will need to re-evaluate those bats which will perform the random walk but not the others, we take a boolean vector which indicates which bats are to be re-evaluated and set this vector using the kernel in parallel. This vector is kept in global memory for access by the forward-propagating kernel which is then called again to re-evaluate the bats that have moved.

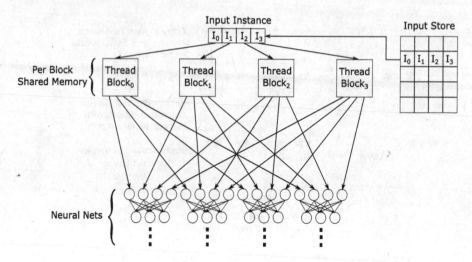

Fig. 2. Alignment of input to all NNs for a simple population of 4 NNs and an input with 4 attributes.

Next we check all bats to see which ones have moved to better positions and update their personal best positions. Again only a single block of threads is launched containing N threads. We also adjust their loudness and pulse rates according to Eqs. (6) and (7).

Finally we obtain the new best bat according to their fitness values. This can be done in parallel using a technique similar to parallel reduction [11], or for small populations we can even transfer the fitness values to host memory and find the best one without significant increase in execution time.

This implementation is designed so that once the entire input and all the networks are moved to device memory, provided they can fit, we do not need any further (significant) memory transfers between host and device for the entire training process. This helps avoid bottlenecks due to memory transfers, and keeps the GPU busy all the time.

3   Results

For all the tests that were conducted, we were able to produce an ideal environment where we needed minimum memory transfers between the GPU and the CPU. Before starting the learning process, we copied all the input as well as initial bats' parameters on to the GPU global memory. We have restricted our work to classification tasks. We have used a mix of easy and difficult to classify datasets. Apart from the Malware dataset, all others are well known benchmark datasets available publicly. We implemented both CPU and GPU versions of BA trained NNs. The tests were run on an Intel Core i7 3610 QM (2.3 GHz) CPU, 6 GB memory and an NVIDIA GT 650M GPU (835 MHz core clock) with 1 GB global memory and CUDA 7.0 toolkit with compute capability 3.0.

Table 1. Accuracies for BA trained NN. Trained to 200 epochs.

Dataset	Instances	Accuracy
Breast Cancer	569	98.24
Wine	178	100.00
Iris	150	98.00
EEG eye state	14980	56.72
Malware	4000	87.90

We briefly describe the datasets used for testing (Table 1). All datasets were obtained from the UCI machine learning repository [10] unless stated otherwise.

Wine Dataset. It consists of 178 instances belonging to 3 classes and having 13 attributes each. It can be learned even by a simple network easily.

Iris Dataset. It consists of 150 instances belonging to 3 classes. It is a very basic dataset which can be learned (at times even 100 %) within a few epochs.

Wisconsin Breast Cancer Diagnostic Dataset. It consists of 569 instances belonging to 2 classes malignant or benign. Each instance has 30 attributes. This is more difficult to learn when compared to the previous two datasets.

EEG Eye State Dataset. It consists of 14980 instances belonging to 2 classes either open or closed eye state. Each instance has 14 attributes. This dataset is much harder to learn then the previous ones.

Malware Dataset. We compiled this dataset ourselves. We obtained malware files from VX heaven[1] and assembled clean files from Windows operating system directories. We filtered the files for size so that the sizes of both clean and malware files were around 10 KB. We then extract features in the form of 16 bit words from the binary files by grouping each block of 16 bits together. We used tf-idf vectors to quantify the features of each file. As there can be upto 2^{16} features, we try and reduce this by using F-scores of the features and then selecting the top 256 features. There were a total of 4000 files with 3000 clean and 1000 malware. This dataset is difficult to classify and to put things in perspective we also learn it using Gradient-Descent (Back-Propagation) and compare accuracies.

The speedups obtained can be somewhat co-related to the size of the network. Larger networks exhibit higher speedups. We can quantify the size of a network by considering the total number of weight elements in the network, which will be directly related to the number of calculations to be made to evaluate the entire population. We display this quantity in Table 2 under the *parameters* column.

For the Wine dataset we use a network with 1 hidden layer with 1024 hidden units and a population size of 128. For the Iris dataset we use network with 2 hidden layers with 256 units each and a population size of 128. We are able to get a speedup of 36× over a serial implementation on this dataset. For the Wisconsin

---

[1] http://vxheaven.org/.

Table 2. Speedups for BA trained NN. The time is average time for one epoch

Dataset	Network	Parameters	Time (CPU).	Time (GPU)	Speedup
Breast Cancer	30-256-2, 128 bats	1048576	84.72 s	5.06 s	16.74×
Wine	13-1024-3, 128 bats	2097152	61.17 s	4.78 s	12.79×
Iris	4-256-256-3, 128 bats	8617984	177.56 s	4.82 s	36.83×
EEG eye	14-256-2, 64 bats	262144	287.32 s	64.79 s	4.43×
Malware	256-256-2, 64 bats	4227072	1379 s	77.1 s	17.88×
Malware	256-640-2, 256 bats	42270720	8881 s	186 s	47.74×

Breast Cancer Dataset we use a network with a single hidden layer of 256 units and a population of 128 bats (Fig. 4). This gives us a speedup similar to that of the Wine dataset as the sizes of both networks are comparable. The EEG eye dataset uses the smallest network in terms of number of parameters and exhibits the smallest speedup. Though the number of parameters do not always dictate the speedups, as can be seen from the difference in speedups between the Wine and Wisconsin Breast Cancer datasets, yet they are a good indicator most of the time.

For the Malware dataset we use 2 different networks. The first one is a large network with a single hidden layer containing 640 units and a population of 256 bats while the second is a smaller network with 256 hidden units and 64 bats. As expected, the larger network obtains higher speedup than the smaller one, with a speedup of 47× over the serial implementation. In this case training with the serial implementation is unfeasible as a single epoch takes around 150 min.

As the Malware dataset has not been studied before we compare the accuracies obtained for our BA trained NNs with that of one trained with gradient descent. Although, we must stress that our work is focused on speeding up the training process rather than obtaining state-of-the-art accuracies, so we refrain from using advanced feature extractors and other methods as in [12] to obtain higher accuracies. The comparison from Fig. 3 reveals that BA outperforms Gradient Descent in this case.

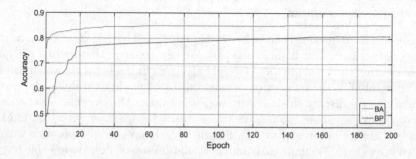

Fig. 3. BAT vs BP on the Malware dataset (256-256-2 network and 64 bats). Average of 10 runs

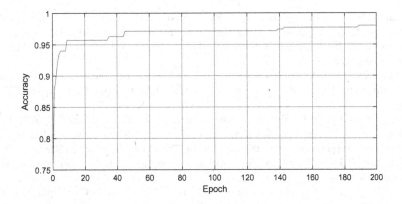

Fig. 4. Accuracy for BA trained NN on Wisconsin Breast Cancer dataset

4 Conclusion and Future Work

The Bat algorithm puts forward a good alternative to traditional gradient based algorithms for Neural Network training. In this paper we make the use of BA in this context much more feasible than before by running the entire process on GPUs. We propose a framework to do the same and test it on various datasets with varied number of instances as well as different difficulties of classification. Our framework is able to minimize the communication between the CPU and GPU to a bare minimum, where communication is needed only to either transfer results at the end or for transferring data before starting. We have also observed that the speedups obtained for various networks are loosely related to the size of those networks.

Further one may explore the possibilities of applying dynamic parallelism to BA as it allows launching of new threads from the GPU itself. This allows us to combine the entire training process in a single kernel launch from the host. A comparative study of parallel BA and other parallel metaheuristics in the context of NN training is needed. Also methods to choose hyper-parameters for metaheuristic trained NNs is required.

References

1. Widrow, B., Rumelhart, D.E., Lehr, M.A.: Neural networks: applications in industry, business and science. Commun. ACM **37**(3), 93–106 (1994)
2. Specht, D.F.: A general regression neural network. IEEE Trans. Neural Netw. **2**(6), 568–576 (1991)
3. Hagan, M.T., Menhaj, M.B.: Training feedforward networks with the Marquardt algorithm. IEEE Trans. Neural Netw. **5**(6), 989–993 (1994)
4. Montana, D.J., Davis, L.: Training feedforward neural networks using genetic algorithms. In: IJCAI, vol. 89, pp. 762–767, August 1989
5. Gupta, J.N., Sexton, R.S.: Comparing backpropagation with a genetic algorithm for neural network training. Omega **27**(6), 679–684 (1999)

6. Gudise, V.G., Venayagamoorthy, G.K.: Comparison of particle swarm optimization and backpropagation as training algorithms for neural networks. In: Proceedings of the 2003 IEEE Swarm Intelligence Symposium, 2003. SIS 3003, pp. 110–117. IEEE, April 2003
7. Khan, K., Sahai, A.: A comparison of BA, GA, PSO, BP and LM for training feed forward neural networks in e-learning context. Int. J. Intell. Syst. Appl. 4(7), 23 (2012)
8. Yang, X.S.: A new metaheuristic bat-inspired algorithm. In: González, J.R., Pelta, D.A., Cruz, C., Terrazas, G., Krasnogor, N. (eds.) NICSO 2010. Studies in Computational Intelligence, vol. 284, pp. 64–74. Springer, Heidelberg (2010)
9. Nvidia, C.U.D.A.: C programming guide version 4.0. NVIDIA Corporation, Santa Clara (2011)
10. Lichman, M.: UCI machine learning repository. University of California, School of Information and Computer Science, Irvine (2013). http://archive.ics.uci.edu/ml
11. Harris, M.: Optimizing parallel reduction in CUDA. NVIDIA DeveloperTechnology 2(4), (2007). http://developer.download.nvidia.com/assets/cuda/files/reduction.pdf
12. Rieck, K., Holz, T., Willems, C., Düssel, P., Laskov, P.: Learning and classification of malware behavior. In: Zamboni, D. (ed.) DIMVA 2008. LNCS, vol. 5137, pp. 108–125. Springer, Heidelberg (2008)

Investigating a Dictionary-Based Non-negative Matrix Factorization in Superimposed Digits Classification Tasks

Somnuk Phon-Amnuaisuk[1]($\boxtimes$) and Soo-Young Lee[2]

[1] Media Informatics Special Interest Group, Centre for Innovative Engineering,
Universiti Teknologi Brunei, Gadong, Brunei
somnuk.phonamnuaisuk@itb.edu.bn
[2] Brain Science Research Center, Korea Advanced Institute
of Science and Technology, Daejeon, Korea
sylee@kaist.ac.kr

Abstract. Human visual system can recognize superimposed graphical components with ease while sophisticated computer vision systems still struggle to recognize them. This may be attributed to the fact that the image recognition task is framed as a classification task where a classification model is commonly constructed from appearance features. Hence, superimposed components are perceived as a single image unit. It seems logical to approach the recognition of superimposed digits by employing an approach that supports construction/deconstruction of superimposed components. Here, we resort to a dictionary-based non-negative matrix factorization (NMF). The dictionary-based NMF factors a given superimposed digit matrix, V, into the combination of entries in the dictionary matrix W. The H matrix from $V \approx WH$ can be interpreted as corresponding superimposed digits. This work investigates three different dictionary representations: pixels' intensity, Fourier coefficients and activations from RBM hidden layers. The results show that (i) NMF can be employed as a classifier and (ii) dictionary-based NMF is capable of classifying superimposed digits with only a small set of dictionary entries derived from single digits.

Keywords: Dictionary-based NMF · Classifying superimposed digits · Restricted Boltzmann Machines

1 Background

The classification task can be seen as a table lookup. A given object is classified according to the similarity of the object to the most similar entry in the table. Therefore, the performance of a classification model depends on (i) the ability to construct a model capable of effectively memorizing/retrieving desired information and (ii) the ability to expressively represent salient properties of a class. It was suggested in [1] that different representations can entangle and hide different

© Springer International Publishing AG 2016
A. Hirose et al. (Eds.): ICONIP 2016, Part III, LNCS 9949, pp. 335–343, 2016.
DOI: 10.1007/978-3-319-46675-0_37

explanatory factors of variations behind the data. Knowledge representation is, therefore, one of the crucial factors that contributes to the success of a classifier. Researchers have been searching for the means to extract a good representation that would effectively generalize the features within the same class and, at the same time, effectively distinguish the features from different classes.

To date, the most successful technique for a digit recognition task is the Convolution Neural Network (CNN) [2–4]. The performance of CNN is far better than other competing techniques such as the Support Vector Machine (SVM) and the multi-layer Artificial Neural Network (ANN). However, the model that is trained to recognize each individual digit 0 to 9 (such as SVM, ANN, CNN, etc.) will not be able to recognize the same collection of digits if they are superimposed together. One may immediately respond that this is so since the model is not trained with superimposed digits. Human, however, can recognize superimposed visual images without the need to be trained with extra superimposed instances.

Although there is no solid theory about how superimposed figures in an image are processed and recognized in humans' visual system, we hypothesize that the representation of individual digits can be separated from superimposed digits. In [5], it has been shown that Non-Negative Matrix Factorization (NMF) technique [6] can successfully reveal the combination of digits from a given superimposed version. In this report, extending from the work done in [5], we are looking into properties of different feature representations. We explore features from (i) pixel intensity; (ii) Fourier coefficients derived from pixel intensity; and (iii) features extracted from the activations of the hidden units of Restricted Boltzmann Machines (AHRBM). Our experiments are carried out with a single digit and a superimposition of two digits. Two different superimposition methods are explored: centred and left/right-shifted which will be elaborated in the later section.

This paper is organized into the following sections: Sect. 2 gives an overview of related works; Sect. 3 discusses our proposed concept and gives the details of the techniques behind it; Sect. 4 provides the output of the proposed approach; and finally, the conclusion and further research are presented in Sect. 5.

2   Related Work

Two common approaches that have been employed to model a representation of digits are (i) the appearance model approach that constructs a digit model based on information derived from the pixel intensity, and (ii) the stroke model approach that constructs a digit model based on how it is drawn i.e., using stroke information. The appearance model is, perhaps, the most common representation paradigm for the image recognition task. In this paradigm, digits are recognized as an image. The stroke model, on the other hand, employs useful information such as the temporal information of stroke sequences in its representation. These two paradigms highlight the fact that different representations can reveal different properties of the domain.

Appearance models are built from pixel intensity of 2D images. This form of representation was popular but could suffer from lacking structural information or redundancy problem. Various techniques such as Gabor filters, Polling, and various transformation techniques (e.g., PCA, NMF), etc., have been explored to increase the expressiveness of the representation (see discussion in [1,7,8]). Currently, the CNN is the state of the art technique in the single digit recognition task. CNN boosts the accuracy of digit classification rate since its convolution layer could represent various bases and its polling layer could handle minor variant in the input.

All techniques discussed so far can be extended to recognize superimposed digits provided that explicit superimposed examples are included in the training set. Without explicitly training the model with superimposed digit examples, it is impossible to recognize superimposed digit using hyperspace partitioning techniques. This is because the structure of the partitions, constructed from the features extracted from individual digits, are totally different from the features extracted from superimposed digits. This is a hard problem since the representation must be able to express meaningful constituents of the whole e.g., this blob of ink is the superimposed image of digits two and five.

The issue of superimposed digit recognition has been somewhat neglected by the community. Some closely relevant works are highlighted here. In [9], the authors explored the concept of selective attention in artificial neural network and exploited it to recognize noisy patterns. In their experiment, a digit is astounded by superimposing another digit on top of it. The feed forward neural network is augmented with a controllable attention gain input. An original digit can be classified correctly if appropriate attention gain is applied to suppress the added noise (i.e., a superimposed digit). The experimental results show that the concept is working. However, scaling this approach to handle variations of handwritten digits could be a challenge since each variation seems to require a different attention gain input. Recent work by [10,11], shows how the authors deal with occlusions and noises in image recognition tasks using Markov random field and restricted Boltzmann machines respectively. The image de-noising and in-painting are obtained from posterior inference from the model. In [5], it was proposed that each individual digit could be extracted from superimposed digits by factoring out the combined pixels into different combinations. There are many plausible combinations of factors for any given matrix. The approach taken in [5] was to constrain the factoring using a dictionary constructed from a training dataset as well as restraining the factorization to only positive combinations using NMF as the factoring technique.

3   Dictionary-Based NMF

Let $V^{m \times n}$ be a matrix where $\mathbf{v}_n$ is a column vector representing the features extracted from a single handwritten digit. Here, the representation of each digit in NMF is obtained by mapping each pixel p_{ij} of a digit $\mathcal{D}^{r \times c}$, where $i = 1..r; j = 1..c$, to v_k, s.t. $k = j + (i - 1)c$. This forms a column vector $\mathbf{v} \in [0, 255]^{784 \times 1}$ (or $\mathbf{v} \in [0, 1]^{784 \times 1}$). In other words, this is a flattening of each row in the matrix $\mathcal{D}$ to form the vector $\mathbf{v}$.

The NMF factorizes a given matrix V into two positive matrices W and H, i.e., $V \approx WH$. Since there could be many good approximations of W and H combinations, it has been shown in [5] that when W is constructed as a dictionary, then H has the interpretation of the aggregation of components in the dictionary. Given a matrix V and the dictionary W, we could compute H by minimizing the cost function below [6]:

$$D_F(V\|WH) = \|V - WH\|_F^2 = \sum_{mn}(V_{mn} - (WH)_{mn})^2 \qquad (1)$$

where $\|.\|_F^2$ denotes the Frobenius norm and V_{mn} stands for the mn^{th} entry of V. In this work, we implement the multiplicative update rule modified from [6]. The matrix W is fixed and the matrix H is iteratively updated as follows:

$$H_{rn} \leftarrow H_{rn}\frac{\sum_m W_{mr}(V_{mn}/(WH)_{mn})}{\sum_{m'} W_{m'r}} \qquad (2)$$

With the constraint imposed using a dictionary and NMF, the non-negative factoring of NMF offers a linear aggregation of the components in the dictionary. In this work, each entry in the dictionary is a complete description of a digit. Hence, the matrix H has the interpretation of combined digits and this allows our dictionary-based NMF to successfully separate superimposed digits.

3.1 Dictionary Constructions

The MNIST handwritten digits[1] are transformed into three different dictionary representations: pixel intensity, Fourier coefficients and RBM activations.

Pixel Intensity. Each image I in the MNIST dataset is a matrix of size 28×28, so it can be flattened to form a vector of size $^{784 \times 1}$. All the training examples are flattened to form 60,000 vectors, for each vector $\mathbf{p}$, its value is the pixel intensity where $p_i \in [0, 255], |\mathbf{p}| = 784$. This constitutes a feature matrix of size $60,000 \times 784$ for the pixel-intensity dictionary construction.

Fourier Coefficients. The pixel intensity input vector $\mathbf{p}$ is transformed into the frequency domain using STFT. Let p_i denote the element i of the vector $\mathbf{p}$, the corresponding frequency domain component F_k can be calculated using

$$\mathbf{F}_k = \sum_{i=0}^{N-1} p_i e^{-j\frac{2knn\pi}{N}} \qquad (3)$$

---

[1] The database is available from http://yann.lecun.com/exdb/mnist/, it consists of 60,000 training examples (roughly 6,000 different handwritten examples for each digit) and 10,000 testing examples (roughly 1,000 different handwritten examples for each digit).

where N is the number of samples in a single window; k is the STFT coefficient index and the magnitude $\|\mathbf{F_k}\|$ is the corresponding frequency domain component. Only half of the coefficients are used as the input feature (i.e., 392 dimensions). This constitutes a feature matrix of size $60,000 \times 392$ for the FFT dictionary construction.

Activations of Hidden Units of RBMs. The input vector of the RBM network is the pixel intensity input vector $\mathbf{p}$ where $p_i \in [0,1], |\mathbf{p}| = 784$. The RBM restricts that their nodes must form a bipartite graph. Here, the input layer has d = 784 nodes and the hidden layer has m = 400 nodes.

Let w_{ij} be the weights connecting visible input node p_i to hidden node h_j, let a_i and b_j be the biases of the input layer and the hidden layer respectively. The energy of a configuration $E(\mathbf{p}, \mathbf{h})$ is defined as

$$E(\mathbf{p}, \mathbf{h}) = -\mathbf{a}^\mathrm{T}\mathbf{p} - \mathbf{b}^\mathrm{T}\mathbf{h} - \mathbf{p}^\mathrm{T}W\mathbf{h} \tag{4}$$

which can be seen as a joint probability as

$$P(\mathbf{p}, \mathbf{h}) = \frac{1}{Z} e^{-E(\mathbf{p}, \mathbf{h})} \tag{5}$$

where Z is a normalizing constant. The conditional probability $P(\mathbf{h}|\mathbf{p})$ is the product of the conditional probability of all hidden nodes

$$P(\mathbf{h}|\mathbf{p}) = \prod_{j=1}^{m=400} P(h_j|\mathbf{p}) \tag{6}$$

The individual activation probabilities of hidden nodes are given by

$$P(h_j = 1|\mathbf{p}) = \sigma \left(b_j + \sum_{i=1}^{d=784} w_{ij}p_i \right) \tag{7}$$

where σ denotes the logistic sigmoid function.

After the RBM network is successfully trained using the 60,000 training samples, the activation output from all the hidden nodes of all the training samples are collected and this constitutes a feature matrix of size $60,000 \times 400$ for the RBM dictionary construction.

Since we know the labels of each training sample, i.e., $\{0,1,2,...,9\}$ in the training dataset, all digits with the same class label are grouped together and then clustered into C clusters using the k-means clustering technique. By using the k cluster centers instead of the original samples as dictionary entries, the size of the dictionary can be much smaller than 60,000. In these experiments, four values of $k \in \{10, 40, 70, 100\}$ are examined.

4   Experimental Results and Discussion

The three dictionary sets are built from the pixel-intensity of single digit examples, FFT and activations from RBM's hidden layers (AHRBM). FFT and

Fig. 1. Examples of test dataset used in this experiment. The last three rows show different superimpositions: centered, left-shifted and right-shifted.

AHRBM features are derived from the pixel-intensity of those single digit examples. For the test dataset, the following variations are applied to the original test dataset: (i) shift a digit from its original position to the left or right by 6 pixels (approximately 20 %), (ii) superimpose two digits, both are positioned at their original positions, and (iii) superimpose two digits, one at the original and one at a shifted position. Figure 1 shows examples of the test dataset. Figure 2 summarizes the experimental setup in this work. In total, 48 test datasets are prepared where there are 16 datasets for each dictionary representation (pixel-intensity, FFT and RBM features). Each test dataset has 10,000 test samples. Instead of taking 10,000 samples as the matrix V of NMF, each test dataset is randomly partitioned into 10 equal groups and the classification H is obtained by factoring $V \approx WH$. The classification accuracy reported here is the average of the 10 groups.

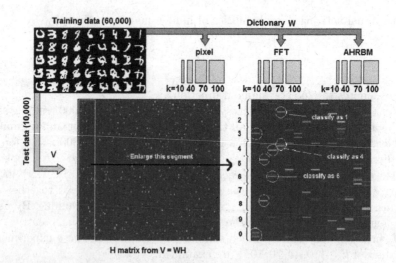

Fig. 2. Experimental setup: 60,000 training examples are employed to create three different dictionaries, each with four different sizes k = 10, 40, 70 and 100 respectively; 10,000 test digits are employed to create single digits and superimposed digits.

Table 1. Summary of classification performance obtained from different dictionary types and different superimposition setup

	k = 10	k = 40	k = 70	k = 100
Single digits				
Pixel	90.04	91.72	92.08	92.74
FFT	74.16	80.40	81.16	81.76
AHRBM	**91.59**	**93.40**	**93.48**	**93.32**
Shifted single digits				
Pixel	20.00	18.22	18.32	19.14
FFT	**75.48**	**81.26**	**81.98**	**81.64**
AHRBM	22.94	23.74	24.38	24.14
Superimposed digits				
Pixel	**80.50**	**82.72**	**83.12**	**83.83**
FFT	45.87	47.86	47.43	47.29
AHRBM	56.90	58.90	58.75	59.13
Shifted superimposed digits				
Pixel	45.07	47.47	48.78	49.42
FFT	**48.79**	**48.66**	**49.25**	**49.70**
AHRBM	46.56	45.62	46.42	48.40

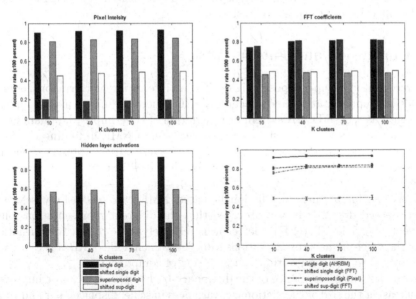

Fig. 3. Summary of all 48 experiments from three dictionaries representations, four k clustering choices and four experimental conditions: single digit, shifted single digit, superimposed digit and shifted superimposed digit. The bottom right plot summarizes the best performance from each category where error bars are also included in the plots

Table 1 summarizes all the results in this experiment. The bold font entries signify the best performance in those categories. Figure 3 graphically summarize the results from the pixel intensity (top left pane), FFT coefficients (top right pane) and AHRBM (bottom left pane). The black bars show the classification results from single digits; the dark gray bars are the results from the shifted single digit; the light gray bars are the results from the superimposition of two digits; and finally the white bars are the results from the superimposition of two digits where one of the digits is shifted to the left or right. In general, the higher the number of clusters for each digit, the higher the performance is.

The results from the pixel-intensity and AHRBM show similar patterns. Superimposed digits can be successfully recognized with an average of 10 % decrease in its performance when the best performance between single digits and superimposed digits categories are compared. This means that the features from pixel-intensity and AHRBM are robust to the composition/factoring process. Both the pixel-intensity and AHRBM perform poorly when the test digits are shifted from its original position. They appear to be very sensitive to spatial translation.

The results from the FFT coefficients show a different behavior. FFT appears to be robust to spatial translation, but less robust to the composition/factoring process. The bottom left pane summarizes the main points in this experiment. The plot summarizes the best performances from the three dictionaries representations. It shows that NMF can be successfully employed in the recognition of superimposed digits, digit-intensity is robust to composition/factoring and FFT coefficient is robust to spatial translation.

5   Conclusion and Future Work

The recognition of superimposed digits offers an interesting platform for studying knowledge representation and machine vision. There are many useful applications such as a CAPTCHA system or recognition of superimposed alphanumeric fonts. In this paper, we investigate the application of NMF in finding the compositions of superimposed digits. Three different dictionaries representations are employed in the dictionary-based NMF for the classification of superimposed digits.

It is conclusive that the dictionary-based NMF can successfully separate superimposed digits. It is also observed that the different representations offer different properties. The FFT coefficient is relatively more robust to translation while the pixel-intensity or activations from the hidden node of RBM is relatively more robust to the composition and factoring of components.

Although the performance of the dictionary-based approach can be improved by increasing the size of the dictionary; variances in scale, displacement and rotation can be handled by augmenting the training examples to increase the dictionary size. The dictionary-based NMF will suffer from computational expenses if the size of the matrix grow too large. In future work, we plan to look into a representation scheme that can be robust to composition/factoring process as well as variances in translation, scale and rotation but with a minimum increment in the dictionary size.

Acknowledgments. We wish to thank anonymous reviewers for their comments, which help improve this paper. We would like to thank the GSR office for their financial support given to this research.

References

1. Bengio, Y., Courville, A., Vincent, P.: Representation learning: a review and new perspectives. IEEE Trans. Pattern Anal. Mach. Intell. **35**(8), 1798–1828 (2013). IEEE
2. Bottou, L., Cortes, C., Denker, J.S., Drucker, H., Guyon, I., Jackel, L.D., LeCun, Y., Müller, U.A., Säckinger, E., Simard, P., Vapnik, V.: Comparison of classifier methods: a case study in handwritten digit recognition. In: Proceedings of the 12th IAPR International. Conference on Pattern Recognition, vol. 2, pp. 77–82. Conference B: Computer Vision & Image Processing (1994)
3. Ciresan, D.C., Meier, U., Masci, J., Gambardella, L.M., Schmidhuber, J.: Flexible, high performance convolutional neural networks for image classification. In: Proceedings of the Twenty-Second International Joint Conference on Artificial Intelligence (IJCAI 2011), pp. 1237–1242 (2011)
4. Lecun, Y., Bengio, Y., Hinton, G.: Deep learning. Nature **521**, 436–444 (2015)
5. Phon-Amnuaisuk, S.: Applying non-negative matrix factorization to classify superimposed handwritten digits. Procedia Comput. Sci. **24**(2013), 261–267 (2013)
6. Lee, D.D., Seung, H.S.: Learning the parts of objects by non-negative matrix factorization. Nature **401**, 788–791 (1999)
7. Sohn, K., Lee, H.: Learning invariant representations with local transformations. In: Proceedings of the 29th International Conference on Machine Learning, ICML 2012, Edinburgh (2012)
8. Song, H.A., Kim, B.K., Xuan, T.L., Lee, S.Y.: Hierarchical feature extraction by multi-layer non-negative matrix factorization network for classification task. Neurocomputing **165**, 63–74 (2015)
9. Lee, S.Y., Mozer, M.C.: Robust recognition of noisy and superimposed patterns via selective attention. In: Proceedings of the International Conference on Neural Information Processing Systems (NIPS 1999), pp. 31–37 (1999)
10. Zhou, Z., Wagner, A., Mobahi, H., Wright, J., Ma., Y.: Face recognition with contiguous occlusion using markov random fields. In: Proceedings of the International Conference on Computer Vision (ICCV 2009), pp. 1050–1057. IEEE (2009)
11. Tang, Y., Salakhutdinov, R., Hinton, G.: Robust Boltzmann machines for recognition and denoising. In: Proceedings of the IEEE Conference on Computer Vision and Pattern Recognition (CVPR 2012), pp. 2264–2271. IEEE (2012)

A Swarm Intelligence Algorithm
Inspired by Twitter

Zhihui Lv, Furao Shen[✉], Jinxi Zhao, and Tao Zhu

National Key Laboratory for Novel Software Technology, Department of Computer
Science and Technology, Nanjing University, Nanjing, China
frshen@nju.edu.cn

Abstract. For many years, evolutionary computation researchers have
been trying to extract the swarm intelligence from biological systems in
nature. Series of algorithms proposed by imitating animals' behaviours
have established themselves as effective means for solving optimization
problems. However these bio-inspired methods are not yet satisfactory
enough because the behaviour models they reference, such as the foraging birds and bees, are too simple to handle different problems. In this
paper, by studying a more complicated behaviour model, human's social
behaviour pattern on Twitter which is an influential social media and
popular among billions of users, we propose a new algorithm named Twitter Optimization (TO). TO is able to solve most of the real-parameter
optimization problems by imitating human's social actions on Twitter:
following, tweeting and retweeting. The experiments show that, TO has
a good performance on the benchmark functions.

Keywords: Swarm Intelligence · Social Media · Twitter Optimization ·
Particle Swarm Optimization

1 Introduction

In recent years, bio-inspired algorithms (BIAs) have been proved to be effective for solving optimization problems by mimicking the biological behaviors
and nature phenomenon. With the characteristics of low computing cost, excellent efficiency and superb performance, these algorithms [1–3] became prevalent
among scientists and engineers. However, there are still some problems such as
premature and curse of dimensionality to be solved in this field.

As the most intelligent bio-system, human society is considered to be very
worthy for investigation. By watching human's daily routine, we surprisingly
found that, we ourselves, as individuals of human society, were actually performing some optimization tasks unconsciously on Internet, or more specifically,
on Twitter. This is because when we use Twitter, we tend to post the Tweet
(a Twitter term similar to message) that is considered to be more valuable. For
example, if someone receives two Tweets, separately about local weather and
presidential election, he would be more willing to share the latter message on

© Springer International Publishing AG 2016
A. Hirose et al. (Eds.): ICONIP 2016, Part III, LNCS 9949, pp. 344–351, 2016.
DOI: 10.1007/978-3-319-46675-0_38

Twitter because it is noteworthy. Similar filtering behaviour happens on every user of Twitter. Everyone receives the Tweets filtered by others and posts the Tweets filtered by himself. Finally the Tweets which can widely spread among the users, have been filtered millions of times and would be considered as the most valuable Tweets at that moment. This explains why Twitter users can always catch up with the hotspot. Inspired by this phenomenon, we proposed a new algorithm named Twitter Optimization (TO).

To create an effective algorithm by the phenomenon mentioned above, TO introduces three metaphors about Twitter. Firstly, every Tweet is considered as a solution vector x composed of D real-valued parameters. And for a minimize optimization mission of objective function F, the smaller the value $F(x)$ is, the more valuable the Tweet is. Secondly, when a person posts a Tweet, he will push the content to all his followers (a Twitter term represents a unidirectional relationship). This means he produces a solution and shares it to the specific crowd connected to him. Finally, each person needs to retweet (a Twitter term means reposting) the most valuable Tweet from the people he followed, which implies he is helping the better solution to spread further. The realization of the three metaphors makes TO tend to find the global optimum.

2 Twitter Optimization

As a Twitter mimicking algorithm, Algorithm 1 shows the TO framework described with Twitter terms, such as Tweet and Retweet. The realization of each term will be explained later. Besides, the goal of TO is to find the solution to minimize the target function, and it is transformed into finding the most valuable Tweet on Twitter.

2.1 Term Explanation

To make a further interpretation about the details of TO, we will explain each operations by listing the equations and parameters. The significance of every equation will be illustrated.

Tweet. Every Tweet is treated as a solution vector of the objective function. Tweet a is more valuable than Tweet b when $F(solution_a) < F(solution_b)$. Where F is the target function to be optimized, $solution_x$ is the math representation of Tweet x's content.

Hottest Tweet. Hottest Tweet is the optimal solution found by TO and its final value will be the result of TO.

Following. Following is a kind of relationship on Twitter as shown in Fig. 1(a). If user A follows user B, A will be able to receive B's latest Tweet. This is similar to the swarm communication topology in SPSO [4]. But the difference is, TO has a dynamic topology which is continuously changing. Firstly, at the beginning of the algorithm, TO randomly initializes the following relationship of every user to generate a directed graph G. And each user is only allowed to

---

Algorithm 1. Twitter Optimization

---

procedure TWITTER

    Initial M persons i.e. M Twitter users.

    Randomly initial $Following_i$ for every person i where $Following_i$ is a set consists of the persons followed by person i.

    Everyone randomly tweets and retweets.

    Find the most valuable Tweet as the hottest Tweet.

    for Each round **do**

        for Each person i **do**

            Receive the latest Tweet tweeted and retweeted by person p, where person p belongs to $Following_i$.

            Choose the most valuable Tweet a and the least valuable Tweet b from what he received, then:

            1. Retweets Tweet a and replace the hottest Tweet with a if a is more valuable.

            2. Follow a's oral author (the source author of a who tweeted a first).

            3. Unfollow b's oral author or retweeter.

            if His own Tweet hasn't been retweeted for several rounds **then**

                Tweet a new Tweet b and replace the hottest Tweet with b if b is more valuable.

            end if

        end for

    end for

end procedure

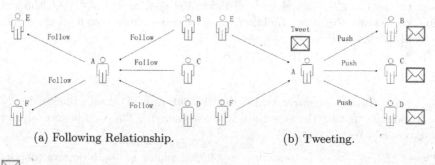

(a) Following Relationship.   (b) Tweeting.

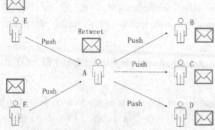

(c) Choosing the best to Retweet.

Fig. 1. Behaviours on Twitter.

follow exactly F other users (F is a constant parameter). Then, in each round, each user will optimize his following set by following and unfollowing. He will choose the most valuable Tweet he received this round, and follow its oral author if he hasn't follow him yet. Once the following operation has been done, the user will unfollow a user in order to make his following set number remain to F. He will choose the least valuable Tweet and unfollow its sender. In other words, the following and unfollowing operation makes each user willing to follow the people who were spreading the valuable Tweet, i.e. better solution. Therefore, the whole topology becomes more efficient, and TO can make a fast-converging to the global optima.

Randomly tweeting for initialization. Tweeting means generating a new solution and sharing it to followers as shown in Fig. 1(b). During the initialization process of TO, each user randomly tweets to generate a random solution vector by the following Eq. (1):

$$x_i^k = min_i + (max_i - min_i) * random(-1, 1) \tag{1}$$

where x^k is the current solution of user k and x_i^k is the value of x^k in i-th dimension, min_i and max_i are bounds of solution value in i-th dimension. Note that the algorithm has no priori knowledge about the target function. The whole initialization process is totally stochastic.

Randomly retweeting for initialization. Every person k randomly chooses a person p from $following_k$ and retweets the Tweet of p for an initialization:

$$y_i^k = x_i^p \tag{2}$$

where y^k is the solution which user k is spreading. And x^p is the current solution found by user p.

Retweeting. Retweeting makes a user able to share the valuable Tweet to his followers as shown in Fig. 1(c). This behaviour can help the better solution spreading further. Besides, TO adds some updating steps during retweeting. When user k decides to retweets the Tweet of user p, user k randomly chooses one of the two operations below to update the Tweet he retweets.

Type 1 comment:

$$x_{i\ new}^p = x_i^p + Character^k * random(-1, 1) \tag{3}$$

Type 2 participation:

$$x_{i\ new}^p = x_i^k \tag{4}$$

When a user retweets a Tweet and chooses to *comment*, he will offer his idea about the Tweet. This idea can possibly improve the quality of the oral Tweet.

We use Eq. (3) to simulate this behaviour. After the Eq. (3) has been calculated for each dimension i, a better solution may be generated. x^p is the oral Tweet, i.e. solution, of user p and x_{new}^p is the new one. If $F(x_{new}^p) < F(x^p)$, the solution will be replaced by the new one. Meanwhile, *Character* is ranging from 10^{-6} to 1 randomly for each user, representing the different updating step sizes of different users.

When a user retweets a Tweet and chooses to *participate* in the related activity, he will contribute his own part to the activity. For example, the ice bucket challenge is an influential activity which is organized on Twitter. People from all over the world take up the challenge. We use Eq. (4) to simulate the phenomenon. Where we randomly choose only one dimension i and replace $x_{i\ new}^p$ with x_i^k, similar to the hybrid model in GA [5]. If $F(x_{new}^p) < F(x^p)$, the oral solution will be replaced by the new one.

Tweeting a new Tweet. This operation means discarding the original solution and regenerate a new one. At the start of each round, every user's follower number will be evaluated and everyone will have a corresponding label. If the follower number of a user ranks up to top 1 % of the whole crowd, he will be labeled as a *celebrity*. Otherwise, the user will be labeled as an *averageman*. For a *celebrity*, if his solution hasn't been updated during the last W times retweeting, he will regenerate a new one. Where W is the follower number of user k in the current round. For an *averageman*, he will regenerate a new solution every round. The *celebrity* will choose the most valuable Tweet whose author is an *averageman* as his new Tweet. And an *averageman* will regenerate a new solution by Eq. (5).

$$x_i^k = bestx_i + \delta * random(-1, 1) \tag{5}$$

Where $bestx$ is the best solution ever found, δ is the exploring radius. Equation (5) generates a random solution near around the best solution. TO algorithm will finally stop when the number of iterations is greater than the upper bound N. For each iteration, the target function will be evaluated $O(M)$ times (tweet solution and retweet solution for every user to calculate fitness at most twice). So the total function evaluation number is $O(MN)$.

3   Experiment

3.1   Benchmark Functions for Comparison

In this section, eight benchmark functions are used to evaluate the performance of TO. Also, TO is compared with two typical methods, CPSO [6] and SPSO [4] (two variants of PSO) on the benchmark functions. The parameters of TO were set as: $F = 10$, $M = 30$, $N = evaluation\ number/60$, $\delta = 1$. And the parameters of CPSO and SPSO were set as [7]. The solution space in our experiment has 30 dimensions and the setting ranges of each dimension are [−100, 100]. Standard benchmark functions are listed in the Table 1.

Table 1. Eight benchmark functions to be tested

Function	Expression	Dimension	Optimum
Sphere	$F_1 = \sum_{i=1}^{D} x_i^2$	30	0
Rosenbrock	$F_2 = \sum_{i=1}^{D-1}(100(x_{i+1} - x_i^2)^2 + (x_i - 1)^2)$	30	0
Rastrigrin	$F_3 = \sum_{i=1}^{D}(x_i^2 - 10cos(2\pi x_i) + 10)$	30	0
Griewank	$F_4 = 1 + \sum_{i=1}^{D} \frac{x_i^2}{4000} - \prod_{i=1}^{D} cos(\frac{x_i}{\sqrt{i}})$	30	0
Ellipse	$F_5 = \sum_{i=1}^{D} 10^{4\frac{i-1}{D-1}} x_i^2$	30	0
Cigar	$F_6 = x_1^2 + \sum_{i=2}^{D} 10^4 x_i^2$	30	0
Tablet	$F_7 = 10^4 x_1^2 + \sum_{i=2}^{D} x_i^2$	30	0
Schwefel	$F_8 = \sum_{i=1}^{D}((x_1 - x_i^2)^2 + (x_i - 1)^2)^2$	30	0

Table 2. Statistical mean and standard deviation of solutions found by TO, CPSO and SPSO on eight benchmark functions over 20 independent runs

Function	Function evaluations	TO's Mean (StD)	CPSO's mean (StD)	SPSO's mean (StD)
Sphere	500000	0.000000	0.0000000	1.909960
		(0.000000)	(0.000000)	(2.594634)
Rosenbrock	600000	18.6452	33.403191	410.522522
		(6.9665)	(42.513450)	(529.389139)
Rastrigrin	500000	0.000000	0.053042	42.912843
		(0.000000)	(0.370687)	(2.177754)
Griewank	200000	0.000000	0.632403	2.177754
		(0.000000)	(0.327648)	(0.294225)
Ellipse	500000	0.000000	0.000000	53.718807
		(0.000000)	(0.000000)	(68.480173)
Cigar	600000	0.000000	0.000000	0.002492
		(0.000000)	(0.000000)	(0.005194)
Tablet	500000	0.000000	0.000000	1.462832
		(0.000000)	(0.000000)	(1.157021)
Schwefel	600000	0.000000	0.095099	0.335996
		(0.000000)	(0.376619)	(0.775270)

3.2 Parameter Selection

As shown in the Fig. 2, different values of M have different experiment results. When the population number M equals 30, TO gets it's best result on most of the functions.

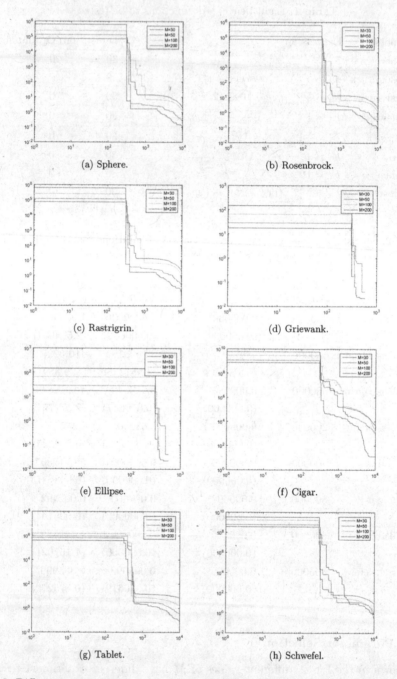

Fig. 2. Different converging speed for different M. The horizontal axis stands for function evaluation times, the vertical axis stands for the best value found by TO and the optimum is zero. $F = 10$, $\delta = 1$, $N = 10000/60$.

3.3 Experiment Results and Analysis

As shown in the Table 2, TO obtains the best results among the three algorithms on the eight benchmark functions. $F1$, $F3$, and $F5$–$F8$ are unimodal functions. The results of these functions show the fast-converging ability of TO. $F2$ and $F4$ are multimodal functions with many local optima. The performance on the two functions shows the global search ability of TO in avoiding premature convergence.

All in all, TO has the following advantages: Firstly, TO's dynamic topology makes it hard to get stuck into a local convergence; Secondly, the adaptive updating step size can fit different solution spaces and different stages of the process; Thirdly, the replacement of one dimension helps the algorithm to jump out a local convergence especially when the problem is dimensionally independent; Finally, the multiple updating methods make TO able to handle different problems.

4 Conclusion

By imitating human's social behaviours on Twitter, an optimization algorithm named TO is proposed. Series of Twitter terms are defined as updating equations for exploration and exploitation. The experiment demonstrates that TO has a better convergence accuracy compared to CPSO and SPSO. It can be concluded that TO has a potential in solving optimization problems.

Acknowledgement. The authors would like to thank the anonymous reviewers for their time and valuable suggestions. This work is supported in part by the National Science Foundation of China under Grant Nos. (61375064, 61373001) and Jiangsu NSF grant (BK20131279).

References

1. Kennedy, J., Eberhart, R.C.: Particle swarm optimization. Proc. IEEE Int. Conf. Neural Netw. **4**, 1942–1948 (1995)
2. Dorigo, M.: Optimization, learning and natural algorithms. PhD thesis. Politecnico di Milano, Italy (1992)
3. Dervis Karaboga, D.: An idea based on honey bee swarm for numerical optimization, Technical report-TR06, Erciyes University, Engineering Faculty, Computer Engineering Department (2005)
4. Bratton, D., Kennedy, J.: Defining a standard for particle swarm optimization. In: Proceedings of IEEE Swarm Intelligence Symposium, pp. 120–127 (2007)
5. Hassan, R., Cohanim, B., de Weck, O.: A comparison of particle swarm optimization and the genetic algorithm. Vanderplaats Research and Development (2005)
6. Tan, Y., Xiao, Z.M.: Clonal particle swarm optimization and its applications. In: Proceedings of IEEE Congress on Evolutionary Computation, pp. 2303–2309 (2007)
7. Tan, Y., Zhu, Y.: Fireworks algorithm for optimization. In: Tan, Y., Shi, Y., Tan, K.C. (eds.) ICSI 2010. LNCS, vol. 6145, pp. 355–364. Springer, Heidelberg (2010). doi:10.1007/978-3-642-13495-1_44

Collaborative Filtering, Matrix Factorization and Population Based Search: The Nexus Unveiled

Ayangleima Laishram[(✉)], Satya Prakash Sahu, Vineet Padmanabhan, and Siba Kumar Udgata

School of Computer and Information Sciences,
University of Hyderabad, Hyderabad, India
{ayang.cs,spsahu}@uohyd.ac.in, {vineetcs,udgatacs}@uohyd.ernet.in

Abstract. Collaborative Filtering attempts to solve the problem of recommending m items by n users where the data is represented as an $n \times m$ matrix. A popular method is to assume that the solution lies in a low dimensional space, and the task then reduces to the one of inferring the latent factors in that space. *Matrix Factorization* attempts to find those latent factors by treating it as a matrix completion task. The inference is done by minimizing an *objective function* by gradient descent. While it's a robust technique, a major drawback of it is that gradient descent tends to get stuck in local minima for non-convex functions. In this paper we propose four frameworks which are novel combinations of population-based heuristics with gradient descent. We show results from extensive experiments on the large scale *MovieLens* dataset and demonstrate that our approach provides better and more consistent solutions than gradient descent alone.

1 Introduction

A growing number of interesting applications like recommender systems (RS) make use of collaborative filtering (CF) for making automatic predictions about the interests of a user by collecting preferences from many users (collaborating). In CF, there is a partially observed $n \times m$ user-item rating matrix Y with n users and m items (it could be movies, books, etc.). The goal is to predict unobserved preference of users for items. A RS based on CF for movie recommendation should be able to make predictions about which movie a user would like given a partial list of that user's tastes. Though these predictions are specific to the user. CF can be visualized as a matrix completion task where the goal is to predict the unknown ratings for a user given the ratings that he/she already provided. Matrix factorization (MF) is a key technique employed for matrix completion wherein the objective is to learn low-rank latent factors U (for users) and V (for items) so as to simultaneously approximate the observed entries (in Y) under some *loss* measure and predict the unobserved entries by computing $UV^{\top}$. Among all the matrices that fit the given data, the one with lowest rank

© Springer International Publishing AG 2016
A. Hirose et al. (Eds.): ICONIP 2016, Part III, LNCS 9949, pp. 352–361, 2016.
DOI: 10.1007/978-3-319-46675-0_39

is preferred. In MF model, one starts by initializing the two matrices (U and V) with some values and then calculates how *different* their product is to Y and thereafter tries to minimize this difference iteratively. Minimizing the difference is akin to minimizing some loss function wherein methods like gradient/stochastic gradient descent is used.

In the current work we demonstrate that if gradient search is combined with either swarm search or evolutionary search in a suitable way, it is possible to obtain more accurate predictions. The gradient search is a popular method for finding a good solution, where if it is initialized closed to the optimum solution (a property called exploitation). The swarm search or evolutionary search has the ability of finding a good solution in large solution space [9]. By combining the exploratory nature of the population based algorithms with the exploitative nature of gradient search, good solutions of the problem are found more quickly and reliably.

There has been some research on combining PSO for *usage-based* RS [2,3], profile matching [13] as well as enhancing scalability [4,14] in RS. None of these works are based on MF techniques. Though SwarmRankCF [7] uses MF techniques, the factorization is achieved via SVD (singular value decomposition) which is not suitable when the matrix to be factored have missing values. A good survey on how different computational intelligence techniques can handle various challenges of RS is outlined in [1] and remain silent on any work combining PSO and MMMF. Two works that are closely related to this paper is the one that combines GA with MF [10] and PSO with MMMF [6]. Of this [10] does not use MMMF as the underlying factorization method and [6] is our own work wherein we have used only a basic variant of PSO. Since both these works have shown beneficial performance, our main objective is to use different variants of optimization techniques with MMMF and observe the corresponding improvements.

The rest of the paper is structured as follows. In Sect. 2 we briefly describe Matrix factorization and one of its variants called MMMF. In Sect. 3 we outline the problem statement and in Sect. 4 we talk about proposed methods. Experimental results are described in Sect. 5 and we conclude with Sect. 6.

2   Matrix Factorization (MF)

Let Y be a $n \times m$ user-item rating matrix. Each entry $y_{ij} \in \{0, 1, 2, \ldots, R\}$ defines the preference of ith user for j^{th} item, while $y_{ij} = 0$ indicates that preference of ith user for j^{th} item is not available (un-sampled entry). Given a partially observed rating matrix $Y \in \mathbb{R}^{n \times m}$, MF aims at determining two matrices $U \in \mathbb{R}^{n \times f}$ and $V \in \mathbb{R}^{m \times f}$ such that $Y \approx UV^{\top}$ where the inner dimension f is called the *numerical rank* of the matrix. The numerical rank is much smaller than m and n, and hence, factorization allows the matrix to be stored inexpensively. $X = UV^{\top}$ is called as low-rank approximation of Y. U_i, the ith row of U, represents the latent factor of ith user and V_j represents the latent factor of jth item. In low-rank MF, the objective is to select U and V among all the matrices that fit the given data, the one with lowest rank is preferred. The intuition behind MF based CF approach is to fit a linear model. In this approach, user $i(1 \leq i \leq n)$ is represented by $u_i \in \mathbb{R}^f$, the ith column of matrix $U^\top$, while item $j(1 \leq j \leq m)$ is modeled by $v_j \in \mathbb{R}^f$, the jth column of matrix $V^\top$. For example, in a movie rating system, each element of v_j can be a feature of movie j, such as whether it is an action movie, or comedy movie etc. Whether user i likes these features in a movie is indicated by the corresponding elements in u_i. Therefore the final rating given by user i to movie j will be a linear combination of these factors, i.e., $y_{ij} = \sum_{i=1}^{k} u_{il} v_{jl} = u_i v_j^\top$. In most applications, additional constraints are also enforced on the factors U and V. In this paper we use *Maximum Margin Matrix Factorization (MMMF)* [11,12] as it is more suitable for non-binary ratings when compared to other MF techniques like *Regularized matrix factorisation(RMF)* [15] and *Non-negative matrix factorization(NMF)* [15] which are more appropriate for binary ratings. In MMMF, `hinge loss` function is used, to get discrete ordinal rating and the values of U and V are restricted to be discrete instead of constraining dimensions of U and V. A set of thresholds $\Theta_{ib}(1 \leq b \leq R - 1)$ are also learned for every user i to generate a discrete rating $[1, \ldots, R]$.

$$J(U, V, \Theta) = \lambda(||U||_F^2 + ||V||_F^2) + \sum_{b=1}^{R-1} \sum_{ij \in S} h(T_{ij}^b(\Theta_{ib} - u_i.v_j^\top)) \tag{1}$$

where, λ is regularization parameter, $||.||_F$ is the Frobenius norm [11], $S = \{ij|y_{ij} > 0\}$, $h(.)$ is smoothed hinge loss function. T_{ij}^b as defined in [11] is given as follows:

$$T_{ij}^b = \begin{cases} +1 & \text{if } b \geq y_{ij} \\ -1 & \text{if } b < y_{ij} \end{cases}, \quad h(z) = \begin{cases} \frac{1}{2}(1-z), & if\, z<0 \\ 0, & if\, z>1 \\ \frac{1}{2}(1-z)^2, & otherwise \end{cases}$$

The factor matrices U, V and Θ are determined iteratively with gradient descent, with updates as shown in Eq. (2), where c is a constant.

$$U_{t+1} = U_t - c\frac{\delta J}{\delta U}, \quad V_{t+1} = V_t - c\frac{\delta J}{\delta V}, \quad \Theta_{t+1} = \Theta_t - c\frac{\delta J}{\delta \Theta} \tag{2}$$

From the loss function in Eq. (1) it can be seen that MMMF does not require $u_i^\top$ to be close to y_{ij} (which is a restriction for RMF). Instead, as in maximum margin classification approaches like SVM, it just expects that compared with Θ_{ib}, $u_i v_j^\top$ should be as small as possible for $b \geq y_{ij}$, and as large as possible for $b \leq y_{ij}$, without caring too much about the specific values of $u_i v_j^\top$.

3   Problem Statement

In this section we describe how population based heuristic techniques like particle swarm optimisation (PSO) and genetic algorithm (GA) can be used to augment gradient descend based MMMF so as to improve the accuracy of the predictions in less number of iterations. We consider here MMMF as our base matrix factorization method, because MMMF has been demonstrated to be the most suitable for discrete ordinal ratings [11]. For our given problem, we represent the candidate solutions (particle/chromosome) to U and V jointly in a single matrix P. The columns of P are user latent vectors $[U_{1,1}, \cdots, U_{f,1}]^{\top}$ to $[U_{1,n}, \cdots, U_{f,n}]^{\top}$ concatenated with columns of item latent vectors $[V_{1,1}, \cdots, V_{f,1}]^{\top}$ to $[V_{1,m}, \cdots, V_{f,m}]^{\top}$ as shown in (3).

$$\text{Potential Solution, } P = \begin{bmatrix} U_{11} & \cdots & U_{1n} & V_{11} & \cdots & V_{1m} \\ \cdots & \cdots & \cdots & \cdots & \cdots & \cdots \\ U_{f1} & \cdots & U_{fn} & V_{f1} & \cdots & V_{fm} \end{bmatrix} \tag{3}$$

Our aim is to minimize the loss with regards to Θ and $\mathcal{J}$ for each Potential Solution P. For our experiments we initialize the particles $\{P_1, ..., P_D\}$ in a randomly selected neighbourhood. One of the particle's components U, V are themselves initialized at a position in the search space that the traditional MMMF achieves as the best solution and the rest are generated randomly in the proximity of that solution. We do so with the assumption that a global solution lies somewhere in that vicinity as the randomly generated positions are less accurate in terms of loss function. As an example, let us consider a 3×3 partially observed rating matrix Y and let Y' be the test data by randomly hiding some known ratings as unknown. Our objective is to predict the unknown ratings such that the loss of the predicted rating of a test data with respect to the actual rating is minimum.

$$Y = \begin{bmatrix} 5 & 3 & 3 \\ 4 & 0 & 0 \\ 0 & 1 & 1 \end{bmatrix} \qquad Y' = \begin{bmatrix} 5 & 3 & 0 \\ 4 & 0 & 0 \\ 0 & 1 & 0 \end{bmatrix}$$

The optimal factor matrices U and V are obtained after iteratively minimizing the objective function $\mathcal{J}$ given in Eq. (1). For simplicity, let us take the latent factor f as 2. Each row of the U matrix represents a user-feature vector. Similarly, each column of the V matrix represents a movie-feature vector. The dot product of the two matrices U and V gives the predicted rating matrix X which is a fully observed matrix. Now we observe in this example that the predicted matrix X that is achieved by the heuristic technique is close to the original matrix Y.

$$U = \begin{bmatrix} 2 & -.03377 \\ 2.0009 & 4.077 \\ 1.002 & -0.0117 \end{bmatrix} \quad V^{\top} = \begin{bmatrix} 3.2438 & 1.1221 & 1.0981 \\ -0.5493 & -1.5493 & -2.5493 \end{bmatrix}$$

$$X = UV^{\top} = \begin{bmatrix} 6.6731 & 2.7674 & 3.0571 \\ 4.2510 & -4.0713 & 3.0571 \\ 3.2567 & 1.1425 & 1.1301 \end{bmatrix}$$

4   Proposed Methods

In this section we propose four methods GDPSO-MMMF, PSOm-MMMF, HPSO-MMMF and GA-MMMF.

Gradient Descent PSO based MMMF (GDPSO-MMMF): GDPSO-MMMF replaces the velocity of the particles in traditional PSO with the gradient direction leading to local best particle by using Eq. (4). Thus, the search direction is now inclined towards swarm search stimulated by the global best particle [6]. The position update rule for the U component of the particle in the swarm is given by Eq. (4) where U_{best} represents local best particle in a neighborhood.

$$U_i^{t+1} = U_i^t - c(\delta(\lambda \frac{\partial J}{\partial U}) - (1 - \delta)(U_{best}^t - U_i^t) \tag{4}$$

V and Θ are updated similarly. The objective function J is iteratively updated according to the current global best. The greater the explored search space, the better minimization of J we can hope to achieve. The proposed algorithm is described in Algorithm 1. The algorithm terminates when the same optimal solution occurs for some consecutive iterations and those set of consecutive iterations themselves occur a fixed number of times, which we set at 30.

Algorithm 1. GDPSO-MMMF

Input: Matrix Y, partially observed known rating matrix.
Output: Matrix X, fully observed predicted rating matrix.
Initialization of Swarm (Initializing U, V and Θ matrices);
while *(stopping criteria is not met)* do
  Evaluate fitness of the particles in swarm using J function given in Eq. (1);
  for *each neighborhood* do
    Find local best;
    Update several particles with some probability, ρ, using Eq. (4);
  end
  Find global best (gb) particle of the swarm;
end

PSO with mutation based MMMF (PSOm-MMMF): In some cases, the optimal solution obtained by GDPSO-MMMF is displaced outside the radius of the swarm. The new optimum then, cannot be found due to the loss of diversity in the swarm. In that scenario, we have to increase the swarm diversity by re-randomizing some particles of the swarm. That re-randomization does not only

---

Algorithm 2. PSOm-MMMF

Input: Matrix Y, partially observed known rating matrix.
Output: Matrix X, complete predicted rating matrix.
Initialization of Swarm (Initializing U, V and Θ matrices);
while *(True)* do
  Evaluate particles in swarm using J function given in Eq. (1);
  for *each neighborhood* do
    Find local best;
    Update several particles with some probability, ρ, using Eq. (4);
  end
  Find global best (gb) particle of the swarm;
  Evaluate *SwarmRadius*;
  if *gb repeats for MaxSame number of consecutive times and SwarmRadius is less than a threshold,* ThresRadius **then**
    Re-randomize the position and velocity of some particles with the mutation probability ρ ;
    Update the new particles using Eq. (4);
  end
end

---

increase the diversity but also preserves the memory of the swarm by keeping some good existing particles near the changed optimum [5]. For PSO, mutation is proposed as a method of re-randomizing new particles in the swarm as mentioned in [5]. We use mutation with GDPSO-MMMF to enhance the exploratory power and to prevent premature convergence. The proposed algorithm PSOm-MMMF, extends GDPSO-MMMF. For PSOm-MMMF, the radius of the swarm, *SwarmRadius*, is given by Eq. (5) as defined in [5].

$$SwarmRadius = max_{j=1...k}\left(\sqrt{\sum_{d=1}^{D}(P_d - lbest_j)^2}\right) \tag{5}$$

where k is the number of neighborhoods, D is the number of particles in the swarm and *SwarmRadius* is the maximum Euclidean distance between all the particles, P, and the optimal position of the neighborhood, $lbest_j$. The *ThresRadius* as given in Algorithm 2 makes sure that the particles do not leave the interesting region of the search space. The algorithm terminates when the mutation occurs for a specific number of times, *MaxSame*, which is fixed when there is no improvement i.e. 5. The mutation probability ρ is fixed at 0.3 and *ThresRadius* at 100.

Hierarchical PSO based MMMF (HPSO-MMMF): In HPSO-MMMF we incorporate the neighbourhood information regarding the solution space, by arranging the particles structurally rather than assigning particles randomly as in GDPSO-MMMF and PSOm-MMMF. The particles are arranged in a hierarchical tree structure where a parent with their immediate children forms a neighborhood with the parent being the best particle in each neighborhood. Each local best particle of a neighborhood also belongs to an immediate upper level neighborhood as a child [8]. In each iteration, rearrangement of the tree guarantees to push the global best particle to the root while still maintaining the constraint that the local best particle to be the parent in each neighborhood. By doing so, this model ensures that the best particle of a neighborhood influences the other neighborhoods with which it is associated and the best particle of the swarm globally influences all the particles. Thus, it is possible to reach the optimal solution faster. The proposed algorithm is described in Algorithm 3. The algorithm terminates when either maximum iterations surpass a user defined threshold or the `global best` does not improve for a certain number of iterations.

Genetic Algorithm based MMMF (GA-MMMF): In GA-MMMF we combine GA with MMMF with a motivation to exploit the principle of GA i.e. survival of the fittest. A chromosome is represented as in (3). Value encoding is used to encode the chromosome. Real values in V part of a chromosome represent the degree of a feature that the movie has and real values in U represent the degree of a feature that the user likes. For instance, music in the movie is enjoyable to what degree, etc. [10]. In *Crossover*, two points are randomly generated in the chromosome. One of the points lies at the beginning part of the chromosome (in the U component) and the other one lies at the end part of it (in V component).

Algorithm 3. HPSO-MMMF

Input: Matrix Y, partially observed known rating matrix.
Output: Matrix X, complete predicted rating matrix.
Initialize swarm (U, V and θ matrices);
Create tree with the particles;
while *(True)* **do**
  Evaluate fitness value for each particle using Eq. (1);
  Rearrange the tree with parents having better position than their children in a
    neighborhood;
  Update U_{best}^t, V_{best}^t, θ_{best}^t;
  Update U^{t+1}, V^{t+1}, θ^{t+1} using (4);
end

Algorithm 4. GA-MMMF

Input: Matrix Y, partially observed known rating matrix.
Output: Matrix X, complete predicted rating matrix.
Initialize population (U, V and θ matrices) ;
Evaluate the fitness values of the chromosomes using Eq. (1);
while *(True)* **do**
  Select parents using tournament selection operator;
  Reproduce offsprings using crossover and mutation operator. Apply gradient descend
    given in Eq. (2) in mutation;
  Selection for the survival of the fittest;
end

In *Mutation*, some random chromosomes are selected with some mutation rate
from the population. We update the selected chromosomes with gradient descent
using Eq. (2). The proposed algorithm is described in Algorithm 4. It terminates
when the maximum number of generations is reached which is fixed at 200.

5 Experimental Results

We conducted experiments on MovieLens data set consisting 100,000 ratings in
the range of 1 to 5, where 1 shows the least preferable while 5 shows the most
preferable. Each user has rated at least 20 movies. Unavailable user rating for
any movies is denoted with 0. We also observed performances of smaller dataset,
such as 10k and 50k ratings datasets, extracted from the 100k ratings MovieLens
dataset. The data set is collected by University of Minnesota under the project
name GroupLens[1]. The configuration of the system that executed the programs
is Microsoft Windows 8, 64 bit Intel Core i3 and the tool used was Matlab.
In our experiments we split the data into 80 % training set and 20 % test set
randomly. The population size D was fixed at 50. The crossover and mutation
rate in GA-MMMF was 0.7 and 0.3 respectively. We experimented with the
value of f as 10, 50 and 100. Then we fixed the value of f at 100 which gave the
most accurate result. The length of a particle/chromosome in the 100k dataset is
$943 + 1682 = 2625$ and its width is 100. The number of particles to be updated in
the proposed Algorithms 1 and 2 are determined by multiplying the probability

---

[1] http://grouplens.org/datasets/movielens/.

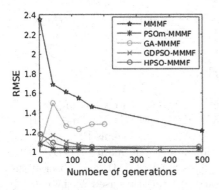 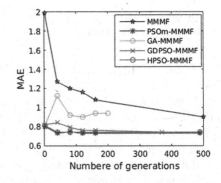

Fig. 1. RMSE and MAE for MMMF, GDPSO-MMMF, PSOm-MMMF, HPSO-MMMF and GA-MMMF for 100k data

Table 1. RMSE and MAE comparisons of basic MMMF, GDPSO-MMMF, PSOm-MMMF, HPSO-MMMF and GA-MMMF for different dataset sizes.

Size	MMMF		GDPSO-MMMF		PSOm-MMMF		HPSO-MMMF		GA-MMMF	
	RMSE	MAE	RMSE	MAE	RMSE	MAE	RMSE	MAE	RMSE	MAE
10k	1.82	1.07	1.033	0.75	1.06	0.78	1.13	0.78	1.21	0.87
50k	1.7	1.32	1.02	0.74	1.03	0.73	1.06	0.74	1.14	0.82
100k	1.645	1.266	1.08	0.785	1.03	0.751	1.05	0.74	1.27	0.94

ρ with the population size. Among all the properties of Recommender Systems we are concerned about the prediction accuracy. For analyzing the performance of our approaches we have used the popular metrics for evaluating predicted ratings, namely Root Mean Square Error (RMSE) and Mean Absolute Error (MAE). Smaller value of these metrics indicate better performance.

Figure 1 presents the graphs comparing the results of all the proposed algorithms. We can see that the RMSE and MAE obtained by the heuristic search based MMMF outperform the traditional MMMF. Among the heuristic based MMMF models, the results obtained by the GDPSO-MMMF, PSOm-MMMF and HPSO-MMMF are better than the GA-MMMF. We can see from Fig. 1 that the PSOm-MMMF starts showing a lower error rate from early generations. All the three variants of PSO model ultimately converge to a similar optimal solution. Similar behaviour is observed on the 10k and 50k MovieLens dataset with all the proposed algorithms. Table 1 shows the average RMSE and MAE results on the 10k, 50k and 100k MovieLens dataset with the proposed algorithms. It is also shown in the that the results of heuristic based algorithms exhibit better performance than the result of the basic MMMF. The robustness of the proposed PSO based algorithms are shown by preserving gb of the previous generation if the current gb has larger value of loss function and that of GA-MMMF by selecting only the fittest chromosomes for next generation. All our proposed algorithms outperform the result of an earlier work [6] which also employed a similar approach.

6   Conclusions and Future Works

We have presented how combining different heuristic search techniques with MMMF affect the accuracy metrics like RMSE and MAE. As some heuristic techniques have already been combined with MF for RS, such as PSO in [6] and GA in [10], and have shown beneficial performance, our main objective is to use different variants of optimization techniques with MMMF and observe the corresponding improvements. The significance of heuristic based MMMF models is getting promising solutions faster. Because of the property of PSO having multiple possible solutions in a search space, it has the potential to search different regions simultaneously and gathering information from each other makes it possible to guide themselves to find a shorter path in getting an optimal solution. Similarly, GA also maintains multiple possible solutions out of which only the fittest chromosomes evolve, resulting in having the potential of getting an optimal solution faster. We are convinced that when we apply optimization techniques to an underlined algorithms, the result can be improved and it is shown in this paper with our experimental results. Since we have seen that the PSO based algorithms take less iterations, yet obtain an encouraging result. PSO based MMMF model can be explored more by applying different neighborhood structures. There are several other popular heuristic techniques that can be exploited and eventually may lead to the correct predictions.

References

1. Abbas, A., Zhang, L., Khan, S.U.: A survey on context-aware recommender systems based on computational intelligence techniques. Computing **97**(7), 667–690 (2015)
2. Alam, S., Dobbie, G., Riddle, P.: Towards recommender system using particle swarm optimization based web usage clustering. In: Cao, L., Huang, J.Z., Bailey, J., Koh, Y.S., Luo, J. (eds.) PAKDD 2011. LNCS (LNAI), vol. 7104, pp. 316–326. Springer, Heidelberg (2012). doi:10.1007/978-3-642-28320-8_27
3. Alam, S., Dobbie, G., Riddle, P., Koh, Y.S.: Hierarchical PSO clustering based recommender system. In: 2012 IEEE Congress on Evolutionary Computation (CEC), pp. 1–8. IEEE (2012)
4. Bakshi, S., Jagadev, A.K., Dehuri, S., Wang, G.-N.: Enhancing scalability and accuracy of recommendation systems using unsupervised learning and particle swarm optimization. Appl. Soft Comput. **15**, 21–29 (2014)
5. Barla-Szabó, D.: A study of gradient based particle swarm optimisers. Ph.D thesis, University of Pretoria, Pretoria, (2010)
6. Devi, V.S., Rao, K.V., Pujari, A.K., Padmanabhan, V.: Collaborative filtering by PSO-based MMMF. In: 2014 IEEE International Conference on Systems, Man and Cybernetics (SMC), pp. 569–574. IEEE (2014)
7. Diaz-Aviles, E., Georgescu, M., Nejdl, W.: Swarming to rank for recommender systems. In: Proceedings of the Sixth ACM Conference on Recommender Systems, RecSys 2012, pp. 229–232. ACM, New York (2012)

8. Ghosh, P., Zarfar, H., Das, S., Abraham, A.: Hierarchical dynamic neighborhood based PSO for global optimization. In: 2011 IEEE Congress on Evolutionary Computation (CEC), pp. 757–764. IEEE (2011)
9. Kennedy, J., Eberhart, R.C.: Swarm Intelligence. Morgan Kaufmann, San Francisco (2001)
10. Navgaran, D.Z., Moradi, P., Akhlaghian, F.: Evolutionary based matrix factorization method for collaborative filtering systems. In: 2013 21st Iranian Conference on Electrical Engineering (ICEE), pp. 1–5, (2013)
11. Rennie, J.D.M., Srebro, N.: Fast maximum margin matrix factorization for collaborative prediction. In: Proceedings of the 22nd International Conference on Machine Learning, ICML 2005, pp. 713–719. ACM, New York (2005)
12. Srebro, N., Rennie, J., Jaakkola, T.S.: Maximum-margin matrix factorization. In: Advances in Neural Information Processing Systems, pp. 1329–1336 (2004)
13. Ujjin, S., Bentley, P.J.: Particle swarm optimization recommender system. In: Proceedings of the 2003 IEEE Swarm Intelligence Symposium, SIS 2003, pp. 124–131. IEEE (2003)
14. Wasid, M., Kant, V.: A particle swarm approach to collaborative filtering based recommender systems through fuzzy features. Procedia Comput. Sci. 54, 440–448 (2015)
15. Mingrui, W.: Collaborative filtering via ensembles of matrix factorization. In: KDDCup 2007, pp. 43–47 (2007)

Adaptive Hausdorff Distances and Tangent Distance Adaptation for Transformation Invariant Classification Learning

Sascha Saralajew[1], David Nebel[2], and Thomas Villmann[2,3(✉)]

[1] Electrical/Electronics Engineering - Driver Assistance Platform/Systems,
Dr. Ing. h.c. F. Porsche AG, Weissach, Germany
[2] Computational Intelligence Group, University of Applied Sciences Mittweida,
Mittweida, Germany
thomas.villmann@hs-mittweida.de
[3] Institut für Computational Intelligence und Intelligente Datenanalyse Mittweida
(CIID) e.V., Mittweida, Germany

Abstract. Tangent distances (TDs) are important concepts for data manifold distance description in machine learning. In this paper we show that the Hausdorff distance is equivalent to the TD for certain conditions. Hence, we prove the metric properties for TDs. Thereafter, we consider those TDs as dissimilarity measure in learning vector quantization (LVQ) for classification learning of class distributions with high variability. Particularly, we integrate the TD in the learning scheme of LVQ to obtain a TD adaption during LVQ learning. The TD approach extends the classical prototype concept to affine subspaces. This leads to a high topological richness compared to prototypes as points in the data space. By the manifold theory of TDs we can ensure that the affine subspaces are aligned in directions of invariant transformations with respect to class discrimination. We demonstrate the superiority of this new approach by two examples.

1 Introduction

Classification learning of noisy or corrupted data is one of the major topics in machine learning. Alternatively, the data may show systematic variations like drifts, different settings during data measuring etc. Those data distortions will contribute to decreased classification performance, if the model is unable to handle these adequately.

Learning vector quantization (LVQ, [1]) or support vector machines (SVM, [2]) are examples of prototype-based classifiers, which have a high classification performance in general. Thereby, LVQs aim is to distribute the prototypes to become class representatives. The prototypes of SVM, denoted as support vectors, are data determining the class borders. Both approaches are able to process noisy data: The LVQ prototypes are weighted averages of the data [3], whereas slack variables can be used for SVM to deal with data noise [4].

© Springer International Publishing AG 2016
A. Hirose et al. (Eds.): ICONIP 2016, Part III, LNCS 9949, pp. 362–371, 2016.
DOI: 10.1007/978-3-319-46675-0_40

Systematic disturbances in data, however, differ from data noise essentially. Mathematically, those variations in data can often be described by continuous data transformations of the original data patterns [5]. If these variations are known beforehand, respective data preprocessing methods can be applied in advance to remove their impact onto the classification system. Transformation invariant feature extraction methods such as scale invariant feature transform (SIFT, [6]) are well-known techniques for image processing. The situation becomes more difficult, if those assumptions cannot be made. For example, measurement data can be affected by drifts, which are caused by the measuring device due to temperature or pollution or due to measures from different devices. In such cases, no specific transformation can be assumed and, hence, classification of respective data may be difficult. One way to overcome this situation is the application of tangent distances (TDs) if a continuous disturbance transformation can be presumed. In that case the given data points are taken as representations of manifolds [7].

The purpose of this paper has two key points: First we show how TDs can be related to Hausdorff distances. Second, we consider *adaptive* TDs in classification using LVQ. Thereby, we are able to determine optimal TDs for class separation depending on the specific classification task. The resulting approach shows similarities to matrix learning in LVQ which was developed to adapt bilinear forms for optimum distance learning in LVQ [8].

The outline of the paper is as follows: We start with the mathematical description of Hausdorff distances and relate them to TDs. Thereafter we present an approach for adaptive TDs for classification learning by LVQ. Numerical experiments demonstrate the capacity of the TD based LVQ models.

2   Hausdorff Distance and Tangent Distance

The Hausdorff distance was developed in order to compare sets. Suppose, a metric space $(\mathbb{M}, \rho)$ with the metric ρ. Then the Hausdorff distance between non-empty sets $X, Y \subseteq \mathbb{M}$ is defined as:

$$d_H(X, Y) = \max \left\{ \sup_{\mathbf{x} \in X} \inf_{\mathbf{y} \in Y} \rho(\mathbf{x}, \mathbf{y}), \sup_{\mathbf{y} \in Y} \inf_{\mathbf{x} \in X} \rho(\mathbf{x}, \mathbf{y}) \right\}$$

If X, Y are non-empty compact sets, the pair $(\mathcal{F}(\mathbb{M}), d_H)$ constitutes a metric space [9], where $\mathcal{F}(\mathbb{M})$ is the set of all non-empty compact subsets of $\mathbb{M}$. Further, if X, Y are not compact, the Hausdorff distance $d_H(X, Y)$ may become infinite. Thus, the pair $(\mathcal{P}(\mathbb{M}), d_H)$ with $\mathcal{P}(\mathbb{M})$ being the power set of $\mathbb{M}$ without the empty set defines an *extended* semi-metric space [10,11], whereby the extension consists in the admission of the ∞-value, i.e. $d_H : \mathcal{P}(\mathbb{M}) \times \mathcal{P}(\mathbb{M}) \longrightarrow \mathbb{R}_+ \cup \{\infty\}$. A semi-metric violates the definiteness property of a usual metric, i.e. the property $d_H(X, Y) = 0$ does not imply $X = Y$.

In the next step we introduce TDs [12] and relate them to Hausdorff distances. Again, we require that $(\mathbb{M}, \rho)$ is a metric space. Additionally, we assume that $\mathbb{M}$ is a vector space over a field K. The TD is defined as

$$\tau\left(\mathbf{v}, \mathcal{W}\right) = \inf\left\{\rho\left(\mathbf{v}, \mathbf{w}'\right) | \mathbf{w}' \in \mathcal{W}\right\} \tag{1}$$

where $\mathcal{W} \subseteq M$ is an affine subspace of M and $\mathbf{v} \in M$. The affine subspace $\mathcal{W}$ can be rewritten as $\mathcal{W} := \mathcal{A}\left(\mathbf{w}, U_{\mathbf{w}}\right) = \left\{\mathbf{w} + \mathbf{u} | \mathbf{u} \in U_{\mathbf{w}}\right\}$ with respect to the linear subspace (also called direction of the affine subspace) $U_{\mathbf{w}} \subseteq M$ and the translation $\mathbf{w} \in M$. Correspondingly to $\mathcal{A}\left(\mathbf{w}, U_{\mathbf{w}}\right)$, we define a parallel affine subspace $\mathcal{V} := \mathcal{A}\left(\mathbf{v}, U_{\mathbf{w}}\right) = \left\{\mathbf{v} + \mathbf{u} | \mathbf{u} \in U_{\mathbf{w}}\right\} \subseteq M$ with the translation $\mathbf{v} \in M$. Further, we assume a translation invariant metric ρ, i.e. $\rho\left(\mathbf{x}, \mathbf{y}\right) = \rho\left(\mathbf{x} + \mathbf{z}, \mathbf{y} + \mathbf{z}\right)$ for all $\mathbf{x}, \mathbf{y}, \mathbf{z} \in M$ is valid. Then we can state the following lemma:[1]

Lemma 1. *Let ρ be a translation invariant metric. Then the following holds:*

$$d_H\left(\mathcal{V}, \mathcal{W}\right) = \tau\left(\mathbf{v}, \mathcal{W}\right)$$

Proof. By the definition of d_H we have

$$d_H\left(\mathcal{V}, \mathcal{W}\right) = \max\left\{\sup_{\mathbf{x} \in \mathcal{A}(\mathbf{v}, U_{\mathbf{w}})} \inf_{\mathbf{y} \in \mathcal{A}(\mathbf{w}, U_{\mathbf{w}})} \rho\left(\mathbf{x}, \mathbf{y}\right), \sup_{\mathbf{y} \in \mathcal{A}(\mathbf{w}, U_{\mathbf{w}})} \inf_{\mathbf{x} \in \mathcal{A}(\mathbf{v}, U_{\mathbf{w}})} \rho\left(\mathbf{x}, \mathbf{y}\right)\right\}$$

$$= \max\left\{\sup_{\mathbf{x} \in U_{\mathbf{w}}} \inf_{\mathbf{y} \in U_{\mathbf{w}}} \rho\left(\mathbf{v} + \mathbf{x}, \mathbf{w} + \mathbf{y}\right), \sup_{\mathbf{y} \in U_{\mathbf{w}}} \inf_{\mathbf{x} \in U_{\mathbf{w}}} \rho\left(\mathbf{v} + \mathbf{x}, \mathbf{w} + \mathbf{y}\right)\right\}$$

$$= \max\left\{\sup_{\mathbf{x} \in U_{\mathbf{w}}} \inf_{\mathbf{y} \in U_{\mathbf{w}}} \rho\left(\mathbf{v}, \mathbf{w} + \mathbf{y} - \mathbf{x}\right), \sup_{\mathbf{y} \in U_{\mathbf{w}}} \inf_{\mathbf{x} \in U_{\mathbf{w}}} \rho\left(\mathbf{v}, \mathbf{w} + \mathbf{y} - \mathbf{x}\right)\right\}$$

For the following simplification we consider

$$\sup_{\mathbf{x} \in U_{\mathbf{w}}} \inf_{\mathbf{y} \in U_{\mathbf{w}}} \rho\left(\mathbf{v}, \mathbf{w} + \mathbf{y} - \mathbf{x}\right)$$

only. By choosing an arbitrarily $\mathbf{x}_0 \in U_{\mathbf{w}}$ we have to compute:

$$\inf_{\mathbf{y} \in U_{\mathbf{w}}} \rho\left(\mathbf{v}, \mathbf{w} + \mathbf{y} - \mathbf{x}_0\right)$$

Due to the subspace properties and $\mathbf{y}, \mathbf{x}_0 \in U_{\mathbf{w}}$ it follows that $\mathbf{y}_0 := \mathbf{y} - \mathbf{x}_0 \in U_{\mathbf{w}}$ and further:

$$\inf_{\mathbf{y}_0 \in U_{\mathbf{w}}} \rho\left(\mathbf{v}, \mathbf{w} + \mathbf{y}_0\right) = \tau\left(\mathbf{v}, \mathcal{W}\right)$$

Because $\mathbf{x}_0 \in U_{\mathbf{w}}$ was arbitrarily chosen it follows that

$$\forall \mathbf{x} \in U_{\mathbf{w}} : \inf_{\mathbf{y} \in U_{\mathbf{w}}} \rho\left(\mathbf{v}, \mathbf{w} + \mathbf{y} - \mathbf{x}\right) = \tau\left(\mathbf{v}, \mathcal{W}\right)$$

which implies:

$$\sup_{\mathbf{x} \in U_{\mathbf{w}}} \inf_{\mathbf{y} \in U_{\mathbf{w}}} \rho\left(\mathbf{v}, \mathbf{w} + \mathbf{y} - \mathbf{x}\right) = \tau\left(\mathbf{v}, \mathcal{W}\right)$$

---

[1] The following statements remain also true if we assume that $(M, +)$ is additional a group instead of a vector space; $U_{\mathbf{w}}$ is a subgroup instead of a subspace and $\mathcal{V}, \mathcal{W}$ are left cosets instead of affine subspaces.

Analogously, we find

$$\sup_{\mathbf{y} \in U_{\mathbf{w}}} \inf_{\mathbf{x} \in U_{\mathbf{w}}} \rho\left(\mathbf{v}, \mathbf{w} + \mathbf{y} - \mathbf{x}\right) = \tau\left(\mathbf{v}, \mathcal{W}\right)$$

which completes the proof.   □

This lemma leads to the following corollary:

Corollary 2. *Let ρ be a translation invariant metric. If $\mathcal{W} \in \mathcal{F}(\mathbb{M})$ is valid then $d_H\left(\mathcal{V}, \mathcal{W}\right)$ induces the metric properties for $\tau\left(\mathbf{v}, \mathcal{W}\right)$. If $\mathcal{W} \in \mathcal{P}(\mathbb{M})$ is valid then $d_H\left(\mathcal{V}, \mathcal{W}\right)$ induces the extended semi-metric properties for $\tau\left(\mathbf{v}, \mathcal{W}\right)$.*

According to these statements, the TD can be seen as a natural instantiation of the Hausdorff distance with underling translation invariant metric ρ.

3   Adaptive Tangent Distances in LVQ

3.1   Standard LVQ

LVQ as introduced by Kohonen [13] is a heuristic adaptive classifier motivated by Bayes classification and unsupervised vector quantization. It is one of the most popular classifiers in machine learning. The generalized LVQ (GLVQ) provides a differentiable cost function approximating the classification error such that a stochastic gradient descent learning (SGDL) can be applied [14]. For GLVQ we assume the ordinary vector space $\mathbb{R}^n$ over $\mathbb{R}$ with the usual vector addition and scalar multiplication. Further, we assume a set $V = \{\mathbf{v}_i \in \mathbb{R}^n, i = 1, ..., m\}$ of data vectors with class labels $c\left(\mathbf{v}_i\right) \in \mathcal{L}$ where $\mathcal{L} = \{1, 2, ..., C\}$ is the set of classes. Further, the set of prototypes $W = \{\mathbf{w}_j \in \mathbb{R}^n, j = 1, ..., M\}$ with labels $c\left(\mathbf{w}_j\right) \in \mathcal{L}$ is required such that each class is represented by at least one prototype. For comparison of data and prototypes, a differentiable non-negative dissimilarity measure $d : \mathbb{R}^n \times \mathbb{R}^n \rightarrow \mathbb{R}^+$ is assumed, usually chosen to be the squared Euclidean distance. After training, a data vector is assigned to a prototype by the winner takes all rule

$$s\left(\mathbf{v}_i\right) = \mathrm{argmin}_{j=1,...,M} d\left(\mathbf{v}_i, \mathbf{w}_j\right). \tag{2}$$

In GLVQ, we denote for given data vector $\mathbf{v}_i$ the prototype $\mathbf{w}_j$ with the smallest dissimilarity $d\left(\mathbf{v}_i, \mathbf{w}_j\right)$ and correct class label $c\left(\mathbf{w}_j\right) = c\left(\mathbf{v}_i\right)$ by $\mathbf{w}^+ = \mathbf{w}^+\left(\mathbf{v}_i\right)$, whereas $\mathbf{w}^- = \mathbf{w}^-\left(\mathbf{v}_i\right)$ is the best matching prototype with incorrect class label, i.e. $c\left(\mathbf{w}_j\right) \neq c\left(\mathbf{v}_i\right)$. Respectively, we define the quantities $d^+\left(\mathbf{v}_i\right) = d\left(\mathbf{v}_i, \mathbf{w}^+\right)$ and $d^-\left(\mathbf{v}_i\right) = d\left(\mathbf{v}_i, \mathbf{w}^-\right)$. Further, we introduce a classifier function

$$\mu\left(\mathbf{v}_i\right) = \frac{d^+\left(\mathbf{v}_i\right) - d^-\left(\mathbf{v}_i\right)}{d^+\left(\mathbf{v}_i\right) + d^-\left(\mathbf{v}_i\right)} \in [-1, 1] \tag{3}$$

resulting in negative values in case of a correct classification. Then the GLVQ cost function to be minimized is $E_{GLVQ}\left(W, V\right) = \sum_{i=1}^{m} f\left(\mu\left(\mathbf{v}_i\right)\right)$ where f is usually defined as $f_\sigma(x) = (1 + \exp(-x \cdot \sigma))^{-1}$ depending on the parameter σ. Prototype adaptation takes place as a SGDL updating both, $\mathbf{w}^+$ and $\mathbf{w}^-$, at the same time according to

$$\Delta \mathbf{w}^\pm \propto \pm \xi_\mu^\pm (\mathbf{v}_i, f, d) \frac{\partial d(\mathbf{v}_i, \mathbf{w}^\pm)}{\partial \mathbf{w}^\pm} \tag{4}$$

with

$$\xi_\mu^\pm (\mathbf{v}_i, f, d) = \frac{2 d^\mp (\mathbf{v}_i) \cdot f'(\mu(\mathbf{v}_i))}{(d^+(\mathbf{v}_i) + d^-(\mathbf{v}_i))^2} \tag{5}$$

being a local scaling factor. After training, an unknown data point $\mathbf{v}$ is classified according to the winner rule (2) and class assignment according to $c(\mathbf{w}_{s(\mathbf{v})})$.

There exist several variants of this basic GLVQ scheme. One of the most prominent extensions are adaptive dissimilarities. In that case the quadratic form

$$d_{\mathbf{\Omega}_j}(\mathbf{v}_i, \mathbf{w}_j) = (\mathbf{\Omega}_j(\mathbf{v}_i - \mathbf{w}_j))^2 \tag{6}$$

comes into play as dissimilarity measure. The respective GLVQ-variant is called local matrix GLVQ (GMLVQ, [8]). This method denotes by $\mathbf{\Omega}^+$ and $\mathbf{\Omega}^-$ the local matrices $\mathbf{\Omega}_j$ for $\mathbf{w}^+$ and $\mathbf{w}^-$, respectively. Then an accompanying adaptation reads as

$$\Delta \mathbf{\Omega}^\pm \propto \pm 2 \xi_\mu^\pm (\mathbf{v}_i, f, d_{\mathbf{\Omega}^\pm}) \cdot \mathbf{\Omega}^\pm (\mathbf{v}_i - \mathbf{w}^\pm)(\mathbf{v}_i - \mathbf{w}^\pm)^T \tag{7}$$

resulted as a SGDL for $E_{GLVQ}(W, V)$ but now with respect to $\mathbf{\Omega}_j$ if it belongs to either $\mathbf{w}^+$ or $\mathbf{w}^-$ [8], with subsequent regularization to ensure stability [3]. If $\mathbf{\Omega} = \mathbf{\Omega}_j$ for all prototypes $\mathbf{w}_j$, standard GMLVQ is obtained. For an overview of further GLVQ variants we refer to [15].

3.2 Adaptive Tangent Distances for GLVQ – GTLVQ

Now we consider adaptive TDs as differentiable dissimilarity measures for GLVQ and explain how the resulting generalized tangent LVQ (GTLVQ) is related to a local GMLVQ. We denote the resulting LVQ algorithm as generalized tangent LVQ (GTLVQ) [12].

We assume the prototype $\mathcal{W}_j \subseteq \mathbb{R}^n$ as affine subspace $\mathcal{W}_j := \mathcal{W}_j(\theta) = \mathbf{w}_j + \mathbf{W}_j \theta$, where $\theta \in \mathbb{R}^{r_j}$ with $r_j \ll n$ is the parameter vector. The matrix $\mathbf{W}_j \in \mathbb{R}^{n \times r_j}$ is a basis of the subspace. Further, the value r_j defines the dimension of the affine subspace for the prototype $\mathcal{W}_j$. Since the complete affine subspace of a prototype $\mathcal{W}_j$ is assigned to exactly one class, it is sufficient to consider a single point of the affine subspace to evaluate the corresponding class. Therefore, we still use the notation $c(\mathbf{w}_j)$ of the classical GLVQ concept to describe the class of the prototype $\mathcal{W}_j$. Now the TD (1) is equivalent to

$$\tau(\mathbf{v}_i, \mathcal{W}_j) = \inf \{\rho(\mathbf{v}_i, \mathbf{w}_j + \mathbf{W}_j \theta) | \theta \in \mathbb{R}^{r_j}\} \tag{8}$$

If ρ is chosen to be the Euclidean metric d_E the Eq. (8) becomes

$$D(\mathbf{v}_i, \mathbf{w}_j, \mathbf{W}_j) = \min \left\{ (\mathbf{v}_i - \mathbf{w}_j - \mathbf{W}_j \theta)^2 | \theta \in \mathbb{R}^{r_j} \right\} \tag{9}$$

where the necessary minimum condition corresponds to the explicit solution

$$\theta = \mathbf{W}_j^T \left(\mathbf{v}_i - \mathbf{w}_j \right) \tag{10}$$

under the assumption of an orthonormal basis $\mathbf{W}_j$, i.e. $\mathbf{W}_j^T \mathbf{W}_j = \mathbf{I}$. The orthonormal property also ensures positive definiteness according to $\frac{\partial^2 D(\mathbf{v}_i, \mathbf{w}_j, \mathbf{W}_j)}{\partial \theta \partial \theta} = \mathbf{I}$ as the sufficient minimum condition.

Thus, we can substitute the solution (10) into (9) and obtain

$$d_{\mathbf{H}_j} \left(\mathbf{v}_i, \mathbf{w}_j \right) := D \left(\mathbf{v}_i, \mathbf{w}_j, \mathbf{W}_j \right) = \left(\mathbf{H}_j \left(\mathbf{v}_i - \mathbf{w}_j \right) \right)^2 \tag{11}$$

where $\mathbf{H}_j = \left(\mathbf{I} - \mathbf{W}_j \mathbf{W}_j^T \right)$ is an orthogonal projector. We define (11) as Euclidean TD.

At this point we remark that (11) is of the form (6) used in local GMLVQ and, therefore, denoted by $d_{\mathbf{H}_j} \left(\mathbf{v}_i, \mathbf{w}_j \right)$ accordingly. In consequence of this observation, we can apply the derived dissimilarity measure $d_{\mathbf{H}_j} \left(\mathbf{v}_i, \mathbf{w}_j \right)$ in local GMLVQ. With the definitions $\mathbf{H}^+ = \mathbf{H}_j$, $\mathbf{W}^+ = \mathbf{W}_j$ and $\mathbf{w}^+ = \mathbf{w}_j$ if $\mathcal{W}_j$ is the best matching prototype of the correct class $c \left(\mathbf{w}_j \right) = c \left(\mathbf{v}_i \right)$ with respect to $d_{\mathbf{H}_j} \left(\mathbf{v}_i, \mathbf{w}_j \right)$ and analogous definitions for $\mathbf{H}^-$, $\mathbf{W}^-$ and $\mathbf{w}^-$, we get the SGDL update rule for the affine subspace translations

$$\Delta \mathbf{w}^\pm \propto \mp 2\varepsilon_t \cdot \xi_\mu^\pm \left(\mathbf{v}_i, f, d_{\mathbf{H}^\pm} \right) \cdot \mathbf{H}^\pm \left(\mathbf{v}_i - \mathbf{w}^\pm \right) \tag{12}$$

with scaling factor $\xi_\mu^\pm \left(\mathbf{v}_i, f, d_{\mathbf{H}^\pm} \right)$ according to (5) and the translation learning rate $0 < \varepsilon_t \ll 1$. The SGDL adaptation for the subspace bases $\mathbf{W}^\pm$ and further for the orthogonal projectors $\mathbf{H}^\pm$ is realized by the adaptation

$$\Delta \mathbf{W}^\pm \propto \mp 2\varepsilon_b \cdot \xi_\mu^\pm \left(\mathbf{v}_i, f, d_{\mathbf{H}^\pm} \right) \cdot \left(\mathbf{v}_i - \mathbf{w}^\pm \right) \left(\mathbf{v}_i - \mathbf{w}^\pm \right)^T \mathbf{W}^\pm \tag{13}$$

looking similar to the matrix update (7) of GMLVQ, and whereby $0 < \varepsilon_b \ll 1$ is the basis learning rate followed by a renormalization of both new bases $\mathbf{W}^+$ and $\mathbf{W}^-$ to ensure the assumed property $\mathbf{W}_j^T \mathbf{W}_j = \mathbf{I}$.

We emphasize that the difference between GMLVQ and the proposed GTLVQ relies on the properties of the projection matrices $\mathbf{\Omega}_j$ and $\mathbf{H}_j$, namely, $\mathbf{H}_j$ is required to be an orthogonal projector. Moreover, the classical concept that a prototype j is a point of the data space $\mathbb{R}^n$ is extended to be an affine subspace which leads to a high topological richness compared to the classical prototype concept. The same result can be obtained by starting the construction of the TD from a manifold definition of the prototypes. This way leads to powerful interpretations of the TD. For example, that the basis vectors of $\mathbf{W}_j$ approximate invariant transformations which are not necessary for class discrimination [7,12, 16,17]. The name TD is inspired by this way of construction, because the affine subspace $\mathcal{W}_j$ is assumed as linear approximation of the prototype manifold j. Hence, the affine subspace $\mathcal{W}_j$ spans a tangent space at the assumed prototype manifold. We refer to [12] for a detailed explanation.

4   Simulations and Numerical Results

We tested the GTLVQ approach for an artificial data set (AD) and a real world data set using the Euclidean TD (11). The comparison is mainly made with respect to the standard GLVQ to concentrate on the effect of TDs while applying the same learning algorithm for the prototypes. The prototype initialization for GTLVQ during the experiments was done using the method of [12].

The AD is a binary non-linear classification problem in $\mathbb{R}^2$, see Fig. 1 (left). We trained several GLVQ classifiers with a different number of prototypes for each class. The results are collected in Table 1 and the first result of it is visualized in Fig. 1 (right). With a greater number of prototypes the GLVQ performance is improved as well as its model complexity is increased. Additionally, a SVM classifier with Gaussian kernel was applied using the LIBSVM toolbox [18]. The SVM-approach yields a perfect classification based on 16 support vectors.

In the next step we trained a GTLVQ model with only 2 prototypes per class. The inner class prototypes are points, i.e. for these prototypes the r-value was set to be zero. Each prototype assigned to the outer class was allowed to learn an affine subspace of $r_j = 1$. More precisely, the prototypes are lines. We achieve a perfect classification for this configuration, see Fig. 2. This result is an improvement over the best GLVQ model considered here, but with significant lower model complexity. Further, GTLVQ yields this optimum result with lower

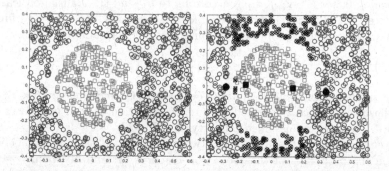

Fig. 1. AD visualization (left) and GLVQ (2+2) classification results (right). Prototypes are depicted as solid symbols. Misclassified data points are marked by a cross.

Table 1. Classification results for the AD. GLVQ $x + y$ has to be interpreted as GLVQ with x prototypes for the inner class and y prototypes for the outer class. The numbers inside the brackets for GTLVQ refer to the r-value regarding the affine subspace dimensionality.

	GLVQ 2+2	GLVQ 3+3	GLVQ 3+6	SVM 16 SV	GTLVQ 2(0)+2(1)
Accuracy	84.99 %	89.35 %	97.68 %	100.0 %	100.0 %

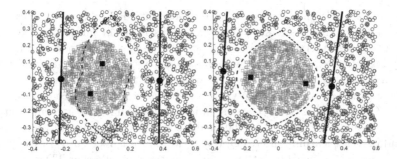

Fig. 2. Prototypes for GTLVQ before (left) and after training (right). The dashed line displays the classification border of the respective GTLVQ model. We observe a perfect classification after GTLVQ training.

Table 2. Classification results for the OJ. The interpretation is as for Table 1.

	GLVQ 2+2	GLVQ 14+14	GMLVQ 2+2	GMLVQ 14+14	GTLVQ 2(6)+2(6)
Train	83.2 %	86.7 %	84.0 %	86.7 %	96.0 %
Test	71.3 %	73.5 %	86.8 %	79.4 %	89.0 %

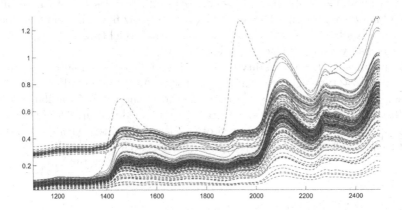

Fig. 3. OJ dataset - absorbance spectra of juices samples. The two class distributions (solid lines and dashed lines) show a high variability such that the data can be supposed to form a manifold.

complexity compared also to SVM. Hence, the adaptive TD learning provides a high variability for LVQ learning while keeping the model complexity moderate.

The real world data set is a collection of near infrared absorbance spectra of orange juices (OJ) already described used in [19]. The absorbances are determined as $(\log(1/R))$ values where R is the light reflectance on the sample surface. Each spectrum consists of 700 absorbance bands at wavelengths between 1100 nm and 2500 nm. The two classes are defined as high/ low sugar (saccharose) concentration in the juice samples. The dataset contains 149 training spectra and 67 test spectra. The classes show a high class distribution variability, see Fig. 3. Particularly, we observe a vertical shift in the data which is more pronounced for the higher bands. Therefore, we can suppose a systematic change such that the data is roughly generated by an invariant data transformation.

We applied GLVQ and GMLVQ with different settings and compared them to GTLVQ, see Table 2. GLVQ with standard Euclidean metric is clearly outperformed although independent of the model complexity (number of prototypes). If we add matrix learning in GLVQ, the performance is increased. However, the GTLVQ result is noticeably better. Hence, we can conclude that TD adaptation in GTLVQ is more appropriate for class distribution learning if an underlying invariant data transformation can be assumed.

5   Conclusion

In this contribution we considered adaptive Hausdorff distance for machine learning. We have shown that for translation invariant underlying metrics the TD is equivalent to the Hausdorff distance, which proves the metric properties for TD. The TD can be easily incorporated in GLVQ learning for classification. Particularly, the TD can be automatically adapted during learning such that a classification optimal TD is obtained. We demonstrated the efficiency of the resulting GTLVQ algorithm for two data sets. The results show first, the power to model nonlinear class boarders and second, the ability to model transformations which are unnecessary for class discrimination.

We emphasize that the ability to model invariant transformations is a result of the new prototype concept. Namely, that the prototypes are affine subspaces. Further, the TD concept is not restricted to the Euclidean TD. Similar TDs can be constructed for example using kernel distances. Moreover, the TD concept can be easily transferred to other prototype-based classifiers with differentiable cost function. These will be subject of further research.

References

1. Kohonen, T.: Self-Organizing Maps. Springer Series in Information Sciences, vol. 30. Springer, Heidelberg (1995). Second Extended Edition 1997
2. Schölkopf, B., Smola, A.: Learning with Kernels. MIT Press, Cambridge (2002)
3. Biehl, M., Hammer, B., Schleif, F.-M., Schneider, P., Villmann, T.: Stationarity of matrix relevance LVQ. In: Proceedings of the International Joint Conference on Neural Networks 2015 (IJCNN), pp. 1–8. IEEE Computer Society Press, Los Alamitos (2015)
4. Xu, H., Caramanis, C., Mannor, S.: Robustness and regularization of support vector machines. J. Mach. Learn. Res. 10, 1485–1510 (2009)
5. Decoste, D., Schölkopf, B.: Training invariant support vector machines. Mach. Learn. 46, 161–190 (2002)
6. Lowe, D.G.: Distinctive image features from scale-invariant keypoints. Int. J. Comput. Vis. 60(2), 91–110 (2004)

7. Simard, P., LeCun, Y., Denker, J.S.: Efficient pattern recognition using a new transformation distance. In: Hanson, S.J., Cowan, J.D., Giles, C.L. (eds.) Advances in Neural Information Processing Systems 5, pp. 50–58. Morgan-Kaufmann, San Mateo (1993)
8. Schneider, P., Hammer, B., Biehl, M.: Adaptive relevance matrices in learning vector quantization. Neural Comput. **21**, 3532–3561 (2009)
9. Henrikson, J.: Completeness and total boundedness of the Hausdorff metric. MIT Undergrad. J. Math. **1**, 69–79 (1999)
10. Pekalska, E., Duin, R.P.W.: The Dissimilarity Representation for Pattern Recognition: Foundations and Applications. World Scientific, Singapore (2006)
11. Villmann, T., Kaden, M., Nebel, D., Bohnsack, A.: Similarities, dissimilarities and types of inner products for data analysis in the context of machine learning. In: Rutkowski, L., Korytkowski, M., Scherer, R., Tadeusiewicz, R., Zadeh, L.A., Zurada, J.M. (eds.) ICAISC 2016. LNCS (LNAI), vol. 9693, pp. 125–133. Springer, Heidelberg (2016). doi:10.1007/978-3-319-39384-1_11
12. Saralajew, S., Villmann, T.: Adaptive tangent distances in generalized learning vector quantization for transformation and distortion invariant classification learning. In: Proceedings of the International Joint Conference on Neural Networks 2016 (IJCNN), pp. 1–8, Vancouver, Canada, (2016)
13. Kohonen, T.: Improved versions of learning vector quantization. In: Proceedings of the IJCNN-90, International Joint Conference on Neural Networks, San Diego, vol. I, pp. 545–550. IEEE Service Center, Piscataway (1990)
14. Sato, A., Yamada, K.: Generalized learning vector quantization. In: Touretzky, D.S., Mozer, M.C., Hasselmo, M.E. (eds.) Advances in Neural Information Processing Systems 8, Proceedings of the 1995 Conference, pp. 423–429. MIT Press, Cambridge (1996)
15. Kaden, M., Lange, M., Nebel, D., Riedel, M., Geweniger, T., Villmann, T.: Aspects in classification learning - review of recent developments in learning vector quantization. Found. Comput. Decis. Sci. **39**(2), 79–105 (2014)
16. Schwenk, H., Milgram, M.: Learning discriminant tangent models for handwritten character recognition. In: Fogelman-Soulié, F., Gallinari, P. (eds.) International Conference on Artificial Neural Networks, volume II, pp. 985–988. EC2 and Cie, Paris (1995)
17. Keysers, D., Macherey, W., Ney, H., Dahmen, J.: Adaptation in statistical pattern recognition using tangent vectors. IEEE Trans. Pattern Anal. Mach. Intell. **26**(2), 269–274 (2004)
18. Chang, C.-C., Lin, C.-J.: LIBSVM : a library for support vector machines. ACM Trans. Intell. Syst. Technol. **2**(3:27), 1–27 (2011)
19. Rossi, F., Lendasse, A., François, D., Wertz, V., Verleysen, M.: Mutual information for the selection of relevant variables in spectrometric nonlinear modelling. Chemometrics Intell. Lab. Syst. **80**, 215–226 (2006)

Data Mining

Semi-supervised Classification by Nuclear-Norm Based Transductive Label Propagation

Lei Jia[1,2], Zhao Zhang[1,2(✉)], and Yan Zhang[1,2]

[1] School of Computer Science and Technology & Joint International Research Laboratory of Machine Learning and Neuromorphic Computing, Soochow University, Suzhou 215006, China
cszzhang@gmail.com
[2] Collaborative Innovation Center of Novel Software Technology and Industrialization, Nanjing 210023, China

Abstract. In this paper, we propose a new transductive label propagation method, *Nuclear-norm based Transductive Label Propagation* (N-TLP). To encode the neighborhood reconstruction error more accurately and reliably, we use the nuclear norm that has been proved to be more robust to noise and more suitable to model the reconstruction error than both L1-norm or Frobenius norm for characterizing the manifold smoothing degree. During the optimizations, the Nuclear-norm based reconstruction error term is transformed into the Frobenius norm based one for pursuing the solution. To enhance the robustness in the process of encoding the difference between initial labels and predicted ones, we propose to use a weighted L2,1-norm regularization on the label fitness error so that the resulted measurement would be more accurate. Promising results on several benchmark datasets are delivered by our N-TLP compared with several other related methods.

Keywords: Classification · Label propagation · Nuclear-norm · Robustness

1 Introduction

Semi-supervised learning (SSL) is a technique that can use both labeled and unlabeled data for learning [3], so it overcomes the shortcoming of fully supervised learning in reality. That is, supervised learning needs class information of all training data, but the number of labeled samples is usually limited and more importantly the acquisition of labeled data by a skilled human agent is costly. Most SSL methods, such as [1, 3, 4, 6, 8, 9, 11–14], are based on the clustering and manifold assumptions [4], i.e., nearby points are most likely to have the same label and points on the same structure are also likely to have the same label.

Label propagation (LP), as a kind of SSL algorithm, has attracted considerable attention in recent years. Typical LP methods include SSL Using Gaussian Fields and Harmonic Functions GFHF [14], *Learning with Local and Global Consistency* (LLGC) [13] and *Linear Neighborhood Propagation* (LNP) [6]. The aforementioned methods are all transductive models, i.e., they produce the prediction results of unlabeled

A. Hirose et al. (Eds.): ICONIP 2016, Part III, LNCS 9949, pp. 375–384, 2016.
DOI: 10.1007/978-3-319-46675-0_41

training samples directly by balancing the manifold smoothness term and label fitness term. More recently, several extensions of LNP have been presented, such as *Prior Class Dissimilarity based LNP* (CD-LNP) [9] and *Nonnegative Sparse Neighborhood Propagation (SparseNP)* [12]. CD-LNP improves LNP by enhancing the assigned weights, i.e., replacing the original Euclidean distances by using a new distance that can enhance the inter-class dispersion over the intra-class dispersion. In contrast, SparseNP improves LNP by reducing the mixed signs in the predicted soft labels and enhancing the robustness to noise and outliers for accurate label estimation.

In this paper, we also consider the problem of enhancing the label prediction performance of LNP via enhancing the robustness and distance metric of the learning model to noise. More specifically, a novel SSL method termed *Nuclear-norm based Transductive Label Propagation* (N-TLP) is proposed for representation and classification. Our N-TLP is built based on the formulation of LNP, but it differs from LNP in two aspects. First, previous LNP defines the neighborhood reconstruction error using Frobenius norm that is proved to be sensitive to noise and outliers in data, while our presented N-TLP applies a more reliable Nuclear-norm [2, 5] as a distance metric to measure the neighborhood reconstruction error, since Nuclear-norm has been prove to be more suitable and reliable for encoding the reconstruction error as a distance metric. In addition, to encode the label fitness error more accurately, we also use a robust weighted L2,1-norm regularization [5, 7, 11] so that the accuracy and robustness in the procedure of characterizing the difference between the initial labels and predicted labels can be enhanced.

We organized the paper as follows. In Sect. 2, we briefly review the related LNP. Section 3 presents N-TLP. Section 4 describes the settings and evaluation results. Finally, the concluding remark is concluded in Sect. 5.

2 Related Work

LNP approach [6] is closely related to our model. Given a collection of samples $X = [x_1, x_2, \ldots, x_N] \in \mathbb{R}^{n \times N}$ and a label set $L = \{1, 2, \ldots, c\}$, where n is the original dimensionality of each sample x_i, N is number of samples. For SSL, the first l points $x_i (i \leq l)$ are considered as labeled, and the remaining points are treated as unlabeled. Graph construction [6, 14] is a core part of LNP, and LNP uses reconstruction weights for encoding the similarities among samples, that is, LNP assumes that all neighborhoods of each sample are linear, so each data point can be reconstructed by using a linear combination of its neighbors. The problem of gaining the reconstruction weights over each sample x_i can be minimized as

$$\varepsilon = \left\| x_i - \sum\nolimits_{j:x_j \in \mathbb{N}(x_i)} w_{i,j} x_j \right\|^2, \ S.t. \ \sum\nolimits_{j:x_j \in \mathbb{N}(x_i)} w_{i,j} = 1, w_{i,j} \geq 0, \tag{1}$$

where $\mathbb{N}(x_i)$ is the K-neighbor set of sample x_i, $x_{i,j}$ is the j-th neighbor of x_i and $w_{i,j}$ denotes the contribution of each $x_{i,j}$ for reconstructing x_i. After the reconstruction weights of points are all computed, a sparse weight matrix [8] W can be got as

$$W(i,j) = [w_{i,j}] \in \mathbb{R}^{N \times N}. \tag{2}$$

In each propagation step, by letting each data object absorb a fraction of label information from its neighbors and retain some label information of its initial sate [9-15], the predicted labels $F = [f_1, f_2, \ldots, f_{l+u}] \in \mathbb{R}^{c \times N}$ of LNP can be obtained by

$$F = (1 - \alpha)(I^N - \alpha W)^{-1} Y, \tag{3}$$

where $Y = [y_1, y_2, \ldots, y_{l+u}] \in \mathbb{R}^{c \times N}$ is the initial label matrix of all samples and I^N is an identify matrix in $\mathbb{R}^N$. Note that $y_{i,j} = 1$ if x_j is labeled as $i (1 \leq i \leq c)$ and else $y_{i,j} = 0$. $0 < \alpha < 1$ is a control parameter for trading-off the terms. Finally, the label of each data object can be assigned as $\arg \max_{i \leq c} F_{i,j}$, that is, the largest entry in each label vector f_i determines the final hard label of each point x_i. According to [6], the criterion of linear label propagation approach can be formulated as

$$\underset{F}{Min} \ J = \sum_{i=1}^{l+u} \left\| f_i^{\mathrm{T}} - \sum_{j: x_j \in \mathbb{N}(x_i)} w_{i,j} f_j^{\mathrm{T}} \right\|_2^2 + \sum_{i=1}^{l+u} \mu_i \| f_i^{\mathrm{T}} - y_i^{\mathrm{T}} \|_2^2, \tag{4}$$

where $\| \bullet \|_2$ is the L2-norm of a vector. By applying the matrix expression over manifold smoothing term, the above objective function can be rewritten as

$$\underset{F}{Min} \ J = \| F^{\mathrm{T}} - W F^{\mathrm{T}} \|_F^2 + \sum_{i=1}^{l+u} \mu_i \| f_i^{\mathrm{T}} - y_i^{\mathrm{T}} \|_2^2. \tag{5}$$

where $\| \bullet \|_F$ is the Frobenius norm of a matrix.

3   Nuclear-Norm Based Transductive Label Propagation

3.1   Notations

In this section, we introduce some important notations used in this paper. For a matrix $M \in \mathbb{R}^{u \times v}$, its $l_{r,p}$-norm can be defined as

$$\| M \|_{r,p} = \left(\sum_{i=1}^u \left(\sum_{j=1}^v |M_{i,j}|^r \right)^{p/r} \right)^{1/p}. \tag{6}$$

When $p = r = 2$, it is the commonly used Frobenius norm (or, simply F-norm); when $p = 1$ and $r = 2$, it is equivalent to the L2,1-norm.

3.2 Proposed Formulation

We mainly present an enhanced version N-TLP, to improve the performance of LNP for predicting the unknown labels of unlabeled data. The improvements are twofold. First, we use a more reliable nuclear-norm [2] as a distance metric to measure the neighborhood reconstruction error. Second, we use a robust weighted L2,1-norm [5, 7, 11] can be enhanced. These lead to the following problem for our N-TLP:

$$\underset{F}{Min} \ \widehat{J} = \left\| F^{\mathrm{T}} - WF^{\mathrm{T}} \right\|_* + \sum_{i=1}^{l+u} \mu_i \left\| f_i^{\mathrm{T}} - y_i^{\mathrm{T}} \right\|_2, \tag{7}$$

where $\left\| F^{\mathrm{T}} - WF^{\mathrm{T}} \right\|_*$ is the nuclear-norm based neighborhood reconstruction error and $\sum_{i=1}^{l+u} \mu_i \left\| f_i^{\mathrm{T}} - y_i^{\mathrm{T}} \right\|_2$ is the weighted L2,1-norm based label fitting term that ensures the predicted soft labels of labeled samples are similar to initial labels. Based on the matrix expressions and suppose that V denotes a diagonal matrix with entries being $v_{ii} = 1/2 \left\| f_i^{\mathrm{T}} - y_i^{\mathrm{T}} \right\|_2$, $i = 1, 2, \ldots, l+u$, the problem (7) can be reformulated as

$$\underset{F}{Min} \ \widehat{J} = \left\| F^{\mathrm{T}} - WF^{\mathrm{T}} \right\|_* + tr\left(\left(F^{\mathrm{T}} - Y^{\mathrm{T}} \right)^{\mathrm{T}} UV \left(F^{\mathrm{T}} - Y^{\mathrm{T}} \right) \right), \tag{8}$$

when $\left\| f_i^{\mathrm{T}} - y_i^{\mathrm{T}} \right\|_2 \neq 0$, $i = 1, 2, \ldots, l+u$, $F = [f_1, \cdots f_N] \in \mathbb{R}^{c \times N}$ is the predicted soft label matrix, U is a diagonal matrix with the adjustable parameter μ_i as its diagonal elements. As the labels of x_i is given, $\mu_i = 10^{10}$, and else $\mu_i = 0$. Note that by imposing the nuclear-norm and the weighted L2,1-norm on the manifold smoothing term and the label fitness error term respectively, the resulted results would be potentially enhanced, which will be verified by simulations. Next, we show its optimization.

3.3 Solution for N-TLP

In this section, we discuss how to solve the objective function in (6). Since the model involves several variables that depends on each other, so we optimize (8) using an alternate manner. We first fix V to update F. We first show how to optimize the term $\left\| F^{\mathrm{T}} - WF^{\mathrm{T}} \right\|_*$. We know the nuclear norm of a matrix is the sum of all singular values of the matrix. Motivated by [2], we convert a nuclear norm based problem to F-norm (L2-norm) based one for optimization. Let us give a lemma firstly.

Lemma 1 [2]: For a matrix $X \in \mathbb{R}^{p \times q}$, one can easily express the nuclear-norm as

$$\|X\|_* = \left\| (XX^T)^{-1/4} X \right\|_F^2. \tag{9}$$

That is, Lemma 1 represents the nuclear norm in the form of F-norm, and provides a base for solving the nuclear norm based term $\|F^T - WF^T\|_*$. In Lemma 1, the α-th power of matrix X of rank γ can be defined as [2]:

$$X^\alpha = U \sum{}^\alpha V^T, \quad \sum{}^\partial = diag\left(\alpha_1^\alpha, \ldots, \alpha_r^\alpha\right), \tag{10}$$

where $U \sum^\alpha V^T$ is the singular value decomposition of X, $\sum = diag(\alpha_1, \ldots, \alpha_r)$. According to Lemma 1, the first term, i.e., $\|F^T - WF^T\|_*$ can be rewritten as

$$\left\| F^T - WF^T \right\|_* = \left\| G(F^T - WF^T) \right\|_F^2, \tag{11}$$

where G is the weight matrix and can be defined as

$$G = \left((F^T - WF^T)(F^T - WF^T)^T \right)^{-1/4}. \tag{12}$$

Then, we can express the formulation in (11) as

$$\begin{aligned}
\left\| F^T - WF^T \right\|_* &= \left\| G(F^T - WF^T) \right\|_F^2 = tr\left(G(F^T - WF^T)(F^T - WF^T)^T G^T \right) \\
&= tr\left(G(I - W)F^T F(I - W^T)G^T \right) = tr\left(F(I - W^T)G^T G(I - W)F^T \right).
\end{aligned} \tag{13}$$

Thus, the objective function of N-TLP can be re-defined as

$$\underset{F}{Min} \; \hat{J} = tr\left(F(I - W^T)G^T G(I - W)F^T \right) + tr\left((F^T - Y^T)^T UV(F^T - Y^T) \right). \tag{14}$$

By taking derivative of $\hat{J}$ with respect to F, soft label matrix F can be inferred as

$$\frac{\partial \hat{J}}{\partial F} = 0 \Rightarrow F = YUV\left((I - W^T)G^T G(I - W) + UV \right)^{-1}. \tag{15}$$

After F is updated, we can update both V and G. That is, given $F = F_{k+1}$, update G and V by the following steps:

$$V_{k+1} = diag\left(1/2 \|h_{k+1}^i\|_2 \right), \tag{16}$$

$$G_{k+1} = \left(\left(F^{\mathrm{T}} - WF^{\mathrm{T}} \right)_{\varepsilon_i^k} \left(F^{\mathrm{T}} - WF^{\mathrm{T}} \right)^{\mathrm{T}} \right)_{\varepsilon_i^k}^{-1/4}, \tag{17}$$

where $h^i = (F^{\mathrm{T}} - Y^{\mathrm{T}})^i$. Let $X = F^{\mathrm{T}} - WF^{\mathrm{T}}$, then weight matrix G can be rewritten as $G = (XX^{\mathrm{T}})^{-1/4}$. Note that when some of the singular values of XX^{T} become small, the computation of G would become ill conditioned. To solve the instability issue, we replace x by using its $\varepsilon-$stabilization X_ε [2] similarly, defined as

$$X_\varepsilon = U \sum\nolimits_\varepsilon V^{\mathrm{T}}, \quad \sum\nolimits_\varepsilon = diag(\max\{\sigma_i, \varepsilon\} i = 1 : r). \tag{18}$$

But for a fixed ε, we would no longer expect the method to converge to the nuclear norm solution of (6), so we similarly select $\varepsilon_i^k = \min\{\varepsilon_i^{k-1}, \sigma_K(X_i^k)\}$ at step k. Then, one may hope for the stability and convergence toward then solution of (6). As a result, we can update the weight matrix G by

$$G = \left(\left(F^{\mathrm{T}} - WF^{\mathrm{T}} \right)_{\varepsilon_i^k} \left(F^{\mathrm{T}} - WF^{\mathrm{T}} \right)_{\varepsilon_i^k}^{\mathrm{T}} \right)^{-1/4}. \tag{19}$$

After convergence of the algorithm, we can get the optimal $F^* = F_{k+1}$, where the position corresponding to the biggest element in the label vector f_i determines the class assignment of each x_i. That is, the hard label of each test sample x_i can be assigned as $\arg\max_{i \leq c}(f_i)_i$, where $(f_i)_i$ is the i-th entry of the estimated soft label vector f_i. For complete presentation of the model, we summarize the procedure of N-TLP in Algorithm 1, where "convergence check" means that the predicted labels of the data will not change drastically during the iterations.

4   Experiments

We evaluate our proposed N-TLP by comparing with the other related LP methods, including GFHF [14], LLGC [13] SLP [10], LNP [6], CD-LNP [9], LapLDA, and SparseNP [11]. Two popular image databases, i.e., Georgia Tech face (GTF) database and Synthetic Control Chart Time Series (SCCTS) database are used. The parameter α in LNP, CD-LNP and SLP is set as 0.99, parameter β in CD-LNP is set to be the average Euclidean between all pairs of data points. In each simulation, we average the results over 10 times run. All the experiments are carried out on a PC with Pentium (R) Dual-Core CPU E5500 @ 2.80 GHz 2.80 GHz 2.00G.

4.1   Face Recognition

We test our N-TLP for recognizing face images. The GTF database that includes 50 persons with each person having 15 face images is used. All face images are resized into $32 * 32$ pixels for the consideration of efficiency, that is, each image can be represented by using a 1024-dimensional vector.

Algorithm 1: Proposed N-TLP framework

Inputs : Training data $X = [x_1, x_2, ..., x_{l+u}] \in \mathbb{R}^{n \times N}$, Label set $Y = [y_1, y_2, ..., y_{l+u}] \in \mathbb{R}^{c \times N}$;

Initialization : $V_0 = G_0 = I^N$; $F_0 = Y$; *gama=kesi=2; K=4;*

While not converged do

1. Fix G and V, update the soft label matrix F by
    $$F_{k+1} = Y U V_k \left((I - W^T) G^T G (I - W) + U V_k + \beta I \right)^{-1} ;$$

2. Fix F and G, update V by $V_{k+1} = diag \left(1/2 \| h^i \|_2 \right)$, where $h^i = (F^T - Y^T)^i$, $i = 1, 2 \cdots N$;

3. Fix F and V, update $G = \left((F^T - WF^T)_{\varepsilon_i^k} (F^T - WF^T)^T_{\varepsilon_i^k} \right)^{-1/4}$;

4. Convergence check: if $\| F_{k+1} - F_k \|_F^2 \le 1e^{-5}$, stop; else $k = k+1$.

5. ***End while***

Output: The optimal predicted soft label matrix $F^* = F_{k+1}$.

Figure 1 shows the classification result of each algorithm. The parameter k of k-neighborhood in computing weight matrix is set to 7 for each algorithm. Figure 1(a) shows the results on the original face data, Fig. 1(b) shows the results on the face data under fixed random noise concentration, and Fig. 1(c) shows the recognition result on the face data under different noise concentrations. We have the following findings. First, we can see from Fig. 1(a) that our N-TLP method outperforms its competitors in most cases. Second, from the results in Fig. 1(b) where noise with the variance being 15 is added, we can find that our presented N-TLP model can deliver more robust and promising results than the others. Third, in Fig. 1(c), we add various noise concentrations with variance being from 0 to 54 with interval 6 to the original images and observe change trends. It can be observed that as the noise concentration is increased, the performance of each method is decreased. But it is important to notice that our N-TLP method goes down slower than the others, i.e., the robustness of our N-TLP algorithm against noise is better than others due to the used reliable nuclear-norm regularization on the neighborhood reconstruction error and the robust weighted L2,1-norm on the label fitness error term.

Table 1 describes the result of each approach according to Fig. 1, where we report the mean accuracy (%) with standard deviation (STD) and the highest accuracy (%) for each model. We can obtain the similar conclusions from the numerical results in

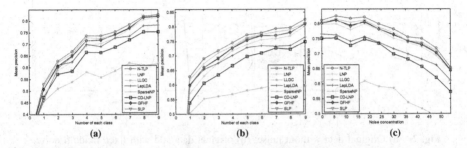

Fig. 1. (a) Original face data without noise; (b) face data under fixed random noise; (c) face data under different noise concentrations.

Table 1, that is, our proposed N-TLP algorithm can achieve the comparable or even higher accuracies than others.

4.2   Synthetic Control Chart Time Series Classification

We also choose a SCCTS database to evaluate performance of our N-TLP and compare with other algorithms. SCCTS database contains 600 training images, with number of 100 per class, and balance equal to 1 ("Balance" is defined as the ratio between the number of samples in the smallest class to the number of sample in the largest class). The method of adding noise is the same as above-mentioned way.

Table 1. Performance comparison of each algorithm on Georgia Tech face database.

Method	Result					
	GTF database (Without noise)		GTF database (Fixed noise)		GTF database (Different noise)	
	Mean ± std (%)	Best (%)	Mean ± std (%)	Best (%)	Mean ± std (%)	Best (%)
SparseNP	67.17 ± 0.458	82.33	73.99 ± 0.345	81.26	76.36 ± 0.551	81.54
SLP	65.92 ± 0.224	82.16	73.02 ± 0.258	81.14	75.91 ± 0.427	81.43
LNP	65.10 ± 0.38	81.50	71.39 ± 0.347	80.23	75.09 ± 0.674	80.06
LLGC	53.78 ± 0.434	61.84	57.21 ± 0.536	59.66	56.10 ± 0.248	63.60
LapLDA	64.63 ± 0.814	78.34	70.34 ± 0.851	76.46	71.38 ± 0.431	76.34
GFHF	70.91 ± 0.741	82.00	72.93 ± 0.641	81.03	75.84 ± 0.674	81.26
CD-LNP	61.80 ± 0.531	75.50	67.29 ± 0.573	74.91	69.49 ± 0.512	75.37
N-TLP	**67.92 ± 0.213**	**82.84**	**75.05 ± 0.341**	**82.57**	**77.44 ± 0.346**	**82.69**

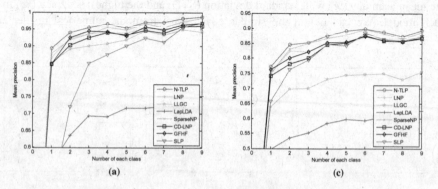

Fig. 2. (a) Original data without noise; (b) original data add with fixed random noise.

Table 2. Performance comparison of each algorithm on SCCTS database.

Method	Result					
	SCCTS database (Without noise)		SCCTS database (Fixed noise)		SCCTS database (Different noise)	
	Mean ± std (%)	Best (%)	Mean ± std (%)	Best (%)	Mean ± std (%)	Best (%)
SparseNP	94.60 ± 0.513	98.33	85.66 ± 0.359	88.61	75.94 ± 0.312	97.82
SLP	83.79 ± 0.246	96.88	82.03 ± 0.294	87.38	75.15 ± 0.457	97.04
LNP	93.76 ± 0.841	96.51	83.99 ± 0.323	87.06	73.65 ± 0.743	96.62
LLGC	91.53 ± 0.144	94.26	72.16 ± 0.419	75.63	61.50 ± 0.557	94.21
LapLDA	69.00 ± 0.932	75.28	57.87 ± 0.391	62.57	54.57 ± 0.657	75.32
GFHF	93.61 ± 0.467	96.82	83.91 ± 0.701	87.08	74.17 ± 0.447	97.22
CD-LNP	92.78 ± 0.364	96.02	83.09 ± 0.512	86.60	73.09 ± 0.146	96.44
N-TLP	**95.72 ± 0.355**	**98.61**	**86.51 ± 0.346**	**89.24**	**76.65 ± 0.241**	**98.29**

Figure 2 shows the evaluation results of each algorithm. The parameter k of k-neighborhood in computing weight matrix is set to 7 for each algorithm. Figure 1(a) shows the results on the original SCCTS database, Fig. 1(b) shows the results on the SCCTS data under fixed random noise concentration. We find that our N-TLP gains a promising and stable superiority performance, where horizontal axis denotes the level of noise and vertical axis is the mean accuracy averaged over 10 times results. From Table 2, we include random noise to corrupt testing images for evaluating the robustness property of each algorithm against corruptions, and our proposed N-TLP can always deliver a better performance.

5   Concluding Remarks

This paper presented a semi-supervised classification model by nuclear-norm based transductive label propagation. Our method improves the existing learning algorithm by enhancing the reliability and robustness of the formulation to noise. The reliability of the model is brought by the regularized nuclear norm on the neighborhood reconstruction error, while the robustness of the model is attributed by the imposed weighted L2,1-norm on the label fitness error. In future, we will explore to extend the transductive setting to the inductive scenario for handling the outside data.

Acknowledgments. This work is partially supported by the National Natural Science Foundation of China (61402310,61672365, 61373093, 61672364), Major Program of Natural Science Foundation of Jiangsu Higher Education Institutions of China (15KJA520002), Special Funding of China Postdoctoral Science Foundation (2016T90494), Postdoctoral Science Foundation of China (2015M580462), Postdoctoral Science Foundation of Jiangsu Province of China (1501091B), Natural Science Foundation of Jiangsu Province of China (BK20140008 and BK20141195), and the Graduate Student Innovation Project of Jiangsu Province of China (SJZZ16_0236).

References

1. Rohban, M.H., Rabiee, H.R.: Supervised neighborhood graph construction for semi-supervised classification. Pattern Recogn. **45**, 1363–1372 (2012)
2. Zhang, F., Yang, J., Qian, J.: Nuclear norm-based 2-DPCA for extracting features from images. Proc. IEEE Trans. Neural Netw. Learn. Syst. **26**(10), 2247–2260 (2015)
3. Blum, A., Chawla, S.: Learning from labeled and unlabeled data using graph mincuts. In: Proceedings of the ICML, pp. 19–26 (2001)
4. Chapelle, O., Weston, J.: Cluster kernels for semi-supervised learning. Adv. Neural Inf. Process. Syst. **15**, 15–17 (2003)
5. Hou, C., Nie, F., Li, X.: Joint embedding learning and sparse regression: a framework for unsupervised feature selection. Proc. IEEE Trans. Cybern. **44**(6), 793–804 (2013)
6. Wang, F., Zhang, C.: Label propagation through linear neighborhoods. In: ICML, pp. 985–992 (2006)
7. Yang, Y.: L21-norm regularized discriminative feature selection for unsupervised learning. In: Proceedings of the AI, pp. 1589–1594 (2011)
8. Wang, J.: Locally Linear Embedding, pp. 203–220. Springer, Heidelberg (2012)
9. Zhang, C., Wang, S., Li, D.: Prior class dissimilarity based linear neighborhood propagation. Knowl.-Based Syst. **83**, 58–65 (2015)
10. Nie, F., Xiang, S., Liu, Y.: A general graph-based semi-supervised learning with novel class discovery. Neural Comput. Appl. **19**, 549–555 (2010)
11. Yang, S.Z., Hou, C.P., Nie, F.P., Wu, Y.: Unsupervised maximum margin feature selection via L2,1-norm minimization. Neural Comput. Appl. **21**(7), 1791–1799 (2012)
12. Zhang, Z., Zhang, L., Zhao, M.B., Jiang, W.M., Liang, Y.C., Li, F.Z.: Semi-supervised image classification by nonnegative sparse neighborhood propagation. In: Proceedings of the ACM-ICMR, pp. 139–146 (2015)
13. Zhou, D., Bousquet, O., Lal, T.N., Weston, J., Scholkopf, B.: Learning with local and global consistency. Adv. Neural Inf. Process. Syst. **17**(4), 321–328 (2004)
14. Zhang, Z., Zhang, Y., Li, F., Zhao, M., Zhang, L., Yan, S.: Discriminative Sparse Flexible Manifold Embedding with Novel Graph for Robust Visual Representation and Label Propagation. Pattern Recogn. **61**, 492–510, (2017)

Effective and Efficient Multi-label Feature Selection Approaches via Modifying Hilbert-Schmidt Independence Criterion

Jianhua Xu$^{(\boxtimes)}$

School of Computer Science and Technology,
Nanjing Normal University, Nanjing 210023, Jiangsu, China
xujianhua@njnu.edu.cn

Abstract. Hilbert-Schmidt independence criterion (HSIC) is a nonparametric dependence measure to depict all modes of dependencies between two sets of variables via matrix trace. When this criterion with linear feature and label kernels is directly applied to multi-label feature selection, an efficient feature ranking is achieved using diagonal elements, which considers only feature-label relevance. But non-diagonal elements essentially characterize feature-feature conditional redundancy. In this paper, two novel criteria are defined by all matrix elements. For a candidate feature, we both maximize its relevance and minimize its average or maximal redundancy. Then an efficient hybrid strategy combining simple feature ranking and sequential forward selection is implemented, where the former sorts all features in descending order using their relevance and the latter finds out the top discriminative features with relevance maximization and redundancy minimization. Experiments on four data sets illustrate that our proposed methods are effective and efficient, compared with several existing techniques, according to classification performance and computational efficiency.

Keywords: Multi-label classification · Feature selection · Feature ranking · Hilbert-Schmidt independence criterion · Sequential forward selection

1 Introduction

Traditional supervised classification solves problems where one instance has one label only, which is regarded as single-label classification, including binary and multi-class cases. However, in many real-world applications, one instance is possibly associated with multiple labels simultaneously. For example, a piece of news belongs to politics, economics, and local conflict [1]. Such a learning task is referred to as multi-label classification [2,3].

This study was supported by the Natural Science Foundation of China under Grant 61273246.

For multi-label classification, nowadays various technological innovations allow us to conveniently collect massive amount of data with a large number of features, which involve some noisy, redundant and irrelevant features, and thus increase computation complexity and possibly deteriorate classification performance. To confront such a problem, dimensionality reduction, including feature extraction (FE) and feature selection (FS), becomes an indispensable pre-processing step [4]. In this paper, we focus on multi-label FS techniques.

Multi-label FS aims to select a small subset of features that minimizes feature-feature redundancy and maximizes feature-label relevance simultaneously, whose methods could still be divided into three groups: filter, wrapper and embedded techniques, according to the same partition as in single-label case [5].

The wrapper techniques iteratively apply a proper search strategy (e.g., genetic algorithm) to determine one or more small subsets of features and evaluate their corresponding classification performance using some classifier (e.g., naive Bayes and multi-label kNN classifiers) [6,7]. The embedded ones directly incorporate FS as a part of the classifier training process (such as linear multi-label support vector machine) [8]. Such two kinds of methods have an extremely high time complexity due to their complicated optimization procedures.

The filter techniques separate FS from classifier training procedure, which rely on a measure of general characteristics in the training data (e.g., distance, consistency, dependency and correlation) and a proper search strategy (e.g., simple ranking and sequential forward selection (SFS)), and further could be divided into two sub-groups: decomposition and extension-based methods.

The decomposition-based FS techniques partition a multi-label data set into either one or more single-label data subsets, evaluate the importance of features on each data subset using an existing single-label FS technique, and finally integrate feature importance over all subsets via aggregation ways. One of the widely-used decomposition tricks is one-versus-rest (OVR) or binary relevance (BR), which divides a C-class multi-label data set into C independent binary subsets [2,3]. This trick is integrated with the χ^2 statistics [1], information gain (IG) [9] and ReliefF [9], respectively. Since each label is processed independently, the label correlations are exploited insufficiently.

The extension-based techniques generalize some existing multi-class FS method to deal with multi-label FS problems directly. Stemmed from multi-class and regression ReliefF methods [10], two multi-label versions (ReliefF-ML and RRliefF-ML) are introduced in [11]. Multivariate mutual information (MI) is an effective measure to describe non-linear correlation among variables. But its estimation depends on high-dimensional joint probability. In PMU [12], the score function based on MI between feature subset and labels is approximated by using fixed second-order interaction, which is further improved by considering any-degree interaction in [13]. In MDMR [14], MI is used to calculate feature-label dependency and feature-feature redundancy. Such three methods utilize SFS strategy to find out the optimal subset of features. But in [15], via selecting some higher entropy features and labels, and limiting the degree of label combinations of interaction, a fast feature ranking method based on MI (FIMF) is proposed.

These extension-based methods have a slightly higher computational complexity. But they consider as much feature-label relevance and feature-feature redundancy as possible, which improves the effectiveness of multi-label FS methods. Therefore in this paper we still aim at designing and implementing novel extension-type filter FS approaches for multi-label classification.

Hilbert-Schmidt independence criterion (HSIC) [16] is a nonparametric dependence measure to consider all modes of dependencies between two sets of variables using the trace of matrix, i.e., the sum of diagonal elements. If HSIC with linear feature and label kernels is directly applied to multi-label FS, a simple feature ranking could be realized according to [17] analysis. But such an efficient approach only considers linear feature-label relevance. In this paper, we point out that the non-diagonal elements essentially depict feature-feature linear conditional redundancy, and then define two novel multi-label FS criteria by all elements in HSIC matrix. For a candidate feature, we simultaneously maximize its relevance and minimize its average or maximal redundancy. Via combining simple feature ranking with SFS, an efficient hybrid search procedure is constructed, in which the former sorts all features in descending order according to their relevances and then the latter finds out the top discriminative features through relevance maximization and redundancy minimization. Experimental results from four data sets demonstrate that, although our novel FS techniques averagely spend slightly more computational time than original ranking HSIC method, our methods achieve the best performance according to six instance-based measures, compared with four existing FS approaches based on original ranking HSIC, FoHSIC [17], and FIMF [15] and ReliefF-ML [11].

2   Preliminaries

Let an *i.i.d.* training set with L instances be,

$$\{(\mathbf{x}_1, \mathbf{y}_1), ..., (\mathbf{x}_i, \mathbf{y}_i), ..., (\mathbf{x}_L, \mathbf{y}_L)\}, \tag{1}$$

where $\mathbf{x}_i \in R^D$ and $\mathbf{y}_i \in \{0,1\}^C$ indicate that the i-th instance is described using D-dimensional feature vector and C-dimensional binary label one. If $\mathbf{x}_i$ belongs to the k-th class, $y_{ik} = 1$ (i.e., relevant label), and 0 otherwise. For the convenience of representation, we convert the above (1) into two matrices,

$$\begin{aligned}
\mathbf{X} &= [\mathbf{x}_1, ..., \mathbf{x}_i, ..., \mathbf{x}_L]^T = [\mathbf{x}_{(1)}, ..., \mathbf{x}_{(k)}, ..., \mathbf{x}_{(D)}], \\
\mathbf{Y} &= [\mathbf{y}_1, ..., \mathbf{y}_i, ..., \mathbf{y}_L]^T = [\mathbf{y}_{(1)}, ..., \mathbf{y}_{(k)}, ..., \mathbf{y}_{(C)}],
\end{aligned} \tag{2}$$

where the i-th rows $(\mathbf{x}_i^T)$ and $(\mathbf{y}_i^T)$ in $\mathbf{X}$ and $\mathbf{Y}$ correspond to the i-th instance feature and label vectors, and the k-th columns $(\mathbf{x}_{(k)})$ and $(\mathbf{y}_{(k)})$ represent the k-th feature and class label vector.

The multi-label classification is to learn a classifier: $f(\mathbf{x}) : R^D \rightarrow \{0,1\}^C$, which could predict the relevant labels for unseen instances.

The multi-label feature selection (FS) is to choose d features from the original D ones $(d < D)$ to remain these relevant features and remove those redundant ones. Formally, FS task is depicted as,

$$\mathbf{x}' = \mathbf{x} \circ \mathbf{p}, \tag{3}$$

where $\circ$ indicates Hadamard product of two vectors, and $\mathbf{p} = [p_1, ..., p_d]^T \in \{0, 1\}^D$ indicates a D-dimensional binary vector where 1 means that the corresponding feature is chosen. In many feature ranking techniques, instead, we usually search for a real vector $\mathbf{p}$ whose each component reflects the feature importance.

3   Effective and Efficient Multi-label Feature Selection Methods via HSIC

In this section, we briefly review Hilbert-Schmidt independence criterion, then apply it to define two novel criteria for multi-label FS, and construct an efficient hybrid search procedure finally.

3.1   Hilbert-Schmidt Independence Criterion

Hilbert-Schmidt independence criterion (HSIC) is a non-parametric dependence measure which considers all modes of dependencies between all variables. With linear feature and label kernels, its empirical estimator is described as a trace of matrix,

$$\text{HSIC} = \frac{1}{(L-1)^2} tr\left(\mathbf{X}^T \mathbf{H} \mathbf{Y} \mathbf{Y}^T \mathbf{H} \mathbf{X}\right), \tag{4}$$

where the centralized matrix $\mathbf{H} = \mathbf{I} - \mathbf{u}\mathbf{u}^T/L$, $\mathbf{I}$ is an identity matrix of size L, and $\mathbf{u}$ denotes a all-one column vector. According to (4), we only need to centralize the matrix $\mathbf{X}$,

$$\bar{\mathbf{X}} = \left[\bar{\mathbf{x}}_{(1)}, ..., \bar{\mathbf{x}}_{(k)}, ..., \bar{\mathbf{x}}_{(D)}\right] = \mathbf{H}\mathbf{X}, \tag{5}$$

and have the following HSIC form after omitting $(L-1)^{-2}$,

$$\text{HSIC} = tr\left(\bar{\mathbf{X}}^T \mathbf{Y} \mathbf{Y}^T \bar{\mathbf{X}}\right) = tr\left(\mathbf{G}\right), \tag{6}$$

where $\mathbf{G} = \bar{\mathbf{X}}^T \mathbf{Y} \mathbf{Y}^T \bar{\mathbf{X}} = [G_{ij} | i, j = 1, ..., D]$.

3.2   Two Novel Feature Selection Criteria

According to [17] analysis, if applying to feature selection, the above HSIC (6) could be simplified into,

$$\text{HSIC} = \sum_{k=1}^{D} G_{kk} = \sum_{k=1}^{D} \bar{\mathbf{x}}_{(k)}^T \mathbf{Y} \mathbf{Y}^T \bar{\mathbf{x}}_{(k)} = \sum_{k=1}^{D} \sum_{j=1}^{C} \left(\mathbf{y}_{(j)}^T \bar{\mathbf{x}}_{(k)}\right)^2, \tag{7}$$

where $G_{kk} = \sum_{j=1}^{C} \left(\mathbf{y}_{(j)}^T \bar{\mathbf{x}}_{(k)} \right)^2 \geq 0$ represents the relevance between the k-th feature and all labels. Further, this criterion (7) shows that FS can be executed via a simple feature ranking, i.e., we select the top d features according to the sorted diagonal elements of $\mathbf{G}$ in descending order. But, this strategy only considers the linear relevance between each feature and all labels (i.e., these diagonal elements), and neglects the possible redundancy among features (i.e., those non-diagonal elements).

In [17] two complicated feature selection approaches are proposed, i.e., FoHSIC and BaHSIC, which combine SFS and SBS strategies with non-linear kernels to depict non-linear dependence. But in their experiments, the RBF feature and delta label kernels are validated for multi-class data sets with 256 feature at most. In this study, we investigate whether the linear kernels are suitable for high dimensional features and the non-diagonal elements could be utilized to improve the performance of HSIC-type FS techniques further.

Now we analyze the non-diagonal elements $(G_{ij}, i \neq j)$ of $\mathbf{G}$,

$$G_{ij} = \bar{\mathbf{x}}_{(i)}^T \mathbf{Y}\mathbf{Y}^T \bar{\mathbf{x}}_{(j)} = \left(\mathbf{Y}^T \bar{\mathbf{x}}_{(i)} \right)^T \left(\mathbf{Y}^T \bar{\mathbf{x}}_{(j)} \right), \tag{8}$$

which indicates that the dependence between $\bar{\mathbf{x}}_{(i)}$ and $\bar{\mathbf{x}}_{(j)}$ is described after such two feature vectors are transformed from the L-dimensional space into a C-dimensional space via linear combination based on binary label matrix $\mathbf{Y}$. Moreover, we provide two special cases.

Case 1: when $\bar{\mathbf{x}}_{(i)} = \bar{\mathbf{x}}_{(j)}$, we have $G_{ij} = G_{ii} = G_{jj}$.

Case 2: if $\bar{\mathbf{x}}_{(i)} = -\bar{\mathbf{x}}_{(j)}$, then $G_{ij} = -G_{ii} = -G_{jj}$.

Therefore, the magnitude of G_{ij} could reflect the linear dependence between the i-th and j-th features, given the binary label matrix $\mathbf{Y}$. In this study, we regarded this linearly conditional dependence as feature-feature redundancy.

Assume F_1 and F_0 to be two subsets of selected and unselected features respectively. To consider both feature-label relevance and feature-feature redundancy, it is needed to find out a candidate feature in F_0 whose relevance is maximal,

$$\max_{j \in F_0} G_{jj}, \tag{9}$$

and whose average or maximal redundancy is minimal,

$$\min_{j \in F_0} \frac{1}{|F_1|} \sum_{i \in F_1} |G_{ij}|, \tag{10}$$

or

$$\min_{j \in F_0} \max_{i \in F_1} |G_{ij}|, \tag{11}$$

where $|F_1|$ denotes the size of F_1 and $|G_{ij}|$ is the absolute value of G_{ij}. Such two indexes measure the redundancy between a candidate feature and those in the subset of selected features using average and maximum ways. Combining (9) with (10) and (11), we define two novel criteria to choose a candidate feature k from F_0,

---

Algorithm 1. An hybrid multi-label feature selection
procedure via modifying HSIC

Input: X: feature matrix of size $L \times D$
      Y: label matrix of size $L \times C$
      n: the number of selected features
Procedure
1: For $k = 1$ to D
    Calculate the diagonal elements G_{kk}
2: Sort G_{kk} in descending order
3: Choose the top $2n$ features to construct
    $F_1 = \{1\}$ and $F_0 = \{2, ..., 2n\}$
4: Estimate the linear kernel matrix of size $2n \times 2n$
5: For all $k \in F_0$
5-1: For all $j \in F_0$
    Calculate $G_{jj} / \max_{i \in F_1} |G_{ij}|$ or $|F_1| G_{jj} / \sum_{i \in F_1} |G_{ij}|$
5-2: Find out the optimal candidate feature k_{opt} using (12) or (13)
5-3: Let $F_1 \leftarrow \{F_1, k_{opt}\}$ and $F_0 \leftarrow \{F_0\} \backslash \{k_{opt}\}$
Output: the top n features in F_1

$$\text{average(avg)} : k = \arg\max_{j \in F_0} \frac{|F_1| G_{jj}}{\sum_{i \in F_1} |G_{ij}|}, \tag{12}$$

and

$$\text{maximum(max)} : k = \arg\max_{j \in F_0} \frac{G_{jj}}{\max_{i \in F_1} |G_{ij}|}. \tag{13}$$

In the next subsection, we will utilize such two criteria to construct an efficient hybrid search procedure.

3.3 An Efficient Hybrid Search Procedure

According to (7), it is only needed to calculate the D diagonal elements of **G** and then to sort them in descending order. The FS procedure could be implemented via a simple feature ranking and thus is efficient since its time complexity is $O(CL^2)$. But this strategy only considers the relevance and neglects the redundancy. When our two new criteria (12) and (13) are combined with SFS directly, it is time consuming for high-dimensional features since the time complexity is $O(CL^3)$. Therefore, we construct a hybrid search strategy. For a given number of selected features (i.e., n), we first choose the $2n$ higher relevant features according to the original criterion (7) and then use SFS to select the n discriminative features using (12) and (13). The detailed algorithm is listed in Algorithm 1.

In Algorithm 1, when its SFS steps (4 and 5) are omitted, its simplified version is referred to as HSIC-rank in this study. According to the average and maximum rules in Algorithm 1, we could derive two multi-label FS methods, simply HSIC-avg and HSIC-max, both of which require $O(Cn^3)$ for Step 4 and $O(Ln^2)$ for Step 5. Therefore, our Algorithm 1 is very efficient for $n << L$.

4   Experiments

In this section, we experimentally evaluate our two HSIC-type feature selection methods and compare them with four existing methods (i.e., ReliefF-ML [11], FIMF [15], FoHSIC [17], and HSIC-rank).

4.1   Four Data Sets and Six Performance Measures

To evaluate the performance of our multi-label FS methods, we choose four indicative data sets: Bibtex, Enron, Langlog and TMC2007-500 for multi-label text categorization, as shown in Table 1. It is noted that all these sets only consist of 0/1 binary features since FIMF [15] only deals with discrete features. Table 1 also list some useful statistics of these data sets, such as, the number of instances, features and labels, and label cardinality and density. These four data sets are downloaded from [18] and their unlabeled instances are removed.

Table 1. Four data sets used in our experiments

Data set	#Instances	#Features	#Labels	Label cardinality	Label density
Bibtex	6995	1836	159	2.40	0.015
Enron	1702	1001	53	3.38	0.064
Slashdot	3782	1079	22	1.18	0.054
TMC2007-500	28596	500	22	2.16	0.098

In order to compare different feature selection techniques, we need to exploit the classification performance of some multi-label classifier. In this paper, six instance-based measures [2,3], i.e., Hamming loss, accuracy, F1, recall, precision and subset accuracy, are evaluated by ML-kNN with $k = 15$ [19]. Additionally, the 10-fold cross validation is executed to report experimental results in mean and standard deviation format.

To sort the performance of multiple feature selection methods for some given data set, we rank all methods for some measure, in which the best performing method gets the rank of 1, the second best rank 2, and in case of ties average ranks are assigned. Then the average rank of each method over all six measures are calculated. This performance comparison is recommended in [20].

4.2   Key Parameter Settings for Compared Methods

In this study, we compare HSIC-avg and HSIC-max with HSIC-rank, ReliefF-ML [11], FIMF [15], and FoHSIC [17], whose basic settings are stated as follows. For ReliefF-ML, the number of nearest neighbor instances is 10 based on Euclidean distance and the number of instances to be selected to estimate the importance of features is fixed to be $0.1L(L \leq 5000)$, $0.05L(5000 \leq L \leq 10000)$

Table 2. Six instance-based measures with 30 features on four data sets

Method	Hamming ↓ loss	Accuracy ↑	F1 ↑	Precision ↑	Recall ↑	Subset ↑ accuracy	Rank ↓
			Bibtex				
ReliefF-ML	1.51(6) ±0.06	1.34(6) ±0.59	1.77(6) ±0.74	3.20(6) ±1.30	1.35(6) ±0.59	0.45(6) ±0.36	6.00
FIMF	1.41(5) ±0.06	9.12(5) ±0.75	11.66(5) ±0.87	19.95(5) ±1.31	9.14(5) ±0.75	3.71(5) ±0.61	5.00
FoHSIC	1.36(3) ±0.06	14.83(3) ±1.51	18.53(3) ±1.72	29.28(3) ±2.22	15.33(3) ±1.58	6.53(3) ±1.15	3.00
HSIC-rank	1.38(4) ±0.06	12.58(4) ±1.04	15.62(4) ±1.13	24.94(4) ±1.57	12.78(4) ±1.03	5.92(4) ±0.95	4.00
HSIC-avg	**1.35**(1.5) ±0.06	**17.77**(1) ±1.87	**21.42**(1) ±2.02	**31.36**(1) ±2.22	**18.28**(1) ±1.96	**9.41**(1) ±1.60	**1.08**
HSIC-max	**1.35**(1.5) ±0.06	16.76(2) ±0.98	20.18(2) ±1.11	29.70(2) ±1.76	17.25(2) ±0.96	9.14(2) ±1.21	1.92
			Enron				
ReliefF-ML	5.21(4.5) ±0.31	33.51(4) ±3.55	44.24(3) ±3.81	59.92(3) ±4.28	38.20(4) ±3.98	5.29(4) ±2.76	3.75
FIMF	5.21(4.5) ±0.24	33.82(3) ±3.65	44.17(4) ±3.69	59.69(4) ±4.59	38.30(3) ±3.85	6.58(3) ±3.52	3.58
FoHSIC	5.74(6) ±0.26	22.10(6) ±3.31	30.93(6) ±4.78	47.97(6) ±8.70	25.16(6) ±4.01	1.53(6) ±1.25	6.00
HSIC-rank	5.15(3) ±0.36	32.38(5) ±4.94	42.83(5) ±5.28	57.97(5) ±6.18	36.98(5) ±5.30	4.64(5) ±3.92	4.67
HSIC-avg	**5.00**(1.5) ±0.29	39.18(2) ±3.07	50.10(2) ±3.16	**65.00**(1) ±4.32	**44.37**(1.5) ±3.29	9.46(2) ±2.42	1.67
HSIC-max	**5.00**(1.5) ±0.33	**39.50**(1) ±3.65	**50.15**(1) ±3.93	64.45(2) ±3.52	**44.37**(1.5) ±4.17	**10.16**(1) ±2.79	**1.33**
			Slashdot				
ReliefF-ML	4.70(4) ±0.13	22.09(6) ±2.88	22.70(6) ±2.90	23.88(6) ±2.98	22.19(6) ±2.86	20.31(6) ±2.89	5.67
FIMF	4.69(3) ±0.17	27.34(4) ±2.26	28.10(4) ±2.21	29.52(4) ±2.11	27.48(4) ±2.27	25.14(4) ±2.52	3.83
FoHSIC	4.75(6) ±0.16	25.16(5) ±2.56	25.91(5) ±2.65	27.38(5) ±2.78	25.25(5) ±2.63	22.95(5) ±2.41	5.17
HSIC-rank	4.63(2) ±0.17	27.86(3) ±2.51	28.71(3) ±2.46	30.22(3) ±2.37	28.09(3) ±2.57	25.36(3) ±2.71	2.83
HSIC-avg	**4.59**(1) ±0.17	**29.06**(1) ±2.55	**29.92**(1) ±2.55	**31.59**(1) ±2.59	**29.17**(1) ±2.51	**26.57**(1) ±2.55	**1.00**
HSIC-max	4.71(5) ±0.19	28.69(2) ±2.57	29.55(2) ±2.55	31.02(2) ±2.52	28.98(2) ±2.60	26.20(2) ±2.79	2.50
			TMC2007-500				
ReliefF-ML	8.14(6) ±0.19	38.19(6) ±1.78	46.81(6) ±2.05	58.51(6) ±2.00	44.19(6) ±2.39	14.35(6) ±1.30	6.00
FIMF	7.43(2) ±0.13	45.10(2) ±1.34	54.24(2) ±1.32	63.83(3) ±0.67	53.45(2) ±1.99	**18.71**(1.5) ±1.33	2.08
FoHSIC	7.98(5) ±0.24	40.63(5) ±2.24	49.71(5) ±2.36	60.10(5) ±2.56	48.38(5) ±2.75	14.97(5) ±2.49	5.00
HSIC-rank	7.57(4) ±0.17	43.25(4) ±1.11	52.46(4) ±1.12	63.75(4) ±0.63	50.55(4) ±1.90	17.12(4) ±1.14	4.00
HSIC-avg	7.48(3) ±0.17	44.63(3) ±1.18	53.72(3) ±1.27	64.02(2) ±1.18	52.50(3) ±1.95	18.62(3) ±1.23	2.83
HSIC-max	**7.42**(1) ±0.17	**45.49**(1) ±1.50	**54.75**(1) ±1.70	**64.61**(1) ±1.51	**53.78**(1) ±2.14	**18.71**(1.5) ±1.33	**1.08**

and $0.01L(L \geq 10000)$. In FIMF, we validate its recommended settings that the number interaction terms $b = 2$ and the size of the highest entropy label set $|Q| = 10$. As to FoHSIC, with delta label kernel, the width of feature RBF kernel is the dimensions of features after each feature is normalized into the zero mean and unit standard deviation, and 1000 instances at most are randomly

Table 3. Computational time (seconds) for feature selection on four data sets

Method	Bibtex	Enron	Slashdot	TMC2007-500	Average time
ReliefF-ML	189.13 ± 2.24	10.23 ± 0.18	30.71 ± 0.38	56.29 ± 0.73	71.59
FIMF	3.55 ± 0.11	0.83 ± 0.03	0.85 ± 0.05	1.50 ± 0.06	1.68
FoHSIC	831.74 ± 4.00	826.17 ± 2.37	811.19 ± 3.07	841.53 ± 3.41	827.66
HSIC-rank	$\mathbf{0.44 \pm 0.02}$	$\mathbf{0.06 \pm 0.02}$	$\mathbf{0.10 \pm 0.01}$	$\mathbf{0.11 \pm 0.01}$	0.18
HSIC-avg	0.59 ± 0.05	0.19 ± 0.02	0.24 ± 0.02	0.25 ± 0.01	0.32
HSIC-max	0.60 ± 0.03	0.22 ± 0.02	0.25 ± 0.01	0.27 ± 0.02	0.34

selected. When the $n=100$ features will be selected, the $2n$ top features are identified firstly to speed up FIMF, FoHSIC and our HSIC-avg and HSIC-max.

4.3 Performance Comparison

In this sub-section, we experimentally evaluate six FS methods: ReliefF-ML, FIMF, FoHSIC, HSIC-rank, HSIC-avg and HSIC-max. The experimental results are listed in Table 2, where the top 30 features are selected. The best measure value is highlighted in boldface and the ranks are denoted by 1–6 in the brackets. The average rank of each method over six measures is listed in the last column.

According to experimental results from Bibtex, our HSIC-avg and HSIC-max perform the best, and ReliefF-ML the worst. On Enron, our HSIC-max and HSIC-avg achieve the top performance, and FoHSIC obtains the lowest rank. As to Shashdot, our HSIC-avg and HSIC-max are superior to the other four method, and ReliefF-ML is inferior. From TMC2007-500, HSIC-max works the best and FIMF obtains the second rank, and ReliefF-ML performs the worst.

To evaluate these methods comprehensively, we calculate their overall average ranks over four data sets and sort these methods as HSIC-avg(1.65), HSIC-max(1.71), FIMF(3.62), HSIC-rank(3.88), FoHSIC(4.79) and ReliefF-ML(5.36). With 30 features only, our two HSIC methods perform the best.

Further we compare the computational time for six FS methods, as shown in Table 3. According to the average time over four data sets in ascending order, six methods are sorted as HSIC-rank, HSIC-avg, HSIC-max, FIMF, ReliefF-ML and FoHSIC. HSIC-rank runs the fastest, and our HSIC-avg and HISC-max achieve the second and third top ranks. Although we speed up FoHSIC with 200 higher relevant features and 1000 randomly selected instances, FoHSIC is still time consuming. In short, our two methods improve the classification performance as the cost of slightly more time. Our computational platform is a HP workstation with two Intel Xeon5675 CPUs and 32G RAM, and Matlab 2010a.

5 Conclusions

In this paper, we proposed two novel feature selection approaches for multi-label classification. After analyzing Hilbert-Schmidt independence criterion, we utilize the non-diagonal elements to characterize feature-feature conditional redundancy, and then define two novel multi-label feature selection rules. Combining simple feature ranking and sequential forward selection, an efficient feature selection procedure is implemented. Experimental results on four data sets illustrate that, (a) our methods consume slightly more computational time than original ranking HSIC algorithm, and (b) according to six instance-based measures, our two methods perform the best, compared with four state-of-the-art methods based on original ranking HSIC, kernel HSIC with sequential forward selection, ReliefF and mutual information. Therefore our novel approaches are effective and efficient comprehensively.

References

1. Lewis, D.D., Yang, Y., Rose, T.G., Li, F.: RCV1: a new benchmark collection for text categorization research. J. Mach. Learn. Res. **5**, 361–397 (2004)
2. Zhang, M.L., Zhou, Z.H.: A review on multi-label learning algorithms. IEEE Trans. Knowl. Data Eng. **26**(8), 1338–1351 (2014)
3. Gibaji, E., Ventura, S.: A tutorial on multilabel learning. ACM Comput. Surv. **47**(3), 52:1–52:38 (2015)
4. Sun, L., Ji, S., Ye, J.: Multi-Label Dimensionality Reduction. CRC Press, Boca Raton (2014)
5. Tang, J., Alelyani, S., Liu, H.: Feature selection for classification: a review. In: Aggarwal, C.C. (ed.) Data Classification: Algorithms and Applications, Chap. 2, pp. 37–64. CRC Press, Boca Raton (2014)
6. Zhang, M.L., Pena, J.M., Robles, V.: Feature selection for multi-label naive Bayes classification. Inform. Sci. **179**(19), 3218–3229 (2009)
7. Yin, J., Tao, T., Xu, J.: A multi-label feature selection algorithm based on multi-objective optimization. In: Proceedings of 2015 International Joint Conference on Neural Networks (IJCNN 2015), pp. 1–7. IEEE Press, New York (2015)
8. Gu, Q., Li, Z., Han, J.: Correlated multi-label feature selection. In: Proceedings of ACM International Conference on Information and Knowledge Management (CIKM2011), pp. 1087–1096. ACM, New York (2011)
9. Spolaor, N., Cherman, E., Monard, M., Lee, H.: A comparison of multi-label feature selection methods using the problem transformation approach. Electron. Notes Theor. Comput. Sci. **292**, 135–151 (2013)
10. Robnik-Sikonja, M., Kononenko, I.: Theoretical and emperical analysis of ReliefF and RReliefF. Mach. Learn. **53**(1), 23–69 (2003)
11. Reyes, O., Morell, C., Ventura, S.: Scalable extensions of the relieff algorithm for weighting and selecting features on the multi-label learning context. Neurocomputing **161**, 168–182 (2015)
12. Lee, J., Kim, W.: Feature selection for multi-label classification using multivariate mutual information. Pattern Recogn. Lett. **34**(3), 349–357 (2013)
13. Lee, J., Kim, D.W.: Mutual information-based multi-label feature selection using interaction information. Expert Syst. Appl. **42**(4), 2013–2025 (2015)
14. Lin, Y., Hu, Q., Liu, J., Duan, J.: Multi-label feature selection based on max-dependency and min-redundancy. Neurocompting **168**, 92–103 (2015)
15. Lee, J., Kim, D.W.: Fast multi-label feature selection based on information-theoretic feature ranking. Pattern Recogn. **48**(9), 2761–2771 (2015)

16. Gretton, A., Bousquet, O., Smola, A.J., Schölkopf, B.: Measuring statistical dependence with Hilbert-Schmidt norms. In: Jain, S., Simon, H.U., Tomita, E. (eds.) ALT 2005. LNCS (LNAI), vol. 3734, pp. 63–77. Springer, Heidelberg (2005)
17. Song, L., Smola, A., Gretton, A., Bedo, J., Borgwardt, K.: Feature selection via dependency maximization. J. Mach. Learn. Res. **13**, 1393–1434 (2012)
18. Multi-label data sets. http://mulan.sourceforge.net/datasets-mlc.html
19. Zhang, M.L., Zhou, Z.H.: ML-kNN: a lazy learning approach to multi-label learning. Pattern Recogn. **40**(5), 2038–2048 (2007)
20. Brazdil, P.B., Soares, C.: A comparison of ranking methods for classification algorithm selection. In: Lopez de Mantaras, R., Plaza, E. (eds.) ECML 2000. LNCS (LNAI), vol. 1810, pp. 63–74. Springer, Heidelberg (2000)

Storm Surge Prediction for Louisiana Coast Using Artificial Neural Networks

Qian Wang[1], Jianhua Chen[1(✉)], and Kelin Hu[2]

[1] Division of Computer Science and Engineering, Louisiana State University,
Baton Rouge, LA 70803-4020, USA
Jianhua@csc.lsu.edu
[2] Center for Computation and Technology, Louisiana State University,
Baton Rouge, LA 70803-4020, USA

Abstract. Storm surge, an offshore rise of water level caused by hurricanes, often results in flooding which is a severe devastation to human lives and properties in coastal regions. It is imperative to make timely and accurate prediction of storm surge levels in order to mitigate the impacts of hurricanes. Traditional process-based numerical models for storm surge prediction suffer from the limitation of high computational demands making timely forecast difficult. In this work, an Artificial Neural Network (ANN) based system is developed to predict storm surge in coastal areas of Louisiana. Simulated and historical storm data are collected for model training and testing, respectively. Experiments are performed using historical hurricane parameters and surge data at tidal stations during hurricane events from the National Oceanic and Atmospheric Administration (NOAA). Analysis of the results show that our ANN-based storm surge predictor produces accurate predictions efficiently.

1 Introduction

1.1 Motivation and Related Works

Tropical cyclones often generate storm surges, wind waves, and flooding when they approach coasts or make landfall, which is a major threat to human life and property in coastal regions throughout the world. In the U.S., this kind of devastation was no more evident than during Hurricanes Katrina and Rita in 2005 and Gustav and Ike in 2008. Over 1600 people lost their lives and several major coastal populations were crippled for months after the hurricanes passed. Obviously, mitigating the impacts of hurricanes requires a timely and accurate prediction of storm surge.

Currently, numerous numerical models have been applied to storm surge modeling, including storm surge models such as SLOSH [1] and ADCIRC [2], and general process-based hydrodynamic and transport models such as FVCOM [3], ROMS [4] and Delft3D [5]. The numerical models are normally based on the two-dimensional or three-dimensional shallow water NavierStokes equations.

© Springer International Publishing AG 2016
A. Hirose et al. (Eds.): ICONIP 2016, Part III, LNCS 9949, pp. 396–405, 2016.
DOI: 10.1007/978-3-319-46675-0_43

Running such models is a computationally intensive task which makes timely prediction difficult. Quite a number of research works have investigated alternative approaches for storm surge prediction using machine learning techniques, in particular, using ANN. Lee [6] developed a back propagation neural network model combined with harmonic analysis to predict the storm surge and surge deviation in Taiwan. The performance of the model were compared with two numerical methods and the results showed the capability of the ANN model in storm surge forecasting. You and Seo [7] developed a cluster neural network model to predict the storm surges in all Korean coastal regions. They showed that the values predicted by their model were closer to observed data than values using harmonic analysis. De Oliveira et al. [8] applied a neural network model to predict coastal sea level variations related to meteorological events along the southeastern coast of Brazil. They used the pressure, wind values and tide gauge time series as inputs to the model to analyze the relationship between these variables and the storm surge event, and showed that the model can be efficient in predicting the non-tidal residuals and effectively complement the standard harmonic analysis model. Bajo and Umgiesser [9] developed an operational surge forecast system based on combination of a hydrodynamic model and an artificial neural network for the city of Venice. They applied a post-processing method which used both observed data and forecast results of the hydrodynamic model to train an ANN.

1.2 Our Approach

In this paper we present our work for storm surge prediction for Louisiana coast using ANN. In contrast to most existing works in which the training data were historical hurricane and storm surge data, we actually utilized simulated hurricane and storm surge data for Louisiana coast region generated by the numerical hydrodynamic model ADCIRC. This use of simulated data has the advantage of providing sufficient training data for learning the weights in ANN, because usually we do not have a lot of historical hurricane data to work with. To verify the validity of our ANN, historical data from 6 hurricanes in the Louisiana coast were used for testing the system's performance.

2   Study Area and the ADCIRC Model

The Louisiana coast is extremely susceptible to the impacts of frequent tropical storms and hurricanes because of its unique geometry and coastal landscapes. When tropical cyclones approach the coast or make landfall, they often generate storm surges, wind waves, and changes in salinity and morphology in estuaries. To mitigate those negative impacts of hurricanes, the ability to accurately predict and model these phenomena is important. In recent years, six major historical hurricanes have affected the Louisiana coast. The Louisiana coast and tracks of these six major hurricanes are shown in Fig. 1. For historical hurricanes, best track data available from the National Hurricane Center (NHC) provide locations of hurricane eye and other hurricane properties such as central pressure,

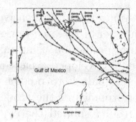

Fig. 1. Louisiana Coast and Tracks of Six Historical Hurricanes in the area

maximum wind speed and its radius. We exploit this information to generate input for testing the ANN system which will be discussed later in the paper. Observations of water level at tidal stations were collected from the National Oceanic and Atmospheric Administration (NOAA) website.

ADCIRC (ADvanced multi-dimensional CIRCulation model for shelves, coasts and estuaries) [10] is a system of computer programs for solving time dependent, free surface circulation and transport problems in two and three dimensions. These programs utilize the finite element method in space and therefore can be run on highly flexible, irregularly spaced grids. ADCIRC is a highly developed computer program for solving the equations of motion for a moving fluid on the rotating earth. These equations have been formulated using the traditional hydrostatic pressure and Boussinesq approximations and have been discretized in space using the finite element method (FEM) and in time using the finite difference method (FDM). Due to space restrictions we will not include the governing equations in ADCIRC here. Please refer to [11] for information about such.

3   Methodology

3.1   Feature Selection

The selection of input variables is very important for an efficient and accurate ANN system. The local storm surge is controlled by hurricane winds and local features such as geometry and bathymetry. The effects of local features are embedded in the ANN model through data training. Hurricane properties should be considered in input variables. As such, six variables are selected as input. Disx (km) and Disy (km) describe the relative location of local stations to the hurricane center in a moving frame of reference whose origin is the hurricane center and the direction of y-axis is the hurricane moving direction. In this way, the input reflected the change of hurricane winds due to the distance to the center in different quadrants. The rest four variables use hurricane properties, that is, forward speed Vf (m/s), minimum central pressure Pm (mb), maximum wind speed Vm (kt) and hurricane radius Rm (nm).

3.2   Data Pre-processing

Two datasets are needed to prepare for the model training and verification, that is, training set and testing set. The testing dataset is obtained by the historical observed data. Six major historical hurricanes that affected the Louisiana coast as mentioned earlier are selected, which are Hurricanes Dennis, Katrina and Rita in 2005, Hurricanes Gustav and Ike in 2008 and Hurricane Isaac in 2012. The six input variables are generated by hurricane track data extracted from Best Track Data provided by the National Hurricane Center. These Best Track data include locations of hurricane center, central pressure, maximum wind speed and its radius. Central pressure (Pm), maximum wind speed (Vm) and its radius (Rm) can be used directly as input variables. The forward speed (Vf) is calculated from the series of hurricane center locations (mostly every 6 hours from NHC). We used a frame of reference moving north at the hurricane forward speed. In this moving frame of reference, the relative location of the station to the hurricane center can be described by its coordinates (x, y) in x- and y- directions, that is, (Disx, Disy). They are interpolated to meet with the hourly surge data. The output variable surge is calculated by

$$surge = ObservedWaterLevel - PredictedTidalLevel$$

Observations of water level and tidal level at tidal stations are collected from the National Oceanic and Atmospheric Administration (NOAA) website. Then the surge data can be calculated. The training dataset is obtained from simulated data. 153 idealized hurricanes that affect the Louisiana coast are selected for the model training. Their tracks are shown in Fig. 2. These hurricanes are defined in a past FEMA and USACE project [14]. Storm parameters for each hurricanes can be found in a Table 11 in [15] and in an appendix of [16]. Maximum wind speed (Vm), its radius (Rm) and forward speed (Vf) can be varying within a same track type (as shown in Fig. 2), which resulted in multiple hurricane scenarios. A best-track-like data file can be achieved and then the input data can be obtained through a similar way to handling testing input. Furthermore, the corresponding output surge data are simulated by ADCIRC model, which we will talk about in detail in the next subsection.

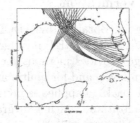

Fig. 2. Tracks of 153 Idealized Hurricanes that Affect Louisiana Coast

3.3 ADCIRC Modeling

Storm parameters for each track of the 153 idealized hurricanes include a series of hurricane center locations (longitudes and latitudes), central pressure, radius for maximum wind, Holland B parameter and forward speed. Maximum wind speed and radii for specified winds (34-, 50- and 64- knot) can be derived from these provided parameters. Thus, a best-track-like data file can be obtained for each storm as wind input file for ADCIRC modeling. An unstructured mesh with about 1M nodes and 2M elements from the Louisiana Coastal Protection and Restoration Authority (CPRA) is employed in ADCIRC modeling. Hurricane-induced water levels (storm surge) are calculated by ADCIRC. Tides were not considered, namely the output by the ADCIRC system is the predicted storm surge level. The time step size in the simulation is set as 1 s. Before running idealized cases, this ADCIRC model was first verified with Hurricane Katrina. Figure 3 shows the comparison of water level (storm surge) at Station PSTL1. The agreement is quite good, especially for the most concerned value, the maximum storm surge.

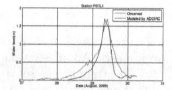

Fig. 3. Comparison of Water Levels at Station PSTL1 during Hurricane Katrina

A typical idealized hurricane event varies from 3 days to 8 days depending on its forward speed. All 153 cases are run on a LSU HPC cluster called QB2 (http://www.hpc.lsu.edu/resources/hpc/system.php?system=QB2). The computation time is about 1 h for a 4-day hurricane event when running with 2000 processors (100 nodes) on QB2.

3.4 ANN Training and Verification

With the simulated training set obtained by ADCIRC modeling, we start to train the ANN system. We used the NNET package [12,13] embedded in language R. The NNET package provides methods for using feed-forward neural networks with a single hidden layer, and for multinomial log-linear models. The optimization that the NNET package applied is done via the Broyden Fletcher Goldfarb Shanno (BFGS) algorithm, which is a gradient-based optimizer. In NNET package, the training process will stop when it reaches convergence. The "activation function" of the hidden layer units is taken to be the logistic function. And the output units can be chosen as linear or logistic. The sum of squared errors of prediction (SSE) is selected to use as the fitting criterion.

SSE is given by $SSE = \Sigma_{i=1}^{n}(y_i - f(x_i))^2$, where x_i is the i-th input data vector, y_i is the corresponding desired output, and $f(x_i)$ is the network predicted output value for the input x_i. Also, we use the parameter "weight decay" to help the optimization process and to avoid over-fitting. The root mean square error (RMSE) is taken to evaluate the training performance: RMSE is given by $RMSE = \sqrt{\Sigma_{i=1}^{n}(y_i - f(x_i))^2/n}$.

After training has finished, the historical data of six hurricanes are applied for model verification. The evaluation for the test performance consists of several error besides the root mean square error, which will be discussed in next section.

4   Experimental Results

We select two NOAA tidal stations as stations of interest in this experiments: Pilots Station East, Sw Pass, LA (Station PSTL1, ID8760922,2855.9' N,89 24.4' W) and Grand Isle, LA (Station GISL1, ID 8761724, 2915.8' N, 89 57.4' W) as shown in Fig. 4. The training set and testing set have been processed as discussed earlier. Here we report several experiments to show the model performance. All the experiments were executed on a LSU HPC cluster called SuperMike-II.

We choose different values of the parameter "weight decay" and different numbers of hidden nodes. For each combination of the parameter setting, the model is trained for 50 times, and each time the training terminates at about 1000 epochs. After each training, we apply the model to make predictions on the test set and record both the training error and testing error. Then we calculate the mean error for both over the 50 times. The mean error is used for evaluating the model training performance. These experiments are implemented for both two stations. The computation time is about half hour for each 50-times training process when running on SuperMike-II. The following Fig. 5 shows the error vs. weight decay values for various choices of number of hidden node, averaged over the 50 training runs, for the station PSTL1. It is apparent that the mean training error decreases a little bit around decay = 0 and almost keeps flat as the value of decay increases. The testing error drops a lot at the beginning and slows down the decrease as the value of decay increases. The number of hidden nodes does not seem to impact the error patterns significantly. For efficiency consideration, we set the number of hidden nodes at 10, and value of decay at 0.01. Now we apply the model to make prediction on testing data and compare the output with the observed water level.

Fig. 4. Tidal stations PSTL1 and GISL1

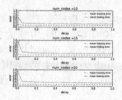

Fig. 5. ANN performance (averaged over 50 runs) for station PSTL1

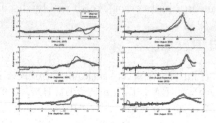

Fig. 6. ANN testing performance for station PSTL1

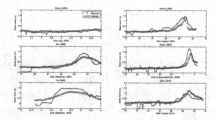

Fig. 7. ANN testing performance for station GISL1

Comparison of ANN results with observations during six historical hurricanes at two NOAA stations (PSTL1 and GISL1) are shown in Figs. 6 and 7, respectively. Note that these results are from one of the 50 training/testing runs described earlier. Among six hurricanes, the ANN results reproduced very well the time series of water level due to Hurricanes Katrina (2005) and Gustav (2008) at Station PSTL1. Both hurricanes passed near stations and induced high storm surge (over 1.5 m), which indicates that the model gets good performance with strong hurricane events. The overall agreement at Station PSTL1 is better than that at Station GISL1. The reason is that station PSTL1 locates more offshore and gets less influence by local geometries. Compared to hurricane-related parameters, local effects would contribute more to storm surges at Station GISL1 which locates near the Grant Isle. Hurricane Dennis (2005) passed to the far east of both stations and induced low storm surge. Hurricane Rita (2005) was very strong but passed to the far west of both stations. Rita caused larger surges at Station GISL1 (which locates more west) than Station PSTL1. The model overestimated surge at both stations. Hurricane Ike (2008) was a super-large storm in size. It induced around 1 m surge at both stations. The ANN model performed well at Station PSTL1. Hurricane Isaac (2012) was a weak but slow-moving storm. It caused over 1 m of surge at both stations while the model showed some underestimation. Also, we find the coverage of the training data is still very limited, both in spatial distributions of hurricane tracks and in variations of hurricane parameters. For example, the tracks of Dennis, Rita and Ike are out of the coverage by 153 idealized tracks. For Hurricane Isaac, its small forward speed may lie beyond the range of input training data. As such, errors would be large when predicting storm surge by those kind of hurricanes.

In coastal areas, people concern primarily about the maximum surge height and the arrival time when a hurricane is approaching. Thus, for practical applications, in addition to the normal test error RMSE, we introduce three more definitions of error below to evaluate the performance of the ANN model.

$$max\text{-}error = max(surge_{predicted}) - max(surge_{observed})$$
$$relative\ max\text{-}error = \frac{max(surge_{predicted}) - max(surge_{observed})}{max(surge_{observed})}$$
$$time_{error} = hour(surge_{predicted} = max) - hour(surge_{observed} = max)$$

The results for both tidal stations are shown in Tables 1 and 2. These are obtained from one of the 50 training/testing runs. For Hurricanes Katrina and Gustav, the error of the maximum surge at Station PSTL1 is less than 0.3 m with about 2 h of phase difference, which means the model performed very well at this site. Except for the case of Gustav at Station GISL1, the absolute error of maximum surge at both stations for all hurricane events is less than 0.36 m (1 foot), which is significant considering the error of high water marks calculated by the ADCIRC model is 0.36–0.4 m during Hurricane Katrina [17]. As for the arrival time, except for the Ike case, the time error is less than 4 h. The accuracy of these information is of great importance for emergency evacuation purpose during a hurricane event. Note that the ANN predictions can be done (and updated timely) in seconds by a personal computer, while predictions by process-based numerical models (e.g. ADCIRC) need lots of computational resources (e.g. hundreds of nodes on LSU HPC SuperMike II) to be finished within several hours. The good performance of this ANN model shows its potential for practical applications.

Table 1. Error for six hurricanes at Station PSTL1

	Dennis	Katrina	Rita	Gustav	Ike	Isaac
Test-error (m)	0.1576	0.1761	0.1534	0.1591	0.1558	0.172
Max-error (m)	−0.0271	−0.2563	0.3576	−0.0189	0.0682	−0.317
Relative max-error	−5.07 %	−14.35 %	48.59 %	−1.48 %	8.38 %	−32.2 %
Time-error (h)	−2	−2	−4	−2	−7	1

Table 2. Error for six hurricanes at Station GISL1

	Dennis	Katrina	Rita	Gustav	Ike	Isaac
Test-error (m)	0.098	0.2375	0.2438	0.233	0.3438	0.22
Max-error (m)	0.0731	0.1886	0.3587	0.4962	0.3509	−0.3103
Relative max-error	46.26%	15.13%	36.8%	37.87%	31.84%	−24.61%
Time-error (h)	−2	−2	−4	−2	−7	1

5   Conclusions and Future Works

We develop an ANN model to make prediction of storm surge in coastal areas of Louisiana. Due to the insufficiency of historical hurricane data in this area, we introduce idealized hurricanes and apply the hydrodynamic model ADCIRC to generate simulated water level data induced by these hurricanes. Experiments are implemented at two NOAA tidal stations and the results show the accuracy and efficiency of the ANN model for storm surge prediction purpose. We can conclude, first, that the selection of hurricane-related input parameters for ANN model is reasonable. Secondly, the ANN model is more suitable for strong hurricane events. Also, the ANN model is more suitable for offshore locations with less local influence. For inside stations, more data is needed to train the model to gain better prediction results.

The ANN model get good results for some hurricanes, while the error for other hurricanes is still large. This situation can be improved by including more data for model training. For the limitation of the current coverage of training data, we may expand the idealized tracks to cover more areas. In addition, for each track, the ranges of hurricane parameters can also be extended. We can also apply this ANN model to more tidal stations. For those inland stations, some parameters that can reflect local effects to storm surge should be added as input variables. These will be addressed in future work.

References

1. Jelesnianski, C.P., Chen, J., Shaffer, W.A.: SLOSH: Sea, Lake, and Overland Surges from Hurricanes. Silver Springs, Maryland (1992)
2. Luettich, R.A., Westerink, J.J., Scheffner, N.W.: ADCIRC: an advanced three-dimensional circulation model for shelves, coasts and estuaries, Report 1: theory and methodology of ADCIRC-2DDI & ADCIRC-3DL (1992)
3. Chen, C., Liu, H., Beardsley, R.C.: An unstructured grid, finite-volume, three-dimensional, primitive equations ocean model: application to coastal ocean and estuaries. J. Atmos. Ocean. Tech. 20(1), 159–186 (2003)
4. Shchepetkin, A.F., McWilliams, J.C.: The regional oceanic modeling system (ROMS): a split-explicit, free-surface, topography-following coordinate oceanic model. Ocean Model 9(4), 347–404 (2004)
5. Lesser, G.R., Roelvink, J.A., Van Kester, J.A.T.M., Stelling, G.S.: Development and validation of a three-dimensional morphological model. Coast. Eng. 51(8), 883–915 (2004)

6. Lee, T.: Prediction of storm surge and surge deviation using a neural network. J. Coast. Res. **24**, 76–82 (2008)

7. You, S., Seo, J.: Storm surge prediction using an artificial neural network model and cluster analysis. Nat. Hazards **51**(1), 97–114 (2009)

8. De Oliveira, M.M.F., Ebecken, N.F.F., De Oliveira, J.L.F., De Azevedo Santos, I.: Neural network model to predict a storm surge. J. Appl. Meteorol. Climatol. **48**, 143–155 (2009)

9. Bajo, M., Umgiesser, G.: Storm surge forecast through a combination of dynamic and neural network models. Ocean Model **33**, 1–9 (2010)

10. The ADvanced CIRCulation model (ADCIRC). http://www.adcirc.org/

11. Luettich, R., Westerink, J.: Formulation and numerical Implementation of the 2D/3D ADCIRC finite element model version 44.XX (2004). http://www.unc.edu/ims/adcirc/adcirc_theory_2004_12_08.pdf

12. Ripley, B.D.: Pattern Recognition and Neural Networks. Cambridge University Press, Cambridge (1996)

13. https://cran.r-project.org/web/packages/nnet/nnet.pdf

14. http://www3.nd.edu/~coast/femaIDS2.html

15. http://www3.nd.edu/~coast/reports_papers/SELA_2007_IDS_2_FinalDraft/Tables

16. http://www3.nd.edu/~coast/reports_papers/SELA_2007_IDS_2_FinalDraft/App

17. Bunya, S., Dietrich, J.C., Westerink, J.J., Ebersole, B.A., Smith, J.M., Atkinson, J.H., Jensen, R., Resio, D.T., Luettich, R.A., Dawson, C., Cardone, V.J., Cox, A.T., Powell, M.D., Westerink, H.J., Roberts, H.J.: Observation of strains: a high-resolution coupled riverine flow, tide, wind, wind wave, and storm surge model for southern louisiana and mississippi. Part I: model development and validation. Mon. Wea. Rev. **138**, 345–377 (2011). doi:10.1175/2009MWR2906.1

Factorization of Multiple Tensors for Supervised Feature Extraction

Wei Liu[(✉)]

Faculty of Engineering and Information Technology, Advanced Analytics Institute,
University of Technology Sydney, Sydney, Australia
Wei.Liu@uts.edu.au

Abstract. Tensors are effective representations for complex and time-varying networks. The factorization of a tensor provides a high-quality low-rank compact basis for each dimension of the tensor, which facilitates the interpretation of important structures of the represented data. Many existing tensor factorization (TF) methods assume there is one tensor that needs to be decomposed to low-rank factors. However in practice, data are usually generated from different time periods or by different class labels, which are represented by a sequence of multiple tensors associated with different labels. When one needs to analyse and compare multiple tensors, existing TF methods are unsuitable for discovering all potentially useful patterns, as they usually fail to discover either common or unique factors among the tensors: (1) if each tensor is factorized separately, the factor matrices will fail to explicitly capture the common information shared by different tensors, and (2) if tensors are concatenated together to form a larger "overall" tensor and then factorize this concatenated tensor, the intrinsic unique subspaces that are specific to each tensor will be lost. The cause of such an issue is mainly from the fact that existing tensor factorization methods handle data observations in an *unsupervised* way, considering only features but not labels of the data. To tackle this problem, we design a novel probabilistic tensor factorization model that takes both features and class labels of tensors into account, and produces informative *common and unique factors of all tensors* simultaneously. Experiment results on feature extraction in classification problems demonstrate the effectiveness of the factors discovered by our method.

Keywords: Feature extraction · Tensor factorization · Supervised learning

1 Introduction

In this paper we study the problem of probabilistically factorizing a sequence of multiple tensors for feature extraction from multi-mode tensor data. Various types of tensor factorization methods have been proposed in the literature, including Tucker decomposition [11], CP [3] (also known as PARAFAC),

© Springer International Publishing AG 2016
A. Hirose et al. (Eds.): ICONIP 2016, Part III, LNCS 9949, pp. 406–414, 2016.
DOI: 10.1007/978-3-319-46675-0_44

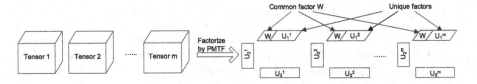

Fig. 1. Factorize all the tensors by our method PMTF, where matrix factors discover both common and unique patterns

non-negative tensor factorization [12], and probabilistic tensor factorization [13]. These methods and their later variants can be considered as higher-order generalizations of matrix factorizations. Most of these existing methods are restricted to decomposing a single instance of a tensor object in an unsupervised manner. This raises the question of what strategy should be used when dealing with multiple tensor objects which are associated with class labels. Given a tensor of M modes, existing TF methods decompose the tensor into M low-rank matrix factors, each of which explains a compact basis of each mode of the tensor. Two common approaches for using TF to factorize a sequence of m tensors are: (*option 1*) decompose each tensor separately – this approach generates a low-rank factor matrix for each mode of each tensor, and does not necessarily identify a potentially important "common factor" matrix that these tensors may share; or (*option 2*) concatenate all tensors along a certain mode to form one big tensor and then decompose it – in contrast to the first option, this strategy may discard possible "unique factor" matrices in the concatenated dimension, and only produces factors that are a consensus of the original tensors. Although it is possible to treat these tensors as a data stream and use sliding windows to analyse them incrementally [8,10], the actual decomposition on each element within each window is still limited to the above two options.

In this research we propose a novel strategy for probabilistically analysing multiple tensors, and introduce the concepts of *common factors* and *unique factors* along each mode of all tensors. As demonstrated in Fig. 1, the common space (denoted by $\mathbf{W}$) along the first dimension of all tensors occupies a fraction of the factor matrix, while the remaining fraction is preserved for each tensor independently so that any unique patterns that are discriminative to each tensor are also preserved. We will show that when the tensors are associated with class labels, the factorization of both common and unique factors is especially beneficial for feature extraction (dimension reduction) of classification tasks.

The strategy of decomposing common and unique factors provides a flexible choice on the sizes of common and unique spaces, such that the preceding two options become special cases of our proposed approach. When the common space is empty (i.e., when the size of $\mathbf{W}$ in Fig. 1 is zero) we obtain *option 1*, and when it is set to the full size of the factor matrix (instead of a fraction) we obtain *option 2*.

In summary, we make the following contributions in this paper:

1. We introduce the concepts of common factors and unique factors in decomposing a sequence of tensors, and formulate the problem of approximating low-rank representations of tensors as simultaneously optimizing the approximation of both common and unique factors;
2. We propose a PMTF (probabilistic multiple tensor factorization) model, which incorporates both the common and unique factor matrices inherently in the factorization process;
3. We perform empirical evaluations of PMTF on feature extraction for graph classification, which demonstrates the power and effectiveness of our method.

2 Related Work

Factorization methods for tensors that are essentially higher order generalizations of those for matrices have been studied, such as the probabilistic tensor factorization (PTF) method [9,13]. As a multi-dimensional generalization of matrix factorization, PTF is more attractive than matrix factorization not only because it considers more dimensions of information, but also because it usually allows for a unique decomposition of a data set into factors under mild conditions, which are usually satisfied by real data [6]. The field of multi-task learning [14] is also related to our research, however there is no existing work in the multi-task learning domain that studied the problem for tensor factorizations. Coupled tensor and/or matrix factorization methods [5] are also closely related to this research. However, no existing coupled factorization methods address the problem of discovering both shared and unique factors simultaneously. It has also been proposed to perform coupled tensor factorizations [1,7] by using generalised learning models. However, these papers only consider the case when the decomposed factors are all the same in the shared mode, and did not address how to discover discriminant factors from the shared identical mode between coupled matrices or tensors.

Different from all the above literature, in this research we propose the first method that simultaneously decomposes tensors into both common and unique factors, incorporating their class labels (or data generation sources) in the factorization processes, which significantly improve the effectiveness of the extracted features.

3 Tensor Factorization

Tensors are multidimensional (aka multi-mode) arrays. We denote tensors with 3 or more modes by calligraphic font (e.g., $\mathcal{X}$), denote matrices (tensors with 2 modes) by boldface uppercase letters (e.g., $\mathbf{U}$), and denote vectors (tensors with 1 mode) by boldface lower letters (e.g., $\mathbf{u}$). In the following, we first briefly introduce preliminaries of PTF, and then elaborate the proposed PMTF model. We give the definition of a standard tensor factorization as follows:

3.1   Probabilistic Tensor Factorization

Given a M-mode tensor $\mathcal{X} \in \mathbb{R}^{n_1 \times \cdots \times n_M}$ and the desired low rank r, probabilistic tensor factorization (PTF) method decompose $\mathcal{X}$ into M matrix factors $\mathbf{U}_d \in \mathbb{R}^{n_d \times r}, (d = 1, 2, 3, ..., M)$, such that $\mathcal{X} \approx \sum_{j=1}^{r} \mathbf{u}_1^j \otimes \mathbf{u}_2^j .... \otimes \mathbf{u}_M^j$, where $\mathbf{u}_d^j$ represents the jth column of $\mathbf{U}_d$, and $\otimes$ represents outer products. Taking 3-mode tensor as an example, the element-wise expression of the decomposition can be written as $\mathcal{X}_{i,j,k} \approx \sum_{d=1}^{r} (\mathbf{U}_1)_{r,i} (\mathbf{U}_2)_{r,j} (\mathbf{U}_3)_{r,k} \equiv < \mathbf{u}_1^i, \mathbf{u}_2^j, \mathbf{u}_3^k >$. For ease of interpretations, we use $\mathbf{U}_1, \mathbf{U}_2, ... \mathbf{U}_M$, or $\mathbf{U}_d|_{d=1}^{M}$ (or simply $\mathbb{U}$) to represent the operation $\sum_{j=1}^{r} \mathbf{u}_1^j \otimes \mathbf{u}_2^j .... \otimes \mathbf{u}_M^j$ in the rest of the paper. Moreover, we will use the example of 3-mode tensor to represent the more generic cases of M modes.

Standard PTF method [13] assumes Gaussian distributions on the likelihood of tensor observations given matrix factors:

$$p(\mathcal{X}|\mathbf{U}_1, \mathbf{U}_2, \mathbf{U}_3, \sigma^2) = \prod_{i=1}^{n_1} \prod_{j=1}^{n_2} \prod_{k=1}^{n_3} \left[\mathcal{N}(\mathcal{X}_{ijk}| < \mathbf{u}_1^i, \mathbf{u}_2^j, \mathbf{u}_3^k >, \sigma^2) \right]^{I_{i,j,k}}, \qquad (1)$$

where $\mathcal{X}$ is a three-mode tensor, $\mathbf{U}_1, \mathbf{U}_2, \mathbf{U}_3$ are respectively the tensors factor matrices in each mode, the inner product of column vectors $< \mathbf{u}_1^i, \mathbf{u}_2^j, \mathbf{u}_3^k >$ is the mean of the Gaussian distribution which has variance σ^2, and the binary indicator $I_{i,j,k}$ equals 1 if value $\mathcal{X}_{i,j,k}$ is observed and equals 0 otherwise.

3.2   Common and Unique Subspaces

Without loss of generality, we define and solve the problem of learning common and unique factors from multiple tensors by using the scenario of two tensors (e.g., $m = 2$ in the example of Fig. 1). This scenario corresponds to the case of binary classes, where each tensor contains instances from a class label. We omit the lengthy derivations for $m > 2$ scenarios due to their close theoretical similarity to the $m = 2$ scenario. Assume it is the first mode of tensors that we want to derive both common and unique factors, the probabilistic learning problem can be defined as follows:

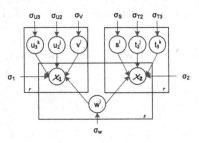

Fig. 2. Graphical representations on using PMTF to factorize two tensors $\mathcal{X}_1$ and $\mathcal{X}_2$. PMTF unveils both the common factor matrix $\mathbf{W}$, and unique factors $\mathbf{V}$ and $\mathbf{S}$ (shown by their column vectors w^i, v^i, and s^i respectively)

Definition 1. *(Probabilistic multiple tensor factorization (PMTF)): Given two M-mode tensors $\mathcal{X}_1$ and $\mathcal{X}_2$, PMTF probabilistically decomposes each of them as the product of $M + 1$ factor matrices so that "$\mathcal{X}_1 \approx [\mathbf{W}|\mathbf{V}] \otimes \mathbf{U}_2 \otimes \mathbf{U}_3 ... \otimes \mathbf{U}_M$" and "$\mathcal{X}_2 \approx [\mathbf{W}|\mathbf{S}] \otimes \mathbf{T}_2 \otimes \mathbf{T}_3 ... \otimes \mathbf{T}_M$" hold simultaneously.*

In this definition, we use "[$\mathbf{W}|\mathbf{V}$]" and "[$\mathbf{W}|\mathbf{S}$]" to represent the first matrix factor of each tensor, where $\mathbf{W}$ is the common factor, and $\mathbf{V}$ and $\mathbf{S}$ are respectively the unique factors of the two tensors. Using the previous illustration in Fig. 1, $\mathbf{V}$ and $\mathbf{S}$ are equivalent to $\mathbf{U}_1^1$ and $\mathbf{U}_1^2$. Since the common factor is located in the first dimension of the tensors, $\mathbf{W}$ is of size $n_1 \times s$, while $\mathbf{V}$ and $\mathbf{S}$ are of size $n_1 \times (r - s)$, where s is the desired cardinality of the common factor matrix ($0 \leq s \leq r$). The concatenation [$\mathbf{W}|\mathbf{V}$] is of size $n_1 \times r$, aligning with $\mathbf{U}_2$ and $\mathbf{U}_3$ which are of size $n_2 \times r$ and $n_3 \times r$ respectively. By using 3-mode tensors as an example, a graphical illustration representing the model of PMTF is shown in Fig. 2.

The conditional probability of tensor data observations is modelled from:

$$
\begin{aligned}
p(\mathcal{X}_1|\mathbf{W}, \mathbf{V}, \mathbf{U}_2, \mathbf{U}_3, \sigma_1^2) = \prod_{i,j,k} & \left[\mathcal{N}((\mathcal{X}_1)_{ijk}| < \mathbf{w}^i, (\mathbf{u}_2^j)_w, (\mathbf{u}_3^k)_w >, \sigma_1^2) \right]^{(I_1)_{i,j,k}} \\
\times \prod_{i,j,k} & \left[\mathcal{N}((\mathcal{X}_1)_{ijk}| < \mathbf{v}^i, (\mathbf{u}_2^j)_v, (\mathbf{u}_3^k)_v >, \sigma_1^2) \right]^{(I_1)_{i,j,k}} ,
\end{aligned}
\tag{2}
$$

and

$$
\begin{aligned}
p(\mathcal{X}_2|\mathbf{W}, \mathbf{S}, \mathbf{K}_2, \mathbf{K}_3, \sigma_2^2) = \prod_{i,j,k} & \left[\mathcal{N}((\mathcal{X}_2)_{ijk}| < \mathbf{w}^i, (\mathbf{t}_2^j)_w, (\mathbf{t}_3^k)_w >, \sigma_2^2) \right]^{(I_2)_{i,j,k}} \\
\times \prod_{i,j,k} & \left[\mathcal{N}((\mathcal{X}_2)_{ijk}| < \mathbf{s}^i, (\mathbf{t}_2^j)_s, (\mathbf{t}_3^k)_s >, \sigma_2^2) \right]^{(I_2)_{i,j,k}} ,
\end{aligned}
\tag{3}
$$

where $(I_1)_{i,j,k}$ and $(I_2)_{i,j,k}$ contain binary indicators that respectively represent whether the entries at $\{i, j, k\}$ position of $\mathcal{X}_1$ and $\mathcal{X}_2$ are observed. Factor matrices of both tensors are modelled by Gaussian priors:

$$
p(\mathbf{W}|\sigma_\mathbf{W}) = \prod_{i=1}^{n_1} \mathcal{N}(\mathbf{w}^i|0, \sigma_\mathbf{W}\mathbf{I}_w), \quad p(\mathbf{V}|\sigma_\mathbf{V}) = \prod_{i=1}^{n_1} \mathcal{N}(\mathbf{v}^i|0, \sigma_\mathbf{V}\mathbf{I}_v),
$$

$$
p(\mathbf{S}|\sigma_\mathbf{S}) = \prod_{i=1}^{n_1} \mathcal{N}(\mathbf{s}^i|0, \sigma_\mathbf{S}\mathbf{I}_s), \quad p(\mathbf{U}_d|\sigma_{\mathbf{U}_d}) = \prod_{i=1}^{n_d} \mathcal{N}(\mathbf{u}_d^i|0, \sigma_{\mathbf{U}_d}\mathbf{I}), \quad p(\mathbf{T}_d|\sigma_{\mathbf{T}_d}) = \prod_{i=1}^{n_d} \mathcal{N}(\mathbf{t}_d^i|0, \sigma_{\mathbf{T}_d}\mathbf{I}),
$$

where $d = 2$ and 3, $\mathbf{I}_w$, $\mathbf{I}_v$, $\mathbf{I}_s$, and $\mathbf{I}$ are respectively identity matrices of size s by s, $r - s$ by $r - s$, $r - s$ by $r - s$, and r by r. The log-posterior probability of the factor matrices is then:

$$
\begin{aligned}
\ln p(&\mathbf{W}, \mathbf{V}, \mathbf{S}, \mathbf{U}_d, \mathbf{T}_d|\mathcal{X}_1, \mathcal{X}_2, \Theta) \\
&= \ln p(\mathcal{X}_1|\mathbf{W}, \mathbf{V}, \mathbf{U}_2, \mathbf{U}_3, \sigma_1^2) + \ln p(\mathcal{X}_2|\mathbf{W}, \mathbf{S}, \mathbf{K}_2, \mathbf{K}_3, \sigma_2^2) + \ln p(\mathbf{W}|\sigma_\mathbf{W}) + \ln p(\mathbf{V}|\sigma_\mathbf{V}) \\
&\quad + \ln p(\mathbf{S}|\sigma_\mathbf{S}) + \ln p(\mathbf{U}_d|\sigma_{\mathbf{U}_d}) + \ln p(\mathbf{T}_d|\sigma_{\mathbf{T}_d}) + C'
\end{aligned}
$$

where $\Theta = \{\sigma_1, \sigma_2, \sigma_W, \sigma_V, \sigma_S, \sigma_{U_d}, \sigma_{T_d}\}$, $d = 2$ and 3, C' is a constant that is not dependent on any of the parameters. By making use of the probability density function of Gaussian distribution, maximizing the above function is equivalent to minimizing the following sum of squared error:

$$\min \sum_{i,j,k} (I_1)_{ijk} \left(((\mathcal{X}_1)_{ijk} - <\mathbf{w}^i, \mathbf{u}_2^j, \mathbf{u}_3^k>)^2 + ((\mathcal{X}_1)_{ijk} - <\mathbf{v}^i, \mathbf{u}_2^j, \mathbf{u}_3^k>)^2 \right)$$

$$+ \sum_{i,j,k} (I_2)_{ijk} \left(((\mathcal{X}_2)_{ijk} - <\mathbf{w}^i, \mathbf{u}_2^j, \mathbf{u}_3^k>)^2 + ((\mathcal{X}_2)_{ijk} - <\mathbf{s}^i, \mathbf{u}_2^j, \mathbf{u}_3^k>)^2 \right)$$

$$+ \sum_i \frac{\lambda_\mathbf{W} \|\mathbf{w}^i\|_2^2}{2} + \sum_i \frac{\lambda_\mathbf{V} \|\mathbf{v}^i\|_2^2}{2} + \sum_i \frac{\lambda_\mathbf{S} \|\mathbf{s}^i\|_2^2}{2} + \sum_j \frac{\lambda_{\mathbf{U}_2} \|\mathbf{u}_2^j\|_2^2}{2} + \sum_k \frac{\lambda_{\mathbf{U}_3} \|\mathbf{u}_3^k\|_2^2}{2} \qquad (4)$$

$$+ \sum_j \frac{\lambda_{\mathbf{T}_2} \|\mathbf{t}_2^j\|_2^2}{2} + \sum_k \frac{\lambda_{\mathbf{T}_3} \|\mathbf{t}_3^k\|_2^2}{2}$$

where $\lambda_\mathbf{W} = \sigma_\mathbf{W}/\sigma_1$, $\lambda_\mathbf{V} = \sigma_\mathbf{V}/\sigma_1$, $\lambda_\mathbf{S} = \sigma_\mathbf{S}/\sigma_2$, $\lambda_{\mathbf{U}_d} = \sigma_{\mathbf{U}_d}/\sigma_1$, $\lambda_{\mathbf{T}_d} = \sigma_{\mathbf{T}_d}/\sigma_2$ (d=2,3). The objective function in Eq. 4 is convex with respect to each matrix factor and can be minimized by gradient descent or block coordinate decent algorithms, which both iteratively update one parameter at a time. In our experiments, we alternate between optimizing the hyperparameters and updating the columns of matrix factors with the hyperparameters fixed.

3.3 Applying PMTF for Supervised Feature Extraction

Given a set of graphs, each of which is associated with a class label, the graph classification task is to predict the class of a new graph. Similar to general factorization-based dimension reduction methods such as PCA (where each Principle Component is used as a new feature), we apply our factorization method PMTF as a feature extraction method for classification problems, by making use of *the features (i.e., column vectors) defined by the new low-dimension feature spaces W, V and S*, in comparison to the *features defined by column vectors of matrix factors from standard factorization methods*. The performance of the features selected by our method for graph classification is evaluated in the next section.

4 Experiments and Analysis

We implement PMTF and PTF in Matlab by using the Tensor Toolbox [2]. This toolbox also contains an implementation of CP tensor decomposition, which we use in the evaluation. All experiment results presented in this section are from 5-fold cross validation with 10 repeated runs.

4.1 Data Sets

We apply PMTF to graph feature extraction and classification problem on chemical compound data sets, where each chemical compound is treated as a graph. We use bioassays of anti-cancer activity and kinase inhibition (AID)[1]: the task is to predict whether a compound is positive or negative in anti-cancer activities or in kinase inhibition activities. Details of these chemical compound data sets are reported in Table 1.

---

[1] http://pubchem.ncbi.nlm.nih.gov.

Table 1. Statistics of chemical compound data sets

Name	#graphs	Descriptions	Name	#graphs	Descriptions
AID83	27784	Breast Cancer	AID81	40700	Colon Cancer
AID123	40152	Leukemia	AID1	40460	Lung Cancer
AID33	40209	Melanoma	AID47	40447	Nerve Cancer
AID109	40691	Ovarian Cancer	AID41	27585	Prostate Cancer
AID145	40164	Renal Cancer	AID1481	217968	ATPase Inhibition
AID1416	217968	PERK Inhibition	AID1446	217968	Janus Kinase

Table 2. Comparisons of different methods in their effectiveness of feature extraction for graph classification using logistic regression

Data sets	AUC from logistic regression					AUC from SVMs with quadratic kernels				
	CP	PTF	GTF	PMTF	Best s	CP	PTF	GTF	PMTF	Best s
AID83	0.615	0.568	0.511	**0.727**	17	0.535	0.582	0.586	**0.610**	18
AID81	0.608	0.757	0.739	**0.762**	16	0.642	0.642	0.699	**0.738**	11
AID123	0.686	0.761	0.695	**0.778**	12	0.604	0.601	0.637	**0.646**	7
AID1	0.782	0.806	**0.886**	0.884	9	0.798	0.738	0.725	**0.810**	7
AID33	0.675	0.595	0.711	**0.806**	11	0.711	0.765	0.762	**0.822**	15
AID47	0.622	0.604	0.801	**0.829**	10	**0.792**	0.774	0.752	0.791	9
AID109	0.797	0.628	0.741	**0.799**	14	0.711	0.761	0.757	**0.808**	13
AID41	0.680	0.554	0.728	**0.741**	7	0.692	0.696	0.722	**0.800**	16
AID145	0.733	0.674	0.762	**0.891**	6	0.807	**0.893**	0.799	0.891	11
AID1481	0.603	0.590	0.660	**0.799**	9	0.720	0.696	0.705	**0.780**	18
AID1416	0.674	0.646	0.721	**0.807**	13	0.780	0.652	0.697	**0.797**	9
AID1446	0.865	0.749	0.758	**0.901**	5	0.889	0.808	0.869	**0.904**	6
Frd. test	✓ 0.006	✓ 0.018	✓ 0.051	Base	–	✓ 0.045	✓ 0.036	✓ 0.002	Base	–

4.2 Feature Extraction for Graph Classification

In each data set we construct two tensors, one for each class, where all unique types of atoms found in a data set are converted to the labels of vertices, and the lengths of bonds between atoms are weights of the edges. So an entry of a tensor is a count, which tells that for a certain compound, how many edges (bonds) connect certain types of atoms and have certain edge weights (lengths).

We compare the accuracy of classification on data points projected into the new low-dimension feature space produced by PTF, PMTF and CP decomposition, where we vary the settings of low ranks (r) from 5 to 20. In PTF, all training data instances are factorized together, so it only discovers common factors of both classes. In CP, tensors belonging to differents classes are factorized separately, hence it only find unique factors of the classes. We also include the GTF (Generalised Coupled Tensor factorization) method [15] in our evaluations, which is built on generalised linear models and produces common factors on the common mode of tensors.

To test the distinctness of the new data points from different classes under the new low-dimension feature space, we use two types of classifiers to learn from the new data points, a linear classifier – logistic regression, and a non-linear classifier – support vector machines (SVMs) with quadratic kernels (i.e., the kernel between two *vectorized* data samples x_i and x_j is: $k(x_i, x_j) = (1+x_i^T x_j)^2$).

The Friedman test is reported as one of the most appropriate methods for validating multiple classifiers among multiple data sets [4]. To confirm the significance of the superiority of PMTF, we perform Friedman tests on the sequences of AUC values across all data sets, where p–values that are lower than 0.05 reject the hypothesis with 95 % confidence that the classifiers in the comparison are not statistically different. In Table 2 we report the performance of the classifiers on different factorization methods when the rank is 20. In each data set the AUC value of the best performing method is put in boldface font. To show the diversity of the data sets, we also present the best s values which are optimized from the training set of cross validation. From the low p–values shown in the bottom of Table 2, it is easy to see that the low-rank spaces produced by PMTF are significantly better than the other corresponding methods in distinguishing the two class labels on each data set.

5   Conclusions and Future Work

In this research we focus on the problem of probabilistically factorizing a sequence of labelled tensors in order to improve tensor feature extraction for supervised learning. We formulate this problem into the task of discovering common and unique factors from multiple tensors. The proposed PMTF model is a generic tensor factorization method that can potentially be applied to many practical problems. We have applied PMTF to the problem of feature extraction (dimension reduction) for graph classification. Empirical results demonstrate the superiority of the factors discovered by PMTF over other existing methods. In future, we plan to apply this research to other data represented in multi-mode forms (such as images and videos).

References

1. Acar, E., Kolda, T.G., Dunlavy, D.M.: All-at-once optimization for coupled matrix and tensor factorizations. MLG workshop (2011)
2. Bader, B., Kolda, T.: Efficient Matlab computations with sparse and factored tensors. SIAM J. Sci. Comput. **30**(1), 205–231 (2007)
3. Carroll, J., Chang, J.: Analysis of individual differences in multidimensional scaling via an n-way generalization of "Eckart-Young" decomposition. Psychometrika **35**(3), 283–319 (1970)
4. Demšar, J.: Statistical comparisons of classifiers over multiple data sets. J. Mach. Learn. Res. **7**, 1–30 (2006)
5. Ermiş, B., Acar, E., Cemgil, A.T.: Link prediction in heterogeneous data via generalized coupled tensor factorization. Data Min. Knowl. Discov. **29**(1), 203–236 (2015)

6. Liu, W., Chan, J., Bailey, J., Leckie, C., Ramamohanarao, K.: Mining labelled tensors by discovering both their common and discriminative subspaces. In: SIAM International Conference on Data Mining (SDM13), pp. 614–622 (2013)
7. Liu, W., Kan, A., Chan, J., Bailey, J., Leckie, C., Pei, J., Kotagiri, R.: On compressing weighted time-evolving graphs. In: Proceedings of the 21st ACM International Conference on Information and Knowledge Management (CIKM 2012), pp. 2319–2322 (2012)
8. Ristanoski, G., Liu, W., Bailey, J.: Discrimination aware classification for imbalanced datasets. In: Proceedings of the 22nd ACM International Conference on Information and Knowledge Management (CIKM 2013), pp. 1529–1532 (2013)
9. Shashua, A., Hazan, T.: Non-negative tensor factorization with applications to statistics and computer vision. In: Proceedings of the 22th International Conference on Machine Learning (ICML), pp. 792–799 (2005)
10. Sun, J., Papadimitriou, S., Yu, P.: Window-based tensor analysis on high-dimensional and multi-aspect streams. In: Proceedings of IEEE International Conference on Data Mining (ICDM), pp. 1076–1080 (2006)
11. Tucker, L.: Some mathematical notes on three-mode factor analysis. Psychometrika **31**(3), 279–311 (1966)
12. Welling, M., Weber, M.: Positive tensor factorization. Pattern Recogn. Lett. **22**(12), 1255–1261 (2001)
13. Xiong, L., Chen, X., Huang, T., Schneider, J., Carbonell, J.: Temporal collaborative filtering with bayesian probabilistic tensor factorization. In: Proceedings of SIAM Conference on Data Mining (SDM) (2010)
14. Xu, J., Tan, P.N., Luo, L.: Orion: Online regularized multi-task regression and its application to ensemble forecasting. In: Proceedings of the 2014 IEEE International Conference on Data Mining (ICDM), pp. 1061–1066 (2014)
15. Yılmaz, Y., Cemgil, A., Simsekli, U.: Generalised coupled tensor factorisation. In: Advances in Neural Information Processing Systems (NIPS) (2011)

A Non-linear Label Compression Coding Method Based on Five-Layer Auto-Encoder for Multi-label Classification

Jiapeng Luo, Lei Cao, and Jianhua Xu$^{(\boxtimes)}$

School of Computer Science and Technology, Nanjing Normal University,
Nanjing 210023, Jiangsu, China
luojiapeng1993@gmail.com, caolei99@hotmail.com, xujianhua@njnu.edu.cn

Abstract. In multi-label classification, high-dimensional and sparse binary label vectors usually make existing multi-label classifiers perform unsatisfactorily, which induces a group of label compression coding (LCC) techniques particularly. So far, several linear LCC methods have been introduced via considering linear relations among labels. In this paper, we extend traditional three-layer auto-encoder to construct a five-layer one (i.e., five-layer symmetrical neural network), and then apply the training principle in extreme learning machine to determine all network weights. Therefore, a non-linear LCC approach is proposed to capture non-linear relations of labels, where the first three-layer network is regarded as a encoder and the last two layers act as a decoder. The experimental results on three benchmark data sets show that our proposed method performs better than four existing linear LCC methods according to five performance measures.

Keywords: Multi-label classification · Label compression coding · Neural network · Auto-encoder · Extreme learning machine

1 Introduction

Traditional supervised classification deals with problems where every instance is only associated with a single class label and thus the classes are mutually exclusive. However, in many real world applications from text categorization, scene annotation and music emotion classification, any instance possibly belongs to several labels simultaneously. For instance, a sunset image could be annotated by sun, sky and mountain at the same time. Such a classification issue is referred to as multi-label classification [1].

In the recent decade, multi-label classification has attracted a lot of attention, thus a variety of multi-label methods have been proposed. In [1–3], these existing methods are reviewed comprehensively and are partitioned mainly into two

This work was supported by the Natural Science Foundation of China (NSFC) under Grant 61273246.

© Springer International Publishing AG 2016
A. Hirose et al. (Eds.): ICONIP 2016, Part III, LNCS 9949, pp. 415–424, 2016.
DOI: 10.1007/978-3-319-46675-0_45

main groups: problem transformation and algorithm extension. The former mainly converts a multi-label problem into either one or more single-label ones to fit existing single-label classifiers, for example, binary relevance transformation with support vector machine [4], whereas the latter generalizes a multi-class classifier to handle multi-label classification problem directly, for instance, multi-label k-nearest neighbour method [5] and multi-label core vector machine [6].

Nowadays, many multi-label applications confront with high-dimensional and sparse label vectors. That is, the number of class labels is large and the number of relevant labels from each instance is very small. Such a special issue makes the aforementioned classifiers work badly. Therefore dimensionality reduction originally in feature space is also applied to label space, to induce a special kind of label compression coding (LCC) methods [7].

The LCC techniques consist of slightly complicated training and testing procedures, which are embedded encoding and decoding steps respectively. In the training phase, for a training data set, there are two tasks: (a) to learn or choose an encoding way from high-dimensional binary label space to low-dimensional real or binary codeword space and a decoding way reversely, and (b) after converting high-dimensional binary label vectors into low-dimensional codewords, a multi-output regressor or a multi-class classifier is learned. In the testing step, for each testing instance, it is needed to predict its short real or binary codeword, and to recover its long binary label vector using decoding way. It is observed that the crucial work is to determine proper encoding and decoding ways. So far, several linear encoding methods and their corresponding decoding ones have been introduced.

The first LCC method is based on compressive sensing (CS) [7]. Under the label vector sparsity assumption, a compressed real codeword is constructed by a random Hadamard matrix in coding stage, and a 1-norm penalized minimization problem is solved for each testing instance in decoding stage. Compressed labeling technique (CL) [8] encodes original binary label vectors into shorter binary codewords using Gaussian random matrix and sign operation, and decodes binary label vectors from predicted codewords using labelset distilling strategy based on recursive clustering and subtraction. Such two methods do not utilize any label information to construct encoding way. Additionally their decoding is time consuming due to complicated decoding phase.

Principal label space transformation (PLST) [9] constructs coding and decoding matrices using eigenvalue vectors from principal component analysis (PCA) on binary label data only. In [10], Boolean matrix decomposition (BMD) on label data is used to define binary encoding and decoding matrices. These two method only apply label information to encoding and decoding ways,

In order to take feature information into account too, conditional principal label space transformation (CPLST) [11] and feature-aware implicit label space encoding (FaIE)[12] exploit feature and label data simultaneously to build the matrix in eigenvalue problems, and then construct coding and decoding matrices using eigenvalue vectors. In LCCMD [13], according to Hilbert-Schmidt independence criterion (HSIC), global feature-label dependence is maximized to build encoding and decoding matrices. In MDLR [14], local feature-label dependence is considered. It is illustrated that the classification performance could be improved further.

From the above analysis, we find out that these existing LLC methods use a linear encoding way to capture possible linear correlation between labels. It is an open problem to design and implement proper non-linear encoding and decoding techniques to depict possible non-linear correlation among labels. In deep learning, an auto-encoder typically is a three-layer symmetrical neural network, which is divided into encoder and decoder [15–18]. In this study, we extend this traditional auto-encoder to construct a five-layer auto-encoder. Further, we apply the training principle in extreme learning machine [19–21] to train this auto-encoder efficiently. Thus a non-linear LCC method is proposed in this paper. Our experiments on three data sets show that our proposed method performs better than four existing linear LCC techniques including LCCMD, PLST, CPLST and CS, according to five instance-based measures.

2   Preliminaries

Assume an *i.i.d.* training set with L instances to be,

$$\{(\mathbf{x}_1, \mathbf{y}_1), ..., (\mathbf{x}_i, \mathbf{y}_i), ..., (\mathbf{x}_L, \mathbf{y}_L)\}, \tag{1}$$

where $\mathbf{x}_i \in R^D$ and $\mathbf{y}_i \in \{+1, -1\}^C$ indicate that the i-th instance is described using D-dimensional feature vector and C-dimensional binary label one. If $\mathbf{x}_i$ belongs to the k-th class, $y_{ik} = 1$ (i.e., relevant label), and -1 (irrelevant label) otherwise. Additionally, we formulate the above (1) as two matrices,

$$\begin{aligned} \mathbf{X} &= [\mathbf{x}_1, ..., \mathbf{x}_i, ..., \mathbf{x}_L]^T, \\ \mathbf{Y} &= [\mathbf{y}_1, ..., \mathbf{y}_i, ..., \mathbf{y}_L]^T, \end{aligned} \tag{2}$$

where the i-th rows $(\mathbf{x}_i^T)$ and $(\mathbf{y}_i^T)$ in $\mathbf{X}$ and $\mathbf{Y}$ is to describe the i-th instance.

The general multi-label classification learns a classifier: $\mathbf{f}(\mathbf{x}) : R^D \to \{+1, -1\}^C$ directly, which could predict the relevant labels for unseen instances.

In label compression coding (LCC) methods, at its training phase, after we design a proper encoder Ψ and its corresponding decoder Ψ^{-1}, the i-th binary label vector $\mathbf{y}_i$ is converted into a lower c-dimensional real or binary codeword $\bar{\mathbf{y}}_i = \Phi(\mathbf{y}_i)$, where $c < C$. Then, a multi-output regressor or multi-class classifier is trained, i.e., $\mathbf{f}(\mathbf{x}) : R^D \to R^c$ or $\{+1, -1\}^c$. At the testing procedure, for a testing instance $\mathbf{x}$, we predict its codeword $\bar{\mathbf{y}} = \mathbf{f}(\mathbf{x})$ and then recover its binary label vector $\mathbf{y} = \Psi^{-1}(\bar{\mathbf{y}})$. Therefore, the critical task is to build an encoder Φ and its corresponding decoder Φ^{-1}. In the previous work [9–14], linear encoding and decoding forms (i.e., orthonormal matrix and its inverse one) are constructed. In this study, we will implement a non-linear encoding and decoding version.

3 Proposed Method

In this section, we extend traditional three-layer auto-encoder to construct a five-layer one which will be divided into a non-linear encoder and its corresponding decoder. To achieve an efficient training procedure, we apply training principle for extreme learning machine to determine all network weights. Finally, an entire non-linear LCC technique based on five-layer auto-encoder is presented.

3.1 Five-Layer Auto-Encoder

Traditional auto-encoder is a three-layer feed forward symmetric neural network [15–18], which is trained to encode the input $\mathbf{y}$ into some representation $\Psi(\mathbf{y})$ so that the input $\mathbf{y}$ could be decoded from $\Psi^{-1}(\mathbf{y})$. The target output of this auto-encoder is the auto-encoder input itself. The activation functions at the hidden and output layers are linear and/or sigmoid forms. When the linear functions and least-square error criterion are used, the auto-encoder could play the same role as principal component analysis (PCA) does.

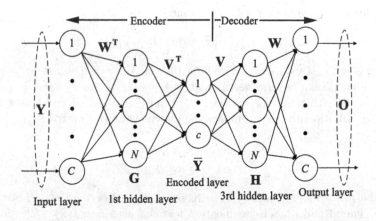

Fig. 1. Five-layer auto-encoder where the first three layers act as an encoder and the last two layers as a decoder

In this study, we generalize such a three-layer auto-encoder to build a five-layer auto-encoder (i.e., a five-layer symmetric feed forward neural network), as shown in Fig. 1, where the first three layers act as an encoder and the last two layers are regarded as a decoder. Note that both its architecture and weight settings are symmetrical.

The input and output layers have the same C (the number of class labels) nodes. The number of nodes at the first and third hidden layers is equal, i.e., N. The encoded layer, that is, the second hidden layer, consists of c nodes, which is also referred to as the length of codewords.

Algorithm 1. Training procedure for five-layer auto-encoder
___
Input:
    $\mathbf{Y} \in [+1, -1]^{L \times C}$: the binary label matrix,
    N: the number of nodes at the first hidden layer,
    c: the number of nodes at the encoded layer,
    K: the maximal number of iterations,
    ε: the error tolerance
Procedure:
    1: Generate $\mathbf{V}$ and $\mathbf{W}^{(0)}$ randomly,
    2: For $p = 1, 2, ..., K$,
        2-1: Calculate the output matrices $\mathbf{G}^{(p)}$, $\bar{\mathbf{Y}}^{(p)}$ and $\mathbf{H}^{(p)}$ using (4),
        2-2: Obtain $\mathbf{W}^{(p)}$ according to (5),
        2-3: if the stopping condition (6) is satisfied, stop the iterative procedure
Output:
    $\mathbf{W}$ and $\mathbf{V}$: two weight matrices

Assume $\mathbf{V} \in R^{c \times N}$ to be a weight matrix between the encoded layer and the third hidden. The weight matrix between the first hidden layer and the encoded one is set to $\mathbf{V}^T$. Let $\mathbf{W} \in R^{N \times C}$ be a weight matrix between the third hidden layer and the output layer. We assign $\mathbf{W}^T$ to be a weight matrix between the input layer and the first hidden layer.

To apply the training principle in extreme learning machine effectively, the linear activation function ($\phi(x) = x$) is used at the encoded layer and output layer, and the tangent sigmoid one ($\phi(x) = 1/(1 + e^{-x})$) is adopted at the first and third hidden layers.

3.2 An Efficient Training Algorithm for Five-Layer Auto-Encoder

In the above five-layer auto-encoder, there are two weight matrices needed to be learned, i.e., $\mathbf{W}$ and $\mathbf{V}$. The widely-used back-propagation technique [15] is time consuming. Therefore, we apply the training principle in extreme learning machine (ELM) [19–21] to determine these two matrices.

The architecture of ELM is a single hidden layer feed forward neural network, i.e., a three-layer neural network, where the hidden and output layers are non-linear and linear respectively. Traditionally, this network is trained by back propagation algorithm, which has a higher time complexity. In ELM, the weight matrix between the input layer and the hidden layer is generated randomly, and the weight matrix between the hidden layer and output is trained via least-square error criterion. Due to its linear output layer, ELM only needs to solve a linear set of equations. Therefore, the training procedure of ELM is extremely efficient.

For the aforementioned five-layer auto-encoder, we generate the weight matrix $\mathbf{V}$ randomly, which is fixed during the entire training procedure. The weight matrix $\mathbf{W}$ is initialized randomly and sequentially is updated iteratively. For our binary label matrix $\mathbf{Y}$, let its output matrices at the three hidden and

---

Algorithm 2. The pseudo-code for LCCAE

===

Training procedure for LCCAE

Input:

   $\mathbf{X} \in R^{L \times D}$: the feature matrix

   $\mathbf{Y} \in \{+1, -1\}^{L \times C}$: the binary label matrix,

   N: the number of nodes at the first hidden layer,

   c: the number of nodes at the encoded layer,

   K: the maximal number of iterations,

   ε: the error tolerance

Procedure:

   1. Obtain $\mathbf{W}$ and $\mathbf{V}$ using **Algorithm 1**,

   2. Estimate the codeword matrix $\bar{\mathbf{Y}}$ using the encoder,

   3. Learn a multi-output linear regressor $\mathbf{f}(\mathbf{x}) : R^D \to R^c$,

Output:

   $\mathbf{W}$ and $\mathbf{V}$: two weight matrics,

   $\mathbf{f}(\mathbf{x})$: the multi-output regressor

===

Testing procedure for LCCAE

Input:

   $\mathbf{W}$ and $\mathbf{V}$: two weight matrics,

   $\mathbf{f}(\mathbf{x})$: the multi-output regressor,

   $\hat{\mathbf{X}} \in R^{M \times D}$: the feature matrix of testing instances

Procedure:

   1. For $i = 1$ to M,

      1-1. Predict the its codeword using $\mathbf{f}(\mathbf{x})$,

      1.2. Calculate its real label vector according to the decoder network,

      1.3. Determine its binary label vector using sign operator

Output:

   $\hat{\mathbf{Y}} \in \{+1, -1\}^{M \times C}$: the predicted binary label matrix for testing instances

---

output layers be $\mathbf{G} \in R^{L \times N}, \bar{\mathbf{Y}} \in R^{L \times c}, \mathbf{H} \in R^{L \times N}$ and $\mathbf{O} \in R^{L \times C}$, which could be calculated as follows,

$$\mathbf{G} = \phi(\mathbf{Y}\mathbf{W}^T), \bar{\mathbf{Y}} = \mathbf{G}\mathbf{V}^T, \mathbf{H} = \phi(\bar{\mathbf{Y}}\mathbf{V}), \mathbf{O} = \mathbf{H}\mathbf{W}, \tag{3}$$

where $\bar{\mathbf{Y}}$ denotes our real codewords, and ϕ represents the tangent sigmoid activation function. In this case, we have the following linear set of equations,

$$\mathbf{H}\mathbf{W} = \mathbf{Y}. \tag{4}$$

According to least-square error criterion, the weight matrix $\mathbf{W}$ becomes,

$$\mathbf{W} = \mathbf{H}^+\mathbf{Y}. \tag{5}$$

where $\mathbf{H}^+$ is the Moore-Penrose generalized inverse of matrix $\mathbf{H}$. This optimization procedure is repeated until the maximal number of iterations (K) is reached or the following stopping condition is satisfied,

$$\left\| \mathbf{W}^{(p)} - \mathbf{W}^{(p-1)} \right\|_F \leq \varepsilon, \tag{6}$$

where $p = 1, 2, ..., K$ denotes the iterative index, $\|.\|_F$ is the Frobenius norm of matrix and ε represents an pre-defined error tolerance. Finally we summarize this training procedure for five-layer auto-encoder in the **Algorithm 1**.

3.3   A Non-linear Label Compression Coding Method for Multi-label Classification

In the above two sub-sections, we introduce a five-layer auto-encoder and its efficient training procedure. In this sub-section, we summarize our entire multi-label classification algorithm as the **Algorithm 2**, which is referred to as a non-linear label compression coding methods based on five-layer auto-encoder for multi-label classification, simply LCCAE.

4   Experiments

In this section, we experimentally evaluate our LCCAE and compare it with four existing methods including LCCMD [13], PLST [9], CPLST [11] and CS [7].

4.1   Experimental Settings

To verify the effectiveness of our proposed method, we download three indicative data sets: Cal500, Corel5K and Delicious from [22], as shown in Table 1, which cover three different application domains: music, image and text. Table 1 also lists some useful statistics of these data sets, such as, the number of training instances, testing instances, features and class labels, and label cardinality and density.

In this study, we evaluate and compare the performance of our proposed method LCCAE and four existing approaches according to five multi-label performance measures (Hamming loss, accuracy, F1, precision and recall). Please refer to their detailed definitions in [2,3]. Among these measures, except for Hamming loss, the higher the other measures are, the better the classifier performs.

Additionally, we use linear ridge regression as a baseline method in all LCC techniques. The length of codeword (i.e., the number of nodes at the encoded layer) is set to be $c = C/2$. For our LCCAE, two stopping conditions are set to be $K = 100$ and $\epsilon = 10^{-6}$, and the number of nodes (N) at the first hidden layer is tuned, which is denoted by #Nodes in Table 1. For CS, its Hadamard matrix

Table 1. Some useful statistics from three data sets in our experiments

Data set	Domain	#Train	#Test	#Feature	#Label	Label cardinality	Label density	#Nodes
Cal500	Music	300	202	68	174	26.044	0.150	67
Corel5k	Image	4500	500	499	374	3.522	0.009	101
Delicious	Text	12920	3185	500	983	19.020	0.019	380

is used as the random coding matrix, and reconstructed algorithm is orthogonal matching pursuit algorithm (OMP) [7].

To sort the performance of multiple classification methods for some given data set, we rank all methods for some measure, in which the best performing method gets the rank of 1, the second best rank 2, and in case of ties average ranks are assigned. Then the average rank of each method over all five measures and overall average rank over all data sets are calculated. This performance comparison is recommended in [23].

4.2   Performance Comparison

In this sub-section, on three data sets in Table 1, we compare our LCCAE with LCCMD, PLST, CPLST, and CS. The original linear regression method without label coding phase is also considered, denoted by RR simply. The experimental results are listed in Table 2, where the best value among each measure is

Table 2. Experimental results from three data sets

Measure	LCCAE	LCCMD	PLST	CPLST	CS	RR
Cal500						
Hamming loss ↓	15.36(2)	16.00(5)	15.87(3)	16.00(5)	**14.99**(1)	16.00(5)
Accuracy ↑	**23.44**(1)	22.62(4)	22.87(2)	22.62(4)	21.78(6)	22.62(4)
F1 ↑	**36.95**(1)	36.01(5)	36.29(2)	36.01(5)	35.09(3)	36.01(5)
Recall ↑	**31.08**(1)	30.23(4)	30.39(2)	30.23(4)	27.39(6)	30.23(4)
Precision ↑	49.36(2)	48.47(5)	48.96(3)	48.47(5)	**52.96**(1)	48.47(5)
Average rank↓	**1.4**	4.6	2.4	4.6	3.4	4.6
Corel5K						
Hamming loss ↓	1.03(6)	**0.94**(3)	**0.94**(3)	**0.94**(3)	**0.94**(3)	**0.94**(3)
Accuracy ↑	**11.30**(1)	7.42(3)	7.31(6)	7.40(4)	7.43(2)	7.37(5)
F1 ↑	**16.21**(1)	10.91(2)	10.76(6)	10.88(4)	10.90(3)	10.83(5)
Recall ↑	**13.83**(1)	7.95(2.5)	7.85(6)	7.95(2.5)	7.93(4)	7.90(5)
Precision ↑	**24.28**(1)	20.45(3)	20.10(6)	20.32(4.5)	20.62(2)	20.32(4.5)
Average rank ↓	**2.0**	2.7	5.4	3.6	2.8	4.5
Delicious						
Hamming loss ↓	1.82(4)	1.82(4)	1.82(4)	1.82(4)	1.82(4)	1.81(1)
Accuracy ↑	**11.38**(1)	10.61(3.5)	10.61(3.5)	10.61(3.5)	10.37(6)	10.61(3.5)
F1 ↑	**17.76**(1)	16.51(3)	16.50(5)	16.51(3)	16.20(6)	16.51(3)
Recall ↑	**12.99**(1)	11.93(3.5)	11.93(3.5)	11.93(3.5)	11.68(6)	11.93(3.5)
Precision ↑	**44.47**(1)	42.16(3.5)	42.16(3.5)	42.16(3.5)	42.05(6)	42.16(3.5)
Average rank ↓	**1.6**	3.5	3.9	3.5	5.6	2.9
Overall rank ↓	**1.67**	3.60	3.90	3.9	3.93	4.00

highlighted in boldface and the ranks are denoted by 1–6 in the brackets. The average rank of each method over five measures is shown too.

For Cal500, our LCCAE works the best on accuracy, F1 and recall, and obtains the second top rank on Hamming loss and precision. According to the average rank, six methods are sorted as LCCAE, PLST, CS, and LCCMD, CPLST and RR, where the last three techniques perform equally.

As to Corel15K, our LCCAE is superior to the other five methods on all measures but Hamming loss. From the average rank, our LCCAE performs the best and PLST the worst.

On Delicious, our LCCAE achieves the best measure values on all measures but Hamming loss. According to the average rank, our LCCAE obtains the top rank and CS the low one.

Additionally, we note that the Hamming loss is insensitive when the number of labels is very high, since the difference among six techniques from three data sets is insignificant.

To evaluate these methods comprehensively, we calculate the overall average rank over three data sets, as shown in the last row in Table 2, and then sort them as LCCAE(1.67), LCCMD(3.60), PLST and CPLST (3.90), CS(3.93) and RR (4.00). It is observed that our LCCAE performs the best and PLST, CPLST and CS with a half of the number of original labels achieve the approximate performance to RR without any label compression trick. Therefore the label compression coding trick indeed improves the performance of multi-label classification applications with high dimensional and sparse label vectors.

5   Conclusion

In this paper, to deal with high dimensional and sparse label vectors, we propose a nonlinear label compression coding technique based on five-layer auto-encoder for multi-label classification, to capture the non-linear correlations among labels. The classical three-layer auto-encoder is generalized to build a five-layer ones, and then the training principle in extreme learning machine is applied to determine all network weights efficiently. Our experiments on three benchmark data sets illustrate that our proposed method is superior to four existing approaches including principal label space transformation and its condition version, label compression coding via maximizing dependence, and compressive sensing. In future, we will validate more benchmark data sets and search for more efficient training procedure.

References

1. Tsoumakas, G., Katakis, I.: Multi-label classification: an overview. Int. J. Data Warehouse. Min. **3**(3), 1–13 (2007)
2. Zhang, M.L., Zhou, Z.H.: A review on multi-label learning algorithms. IEEE Trans. Knowl. Data Eng. **26**(8), 1338–1351 (2014)

3. Gibaji, E., Ventura, S.: A tutorial on multilabel learning. ACM Comput. Surv. **47**(3), 52:1–52:38 (2015)
4. Boutell, M.R., Luo, J., Shen, X., Brown, C.M.: Learning multi-label scene classification. Pattern Recogn. **37**(9), 1757–1771 (2004)
5. Zhang, M.L., Zhou, Z.H.: ML-kNN: a lazy learning approach to multi-label learning. Pattern Recogn. **40**(5), 2038–2048 (2007)
6. Xu, J.: Multi-label core vector machine with a zero label. Pattern Recogn. **47**(7), 2542–2557 (2014)
7. Hsu, D., Kakade, S.M., Langfors, J.L., Zhang, T.: Multi-label prediction via compressed sensing. In: Proceedings of the 24th Conference on Neural Information Processing Systems (NIPS2009), pp. 772–780. MIT Press, Cambridge (2010)
8. Zhou, T., Tao, D., Wu, X.: Compressed labeling on distilled labelsets for multi-label learning. Mach. Learn. **88**(1/2), 69–126 (2012)
9. Tai, F., Lin, H.T.: Multi-label classification with principal label space transformation. Neural Comput. **24**(9), 2508–2542 (2012)
10. Wicker, J., Pfahringer, B., Kramer, S.: Multi-label classification using Boolean matrix decomposition. In: Proceedings of the 27th Annual ACM Symposium on Applied Computing (SAC2012), pp. 179–186. ACM, New York (2012)
11. Chen, Y., Lin, H.T.: Feature-aware label ppace dimension reduction for multi-label classification. In: Proceedings of the 27th Conference on Neural Information Processing Systems (NIPS2012), pp. 1538–1546. MIT Press, Cambridge (2012)
12. Lin, Z., Ding, G., Hu, M., Wang, J.: Multi-label classification via feature-aware implicit label space encoding. In: Proceedings of the 31th International Conference on Machine Learning (ICML2014), pp. 325–333. Microtome Publishing, Brookline (2014)
13. Cao, L., Xu, J.: A label compression coding approach through maximizing dependence between features and labels for multi-label classification. In: Proceedings of the 2015 International Joint Conference on Neural Networks (IJCNN2015), pp. 1–8. IEEE Press, New York (2015)
14. Zhang, J.J., Fang, M., Wang, H., Li, X.: Dependence maximization based label space dimension reduction for multi-label classification. Eng. Appl. Artif. Intell. **45**, 453–463 (2015)
15. Rumelhart, D.E., Hinton, G.E., Williams, R.J.: Leraning representations by back-propagating error. Nature **323**, 533–536 (1986)
16. Bengio, Y.: Learning deep architectures for AI. Found. Trends Mach. Learn. **2**(1), 1–127 (2009)
17. Bengio, Y., Courville, A., Vincent, P.: Representation learning: a review and new perspectives. IEEE Trans. Pattern Anal. Mach. Intell. **35**(8), 1798–1828 (2013)
18. LeCun, Y., Bengio, Y., Hinton, G.: Deep learning. Nature **521**, 436–444 (2015)
19. Huang, G.B., Zhu, Q.Y., Siew, C.K.: Extreme learning machine: theory and applications. Neurocomputing **70**(1–3), 489–501 (2006)
20. Huang, G.B., Wang, D.H., Lan, Y.: Extreme learning machine: a survey. Int. J. Mach. Learn. Cybern. **2**(2), 107–122 (2011)
21. Huang, G., Huang, G.B., Song, S., You, K.: Trends in extreme learning machines: a review. Neural Netw. **61**, 32–48 (2015)
22. Multi-label data sets. http://mulan.sourceforge.net/datasets-mlc.html
23. Brazdil, P.B., Soares, C.: A comparison of ranking methods for classification algorithm selection. In: Lopez de Mantaras, R., Plaza, E. (eds.) ECML 2000. LNCS (LNAI), vol. 1810, pp. 63–74. Springer, Heidelberg (2000)

Fast Agglomerative Information Bottleneck Based Trajectory Clustering

Yuejun Guo[1,2], Qing Xu[1(✉)], Yang Fan[1], Sheng Liang[1], and Mateu Sbert[1,2]

[1] School of Computer Science and Technology, Tianjin University, Tianjin, China
qingxu@tju.edu.cn
[2] Graphics and Imaging Laboratory, Universitat de Girona, Girona, Spain

Abstract. Clustering is an important data mining technique for trajectory analysis. The agglomerative Information Bottleneck (aIB) principle is efficient for obtaining an optimal number of clusters without the direct use of a trajectory distance measure. In this paper, we propose a novel approach to trajectory clustering, fast agglomerative Information Bottleneck (faIB), to speed up aIB by two strategies. The first strategy is to do "clipping" based on the so-called feature space, calculating information losses only on fewer cluster pairs. The second is to select and merge more candidate clusters, reducing iterations of clustering. Remarkably, faIB considerably runs above 10 times faster than aIB achieving almost the same clustering performance. In addition, extensive experiments on both synthetic and real datasets demonstrate that faIB performs better than the clustering approaches widely used in practice.

Keywords: aIB · faIB · Trajectory clustering · Speedup

1 Introduction

Clustering is one of the most important techniques to analyze the important trajectory data [8,10]. Over the past few years, a lot of approaches to trajectory clustering have been developed and demonstrated to work well and, we advise the readers of interest to refer to a related survey [13].

Basically, all the trajectory clustering methods can be approximately divided into supervised and unsupervised categories [10], and the latter type has become popularly applied in many situations [13]. k-means and its modified versions, such as k-medoids [3], which are the most classical schemes, can be easily used. Some of the most widely used methods in practice, such as Density-Based Spatial Clustering of Applications with Noise (DBSCAN) [6] and agglomerative hierarchical clustering (aHC) [1], can do the trajectory clustering very efficiently. Some state-of-the-art methods based on probabilistic representations of trajectory data

This work has been funded by Natural Science Foundation of China (61471261, 61179067, U1333110), and by grants TIN2013-47276-C6-1-R from Spanish Government and 2014-SGR-1232 from Catalan Government (Spain).

A. Hirose et al. (Eds.): ICONIP 2016, Part III, LNCS 9949, pp. 425–433, 2016.
DOI: 10.1007/978-3-319-46675-0_46

[3] can manage the complex trajectory data very well. The latest approaches capable of online updating of trajectory clusters [8] have been paid attractions.

Unfortunately, there exist two obvious but difficult issues for trajectory clustering methods. First, to generally determine an optimal number of the trajectory clusters remains a big challenge [5]. Second, to define an appropriate distance measure between trajectories is still open [11]. Importantly, a very recent technique has emerged to adaptively obtain an optimal number of the clusters without the direct use of a distance measure for trajectories [5], based on utilizing the agglomerative Information Bottleneck (aIB) [4] tool in the powerful information theory literature. In this paper, we pay much attention to the fast run of the clustering procedure. It is worthy to point out that an efficient and effective method for trajectory clustering is quite meaningful in many applications, such as the centralized traffic management systems, for a quickly and accurately strategic decision coping with the potential threat.

In this paper, a novel and fast unsupervised clustering method, called as faIB, is proposed to speed up the aIB based trajectory clustering and we summarize the contributions as follows. Firstly, we obtain the trajectory representations using Discrete Fourier Transform (DFT). And then we take advantage of the aIB principle from information theory and, inspired by the two observations in the iterative process of aIB, we improve the aIB by utilizing two strategies to speed up the trajectory clustering. The first strategy is the "clipping" based on the relevant variable, the so-called feature space F (see more details in Sect. 3) and, the second is the selecting more candidate clusters. We conduct extensive experiments to demonstrate our proposed approach.

2   Data Preprocessing Based on DFT

For a trajectory $T_i = \{x_1, y_1, x_2, y_2, \ldots, x_m, y_m\}, i = 1, 2, \ldots, n$, consisting of planar position values, where m and n are respectively the total numbers of the sampling points on T_i and of the trajectories in the dataset, we obtain the Fourier coefficient for its each sampling point [6]

$$R_t = \frac{1}{\sqrt{m}} \sum_{k=1}^{m} (x_k + \mathrm{j}y_k)e^{-\mathrm{j}kt2\pi/m}, t = 1, 2, \ldots, m. \tag{1}$$

Generally, low frequency coefficients, distributing at both ends of the sequence of all the coefficients, provide a global information of the trajectory. As a result, a limited $2u$ number of low frequency coefficients, being the former and latter u coefficients, can be used to represent the original trajectory. By the inverse DFT

$$r_t = \frac{1}{\sqrt{2u}} \left[\sum_{k=1}^{u} R_k e^{\mathrm{j}kt\pi/u} + \sum_{k=m-u+1}^{m} R_k e^{\mathrm{j}kt\pi/u} \right], t = 1, 2, \ldots, m, \tag{2}$$

the reconstructed version of T_i, $T_i^u = \{x_1^u, y_1^u, x_2^u, y_2^u, \ldots, x_t^u, y_t^u, \ldots, x_m^u, y_m^u\}$, is obtained. Here x_t^u and y_t^u are the real and imaginary parts of r_t, respectively.

Based on the distance between T_i and T_i^u,

$$Dis(T_i, T_i^u) = \sqrt{\frac{1}{m} \sum_{t=1}^{m} [(x_t - x_t^u)^2 + (y_t - y_t^u)^2]}, \qquad (3)$$

an optimum number u_i^{opt} of coefficients for T_i is obtained by

$$u_i^{opt} = \underset{u \in \{1,2,\ldots,\lfloor \frac{m}{2} \rfloor\}}{\text{argmin}} [G1 + G2], \qquad (4)$$

where $G1 = \frac{Dis(T_i, T_i^u)}{Dis(T_i, T_i^1)}$ and $G2 = \frac{\int_1^u Dis(T_i, T_i^z)\, dz}{\int_1^{\lfloor \frac{m}{2} \rfloor} Dis(T_i, T_i^z)\, dz}$ (the denominators are the normalization factors). As exemplified in Fig. 1, the distance between T_i and T_i^u decreases with the increasing number of coefficients used (the green plot $G1$), while the area resulted by the distance plot increases gradually (the pink area covered by $G2$). Thus, u_i^{opt} can be determined by the minimization of $G1 + G2$. For simplicity, an optimal number u^{opt} of coefficients usable for all the trajectories can be set as the maximum value of $\{u_i^{opt}\}, i = 1, 2 \ldots, n$, achieving the least possible error by the DFT on trajectory data using the smallest possible number of coefficients.

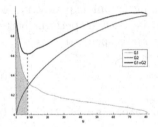

Fig. 1. Determination of the optimal number of coefficients for a trajectory. (Color figure online)

Fig. 2. A feature space F of the real trajectory data.

3 Trajectory Clustering Based on FaIB

According to the Information Bottleneck (IB) principle [4,5], given a dataset X, an optimal number of clusters C, which furthest compact the dataset X and at the same time maximize the information shared between C and the so-called feature space F, are obtained by

$$\min\{I(C; X) - \beta I(C; F)\}, \qquad (5)$$

where β denotes the Lagrange multiplier controlling the trade-off between data compression and information keeping. In this paper, we improve on an extension of IB, namely the aIB which employs a greedy manner to iteratively merge two candidate clusters [4], leading to the faIB technique for the purpose of considerably speeding up the aIB driven trajectory clustering [5].

3.1 The Feature Space and Conditional Probability Distribution

First of all, obtaining a feature space F for the trajectory dataset X is important for the speedup of our faIB approach and, this is usually easy and direct. As illustrated by a real data example in Fig. 2 (see details on this data in Sect. 4), there exist several paths in a small square and the trajectories of pedestrians passing through this square happen to these paths. Hence, in this case, the feature space F can be obtained by partitioning the trajectories according to the paths to which they happen.

Obviously, the feature space F has a clear correlation with the dataset X, namely reflecting the feature information of X, which is the key to build the channel between X and F and to calculate the necessary conditional probability $p(f_i|x), x \in X, f_i \in F$ ($F = \{f_1, f_2, \ldots, f_m\}$). For a given trajectory x, the distances, $D = \{d_1, d_2, \ldots, d_m\}$, of it with the centers of F are calculated with the traditional and fast distance measure, Euclidean distance, which is usually used in the frequency domain [6]. Here, the center of f_i is identified as the one that results in the minimum sum distances between it and the others within f_i. We design the conditional probability to be inverse related with this distance. In fact, the larger the distance between x and $f_{icenter}$ is, the smaller the probability $p(f_i|x), i \in \{1, 2, \ldots, m\}$ is. To this end, we use the exponential function to do the transformation for each distance d_i, namely e^{-d_i}. Notice that, the exponent function can be used to achieve this well, according to our experimentation. In fact, some other mathematical functions could be tried to perform better. Finally, the conditional probability that relates x to y_i is specified as

$$p(f_i|x) = \frac{e^{-d_i}}{\sum_{k=1}^{m} e^{-d_k}}, \tag{6}$$

where $\sum_{k=1}^{m} e^{-d_k}$ is a normalization factor. In this paper, we use $p(x) = 1/n$, where n is the total number of trajectories in X.

As mentioned above, in each iteration of the aIB procedure [4], two candidate clusters are merged if the mutual information loss induced by merging them is minimum. Initially, each trajectory is specified into a single cluster, namely $C = X$. Given two clusters $c_1, c_2 \in C$, the information loss caused by the merging of c_1 and c_2 is, as [4] indicated,

$$\Delta I(c_1, c_2) = I(C_{bef}) - I(C_{aft}) = \sum_{f_i \in F, k=1,2} p(c_k, f_i) \log \frac{p(f_i|c_k)}{p(f_i|c_{merge})} \tag{7}$$
$$= \sum_{f_i \in F, k=1,2} p(c_k) D_{KL}(p(f_i|c_k)||p(f_i|c_{merge})).$$

Here, $I(C_{bef})$ and $I(C_{aft})$ indicate the clusters before and after merging, respectively. $D_{KL}(v||w)$ is the Kullback-Leibler divergence [4]. c_{merge} is the new cluster generated by merging c_1 and c_2 and, the updated conditional probability of c_{merge} is computed by

$$p(f_i|c_{merge}) = \frac{p(c_1)}{p(c_{merge})} p(f_i|c_1) + \frac{p(c_2)}{p(c_{merge})} p(f_i|c_2) \tag{8}$$

where $p(c_{merge}) = p(c_1) + p(c_2)$.

3.2 The Speeding Up Strategies

We speed up the aIB based trajectory clustering by using two strategies from two perspectives: the clipping based on feature space for calculating information losses on fewer cluster pairs, and the selecting more candidate clusters for merging to reduce iterations. The two strategic methods are utilized in sequence, and we call this the faIB in our paper.

Strategy 1: clipping based on feature space. Based on lots of experiments on different datasets, we have found that the trajectories in a same final cluster also usually belong to a subset of the feature space. Actually, this is because that, based on a feature space F, the aIB principle obtains an optimal number of clusters through maximizing the mutual information of F and C, leading to a high similarity between them. Considering this, we limit the trajectories in the two candidate clusters to come from the "connected" subset f_j^{con} of F (assuming $F = \{f_j^{con}\}, j = 1, 2, \ldots, n_c$, here n_c is the number of "connected" subsets in F), instead of selecting the candidate clusters with respect to the whole set of F as aIB does. In this case, the calculation of information loss caused by merging each two clusters which include the trajectories, respectively, belonging to f_j^{con} and $F - f_j^{con}$ is "clipped" out. Thus, we call this strategy as "clipping" based on the feature space F. In this paper, if the elements in F, f_1 and f_2, are adjacent, they are considered as "connected" and $\{f_1, f_2\} \in f_j^{con}$. Note that, we deem that f_i $(f_i \in F)$ is "connected" with itself. Therefore, the information loss resulted from merging c_1 and c_2 is defined by faIB as

$$\Delta I'(c_1, c_2) = \sum_{f_i \in f_j^{con}, k=1,2} p(c_k) D_{KL}(p(f_i|c_k)||p(f|c_{merge})), \qquad (9)$$

where the trajectories in c_1 and c_2 belong to f_j^{con}. Apparently, the "clipping" can very effectively reduce the computational complexity by calculating the information loss on fewer cluster pairs. For example, as shown in Fig. 3, we only calculate the information losses on four cluster pairs, $\{(c_1, c_2), (c_1, c_3), (c_2, c_3), (c_4, c_5)\}$, by faIB in an iteration, which is much fewer than those on ten pairs, $\{(c_1, c_2), (c_1, c_3), (c_1, c_4), (c_1, c_5), (c_2, c_3), (c_2, c_4), (c_2, c_5), (c_3, c_4), (c_3, c_5), (c_4, c_5)\}$, by aIB.

Strategy 2: selecting more candidate clusters. It is a common observation that the information loss caused by merging only two candidate clusters in each iteration changes very small, until the iteration number reaches a "break point", which in essential indicates an optimal number of final clusters as well as a termination condition for the iteration. Thus, we propose to merge more than two candidate clusters in an iteration, to terminate the iterative clustering procedure as fast as possible. Importantly, only the candidate clusters that belong to a same final cluster should be merged, otherwise an error will occur to the clustering. For this purpose, in an iteration of faIB, we propose that each pair of the candidate clusters $\{c_i, c_j\}$ $(i \neq j)$ to be merged should meet $\Delta I'(c_i, c_j) \leq \lambda \cdot \min \Delta I'$, where $\min \Delta I'$ is the minimum information loss by merging two clusters in this iteration, and λ is a predefined value to ensure the clusters to be merged are

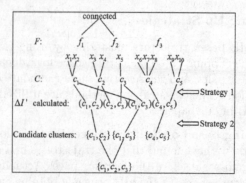

Fig. 3. An example of the speeding up in an iteration in the procedure of faIB. In this example, the trajectory data $X = \{x_1, x_2, \ldots, x_{10}\}$, and the feature space $F = \{f_1, f_2, f_3\} = \{\{x_1, x_2, x_3\}, \{x_4, x_5\}, \{x_6, x_7, x_8, x_9, x_{10}\}\} = \{f_1^{con}, f_2^{con}\}$, here the connected subsets $f_1^{con} = \{f_1, f_2\}$ and $f_2^{con} = f_3$.

from the same final cluster. Generally, a good λ is experimentally set between 1 and 1.1. Finally, if the candidate clusters include the same element then we merge them for speeding up. For example, $\{c_1, c_2\}$ and $\{c_1, c_3\}$ are combined as $\{c_1, c_2, c_3\}$ (see the bottom of Fig. 3). As a result, in contrast to the traditional aIB, faIB speeds up the trajectory clustering by, if possible, merging more than two candidate clusters in each iteration.

4   Experimental Results and Discussions

Extensive experiments on 3 real (edb[1] [9], cross and lab[2] [12]) and 49 synthetic[3] [14] datasets are carried to evaluate the performance of our proposed approach. We compare faIB with aIB, the method in [6] and popular aHC [1]. Additionally, we employ two widely used distance measures, Euclidean (EU) and Hausdorff (Haus) distances [7], for both the method in [6] and aHC. All the parameters in [6] and aHC are tuned to obtain their best possible results. The experiments are conducted on a Windows PC with Intel(R) Core(TM) 4 Duo processor i5-3570, 3.40 GHz and 4 GB RAM.

For reasons of space, only the visual results on the real dataset edb are presented, as shown in Fig. 4. In this experimentation, Fig. 4(a) and (b) depict almost the same clusters attained by aIB and faIB algorithms, respectively. Significantly, the runtime by faIB is almost 10 times smaller than that by aIB, as shown in Table 1. Compared with [6] (Fig. 4(c) and (d)) and AHC (Fig. 4(e) and (f)), obviously, our results are better, because [6] and AHC cannot give the correct clusters. For instance, the trajectories visualized in red and yellow colors by faIB (Fig. 4(b)) have two directions in this path and should be partitioned,

---

[1] http://homepages.inf.ed.ac.uk/rbf/forumtracking.
[2] http://cvrr.ucsd.edu/bmorris/datasets/dataset_trajectory_analysis.html.
[3] http://avires.dimi.uniud.it/papers/trclust.

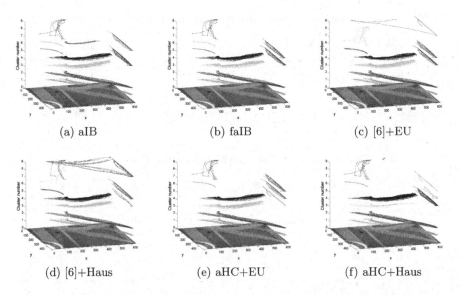

(a) aIB   (b) faIB   (c) [6]+EU

(d) [6]+Haus   (e) aHC+EU   (f) aHC+Haus

Fig. 4. Comparative clustering results of the real data by six approaches. aIB and faIB obtain almost the same clusters. (Color figure online)

however, [6] mixes them in red color (Fig. 4(c)). Similarly, these trajectories in Fig. 4(d) is also with the wrong "mix" problem. Moreover, there are several trajectories in Fig. 4(c) and (d) incorrectly grouped into the cluster in dark blue color. This is due to that the clustering algorithm used in [6], DBSCAN, has to use an explicit measure of the trajectory distance which is usually very difficult to compute [11]. Additionally, in Fig. 4(e) and (f), the trajectories in yellow color essentially have different directions but are mixed together by aHC. Likewise, this mistaken separation problem also happens to the trajectories, which should be two clusters (see blue and rose trajectories in Fig. 4(b)). This is because that aHC highly relies on the distance between trajectories and the distance is very sensitive to the various trajectory lengths. By contrast, the approaches based on IB avoid the direct use of the trajectory distance, achieving largely better performance (Table 2).

Table 1 shows the runtimes by aIB and faIB for 52 datasets. Due to space limit, the 49 synthetic datasets are given by 5 mean results of 5 sets of data. For example, "TS(1-10)" means the average results of the datasets from 1 to 10. Apparently, faIB reduces the iterations and largely improves the run speed of aIB. In addition, we make use of two metrics, Precision and Recall [2], to measure the clustering quality quantitatively. As demonstrated in Table 2, aIB and faIB achieve better performances than [6] and aHC, which is consistent with the visual clustering results in Fig. 4.

Table 1. Comparisons of runtime and iteration number by aIB and faIB.

	faIB		aIB			faIB		aIB	
	Time (s)	Iterations	Time (s)	Iterations		Time (s)	Iterations	Time (s)	Iterations
corss	1.12	129	23.194	439	lab	0.16	76	1.035	107
edb	0.301	123	2.83	199	TS(1–10)	0.361	191	5.751	244
TS(11–20)	0.330	194	5.342	244	TS(21–30)	0.383	191	5.707	244
TS(31–40)	0.342	192	5.355	244	TS(41–49)	0.342	192	5.264	244

Table 2. Comparison results on Precision and Recall of six approaches.

	faIB		aIB		[6]+EU		[6]+Haus		aHC+EU		aHC+Haus	
	Precision	Recall	Precision	Recall	Precision	Recall	Precision	Recall	Precision	Recall	Precision	Recall
corss	0.980	0.983	0.979	0.984	0.750	0.745	0.735	0.691	0.695	0.689	0.742	0.733
edb	1.000	1.000	1.000	1.000	0.846	0.862	0.720	0.680	0.846	0.870	0.552	0.567
lab	0.877	0.890	0.876	0.890	0.804	0.808	0.553	0.451	0.719	0.730	0.625	0.565
TS(1–10)	0.998	0.992	0.998	0.987	0.920	0.900	0.894	0.877	0.951	0.936	0.886	0.901
TS(11–20)	0.990	0.993	0.990	0.993	0.906	0.890	0.886	0.867	0.962	0.956	0.833	0.865
TS(21–30)	0.990	0.989	0.988	0.985	0.934	0.907	0.847	0.864	0.933	0.920	0.879	0.869
TS(31–40)	0.997	0.991	0.997	0.990	0.956	0.951	0.864	0.893	0.904	0.886	0.903	0.892
TS(41–49)	1.000	1.000	1.000	1.000	0.969	0.965	0.869	0.886	0.956	0.942	0.888	0.869

5   Conclusions

In this paper, a fast trajectory clustering approach based on the aIB principle, called faIB, has been developed, and extensive experiments on real and synthetic datasets have demonstrated the effectiveness of the new technique. Thanks to the two proposed strategies for speeding up, faIB can significantly increase the clustering speed above 10 times than aIB and meanwhile keeping the clustering performance of aIB. Moreover, compared with the state-of-the-art method in [6] and the popularly applied aHC in many practical scenarios, our approach behaves much better. Due to its fast running, we believe that the proposed approach should be especially meaningful for dealing with large scale data.

For the future work, online learning is also important for trajectory clustering [8], we will extend our scheme as an online learning version. Additionally, we intend to design a graphical interface for deeply understanding the clustering procedure, allowing the users to observe and analyze the (intermediate) results of interest.

References

1. Aggarwal, C.C., Reddy, C.K.: Data Clustering: Algorithms and Applications. CRC Press, Boca Raton (2013)
2. Anjum, N., Cavallaro, A.: Multifeature object trajectory clustering for video analysis. IEEE Trans. Circ. Syst. Video Technol. **18**(11), 1555–1564 (2008)
3. Calderara, S., Prati, A., Cucchiara, R.: Mixtures of von mises distributions for people trajectory shape analysis. IEEE Trans. Circ. Syst. Video Technol. **21**(4), 457–471 (2011)

4. Goldberger, J., Gordon, S., Greenspan, H.: Unsupervised image-set clustering using an information theoretic framework. IEEE Trans. Image Process. **15**(2), 449–458 (2006)
5. Guo, Y., Xu, Q., Yang, Y., Liang, S., Liu, Y., Sbert, M.: Anomaly detection based on trajectory analysis using kernel density estimation and information bottleneck techniques. Technical report 108, University of Girona (2014)
6. Annoni Jr., R.A., Forster, C.H.Q.: Analysis of aircraft trajectories using fourier descriptors and kernel density estimation. In: Proceedings of the 15th International IEEE Conference on Intelligent Transportation Systems, pp. 1441–1446 (2012)
7. Junejo, I.N., Javed, O., Shah, M.: Multi feature path modeling for video surveillance. In: Proceedings of the 17th International Conference on Pattern Recognition, vol. 2, pp. 716–719. IEEE (2004)
8. Laxhammar, R., Falkman, G.: Online learning and sequential anomaly detection in trajectories. IEEE Trans. Pattern Anal. Mach. Intell. **36**(6), 1158–1173 (2014)
9. Majecka, B.: Statistical models of pedestrian behaviour in the forum. Master's thesis, School of Informatics, University of Edinburgh (2009)
10. Mitsch, S., Müller, A., Retschitzegger, W., Salfinger, A., Schwinger, W.: A survey on clustering techniques for situation awareness. In: Ishikawa, Y., Li, J., Wang, W., Zhang, R., Zhang, W. (eds.) APWeb 2013. LNCS, vol. 7808, pp. 815–826. Springer, Heidelberg (2013)
11. Morris, B., Trivedi, M.: Learning trajectory patterns by clustering: experimental studies and comparative evaluation. In: Proceedings of the IEEE Conference on Computer Vision and Pattern Recognition, pp. 312–319 (2009)
12. Morris, B.T., Trivedi, M.M.: Trajectory learning for activity understanding: unsupervised, multilevel, and long-term adaptive approach. IEEE Trans. Pattern Anal. Mach. Intell. **33**(11), 2287–2301 (2011)
13. Morris, B.T., Trivedi, M.M.: Understanding vehicular traffic behavior from video: a survey of unsupervised approaches. J. Electron. Imaging **22**(4), 041113–041113 (2013)
14. Piciarelli, C., Micheloni, C., Foresti, G.L.: Trajectory-based anomalous event detection. IEEE Trans. Circ. Syst. Video Technol. **18**(11), 1544–1554 (2008)

Anomaly Detection Using Correctness Matching Through a Neighborhood Rough Set

Pey Yun Goh[✉], Shing Chiang Tan, and Wooi Ping Cheah

Multimedia University, Jln. Ayer Keroh Lama, 75450 Melaka, Malaysia
{pygoh, sctan, wpcheah}@mmu.edu.my

Abstract. Abnormal information patterns are signals retrieved from a data source that could contain erroneous or reveal faulty behavior. Despite which signal it is, this abnormal information could affect the distribution of a real data. An anomaly detection method, i.e. Neighborhood Rough Set with Correctness Matching (NRSCM) is presented in this paper to identify abnormal information (outliers). Two real-life data sets, one mixed data and one categorical data, are used to demonstrate the performance of NRSCM. The experiments positively show good performance of NRSCM in detecting anomaly.

Keywords: Neighborhood · Rough set · Anomaly detection · Outlier detection

1 Introduction

Anomaly discovery is a fundamental research problem in data mining [1]. Information due to anomaly could be either useful or meaningless. Anomaly is inherently embedded in a data set. We should accept the existence of anomaly but at the same time, finding ways to reduce its impacts. Various kinds of anomaly detection methods have been proposed. Some researchers worked on the methods, such as clustering methods [2], distant based methods [3]; some focus on the nature of data, such as high dimensional data [4] and categorical data [5]. We approach anomaly detection by resorting to the Rough Set Theory (RST). It is a famous method used for feature selection [6]. Pawlak [7], the founder of RST, applied it to find the uncertain boundary in information, which covers upper approximation (rough information when boundary is not empty) and lower approximation (certain information if boundary is empty). However, the application of RST for anomaly detection is relatively few [8]. Some examples of the related research work by using RST can be seen in [8–10]. Authors in [8] suggested two versions of anomaly detection method to deal with spatiotemporal data. All input set are considered in the first version. The second version selects a subset, called a kernel set to represent the original input set. Anomaly will be computed as upper and lower approximations which are determined through a threshold. This threshold is generated automatically based on a spatiotemporal weight. According to the authors, the second version is better in term of computational time. Another two proposed methods were explained in [10]. The first version is called sequence-based outlier detection (SEQ) through RST. The second version is called distance-based outlier detection (DIS) through RST. SEQ examines the abnormal degree of an object by identifying the

A. Hirose et al. (Eds.): ICONIP 2016, Part III, LNCS 9949, pp. 434–441, 2016.
DOI: 10.1007/978-3-319-46675-0_47

sequence change of feature subsets and the equivalence classes. For DIS, [10] recommended a distance outlier factor (DOF) to determine the magnitude of anomaly for each sample. Both SEQ and DIS are applicable to process categorical data only. The reason is the original RST can only deal with categorical data. However, the neighborhood rough set (NRS) extends its capability to deal with numerical data. Thus, we proposed to use NRS for anomaly detection. An almost similar research can be seen in [9]. The authors proposed a Neighborhood Entropy-based Outlier Detection (NEOD) algorithm. They defined the relative neighborhood entropy to examine uncertainty information in the neighborhood. Such entropy is then used to guide the calculation of neighborhood entropy outlier factor. Our proposed NRS model is different from the method in [9]. We propose correctness matching and rank order to substitute inclusion degree in the NRS model. The reported results are encouraging to show the effectiveness of the proposed NRSCM in detecting anomaly.

The arrangement of each section in this paper is as follows: an overview of NRS and NRSCM are explained in next section. Experiments and discussion are presented in Sect. 3. Last but not least, concluding remarks as well as the potential work in future are described in Sect. 4.

2 The Methods

2.1 Overview of Neighborhood Rough Set (NRS) Model

The NRS model was proposed by Hu et al. [6] for feature selection. They extended the usage of RST, which is effective in handling categorical data to handle numerical data through a neighbourhood concept.

Assumed that the data is an information system, denoted by $InfSys = <Uni, F>$. Uni is the nonempty universe that represents a finite set of data $\{x_1, x_2, x_3, \ldots x_n\}$. Descriptions about data are presented through features F, i.e. $\{f_1, f_2, f_3, \ldots f_m\}$. There are two groups of features, information features IF and decision feature DF. As such, a complete description about a data set is $F = IF \cup DF$. The neighbourhood granule of data, x_i in B, i.e. the feature space, where $B \subseteq IF$, is denoted by

$$\delta_B(x_i) = \{x_j | x_j \in Uni, \Delta^B(x_i, x_j) \leq \delta\} \tag{1}$$

The distance measure, Δ, with $\forall x_1, x_2, x_3 \in Uni$ usually fulfills the following conditions:

(a) $\Delta(x_1, x_2) \geq 0$, $\Delta(x_1, x_2) = 0$ if and only if $x_1 = x_2$;
(b) $\Delta(x_1, x_2) = \Delta(x_2, x_1)$;
(c) $\Delta(x_1, x_3) \leq \Delta(x_1, x_2) + \Delta(x_2, x_3)$.

A general distant metric, i.e. Minkowsky distance is applied:

$$\Delta_P(x_1, x_2) = \left(\sum_{i=1}^{m} |D(x_1, f_i) - D(x_2, f_i)|^P \right)^{1/P} \tag{2}$$

where $x_1, x_2 \in Uni$; $F = \{f_1, f_2, \ldots, f_m\}$; m = dimensional space; $D(x, f_i)$ denotes the value of data x in the i-th feature f_i; P represents the mode of the distant function. Manhattan distance is implemented when $P = 1$; Euclidean distance is implemented when $P = 2$; and Chebychev distance is implemented when $P = \infty$. The model is designed to deal with both categorical and numerical features according to the following definitions:

Definition 1. Assumed B_1 = numerical features, where $B_1 \subseteq F$ and B_2 = categorical features, where $B_2 \subseteq F$. The neighborhood granule of $\delta_{B_1}(x)$, $\delta_{B_2}(x)$ and $\delta_{B_1 \cup B_2}(x)$ are:

(a) $\delta_{B_1}(x) = \{x_i | \Delta_{B_1}(x, x_i) \leq \delta, x_i \in Uni\}$;

(b) $\delta_{B_2}(x) = \{x_i | \Delta_{B_2}(x, x_i) = 0, x_i \in Uni\}$;

(c) $\delta_{B_1 \cup B_2}(x) = \{x_i | \Delta_{B_1}(x, x_i) \leq \delta \wedge \Delta_{B_2}(x, x_i) = 0, x_i \in Uni\}$, where $\wedge$ represents an "and" operator.

Definition 1(a) identifies the neighborhood of numerical features; Definition 1(b) identifies the neighborhood of categorical features; and Definition 1(c) deals with a mixture of numerical and categorical features. The relation matrix, which presents the neighborhood relation $Neigh$ on the Uni, is as below:

$$Mat(Neigh) = (R(x_i, x_j))_{n \times n} \tag{3}$$

where $x_i, x_j \in Uni$, $R(x_i, x_j) = 1$ when $\Delta(x_i, x_j) \leq \delta$ or else $R(x_i, x_j) = 0$. $Neigh$ can be considered as similarity relations as it fulfils both conditions of reflexivity and symmetry, i.e. $R(x_i, x_j) = 1$ and $R(x_i, x_j) = R(x_j, x_i)$. These conditions reflect that the data having an arbitrary similarity level in distant are grouped together in the same neighborhood and their neighborhood granules are close to each other. The neighborhood is approximated according to Definition 2, as follows:

Definition 2. Assumed $<Uni, Neigh>$ is a neighborhood approximation space, with Uni = universe (a set of data); $Neigh$ = neighborhood relation. In order to define the upper and lower approximations of X in $< Uni, Neigh >$, with $X \subseteq Uni$,

$$\begin{aligned}\overline{Neigh}X &= \{x_i | \delta(x_i) \cap X \neq \emptyset, x_i \in Uni\}, \\ \underline{Neigh}X &= \{x_i | \delta(x_i) \subseteq X, x_i \in Uni\}\end{aligned} \tag{4}$$

Therefore, $\underline{Neigh}X \subseteq X \subseteq \overline{Neigh}X$. The boundary region of X, namely $BNRX$, can be defined as

$$BNRX = \overline{Neigh}X - \underline{Neigh}X \tag{5}$$

The roughness degree of set X in $< Uni, Neigh >$ is presented through the size of boundary region. For practical use of the model, an inclusion degree is introduced. This inclusion degree generalizes the idea of upper and lower approximation and the model is more robust against noisy data. Assumed $A \subseteq Uni$ and $DF \subseteq Uni$, where A could represent an information feature, or a combination of several information features. For this example, we assume A includes only a single feature and DF is the decision feature. The inclusion degree of A in DF with respect to A is defined as:

$$Inclusion(A, DF) = \frac{\text{Card}(A \cap DF)}{\text{Card}(A)}, \text{ where } A \neq \emptyset \tag{6}$$

For example, $A = \{x_2, x_4, x_5, x_6, x_7, x_8\}$, $DF = \{x_1, x_4, x_6\}$, the cardinality of $A \cap DF$ is 2, cardinality of A is 6. Therefore, $Inclusion(A, DF) = 0.33$. Then, the upper and lower approximation of a subset X could be defined as below:

Definition 3. Assumed subset $X \subseteq Uni$ in $(Uni, F, Neigh)$, the variable correctness, i.e. upper and lower approximations of X are:

$$\overline{Neigh^{id}}X = \{x_i | Inclusion(\delta(x_i), X) \geq 1 - id, x_i \in Uni\},$$
$$\underline{Neigh^{id}}X = \{x_i | Inclusion(\delta(x_i), X) \geq id, x_i \in Uni\} \tag{7}$$

where $1 \geq id \geq 0.5$, id is the inclusion degree. If the id is set as minimum 0.5, with the above example, i.e. $Inclusion(A, DF) = 0.33$ does not fulfill the condition of both upper and lower approximations. Therefore, A is not selected as the information feature to predict the decision.

2.2 The Proposed Method

The proposed model is designed to analyze whether or not a sample input is an anomaly. We apply the neighborhood relation as in the original NRS [6]. We adopt Euclidean distance as a metric for distance calculation. We modify the inclusion degree, which is initially proposed for feature selection in [6]. Our main intention is to examine the abnormal degree of each sample input. Thus we proposed a metric to measure the correctness matching degree between a sample input neighborhood with the decision attributes' neighborhood. In this case, we consider the matching of elements with relationship as well as the complement sets, i.e. elements without relationship. Assume a heterogeneous data set is available in Table 1:

Table 1. A heterogeneous data set

Data, X	Numerical feature, B_1	Categorical feature, B_2	Decision feature, D
x_1	0.81	3	No
x_2	0.95	1	Yes
x_3	0.84	2	Yes
x_4	0.41	2	No
x_5	0.92	1	Yes
x_6	0.30	1	No

Let the neighborhood parameter, $\delta = 0.1$. We obtain the following information for the equivalence classes and we consider also those instances which contain no relation. For example, consider x_1 in B_1, it has Relationship (Rel) with $\{x_1, x_3, x_5\}$, i.e. $Relx_{11} = \{x_1, x_3, x_5\}$ but contains No Relationship (NoRel) with $\{x_2, x_4, x_6\}$, NoRelx$_{11} = \{x_2, x_4, x_6\}$; in B_2, $Relx_{12} = \{x_1\}$ but NoRelx$_{12} = \{x_2, x_3, x_4, x_5, x_6\}$. There are two groups of decision class, $1 = $ No and $2 = $ Yes. Thus, DF_1 has Rel with $\{x_1, x_4, x_6\}$ but not $\{x_2, x_3, x_5\}$, i.e., $DF_1\text{Rel} = \{x_1, x_4, x_6\}$ and $DF_1\text{NoRel} = \{x_2, x_3, x_5\}$ whereas $DF_2\text{Rel} = \{x_2, x_3, x_5\}$ and $DF_2\text{NoRel} = \{x_1, x_4, x_6\}$. The Correctness Matching (CM) for an input data x_i per each feature B_j relevant to a class c is defined as follows:

$$CM_{ij} = \text{Card}(\text{Rel}x_{ij} \cap DF_c\text{Rel}) + \text{Card}(\text{NoRel }x_{ij} \cap DF_c\text{NoRel}) \tag{8}$$

CM reflects the ability of an input data to approximate the decision feature. This includes the cardinality of intersection between input elements and decision feature that have Rel as well as the cardinality of elements that contain NoRel. For example, decision feature for x_1 is No. With Eq. 8, cardinality of $\text{Rel}x_{11}$ intercept $DF_1\text{Rel}$ is 1 for NoRel is 2. Thus, $CM_{11} = 1{+}2$. The probability of CM is then computed as follows:

$$\text{Prob}(CM_{ij}) = \frac{CM_{ij}}{n} \tag{9}$$

where n is total number of input. We then average the $\text{Prob}(CM)$ score for each feature B_j using equation below:

$$E(CM_j) = \frac{\sum_{i=1}^{n} \text{Prob}(CM_{ij})}{n} \tag{10}$$

For example, for each data under $B_{j=1}$ from Table 1, by applying Eq. (9), $\text{Prob}(CM_{11})$ for $x_1 = 0.50$; and the following x_2 to x_6 has 0.83; 0.50; 0.67; 0.83; and 0.67 respectively. $E(CM_1)$ is equal to 0.67. By applying similar process for $B_{j=2}$, $E(CM_2)$ is 0.56. We rank each feature according to $E(CM_j)$. The higher the $E(CM_j)$, then the feature is positioned at a higher rank. In this case, $B_{j=1}$ is ranked as 2, $B_{j=2}$ is ranked as 1. The purpose of doing this is to include the details where some features are important but some features are not. Finally, the degree of CM (CMD) is calculated:

$$CMD_i = \frac{\sum_{j=1}^{M} (rank_j * CM_{ij})}{\sum_{j=1}^{M} rank_j} \tag{11}$$

With rank order, the CMD of x_1 to x_6 are 0.44, 0.78, 0.50, 0.61, 0.78 and 0.67 respectively. As we look for potential anomaly, therefore the upper and lower approximations are re-defined as:

$$\begin{aligned}
\underline{Neigh}^{id}Out &= \{x_i | CMD(\delta(x_i), Out) \le id, x_i \in Uni\}, \\
\overline{Neigh}^{id}Out &= \{x_i | CMD(\delta(x_i), Out) \le 1 - id, x_i \in Uni\}
\end{aligned} \tag{12}$$

If the parameter of inclusion degree is set as $id = 0.5$, then x_1 is an anomaly.

3   Experiment and Discussion

The performance results from the experiment are presented in two sub-sections. Each section includes a benchmark comparison between NRSCM and other methods. In Sect. 3.1, Annealing is used to demonstrate the performance of NRSCM. Annealing contains both categorical data and numerical data. Settings of neighborhood parameter are reported in Table 2. In Sect. 3.2, Wisconsin Breast Cancer (WisBC) is used to examine the performance of NRSCM. This data set is a categorical data. Parameter setting for neighborhood is 0. Both Annealing and WisBC data are publicly available from Machine Learning Repository of University of California, Irvine [11]. Computer that is used to run the experiments has these specifications: Windows 7 as the operating system, RAM with 4.0 GB and CPU Intel Core (TM), i.e. i5-2410M.

Table 2. Neighborhood parameter setting of annealing

Feature	4	5	9	33	34	35	37	Others
Neighbourhood	5	10	200	0.2	300	800	50	0

3.1   Annealing Data Set

The annealing data set has 798 samples, 38 attributes and 5 classes. As class 3 is the majority class, so other class labels are deemed as anomaly. We followed the neighborhood parameters setting as stated by Li [9]. Table 2 shows the settings. We evaluate the effectiveness of the anomaly detection by using the evaluation method proposed by Aggarwal [12]. Minority classes would be presented with high percentage if the anomaly detection method works well. NRSCM achieves a good result as compared to the performance of Neighborhood Entropy Outlier Detection (NEOD) and Distance-based method (DIS), which are RST-based and another method, K-Nearest Neighbor (KNN) that is none RST-based. The results are presented in Table 3. For top ratio number of objects ranges from 10 % to 30 %, NRSCM identifies all the minority classes from 80(42 %) out of 80 objects, 105(55 %) out of 105 objects, 140(74 %) out of 140 objects, 175(92 %) out of 175 objects and 190(all minority objects) out of 209 objects. Unlike other methods, only a smaller portion of anomalies are detected. This result reflects that NRSCM can deal well with mixed data effectively.

Table 3. Performance comparison through annealing

Top ratio (number of objects)	Number of rare class included (coverage)			
	NRSCM	NEOD	KNN	DIS
10 % (80)	80(42 %)	51(27 %)	21(11 %)	33(17 %)
15 % (105)	105(55 %)	67(35 %)	30(16 %)	44(23 %)
20 % (140)	140(74 %)	81(43 %)	41(22 %)	61(32 %)
25 % (175)	175(92 %)	84(44 %)	58(31 %)	77(41 %)
30 % (209)	190(100 %)	92(48 %)	62(33 %)	84(44 %)

3.2   Wisconsin Breast Cancer Data Set

WisBC is a categorical data set that has 699 samples with 9 features. For benchmarking purposes, an unbalanced distribution is formed according to [13], i.e. the dataset contains 92 %(444) benign samples and 8 %(39) malignant samples. A Replicator Neural Network (RNN) is added for comparison, as this distribution method is proposed by [13]. The results are shown in Table 4. NRSCM can perform either better than or comparable to the other methods, i.e. NEOD, DIS, KNN and RNN. The results show that when the top ratio number of objects is among 1 % to 6 %, NRSCM achieves better results as compared to other anomaly detection methods. For the remaining top ratios, NRSCM is better than DIS and RNN but slightly weaker than NEOD and KNN in anomaly detection. In this data set, none of the anomaly detection method is the best.

Table 4. Performance comparison through WisBC

Top ratio (number of objects)	Number of minority classes included (Coverage)				
	NRSCM	NEOD	DIS	KNN	RNN
1 %(4)	4(10.3 %)	4(10.3 %)	4(10.3 %)	3(7.7 %)	3(10.3 %)
2 %(8)	8(20.5 %)	7(17.8 %)	5(12.8 %)	6(15.4 %)	6(20.5 %)
4 %(16)	15(38.4 %)	13(33.3 %)	11(35.9 %)	11(35.9 %)	11(35.9 %)
6 %(24)	21(53.8 %)	20(51.3 %)	18(46.2 %)	18(46.2 %)	18(46.2 %)
8 %(32)	25(64.1 %)	27(69.2 %)	24(61.5 %)	27(69.2 %)	25(64.1 %)
10 %(40)	31(76.9 %)	32(82.1 %)	29(74.4 %)	32(82.1 %)	30(76.9 %)
12 %(48)	36(92.3 %)	36(92.3 %)	36(92.3 %)	38(97.4 %)	35(89.7 %)
14 %(56)	39(100 %)	39(100 %)	39(100 %)	39(100 %)	36(92.3 %)
16 %(64)	39(100 %)	39(100 %)	39(100 %)	39(100 %)	36(92.3 %)
18 %(72)	39(100 %)	39(100 %)	39(100 %)	39(100 %)	38(97.4 %)
20 %(80)	39(100 %)	39(100 %)	39(100 %)	39(100 %)	38(97.4 %)
28 %(112)	39(100 %)	39(100 %)	39(100 %)	39(100 %)	39(100 %)

4   Conclusion

In this study, an anomaly detection method, NRSCM is proposed. The method provides good results in identifying anomaly. Experiments show that NRSCM can work well with mixed data sets as well as categorical data. In the future, the functionality of NRSCM should be upgraded with an ability to automate the selection of inclusion degree and neighborhood setting. It is also an interesting work to extend NRSCM to incorporate a prediction capability.

References

1. Chen, D., Lu, C.-T., Kou, Y., Chen, F.: On detecting spatial outliers. GeoInformatica **12**(4), 455–475 (2008)
2. Goldstein, M., Dengel, A.: Histogram-based outlier score (HBOS): a fast unsupervised anomaly detection algorithm. In: 35th German Conference on AI (KI-2012), pp. 59–63 (2012)
3. Kontaki, M., Gounaris, A., Papadopoulos, A.N., Tsichlas, K., Manolopoulos, Y.: Continuous monitoring of distance-based outliers over data streams. In: IEEE 27th International Conference on Data Engineering (ICDE), pp. 135–146. IEEE, Hannover (2011)
4. Ye, M., Li, X., Orlowska, M.E.: Projected outlier detection in high-dimensional mixed-attributes data set. Expert Syst. Appl. **36**(3), 7104–7113 (2009). Part 2
5. Suri, N.N.R.R., Murty, M.N., Athithan, G.: Detecting outliers in categorical data through rough clustering. Natural Comput. **15**, 1–10 (2015)
6. Hu, Q., Yu, D., Liu, J., Wu, C.: Neighborhood rough set based heterogeneous feature subset selection. Inf. Sci. **178**(18), 3577–3594 (2008)
7. Pawlak, Z.: Rough sets. Int. J. Comput. Inf. Sci. **11**(5), 341–356 (1982)
8. Albanese, A., Pal, S.K., Petrosino, A.: Rough sets, kernel set, and spatiotemporal outlier detection. IEEE Trans. Knowl. Data Eng. **26**(1), 194–207 (2014)
9. Li, X., Rao, F.: Outlier detection using the information entropy of neighborhood rough sets. J. Inf. Comput. Sci. **9**(12), 3339–3350 (2012)
10. Jiang, F., Sui, Y., Cao, C.: Some issues about outlier detection in rough set theory. Expert Syst. Appl. **36**(3), 4680–4687 (2009). Part 1
11. Bache, K., Lichman, M.: UCI Machine Learning Repository, University of California, School of Information and Computer Science. http://archive.ics.uci.edu/ml
12. Aggarwal, C.C., Yu, P.S.: Outlier Detection for High Dimensional Data. SIGMOD Rec. **30**(2), 37–46 (2001)
13. Hawkins, S., He, H., Williams, G.J., Baxter, R.A.: Outlier detection using replicator neural networks. In: Kambayashi, Y., Winiwarter, W., Arikawa, M. (eds.) DaWaK 2002. LNCS, vol. 2454, pp. 170–180. Springer, Heidelberg (2002)

Learning Class-Informed Semantic Similarity

Tinghua Wang[(✉)] and Wei Li

School of Mathematics and Computer Science, Gannan Normal University,
Ganzhou 341000, People's Republic of China
wthgnnu@163.com

Abstract. Exponential kernel, which models semantic similarity by means of a diffusion process on a graph defined by lexicon and co-occurrence information, has been successfully applied to the task of text categorization. However, the diffusion is an unsupervised process, which fails to exploit the class information in a supervised classification scenario. To address the limitation, we present a class-informed exponential kernel to make use of the class knowledge of training documents in addition to the co-occurrence knowledge. The basic idea is to construct an augmented term-document matrix by encoding class information as additional terms and appending to training documents. Diffusion is then performed on the augmented term-document matrix. In this way, the words belonging to the same class are indirectly drawn closer to each other, hence the class-specific word correlations are strengthened. The proposed approach was demonstrated with several variants of the popular 20Newsgroup data set.

Keywords: Semantic similarity · Class information · Exponential kernel · Text categorization · Support vector machine (SVM) · Kernel method

1 Introduction

The widespread and increasing availability of massive textual data stimulates the development of text categorization field, which aims to automatically classify unlabeled documents into predefined categories according to some criteria of interest. Pioneered by [1], kernel methods [2] in general and support vector machines (SVM) in particular have been heavily used for text categorization tasks, typically showing good results [3–9]. In kernel methods, the kernel can be considered as an interface between the input data and the learning algorithm, and is the key component to ensure the good performance of kernel methods. In text categorization, the widely used kernel is the "Bag of Words" (BOW) kernel [2], which encodes the input documents as vectors whose dimensions correspond to the words or terms occurring in the corpus. Despite its ease of use, this kernel suffers from well-known limitations, mostly due to its inability to exploit semantic similarity between terms: documents sharing terms that are different but semantically related will be considered as unrelated. To address this problem, a number of attempts have been made to incorporate semantic knowledge into the BOW kernel, resulting in the so-called semantic kernels [2]. For example: the semantic kernels that use the external semantic knowledge like WordNet and Wikipeida were proposed to improve the kernel-based text categorization systems [3, 4]. In the absence

© Springer International Publishing AG 2016
A. Hirose et al. (Eds.): ICONIP 2016, Part III, LNCS 9949, pp. 442–449, 2016.
DOI: 10.1007/978-3-319-46675-0_48

of external semantic knowledge, corpus-based statistical approaches are applied to capture the semantic relations between terms, resulting in the corpus-based semantic kernels [5–9].

In this paper, we only consider the exponential kernel [6, 8, 10], which is one of the corpus-based semantic kernels, for text categorization. Exponential kernel models semantic similarities by means of a diffusion process on a graph defined by lexicon and co-occurrence information and virtually exploits high order co-occurrences to infer semantic similarity between terms. However, the diffusion is an unsupervised process, which fails to exploit the class information in a classification scenario and may not be optimal for text categorization systems. Chakraborti et al. [11, 12] introduced a simple approach called "sprinkling" to incorporate class labels of documents into Latent Semantic Indexing (LSI). In sprinkling, a set of class-specific artificial terms are appended to the representations of documents of the corresponding class. LSI is then applied on the sprinkled term-document space resulting in a concept space that better reflects the underlying class distribution of documents. Recently, this approach was also applied to sprinkle Latent Dirichlet Allocation (LDA) topics for weakly supervised text categorization [13]. The inherent reason for this approach is that the sprinkled term can add contribution to exploit the class information of text documents in a classification procedure. Motivated by the above work, this paper presents a class-informed exponential kernel to make use of the class knowledge of training documents in addition to the co-occurrence knowledge. The basic idea is to construct an augmented term-document matrix by encoding class information as additional terms and appending them to training documents. Diffusion is then performed on the augmented term-document matrix to learn the semantic matrix, which is the key component of semantic kernels. In this way, the words belonging to the same class are indirectly drawn closer to each other, hence the class-specific word correlations are strengthened. Our evaluation on the variants of popular 20Newsgroup data set shows the proposed kernel significantly outperforms both the BOW kernel and exponential kernel.

2 A Review on Exponential Kernel

Let $X = (x_1, \ldots, x_l)$ be a set of documents. Consider that we are also given a dictionary V consisting of n words. The BOW model (also called vector space model, VSM) [1, 2] of the document x is defined as follows:

$$\phi : x \rightarrow \phi(x) = (tf(t_1, x), \ldots, tf(t_n, x))^{\mathrm{T}} \in \mathbf{R}^n \tag{1}$$

where $tf(t_i, x)$, $1 \leq i \leq n$, is the frequency of the occurrence of the word t_i in the document x, and R denotes the set of real numbers. The BOW kernel is given by

$$k(x_i, x_j) = <\phi(x_i), \phi(x_j)> = \sum_{t \in V} tf(t, x_i) tf(t, x_j) \tag{2}$$

BOW model is probably one of the simplest constructions used in text processing. In this model, the feature vectors $\phi(x)$ are typically sparse with a small number of non-zero entries for those words occurring in the documents. Two documents that use semantically related but distinct words will therefore show no similarity. Ideally, semantically similar documents should be mapped to nearby positions in the feature space. In order to address this problem, a transformation of the feature vector of the type $\bar{\phi}(x) = S\phi(x)$ is required, where S is a semantic matrix indexed by pairs of words with the entries $S_{i,j}$ $(1 \leq i \leq n)$ indicating the strength of their semantic similarity. Using this transformation, the semantic kernels [2] take the form of

$$k(x_i, x_j) = \bar{\phi}(x_i)^T \bar{\phi}(x_j) = \phi(x_i)^T S^T S \phi(x_j) \qquad (3)$$

In practice, the problem of how to infer semantic similarities between terms from a corpus remains an open issue. Kandola et al. [6] proposed a semantic kernel named exponential kernel given by:

$$K(\lambda) = K_0 \exp(\lambda K_0) \qquad (4)$$

where K_0 is the kernel matrix of the BOW kernel, $\lambda \in [0, +\infty)$ is a decay factor. Let D be the feature example (term-by-document) matrix in the BOW representation, then $K_0 = D^T D$. Let $G = DD^T$, it is easy to prove that $K(\lambda)$ corresponds to a semantic matrix $\exp(\lambda G/2)$ [2, 6], i.e.

$$S = \exp(\lambda G/2) = \left(\sum_{d=0}^{\infty} \frac{\lambda^d}{d!} G^d \right)^{1/2} = \left(I + \lambda G + \frac{\lambda^2}{2!} G^2 + \ldots + \frac{\lambda^2}{d!} G^d + \ldots \right)^{1/2} \qquad (5)$$

where I denotes the identity matrix. In fact, noting that S is a symmetric positive semi-definite matrix since G is symmetric, we have

$$K(\lambda) = D^T S^T S D = D^T \exp(\lambda G) D$$

$$= \sum_{d=0}^{\infty} \frac{\lambda^d}{d!} D^T G^d D = K_0 \left(\sum_{d=0}^{\infty} \frac{\lambda^d}{d!} K_0^d \right) = K_0 \exp(\lambda K_0) \qquad (6)$$

Geometrically, exponential kernel models semantic similarities by means of a diffusion process on a graph defined by lexicon and co-occurrence information [6, 8, 10]. Specifically, such a graph has nodes indexed by all the terms in the corpus, and the edges are given by the co-occurrence between terms in documents of the corpus. A diffusion process on the graph can capture high order co-occurrences between indirectly connected terms. Noting that $\left[G^d \right]_{i,j}$ in (5) is the number of dth-order co-occurrence paths between terms t_i and t_j in the graph (the identity matrix I, i.e. G^0, can be regarded as the indication of the zero-order correlation between terms, meaning only the similarity between a term and itself equals 1 and 0 for other cases), and the semantic matrix S combines all the order co-occurrence paths with exponentially decaying weights, we can easily find that the semantic similarity between two terms is measured by the number of the co-occurrence paths between them, and the semantic matrix S essentially exploits the higher order correlation between terms. Intuition shows that the higher the co-occurrence order is, the less similar the semantics becomes. The parameter λ is used to control the decaying speed for increasing orders. To summarize, exponential kernel takes all possible paths connecting two nodes into account, and propagates the similarity between two remote terms (or documents) in an elegant way.

3 Class-Informed Semantic Similarity and Kernel

As mentioned before, the elements of the semantic matrix $\mathbf{S}$ give the strength of the semantic similarities between terms. Exponential kernel essentially exploits the high order correlations to refine the similarity measure by performing a diffusion process on a graph defined by lexicon and co-occurrence information. However, the diffusion is an unsupervised process, which fails to exploit the class information in a classification scenario and may not be optimal for text categorization systems. In a supervised classification scenario, we have class knowledge of training corpus in addition to the co-occurrence information. To address this issue, motivated by the Sprinkled LSI [11–13], we incorporate class knowledge into the semantic similarity learning process to obtain the class-informed exponential kernel. Specifically, we first generate a set of artificial terms corresponding to the class labels and sprinkling (appending) these terms to the training documents. Diffusion is then carried out on the augmented term-document matrix, and the semantic matrix with sprinkled terms is obtained. Obviously, such semantic matrix has the same dimensionality as the augmented matrix. In order to make the training document representations compatible with test documents, the sprinkled dimensions are removed. Figure 1 shows the process of learning the semantic matrix for the class-informed exponential kernel ($\lambda = 0.2$) with a toy corpus. As shown if Fig. 1(a), this corpus comprises three documents x_1, x_2 and x_3, which consist of a set of terms $\{t_1, t_2\}, \{t_2, t_3\}$ and $\{t_3, t_4\}$, respectively. These documents belong to two different classes, i.e., document x_1 and x_2 belong to one class, while x_3 belongs to the other class. As shown in Fig. 1(b), instead of performing diffusion directly on the term-document matrix constructed from this corpus, sprinkling augments the feature set with artificial terms corresponding to the class labels (we here set 1 when the documents are from the same class, 0 when they are not in the same class). These terms act as carriers of class knowledge. The "unsprinkle" functions as removing sprinkled dimensions and "normalization" as ensuring the semantic similarity between a term and itself equals 1. For a semantic matrix $\mathbf{S}$, its normalized version $\bar{\mathbf{S}}$ can be obtained by $\bar{\mathbf{S}}_{i,j} = \mathbf{S}_{i,j} / \sqrt{\mathbf{S}_{i,i} \mathbf{S}_{j,j}}$. Conceptually, the learned semantic matrix $\bar{\mathbf{S}}$ encodes the class-informed semantic similarities between terms, and then the class-informed exponential kernel can be obtained by (3), where $\mathbf{S}$ is replaced by $\bar{\mathbf{S}}$.

Why does the class-informed exponential kernel work? For one thing, just the same as the exponential kernel, the class-informed exponential kernel takes advantage of high order correlations between terms to infer semantic similarities. Benefits of using high order correlations between terms can be found in Fig. 1. Using a traditional similarity measure which is based on the co-occurrence information (e.g. inner product), the similarity value between t_1 and t_3 will be zero since they do not co-occur in any of documents. But this measure is misleading since these two terms share a second-order correlation through t_2: t_1 co-occurs with t_2 in document $\boldsymbol{x}_1$, and t_2 co-occurs with t_3 in document $\boldsymbol{x}_2$. This shows that, by using higher order correlations between terms, we obtain a non-zero similarity value (i.e., 0.0108) between t_1 and t_3, which is not possible in the BOW representation. The similarity between t_1 and t_3 becomes more eminent interconnecting more terms like t_2. Besides, the similarity values between terms t_1 and t_2 (they share a first-order correlation), t_1 and t_3 (they share a second-order correlation), and t_1 and t_4 (they share a third-order correlation) are decreasing (i.e., $0.1094 > 0.0108 > 0.0004$). This is because the higher the co-occurrence order is, the less similarity values becomes. On the other hand, sprinkling aims to make explicit any implicit correlation between terms indicative of underlying classes. When co-occurrences are mined on the augmented term-document matrix which encodes class knowledge, terms belonging to the same class are drawn closer to each other, hence the class-specific word correlations are strengthened. For example, in Fig. 1(b), the similarity value between t_1 and t_3 is larger than that between t_2 and t_4 (i.e., $0.0108 > 0.0055$) rather than equal in the exponential kernel. This happens because t_1 and t_3 are in the same class while t_2 and t_4 are not, and the sprinkled terms boost the second-order correlations between terms related to the same class.

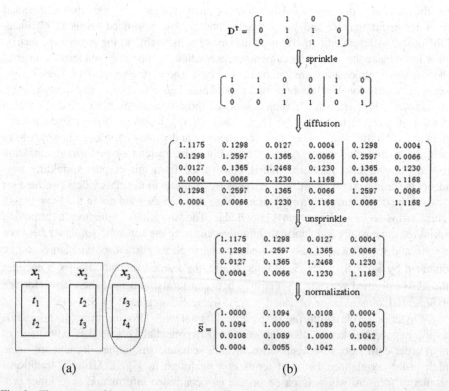

Fig. 1. The process of learning semantic matrix for class-informed exponential kernel with a toy corpus comprising three documents belonging to two different classes

To summarize, the exponential kernel simultaneously exploits high order co-occurrences and class knowledge in an elegant way.

Finally, while in the above example we used one sprinkled term per class, in principle we can sprinkle more terms per class to boost the contribution of class knowledge to the diffusion process. One important question is how to decide the number of sprinkled terms for each class. An empirical solution is to use as many as sprinkled terms which yield the best cross-validation classification performance. While sprinkled terms help in emphasizing class knowledge, using too many of them (over-sprinkling) may fail to preserve interesting variations in the original term-document structure. Therefore, we should impose a good trade-off between under- and over-sprinkling in practice [12].

4 Experimental Evaluation

This experiment evaluates the performance of the proposed method on textual data sets which are shown in Table 1. It presents, for each data set, the number of samples, the number of features and the number of classes. These data sets are variants of the 20Newsgroup data set (http://www.ai.mit.edu/people/jrennie/20Newsgroups/), which is a collection of approximately 20000 newsgroup documents, partitioned (nearly) evenly across 20 different newsgroups. We used four basic subgroups "SCIENCE", "COMP", "POLITICS" and "RELIGION" from the 20Newsgroup data set. The documents are evenly distributed to the classes (each class including 500 documents). All data sets were preprocessed using the Text Mining Infrastructure (TMI) [14]. The preprocessing includes sentence boundary determination, stop word removal and stemming. We used the stemmer and stop word list embedded in the TMI.

Table 1. Statistics of selected four data sets

Data set	#samples	#features	#classes
SCIENCE	2000	2225	4
COMP	2500	2478	5
POLITICS	1500	2478	3
RELIGION	1500	2125	3

For each data set in Table 1, after the proper preprocessing, we partitioned it into a training set and a test set by stratified sampling: 70 % of the data set is used for training and the rest for prediction. We then used the LIBLINEAR package [15] to train and test the SVM. We consider three types of kernels, i.e., BOW kernel, exponential kernel and class-informed exponential kernel for comparison. These kernels are embedded in the SVM classifier individually. The parameters of the SVM were optimized by five-fold cross-validation on the training set. For the BOW kernel, there is only one parameter C that needs to be optimized. For the exponential kernel, there are two parameters C and λ that need to be optimized. For the class-informed exponential kernel, besides the regularization parameter C and decay factor λ, the number of the sprinkled terms per class (denoted by μ) is also needed to be optimized. We performed grid-search in one dimension (C), two dimensions (C and λ) and three dimensions (C, λ and μ) to choose parameters from the sets $C = \{2^{-2}, 2^0, \ldots, 2^{10}\}$, $\lambda = \{2^{-1}, 2^{-2}, \ldots, 2^{-10}\}$ and $\mu = \{2^1, 2^2, \ldots, 2^5\}$, respectively.

The average classification accuracies with standard deviations over 10 trials are summarized in Table 2. The bold font indicates the best performance. From this table, we find that the exponential and class-informed exponential kernels produce significantly better classification performances than the BOW kernel baseline. More importantly, for all data sets we see that the class-informed exponential kernel achieves significant performance improvement over the exponential kernel. It should be noted that the performance differences are statistically significant with the significant level 0.05 in light of the pairs t-tests on all four data sets. Since whether or not the class information is taken into consideration is the only difference between the exponential kernel and class-informed exponential kernel, these results imply that the class knowledge has a conspicuous impact on the kernel construction for text categorization and demonstrate the effectiveness of the proposed class-informed exponential kernel with application to text categorization.

Table 2. Classification accuracy of different kernels on four data sets

Data set	Classification accuracy (%)		
	BOW	Exponential	Class-informed exponential
SCIENCE	90.61 ± 0.83	92.35 ± 0.92	**95.73 ± 0.65**
COMP	79.64 ± 1.59	84.59 ± 1.83	**85.62 ± 1.17**
POLITICS	93.20 ± 1.17	94.50 ± 1.17	**95.86 ± 0.74**
RELIGION	89.68 ± 1.36	90.63 ± 1.15	**92.37 ± 1.22**

5   Conclusion

We have presented a novel class-informed exponential kernel which incorporates class knowledge into the diffusion process of mining high order correlations between terms for text categorization tasks. Since sprinkled terms are essentially class labels, the inclusion of them helps to artificially promote co-occurrences between existing terms and classes. As a result, our approach can be considered as a supervised semantic smoothing kernel which makes use of the class knowledge. Experimental evaluation shows the superior effectiveness of the proposed approach compared with other baseline models. Future work will focus on the theoretical verification of the superior performance of the proposed approach, as well as making comparisons with other newly proposed methods for text categorization.

Acknowledgements. This work is supported in part by the National Natural Science Foundation of China (No. 61562003), the Natural Science Foundation of Jiangxi Province of China (Nos. 20151BAB207029 and 20161BAB202070), the China Scholarship Council (No. 201508360144) and the "Bai Ren Yuan Hang" Project of Jiangxi Province of China in 2015.

References

1. Joachims, T.: Text categorization with support vector machines: learning with many relevant features. In: Nédellec, C., Rouveirol, C. (eds.) ECML 1998. LNCS, vol. 1398. Springer, Heidelberg (1998)
2. Shawe-Taylor, J., Cristianini, N.: Kernel methods for pattern analysis. Cambridge University Press, New York (2004)
3. Bloehdorn, S., Basili, R., Cammisa, M., Moschitti, A.: Semantic kernels for text categorization based on topological measures of feature similarity. In: Proceedings of the 6th IEEE International Conference on Data Mining, Hong Kong, China, pp. 808–812 (2006)
4. Wang, P., Domeniconi, C.: Building semantic kernels for text categorization using Wikipedia. In: Proceedings of the 14th ACM SIGKDD International Conference on Knowledge Discovery and Data Mining, Las Vegas, USA, pp. 713–721 (2008)
5. Cristianini, N., Shawe-Taylor, J., Lodhi, H.: Latent semantic kernels. J. Intell. Inf. Syst. 18(2–3), 127–152 (2002)
6. Kandola, J., Shawe-Taylor, J., Cristianini, N.: Learning semantic similarity. In: Advances in Neural Information Processing Systems, vol. 15, pp. 657–664 (2003)
7. Gliozzo, A.M., Strapparava, C.: Domain kernels for text categorization. In: Proceedings of the 9th Conference on Computational Natural Language Learning, Ann Arbor, USA, pp. 56–63 (2005)
8. Chen, J., Zhong, J., Xie, Y., Cai, C.: Text categorization using SVM with exponential kernel. Appl. Mech. Mater. 519–520, 807–810 (2014)
9. Altınel, B., Caniz, M.C., Diri, B.: A corpus-based semantic kernel for text categorization by using meaning values of terms. Eng. Appl. Artif. Intell. 43, 54–66 (2015)
10. Wang, T., Rao, J., Hu, Q.: Supervised word sense disambiguation using semantic diffusion kernel. Eng. Appl. Artif. Intell. 27, 167–174 (2014)
11. Chakraborti, S., Lothian, R., Wiratunga, N., Watt, S.N.: Sprinkling: supervised latent semantic indexing. In: Lalmas, M., MacFarlane, A., Rüger, S.M., Tombros, A., Tsikrika, T., Yavlinsky, A. (eds.) ECIR 2006. LNCS, vol. 3936, pp. 510–514. Springer, Heidelberg (2006)
12. Chakraborti, S., Mukras, R., Lothian, R., Wiratunga, N., Watt, S., Harper, D.: Supervised latent semantic indexing using adaptive sprinkling. In: Proceedings of the 20th International Joint Conference on Artificial Intelligence, Hyderabad, India, pp. 1582–1587 (2007)
13. Hingmire, S., Chakraborti, S.: Sprinkling topics for weakly supervised text categorization. In: Proceedings of the 52nd Annual Meeting of the Association for Computational Linguistics, vol. 2, Short Paper, Baltimore, USA, pp. 55–60 (2014)
14. Holzman, L.E., Fisher, T.A., Galitsky, L.M., Kontostathis, A., Pottenger, W.M.: A software infrastructure for research in textual data mining. Int. J. Artif. Intell. Tools 14(4), 829–849 (2004)
15. Fan, R.E., Chang, K.W., Hsieh, C.J., Wang, X.R., Lin, C.J.: LIBLINEAR: a library for large linear classification. J. Mach. Learn. Res. 9, 1871–1874 (2008)

Aggregated Temporal Tensor Factorization Model for Point-of-interest Recommendation

Shenglin Zhao[1,2]($\boxtimes$), Michael R. Lyu[1,2], and Irwin King[1,2]

[1] Shenzhen Key Laboratory of Rich Media Big Data Analytics and Application,
Shenzhen Research Institute, The Chinese University of Hong Kong,
Shenzhen, China
[2] Department of Computer Science and Engineering,
The Chinese University of Hong Kong, Hong Kong, China
{slzhao,lyu,king}@cse.cuhk.edu.hk

Abstract. Temporal influence plays an important role in a point-of-interest (POI) recommendation system that suggests POIs for users in location-based social networks (LBSNs). Previous studies observe that the user mobility in LBSNs exhibits distinct temporal features, summarized as periodicity, consecutiveness, and non-uniformness. By capturing the observed temporal features, a variety of systems are proposed to enhance POI recommendation. However, previous work does not model the three features together. More importantly, we observe that the temporal influence exists at different time scales, yet this observation cannot be modeled in prior work. In this paper, we propose an Aggregated Temporal Tensor Factorization (ATTF) model for POI recommendation to capture the three temporal features together, as well as at different time scales. Specifically, we employ temporal tensor factorization to model the check-in activity, subsuming the three temporal features together. Furthermore, we exploit a linear combination operator to aggregate temporal latent features' contributions at different time scales. Experiments on two real life datasets show that the ATTF model achieves better performance than models capturing temporal influence at single scale. In addition, our proposed ATTF model outperforms the state-of-the-art methods.

Keywords: Point-of-interest recommendation · Tensor factorization · Temporal influence · Location-based social network

1 Introduction

Temporal influence plays an important role in a point-of-interest (POI) recommendation system that recommends POIs for users in location-based social networks (LBSNs). Previous studies show that the user mobility in LBSNs exhibits distinct temporal features [3,4,17]. For example, users always stay in the office in the Monday afternoon, and enjoy entertainments in bars at night. The temporal features in users' check-in data could be summarized as follows.

© Springer International Publishing AG 2016
A. Hirose et al. (Eds.): ICONIP 2016, Part III, LNCS 9949, pp. 450–458, 2016.
DOI: 10.1007/978-3-319-46675-0_49

- **Periodicity.** Users share the same periodic pattern, visiting the same or similar POIs at the same time slot [3,17]. For instance, a user always visits restaurants at noon, so do other users. Hence, the periodicity inspires the time-aware collaborative filtering method to recommend POIs.
- **Consecutiveness.** A user's current check-in is largely correlated with the recent check-in [2,4]. Geo et al. [4] model this property by assuming that user preferences are similar in two consecutive hours. Cheng et al. [2] assume that two checked-in POIs in a short term are highly correlated.
- **Non-uniformness.** A user's check-in preference changes at different hours of a day, or at different months of a year, or different days of a week [4]. For example, at noon a user may visit restaurants while at night he/she may have fun in bars.

By capturing the observed temporal features, a variety of systems are proposed to enhance POI recommendation performance [2,4,17]. They do gain better performance than general collaborative filtering (CF) methods [15]. Nevertheless, previous work [2,4,17] cannot model the three features together. Moreover, an important fact is ignored in prior work that the temporal influence exists at different time scales. For example, in day level, you may check-in POIs around your home in the earning morning, visit places around your office in the day time, and have fun at nightclubs in the evening. In week level, you may stay in the city for work in weekdays and go out for vocation in weekends. Hence, to better model the temporal influence, capturing the temporal features at different time scales is necessary.

In this paper, we propose an Aggregated Temporal Tensor Factorization (ATTF) model for POI recommendation to capture the three temporal features together, as well as at different time scales. We construct a user-time-POI tensor to represent the check-ins, and then employ the interaction tensor factorization [14] to model the temporal effect. Different from prior work that represents the temporal influence at single scale, we index the temporal information for latent representation at different scales, i.e., hour, week day, and month. Furthermore, we employ a linear combination operator to aggregate different temporal latent features' contributions, which capture the temporal influence at different scales. Specifically, our ATTF model learns the three temporal properties as follows: (1) periodicity is learned from the temporal CF mechanism; (2) consecutiveness is manifested in two aspects—time in a slot brings the same effect through sharing the same time factor, and relation between two consecutive time slots could be learned from the tensor model; (3) non-uniformness is depicted by different time factors for different time slots at different time scales. Moreover, an aggregate operator is introduced to combine the temporal influence at different scales, i.e., hour, week day, and month, and represent the temporal effect in a whole.

The contributions of this paper are summarized as follows:

- To the best of our knowledge, this is the first temporal tensor factorization method for POI recommendation, subsuming all the three temporal properties: periodicity, consecutiveness, and non-uniformness.

– We propose a novel model to capture temporal effect in POI recommendation at different time scales. Experimental results show that our model outperforms prior temporal model more than 20 %.
– The ATTF model is a general framework to capture the temporal features at different scales, which outperforms single temporal factor model and gains 10 % improvement in the top-5 POI recommendation task on Gowalla data.

2   Related Work

In this section, we first review the literature of POI recommendation. Then we summarize the progress of modeling temporal effect for POI recommendation.

POI Recommendation. Most of POI recommendation systems base on the collaborative filtering (CF) techniques, which can be reported in two aspects, memory-based and model-based. On the one hand, Ye et al. [15] propose the POI recommendation problem in LBSNs solved by user-based CF method, and further improve the system by linearly combining the geographical influence, social influence, and preference similarity. In order to enhance the performance, more advanced techniques are then applied, e.g., incorporating temporal influence [17], and utilizing a personalized geographical model via kernel density estimator [18,19]. On the other hand, model-based CF is proposed to tackle the POI recommendation problem that benefits from its scalability. Cheng et al. [1] propose a multi-center Gaussian model to capture user geographical influence and combine it with social matrix factorization (MF) model [12] to recommend POIs. Gao et al. [4] propose an MF-based model, Location Recommendation framework with Temporal effects (LRT), utilizing similarity between time-adjacent check-ins to improve performance. Lian et al. [9] and Liu et al. [11] enhance the POI recommendation by incorporating geographical information in a weighted regularized matrix factorization model [7]. In addition, some researchers subsume users' comments to improve the recommendation performance [5,8,16]. Other researchers model the consecutive check-ins' correlations to enhance the system [2,10,21].

Temporal Effect Modeling. In 2011, Cho et al. [3] propose the periodicity of check-in data in LBSNs. For instance, people always visit restaurants at noon. Hence, it is reasonable to recommend users restaurants he/she did not visit at noon. In 2013, Yuan et al. [17] improve the similarity metric in CF model by combining the temporal similarity and non-temporal similarity, which enhances the recommendation performance. At the same year, Gao et al. [4] observe the non-uniformness property (a user's check-in preference changes at different hours of a day), and consecutiveness (a user's preference at time t is similar with time $t-1$). Further, Gao et al. propose LRT model to capture the non-uniformness and consecutiveness. Meantime, Cheng et al. [2] propose FPMC-LR model to capture the consecutiveness, supposing strong correlation between two consecutive checked-in POIs.

3   Aggregated Temporal Tensor Factorization (ATTF) Model

3.1   Formulation

The Aggregated Temporal Tensor Factorization (ATTF) model estimates the preference of a user u at a POI l given a specific time label t through a score function $f(u, t, l)$. In our case, the time label t is represented in three scales, month (from Jan. to Dec.), week day (from Monday to Sunday), and hour (from 0 to 23). Denote that $\mathcal{U}$ is the set of users and $\mathcal{L}$ is the set of POIs. In addition, $\mathcal{T}_1$, $\mathcal{T}_2$, and $\mathcal{T}_3$ are the set of months, days of week, and hours respectively. Further we define $\mathcal{T}$ as the set of time label tuples, where a tuple $t := (t_1, t_2, t_3)$. For instance, a time stamp, "2011-04-05 10:12:35", could be represented as $(3, 2, 10)$ since we index the temporal id from zero. Therefore, the goal of ATTF model is to learn the preference score function $f(u, t, l)$, where u denotes a user id, t denotes a time label tuple, and l denotes a POI id. Under this circumstance, the score function subjects to that a high score corresponds to high preference of a user u for the POI l at time t.

Furthermore, we formulate the Aggregated Temporal Tensor Factorization (ATTF) model through interaction tensor factorization. Thus the score of a POI l given user u and time t is factorized into three interactions: user-time, user-POI, and time-POI, where each interaction is modeled through the latent vector inner product. As different candidate POIs share the same user-time interaction, the score function for comparing preference over POIs could be simplified as the combination of interactions of user-POI and time-POI. Formally, the score function for a given time label t, user u, and a target POI l could be formulated as,

$$f(u, t, l) = \left\langle U_u^{(L)}, L_l^{(U)} \right\rangle + \left\langle A\left(T_{1,t_1}^{(L)}, T_{2,t_2}^{(L)}, T_{3,t_3}^{(L)}\right), L_l^{(T)} \right\rangle, \qquad (1)$$

where $\langle \cdot \rangle$ denotes the vector inner product, $A(\cdot)$ is the aggregate operator. Suppose that d is the latent vector dimension, $U_u^{(L)} \in R^d$ is user u's latent vector for POI interaction, $L_l^{(U)}, L_l^{(T)} \in R^d$ are POI l's latent vectors for user interaction and time interaction, $T_{1,t_1}^{(L)}, T_{2,t_2}^{(L)}, T_{3,t_3}^{(L)} \in R^d$ are time t's latent vector representations in three aspects: month, week day, and hour.

Aggregate operator combines several temporal characteristics together. In this paper, we propose a linear convex combination operator. It is formulated as follows,

$$A(\cdot) = \alpha_1 \cdot T_{1,t_1}^{(L)} + \alpha_2 \cdot T_{2,t_2}^{(L)} + \alpha_3 \cdot T_{3,t_3}^{(L)}, \qquad (2)$$

where α_1, α_2, and α_3 denote the weights of each temporal factor, satisfying $\alpha_1 + \alpha_2 + \alpha_3 = 1$, and $\alpha_1, \alpha_2, \alpha_3 >= 0$.

3.2   Inference

We infer the model via BPR criteria [13] that is a general framework to train a recommendation system from implicit feedback. We treat the checked-in POIs as positive and the unchecked as negative. Suppose that the score of $f(u, t, l)$ at positive observations is higher than the negative given u and t, we formulate the relation that user u prefers a positive POI l_i than a negative one l_j at time t as $l_i >_{u,t} l_j$. Further, we extract the set of pairwise preference relations from the training examples,

$$D_S := \{(u, t, l_i, l_j) | l_i >_{u,t} l_j, u \in \mathcal{U}, t \in \mathcal{T}, l_i, l_j \in \mathcal{L}\}. \tag{3}$$

Suppose the quadruples in D_S are independent of each other, then to learn the ATTF model is to maximize the likelihood of all the pair orders,

$$\arg\max_{\Theta} \prod_{(u,t,l_i,l_j) \in D_S} p(l_i >_{u,t} l_j), \tag{4}$$

where $p(l_i >_{u,t} l_j) = \sigma(f(u, t, l_i) - f(u, t, l_j))$, σ is the sigmoid function, and Θ is the parameters to learn, namely $U^{(L)}$, $L^{(U)}$, $L^{(T)}$, $T_1^{(L)}$, $T_2^{(L)}$, and $T_3^{(L)}$. The objective function is equivalent to minimizing the negative log likelihood. To avoid the risk of overfitting, we add a Frobenius norm term to regularize the parameters. Then the objective function is

$$\arg\min_{\Theta} \sum_{(u,t,l_i,l_j) \in D_S} - ln(\sigma(y_{u,t,l_i} - y_{u,t,l_j})) + \lambda_{\Theta} ||\Theta||_F^2, \tag{5}$$

where λ_{Θ} is the regularization parameter.

Algorithm 1. ATTF model learning algorithm

Input: Training tuples $\{(u_i, t_i, l_i)\}_{i=1,...,N}$
Output: $U^{(L)}, T_1^{(L)}, T_2^{(L)}, T_3^{(L)}, L^{(U)}, L^{(T)}$
1: Initialize $U^{(L)}, T_1^{(L)}, T_2^{(L)}, T_3^{(L)}, L^{(U)}, L^{(T)}$
2: **repeat**
3:     Draw (u, t, l_p) uniformly from training tuples
4:     For $s = 1, 2, ..., k$, where k is the number of sampled negative POIs
5:         Draw (u, t, l_p, l_n) uniformly
6:         $\delta \leftarrow 1 - \sigma(f(u, t, l_p) - f(u, t, l_n))$
7:         Update parameters according to Eq. (6)
8: **until** convergence
9: **return** $U^{(L)}, T_1^{(L)}, T_2^{(L)}, T_3^{(L)}, L^{(U)}, L^{(T)}$

3.3   Learning

We leverage the Stochastic Gradient Decent (SGD) algorithm to learn the objective function for efficacy. Due to convenience, we define $\delta = 1 - \sigma(f(u, t, l_p) - f(u, t, l_n))$. As $T_{1,t_1}^{(L)}$, $T_{2,t_2}^{(L)}$, and $T_{3,t_3}^{(L)}$ are symmetric, they have the same gradient form. For simplicity, we use $T_t^{(L)} \in \{T_{1,t_1}^{(L)}, T_{2,t_2}^{(L)}, T_{3,t_3}^{(L)}\}$ to represent any of them, $\alpha \in \{\alpha_1, \alpha_2, \alpha_3\}$ to denote corresponding weight, and $A(\cdot)$ to denote $A(T_{1,t_1}^{(L)}, T_{2,t_2}^{(L)}, T_{3,t_3}^{(L)})$. Then the parameter updating rules are as follows,

$$
\begin{aligned}
U_u^{(L)} &\leftarrow U_u^{(L)} + \gamma \cdot \left(\delta \cdot \left(L_{l_p}^{(U)} - L_{l_n}^{(U)} \right) - \lambda \cdot U_u^{(L)} \right), \\
L_{l_p}^{(U)} &\leftarrow L_{l_p}^{(U)} + \gamma \cdot \left(\delta \cdot U_u^{(L)} - \lambda \cdot L_p^{(U)} \right), \\
L_{l_p}^{(T)} &\leftarrow L_{l_p}^{(T)} + \gamma \cdot \left(\delta \cdot A(\cdot) - \lambda \cdot L_{l_p}^{(T)} \right), \\
L_{l_n}^{(U)} &\leftarrow L_{l_n}^{(U)} - \gamma \cdot \left(\delta \cdot U_u^{(L)} + \lambda \cdot L_n^{(U)} \right), \\
L_{l_n}^{(T)} &\leftarrow L_{l_n}^{(T)} - \gamma \cdot \left(\delta \cdot A(\cdot) + \lambda \cdot L_{l_n}^{(T)} \right), \\
T_t^{(L)} &\leftarrow T_t^{(L)} + \gamma \cdot \left(\delta \cdot \alpha \cdot \left(L_{l_p}^{(T)} - L_{l_n}^{(T)} \right) - \lambda \cdot T_t^{(L)} \right),
\end{aligned}
\tag{6}
$$

where γ is the learning rate, λ is the regularization parameter. To train the model, we use the bootstrap sampling skill to draw the quadruple from D_S, following [13]. Algorithm 1 gives the detailed procedure to learn the ATTF model.

Complexity. The runtime for predicting a triple (u, t, l) is in $O(d)$, where d is the number of latent vector dimension. The updating procedure is also in $O(d)$. Hence training a quadruple is in $O(d)$, then training an example (u, t, l) is in $O(k \cdot d)$, where k is the number of sampled negative POIs. Therefore, to train the model costs $O(N \cdot k \cdot d)$, where N is the number of training examples.

4   Experiments

4.1   Data Description and Experimental Setting

Two real-world datasets are used in the experiment: one is Foursquare data from January 1, 2011 to July 31, 2011 provided in [6] and the other is Gowalla data from January 1, 2011 to September 31, 2011 in [20]. We filter the POIs checked-in by less than 5 users and then choose users who check-in more than 10 times as our samples. Table 1 shows the data statistics. We randomly choose 80 % of each user's check-ins as training data, and the remaining 20 % for test data. Moreover, we use each check-in (u, t, l) in training data to learn the latent features of user, time, and POI. Then given the (u, t), we estimate the score value of different candidate POIs, select the top N candidates, and compare them with check-in tuples in test data. Finally, we exploit the **precision** and **recall** to evaluate the model performance, following [8,21].

Table 1. Data statistics

Source	#users	#POIs	#check-ins	Density
Foursquare	10,180	16,561	867,107	0.0015
Gowalla	3,318	33,665	635,600	0.03

4.2 Model Comparison

To demonstrate the advantage of ATTF in aggregating several temporal latent factors, we propose three single temporal latent factor models as competitors: **TTFM**, **TTFW**, and **TTFH**. They correspondingly model the month, week day, and hour latent factor. We attain their results by setting the corresponding weight as 1, and others as 0 in ATTF.

We compare our ATTF model with state-of-the-art collaborative filtering (CF) methods: **WRMF** [7] and **BPRMF** [13], which are two popular MF models for processing large scale implicit feedback data. In addition, we also compare with state-of-the-art POI recommendation models capturing temporal influence: **LRT** [4] and **FPMC-LR** [2].

4.3 Experimental Results

In the following, we demonstrate the performance comparison on precision and recall for a top-N (N ranges from one to ten) POI recommendation task. We set the latent factor dimension as 60 for all compared models. We leverage grid search method to find the best weights in ATTF model. As a result, ATTF model on Foursquare data gets the best performance when $\alpha_1 = 0.7$, $\alpha_2 = 0.1$, and $\alpha_3 = 0.2$, while ATTF model on Gowalla data gets the best performance when $\alpha_1 = 0.2$, $\alpha_2 = 0.1$, and $\alpha_3 = 0.7$.

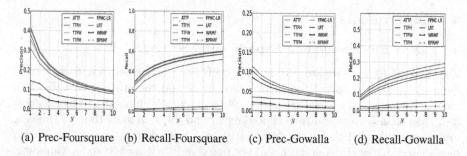

(a) Prec-Foursquare   (b) Recall-Foursquare   (c) Prec-Gowalla   (d) Recall-Gowalla

Fig. 1. Model comparison

Figure 1 demonstrates the experimental results on Foursquare and Gowalla data. We sihe following observations. **(1)** Our proposed models, ATTF, TTFH, TTFW, and TTFM, outperform state-of-the-art CF methods and POI recommendation models. Compared with the state-of-the-art competitor, FPMC-LR, ATTF model gains more than 20 % enhancement on Foursquare data, and more than 36 % enhancement on Gowalla data at Precision@5 and Recall@5. We observe that models perform better on Foursquare dataset than Gowalla dataset, even though it is sparser. The reason lies in Gowalla data contain much more POIs. **(2)** The ATTF model outperforms single temporal factor models. Compared with best single temporal factor model, ATTF gains about 3 % enhancement on Foursquare data, and about 10 % improvement on Gowalla data in a top-5 POI recommender system. **(3)** Our proposed models and FPMC-LR perform much better than other competitors, especially at recall measure. They try to recommend POIs at more specific situations, which is the key to improve performance. Our models recommend a user POIs at some specific time, and FPMC-LR recommends POIs given a user's recent checked-in POIs; while, the other three models give general recommendations.

In addition, different weight assignments on both data give us two interesting insights: (1) When data are sparse, the temporal characteristic on month dominates the POI recommendation performance. Foursquare dataset has high weight on month temporal factor. However, when data are denser, check-ins on hour are not so sparse. Consequently the characteristic on hour of day becomes prominent. (2) We usually pay much attention on characteristics on hour of day and day of week. Our experimental result indicates that the characteristic on month is important, especially for sparse data. The temporal characteristic on day of week is the least significant factor, which always has smallest weight in ATTF model.

5   Conclusion and Future Work

We propose a novel Aggregated Temporal Tensor Factorization (ATTF) model for POI recommendation. The ATTF model introduces time factor to model the temporal effect in POI recommendation, subsuming all the three temporal properties: periodicity, consecutiveness, and non-uniformness. Moreover, the ATTF model captures the temporal influence at different time scales through aggregating several time factor contributions. Experimental results on two real life datasets show that the ATTF model outperforms state-of-the-art models. Our model is a general framework to aggregate several temporal characteristics at different scales. In this work, we only consider three temporal characteristics: hour, week day, and month. In the future, we may add more in this model, e.g., splitting one day into several slots rather than in hour. In addition, we may try nonlinear aggregate operators, e.g., max, in our future work.

Acknowledgments. The work described in this paper was partially supported by the Research Grants Council of the Hong Kong Special Administrative Region, China (No. CUHK 14203314 and No. CUHK 14205214 of the General Research Fund), and 2015 Microsoft Research Asia Collaborative Research Program (Project No. FY16-RES-THEME-005).

References

1. Cheng, C., Yang, H., King, I., Lyu, M.R.: Fused matrix factorization with geographical and social influence in location-based social networks. In: AAAI (2012)
2. Cheng, C., Yang, H., Lyu, M.R., King, I.: Where you like to go next: successive point-of-interest recommendation. In: IJCAI (2013)
3. Cho, E., Myers, S.A., Leskovec, J.: Friendship and mobility: user movement in location-based social networks. In: SIGKDD (2011)

4. Gao, H., Tang, J., Hu, X., Liu, H.: Exploring temporal effects for location recommendation on location-based social networks. In: RecSys (2013)
5. Gao, H., Tang, J., Hu, X., Liu, H.: Content-aware point of interest recommendation on location-based social networks. In: AAAI (2015)
6. Gao, H., Tang, J., Liu, H.: gSCorr: modeling geo-social correlations for new check-ins on location-based social networks. In: CIKM (2012)
7. Hu, Y., Koren, Y., Volinsky, C.: Collaborative filtering for implicit feedback datasets. In: ICDM (2008)
8. Lian, D., Ge, Y., Zhang, F., Yuan, N.J., Xie, X., Zhou, T., Rui, Y.: Content-aware collaborative filtering for location recommendation based on human mobility data. In: ICDM (2015)
9. Lian, D., Zhao, C., Xie, X., Sun, G., Chen, E., Rui, Y.: GeoMF: joint geographical modeling and matrix factorization for point-of-interest recommendation. In: SIGKDD (2014)
10. Liu, X., Liu, Y., Aberer, K., Miao, C.: Personalized point-of-interest recommendation by mining users' preference transition. In: CIKM (2013)
11. Liu, Y., Wei, W., Sun, A., Miao, C.: Exploiting geographical neighborhood characteristics for location recommendation. In: CIKM (2014)
12. Ma, H., Zhou, D., Liu, C., Lyu, M.R., King, I.: Recommender systems with social regularization. In: WSDM (2011)
13. Rendle, S., Freudenthaler, C., Gantner, Z., Schmidt-Thieme, L.: BPR: bayesian personalized ranking from implicit feedback. In: UAI (2009)
14. Rendle, S., Schmidt-Thieme, L.: Pairwise interaction tensor factorization for personalized tag recommendation. In: WSDM (2010)
15. Ye, M., Yin, P., Lee, W.C., Lee, D.L.: Exploiting geographical influence for collaborative point-of-interest recommendation. In: SIGIR (2011)
16. Yin, H., Sun, Y., Cui, B., Hu, Z., Chen, L.: Lcars: a location-content-aware recommender system. In: SIGKDD (2013)
17. Yuan, Q., Cong, G., Ma, Z., Sun, A., Thalmann, N.M.: Time-aware point-of-interest recommendation. In: SIGIR (2013)
18. Zhang, J.D., Chow, C.Y.: iGSLR: personalized geo-social location recommendation: a kernel density estimation approach. In: SIGSPATIAL (2013)
19. Zhang, J.D., Chow, C.Y.: GeoSoCa: exploiting geographical, social and categorical correlations for point-of-interest recommendations. In: SIGIR (2015)
20. Zhao, S., King, I., Lyu, M.R.: Capturing geographical influence in POI recommendations. In: Lee, M., Hirose, A., Hou, Z.-G., Kil, R.M. (eds.) ICONIP 2013, Part II. LNCS, vol. 8227, pp. 530–537. Springer, Heidelberg (2013)
21. Zhao, S., Zhao, T., Yang, H., Lyu, M.R., King, I.: Stellar: spatial-temporal latent ranking for successive point-of-interest recommendation. In: AAAI (2016)

Multilevel–Multigroup Analysis Using a Hierarchical Tensor SOM Network

Hideaki Ishibashi, Ryota Shinriki, Hirohisa Isogai, and Tetsuo Furukawa$^{(\boxtimes)}$

Department of Brain Science and Engineering, Kyushu Institute of Technology,
Kitakyushu, Japan
{ishibashi-hideaki,shinriki-ryota}@edu.brain.kyutech.ac.jp,
{isogai,furukawa}@brain.kyutech.ac.jp

Abstract. This paper describes a method of multilevel–multigroup analysis based on a nonlinear multiway dimensionality reduction. To analyze a set of groups in terms of the probabilistic distribution of their constituent member data, the proposed method uses a hierarchical pair of tensor self-organizing maps (TSOMs), one for the member analysis and the other for the group analysis. This architecture enables more flexible analysis than ordinary parametric multilevel analysis, as it retains a high level of translatability supported by strong visualization. Furthermore, this architecture provides a consistent and seamless computation method for multilevel–multigroup analysis by integrating two different levels into a hierarchical tensor SOM network. The proposed method is applied to a dataset of football teams in a university league, and successfully visualizes the types of players that constitute each team as well as the differences or similarities between the teams.

Keywords: Self-organizing map · Multilevel–multigroup analysis · Tensor SOM

1 Introduction

The aim of this study is to develop a multilevel-multigroup analysis method based on a nonlinear multiway dimensionality reduction. The task is to analyze a set of groups in terms of the probability distributions of their constituent member data. A typical example is the analysis of sports teams belonging to the same league, in which the task is to visualize how each team is different or similar to other teams by comparing the statistical data of their members. In addition, we may like to visualize what types of players constitute those teams. A hierarchical analysis is required to realize this task, with the lower (member) level modeling the data distribution of each group and the upper (group) level analyzing the obtained distribution set.

The simplest method of multigroup analysis is to use the mean value of the constituent member data. In this case, the individual data are averaged over the members for each group to form the feature vector of the group [1] (although

© Springer International Publishing AG 2016
A. Hirose et al. (Eds.): ICONIP 2016, Part III, LNCS 9949, pp. 459–466, 2016.
DOI: 10.1007/978-3-319-46675-0_50

another parameter is sometimes used for the feature vectors [2]). Generally, a parametric model can be employed to represent the data distribution of each group, and the obtained parameters are regarded as the feature vectors of the groups. This approach is convenient in that the employed parametric model sufficiently captures the data distributions. However, if the model is not adequate, the parametric approach produces false results. For example, if the members of a group form some distinct clusters, then the average only represents the intermediate point between clusters, and there may not be any members around the mean point.

The alternative is to use a nonparametric representation such as a histogram. In this case, the obtained histograms are regarded as the feature vectors of the groups. This provides a more flexible representation than the parametric approach, but is only available for low-dimensional cases. To overcome this limitation, a combination of dimensionality reduction and nonparametric modeling is required. Thus, high-dimensional data are mapped to a low-dimensional latent space in advance, and the data distributions in the latent space are estimated by a nonparametric model [3]. This approach is promising, but dimensionality reduction sacrifices the translatability of the results, especially in nonlinear cases.

In this paper, we introduce a nonlinear multiway dimensionality reduction based on the tensor self-organizing map (TSOM). The TSOM is a visualization tool for relational datasets that produces two (or more) maps [4]. Though designed to analyze relational datasets, the TSOM can also visualize ordinary high-dimensional datasets by regarding them as relational ones. In this case, the TSOM produces a map of the target objects and a map of the data components, with the latter visualizing the intrinsic factors underlying the data. Thus, the TSOM enhances the translatability of high-dimensional datasets. By introducing the TSOM to multilevel–multigroup analysis, we achieve both a flexible representation of the member distributions and a translatable result.

Furthermore, the TSOM can be employed to visualize the groups, as the member–group affiliations can be represented by relational data. As a result, the entire architecture becomes a hierarchical network of two TSOMs, one for the member-level analysis and the other for the group-level analysis. Using the TSOM network, multilevel–multigroup analysis can be executed in a consistent and seamless manner.

The remainder of this paper is organized as follows. Section 2 formulates the task, before the algorithm is described in Sect. 3. In Sect. 4, this method is applied to analyze the football teams in a university league. The final section contains the conclusions to this study.

2   Problem Formulation

Suppose that we survey I subjects (members) with J queries. The entire dataset can be represented by a matrix $\mathbf{X} = (x_{ij}) \in \mathbb{R}^{I \times J}$. Suppose that there are K groups, and each member belongs to at least one of the groups. (Duplication is allowed.) This member–group affiliation is represented by a matrix $\mathbf{Y} = (y_{ik}) \in$

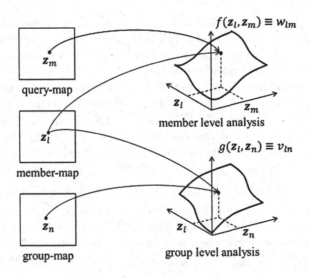

Fig. 1. Architecture of the proposed method.

$\{0, 1\}^{I \times K}$, where $y_{ik} = 1$ if the ith member belongs to the kth group, and otherwise $y_{ik} = 0$.

The aim of the proposed method is to visualize the relationships between the members, the groups, and the queries by mapping them to low-dimensional latent spaces via *member-maps*, *query-maps*, and *group-maps*, respectively. Thus, our task is to estimate the three latent variable sets $\left\{\mathbf{z}_i^{\text{member}}\right\}$, $\left\{\mathbf{z}_j^{\text{query}}\right\}$, $\left\{\mathbf{z}_k^{\text{group}}\right\}$. This task consists of two subtasks, corresponding to the member-level analysis and the group-level analysis.

In the member-level analysis, the subtask is to estimate the latent variables $\left\{\mathbf{z}_i^{(\text{member})}\right\}$, $\left\{\mathbf{z}_j^{(\text{query})}\right\}$, and a smooth nonlinear map f, so that the observed data is approximated as

$$x_{ij} \simeq f(\mathbf{z}_i^{(\text{member})}, \mathbf{z}_j^{(\text{query})}). \tag{1}$$

In the group-level analysis, the subtask is to estimate the latent variable $\left\{\mathbf{z}_k^{(\text{group})}\right\}$ and a nonlinear smooth map g, so that the member distribution of the kth group is represented as

$$p\left(\mathbf{z}^{(\text{member})} \,\middle|\, \mathbf{z}_k^{(\text{group})}\right) = g(\mathbf{z}^{(\text{member})}, \mathbf{z}_k^{(\text{group})}). \tag{2}$$

3 Architecture and Algorithm

3.1 Architecture

In an ordinary SOM, the latent space is discretized to grid nodes, and a reference vector is assigned to every node. In the case of the TSOM of order N, the TSOM has N latent spaces that are discretized to grid nodes, as in the ordinary SOM. However, unlike the conventional SOM, the reference vectors of the TSOM are assigned to all combinations of L nodes. Thus, if the latent spaces are discretized to L nodes, there are L^N reference vectors.

The proposed architecture for multilevel-multigroup analysis consists of two TSOMs of order 2, one for the member-level and the other for the group-level (Fig. 1). Suppose that $\mathbf{z}_l^{(\mathrm{member})}$, $\mathbf{z}_m^{(\mathrm{query})}$, $\mathbf{z}_n^{(\mathrm{group})}$ are the positional vectors in the latent spaces of the discretized nodes l, m, n. In the TSOM for the member-level analysis, reference vectors are assigned to all combinations of $\left\{ (\mathbf{z}_l^{(\mathrm{member})}, \mathbf{z}_m^{(\mathrm{query})}) \right\}$ so that $w_{lm} \equiv f(\mathbf{z}_l^{(\mathrm{member})}, \mathbf{z}_m^{(\mathrm{query})})$. Then, the entire set of reference vectors becomes a matrix $\mathbf{W} \equiv (w_{lm}) \in \mathbb{R}^{L \times M}$. The row vectors $\mathbf{w}_l^{(\mathrm{member})} \equiv (w_{l1}, \dots, w_{lM})$ and the column vectors $\mathbf{w}_m^{(\mathrm{query})} \equiv (w_{1m}, \dots, w_{Lm})$ act as the conventional reference vectors for members and for queries, respectively.

Similarly, reference vectors in the group-level TSOM are assigned to all combinations of $\left\{ (\mathbf{z}_l^{(\mathrm{member})}, \mathbf{z}_n^{(\mathrm{group})}) \right\}$ so that $v_{ln} \equiv g(\mathbf{z}_l^{(\mathrm{member})}, \mathbf{z}_n^{(\mathrm{group})})$. Thus, the entire set becomes a matrix $\mathbf{V} = (v_{ln}) \in \mathbb{R}^{L \times N}$. In this case, the column vector $\mathbf{v}_n^{(\mathrm{group})} \equiv (v_{1n}, \dots, v_{Ln})$ acts like a reference vector for groups.

Besides these reference vectors, the algorithm includes the vectors $\mathbf{u}_i^{(\mathrm{member})} \in \mathbb{R}^L$, $\mathbf{u}_j^{(\mathrm{query})} \in \mathbb{R}^M$, and $\mathbf{u}_k^{(\mathrm{group})} \in \mathbb{R}^N$. These play the role of data vectors for members, queries, and groups, and are calculated iteratively in the algorithm.

3.2   TSOM for Member-Level Analysis

The proposed algorithm consists of two stages: the member-level TSOM learns first, followed by the group-level TSOM. The learning algorithm for the TSOM is the expectation-maximization (EM) algorithm in a broad sense, with the E and M steps iterating alternately.

The learning algorithm of the member-level TSOM can be described as follows.

E Step. At calculation time t, the best matching nodes are determined from the result of the M step at $t - 1$.

$$l_i^\star = \arg\min_l \| \mathbf{u}_i^{(\mathrm{member})} - \mathbf{w}_l^{(\mathrm{member})} \|^2 \tag{3}$$

$$m_j^\star = \arg\min_m \| \mathbf{u}_j^{(\mathrm{query})} - \mathbf{w}_m^{(\mathrm{query})} \|^2. \tag{4}$$

(For the first loop, the best matching nodes are assigned randomly.) The latent variables are then estimated as $\mathbf{z}_i^{(\mathrm{member})} = \mathbf{z}_{l_i^\star}^{(\mathrm{member})}$, $\mathbf{z}_j^{(\mathrm{query})} = \mathbf{z}_{m_j^\star}^{(\mathrm{query})}$.

M-Step. $\mathbf{W}$ is calculated using a kernel smoother:

$$w_{lm} = \sum_{i=1}^I \sum_{j=1}^J R_{li}^{(\mathrm{member})} R_{mj}^{(\mathrm{query})} x_{ij}, \tag{5}$$

where $R_{li}^{(\text{member})}$ and $R_{mj}^{(\text{query})}$ are the neighborhood functions normalized with respect to data points $s.t.$ $\sum_i R_{li}^{(\text{member})} = 1$ and $\sum_j R_{mj}^{(\text{query})} = 1$. In this paper, we use $R_{li} \propto \exp\left[\|\mathbf{z}_l - \mathbf{z}_i\|^2 /2\sigma^2\right]$. After updating $\mathbf{W}$, $\{\mathbf{u}^{(\text{member})}\}$ and $\{\mathbf{u}^{(\text{query})}\}$ are also updated by

$$u_{im}^{(\text{member})} = \sum_{j=1}^{J} R_{mj}^{(\text{query})} x_{ij} \tag{6}$$

$$u_{jl}^{(\text{query})} = \sum_{i=1}^{I} R_{li}^{(\text{member})} x_{ij}. \tag{7}$$

These E and M steps are iterated until the TSOM reaches the steady state with a reduced neighborhood size.

3.3   TSOM for Group-Level Analysis

Once the first stage is finished, the group-level TSOM is trained.

Initialization. Before starting the process, $\{\mathbf{u}_k^{(\text{group})}\}$ are calculated in advance as:

$$u_{kl}^{(\text{group})} = \sum_{i=1}^{I} \tilde{R}_{li}^{(\text{member})} \tilde{y}_{ik}, \tag{8}$$

where $\tilde{y}_{ik} \equiv y_{ik}/Y_k$ and $Y_k \equiv \sum_i y_{ik}$. $\tilde{R}_{li}^{(\text{member})}$ is the neighborhood function normalized with respect to the latent space $s.t.$ $\sum_l \tilde{R}_{li}^{(\text{member})} = 1$. Note that $\mathbf{u}_k^{(\text{group})}$ represents the conditional probability $p(\mathbf{z}^{(\text{member})}|y_{ik} = 1)$, which means the member distribution of the kth group.

E Step. The E step determines the best matching nodes. As $\mathbf{u}_k^{(\text{group})}$ and $\mathbf{v}_n^{(\text{group})}$ represent conditional probabilities, the errors are evaluated by the cross entropy

$$n_k^\star = \arg\max_n \sum_{n=1}^{N} u_{kl}^{(\text{group})} \ln v_{ln}. \tag{9}$$

(For the first loop, the best matching nodes are assigned randomly.) The latent variables are then estimated as $\mathbf{z}_k^{(\text{group})} = \mathbf{z}_{n_k^\star}^{(\text{group})}$.

M Step. In the M step, the reference vector $\mathbf{V}$ is updated by

$$\mathbf{v}_n^{(\text{group})} = \sum_{k=1}^{K} \tilde{R}_{nk}^{(\text{member})} \mathbf{u}_k^{(\text{group})}. \tag{10}$$

Note that $\mathbf{V}$ represents the conditional probability $p(\mathbf{z}^{(\text{member})}|\mathbf{z}^{(\text{group})})$.

These E and M steps are iterated alternately until the group map converges.

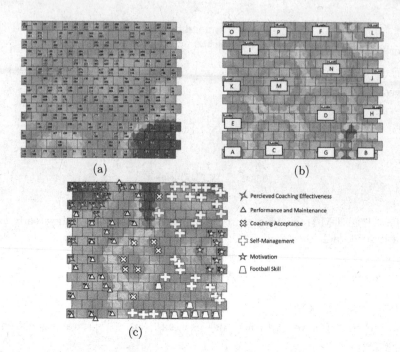

(a)   (b)

(c)

Fig. 2. The maps generated by our algorithm. (a) The member (player) map. (b) The group (team) map. (c) The query map. The color of these maps is represented by U-matrix. (Color figure online)

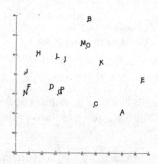

Fig. 3. Result of the team analysis by PCA. The average score of the members is used as the feature vector of each team.

4   Experiments

The proposed method was applied to a questionnaire survey from a university football league in Japan. The survey was taken for the 439 players of 16 universities. The questionnaire consisted of 112 queries asking how players assess themselves and their coaches. The queries were categorized as six types: (1) motivation, (2) self-management skill, (3) football skill, (4) coach acceptance,

(5) performance and maintenance (PM) of coaching, and (6) perceived coaching effectiveness (PCE). Questions (1)–(3) concern self-assessment, and (4)–(6) concern the assessments of the coach by the players. The data were standardized in advance so that the mean and variance were zero and one, respectively.

The results are shown in Fig. 1. In the query map (c), the six categories are separated, suggesting the result is plausible. To compare with the conventional method, the averaged team scores were analyzed by principal component analysis (PCA). The results are shown in Fig. 2. Compared with Fig. 1 (b), the arrangement of the teams is similar, but some teams (such as B and O) are located differently. Figures 3 and 4 show the constitution of teams B and O, indicating that both consist of three clusters of players (Figs. 4a, and 5a). The averaged score distributions are almost equal in the query maps (Figs. 4b, and 5b), but the score distributions of each cluster are quite different (Figs. 4c–e, 5c–e). The proposed method clearly discriminates between such differences. In addition, it is easy to see the properties of clusters intuitively using the interactive graphical interface of TSOM. These are the major advantages of using TSOM in such multilevel–multigroup analysis.

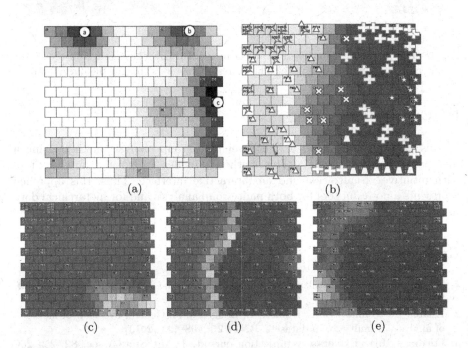

(a)   (b)

(c)   (d)   (e)

Fig. 4. The player distribution of team B in Fig. 1b. (a) The distribution of players from team B in the member map. (b) The averaged score of the members in the query map. (c)–(e) The scores of clusters a, b, c in the query map.

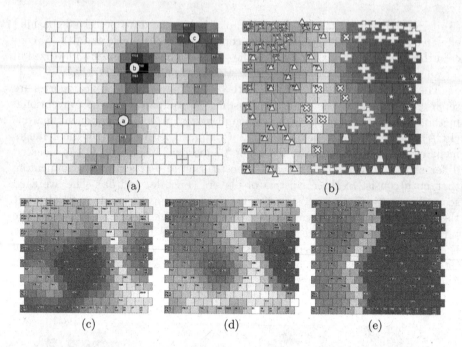

(a)   (b)

(c)   (d)   (e)

Fig. 5. The player distribution of team O in Fig. 1b. (b) The averaged score of the members in the query map. (c)–(e) The scores of clusters a, b, c in the query map.

5   Conclusion

In this paper, we proposed a multilevel–multigroup analysis method using a hierarchical TSOM network. This method provides an effective analysis tool with intuitive visualization and an interactive interface. Thus, this approach represents a powerful tool for both multigroup analysis and for hierarchical data in general.

References

1. Timmerman, M.E.: Multilevel component analysis. Br. J. Math. Stat. Psychol. **59**, 301–320 (2006)
2. Eslami, A., Qannari, E.M., Kohler, A., Bougeard, S.: General overview of methods of analysis of multi-group datasets. RNTI **25**, 108–123 (2013)
3. Friedman, J.H.: Exploratory projection pursuit. J. Am. Stat. Assoc. **82**, 259–266 (1987)
4. Iwasaki, T., Furukawa, T.: Tensor SOM and tensor GTM: nonlinear tensor analysis by topographic mappings. Neural Netw. **77**, 107–125 (2016)

A Wavelet Deep Belief Network-Based Classifier for Medical Images

Amin Khatami$^{(\boxtimes)}$, Abbas Khosravi, Chee Peng Lim, and Saeid Nahavandi

Institute for Intelligent Systems Research and Innovation, Deakin University,
Geelong, VIC 3216, Australia
{skhatami,abbas.khosravi,chee.lim,saeid.nahavandi}@deakin.edu.au

Abstract. Accurately and quickly classifying high dimensional data using machine learning and data mining techniques is problematic and challenging. This paper proposes an efficient and effective technique to properly extract high level features from medical images using a deep network and precisely classify them using support vector machine. A wavelet filter is applied at the first step of the proposed method to obtain the informative coefficient matrix of each image and to reduce dimensionality of feature space. A four-layer deep belief network is also utilized to extract high level features. These features are then fed to a support vector machine to perform accurate classification. Comparative empirical results demonstrate the strength, precision, and fast-response of the proposed technique.

Keywords: Deep belief network · Wavelet transforms · Classification

1 Introduction

Medical images are widely applied to medicine tasks for treatment and research. Manual classification of medical images is time consuming and impractical specially when there is a huge image dataset. Computer vision provides techniques to overcome these difficulties. Image processing is a method in the field of computer vision that extracts meaningful information from raw data. Analysing large datasets of the raw images are usually cumbersome, and always involves treating big data. A huge researches called "data analytics" has being developed widely utilizing statistical and computational intelligence techniques to overcome some difficulties such as reducing the dimensionality of big data to obtain a manageable size of input data. This results in a robust model for any data mining applications. Proposing a reliable classification system in medical approaches requires an accurate, efficient, and effective technique in data processing. Recent efforts have been proposed several data mining techniques to classify medical images, which use statistical methods such as Support vector machine (SVM), type-2 fuzzy logic system, etc. [1,2,4,11]. In more details, Celebi et al. [2] used SVM to classify features extracted from skin lesions in dermoscopy images. Lee and co-workers [1] introduced a classification model based on linear discriminant analysis. They combined the genetic algorithm with the random subspace

© Springer International Publishing AG 2016
A. Hirose et al. (Eds.): ICONIP 2016, Part III, LNCS 9949, pp. 467–474, 2016.
DOI: 10.1007/978-3-319-46675-0_51

method to estimate pulmonary nodule. Fuzzy type II is another classification technique utilized for medical approaches [5,11]. Hassanien et al. applied Fuzzy type II classifier along with a pulse coupled neural networks algorithm for detecting the region of interest [5]. Wavelet analysis is also utilized in this field in order to obtain proper information of data [7,10,11,16]. The authors mostly utilized the wavelet filtering as a pre-processing phase of their works. Processing time and accuracy performance are two main motivations behind this approach. The major goal to propose this article is to develop an efficient and effective technique to classify five different classes of IRMA (the Image Retrieval in Medical Applications) project dataset [14], which consists of 12,677 radiographs with known categories. Figure 1 depicts five classes of study here. The proposed method consists of three stages. We combine a pre-processing and a feature extractor phases along with a SVM classification stage to introduce the efficient system to classify those classes.

The main contributions of this study are: (1) to show the significant effect of using wavelet transforms on the performance of deep classification of X-ray images. (2) to demonstrate that an appropriately designed DBN selects features that allow a fast and precise classification. The rest of the paper is organised as follows. Next section, briefly describes the techniques used in this study. Section 2 presents the flow chart and pseudo-code of the algorithm. Section 3 reports the experimental results, followed by concluding remarks in Sect. 4.

Fig. 1. 5 classes selected form IRMA to be classified

1.1   Technical Route

- Single-level two-dimensional wavelet transformation: The 2D wavelet decomposition diagram proposed by Mallat [9] to find the corresponding density utilizes two Low Pass (LPF) and High Pass Filters. The approximation coefficients is called LL, and the details of the horizontal, vertical, and diagonal coefficient matrixes are called $LH, HL,$ and HH, respectively. Utilizing the low level filter for a picture in frequency domain performs computationally fast compared to that of its image domain.
- Feature extraction: Informative values of initial data called features help scientists to develop robust techniques for classifying the objects of interest in different applications. Several methods depended on applications in the computer vision approach are used to obtain these informative features. Shape feature extractor methods are playing a significant role in classification of

huge image dataset [15]. Deep learning that is a sub-field of machine learning is a tool to find meaningful shape-based features. Deep Belief Network (DBN) is a new sub-field of deep learning which learn to extract a deep hierarchical representation of the training data proposed by Hinton in 2006 [6]. The main reason to choose DBN in this paper as a feature extractor method is to generate robust features in its last layer with the capability to reconstruct visible data from generated features. It results in improving the classification performance [8]. DBN is an unsupervised greedy layer-wise training procedure in which the output of each individual RBM is fed to the visible layer of the next RBM. There is a supervised fine-tuning algorithm at the last stage of the Hinton's proposed technique to compute the derivatives at the top, and then propagate backward through the network.

- Classifiers: A classification problem is involved in determining the classes of new observations among a set of categories. A brief discussion on the two state-of-the-art supervised classifiers are mentioned as follows:
  - Softmax function: General speaking, a softmax layer at the top of the stacked RBMs in a DBN structure uses the Eq. 1, called Softmax function, to value the units in order to categorise the data fed to this layer. It drives those values close to the maximum to one, otherwise to zero.

$$p_j = \frac{exp(-x_j)}{\sum_{i=1}^{k} \exp(x_i)} \tag{1}$$

where p_j is the value for the jth unit, and x_j is the input of jth unit calculated by $x_j = a_j + \sum_i h_i w_{ij}$. At the result, the predicted class in the case of the jth unit is:

$$The\ predeicted\ class = arg_j max(p_j) \tag{2}$$

  - Support vector machine classifier: SVM is another state-of-the-art supervised discriminative classifier method maximizing the margin hyperplane in a converted feature space to categorise a dataset [3]. The output of SVM algorithm is an optimal hyperplane which classifies new observations.

2   Proposed Method

This study contributes to propose a wavelet-based discriminating feature selection system using DBN to extract high level features, following by a classifier to classify the selected features. For classification approach, usually, the softmax function is used at the top of the most of deep learning models. In 2013, Tang [13] demonstrated a consistent advantage of replacing the softmax layer with a linear SVM. Hence, in our proposed paper, we use linear SVM at the top of the model as the classifier. Figure 2 depicts the diagram of the proposed method, and the following pseudo-code explains its steps:

- Step 1 (Wavelet filtering): The low-level filter is utilized to transform the images from image domain to frequency one. It reduces the size of each image, and also the computation time.

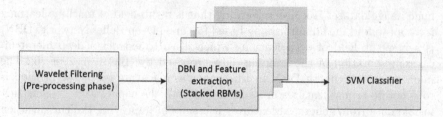

Fig. 2. The structure of proposed model

Table 1. Parameters of RBMs

Parameters	DBN structure			
	RBM 1	RBM 2	RBM 3	RBM 4
Visible units	25600	100	50	25
Latent units	Binary	Binary	Binary	Gaussian
Latent units #	100	50	25	3
Performance	Free energy	Free energy	Free energy	Free energy
Max epoch	50	50	50	50
Learning rate	0.1	0.1	0.1	0.001
Model	Generative	Generative	Generative	Generative

- Step 2 (Initializing): Initialize RBM parameters such as weight, visible and hidden biases, and define the number of visible and hidden units for each RBM.
- Step 3 (Adding layers): Add four RBMs in which each individual RBM has one visible and one hidden layer. Table 1 shows the description of the stacked RBMs used to construct DBN.
- Step 4 (Obtaining the high level features): After training the network, new observations are applied by running a procedure called getfeature function to obtain high level features of the new data. Therefore, the three high-level features for the test data are extracted in this stage by using the optimal parameters of the network. The output of this step is a two-dimensional matrix, $n \times 3$ which n is the number of test data.
- Step 5 (Classifying the high-level features): A 10-fold SVM cross-validation is applied to investigate the performance.

3   Experimental Result

The proposed method is carried out on a dataset called IRMA (the Image Retrieval in Medical Applications) project dataset. All images are classified according to the IRMA code [14]. In this paper, 5 classes of the total including of 10489 images are chosen with respect to their labels. Briefly speaking, all images

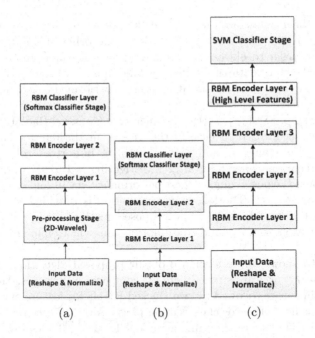

Fig. 3. (a) Model 1. (b) Model 2. (c) Model 3.

Table 2. Comparing accuracy and time consuming among models 1-4

Categories	Perf	Model 1		Model 2		Model 3		Proposed method	
		Train	Test	Train	Test	Train	Test	Train	Test
2-class	acc	0.9805	0.9750	0.9541	0.9378	0.9927	0.9865	0.9929	0.9865
	time(second)	219.7	0.4	6001.1	1.1	2810.7	0.3	688.4	0.08
5-class	acc	0.8304	0.8363	0.8250	0.8200	0.8010	0.8375	0.8964	0.8852
	time(second)	1342.1	0.6	8648.9	1.39	3649.7	1.2	830.9	0.3

are reshaped into 160 × 160 matrixes in which each individual image is demonstrated by a row vector with 25600 dimensions. Accordingly, the input data is a matrix with 10489 × 25600 dimensions (75 % of the total data are considered as the training set). 25600 is the number of features for each individual image that are fed into the system at first. This input matrix are passed through the wavelet filter in order to find the sparse representation of each image. Experimental results show that this pre-processing stage causes a better performance in the case of accuracy and processing time. In order to demonstrate the results, the input matrix is categorized into two groups; 2 classes in group one, and 5 classes in group two.

In the second stage of the proposed algorithm, the pre-processed input data is fed to a DBN to extract the three high level features using four hidden layers. This network starts with 25600 units in the visible layer, and then 100, 50, 25, and 3 units in the hidden layers respectively; 3 units in the last layer force the network to select the three most relevant of the features for representing each image. In other words, the 100 low level features of the pre-processed data are extracted from the first RBM, and then are pushed to the other three stacked RBMs sequentially in order to obtain the three high level features. In the next step, these three extracted features are used as the input data for the classification phase. The next supervised classification stage is performed based on these three high level features.

In the last step, SVM supervised classifier with Gaussian Radial Basis (RBF) kernel function is utilized to classify the images in the dataset. The amount of two parameters called C and $sigma$ plays an important role in terms of maximizing accuracy. Therefore, after investigating a variety value of C and $sigma$, we select the one giving the best performance via 10-fold cross-validation on the training dataset. The following explanation is provided to describe the experimental results achieved by the proposed method along with discussion of the classification accuracy and consumption time via 10-fold cross-validation technique.

In order to show the performance of our proposed algorithm, three other models shown in Fig. 3 are studied here for comparison purposes. Model 1 uses a per-processing stage followed by two stacked RBMs for feature extraction and a softmax classifier kernel for classification phase. Model 2 uses raw input data, and the further stages are the same as model 1. Model 3 uses raw input data followed by four stacked RBMs for feature extraction and a SVM classifier kernel for classification phase. Model 4 is our proposed method consisting of the pre-processing stage followed by four RBMs and the SVM classifier kernel. There is no wavelet filtering stage in the model 2 and 3. Also, the type of classifier is the main difference between model 1 and 4. The details of the DBN structure consisting of the stacked RBMs is shown in Table 1. Two experiments with two and fives classes are considered here. Table 2 reports the comparative results for two experiments evaluated by elapsed time to generate results and obtained accuracy. These outputs are provided by implementing the new model in Matlab 2014. All computation are performed on an Intel Core 7 CPU with 2.7 GHz.

2-class study: This category consists of a separation of the two most frequent classes of the dataset, 3587 chest and 1042 hand x-ray images depicted in Fig. 1. The input data of the first stage of the proposed method in the case of this category is a 4629×25600 matrix fed to DBN to gain the high level features in 3 dimensions. Finally, as described in Table 2, the accuracy for the test data of 2-class category, 0.9865 %, is achieved by a 10-fold cross-validation process on different $sigma$ and C. It is the mean of ten accuracy percentage provided.

5-class study: This category is including of the five most frequent classes of the dataset depicted in Fig. 1. The input data that is a 5968×25600 matrix is passed to the proposed algorithm to provide the 1494×3 matrix to be classified by SVM. The accuracy for the test data of the 5-class category, 0.8852 % , is achieved by a 10-fold cross-validation process on different $sigma$ and C. It is the mean of ten accuracy percentage provided. The functional boxplot is an informative data exploratory visualization tool for descriptive statistics describing the major attributes of a collection of information [12]. Figure 4 shows the box plot of

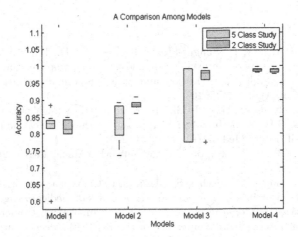

Fig. 4. 10-fold cross-validation results

the outcomes of 10-fold cross-validation on the above categories. It shows the proposed method obtains better results compared to other models studied.

4 Conclusion and Future Work

The diagnosis of medical images into different categories is one of the most troublesome studies in the medical image processing approaches, especially when there is a huge image dataset. In general, achieving high accuracy and precision might be significantly demanding and time-consuming. These difficulties motivate us to add the pre-processing phase to reduce the processing time, and examine different classifiers to categorise the data in order to obtain high accuracy. Experimental results show that the proposed model improves about 0.07 % accuracy performance compared to the model 2 using raw data without pre-processing at the beginning and using the softmax classifier at the end of algorithm for the classification phase. Moreover, it reduces time processing to about 11 times in the training process, and to about 5 times in the generation of the classes corresponding to testing data. In the future, other types of deep learning and classifiers will be tested with different kinds of pre-processing methods.

Acknowledgment. The authors would like to thank Thomas M. Deserno for the IRMA database support.

References

1. Lee, M.C., Boroczky, L., Sungur-Stasik, K., Cann, A.D., Borczuk, A.C., Kawut, S.M., Powell, C.A.: Computer-aided diagnosis of pulmonary nodules using a two-step approach for feature selection and classifier ensemble construction. Artif. Intell. Med. **50**(1), 43–53 (2010)
2. Celebi, M.E., Kingravi, H.A., Uddin, B., Iyatomi, H., Aslandogan, Y.A., Stoecker, W.V., Moss, R.H.: A methodological approach to the classification of dermoscopy images. Artif. Intell. Med. **31**(6), 362–373 (2007)
3. Cortes, C., Vapnik, V.: Support-vector networks. Mach. Learn. **20**(3), 273–297 (1995)
4. Doukas, C., Goudas, T., Fischer, S., Mierswa, I., Chatziioannou, A., Maglogiannis, I.: An open data mining framework for the analysis of medical images: application on obstructive nephropathy microscopy images. In: Annual Conference proceedings: Annual International Conference of the IEEE Engineering in Medicine and Biology Society. IEEE Engineering in Medicine and Biology Society (2010)
5. Hassanien, A.E., Kim, T.: Breast cancer mri diagnosis approach using support vector machine and pulse coupled neural networks. Mach. Learn. **10**(4), 277–284 (2012)
6. Hinton, G., Osindero, S., Teh, Y.W.: A fast learning algorithm for deep belief nets. Neural Comput. **18**, 1527–1554 (2006)
7. Jaleel, J.A., Salim, S., Archana, S.: Mammogram mass classification based on discrete wavelet transform textural features (2014)
8. Kim, Y., Lee, H., Provost, E.M.: Deep learning for robust feature generation in audiovisual emotion recognition (2013)
9. Mallat, S.G.: A theory for multiresolution signal decomposition: the wavelet representation. IEEE Trans. Pattern Anal. Mach. Intell. **11**(7), 674–693 (1989)
10. Meenakshi, R., Anandhakumar, P.: Wavelet statistical texture features with orthogonal operators tumour classification in magnetic resonance imaging brain. Am. J. Appl. Sci. **10**(10), 1154–1159 (2013)
11. Nguyen, T., Khosravi, A., Creighton, D., Nahavandi, S.: Medical data classification using interval type-2 fuzzy logic system and wavelets. Appl. Soft Comput. **30**, 812–822 (2015)
12. Sun, Y., Genton, M.G.: Functional boxplots. J. Comput. Graph. Stat. **20**(2), 316–334 (2011)
13. Tang, Y.: Deep learning using linear support vector machines. arXiv preprint (2013). arXiv:1306.0239
14. Tommasi, T., Caputo, B., Welter, P., Güld, M.O., Deserno, T.M.: Overview of the clef 2009 medical image annotation track (2010)
15. Yang, M., Kpalma, K., Ronsin, J.: A survey of shape feature extraction techniques. Pattern Recogn. **30**(4), 643–658 (2008)
16. Zainuddin, Z., Ong, P.: An effective and novel wavelet, neural network approach in classifying type 2 diabetics. Neural Netw. World **5**(12), 407–428 (2012)

Bayesian Neural Networks Based Bootstrap Aggregating for Tropical Cyclone Tracks Prediction in South China Sea

Lei Zhu[1], Jian Jin[1(✉)], Alex J. Cannon[2], and William W. Hsieh[2]

[1] Department of Computer Science and Technology, East China Normal University,
Shanghai 200241, People's Republic of China
jjin@cs.ecnu.edu.cn
[2] Department of Earth, Ocean and Atmospheric Sciences,
University of British Columbia, Vancouver, BC V6T 1Z4, Canada

Abstract. Accurate forecasting of Tropical Cyclone Track (TCT) is very important to cope with the associated disasters. The main objective in the presented study is to develop models to deliver more accurate forecasts of TCT over the South China Sea (SCS) and its coastal regions with 24 h lead time. The model proposed in this study is a Bayesian Neural Network (BNN) based committee machine using bagging (bootstrap aggregating). Two-layered Bayesian neural networks are employed as committee members in the committee machine. Forecast error is measured by calculating the distance between the real position and forecast position of the tropical cyclone. A decrease of 5.6 km in mean forecast error is obtained by our proposed model compared to the stepwise regression model, which is widely used in TCTs forecast. The experimental results indicated that BNN based committee machine using bagging for TCT forecast is an effective approach with improved accuracy.

1 Introduction

The coastal areas of the South China Sea (SCS) are directly affected by Tropical Cyclones (TCs), which extremely caused damage to people and property when they make landfall. So, delivering the accurate Tropical Cyclone Track (TCT) forecasts is utmost important when it comes to saving human lives and reducing economic loss.

Artificial Neural Networks (ANNs) have been widely applied to TCT forecasts. Some ANN based techniques are capable of producing forecasts with acceptable accuracy [1–4]. The forecasting of TCT is also a time series problem and associated techniques have been proposed [5,6]. According to Chu [7,8], the Bayesian framwork is shown to be suitable to establish the regression model for TC forecasts in the central North Pacific. Another aspect, ensemble techniques were developed and widely used in TCT forecasts with outstanding results obtained [9–12]. Tresp has developed Bayesian committee machine which performances well in prediction [13]. In committee machines, Bagging is

© Springer International Publishing AG 2016
A. Hirose et al. (Eds.): ICONIP 2016, Part III, LNCS 9949, pp. 475–482, 2016.
DOI: 10.1007/978-3-319-46675-0_52

Algorithm 1. The Procedure of Bagging

---

1 **begin**
2 | Initialize the parameter C **repeat**
3 | | Generate a subset $\mathbf{P}_c$ from training set $\mathbf{P}$ using bootstrapping
4 | | Train the c-th model using the training set $\mathbf{P}_c$
5 | **until** *the maximum number of iterations C is reached*;
6 | Averaging the predictions made by C different models to produce the
   | bagging prediction

---

widely used for its simplicity and efficiency in improving the prediction ability of experts [14]. However, bagging has not been applied for short-term (24 h) forecasting of TCs. Hence, we choose bagging to construct the forecast model and Bayesian Neural Networks (BNNs) employed as the committee members. The aim of this research is to deliver accurate short-term forecasts of TCTs in SCS and its coastal regions. We applied our proposed Bayesian neural network based bagging forecasting model to the short term forecasting of TCTs recorded in the 20-year period(1984-2003). The forecasting accuracy of our proposed model will be analysed in comparison with that obtained by stepwise regression.

2 Method

In this study, bootstrap aggregating (Bagging) is used and its committee member are two-layer feed-forward neural networks. Algorithm 1 illustrates the workflow of bagging. The Bayesian approach is used to train the MLP network [16]. In Bayesian approach, we approximate the prior distribution over the weights $\mathbf{w}$ to a Gaussian with zero mean and inverse variance α, under the assumption of small network weights,

$$p(\mathbf{w}) = \mathcal{N}(\mathbf{w}|\mathbf{0}, \alpha^{-1}\mathbf{I}). \tag{1}$$

Once a prior has been constructed, the network weights are then initialized using the values drawn from the prior distribution. In the network training process, the probability distribution over a target t given an input vector $\mathbf{x}$ is assumed to a Gaussian with zero mean and constant inverse variance β,

$$p(t|\mathbf{x}, \mathbf{w}) = \mathcal{N}(t|y(\mathbf{x}, \mathbf{w}), \beta^{-1}). \tag{2}$$

For the dataset of N observations $\mathbf{x_1}, \mathbf{x_2}, \ldots, \mathbf{x_N}$, with the corresponding target dataset $D = t_1, t_2, \ldots, t_N$, the likelihood is given by

$$p(D|\mathbf{w}) = \prod_{n=1}^{N} \mathcal{N}(t_n|y(\mathbf{x}_n, \mathbf{w}), \beta^{-1}). \tag{3}$$

So far, the prior distribution and the likelihood function have been obtained under the assumption that α and β are fixed. However, as a consequence of

---

Algorithm 2. The Training Process of Bayesian Neural Network

---

1 **begin**
2 | Set the number of the input units d, hidden units m, and output units n, respectively. Choose initial values for the hyperparameters α and β. Initialize the network weights using the values drawn from the weights prior $p(\mathbf{w})$ defined by α.
3 | **repeat**
4 | | **repeat**
5 | | | Optimize the weights using scaled conjugate gradients algorithm
6 | | **until** *the maximum iterative times y for scaled conjugate gradients algorithm is reached, or precision γ for weights updating is reached, or convergence precision δ is reached*;
7 | | **repeat**
8 | | | Re-estimate the hyperparameters α and β using the evidence procedure
9 | | **until** *the number of the evidence re-estimations z is reached*;
10 | **until** *the maximum iterative times x for weight optimizations is reached*;
11 | Make predictions using the optimal weights and hyperparameters.

---

the nonlinear dependency of the network output $y(\mathbf{x}, \mathbf{w})$ on $\mathbf{w}$, the posterior distribution $p(\mathbf{w}|\mathbf{D})$ is incalculable. To solve this problem, we use the scaled conjugate gradients algorithm to search a maximum of the posterior, which is denoted by $\mathbf{w}_{\mathbf{MAP}}$, by minimize the sum-of-squares error function. The posterior distribution is then approximated to a Gaussian with mean $\mathbf{w}_{\mathbf{MAP}}$ using Laplace approximation. Finally, the predictive distribution $p(t|\mathbf{x}, \mathbf{D})$ given an input $\mathbf{x}$ is obtained by marginalizing with respect to the approximated posterior,

$$p(t|\mathbf{x}, D) = \int p(t|\mathbf{x}, \mathbf{w})p(\mathbf{w}|D)d\mathbf{w}. \tag{4}$$

To eliminated the uncertainty with the arbitrary hyperparameters, the evidence procedure is used to determine the optimal weights and hyperparameters. Algorithm 2 illustrates the training process of Bayesian neural network.

3   Experiments

3.1   Data Set

The TCT dataset used in our experiment is published by China Meteorological Administration. The objects are the TCs generated in or moved into the South China Sea and last for at least 48 h. A real TCT forms a continuous line, and was sampled every 12 h, starting from the moment when a TC moves in the waters of SCS or generated in the area.

A TCT and its changes are associated with its intensity, accumulation and replenish of energy, and various nonlinear changes in its environmental flow field, which are referred to as variables in this paper. The variables used in our experiment include the climatology and persistence (CLIPER) factors representing changes of TC itself, such as changes in the latitude, longitude, and intensity of a TC at 12 and 24 h before prediction time.

750 samples were available during 1984 and 2003 for longitude and latitude, respectively. Each sample has 16 predictors chosen. The statistical informations of 16 common used predictors (v1 to v16) and the corresponding observations (Lat.t and Lon.t) for latitude and longitude of the TC are listed in Table 1. Our object is to predict the longitude and latitude of the TC center in the next 24 h [15]. The 750-sample data set have the records in chronological order in which the preceding 720 samples form the training set DATA_TRN, and the last 30 samples form the independent testing set DATA_TST.

Table 1. Variable information of the dataset

Var.	Min.	Max.	Mean	Std.	Meaning
v1	105.3	123.0	114.7	4.2	Initial Lon.(E)
v2	−39.8	23.1	−9.1	10.3	Zonal motion at −12 h
v3	−36.1	20.8	−9.3	9.7	Zonal motion at −24 h
v4	106	130	116.8	4.9	Lon.(E) at -12h
v5	−43.9	20.8	−9.6	10.3	Zonal motion from −24 to −12 h
v6	−4.5	2.7	−1.0	1.2	Lon differ between 0 and −12 h
v7	106.0	124.6	115.2	4.3	Lon.(E) at −6 h
v8	106.0	128.3	116.3	4.7	Lon.(E) at −24 h
v9	10.8	23.5	18.7	2.4	Initial Lat.(N)
v10	−17.4	26.6	4.2	6.5	Meridional motion at −12 h
v11	0.0	525.2	50.2	62.0	Squared zonal motion at −24 h
v12	10.0	23.7	18.2	2.4	Lat.(N) at −12 h
v13	−8.2	4.8	−2.1	2.2	Lon differ between 0 and −24 h
v14	10.3	23.8	18.5	2.4	Lat.(N) at −6 h
v15	9.6	24.0	18.0	2.5	Lat.(N) at −24 h
v16	10	60	23.5	9.9	Max surface wind at −6 h
Lat.t	101.4	128.3	112.8	4.7	Current Lon.(E)
Lon.t	12.2	30.0	19.8	2.6	Current Lat.(N)

3.2 Experiment Set Up

In our experiment, 10 parameter values are required to be initialized before training a committee member. When training the committee member, the 16 predictors are used for net-inputs, and the 2 targets are used for net-outputs. So, the parameter d and n, which should be consistent with the dimension of

Table 2. Parameter range of the bayesian neural network

Parameter	Min.	Max.	Meaning
m	1	30	Number of hidden units
α	0.01	1.00	Hyperparameter for the prior
β	1	100	Hyperparameter for the targets
x	10	30	Iterative times of the weight optimizations
y	30	200	Iterative times of the SCG algorithm
z	40	200	Iterative times of the evidence re-estimations

inputs and outputs of each committee member, are set to 16 and 2, respectively. The precision for weight updating and the convergence precision for the error function are both empirically set to 10^{-10}.

The committee machine model requires that each member has some degree of accuracy, which depends on the initial values of the parameters in Table 2. However, we cannot directly determine the initial values for the committee member to guarantee its accuracy due to the unpredictable impact of the initial values of these six parameters. To find proper initial parameter values that bring committee members accuracy, we generate 2000 sets of values which are randomly selected from large value ranges using the Sobol sequence [17,18], where each set of values is a group of six initialized parameters. For each group of parameters, one committee member is then trained using DATA_TRN, that is to say, 2000 committee members were obtained totally. All committee members are used to make prediction on the test set and some committee members with high forecast accuracy are marked and their corresponding initial parameter values are then used to estimate six numeric ranges for six parameters, respectively. The range of values obtained are listed in Table 2.

In this study, the committee machine consists of 30 members. For diversity among the committee members, each committee member is trained using different training set. So, the process of bootstrap sampling is independently executed for 30 times to produce a different training set for each committee member. The training set of each committee member is known as bootstrap dataset, which consisted of 720 samples that were randomly sampled with replacement from the original training set DATA_TRN. Training data not sampled are reserved as the validation set, which will be used for model selection. So far, 30 pairs of training set and validation set are constructed for 30 committee members, respectively.

For each training set, one model is trained to become a member of the committee machine. Before training a model, each of the six model parameters is initialized using the random value selected from its corresponding allowed value range provided by Table 2. However, we cannot make sure that any one of the models trained with these initialized parameters always has enough accuracy to be a member of the committee. To satisfy the requirement for the accuracy of the committee member, we train a number of candidate models with different initialized parameters using the same training set and select the model which best performance in predictive precision as a committee member, according to the distance error of the predictions these candidate models made on the validation set. Here, 200 candidate models are trained for each training set and the most accurate one is chosen based on the validation set as a member of the committee machine. When the construction of the committee machine is finished, the testing set DATA_TST is then used to evaluate the generalization performance and predictive precision of the final model.

When making predictions on the testing set, each committee member uses the same 30 testing samples in the DATA_TST and outputs 30 predictive values of the latitude and longitude. The committee machine then averages the predictions of 30 committee members to give the final bagging prediction. The distance error of a single test data point, Δd, is calculated by

$$\Delta d = \sqrt{\Delta x^2 + \Delta y^2} \times 110(km), \tag{5}$$

where Δx and Δy represent the absolute error of longitude and latitude of a single test data point, respectively. In the evaluation of the committee machine, the performance assessment indicators are the MAE (Mean Absolute Error) of longitude and latitude, and the MDE (Mean Distance Error), the mean of Δd over the testing set.

3.3   Result and Analysis

Figure 1 shows the MAE and the MDE of the committee. As can be seen from the top two panels, the solid lines are both lower than the dashed lines, which means the MAE of the bagging prediction is lower than the average MAE of 30 committee members, both in longitude and latitude. It indicates that the error of committee is reduced by averaging multi version of committee members. In the bottom panel, the solid line is lower than the dashed line, too. That is to say the bagging prediction has lower MDE, compared to the average MDE of 30 committee members. Besides, in all panels, extreme points of jagged line are distributed in a large fluctuation range and most of them are located above the solid line. That means, there are significant difference among the performances, whether measured by MAE of longitude (latitude) or MDE, of members in the committee, however, the ensemble performance of members is generally superior to individual committee member in prediction accuracy.

Stepwise regression analysis is a well-developed method in statistical forecast technologies, and it is an important tool for meteorological statistical forecast. In order to make an objective assessment for prediction performance of committee machine model established in this study, the prediction of stepwise regression is used as a comparison. Results of two methods are both listed in Table 3. As can be seen from the table, the MAE of latitude from the committee is slightly smaller than that from stepwise regression. However, the MAE of longitude obtained by the committee is more smaller than that obtained by stepwise regression analysis.

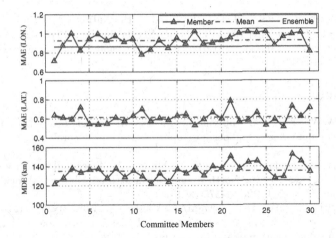

Fig. 1. Experimental results on DATA_TST of the committee. Each node of the gray jagged line represents the result obtained by a member in the committee. The blue dashed line represents the average result of 30 committee members, and the red solid line represents the result of the committee prediction, which is an ensemble of 30 member-predictions.

Table 3. Experimental results of the committee and stepwise regression

Method	MAE (Lon.)	MAE (Lat.)	MDE (km)
Committee	0.8649	0.5503	125.37
Stepwise regression	0.9471	0.5618	131.00

For MDE, there is a difference of 5.63 km between the two approaches. Hence the committee machine is superior to stepwise regression analysis in the performance of predictive precision when applied to the SCS TCT forecasts.

4   Conclusion

In this study, the Bayesian neural network based committee machine using bagging has been applied to the forecasting of tropical cyclone tracks over the South China Sea. Stepwise regression analysis is then used for comparison. The presented results suggest that the committee machine model established in this study has achieved an acceptable performance in tropical cyclone track forecast. The comparison results show that the committee machine model is superior to the stepwise regression model in forecast accuracy. Actually, this improvement in accuracy comes at the cost of time consuming during the network training, but this is acceptable for the reason that the training speed of network is fast enough to give forecasts within 12 h.

Acknowledgment. This work was supported by the National Natural Science Foundation of China (Grant No.61203301), Major Special Funds for Guangxi Natural Science Foundation (No.2011GXNSFE018006) and National Natural Science Foundation of China (No.41065002).

References

1. Hsieh, W.W., Tang, B.: Applying neural network models to prediction and data analysis in meteorology and oceanography. Bull. Am. Meteorol. Soc. **79**, 1855–1870 (1998)
2. Chaudhuri, S., Dutta, D., Goswami, S., Middey, A.: Track and intensity forecast of tropical cyclones over the North Indian Ocean with multilayer feed forward neural nets. Meteorol. Appl. **22**, 563575 (2015)
3. Chaudhuri, S., Dutta, D., Goswami, S., Middey, A.: Intensity forecast of tropical cyclones over North Indian Ocean using multi layer perceptron model: skill and performance verification. Nat. Hazards **65**, 97113 (2013)
4. Ali, M.M., Kishtawal, C.M., Jain, S.: Predicting cyclone tracks in the North Indian ocean: an artificial neural network approach. Geophys. Res. Lett. **34**, L04603 (2007)
5. Egrioglu, E., Yolcu, U., Aladag, C.H., Bas, E.: Recurrent multiplicative 5 neuron model artificial neural network for non-linear time series forecasting. Procedia - Soc. Behav. Sci. **109**, 1094–1100 (2014)
6. Egrioglu, E., Yolcu, U., Aladag, C.H., Bas, E., Kocak, C.: An ARMA type fuzzy time series forecasting method based on particle swarm optimization. Math. Probl. Eng. **2013**, 19 (2013)
7. Chu, P.S., Zhao, X., Ho, C.H., Kim, H.S., Lu, M.M., Kim, J.H.: Bayesian forecasting of seasonal typhoon activity: a track-pattern-oriented categorization approach. J. Clim. **23**, 66546668 (2010)
8. Chu, P.S., Zhao, X.: A bayesian regression approach for predicting seasonal tropical cyclone activity over the central north pacific. J. Clim. **20**, 40024013 (2007)
9. Lee, T.C., Leung, W.M.: Performance of multiple-model ensemble techniques in tropical cyclone track prediction. In: The 35th session of the Typhoon Committee, Chiang Mai, Thailand (2002)
10. Lee, T.C., Wong, M.S.: The use of multiple-model ensemble techniques for tropical cyclone track forecast at the Hong Kong Observatory. In: WMO Commission for Basic Systems Technical Conference on Data Processing and Forecasting Systems (2002)
11. Wang, Q., Liu, J., Zhang, L.: The study on ensemble prediction of typhoon track. J. Meteorol. Sci. **32**, 137–144 (2012)
12. Goerss, J.S.: Tropical cyclone track forecasts using an ensemble of dynamical models. Mon. Weather Rev. **128**, 1187–1193 (2000)
13. Tresp, V.: A Bayesian committee machine. Neural Comput. **12**, 2719–2741 (2000)
14. Breiman, L.: Bagging predictors. Mach. Learn. **24**, 123140 (1996)
15. Jin, L., Huang, X., Shi, X.: A study on influence of predictor multicollinearity on performance of the stepwise regression prediction equation. Acta Meteorol. Sin. **24**, 593–601 (2010)
16. Hsieh, W.W.: Machine Learning Methods In The Environmental Sciences: Neural Networks and Kernels. Cambridge University Press, New York (2009)
17. Sobol, I.M.: On the distribution of points in a cube and the approximate evaluation of integrals. USSR Comput. Math. Math. Phys. **7**, 86–112 (1967)
18. Joe, S., Kuo, F.Y.: Remark on algorithm 659: implementing Sobol's quasirandom sequence generator. ACM Trans. Math. Softw. (TOMS) **29**, 49–57 (2003)

Credit Card Fraud Detection Using Convolutional Neural Networks

Kang Fu, Dawei Cheng, Yi Tu, and Liqing Zhang[✉]

Key Laboratory of Shanghai Education Commission for Intelligent Interaction and
Cognitive Engineering, Department of Computer Science and Engineering,
Shanghai Jiao Tong University, Shanghai, China
{fukang1993,dawei.cheng,tuyi1991,lqzhang}@sjtu.edu.cn

Abstract. Credit card is becoming more and more popular in financial transactions, at the same time frauds are also increasing. Conventional methods use rule-based expert systems to detect fraud behaviors, neglecting diverse situations, extreme imbalance of positive and negative samples. In this paper, we propose a CNN-based fraud detection framework, to capture the intrinsic patterns of fraud behaviors learned from labeled data. Abundant transaction data is represented by a feature matrix, on which a convolutional neural network is applied to identify a set of latent patterns for each sample. Experiments on real-world massive transactions of a major commercial bank demonstrate its superior performance compared with some state-of-the-art methods.

Keywords: Credit card fraud · Convolutional neural network · Imbalanced data

1 Introduction

With the rapid development of economy globalization in recent decades, credit cards are much more popular in commercial transactions. The corresponding problem of the credit card fraud emerges accordingly. Machine learning approaches have been proposed to overcome these challenges. Kokkinaki [4] proposed the decision tree and boolean logic functions to characterize the normal transaction modes so as to detect fraudulent transactions. However, some of the fraudulent transactions similar to the legitimate trading patterns can not be identified. So neural networks and Bayesian networks have been employed. Ghosh [2] used a neural network to detect credit card frauds. Bayesian belief networks and artificial neural networks have been also introduced to tackle the problem [6]. These models have been criticized for being overly complex to detect frauds and there has been a high probability of being over-fitting. In order to reveal the latent patterns of fraudulent transactions and avoid the model over-fitting, we use a convolutional neural network to reduce the feature redundancy effectively.

How to generate features of credit card transactions successfully is one of the major challenges to machine learning approaches. Some aggregation strategies

© Springer International Publishing AG 2016
A. Hirose et al. (Eds.): ICONIP 2016, Part III, LNCS 9949, pp. 483–490, 2016.
DOI: 10.1007/978-3-319-46675-0_53

[1,3,7,8] have been proposed to obtain the customer spending modes in the recent transactions. But these models can not describe the complicated patterns of consumer spending. Therefore, we design a novel trading feature called trading entropy based on the latest consumption preferences for each customer. In order to apply a convolutional neural network(CNN) to credit card fraud detection, we need to transform features into a feature matrix to fit the CNN model.

Besides, extremely imbalanced data is another issue in fraud detection. The random undersampling method for dominated class is a common technique to adjust the ratio of the minority. Unfortunately, it will inevitably dismiss valuable information. In this paper we generate synthetic fraudulent samples from real frauds by a cost based sampling method. Thus, we can get a comparable number of frauds with legitimate transactions for training.

In brief, the main contributions to this paper can be summarized as bellows:

1. We propose a CNN-based framework of mining latent fraud patterns in credit card transactions.
2. We transform transaction data into a feature matrix for each record, by which the inherent relations and interactions in time series can be revealed for the CNN model.
3. By combining the cost based sampling method in characteristic space, the extremely imbalanced sample sets are alleviated, yielding a superior performance of fraud detection.
4. A novel trading feature called trading entropy is proposed to identify more complex fraud patterns.

2 Methodology

In this section, we firstly provide a description of our CNN-based fraud detection framework. Secondly, we propose a novel trading feature. Thirdly, the cost based sampling method is elaborated. At the end of this section, the CNN model is employed to the problem of credit card fraud detection.

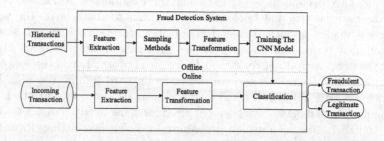

Fig. 1. The illustration of credit card fraud detection system.

2.1   Fraud Detection Framework

Our fraud detection framework is shown as Fig. 1, it consists of training and prediction parts. The training part mainly includes four modules: feature engineering, sampling methods, feature transformation and a CNN-based training procedure. The training part is offline and the prediction part is online. When a transaction comes, the prediction part can judge whether it is fraudulent or legitimate immediately. The detection procedure consists of feature extraction, feature transformation and the classification module.

For feature extraction, we adopt the aggregation strategies from [1,3,7,8]. In our system, we add trading entropy to the collection of traditional features so as to model more complicated consuming behaviors.

In the general process of data mining, we train the model after feature engineering. But a problem is that the data of credit card is extremely imbalanced. We propose a cost based sampling method to generate synthetic frauds.

Besides, in order to apply the CNN model to this problem, we need to transform features into a feature matrix to fit this model.

2.2   Feature Engineering

For traditional features, we can define the average amount of the transactions with the same customer during the past period of time as AvgAmountT. T means the time window length. For example, we can set T as different values: one day, two days, one week and one mouth, then four features of these time windows are generated. Table 1 shows the details of our feature types. Traditional feature types can not describe the complicated patterns of consumer spending. Therefore, we propose a new kind of feature called trading entropy. Assume in all transactions of the same customer during the past period of time before the current transaction, there are K kinds of merchant types, the total amount is TotalAmountT, the sum amount of the i-th merchant type is $AmountT_i (i = 1, 2, \ldots, K)$, the proportion of the i-th merchant type is p_i:

$$p_i = \frac{AmountT_i}{TotalAmountT} \qquad (1)$$

The entropy of the i-th merchant type can be defined as $EntT$:

$$EntT = -\sum_i^K p_i log p_i \qquad (2)$$

The above calculations only use previous transactions while the current transaction is not involved in. Then we add the current transaction to join the above calculation to obtain the current entropy: $NewEntT$. So the trading entropy is defined as $TradingEntropyT$:

$$TradingEntropyT = EntT - NewEntT \qquad (3)$$

If the trading entropy is too large, it has a higher probability of being fraudulent.

2.3 Cost Based Sampling

The cost based sampling method is developed based on the following observation. The fraudulent transactions near the decision boundary have higher probabilities to generate more synthetic fraudulent samples. For the i-th fraud transaction, the number of frauds around i is defined as fd_i, and the number of normal transactions around i is defined as nd_i, the cost of the i-th transaction can be defined as $cost_i$. d_{ij} is the distance between the i-th fraud and j-th transaction. The number of neighborhoods of the i-th fraud can be limited by a transaction function $f(x)$ and a cutoff value. $f(x) = 1$ if $x < 0$ and $f(x) = 0$ otherwise and C is the cutoff.

$$cost_i = \frac{\sum_{j \in legitimate} f(d_{ij} - C)}{\sum_{k \in fraud} f(d_{ik} - C)} \tag{4}$$

After obtaining the cost of each fraudulent transaction, we use k-means algorithm to divide the frauds into some clusters. If we want to generate a new fraud sample, we choose a fraud transaction x_1 as the seed in accordance with the cost. Then we choose another fraud transaction x_2 from the same cluster as x_1. The new synthetic fraud sample can be generated as $newFraud = \alpha \cdot x_1 + (1 - \alpha) \cdot x_2$ where α is randomly generated between 0 and 1.

Table 1. The description of different feature types. The first seven feature types are traditional and the last one is proposed by us.

Feature types	Description
AvgAmountT	Average amount of the transactions during the past period of time
TotalAmountT	Total amount of the transactions during the past period of time
BiasAmountT	The bias of the amount of this transaction and AvgAmountT
NumberT	Total number of the transactions during the past period of time
MostCountryT	The mostly used country during the past period of time
MostTerminalT	The mostly used terminal during the past period of time
MostMerchantT	The mostly used merchant type during the past period of time
TradingEntropyT	The trading entropy gain during the past period of time

2.4 CNN Modeling

In this paper we apply the convolutional neural network to detect the frauds of credit cards, because the CNN model is suitable for training a large size of data and it has the mechanism to avoid the model over-fitting. Convolutional neural networks have been successfully applied to some fields, such as image classification and speech signal processing. But not all kinds of data are suitable for the CNN model. The method of feature transformations is proposed to adapt the CNN model. The features of credit card transactions can be partitioned into several groups. And each group has different features by different time windows.

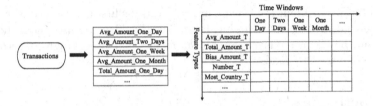

		Time Windows				
		One Day	Two Days	One Week	One Month	...
	Avg_Amount_T					
	Total_Amount_T					
	Bias_Amount_T					
	Number_T					
	Most_Country_T					
	...					

(Transactions → Avg_Amount_One_Day / Avg_Amount_Two_Days / Avg_Amount_One_Week / Avg_Amount_One_Month / Total_Amount_One_Day / ... → Feature Types)

Fig. 2. The illustration of feature transformation. One dimensional features are generated from original attributes of transactions, then they are transformed into feature matrices.

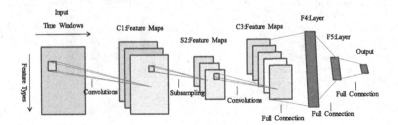

Fig. 3. The illustration of feature matrices. We randomly select four fraudulent and four legitimate samples to generate their heat maps after feature transformation.

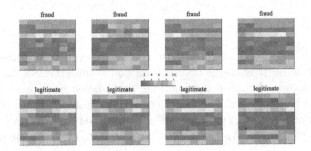

Fig. 4. The structure of our CNN model.

Two features of the same feature type by different time windows have strong relationship. Therefore, in the feature matrix, these two features are set in close positions. The original features are one dimensional. We can reshape them as a feature matrix where the rows have different feature types and the columns have different time windows. The procedure of transforming original features into a feature matrix is shown as Figs. 2 and 3.

These heat maps show the strong local correlation, in both row and column formats. According to the local correlation in the feature matrix, the CNN model can reduce the time complexity of data processing while retaining useful information. Our CNN structure is similar with the LeNet [5]. There are six layers in total. The input is a feature matrix. And the first layer is a convolutional layer, the following is a sub sampling layer. The third layer is also a convolutional layer. And the last three layers are all full connection layers. Figure 4 shows the structure of our CNN model.

3   Experiments

This section is organized as three parts. Firstly, we introduce the dataset. Secondly, we show the importance of trading entropy. Finally, we demonstrate the best accuracy of the CNN model.

3.1   Dataset

To evaluate the proposed model, we use real credit card transaction data from a commercial bank. It contains over 260 million transactions of credit cards in a year. About four thousand transactions are labeled as frauds and the rest are legitimate transactions. The transaction data is divided into two sets. We take the data of the first 11 months as the training set and the data of the next month as the testing set. And we take the F1 score to evaluate the performance of models.

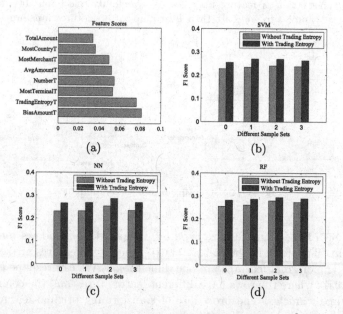

Fig. 5. Feature Evaluations. (a) Feature scores of different feature types. (b) The performances of SVM with trading entropy or not on different sample sets. (c) The performances of the neural network. (d) The performances of random forest.

3.2   Feature Evaluation

To evaluate the importance of trading entropy, we propose the feature score:

$$FeatureScore = \frac{1}{T} \sum_{t=1}^{T} \frac{\|u_t^f - u_t^l\|_1}{\sqrt{S_t^f + S_t^l}} \tag{5}$$

where T is the size of time windows, u_t^f and u_t^l represent the means of the fraud and legitimate samples for a given feature on t-th time window respectively, S_t^f and S_t^l are variances of the fraudulent and legitimate feature respectively. These scores are computed for each feature type. If the score is higher, this feature type is more important. Figure 5(a) shows the feature scores of different feature types. We can observe that the trading entropy ranks highly.

In order to demonstrate the efficiency of trading entropy better, we use different models and various sampling rates to obtain the performances with or without the trading entropy features. The testing results are illustrated in Fig. 5.

3.3   Model Evaluation

As shown above, the proposed trading entropy could significantly improve the classification accuracy. To relieve the problem of the imbalanced dataset, we employ the cost based sampling method to generate different number of frauds. In our experiments, we increase the fraudulent samples to 1, 2 and 3 times of the size of original frauds respectively. The samples of legitimate transactions are randomly undersampled. After feature engineering and data sampling, we evaluate the performance of the CNN model by comparing with other existing models. Figure 6 shows the comparison results. We can find that the cost based sampling method can make use of more legitimate data and alleviate the imbalanced problem. Besides, the CNN model on different sample sets achieves the best performance.

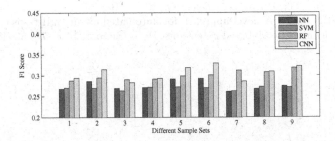

Fig. 6. The performances of different models on various sample sets.

4   Conclusion

In this paper, we introduce a CNN-based method of credit card fraud detection. And the trading entropy is proposed to model more complex consuming behaviors. Besides, we recombine the trading features to feature matrices and use them in a convolutional neural network. Experimental results from the real transaction data of a commercial bank show that our proposed method performs better than other state-of-art methods.

Acknowledgements. The work was supported by the National Natural Science Foundation of China (61272251), the Key Basic Research Program of Shanghai Municipality, China (15JC1400103) and the National Natural Science Foundation of China (91420302).

References

1. Bhattacharyya, S., Jha, S., Tharakunnel, K., Westland, J.C.: Data mining for credit card fraud: a comparative study. Decis. Support Syst. **50**(3), 602–613 (2011)
2. Ghosh, S., Reilly, D.L.: Credit card fraud detection with a neural-network. In: Proceedings of the Twenty-Seventh Hawaii International Conference on System Sciences, 1994, vol. 3, pp. 621–630. IEEE (1994)
3. Khandani, A.E., Kim, A.J., Lo, A.W.: Consumer credit-risk models via machine-learning algorithms. J. Banking Finan. **34**(11), 2767–2787 (2010)
4. Kokkinaki, A.I.: On a typical database transactions: identification of probable frauds using machine learning for user profiling. In: Proceedings of Knowledge and Data Engineering Exchange Workshop, 1997, pp. 107–113. IEEE (1997)
5. LeCun, Y., Bottou, L., Bengio, Y., Haffner, P.: Gradient-based learning applied to document recognition. Proc. IEEE **86**(11), 2278–2324 (1998)
6. Maes, S., Tuyls, K., Vanschoenwinkel, B., Manderick, B.: Credit card fraud detection using bayesian and neural networks. In: Proceedings of the 1st International Naiso Congress on Neuro Fuzzy Technologies, pp. 261–270 (2002)
7. Ravisankar, P., Ravi, V., Rao, G.R., Bose, I.: Detection of financial statement fraud and feature selection using data mining techniques. Decis. Support Syst. **50**(2), 491–500 (2011)
8. Van Vlasselaer, V., Bravo, C., Caelen, O., Eliassi-Rad, T., Akoglu, L., Snoeck, M., Baesens, B.: Apate: A novel approach for automated credit card transaction fraud detection using network-based extensions. Decis. Support Syst. **75**, 38–48 (2015)

An Efficient Data Extraction Framework for Mining Wireless Sensor Networks

Md. Mamunur Rashid[1]([✉]), Iqbal Gondal[1,2], and Joarder Kamruzzaman[1,2]

[1] Faculty of Information Technology, Monash University, Melbourne, Australia
{md.rashid,iqbal.gondal,joarder.kamruzzaman}@monash.edu
[2] ICSL, Federation University, Ballarat, Australia

Abstract. Behavioral patterns for sensors have received a great deal of attention recently due to their usefulness in capturing the temporal relations between sensors in wireless sensor networks. To discover these patterns, we need to collect the behavioral data that represents the sensor's activities over time from the sensor database that attached with a well-equipped central node called sink for further analysis. However, given the limited resources of sensor nodes, an effective data collection method is required for collecting the behavioral data efficiently. In this paper, we introduce a new framework for behavioral patterns called associated-correlated sensor patterns and also propose a MapReduce based new paradigm for extract data from the wireless sensor network by distributed away. Extensive performance study shows that the proposed method is capable to reduce the data size almost 50 % compared to the centralized model.

Keywords: Wireless sensor networks · Data mining · Data extraction · Knowledge discovery · Associated-correlated sensor pattern

1 Introduction

Advances in technologies have allowed the development of a new kind of network, called the wireless sensor network (WSN). These WSNs consists of sensor nodes that are capable of sensing, processing, and transmitting. The main function of each sensor in WSN is to sense the surrounding area and send the detected event to a predefined access point called sink in multi hop fashion. The detected events are send to the sink periodically or if they meet a particular testify [1].

In WSNs time is a very important issue that introduces the possibility of temporal correlations between sensors. These relations can be use for predicting the sources of future events. Recently data mining techniques received a great deal of attention for extracting temporal correlations between sensors. However, the stream nature of the sensor data and the limited resources of wireless sensor networks bring new challenges to the data mining techniques that need to be addressed. Data mining techniques are used to extract patterns that describe the behavior of the network. These behavioral patterns use to enhance the performance of the sensor network and thus improve the overall QoS [2].

© Springer International Publishing AG 2016
A. Hirose et al. (Eds.): ICONIP 2016, Part III, LNCS 9949, pp. 491–498, 2016.
DOI: 10.1007/978-3-319-46675-0_54

Several works have been proposed for utilizing association rules for WSNs [3,4], where the data values of the sensors have targeted. This means the sensor's values are the main object of the rules. Sensor association rules proposed in [5] and associated sensor patterns proposed in [6], where sensor's are the main values regardless of their values. An example of sensor association rules could be $(s_1, s_2 \rightarrow s_3, 90\%, \lambda)$ which means that if sensor s_1 and s_2 detect events within time λ, then there is 90% of chance that s_3 detects events within same time interval. However, association rule mining with real datasets is not so simple. When the minimum support threshold is set low, a huge number of association rules will usually generated, a majority of them are redundant or non-informative. In this paper, we introduce a new type of behavioral pattern called associated-correlated sensor patterns that capture the temporal relations between sensors more accurately.

Two main steps are essential for mining associated-correlated sensor patterns. The first step is deciding how to collect data that describes the sensors activities which is called behavioral data. The second step is focusing how to generate these patterns. In this paper, we discuss the first step and mainly how to collect the behavioral data effectively. Sensor data values can be collected using existing routing protocols such as Direct Diffusion [7] or data gathering algorithms LEACH [8], PEGASIS [9], Chain- Based Three Level Scheme [10]. The problem is that, these methodologies are not designed to collect data for mining purpose. Boukerche et al. [5] proposed two methodologies for extracting the data from WSN for mining sensor associations. These are direct reporting and Decentralized storage. The primary assumption of this work is to attach a flash memory to each sensor to store metadata about the sensors behavior. Sensor behavioral data presents these activities of sensors over time and defines the intervals between the sensor's activities. Due to the limited resources of sensor nodes, collecting this data is a costly process. In most data collecting models, a sensor is not only a data source, but also forwards the data for other sensors. As a result, this data forwarding strategy may incur significant energy consumption. To reduce energy consumption and prolong the network life time of WSNs in this paper we propose a MapReduced based Decentralized data extraction mechanism that not only overcome the problem but also prolong the network lifetime and enhance the performance of the network.

The main contribution of this paper can be summarized as follows. First, a framework of the associated-correlated sensor pattern mining problem in which the sensors themselves are the main object is proposed. Second, an efficient data collection method is presented. Finally, a performance analysis of the Decentralized methodology is presented.

2   Associated-Correlated Sensor Pattern Mining in WSNs: Problem Formulation

Let $S = \{s_1, s_2, ..., s_p\}$ be a set of sensor in a particular wireless sensor network. We assume that the time is divided into equal-sized slots $t = \{t_1, ..., t_q\}$ such that $t_{j+1} - t_j = \lambda, j \in [1, q-1]$ where λ is the size of the each time slot. A set $P = \{s_1, s_2, ..., s_n\} \subseteq S$ is called a pattern of a sensors. A sensor database, SD, is defined to be a set of epochs where each epoch is a tuple $E(E_{ts}, Y)$ such that Y is a pattern of the event detecting sensors that report events within the same time slot and E_{ts} is the epoch's time slot. Let size(E) be the size of E i.e., the number of sensors in E. An epoch $E(E_{ts}, Y)$ supports a pattern Y' if $Y \subseteq Y'$. Pattern Y' is said to be a frequent pattern if $Freq(Y', SD) \geq min\_sup$, where min_sup is a user given *minimum support threshold* in percentage of SD size in number of epochs.

The interestingness measure *all-confidence* denoted by α of a pattern Y is defined as follows:

$$\alpha = \frac{Sup(Y)}{Max\_sensor\_Sup(Y)} \tag{1}$$

Definition 1 (Associated Pattern): A pattern is called an associated pattern, if its *all-confidence* is greater than or equal to the given minimum *all-confidence* threshold. In statistical theory, $s_1, s_2, ..., s_p$ are independent if $\forall K$ and $\forall 1 \leq i_1 < i_2 < ... < i_k \leq n$,

$$P(s_{i_1}, s_{i_2}, ..., s_{i_k}) = P(s_{i_1}), P(s_{i_2}), ..., P(s_{i_k}) \tag{2}$$

If a pattern has two sensor, such as pattern $s_1 s_2$, then *corr-confidence* of the pattern is,

$$\rho(s_1 s_2) = \frac{P(s_1 s_2) - P(s_1)P(s_2)}{P(s_1 s_2) + P(s_1)P(s_2)} \tag{3}$$

If a pattern has more than two sensors, such as pattern $Y = \{s_1, s_2, ..., s_n\}$, then *corr-confidence* of the pattern is,

$$\rho = \frac{P(s_1, s_2, ..., s_n) - P(s_1)P(s_2)...P(s_n)}{P(s_1, s_2, ..., s_n) + P(s_1)P(s_2)...P(s_n)} \tag{4}$$

From Eqs. (3) and (4), we see that ρ has two bounds, i.e., $-1 \leq \rho \leq 1$. Let β be a given minimum *corr-confidence*, if pattern Y has two sensors s_1, s_2 and if $|\rho(s_1 s_2)| > \beta$ then Y is called a correlated pattern (i.e., s_1 and s_2 are called correlated with each other). If pattern Y has more than two event detecting sensors, we define a correlated pattern as follows:

Definition 2 (Correlated Pattern): Pattern Y is called a correlated pattern, if and only if there exists a pattern Y' which satisfies $Y' \subseteq Y$ and $|\rho(Y')| > \beta$.

Definition 3 (Associated-correlated pattern): A pattern is called an associated-correlated pattern if it is not only an associated pattern but also correlated pattern.

Let pattern Y be an associated-correlated pattern, then it must have two members s_1 and s_2 which satisfy the condition that the event detect of sensor s_1 can increase the likelihood of the event detection of sensor s_2.

Problem definition: Given a sensor database SD, $min\_sup(\delta), min\_all\_conf(\alpha)$ and $min\_corr\_conf(\beta)$ constraints, the objective is to discover the complete set of associated-correlated frequents patterns in SD having the value no less than δ, α, and β.

3   Decentralized Data Extraction Methodology

In our model sensors themselves are the main objects regardless of their values. We collect the sensors activity data (e.g., triggering on detection of events) called metadata. We are interested to capture sensor activity data, not the actual value of sensed data, and use this activity data for mining purpose later. The proposed network architecture for data extraction is shown in Fig. 1 which consists of sensors and a well-equipped sink where sensors are deployed in an ad hoc fashion on the area under monitoring. Each sensor node is coupled with a flash memory device that acts as a local storage to keep the detected events during monitoring. It is shown in [12] that energy consumption to maintain a unit of data in a flash memory coupled within a sensor node is very low compared to energy required to transmit this unit of data. All the sensor nodes share a Sensor Decentralized File System (SDFS) which is built upon their local storage. Each sensor node has the ability to download files from SDFS to its local storage and upload files to SDFS.

The decentralized method for data extraction is designed to put more computational load on the sensors. This is obtained by equipping each sensor with an additional storage to store the metadata that exhibit the sensor's activity

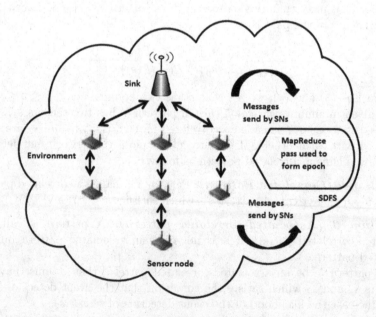

Fig. 1. Network architecture

Fig. 2. Detected events for 80 min historical period

S_1	1	1	1	1	0	0	0	0
S_2	1	1	1	1	1	0	0	1
S_3	0	1	0	1	1	1	0	1
S_4	0	0	0	0	1	1	1	1
S_5	1	1	1	1	0	1	1	0
S_6	1	0	1	0	0	0	1	0
S_7	1	0	0	0	0	0	1	0

Fig. 3. Example of buffer activity

during the monitoring/observation period. The notion behind the decentralized data extraction is to filter out the sensors whose frequencies (i.e., support) are less than min_sup. This will reduce the communication cost when the messages are uploading into the SDFS. The data extraction process is begins by sending the mining parameters from the sink to the sensor nodes in the network. These parameters are include time slot (λ), historical period (T_his) and minimum support min_sup. After getting the parameters, each sensor creates a local buffer. Each buffer has one bit entry for each time slot in the historical period of the data extraction. At first, all the bit entries in the buffer are unset. At the end of the each time slot every node checks whether there is any detected event for the current time slot. If an event is detected, the bit value for the corresponding time slot is set. After end of the historical period, each sensor scans over its local buffer. If the numbers of the bits is greater than or equal to the min_sup, the node will form a message or series of messages depending on the packet size. The message contains sensor ids and the time slot numbers in which the corresponding bits are set. Then the sensor uploads these messages to the SDFS. Note that the messages may be stored in different sensor nodes. This characteristic of our model ensure the data availability in noisy/erroneous environment.

Then a MapReduce pass is employed to merge messages into epochs in parallel. Each mapper takes in an input pair in the form of (key = message ID, value = message), where message = (ts, s) is a message that generated previously. It splits message into a time slot (ts) and a sensor (s), and outputs a key-value pair (key = ts, value = s). After all mapper instances have completed, for each distinct key ts, the MapReduce infrastructure collects its corresponding values as sensors, and feeds reducers with key-value pair (key = ts, value = S), where S is the set of sensors that triggered at the same time slot. The reducer receives the key-value pair, merges all

Table 1. Epochs on SDFS (an example sensor database (SD))

TS	Epoch	TS	Epoch
1	$s_1 s_2 s_5 s_6$	5	$s_2 s_3 s_4$
2	$s_1 s_2 s_3 s_5$	6	$s_3 s_4 s_5$
3	$s_1 s_2 s_5 s_6$	7	$s_4 s_5 s_6$
4	$s_1 s_2 s_3 s_5$	8	$s_2 s_3 s_4$

sensors with the same time slot into a epoch E, and associates its time slot number (TS) with ts. Finally, it outputs the key-value pair (key = TS, value = E).

As an example, let us consider the following simple scenarios. Let $s = \{s_1, s_2, s_3, s_4, s_5, s_6, s_7\}$ be the sensors in a particular sensor network. Let the time slot, $\lambda = 10$ min and the historical period $T_{his} = 80$ min. Suppose that the data extraction process start at time 09:00. Each sensor node will keep a buffer length of 8, one entry for each time slot. A buffer entry is set if there is an event detected in that time slot. The detected event for 80 min historical period is shown in Fig. 2 and the activity buffers for the sensor nodes of our example WSN is shown in Fig. 3. Assume the minimum support is 3. At the end of the historical period (10:20) sensors s_1, s_2, s_3, s_4, s_5 and s_6 will formulate the following messages $(s_1, [1, 1, 1, 1, 0, 0, 0, 0])$, $(s_2, [1, 1, 1, 1, 1, 0, 01])$, $(s_3, [0, 1, 0, 1, 1, 1, 0, 1])$, $(s_4, [0, 0, 0, 0, 1, 1, 1, 1])$, $(s_5, [1, 1, 1, 1, 0, 1, 1, 0])$ and $(s_6, [1, 0, 1, 0, 0, 0, 1, 0])$. Then each sensor sends the message as a set of sensor's id and time slot to the SDFS (e.g., messages send by the sensor s_1 are $m_1(s_1, 1), m_2(s_1, 2), m_3(s_1, 3), m_4(s_1, 4)$). Finally, the MapReduce pass is used to merge these messages into epoch and epochs on SFDS shown in Table 1. Sensor node s_7 does not send any message because the number of set entries is less than the required *min_sup*.

4   Experimental Results

In this section, we presented the simulation results of our proposed data extraction model. To generate synthetic data for WSN, an event generation program was written in Microsoft Visual C++ and run with Windows 7 on a 2.66 GHz machine with 4 GB of main memory. We also used another dataset containing real WSN data from Intel Berkely Research Lab [13], which has been widely used in literature (e.g., [4–6]). From the available Intel dataset, we utilized the sensor data collected over 5 days and 10 days at 30 s interval where the datasets consists of tuples from 54 sensors reporting environmental reading in every 30 s.

The simulator generates events with the several input parameters including the number of nodes, the historical period, slot size and minimum support value for mining. In this simulator, we considered two scenarios consisting of 150 and 300 nodes respectively placed in a grid of 250m × 250m, where the sensors are evenly placed over the area. We used the radio model introduced in [11] and the

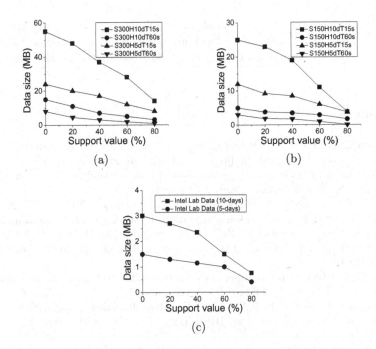

Fig. 4. Data size v/s support values: (a) *S300 data*, (b) *S150 data*, and (c) *Intel Lab Data*

storage model used by the sensor nodes introduced in [12]. We assumed a Toshiba 16 MB NAND flash memory that costs 0.017 uJ to read, write, and erase a byte of data [12]. Event generation by each sensor node was assumed to be uniformly Decentralized over the possible number of slots within the given historical period. In addition, we assumed the nodes were uniformly Decentralized in space in the monitoring area and messages were delivered reliably by acknowledging the number of set bits in each sensor's buffer.

The performance of the proposed data model is evaluated on the basis of the amount of data that are accumulated in the SDFS. Figure 4 shows the data size for different support values accumulated at SDFS. The amount of data at 0 support value refers to the data size accumulated using centralized mechanism. The results indicate that the data size is reduced with increasing support values and this reduction rate is very sharp when the support value exceeds 50 %.

5   Conclusion

In this paper, we have proposed a framework for behavioral patterns called as associated-correlated sensor pattern that can create the time relations and correlations between sensors in a particular wireless sensor network. We also introduce, MapReduce based new paradigm for extract data from the wireless sensor network by decentralized mechanism. The decentralized data collection model tries to prolong the life time of the WSN through optimizing number of exchanged messages. Extensive performance study shows that the decentralized method is able to reduce data size almost 50 % compared to the centralized model.

References

1. Boukerche, A., Pazzi, R.W., Araujo, R.B.: A fast and reliable protocol for wireless sensor networks in critical conditions monitoring applications. In: Proceedings of the 7th ACM International Symposium on Modeling, Analysis and Simulation of Wireless and Mobile Systems, p. 157164 (2004)
2. Tan, P.-N.: Knowledge discovery from sensor data. Sensors **23**(3), 1419 (2006)
3. Loo, K.K., Tong, I., Kao, B.: Online algorithms for mining inter-stream associations from large sensor networks. In: Ho, T.B., Cheung, D., Liu, H. (eds.) PAKDD 2005. LNCS (LNAI), vol. 3518, pp. 143–149. Springer, Heidelberg (2005). doi:10.1007/11430919_18
4. Romer, K.: Decentralized mining of spatio-temporal event patterns in sensor networks. In: EAWMS / DCOSS, pp. 103–116 (2006)
5. Boukerche, A., Samarah, S.A.: Novel algorithm for mining association rules in wireless ad-hoc sensor networks. IEEE Trans. Parallel Distrib. Syst. **19**, 865–877 (2008)
6. Rashid, M.M., Gondal, I., Kamruzzaman, J.: Mining associated patterns from wireless sensor networks. IEEE Trans. Comput. **64**(7), 1998–2011 (2015)
7. Intanagonwiwat, C., Govindan, R., Estrin, D., Heidemann, J., Silva, F.: Directed diffusion for wireless sensor networking. IEEE/ACM Trans. Networking (ToN) **11**(1), 2–16 (2003)
8. Handy, M.J., Haase, M., Timmermann, D.: Low energy adaptive clustering hierarchy with deterministic cluster-head selection. In: 4th International Workshop on In Mobile and Wireless Communications Network pp. 368–372 (2002)
9. Lindsey, S., Raghavendra, C.S.: PEGASIS: power-efficient gathering in sensor information systems. Aerosp. Conf. Proc. **3**, 3–11 (2002)
10. Tan, H., Krpeolu, I.: Power efficient data gathering and aggregation in wireless sensor networks. ACM Sigmod Rec. **32**(4), 66–71 (2003)
11. Heinzelman, W., Chandrakasan, A., Balakrishnan, H.: Energy-efficient communication protocol for wireless microsensor networks. In: Proceedings of the Hawaii Conference on System Sciences, pp. 1–10 (2000)
12. Mathur, G., Desnoyers, P., Chukiu, P., Ganesan, D., Shenoy, P.: Ultra-low power data storage for sensor networks. ACM Trans. Sens. Netw. (TOSN) **5**(4), 1–33 (2009)
13. Madden, S.: Intel berkeley research lab data (2003)

Incorporating Prior Knowledge into Context-Aware Recommendation

Haitao Zheng[✉] and Xiaoxi Mao

Tsinghua -Southampthon Web Science Laboratory, Graduate School at Shenzhen,
The University Town, Shenzhen, China
{zheng.haitao,mxx14}@mails.tsinghua.edu.cn

Abstract. In many recommendation applications, like music and movies recommendation, describing the features of items heavily relies on user-generated contents, especially social tags. They suffer from serious problems including redundancy and self-contradiction. Direct exploitation of them in a recommender system leads to reduced performance. However, few systems have taken this problem into consideration.

In this paper, we propose a novel framework named as prior knowledge based context aware recommender (PKCAR). We incorporate Dirichlet Forrest priors to encode prior knowledge about item features into our model to deal with the redundancy, and self-contradiction problems. We also develop an algorithm which automatically mine prior knowledge using co-occurrence, lexical and semantic features. We evaluate our framework on two datasets from different domains. Experimental results show that our approach performs better than systems without leveraging prior knowledge about item features.

Keywords: Context-aware recommendation · Topic modeling · KBTM · Semi-supervised learning · Collaborative filtering

1 Introduction

Context-aware recommender systems could make recommendations that are both consistent with a user's long term interests and his or her short term interests. It requires additional information describing item features to do that. However, though there are some techniques proposed to directly generate descriptions for multimedia items, in most cases, social tagging remains the major source of item descriptions. Generally social tags are freely edited and rarely elaborately revised, thus they consequently contain redundant, sometimes even self-contradictory tags. Redundant tags have the same meaning but take different textual forms, like "90s" and "1990s", while the self-contradictory tags have opposite meanings but co-occur describing the same item. Table 1 shows an example, where both redundant tags like "fav", "favourites" and self-contradictory tags like "sad", "fun" are included. To deal with the problems described above, we must exploit the relations of the social tags, or more generally, the item features. For example, if a topic model knows that "favorites" and

© Springer International Publishing AG 2016
A. Hirose et al. (Eds.): ICONIP 2016, Part III, LNCS 9949, pp. 499–508, 2016.
DOI: 10.1007/978-3-319-46675-0_55

"favorite" mean the same, they would have a big chance appearing in the same topic. To achieve this, we propose a novel framework named Prior Knowledge based Context-Aware Recommender (PKCAR), which incorporates the Dirichlet Forrest priors proposed in [2] to encode specific feature pairs as *must-link* or *cannot-link* pairs. Feature pairs taking similar meanings are *must-links*, like "fav"and "favourite"; pairs that have opposite meanings are *cannot-links*, like "male artist" and "female artist", and they should have very low chance appearing in the same topic. Besides, since manually annotating these prior knowledge sets needs a lot of human laboring, we propose an automatically knowledge mining algorithm using co-mention, lexical and semantic features. To our best knowledge, we are the first to exploit the relations of items features in a context-aware recommendation system.

Table 1. A sample of songs from the *playlist* dataset

Song	Cleanin Out My Closet
Artist	Eminem
Tags	favorites, american, awesome, loved, fun, favourite, sad, hip hop, rap, memories, great, best, fav, english, hiphop ...

We evaluate our framework on two datasets from different domains. We compare our algorithm with approaches that don't consider the problems of social tagging. The experimental results show that our framework significantly outperforms the others.

2   Related Works

Context-aware recommender systems could be categorized into 3 kinds: contextual pre-filtering, contextual post-filtering, and contextual modeling [1]. Baltrunas et al. introduced an item splitting method which exploits contextual information to split the item ratings, converting the form $U \times I \times C \times R$ to $U \times I \times R$ thus traditional recommending techniques could be used [3]. Said et al. proposed another pre-filtering method which issues user-splitting on the $U \times I \times C \times R$ form [8]. Both two methods suffer from sparsity problem in applications like music recommendation since quite a few features within the context are not always available across all the songs. There also exists quite a few methods for music playlists prediction. The method called Logistic Markov Embedding (LME) proposed by Chen et al. models the playlists as Markov chains in latent space and represents the songs as points in that space [4].

Knowledge-based topic modeling (KBTM) has been receiving attentions from researchers in recent years. Andrzejewski et al. proposed the Dirichlet Forest priors which could code the prior lexical knowledge as must-links and cannot-links [2]. Chen et al. argued that the must-links defined in [2] are transitive thus introducing incorrect prior knowledge when it is from different domains or general domain [5,6]. They proposed two models respectively named as *MDK-LDA* which takes knowledge from different domains as input, and *GK-LDA* which takes general knowledge.

3 Our Approach

Our framework consists of three components. The first one is the knowledge mining algorithm, which could automatically mine *must-link* and *cannot-link* pairs from item descriptions. The second component is our PKCAR model, which uses the prior knowledge sets to form Dirichlet Forrest priors and outputs the posterior probabilities of topics over users, items over topics and features over topics. The third component, a ranking algorithm takes the posterior probabilities, the prior knowledge sets, and a query given by a user to make recommendations which are consistent with the user's profile and the given query. In below we discuss them in detail.

3.1 Mining Prior Knowledge Sets

We use the co-occurrence frequency and lexical features to mine must-links because leveraging semantic features may introduce noises due to the transitivity of *must-links*. For *cannot-links*, we use semantic relations for mining, because *cannot-links* are not transitive. We use the WordNet to determine if two given tags contain opposite meanings. If they do, they are annotated as a pair of *cannot-link*. The details of our mining algorithm is summarized in Algorithm 1.

$$score(< t1, t2 >) = \lambda * \frac{count(< t1, t2 >)}{count(t1, t2)} \tag{1}$$

$$\lambda = \frac{2.0 + a}{1.0 + a} - \frac{count(all)}{count(t1, t2) + a * count(all)} \tag{2}$$

$$count(t1, t2) = count(t1) + count(t2) - count(< t1, t2 >) \tag{3}$$

$count(t1)$ indicates the number of item features lists where $t1$ occurs. $count(< t1, t2 >)$ is the number of item features lists where $t1$ *and* $t2$ co-occur. $count(t1, t2)$ is the number of item features lists where $t1$ *or* $t2$ occurs. $count(all)$ is the number of all items. λ is a penalty factor which ranges from $[1 - (a^2 + a)^{-1}, 1]$. a is supposed to be greater than 0.62. We choose this penalty factor to ensure more popular feature pairs are ranked higher, while keeping the co-occurrence ratio as the main metric for ranking. *is_anonym* is implemented with WordNet, we split the given two tags into two groups of tokens respectively, using WordNet to determine if there exists a pair of tokens that are antonyms. *clean_string* is used to filter out any character that is either a number or letter in a given tag. *knowledge* is a map which takes *must-links* and *cannot-links* as keys, and 1(if the key is a *must-link*) and −1(on the contrary) as values.

Algorithm 1. Mining must-links and cannot-links automatically
Require:
    tag_hash, mapping from tag integer id to tag name
    tags, social tags in integer id to represent items
    n, the number of highest ranked pairs for further mining
    a, a parameter to adjust the penalty factor;
Ensure:
    knowledge, a hashmap of pairs;
 1: Make statistics of the occurrences of each feature pair $< t1, t2 >$ on *tags*, rank the
    pairs with scores computed using the formulation 1, 2 and 3.
 2: $top\_n\_pairs \Leftarrow ranked\_pairs[: n]$
 3: **for** each $pair \in ranked\_pairs$ **do**
 4:    $t1 \Leftarrow pair[0]$ $t2 \Leftarrow pair[1]$
 5:    $t1\_text \Leftarrow tag\_hash[t1]$ $t2\_text \Leftarrow tag\_hash[t2]$
 6:    **if** $is\_antonym(t1\_text, t2\_text)$ **then**
 7:       $knowledge[pair] = -1$
 8:    **else**
 9:       $t1\_cleaned \Leftarrow clean\_string(t1\_text)$
10:       $t2\_cleaned \Leftarrow clean\_string(t2\_text)$
11:       **if** $sub\_string(t1\_cleaned, t2\_cleaned)$**and**$pair \in top\_n\_pairs$ **then**
12:         $knowledge[pair] = 1$
13:       **end if**
14:    **end if**
15: **end for**
16: **return** *knowledge*;

3.2   The PKCAR Model

The graphical representation of our model is shown in Fig. 1. The M in the outermost plate denotes the number of user profiles. The N in the inner plate is the number of items in the mth user profile. And the O in the innermost plate denotes the number of corresponding features describing song $s_{m,n}$. The other variables and parameters in the model are shown in Table 2.

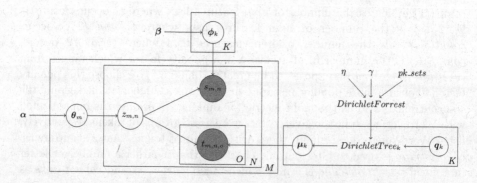

Fig. 1. The graphical representation of our proposed model

Table 2. The descriptions of variables and parameters in our proposed model

θ_m	the topic distribution for user m
$z_{m,n}$	the topic for the nth item in mth user profile
$s_{m,n}$	the nth item in mth user profile
$t_{m,n,o}$	the oth feature for the nth item in mth user profile
ϕ_k	the item distribution for topic k
μ_k	the feature distribution for topic k
q_k	the index vector for topic k
$DirichletTree_k$	the Dirichlet tree distribution for topic k
pk_sets	prior knowledge encoded as must-links and cannot-links
η	hyperparameter controlling the influences of the prior knowledge

The joint distribution of items s, features $\underline{T}$, the index vectors $\underline{Q}$ is computed as follows:

$$p(s, z, \underline{T}, \underline{Q} | \alpha, \beta, \gamma, \eta) = \prod_{z=1}^{K} \frac{\Delta(\boldsymbol{n}_z + \boldsymbol{\beta})}{\Delta(\boldsymbol{\beta})} \times \prod_{z=1}^{K} \prod_{\chi}^{Tree_z} \frac{\Delta(\boldsymbol{b}_\chi + \boldsymbol{n}_z^\chi)}{\Delta(\boldsymbol{b}_\chi)}$$
$$\times \prod_{m=1}^{M} \frac{\Delta(\boldsymbol{n}_m + \boldsymbol{\alpha})}{\Delta(\boldsymbol{\alpha})} \times \prod_{z=1}^{K} \prod_{r=1}^{R} p(q_z^{(r)}) \tag{4}$$

In the equation described above, K is the number of topics, $\boldsymbol{n}_z$ denotes the numbers of each item under the topic z, $Tree_z$ is the Dirichlet tree distribution selected from Dirichlet priors with $\boldsymbol{q}_z$, χ is the internal node within $Tree_z$, the internal nodes are nodes in a Dirichlet tree except the leaf nodes. $\boldsymbol{b}_\chi$ is the weights of branches under node χ, $\boldsymbol{n}_z^\chi$ denotes the numbers of items that are under node χ in $Tree_z$. $\boldsymbol{n}_m$ refers to the numbers of times that each topic is observed with an item of user profile m. R means the number of branches under the root node, $p(q_z^{(r)})$ is computed according to the item nodes within each subtree of the clique $C^{(r)}$, i.e., $p(q_z^{(r)}) \propto |M_{rq}|, q = 1...|C^{(r)}|$.

We use collapsed Gibbs sampling to do inference over the model. Each MCMC iteration consists of a sweep through both z and $\underline{Q}$. Using the joint distribution described above, we could derive the conditional probabilities for collapsed Gibbs sampling as follows.

$$p(z_i = v | \boldsymbol{z}_{-i}, \underline{Q}, s, \underline{T}) \propto (n_{-i,m}^{(v)} + \alpha_v) \times \frac{n_{-i,v}^{(w)} + \beta_w}{\sum_{s=1}^{S} n_{-i,v}^{(s)} + \beta_s}$$
$$\times \prod_{\chi}^{Tree_v(\uparrow i)} \frac{\prod_{br}^{branch_\chi(\uparrow i)} \prod_{h=1}^{dec(br)} (n_{-i,v}^{(br)} + b_\chi^{(br)} + h - 1)}{\prod_{h=1}^{dec(\chi)} (\sum_{br}^{branch_\chi} n_{-i,v}^{(br)} + b_\chi^{(br)} + h - 1)} \tag{5}$$

$$p(q_j^{(r)} = q'|z, \underline{Q}_{-j}, q_j^{(-r)}, \underline{T}) \propto \sum_k^{M_{rq'}} \gamma_k \times \prod_\chi^{Tree_{j,r=q'}} \frac{\Delta(b_\chi + n_z^\chi)}{\Delta(b_\chi)} \qquad (6)$$

Equation 5 is the full conditional distribution for an item with index $i = (m, n)$. $n_{-i,m}^{(v)}$ indicates the number of items in user profile m assigned to topic v excluding item i. $n_{-i,v}^{(w)}$ denotes the number of item w assigned to topic v excluding item i, noting that $s_i = w$ and S is the number of items in vocabulary. $Tree_v(\uparrow i)$ denotes the subset of internal nodes in topic v's Dirichlet tree which are ancestors of all the features associated with item i, i.e., τ_i. χ denotes a internal node. $branch_\chi(\uparrow i)$ indicates the branches under node χ which lead to features in τ_i. $n_{-i,v}^{(br)}$ denotes the number of features that br leads to and assigned to topic v excluding τ_i. $dec(br)$ means the decrease of the number after excluding τ_i, i.e., $n_v^{(br)} - n_{-i,v}^{(br)}$. $dec(\chi)$ indicates the decrease of the number of features that are children of χ excluding τ_i. $b_\chi^{(br)}$ denotes the weight of branch br of node χ. Equation 6 is used for sampling q_j. q' ranges from 1 to $|Q^{(r)}|$. $Tree_{j,r=q'}$ indicates the subset of internal nodes under branch q' in topic j's Dirichlet tree.

After sufficient iterations, we use the final sample to estimate $\underline{\Theta}$, $\underline{\Phi}$ and $\underline{M}$. The estimation of $\underline{\Theta}$ and $\underline{\Phi}$ is the same as in standard LDA, like follows.

$$\hat{\theta}_m^{(k)} = \frac{n_m^{(k)} + \alpha_k}{\sum_{k=1}^K n_m^{(k)} + \alpha_k} \qquad \hat{\phi}_k^{(w)} = \frac{n_k^{(w)} + \beta_w}{\sum_{w=1}^S n_k^{(w)} + \beta_w} \qquad (7)$$

To estimate $\underline{\Phi}$, we use the following equation.

$$\hat{\mu}_k^{(t)} = \prod_\chi^{Tree_k(\uparrow t)} \frac{n_k^{(br(\chi\downarrow t))} + b_\chi^{((br(\chi\downarrow t))}}{\sum_{br}^{branch_\chi} n_k^{(br)} + b_\chi^{(br)}} \qquad (8)$$

In Eq. 8, $br(\chi \downarrow t)$ refers to the unique branch that leads to χ's immediate child and ancestor of t(including t itself).

3.3 Items Ranking Algorithm

Our ranking algorithm is quite similar to the algorithm proposed in [7], despite we incorporate *must-links* here. Before computing a score for each item, the given query will be preprocessed to find out the must-link cliques within the query. Every clique will be regarded as a single feature in the subsequent computation. Let $q = t_1, t_2, ..., t_w$ be the query consisting of w features t_i, combine the features that are within the same must-link clique, reform the query as $q' = c_1, c_2, ..., c_v$ where c_i is a clique of features. The computation of a ranking score for an item s given the reformed query q', user profile p is like the following. With the posterior estimates of $\underline{\hat{\Theta}}$, $\hat{\Phi}$ and $\hat{M}$, we have:

$$p(s|q', p, \underline{\hat{\Theta}}, \underline{\hat{\Phi}}, \underline{\hat{M}}) \propto p(q'|s, p, \underline{\hat{\Theta}}, \underline{\hat{\Phi}}, \underline{\hat{M}}) p(s|p, \underline{\hat{\Theta}}, \underline{\hat{\Phi}}) \qquad (9)$$

$$p(s|p, \hat{\Theta}, \hat{\Phi}) = \sum_{j=1}^{K} p(s|z_j)p(z_j|p) = \sum_{j=1}^{K} \hat{\phi}_{s,j}\hat{\theta}_{j,p} \qquad (10)$$

$$p(q'|s, p, \hat{\Theta}, \hat{\Phi}, \hat{M}) = \prod_{c_i \in q'} \sum_{j=1}^{|c_i|} p(t_j|s, p, \hat{\Theta}, \hat{\Phi}, \hat{M})/|c_i| \qquad (11)$$

$|c_i|$ refers to the number of features in c_i, we use the average of posterior probability for each feature in c_i as its posterior probability. The computation of $p(t_j|s, p, \hat{\Theta}, \hat{\Phi}, \hat{M})$ is similar to [7], so we don't repeat it here.

4 Experiments

4.1 Datasets

Playlist. This dataset[1] is collected by Chen and used in [4,5]. The playlists and tag data are respectively crawled from Yes.com and Last.fm. The original dataset comprises three datasets of various size, *yes_small*, *yes_big* and *yes_complete*. The former two are pruned versions of *yes_complete*. In our experiment, we only use the *yes_complete* for evaluation. It consists of 75,262 distinct songs, 250 tags, 11,137 and 4,773 playlists respectively in training set and test set.

Who-posted-what. This dataset is obtained from CiteULike[2]. Users with less than 4 postings, articles which have appeared less than 5 times are removed. The top 300 most occurring tags and the most popular tag of each article are kept. After pruning, the number of users, articles and tags were 17,707, 80,433 and 2,012 respectively.

4.2 Preprocess

For each user profile in test set, we randomly filter out 20 % items and combine them as the answer set. The filtered profiles form the new test set. Each query for a user profile is formed by randomly picking one from the features of each filtered out item, removing repeated ones. They form the query set. In experiment, we iterate through the user profiles in the test set. For a given user profile p and the respective query q, the recommender generates a list of recommendations, which is then compared to the corresponding answer to evaluate the performance.

4.3 Baselines

Simple Filtering: This method simply filter out all the items that do not contain any feature in the given context q. The rest items are randomly ranked.

---

[1] http://www.cs.cornell.edu/~shuochen/lme/data_page.html.
[2] http://www.citeulike.org/faq/data.adp.

k**NN:** For every user profile in the test set, return items in the k nearest profiles in the training set and remove the duplicates.

LDA: This method is only context-sensitive. We train the LDA model on the features of every item, regarding every item as a document and every feature as a word. Thus we obtain estimates of the distributions of latent topics over each item $\hat{\Theta}$ and the distributions of items over latent topics $\hat{\Phi}$. Given $\hat{\Theta}$, $\hat{\Phi}$, and the given context q, we are able to rank the items using the following formulations:

$$p(q|i) = \prod_{j \in q} p(t_j|i) \qquad p(t_j|i) = (1 - \kappa)p_{bd}(t_j|i) + \kappa p_{cr}(t_j|i) \qquad (12)$$

$$p_{cr}(t_j|i) = \sum_{k=1}^{K} \hat{\theta}_{i,k}\hat{\phi}_{k,j} \qquad p_{bd}(t_j|i) = \frac{c(t_j; i) + \delta p(t_j|c)}{\sum_{t \in i} c(t; i) + \delta} \qquad (13)$$

In the equations above, $p_{bd}(t_j|i)$ is a *Bayesian smoothing* item. This technique is also used in QDCA and PKCAR to rank the items. δ is the smoothing parameter, $c(t_j; i)$ refers to the frequency of feature t_j in features of item i. $p(t_j|c)$ is the probability of feature t_j in the whole corpus.

Query-driven context-aware recommender (QDCA): The approach proposed in [7]. The model is trained on a combination of profiles both in training set and test set. After training, the estimates are used to do recommendation for every user profile in the test set, with the respective query.

4.4   Metrics

To compare the performance across different approaches, we use the *hit ratio* to demonstrate an algorithm's ability to make precise recommendations. The hit ratio $h(N)$ is calculated using Eq. 14. R_N is the top N recommended items for every profile in the test set.

$$h(N) = |R_N \bigcap answer\_set|/|answser\_set| \qquad (14)$$

4.5   Results and Discussion

In order to evaluate the effectiveness of our automatically mined prior knowledge, we manually annotate a knowledge set. The result using manually annotated knowledge set is labeled as *PKCAR1*, and the result using automatically mined knowledge is labeled as *PKCAR2*. The parameters for each method could be seen in Table 3. As seen in Fig. 2, our model using manually annotated prior knowledge performs better over all other methods, the model using automatically mined knowledge performs slightly worse than PKCAR1 but still outperforms other methods. The kNN method performs well on *playlist* dataset because it has many popular items which are shared across many users. The *Who-posted-what* dataset contains very few popular items, leading to reduced kNN performance.

Table 3. Parameters used for each method

Method	Parameters
kNN	Number of neighbors = 10
LDA	topics = 50, iterations = 1500, $\alpha = 5$, $\beta = 0.01$, $\delta = 1000$, $\kappa = 0.7$
QDCA	topics = 50, iterations = 1500, $\alpha = 5$, $\beta = 0.01$, $\gamma = 0.01$, $\delta = 1000$, $\kappa = 0.7$
PKCAR1	topics = 50, iterations = 1500, $\alpha = 5$, $\beta = 0.01$, $\gamma = 0.01$, $\eta = 20$, $\delta = 1000$, $\kappa = 0.7$
PKCAR2	topics = 50, iterations = 1500, $\alpha = 5$, $\beta = 0.01$, $\gamma = 0.01$, $\eta = 20$, $\delta = 1000$, $\kappa = 0.7$, $n = 200$, $a = 1.0$

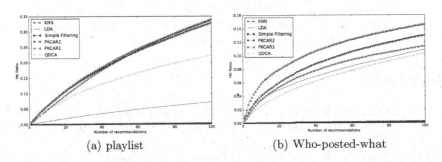

(a) playlist   (b) Who-posted-what

Fig. 2. Hit ratio for different number of recommendation for different datasets

Table 4. A sample of topics generated by our PKCAR model

#Topic 1	#Topic 2	#Topic 3	#Topic 4	#Topic 5
Favourite	love songs	80s	indie	female vocalists
Favorites	love	new wave	alternative	female vocalist
Favourites	easy listening	dance	alternative rock	female
Favorite	romantic	party	indie rock	female vocals
Favorite songs	love song	80's	alternative punk	driving

QDCA doesn't perform well, because it generates too many uninterpretable topics that it could hardly interpret the queries. The problems in social tags also greatly influence the effects of simple filtering, which is hardly better than randomly choosing. In comparison, our PKCAR model could generate much more consistent topics with the exploitation of prior knowledge, as seen in Table 4, leading to good recommendation results.

5 Conclusion

In this paper we propose a query-driven context-aware recommendation framework which exploits relations of item features to deal with the redundancy and self-contradiction problems in social tags. We use the Dirichlet Forrest priors to incorporate relations between item features encoded as *must-links* and *cannot-links* in our model. We also propose an algorithm to automatically mine knowledge sets from given datasets using co-occurrences, lexical and semantic features. Experiments show that we do have a significant performance improvement with the prior knowledge introduced.

Acknowledgements. This research is supported by National Natural Science Foundation of China (Grant No. 61375054 and 61402045), Natural Science Foundation of Guangdong Province (Grant No. 2014A030313745), Tsinghua University Initiative Scientific Research Program (Grant No.20131089256), and Cross fund of Graduate School at Shenzhen, Tsinghua University (Grant No. JC20140001).

References

1. Adomavicius, G., Tuzhilin, A.: Context-aware recommender systems. In: Proceedings of the 2008 ACM Conference on Recommender Systems, RecSys 2008, p. 335 (2008)
2. Andrzejewski, D., Zhu, X., Craven, M.: Incorporating domain knowledge into topic modeling via dirichlet forest priors (2009)
3. Baltrunas, L., Ricci, F.: Context-based splitting of item ratings in collaborative filtering. In: Proceedings of the Third ACM Conference on Recommender Systems, RecSys 2009, p. 245 (2009)
4. Chen, S., Moore, J.L., Turnbull, D., Joachims, T.: Playlist prediction via metric embedding. In: Proceedings of the 18th ACM SIGKDD International Conference on Knowledge Discovery and Data Mining, KDD 2012, p. 714. ACM Press, New York (2012)
5. Chen, S., Xu, J., Joachims, T.: Multi-space probabilistic sequence modeling. In: Proceedings of the 19th ACM SIGKDD International Conference on Knowledge Discovery and Data Mining, KDD 2013, p. 865. ACM Press, New York (2013)
6. Chen, Z., Mukherjee, A., Liu, B., Hsu, M., Castellanos, M., Ghosh, R.: Leveraging multi-domain prior knowledge in topic models, pp. 2071–2077 (2013)
7. Hariri, N., Mobasher, B., Burke, R.: Query-driven context aware recommendation. In: Proceedings of the 7th ACM Conference on Recommender Systems, RecSys 2013, pp. 9–16. ACM Press, New York (2013)
8. Said, A., De Luca, E.W., Albayrak, S.: Inferring contextual user profiles improving recommender performance, vol. 791. CEUR Workshop Proceedings, Chicago (2011)

Deep Neural Networks

Unsupervised Video Hashing by Exploiting Spatio-Temporal Feature

Chao Ma[1], Yun Gu[1], Wei Liu[1], Jie Yang[1(✉)], and Xiangjian He[2]

[1] Institute of Image Processing and Pattern Recognition,
Shanghai Jiao Tong University, Shanghai, China
{sjtu_machao,geron762,liuwei.1989,jieyang}@sjtu.edu.cn
[2] University of Technology, Sydney, Australia
sean@it.uts.edu.au

Abstract. Video hashing is a common solution for content-based video retrieval by encoding high-dimensional feature vectors into short binary codes. Videos not only have spatial structure inside each frame but also have temporal correlation structure between frames, while the latter has been largely neglected by many existing methods. Therefore, in this paper we propose to perform video hashing by incorporating the temporal structure as well as the conventional spatial structure. Specifically, the spatial features of videos are obtained by utilizing Convolutional Neural Network (CNN), and the temporal features are established via Long-Short Term Memory (LSTM). The proposed spatio-temporal feature learning framework can be applied to many existing unsupervised hashing methods such as Iterative Quantization (ITQ), Spectral Hashing (SH), and others. Experimental results on the UCF-101 dataset indicate that by simultaneously employing the temporal features and spatial features, our hashing method is able to significantly improve the performance of existing methods which only deploy the spatial feature.

Keywords: Video hashing · Unsupervised Method · Spatio-temporal feature

1 Introduction

Video retrieval is a very challenging task in the area of computer vision. Most of current video search engines rely on textual keyword matching rather than visual content-based retrieval. One of the bottlenecks for content-based search is the unaffordable computational cost when handling a large collection of video clips. Consequently, hashing is a popular method to solve this problem by encoding high-dimensional feature vectors into short binary codes, so that the hamming distance, which is very efficient to compute, can be used to represent the similarity between different videos. This has enabled significant efficiency gains in both storage and speed.

Recently, great achievements have been made on hashing by incorporating various machine learning techniques. These methods can be divided into

© Springer International Publishing AG 2016
A. Hirose et al. (Eds.): ICONIP 2016, Part III, LNCS 9949, pp. 511–518, 2016.
DOI: 10.1007/978-3-319-46675-0_56

three categories: unsupervised, semi-supervised, and supervised. Unsupervised hashing methods such as Spectral Hashing (SH) [17] mainly utilize data properties like distribution or manifold structure to design effective indexing schemes. Supervised methods such as deep neural network-based method [15] treat the design of hash functions as a special classification problem and utilize supervised (label) information in the training procedure. Some other supervised methods, *e.g.* supervised hashing with kernels [4], take into account the pairwise relationship of samples in the hash function learning procedure. Semi-supervised hashing methods [16] play a tradeoff between supervised information and data properties, which leverage semantic similarity using label data while remaining robustness to overfitting.

Video hashing is different from image hashing because videos not only have the spatial structure within each frame but also have the temporal correlation between frames.

On one hand, Convolutional Neural Network (CNN) can be used to learn spatial structure features, as CNN is effective in learning rich mid-level image descriptors. By utilizing the feature vectors generated by the seventh layer in the trained CNN, the method proposed by Krizhevsky *et al.* [13] achieved the state-of-the-art performance in image retrieval on ImageNet dataset [2].

On the other hand, there are also networks that perfectly learn the correlation between signal sequences. It is widely acknowledged that Recurrent Neural Network (RNN) models are "deep in time", which means that RNN is connectionist models that capture the dynamics of sequences via cycles in the nodes of network. However, a significant limitation of simple RNN models is the "vanishing gradient" effect, *i.e.* practically back propagating an error signal through a long-range temporal interval will become increasingly intractable. To handle the vanishing gradient problem, Hochriter *et al.* [9] introduced the Long-Short Term Memory (LSTM) model which resembled a standard RNN with a hidden layer. Each ordinary node in the hidden layer is replaced by a memory cell that contains a node with a self-connected recurrent edge of fixed weight. This ensures that the gradient can pass across a long-range temporal interval without vanishing or exploding.

Most of recent hashing methods [4,15–17] generate binary codes for each sample independently but pay little attention to developing specific hash functions to index structured data like videos. Recently, Song *et al.* [11] proposed multiple feature based hashing for video near-duplicate detection. Cao *et al.* [1] proposed a submodular hashing framework to index videos. Although these methods achieve satisfactory performance to some extent, the specific temporal structure between video frames is neither considered nor encoded into the binary codes.

In this paper, we propose to perform video hashing by making use of not only the spatial structure within each frame but also the temporal correlation structure between frames. We construct a spatio-temporal feature learning framework by using CNN for spatial feature learning and LSTM for temporal feature learning. We apply our spatio-temporal feature learning framework to many unsupervised hashing methods including Iterative Quantization (ITQ) [7], Locality Sensitive Hashing (LSH) [6], PCA Hashing (PCAH) [7], Spectral Hashing (SH) [17], Density Sensitive Hashing (DSH) [12], and Spherical Hashing (SpH) [8]. We use UCF-101 dataset [14] to compare our hashing method with the existing algorithms that only deploy the spatial features, and the results reveal that our method is significantly superior to the existing methodologies.

The rest of our paper is organized as follows. The details of our approach are described in Sect. 2. Our approach is empirically evaluated on the UCF-101 dataset in Sect. 3. Finally, we conclude the entire paper in Sect. 4.

2   Methodology

2.1   The Recurrent Neural Networks

Figure 1(a) is a simple recurrent net with one input unit, one output unit and one recurrent hidden unit. Such recurrent net can learn complex temporal dynamics by mapping input sequences to a sequence of hidden states. The hidden states are then mapped to the output via the following recurrence equations:

$$h_t = g(W_{hx}x_t + W_{hh}h_{t-1} + b_h)$$
$$z_t = g(W_{hz}h_t + b_z) \tag{1}$$

where $g(\cdot)$ is an element-wise non-linearity function, such as sigmoid or hyperbolic tangent, x_t is the input, $h_t \in \mathbb{R}^N$ denotes the hidden state with N hidden units, and z_t is the output at time t. The weight matrices W_{ij} and biases b_j are the parameters to be learned.

As mentioned in the introduction, although RNN has been proven to be successful on several tasks, a significant drawback of simple RNN models is the vanishing gradient problem. This problem makes it difficult to train RNN to learn long-term dynamics. As a result, Long-Short Term Memory (LSTM) model provides a solution by incorporating memory units that allow the network to decide whether to forget the previous hidden states or to update the hidden states according to the new information. We use the LSTM unit as described in [5] (Fig. 1(b)), which was derived from the LSTM initially proposed in [9]. The formal definition of LSTM with forget gates is formulated as follows:

$$g_t = \phi(W_{gx}x_t + W_{gh}h_{t-1} + b_g)$$
$$i_t = \sigma(W_{ix}x_t + W_{ih}h_{t-1} + b_i)$$
$$f_t = \sigma(W_{fx}x_t + W_{fh}h_{t-1} + b_f)$$
$$o_t = \sigma(W_{ox}x_t + W_{oh}h_{t-1} + b_o)$$
$$s_t = g_t \odot i_t + s_{t-1} \odot f_t$$
$$h_t = \phi(s_t) \odot o_t \tag{2}$$

where $\sigma(x) = (1 + e^{-x})^{-1}$ is the sigmoid non-linearity function that maps real-valued inputs into the interval $[0, 1]$, $\phi(x) = \frac{e^x - e^{-x}}{e^x + e^{-x}} = 2\sigma(2x) - 1$ is the hyperbolic tangent nonlinearity that maps its inputs into the interval $[-1, 1]$, and $\odot$ donates element-wise product.

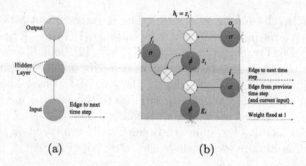

Fig. 1. Recurrent Neural Networks: (a) A simple recurrent net. (b) LSTM memory cell used in this paper.

The advantages of LSTM for modeling sequential data in computer vision are twofold. First, when integrated with current vision systems, the parameters in the LSTM model can be easily fine-tuned in an end-to-end way. Second, LSTM models are not limited to fixed length inputs. It is able to model the sequential data with a varying length such as videos.

2.2 Obtaining Binary Codes of Videos

As CNN has shown impressive performance for spatial feature learning in various tasks, such as image classification [13], we thus adopt CNN to learn the spatial features in our spatio-temporal feature learning framework. Besides, we use the LSTM model to learn the temporal features between frames. Figure 2 shows the proposed method. Our method includes two main components. The first component involves the supervised pre-training to learn the rich mid-level video representation features. We use the CNN+LSTM model proposed by Donahue *et al.* [3] in the Caffe library [10]. The model is then trained on the UCF-101 dataset [14]. The second component is the unsupervised hashing which exploits the spatio-temporal features obtained by the first component. The entire procedure for obtaining binary codes of spatio-temporal features is detailed as follows.

As illustrated in Fig. 2, we pass the visual input v_{it} (the t-th frame from the i-th video) through a few convolutional layers. The architecture used for each CNN layer is similar to the one proposed in [3]. The initial weights of CNN are set to the same values as trained on the ImageNet dataset [2]. The weights are then fine-tuned when the LSTM layer is added. The output of the fc6 layer is chosen as the output of the CNN framework, and it has 4096-dimensional feature. Rectified Linear Units (ReLU) and dropout layers are used to avoid the overfitting problem.

After modeling the spatial features with CNN, we pass the learned spatial features to LSTM layer. The LSTM layer is used to model the temporal correlation structure between frames in videos. Similar to the strategy proposed in [3], we use only one LSTM layer of which the output is a 256-dimensional vector. The reason why we choose only one LSTM layer is that our model requires to

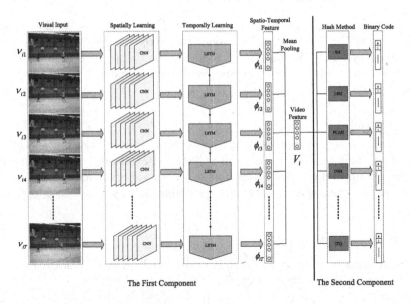

Fig. 2. The proposed video hashing framework consists of two main components. The first component involves the supervised per-training based on the UCF-101 dataset to learn the spatio-temporal features. The second component is unsupervised hashing which exploits the spatio-temporal features generated by the first component.

deal with sequential input while the output is a fixed-length vector, so other LSTM layers are not necessarily needed.

After all the frames have been processed, we obtain a vector representation sequence $\{\phi_{i1}, \phi_{i2}, ..., \phi_{iT}\}$. The spatio-temporal representation of each video is then calculated by:

$$V_i = \frac{1}{T} \sum_{t=1}^{T} \phi_{it} \tag{3}$$

where T is the number of frames in the i-th video.

After obtaining the feature set $V = \{V_1, V_2, \ldots, V_m\} \in \mathbb{R}^d$ (m is the number of videos), the next step for hashing is to look for a group of appropriate hashing functions $h : \mathbb{R}^l \longmapsto \{1, -1\}^1$ with each of them accounting for generating of a single hash bit. Many unsupervised hashing methods can be used here such as Iterative Quantization (ITQ) [7] and Spectral Hashing (SH) [17].

3   Experiments

In this section, we apply our spatio-temporal feature learning framework to many existing unsupervised hashing methods including Iterative Quantization (ITQ) [7], Locality Sensitive Hashing (LSH) [6], PCA Hashing (PCAH) [7], Spectral Hashing (SH) [17], Density Sensitive Hashing (DSH) [12], and Spherical Hashing (SpH) [8]. By testing the proposed framework on the UCF-101 dataset, we compare between the hashing methods with our spatio-temporal feature learning framework and the ones that only use spatial features which are learned by CNN. We start by introducing the dataset and evaluation metrics, and then present the comparison results of our method with the existing approaches on the UCF-101 dataset [14].

3.1 Dataset

UCF-101 Dataset [14] consists of 101 action classes, over $13\,k$ clips and $27\,h$ of video data. This database consists of realistic user-uploaded videos containing camera motion and cluttered background. In our experiments, we select 9,537 videos as the training data, and the remaining 3,783 videos are adopted for testing.

3.2 Evaluation Metrics

The performance of video retrieval is evaluated by Mean Average Precision (MAP). For a query q, the average precision is defined as:

$$AP(q) = \frac{1}{L_q} \sum_{r=1}^{R} P_q(r)\delta_q(r) \tag{4}$$

where L_q is the number of groundtruth neighbors in the retrieval list. $P_q(r)$ is the precision of the top r retrieved results, and $\delta_q(r) = 1$ if the r-th result is the true neighbor and 0 otherwise.

3.3 Results

To show the effectiveness of the proposed spatio-temporal feature learning framework, we first apply it to the unsupervised hashing methods mentioned above and see the results. Then, we compare the results with those of the corresponding methods that only use the spatial features learned by CNN. Figure 3(a) shows the comparison results. It is clear that the methods using the proposed spatio-temporal features obtain significantly larger MAP than the ones that only use spatial features. This is because our spatio-temporal feature learning framework fully takes advantages of not only the spatial structure but also the temporal correlation of video data. The proposed spatio-temporal feature learning framework has a better representation of the video than spatial feature representation.

To further validate the effectiveness of the proposed method using binary codes of shorter bits, we perform the corresponding experiments using the binary hashing codes of 8 bits, 16 bits, 32 bits, and 64 bits, respectively. The corresponding results are shown in Fig. 3(b) and (c). It can be clearly observed that the MAP results of both settings (corresponding to using the proposed spatio-temporal features and only the spatial features) increase as the length of hashing binary codes increase from 8 to 64 bits. As showed in Fig. 3(b) and (c), in all cases, the methods that exploit the proposed spatio-temporal features still outperform the ones that only use spatial features with a noticeable gain regarding MAP.

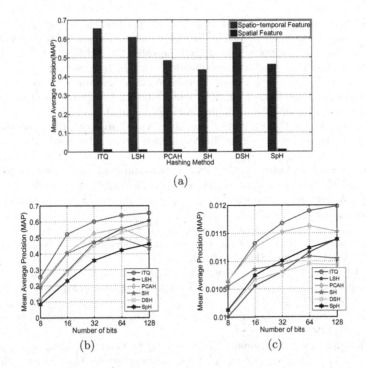

(a)

(b)   (c)

Fig. 3. Results of six unsupervised hashing methods on UCF-101. (a) The performance comparison between using spatio-temporal features and only using the spatial features. The comparison is performed on the 128-bit binary codes. (b) MAP results of the methods using the spatio-temporal features proposed in this paper. (c) MAP results of the methods only using spatial features learned by CNN.

4   Conclusion

In this paper, we have combined CNN and LSTM units into a unified framework, which is both spatially and temporally deep, for video hashing. Due to the utilization of spatio-temporal features, the created binary codes are more representative for video retrieval, and thus encouraging performances can be achieved. Experimental results on UCF-101 dataset have well demonstrated the feasibility and effectiveness of the proposed method.

Acknowledgments. This research is partly supported by 973 PlanChina (No. 2015CB856004) and NSFC, China (No: 61572315).

References

1. Cao, L., Li, Z., Mu, Y., Chang, S.F.: Submodular video hashing: a unified framework towards video pooling and indexing. In: Proceedings of the 20th ACM International Conference on Multimedia, pp. 299–308. ACM (2012)

2. Deng, J., Dong, W., Socher, R., Li, L.J., Li, K., Fei-Fei, L.: ImageNet: a large-scale hierarchical image database. In: IEEE Conference on Computer Vision and Pattern Recognition, 2009. CVPR 2009, pp. 248–255. IEEE (2009)
3. Donahue, J., Anne Hendricks, L., Guadarrama, S., Rohrbach, M., Venugopalan, S., Saenko, K., Darrell, T.: Long-term recurrent convolutional networks for visual recognition and description. In: IEEE Conference on Computer Vision and Pattern Recognition, 2015. CVPR 2015, pp. 2625–2634 (2015)
4. Douze, M., Jégou, H., Schmid, C.: An image-based approach to video copy detection with spatio-temporal post-filtering. IEEE Trans. Multimed. 12(4), 257–266 (2010)
5. Gers, F.A., Schmidhuber, J., Cummins, F.: Learning to forget: continual prediction with LSTM. Neural Comput. 12(10), 2451–2471 (2000)
6. Gionis, A., Indyk, P., Motwani, R., et al.: Similarity search in high dimensions via hashing. In: VLDB, vol. 99, pp. 518–529 (1999)
7. Gong, Y., Lazebnik, S.: Iterative quantization: a procrustean approach to learning binary codes. In: IEEE Conference on Computer Vision and Pattern Recognition, 2011. CVPR 2011, pp. 817–824. IEEE (2011)
8. Heo, J.P., Lee, Y., He, J., Chang, S.F., Yoon, S.E.: Spherical hashing. In: IEEE Conference on Computer Vision and Pattern Recognition, 2012. CVPR 2012, pp. 2957–2964. IEEE (2012)
9. Hochreiter, S., Schmidhuber, J.: Long short-term memory. Neural Comput. 9(8), 1735–1780 (1997)
10. Jia, Y., Shelhamer, E., Donahue, J., Karayev, S., Long, J., Girshick, R., Guadarrama, S., Darrell, T.: Caffe: convolutional architecture for fast feature embedding. arXiv preprint arXiv:1408.5093 (2014)
11. Jiang, Y.G., Ye, G., Chang, S.F., Ellis, D., Loui, A.C.: Consumer video understanding: a benchmark database and an evaluation of human and machine performance. In: Proceedings of the 1st ACM International Conference on Multimedia Retrieval, p. 29. ACM (2011)
12. Jin, Z., Li, C., Lin, Y., Cai, D.: Density sensitive hashing. IEEE Trans. Cybern. 44(8), 1362–1371 (2014)
13. Krizhevsky, A., Sutskever, I., Hinton, G.E.: ImageNet classification with deep convolutional neural networks. In: Advances in Neural Information Processing Systems, pp. 1097–1105 (2012)
14. Soomro, K., Zamir, A.R., Shah, M.: UCF101: a dataset of 101 human actions classes from videos in the wild. arXiv preprint arXiv:1212.0402 (2012)
15. Torralba, A., Fergus, R., Weiss, Y.: Small codes and large image databases for recognition. In: IEEE Conference on Computer Vision and Pattern Recognition, 2008. CVPR 2008, pp. 1–8. IEEE (2008)
16. Wang, J., Kumar, S., Chang, S.F.: Semi-supervised hashing for large-scale search. IEEE Trans. Pattern Anal. Mach. Intell. 34(12), 2393–2406 (2012)
17. Weiss, Y., Torralba, A., Fergus, R.: Spectral hashing. In: Koller, D., Schuurmans, D., Bengio, Y., Bottou, L. (eds.) Advances in Neural Information Processing Systems, vol. 21, pp. 1753–1760. Curran Associates, Inc. (2009)

Selective Dropout for Deep Neural Networks

Erik Barrow[1]([⊠]), Mark Eastwood[1], and Chrisina Jayne[2]

[1] Coventry University, Coventry, West Midlands, UK
ab3065@coventry.ac.uk
[2] Robert Gordon University, Aberdeen, Scotland, UK

Abstract. Dropout has been proven to be an effective method for reducing overfitting in deep artificial neural networks. We present 3 new alternative methods for performing dropout on a deep neural network which improves the effectiveness of the dropout method over the same training period. These methods select neurons to be dropped through statistical values calculated using a neurons change in weight, the average size of a neuron's weights, and the output variance of a neuron. We found that increasing the probability of dropping neurons with smaller values of these statistics and decreasing the probability of those with larger statistics gave an improved result in training over 10,000 epochs. The most effective of these was found to be the Output Variance method, giving an average improvement of 1.17 % accuracy over traditional dropout methods.

Keywords: MNIST · Artificial neural network · Deep learning · Dropout network · Non-random dropout · Selective dropout

1 Introduction

Dropout is an effective method for reducing overfitting in neural networks [1] that works by switching off neurons in a network during training to force the remaining neurons to take on the load of the missing neurons. This is typically done randomly with a certain percentage of neurons per layer being switched off. Dropout has also been previously tested on large well known data sets such as MNIST [1,2], ImageNet [3,4] which is significantly larger than MNIST, and CIFAR [1,5].

We propose a new method of dropout that selectively chooses the best neurons (neurons which will have the biggest positive effect on the network if switched off) to be given a higher probability of being switched off on the assumption that dropout could be made more effective and efficient by not dropping neurons that should be forced to continue to learn. Our main contributions are 3 new methods of selecting these neurons as described in Sect. 2.2.

Recently some other methods of altering the way dropout works have surfaced. These include a method of adaptive dropout that was explored by Ba and Frey, where they change the probability of neurons being switched off using a binary belief network, which runs separately from the neural network being

© Springer International Publishing AG 2016
A. Hirose et al. (Eds.): ICONIP 2016, Part III, LNCS 9949, pp. 519–528, 2016.
DOI: 10.1007/978-3-319-46675-0_57

trained [6]. This allows the network to find the optimum drop rate for any given neuron.

Another Modified dropout method was introduced by Duyckm et al. where they apply optimisation methods to dropout such as simulated annealing [7]. Dropout has also been used in an innovative way by Wan et al. to create a new method called DropConnect, where a random subset of the weights on a neuron are randomly set to zero instead of all of a neurons weights [8].

Our experiments on Selective Dropout will be conducted using the MNIST dataset [2], a dataset consisting of pre-processed handwritten digits. MNIST is a well-known and well-used dataset in the field of deep learning and will provide results that can be easily benchmarked against other research. Number recognition is a widely used machine learning application, with applications such as number plate recognition, and automated bank teller machines. Dropout has also been previously tested on large ImageNet [3,4] which is significantly larger than MNIST.

To test the effectiveness of selective dropout against that of the traditional dropout we will use a Deep Neural Network architecture. Since 2006 Deep learning has been an effective method of machine learning since the discovery of the use of pre-training deep networks with Restricted Boltzmann Machines [9]. All our experiments will be pre-trained by a stack of Restricted Boltzmann Machines in two different hidden layer architectures, these are [500,500,500] & [500,500,1000]. All experiments will be validated using 10 Fold validation to ensure an accurate test score for comparison of methods.

Section 2 outlines the methodology of the traditional dropout method and goes into detail on the various methods of selective dropout used for the experiments in this paper. Section 3 outlines the MNIST dataset used in this experiments, and Sects. 4 and 5 outline the experiments results and conclusions.

2 Dropout Methodology

2.1 Traditional Dropout

The Traditional dropout method [1] is a method used to reduce the error rate of artificial neural networks, working by reducing the amount of overfitting that happens in a network and also helping to prevent the network settling in local minima. The dropout method does this by selecting a proportion of random neurons from the input layer through to the last hidden layer, and effectively switching them off for a training epoch. At the end of an epoch, the neurons are then switched back on and another set of random neurons are switched off. This switching off of neurons causes the remaining neurons that may not have learnt anything useful, to try and learn new features.

Neuron outputs are scaled for the use of the network after training, relative to the amount of neurons being switched off in that particular layer. This prevents issues occurring where neurons are overloaded with more inputs than expected. e.g. A network has 50 % of neurons switched off on each layer during training, however after training all neurons are switched on. Neurons on a receiving layer would get double the expected input if the outputs are not scaled down by 50 %.

If $\mathbf{x}^{(k)}$ are the node outputs at layer k, and p_k is the proportion of nodes retained during dropout for layer k, the node output during prediction is as shown in Eq. 1 [1]:

$$\mathbf{x}^{(k)} = sigmoid(p_k\mathbf{W}^{(k)}\mathbf{x}^{(k-1)} + \mathbf{b}) \tag{1}$$

In Sect. 4 the results of traditional dropout are represented by **TD**.

2.2   Selective Dropout

Selective Dropout is different from traditional dropout in that it attempts to choose the neurons which it switches off by giving a higher drop probability to a neuron. By increasing the chances of switching off a selective set of neurons we could improve the effectiveness of the traditional dropout method by either better test results or faster network convergence. There are multiple ways in which a neuron can be selected, and this paper looks at selecting neurons based on the average change in weights between training iterations, the neurons average weight size, and the variance of a neurons output during a training iteration.

With the statistic calculated from each method, we can use a layers drop probability to calculate an individual drop probability for each neuron in the layer while keeping the mean of all probabilities assigned to those neurons the same as the drop percentage for each layer. The below formula (Eq. 2) comes from the demands of having a probability that is not only proportionate to the calculated statistics, but also gives an average that is equal to the drop probability given to the network.

$$\mathbf{P} = C\mathbf{S}^\alpha \tag{2}$$

Equation 2 shows how this is calculated, where $\mathbf{P}$ is the calculated probability for all the neurons on a given layer, and $\mathbf{S}$ is the given statistics we are scaling probabilities by (Such as $\mathbf{avg_k}$ in Eqs. 6 and 7, or $\mathbf{N\_Variance_k}$ in Eq. 8). C is a constant that allows us to calculate a probability that is proportionate to our statistics ($\mathbf{S}$) and the drop rate given to the network, shown by Eq. 3 below. This allows us to meet our demands that the calculated probability is proportionate to the statistics as well as the given drop rate. N is used in the equation to ensure that the mean of all the probabilities generated is equal to the layers drop probability specified in the network parameters, such that $N = C\sum\mathbf{S}^\alpha$.

$$C = \frac{N}{\sum\mathbf{S}^\alpha} \tag{3}$$

In Eq. 3, α is a constant between 0 and 1 chosen such that $Cmax(\mathbf{S}^\alpha) \leq 1$. This is chosen in numerical decrements starting at 1 until a solution is found. α is used in the equation to ensure that the result for a neurons probability is always less than or equal to 1, this also allows us to work with larger statistics values. It is also worth noting that an α of 0 would result in the traditional dropout method. N is the number of neurons we expect to drop on any given layer, based on the size of the layer and the drop rate provided to the network. Equation 4 below shows how N is calculated, with p_k representing the neuron retention rate given to the layer (k) being calculated, such as the retention rate of 0.6 used for the hidden layers in our experiments.

$$N = size(\mathbf{S}) \times (1 - p_k) \tag{4}$$

Equation 2 will give a higher probability of being dropped to statistics with a high value. However in some of our experiments we need to give a higher probability to statistics with a small value. We can accomplish this with the following equation (Eq. 5). In the equation 1.1 is used to ensure that no statistical values equal zero, as it is preferable that all neurons have some probability of being switched off.

$$\mathbf{S} = 1.1 \times max(\mathbf{S}) - \mathbf{S}; \tag{5}$$

The methods described below can only be applied to the neurons in the hidden layers of a network and cannot be applied to neurons in the input layer, as the input layer either has no incoming weights or the input layer output is the image being inputted, of which the network has no control over. Due to this we use the traditional dropout method on the input layer to select random neurons, and then the remaining hidden layers are run with our selective method to choose selective neurons.

Average Weight Change. The method of Average Weight Change looks at the average of the differences between a neurons weights before and after a training iteration. The motivation of this method is that we can tell whether a neuron is still learning by the amount of change in its weights between epochs/batches and by how active the learning process is by the scale of this change.

The equation below (Eq. 6) shows the method of finding the average weight change for each neuron where $\mathbf{avg}_k$ is the nodes input weight change average at layer k, and $\mathbf{W}_{jk}^{(i)}$ is the weight matrix feeding into layer k at the beginning of iteration i (Before training). n represents the number of weights feeding into a neuron from the nodes in the previous layer.

$$\mathbf{avg}_k = \frac{1}{n} \sum_{j=1}^{n} (|\mathbf{W}_{jk}^{(i)} - \mathbf{W}_{jk}^{(i-1)}|) \tag{6}$$

For neurons switched off during a dropout iteration there should be no change in their weights between iterations, which would cause the average change to be 0. In some selections as detailed below this would cause the neurons to either always be given a high probability of being selected to be switched off, or could cause the dropped neurons to alternate between a high and low probability of being switched off. To overcome this, the average weight change for neurons switched off are replaced with the average from the last time they were switched on. To facilitate this the first iteration of training is performed without any dropout to allow a base average for each neuron to be obtained.

When the vector of the average change in weights for each neuron (as calculated in Eq. 6) has been obtained it can then be used to find the neurons with either the highest change in weights or the lowest change in weights, before then using algorithm (2) to assign a drop probability to these neurons. This allows us to probabilistically switch off a percentage of neurons which have either had a large change in weights or a small change in weights.

Giving a higher probability to neurons with the smallest average change in weights allows us to increase their probability of being switched off, and is much like sorting our statistic in Ascending order and giving high probabilities to the first x neurons in the vector. This could be considered as neurons that have finished learning as their weights are changing the least. In Sect. 4 the results of average weight change dropout (dropping smallest average change) are represented by **WA**.

Giving a higher probability to neurons with a high average change in weights (as if sorting the statistics in Descending order), will allow us to increase the chances of neurons with a high change in weights being switched off. This could be considered as increasing the chances of switching off the neurons that are learning, in order to force those that are not learning to learn. In Sect. 4 the results of average weight change dropout (dropping highest average change) are represented by **WD**.

Average Neuron Weight. This method looks at the average weight for each neuron in a layer. The motivation behind this method is that neurons or weights that have learnt something tend to increase in size and start out very small or close to zero when initialised, allowing us to choose neurons that have or haven't done much learning.

To find the average weight for each neuron the following equation (Eq. 7) is used where $\mathbf{avg}_k$ is the average input weight of a neuron on layer k and $\mathbf{W}_{jk}^{(i)}$ is the weight matrix for the current iteration i before training commences.

$$\mathbf{avg}_k = \frac{1}{n} \sum_{j=1}^{n} (|\mathbf{W}_{jk}^{(i)}|) \tag{7}$$

As with the previous method, the average of the weights can be used to give a drop probability to each neuron relative to the layer drop rate using Eq. 2. As dropout typically already replaces the weights of switched off neurons at the end of an iteration there is no need to keep a base representation for neurons switched off.

By giving the neurons with high average weight values higher drop probabilities (much like sorting the statistics in descending order) we can increase the chances of switching off neurons with high weights attached to them. The idea of this method is that weights are initialised close to zero and that weights that are closer zero will have learnt less and should be kept on to learn while the neurons that have learnt are switched off. Another motivation for this method is regularisation, typical regularisation methods penalise large weights in some way, and this method also does so by preventing neurons with very large weights from increasing. In Sect. 4 the results of average weight dropout (dropping highest average weight) are represented by **AD**.

On the other hand giving a high drop probability to neurons with low average weights (much like sorting the statistic in ascending order) allows us to increase the chances of switching off neurons where the average weight is of a small value. We expect that this could cause neurons with small initial weights to be given a high drop probability for a majority of the training, causing them to spend a majority of training time switched off and rarely being switched back on, in turn leaving the same neurons learning a majority of the time. We still run this experiment to give us some comparison of the effect between giving high probabilities to switching of neurons with high or low weight magnitudes. In Sect. 4 the results of average weight dropout (dropping lowest average weight) are represented by **AA**.

Output Variance. This method looks at the variance of outputs for a neuron over an iteration. The motivation behind this method is that you may be able to tell how much a neuron has learnt by the difference in the outputs it gives between images, i.e. a neuron that always outputs the same number no matter the input could be considered as not having learnt anything useful.

The variance is found by taking the output matrix from the previous iteration as shown by the equation below (Eq. 8) where $\mathbf{N\_Variance}_k$ represents the variance of neurons of layer k and $\mathbf{X}_k^{(i-1)}$ represents the output matrix of layer k for iteration $i - 1$. Variance is calculated along the dimension that holds all the outputs for a neuron over an epoch/batch.

$$\mathbf{N\_Variance}_k = variance(\mathbf{X}_k^{(i-1)}) \tag{8}$$

As with the Average Weight Change method, there is a need to keep a baseline for neurons switched off as switched off neurons would produce no output giving a variance of 0. This baseline is facilitated by running the first iteration with no dropout to allow for an output matrix where every neuron has an output. As with all previous methods the variance vector can be used to find neurons with high output variance or low output variance, and apply a probability to them with Eq. 2.

Giving a high probability to neurons with high output variance (much like ordering the statistic in descending order) will allow us to increase the chance of switching off neurons with high variance. High variance could indicate that the neuron has learnt an interesting feature and no longer needs training and as such can be switched off to allow neurons with a lower variance to learn. In Sect. 4 the results of neuron output variance (dropping highest variance) are represented by **VD**.

Giving a high probability to neurons with a small output variance (much like sorting the statistic in ascending order) allows us to increase the chance of switching off neurons with low variance. Low variance may indicate that a neuron is outputting roughly the same value no matter the input and has learnt no interesting features. We expect that switching off these neurons that have not learnt, will cause the network to learn at a very slow rate and possibly be less efficient than the traditional dropout method. In Sect. 4 the results of neuron output variance (dropping lowest variance) are represented by **VA**.

3   Dataset

The dataset used for this experiment was MNIST's handwritten number dataset [2]. MNIST consists of 60,000 training images and 10,000 test images of hand-written numbers in the range of 0–9, each of which 28×28 in size (784 pixels) and has been normalised and centred. We used the 60,000 training image set for training, validation, and testing to allow for 10 Fold validation. Figure 1 shows a small sample of MNIST images.

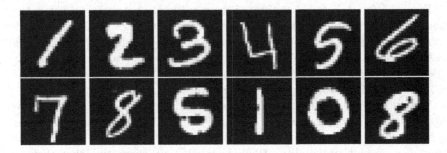

Fig. 1. A sample of images from the MNIST data set

For our experiment, the MNIST dataset was split into 10 folds for data validation. Each fold contains 54,000 images for Training and Validation (at an approximate 90 % and 10 % split), and 6,000 images for testing. The Training images are used for the training of the Deep Neural network with the Validation images used to facilitate early stopping. The remaining test set images are solely used for testing the network after training and have no impact on the training process.

4   Experiments and Results

140 Experiments were conducted over a maximum of 10,000 epochs each. Each method ran 20 experiments consisting of 2 network architectures and 10 different folds of data. Each dropout method used the same pre-initialised weights from their respective architectures/folds Restricted Boltzmann Machine. The results of each fold were combined to get an average result for the testing of the method

Table 1. 10 fold validated experiment results (test set), to 4 decimal places

Method	% Error [500,500,500]	% Error [500,500,1000]
TD - Traditional Dropout	4.7300	4.5933
WD - Weight Change (Descending)	7.0533	7.5783
WA - Weight Change (Ascending)	4.3167	4.2517
AD - Average Weight (Descending)	5.2983	5.6533
AA - Average Weight (Ascending)	4.3217	4.1733
VD - Output Variance (Descending)	14.4633	13.7833
VA - Output Variance (Ascending)	3.5167	3.4667

and architecture. Each network was trained with a drop rate of 20 % for the input layer (using traditional dropout), and 40 % on all remaining hidden layers (using our altered methods). Table 1 shows the results of the experiments for each method under each network architecture.

From the results listed above we can see that 3 of the networks did better than the traditional dropout method, these being Weight Change in Ascending order (**WA**) which was 0.378 % better on average, Average Weight in Ascending order (**AA**) which was 0.414 % better on average, and Output Variance in ascending order (**VA**) which was 1.17 % better on average. We can also see that these methods worked for both sizes of architecture tested.

Several of these results were unexpected and the complete opposite of our expectations in Sect. 2.2. For the Average Weight Change method our experiments performed as expected, with neurons that had little change in weights being given a high probability of being switched off, suggesting that they had finished learning, while leaving neurons that were still learning switched on. As expected, giving a high probability of switching neurons with a large weight change off during training caused the network to be 2.744 % worse off on average than the traditional dropout method.

With the Average Weights method, we expected neurons with larger weights to have learnt more, and that they could be given a higher probability of being switched off, allowing neurons with smaller weights to continue learning. However, this did worse than the traditional dropout method, coming out 0.814 % worse off on average than the traditional dropout method. The method of switching off neurons with smaller weights did better than this, suggesting that leaving on larger weights gave better results. This could possibly indicate that similar to the weight change method, that it is better to leave neurons that are learning the most switched on.

We also expected that increasing the probability of switching off neurons with high Output Variance to perform well, indicating the neurons had learnt an interesting feature and could be switched off. However, this gave the worst results of all the experiments giving an average of being 9.462 % worse than the traditional dropout method. Giving a high probability to neurons with low variance did much better than all the other experiments, suggesting leaving the neurons with high variance switched on is the best method. One possible explanation for this is that the worst neurons that were hindering the network were switched off for a majority of training, leaving the better neurons to continue to learn and refine their weights. It is also a possibility that the method could be acting as a form of network pruning [10] or sparsing [11], causing the network to prune or sparse particular neurons for the majority or all of the training, consistent with what can be seen in biology.

5 Conclusion

To conclude we have found that the three methods we have presented can improve the effectiveness of the dropout method in deep neural networks by probabilistically dropping neurons with smaller values in the statistics given (Eqs. 6, 7, and 8). The best of these given methods was shown to be giving higher drop probability to neurons with a low output variance, yielding an average improvement of 1.17 % over normal dropout.

Dropping neurons based on the smallest change in weights between iterations, and a smaller average weight size, also yielded improved results of 0.378 % and 0.414 % on average. Future work could explore the effects of the networks architecture size on these methods, as well as explore other new statistics that could be used to select neurons to be switched off.

References

1. Srivastava, N., Hinton, G., Krizhevsky, A., Sutskever, I., Salakhutdinov, R.: Dropout: a simple way to prevent neural networks from overfitting. J. Mach. Learn. Res. 15(1), 1929–1958 (2014)
2. LeCun, Y., Bottou, L., Bengio, Y., Haffner, P.: Gradient-based learning applied to document recognition. Proc. IEEE 86(11), 2278–2324 (1998)
3. Krizhevsky, A., Sutskever, I., Hinton, G.E.: ImageNet classification with deep convolutional neural networks. In: Advances in Neural Information Processing Systems, pp. 1097–1105 (2012)
4. Russakovsky, O., Deng, J., Su, H., Krause, J., Satheesh, S., Ma, S., Huang, Z., Karpathy, A., Khosla, A., Bernstein, M., Berg, A.C., Fei-Fei, L.: ImageNet large scale visual recognition challenge. Int. J. Comput. Vis. (IJCV) 115(3), 211–252 (2015)
5. Krizhevsky, A., Hinton, G.: Learning multiple layers of features from tiny images (2009)
6. Ba, J., Frey, B.: Adaptive dropout for training deep neural networks. In: Advances in Neural Information Processing Systems, pp. 3084–3092 (2013)
7. Duyck, J., Lee, M.H., Lei, E.: Modified dropout for training neural network (2014)
8. Wan, L., Zeiler, M., Zhang, S., Cun, Y.L., Fergus, R.: Regularization of neural networks using dropconnect. In: Proceedings of the 30th International Conference on Machine Learning (ICML-13), pp. 1058–1066 (2013)
9. Hinton, G.E., Osindero, S., Teh, Y.W.: A fast learning algorithm for deep belief nets. Neural Comput. 18(7), 1527–1554 (2006)

10. Karnin, E.D.: A simple procedure for pruning back-propagation trained neural networks. IEEE Trans. Neural Netw. **1**(2), 239–242 (1990)
11. Liu, B., Wang, M., Foroosh, H., Tappen, M., Pensky, M.: Sparse convolutional neural networks. In: Proceedings of the IEEE Conference on Computer Vision and Pattern Recognition, pp. 806–814 (2015)

Real-Time Action Recognition in Surveillance Videos Using ConvNets

Sheng Luo[✉], Haojin Yang, Cheng Wang, Xiaoyin Che, and Christoph Meinel

Hasso Plattner Institute, University of Potsdam, Prof.-Dr.-Helmert-Str. 2-3,
14482 Potsdam, Germany
{Sheng.Luo,Haojin.Yang,Cheng.Wang,Xiaoyin.Che,Meinel}@hpi.de

Abstract. The explosive growth of surveillance cameras and its 7 * 24 recording period brings massive surveillance videos data. Therefore how to efficiently retrieve the rare but important event information inside the videos is eager to be solved. Recently deep convolutinal networks shows its outstanding performance in event recognition on general videos. Hence we study the characteristic of surveillance video context and propose a very competitive ConvNets approach for real-time event recognition on surveillance videos. Our approach adopts two-steam ConvNets to respectively recognition spatial and temporal information of one action. In particular, we propose to use fast feature cascades and motion history image as the template of spatial and temporal stream. We conducted our experiments on UCF-ARG and UT-interaction dataset. The experimental results show that our approach acquires superior recognition accuracy and runs in real-time.

Keywords: Real-time application · Event recognition · Surveillance videos · Motion history image

1 Introduction

A sharp decline in the cost of semiconductor components brings explosive growth for surveillance cameras, specially the HD surveillance cameras. IHS forecasts that in the professional market, shipments of HD CCTV cameras will grow from fewer than 0.2 million units in 2012 to over 28 million units in 2016 [7]. The huge amount of surveillance cameras brings massive surveillance video data in the same time. According to a new report from IHS, the amount of daily data generated by new video surveillance cameras installed worldwide in 2015 approaches twice the amount of all user data stored by Facebook. However, the majority parts of these videos are useless and meaningless, and the manually information retrieve is impossible for such huge amount of data. Therefore, how to effectively utilize these data and retrieve the rare and important information inside is eager to be solved. Furthermore, the 7/24 uninterrupted recording period raises higher requirement for processing performance.

© Springer International Publishing AG 2016
A. Hirose et al. (Eds.): ICONIP 2016, Part III, LNCS 9949, pp. 529–537, 2016.
DOI: 10.1007/978-3-319-46675-0_58

Recently a novel range of methods based on deep convolution networks (ConvNets) beats the traditional approaches and shows excellent recognition accuracy on event recognition on general videos. In the same time, the traditional methods such as spatio-temporal interests point, hog, optical flow etc. are mainstream event recognition methods on surveillance videos as well. In this work we intended to apply ConvNets for further improving the action recognition performance on surveillance videos. However, it is well known that deep neural network has high demand of computation resources and relative slow processing speed due to its huge amount of parameters. Therefore, we propose a novel approach which apply ConvNets in combination with fast feature cascades and motion history image. Our approach achieves the real-time performance with state-of-the-art recognition accuracy.

The rest of this paper can be listed as follows: Sect. 2 reviews state-of-the-art ConvNets methods for action recognition. Our real-time two-stream architecture is described in Sect. 3. Section 4 gives the implementation detail and evaluation result. Conclusion and future works are presented in Sect. 5.

2   Related Work

Video is composed of images, image classification naturally becomes the fundamental of event recognition on video. Therefore the outstanding performance of ConvNets on image classification attracted a great deal of researchers to introduce ConvNets into video domain. [3,11] firstly proposed to adopt image CNN in video recognition, they performed convolutions in both space and time dimensions. Then [13] utilized this idea and proposed to use stacked video frames for better clue in time domain, but they didn't achieved a outstanding performance improvement. After that, [16] proposed to incorporate spatial and temporal networks based on this idea and proved that two-stream ConvNets architecture is appropriate for event recognition on videos. In addition, they also compared different temporal expression approaches and proved that multi-frame dense optical flow is more superior for ConvNets to understand motion changes. Since then the two-streaming architecture and multi-frame dense optical flow become the general reference method for successor researchers. The main optimizing directions of the reference methods are increasing the complexity of ConvNets and parameter engineering of the network. For the first one, [10,20] introduced recurrent convolutional architecture into the model and improved the performance with the computational complexity in the same time. Based on that, [18] proposed a regularized feature fusion network for the fusion part, it further increased the classification result and computation cost. Regarding the later direction, [17,19] studied the important parameter and implementation options to improve the performance without higher computation demand. These methods are based on general videos, how to more effectively apply ConvNets on surveillance video is the issue we want to solve in this paper.

3 Framework

Human action can be expressed by temporal and spatial variation, therefore it is reasonable to respectively understand these two parts for recognizing human action. In our framework for human action recognition, we naturally adopt the two-stream architecture which shows in Fig. 1 to analyze the variation.

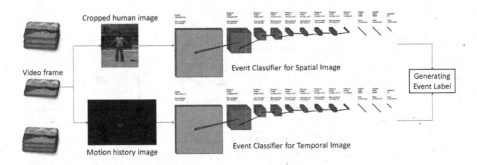

Fig. 1. System architecture

Regarding surveillance video, the main difference from general videos is that the actor inside video is small in most circumstance because the recording camera is quite far away from actor. In other words, comparing to general video, surveillance video contains more similar and irrelevant information. The action irrelevant information not only increases the difficulty for action recognition but also wastes computation resources. Therefore we apply human detector for eliminating the background noise before ConvNets in spatial stream.

For temporal stream, static background makes it easier to separate the motion of the object from camera-induced or distractor motion. Hence we utilize motion history image (MHI) to record the motion change from the actor. Unlike optical flow and stacked optical flow Fig. 2(b), MHI can capture and express the short-term motion changes in a more compact way. Figure 2(c) shows the principle of MHI which express the consecutive motion change in one single image. To create a MHI H_r, binary image difference frames are firstly generated with a defined buffer length. Then consecutive temporal motion information is collapsed into H_r according to their temporal appearances, i.e. the pixel intensity in H_r is a function of the temporal history of motion at that point. Hence, H_r can be presented by utilizing a replacement and a decay operator D [8]:

$$H_r(x,y,t) = \begin{cases} \tau, & D(x,y,t) = 1 \\ max(0, H_r(x,y,t-1) - 1) & otherwise \end{cases} \quad (1)$$

The essence of MHI is to record where the motion is happening and how the motion is occurring in one image. In this way, a consecutive motion scene can be explicitly stored in MHI. However, the common method—optical flow, which is used frequently, not only need to be previously stacked, to explicitly express continuous motion, but also not able to fulfill the real-time requirement. The extracted temporal and spatial image will be input into pre-trained ConvNets models for the classification respectively. Then we fuse the classification results from two ConvNets and print the label to the video if the corresponding probability is over the threshold.

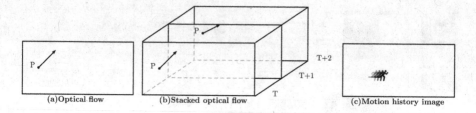

Fig. 2. Comparison of MHI and optical flow

In general, we respectively use fast pedestrian detector and motion history image to retrieve the spatial and temporal clues from the video. Then we apply ConvNets to learn the representative action features. Our approach takes advantage of the speed benefit from shallow features and high accuracy from ConvNets.

4   Evaluation

4.1   Datasets and Pre-trained Model

In order to verify the effectiveness of our method, we conducted experiments on UCF-ARG [2] and UT-Interaction dataset [15]. UCF-ARG dataset contains 10 classes actions which are Boxing, Carrying, Clapping, Digging, Jogging, Open-Close Trunk, Running, Throwing, Walking and Waving. In this work we only adopted the video data taken by the ground camera and put first 3/4 clips into train set and the rest 1/4 into test set, specifically the clips from actors 1–9 are in train set, and test set own the clips of actor 10–12. The UT-Interaction dataset is made up of 6 classes of human-human interactions: shake-hands, point, hug, push, kick and punch. This dataset has been evenly divided into two sub-sets, and we work on set 1 to further verify our method. For every category, the first 8 clips are used for training and the rest 2 clips for testing. In order to recognize the event inside the video, we need to train relevant ConvNets model for two streams.

UCF-ARG: Regarding the temporal stream, we adopt the off-the-shelf implementation from openCV library for motion history image creation. We set the threshold parameter to 15 and use different buffer size 2, 5, 10. Because the MHI method can filter out the background information, the generated MHI images can be directly used as the input for ConvNets.

For spatial stream, we adopt ACF pedestrian algorithm proposed by Piotr et al. [9] to acquire the human actor image. In order to verify the influence of actor image on ConvNets, we conduct an experiment on complete images and actor images. We train ConvNets based on these two kinds of images, the classification result can be found in Table 1. It is obvious that actor image can be better understood by ConvNets and bring higher recognition accuracy, which also proves the importance of the human detection process.

UT-Interaction: Because UT-Interaction dataset provides the ground truth of actor, we extract the actor image from videos of that based on ground truth as the input data of spatial stream. For the training data of temporal stream, we use the same method above. But the actor in video of UT-Interaction is quite big and the motion is

Table 1. Classification result on complete and actor image

Images	Accuracy
Complete image	23.1804 %
Actor image	72.7831 %

easily to be extracted, so we set the threshold parameter to 30 for a better background noise removal. Because there is overlap between actions, we calculate the motion history image based on the complete video and extract the relevant motion part based on ground truth. In this way, we can steadily isolate the background and assure that every image only has one action inside.

Concerning training process, we adopt Caffe framework [12] and finetune our model from robust models instead of training from scratch. Because the training samples of both UCF-ARG and UT-interaction are quite limited, in this case pre-training is an effective way to initialize the parameters for ConvNets. The effectiveness of finetuning has been proved by lots of researchers. We adopt VGG_CNN_S as our pre-training model for both temporal and spatial model.

In order to compare our approach with the mainstream methods based on ConvNets, we also adopt optical flow approach from [1] to extract motion feature. We finetune SFOF model based on single frame optical flow images from VGG_CNN_S model. Like in [17], we modify the original VGG_CNN_S model and finetune SOF model based on stack optical flow images (20 Channels).

4.2 Implementation Detail

Our real-time recognition is based on open source code, so it is easy to implement. Specifically, we adopt the off-the-shelf VeryFast algorithm implemented by Bennenson and collaborators in "Doppia" source code [4] for extracting spatial image. Regarding temporal motion feature, we utilize the high efficiency MHI algorithm implemented in opencv library [6]. According to the limitation of real-time, the whole process should finish in 33 ms, hence the parameters of every process should be well optimized. For every frame from the video, we use VeryFast detector to find the human actor and cropped the key sub-image from the frame. According to the theory of VeryFast algorithm, detection window scale explored highly affect the detection speed and quality. Because the actor is relatively small in video of UCF-ARG data set, here we set min-scale, max-scale to 0.4 and 1.5 to accelerate detection. Through this process, the majority of irrelevant background information are cropped and the whole useful information is input into the ConvNets.

On the other hand, we utilize the MHI detector to calculate the motion history based on the whole frame and generate temporal feature. Buffer size and threshold parameter are important to the purity of action motion and should be the same as in the training model. We set the buffer size to 2 and threshold parameter to 15 for UCF-ARG dataset. Comparing UCF-ARG, the action in UT-interaction is longer and the actor is larger. Therefore, the buffer size is 10 and the threshold parameter is 30 in UT-interaction.

Then the spatial and temporal feature are input into relevant ConvNets model for prediction. After prediction, the two ConvNets model provide two arrays with the probabilities in ten/six classes. Regarding the fusion part, we average these two arrays and get the class with the max probability. We set the threshold for probability to 0.3.

Because of the overlap between actions in UT-interaction dataset, we use pre-cropped video for real-time recognition instead of complete video. Therefore it is not necessary to use human detector for spatial stream in this context. The resolution of complete video in UT-interaction is smaller than video in UCF-ARG. There is no doubt that computation cost on video in UCF-ARG is higher than video in UT-interaction. But for a fair comparison, we only list the ConvNets model performance of UT-interaction dataset.

4.3 Performance

Figure 3 shows the recognition example, the top-left, top-right, bottom-left and bottom-right respectively show the original video, two-stream fusion video, spatial stream and temporal stream. Interestingly, the spatial stream shows more miss-shooting and wrong classification label than temporal stream. The whole recognition is running steadily at 30 frames per second. The recognition video is also online accessible.[1]

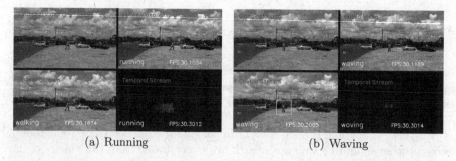

(a) Running   (b) Waving

Fig. 3. Example recognition of our two-stream approach on UCF-ARG test data (a) Running (b) Waving

---

[1] https://youtu.be/IwG5Q0zwOzU.

Table 2. System configuration

CPU	Intel Xeon E5-1607
RAM	8G, DDR3 1600 MHz
GPU	Nvidia GeForce GTX 780 3 GB
OS	Ubuntu 14.04 x64

Table 3. Approach performance

Method	FPS	Mode
Veryfast & MHI	~30	CPU
Veryfast & SFOF	~4-5	GPU
Veryfast & SOF	~0.8	GPU

Table 4. Image-level classification result

	UCF-ARG	UT-interaction
Spatial	74.62 %	43.25 %
MHI2	**85.97%**	70.24 %
MHI5	85.37 %	76.88 %
MHI10	84.62 %	**79.76%**
SFOF	76.31 %	72.22 %
SOF	83.56 %	76.44 %

Table 5. Clip-level classification result on UCF-ARG

Method	Accuracy
Spatial & MHI	**94.87%**
Spatial & SFOF	92.31 %
Spatial & SOF	**95.73%**
Piotr Bilinski [5]	82.05 %
Laptev et al. [14]	80.98 %

To a certain extent, real-time implementation depends on the hardware performance. Table 2 shows our experiment environment which is a normal and affordable computer. Table 3 gives a intuitionistic performance comparison between different approaches on UCF-ARG. There is no doubt that our approach "VeryFast & MHI" can run in real-time on both UCF-ARG and UT-interaction dataset. In contrast, temporal stream with SFOF can only process 4-5 frames per second which is 1/6 of MHI stream. Considering temporal stream with SOF which needs process 20 channels in one time, it has highest computation cost but can not process one frame in one second. Tables 4 and 5 shows the image-level and clip-level classification result of our ConvNets models. From the experiments, we can draw the following conclusions:

1. Motion information is superior for action recognition in surveillance video, SOF can better describe the temporal clue than SFOF.
2. At the image level, MHI significantly outperforms the other settings, which proves that this temporal representation can be well learned by CNNs.
3. At the clip level, MHI can also achieve very competitive accuracy result to SOF, but with much higher processing speed.

5   Conclusion and Future Works

In this paper, we investigate the state of the art ConvNets method for event recognition and analyze the characteristic of surveillance video. Based on that, we propose the real-time ConvNets architecture and implemented real-time event recognition approach on surveillance video. The experiments show that our method fulfill the real-time requirement and acquire high recognition performance for event recognition on surveillance. Although our approach achieves the best classification result in image-level, but the result in clip-level is in different, so we intend to further evaluate our method on a bigger dataset and with different fusion method. Besides, our ConvNets model is trained on the image which only has one action per image, it is not appropriate method for multi-action context. In the next, we intend to enable the action localization for temporal stream and make it suitable for multi-action context.

References

1. Brox, T., Bruhn, A., Papenberg, N., Weickert, J.: High accuracy optical flow estimation based on a theory for warping. In: Pajdla, T., Matas, J.G. (eds.) ECCV 2004. LNCS, vol. 3024, pp. 25–36. Springer, Heidelberg (2004)
2. UCF-ARG Data Set. http://crcv.ucf.edu/data/UCF-ARG.php. Accessed 10 Nov 2015
3. Baccouche, M., Mamalet, F., Wolf, C., Garcia, C., Baskurt, A.: Sequential deep learning for human action recognition. In: Salah, A.A., Lepri, B. (eds.) HBU 2011. LNCS, vol. 7065, pp. 29–39. Springer, Heidelberg (2011)
4. Benenson, R., Mathias, M., Timofte, R., Van Gool, L.: Pedestrian detection at 100 frames per second. In: CVPR (2012)
5. Bilinski, P., Bremond, F.: Statistics of pairwise co-occurring local spatio-temporal features for human action recognition. In: Fusiello, A., Murino, V., Cucchiara, R. (eds.) ECCV 2012 Ws/Demos, Part I. LNCS, vol. 7583, pp. 311–320. Springer, Heidelberg (2012)
6. Bradski, G.: Dr. Dobb's J. Softw. Tools (2000). Article ID 2236121
7. Cropley, J.: Top video surveillance trends for 2016, February 2016. https://technology.ihs.com/api/binary/572252
8. Davis, J.W., Bobick, A.E.: The representation and recognition of human movement using temporal templates. In: IEEE Computer Society Conference on CVPR, 1997, pp. 928–934. IEEE (1997)
9. Dollár, P., Belongie, S., Perona, P.: The fastest pedestrian detector in the west. In: BMVC, vol. 2, p. 7. Citeseer (2010)
10. Donahue, J., Anne Hendricks, L., Guadarrama, S., Rohrbach, M., Venugopalan, S., Saenko, K., Darrell, T.: Long-term recurrent convolutional networks for visual recognition and description. In: CVPR, pp. 2625–2634 (2015)
11. Ji, S., Xu, W., Yang, M., Yu, K.: 3D convolutional neural networks for human action recognition. IEEE Trans. Pattern Anal. Mach. Intell. 35(1), 221–231 (2013)
12. Jia, Y., Shelhamer, E., Donahue, J., Karayev, S., Long, J., Girshick, R., Guadarrama, S., Darrell, T.: Caffe: convolutional architecture for fast feature embedding. In: Proceedings of ACMMM, pp. 675–678. ACM (2014)
13. Karpathy, A., Toderici, G., Shetty, S., Leung, T., Sukthankar, R., Fei-Fei, L.: Large-scale video classification with convolutional neural networks. In: 2014 IEEE Conference on CVPR, pp. 1725–1732. IEEE (2014)
14. Laptev, I., Marszałek, M., Schmid, C., Rozenfeld, B.: Learning realistic human actions from movies. In: 2008 IEEE Conference on CVPR, pp. 1–8. IEEE (2008)
15. Ryoo, M.S., Chen, C.-C., Aggarwal, J.K., Roy-Chowdhury, A.: An overview of contest on semantic description of human activities (SDHA) 2010. In: Ünay, D., Çataltepe, Z., Aksoy, S. (eds.) ICPR 2010. LNCS, vol. 6388, pp. 270–285. Springer, Heidelberg (2010)

16. Simonyan, K., Zisserman, A.: Two-stream convolutional networks for action recognition in videos. In: NIPS, pp. 568–576 (2014)
17. Wang, L., Xiong, Y., Wang, Z., Qiao, Y.: Towards good practices for very deep two-stream convnets. CoRR abs/1507.02159 (2015)
18. Wu, Z., Wang, X., Jiang, Y.G., Ye, H., Xue, X.: Modeling spatial-temporal clues in a hybrid deep learning framework for video classification. In: Proceedings of the 23rd ACM MM, pp. 461–470. ACM (2015)
19. Ye, H., Wu, Z., Zhao, R.W., Wang, X., Jiang, Y.G., Xue, X.: Evaluating two-stream CNN for video classification. In: ICMR 2015, pp. 435–442. ACM (2015)
20. Yue-Hei Ng, J., Hausknecht, M., Vijayanarasimhan, S., Vinyals, O., Monga, R., Toderici, G.: Beyond short snippets: deep networks for video classification. In: 2015 IEEE Conference on CVPR, pp. 4694–4702 (2015)

An Architecture Design Method of Deep Convolutional Neural Network

Satoshi Suzuki[✉] and Hayaru Shouno

University of Electro-Communications, Chofugaoka 1-5-1, Chofu, Japan
sat.suzuki@uec.ac.jp

Abstract. Deep Convolutional Neural Network (DCNN) is a kind of multi layer neural network models. In these years, the DCNN is attracting the attention since it shows the state-of-the-arts performance in the image and speech recognition tasks. However, the design for the architecture of the DCNN has not so much discussed since we have not found effective guideline to construct. In this research, we focus on within-class variance of SVM histogram proposed in our previous work [8]. We try to apply it as a clue for modifying the architecture of a DCNN, and confirm the modified DCNN shows better performance than that of the original one.

1 Introduction

In recent years, the performance of the image classification task was dramatically improved by Deep Convolutional Neural Network (DCNN) models [3]. Most notable work, which is known as the AlexNet that was proposed by Krizhevsky *et al.* shows the highest performance in the image recognition task contest called the ILSVRC (ImageNet Large Scale Visual Recognition Challenge) 2012 [5]. In the contest, the AlexNet shows far and away the best score, which is 16.4 % error rate, compared to the 2nd place, which is 26.1 %. The DCNN has the almost same structure to the Neocognitron proposed by Fukushima [4,6], which can automatically extract feature expression from input data. The DCNN is now going to become a *de facto standard* classification tool in the image classification field, and the related research has been increased [3]. Many of the learning algorithms for the DCNN have been proposed various method since Fukushima proposed the basic structure [7], however, in recent years, the error back propagation (BP) method is typically used [6]. On the other hand, there is few design guidelines with respect to the architecture of DCNNs. So that, if we obtain a good clue for the architecture design, we can reduce the number of trial-and-error times to obtain a good DCNN for the specific purpose. For example, Zeiler *et al.* focused on the weight shapes, which is described as convolution kernel. They reduced the useless kernel, and obtained the improved performance [9]. In this work, we focus on a within-class variance of support vector machine (SVM) histogram representation for each layer, which is proposed in our previous work [8]. In the previous work, we found the shape of the SVM histogram becomes narrower throughout the DCNN pattern transformation. Hence, we try to apply the quantity as a clue of the DCNN architecture design.

A. Hirose et al. (Eds.): ICONIP 2016, Part III, LNCS 9949, pp. 538–546, 2016.
DOI: 10.1007/978-3-319-46675-0_59

2   Method

2.1   Deep Convolutional Neural Network

Deep Convolutional Neural Network (DCNN) has the similar structure to Neocognitron, which has characteristic connection weight structure described as a convolution operation [4]. In this work, we fixed the training method for the DCNN as the BP, which is used as the standard training method in these decades [6]. Generally, the DCNN for the visual task takes 2-dimensional image input, and the input is transformed the representation throughout the layers. The output layer of the DCNN provides class probability for the input image.

The DCNN consists of following four types of layers, that is, "convolution", "rectified linear unit (ReLU)", "normalized", and "polling" layers. The Convolution layer extracts a feature representation from the previous layer with a set of learned filters. The ReLU layer modifies the output of the convolution layer. The response of the ReLU layer is described as a rectified linear function $(relu(x) = \max(x, 0))$. The normalized layer is used as the optional layer, which is for reducing the contrast variance. The pooling layer applies max-pooling operation, which is used for reducing the effect of the local pattern deformation. Figure 1 shows a CaffeNet architecture used in previous work. This CaffeNet is used as the baseline DCNN, this CaffeNet has a similar architecture to the AlexNet [5].

2.2   Layer Response Representation with SVM Histogram

This section explains about the SVM histogram for layer representation in the DCNN [8]. In the previous work, we introduced a linear SVM for each layer in order to observe how representations in the DCNN develops throughout layer transformations. Let us consider the linear SVM which finds a decision boundary. The decision boundary is described as $y(\boldsymbol{x}) = \boldsymbol{w}^t \phi(\boldsymbol{x}) + b = 0$, where $\phi(\boldsymbol{x})$ denotes the layer representation for the input pattern $\boldsymbol{x}$. Here we introduce t as the teacher signal, which is described as $t_n \in \{1, -1\}$ in the case of two-class classification problem where n means the index for the test class pattern. In the feature space, the decision boundary of the linear SVM is obtained by maximizing the margin, $1/||\boldsymbol{w}|| \min_n [t_n(w^T \phi(\boldsymbol{x}_n) + b)]$ [1].

In our previous work, we introduced a distance from the decision boundary for each layer representation $\phi(\boldsymbol{x})$ as a measure of discriminability, which is described as $y(\boldsymbol{x}) = \boldsymbol{w}^T \phi(\boldsymbol{x}) + b$ where $\boldsymbol{w}$ and b are optimized by the linear SVM. Then, we can obtain a test class projection $\{y(\boldsymbol{x}_n)\}$. We analyze the projection $\{y(\boldsymbol{x}_n)\}$ as a histogram in the previous work [8]. Hereafter, we call it as "SVM histogram" for layer representation. Using the SVM histogram for each layer, we can visualize the distribution of each class data for intermediate layer of the DCNN. Here, Fig. 2 shows the overview of the SVM histogram.

According to the previous work, the SVM histogram becomes narrower through the hierarchical development of representation in the DCNN layers. As the result, we concluded the narrower representation of each class plays

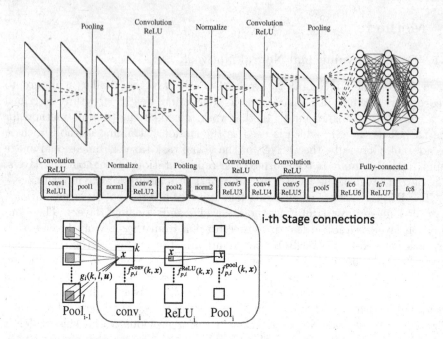

Fig. 1. Top: The summary of a kind of DCNN model CaffeNet. Bottom: The details of the process. In general, DCNN model acquire feature representation by repeating the processing Convolution → ReLU → Pooling.

important role for the class discriminability [8]. Therefore, in this study, we focus on the within-class variance of the SVM histogram as a design guideline for the DCNN architecture.

3   Training Details

In the following, we describe the details of the learning of DCNN, which is carried out in the Sects. 4 and 5. Figures 1 and 5 shows the architectures in this experiment. These models are trained with the ImageNet 2012 training set (1.3 million images, spread over 1000 different classes), and the classification accuracies are evaluated using the 50,000 images [2,5].

Each RGB image was pre-processed by resizing 256×256 [pixel], subtracted the per-pixel mean (across all images) and cropping out the 227×227 [pixel] from the center. In the BP training, stochastic gradient decent with mini-batch of size 256 is used to update the parameters. In addition, the learning rate starts with 10^{-2}, the momentum parameter is set to 0.9, and we anneal the learning rate with $\gamma = 0.1$ for stepsize $100,000$.

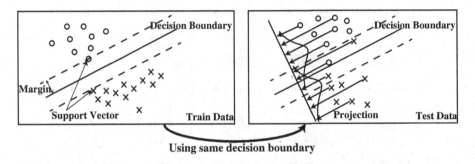

Fig. 2. Linear SVM finds the support vector which maximize the margin from feature space input data, and it make discrimination plane between the support vectors (Left). We create the histogram of the distance to the test data of decision boundary (Right).

4   Experiments of CaffeNet

In this section, we train the CaffeNet with ImageNet 2012, and evaluate the class separation degree by using the SVM histogram.

4.1   Evaluation Data

To evaluate the SVM histogram, we extract 50 images from each 10 classes of "Tench", "Goldfish", "Brambling", "Black swan", "Tusker", "Echidna", "Platypus", "Wallaby", "Koala", and "Wombat" from the ImageNet 2012 evaluation dataset. From the 500 image, we divide them into the 2 dataset. One is used for the linear SVM training, and the other is used for the evaluation of the SVM histograms. Each dataset has 250 images without overlapping.

4.2   Evaluation of Class Separation

As described in the Sect. 4.1, we evaluate the SVM histogram with subset of the ImageNet 2012 dataset. Figure 3 shows the histograms results for the classifier of "Tench" and "Goldfish" classes. In each histogram, the horizontal axis shows the location from the decision boundary, where the origin indicates the decision boundary, and the vertical one shows the frequency of the test class examples. In the figure, each column shows the same layer results, and each row shows same training epochs results. In low layers of CaffeNet, the class separation is not enough, however, we can see that it gradually improves through the hierarchy. Furthermore, the features of the 90 Epochs, which has advanced learning, the within-class variance is smaller than those of the 2 and 20 Epochs. We can guess that within-class variance of DCNN of 90 Epochs is smaller than other two DCNNs.

In addition to the above qualitative evaluation, we also evaluate the result quantitatively in following way. Figure 4 shows the box plot of the within-class variance histogram for the all combinations of SVM histograms. In each graph,

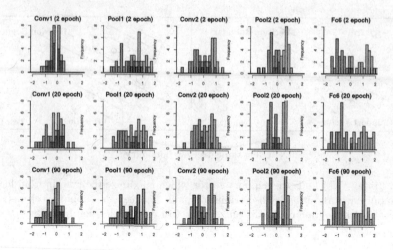

Fig. 3. The SVM histogram of Tench class and Goldfish class in CaffeNet. In the figure, each column shows the same layer results, and each row shows same epochs results. In each graph, the horizontal axis shows the distance from the decision plane, where the origin indicates the decision boundary, and the vertical one shows the frequency of the test examples.

the horizontal axis shows the layer index of extracted feature representation, and the vertical one shows the within-class variance. We can see the within-class variance of SVM histogram is reduced as the learning progresses, and the within-class variance is decreased in Pool2 and 5 layers. Also with respect to the position where the Conv3,4,5 of CaffeNet of 90 Epochs, within-class variance did not changed significantly.

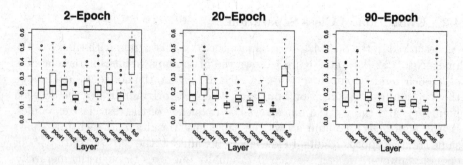

Fig. 4. In the feature representation extracted from CaffeNet, box plots of the within-class variance of SVM histogram which in all of the combination 10 class. The horizontal axis shows layer which extracted feature representation, the vertical axis shows the within-class variance.

4.3   Architecture Modification with SVM Histograms Analysis

From the result in the Subsect. 4.2, we focus on the following 2 points, that is, in the conventional CaffeNet,

(i) Within-class variance is not reduced in the Pool1 layer.
(ii) Within-class variance is not reduced even in Conv3,4,5 layers.

In the following, we modify the architecture design of the CaffeNet by use of the analysis of the within-class variances. We assume the shrinking that within-class variances through the layer development in the DCNN might play an important role in classification task. From the assumption, we can present several solving points, those are

(i) The class separation is almost not proceeding at previous layer Conv1.
(ii) The Pooling processing has not been performed after layers Conv3,4.

Therefore, we eliminate the Pool1 layer for the purpose of improving the degree of separation in a low layer. As a result, the proposed structure of the DCNN becomes repeating the convolution operation twice. Elimination of the pooling layer means that the detected feature location with convolution filter is preserved. In addition, the representation becomes sensitive to the local deformation of the input pattern. Moreover, we insert the Pool3,4 layer after the Conv3,4 layer respectively. Inserting the pooling layer means that feature location information might be lost, however, the representation becomes stable for the local deformation of the input pattern. The conventional CaffeNet, the within-class variance of the convolution 3 and 4 layer looks the same level, so that, we insert the pooling layer in order to reduce the within-class variances. Details of the new architecture are shown in Fig. 5.

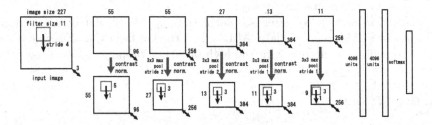

Fig. 5. Overview of DCNN model that we proposed. Remove the Pool1 layer from CaffeNet, and insert a new Pool3,4 layer. In addition, Pool4,5 layer so that the feature map does not become too small have the stride to 1.

5   Experiments of New Architecture

In this section, we compare CaffeNet and our proposal model of Fig. 5 that trained by ImageNet 2012.

5.1   Within-Class Variance of SVM Histogram

From our model training, we obtain the Fig. 6, which is similar to the box plots of Fig. 4. As from design guidelines, within-class variance of DCNN of 90 Epochs reduced in the first of pooling layer, which is described as "Pool2". Moreover, within-class variance changes after layer Conv3, which indicates the advance of pattern separation. These phenomenons suggests the legitimacy of our proposal model.

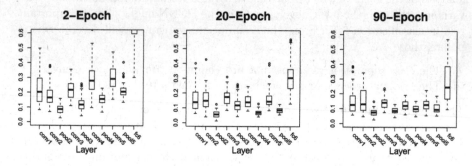

Fig. 6. Within-class variance of SVM histogram in our proposal model. It can be seen that within-class variance has decreased in all Pool layer.

5.2   Accuracy Evaluation

Table 1 shows the accuracy comparison in the Top-1 and Top-5 accuracy of our model and CaffeNet. In DCNN with 90 Epochs training, our model indicates a higher classification accuracy than CaffeNet, so it can be seen that it have been successful on the architecture design.

Further, in each DCNN after 90 Epochs training, we calculated classification accuracy in SVM histogram data. Although in the low layer CaffeNet shows the higher classification accuracy, in the high layer our proposal model shows the higher accuracy than CaffeNet. This is considered suggesting that class separation proceeds, preferable to use some Pool layers at higher layer than used in low layer (Table 2).

Table 1. The accuracy comparison in the Top-1 and Top-5 accuracy of our model and CaffeNet. Bold texts show the highest classification accuracy in the respectively Top-1, Top-5

DCNN		2 epoch	20 epoch	90 epoch
CaffeNet	Top 1	20.96 %	41.01 %	56.99 %
Ours		13.87 %	35.22 %	**58.26 %**
CaffeNet	Top 5	42.60 %	66.96 %	80.13 %
Ours		33.49 %	61.53 %	**81.36 %**

Table 2. Classification accuracy of the test data collected in Subsect. 4.1 with SVM histogram

	Conv1	Pool1	Conv2	Pool2	Conv3	Pool3	Conv4	Pool4	Conv5	Pool5	Fc6
CaffeNet	22%	44.8%	49.6%	62%	61.6%	–	66.8%	–	71.6%	82.4%	85.2%
Ours	24.4%	–	41.2%	54.8%	55.2%	65.2%	68.4%	78%	80.4%	84.8%	88%

6   Summary and Discussion

In this study, we showed the new guidelines in architecture design of DCNN, as a result, we found high classification performance models than conventional CaffeNet for ImageNet 2012 classification. Architecture design of DCNN has not been made so much research, because it is difficult to determine the design guidelines, and the architecture design of DCNN had been a black box. However, by using within-class variance of SVM histogram, it has become possible to determine the guideline in architecture design.

In addition, despite our proposal model is consisted by almost the same techniques as the DCNN of AlexNet [5] and Zeiler *et al.* [9], its classification accuracy is inferior to that they have reported. This is because in their approach, they tune the learning rate manually, our proposal model and CaffeNet used in this work has learned to fix the stepsize 100,000, so it is considered that our model can not beat their reported accuracy.

Acknowledgment. This work is partly supported by MEXT/JSPS KAKENHI Grant number 26120515 and 16H01542.

References

1. Bishop, C.M.: Pattern Recognition and Machine Learning. Springer, New York (2006)
2. Deng, J., Dong, W., Socher, R., Li, L.-J., Li, K., Fei-Fei, L.: ImageNet: a large-scale hierarchical image database. In: CVPR 2009 (2009)
3. Deng, L., Yu, D., Deep learning: methods and applications. Technical report MSR-TR-2014-21, Microsoft Research, May 2014
4. Fukushima, K.: Neocognitron: a self-organizing neural network model for a mechanism of pattern recognition unaffected by shift in position. Biol. Cybern. **36**(4), 193–202 (1980)
5. Krizhevsky, A., Sutskever, I., Hinton, G.E.: ImageNet classification with deep convolutional neural networks. In: Pereira, F., Burges, C.J.C., Bottou, L., Weinberger, K.Q. (eds.) Advances in Neural Information Processing Systems, vol. 25, pp. 1097–1105. Curran Associates Inc. (2012)
6. LeCun, Y., Boser, B., Denker, J.S., Henderson, D., Howard, R.E., Hubbard, W., Jackel, L.D.: Backpropagation applied to handwritten zip code recognition. Neural Comput. **1**(4), 541–551 (1989)
7. Shouno, H.: Recent studies around the neocognitron. In: Ishikawa, M., Doya, K., Miyamoto, H., Yamakawa, T. (eds.) ICONIP 2007, Part I. LNCS, vol. 4984, pp. 1061–1070. Springer, Heidelberg (2008)

8. Shouno, H., Suzuki, S., Kido, S.: A transfer learning method with deep convolutional neural network for diffuse lung disease classification. In: Arik, S., Huang, T., Lai, W.K., Liu, Q. (eds.) ICONIP 2015. LNCS, vol. 9489, pp. 199–207. Springer, Heidelberg (2015). doi:10.1007/978-3-319-26532-2_22
9. Zeiler, M.D., Fergus, R.: Visualizing and understanding convolutional networks. In: Fleet, D., Pajdla, T., Schiele, B., Tuytelaars, T. (eds.) ECCV 2014, Part I. LNCS, vol. 8689, pp. 818–833. Springer, Heidelberg (2014)

Investigation of the Efficiency of Unsupervised Learning for Multi-task Classification in Convolutional Neural Network

Jonghong Kim, Gil-Jin Jang, and Minho Lee[✉]

School of Electronics Engineering, Kyungpook National University,
1370 Sankyuk-Dong, Puk-Gu, Daegu 702-701, South Korea
jonghong89@gmail.com, mholee@gmail.com,
gjang@knu.ac.kr

Abstract. In this paper, we analyze the efficiency of unsupervised learning features in multi-task classification, where the unsupervised learning is used as initialization of Convolutional Neural Network (CNN) which is trained by a supervised learning for multi-task classification. The proposed method is based on Convolution Auto Encoder (CAE), which maintains the original structure of the target model including pooling layers for the proper comparison with supervised learning case. Experimental results show the efficiency of the proposed feature extraction method based on unsupervised learning in multi-task classification related with facial information. The unsupervised learning can produce more discriminative features than those by supervised learning for multi-task classification.

Keywords: Unsupervised learning · Convolutional Neural Networks · Auto-encoder · Feature extraction · Deep learning

1 Introduction

Humans use their past experiences or prior knowledge when they encounter new cognitive task. When a new task is given, we usually search for the past experience that is related to the given task to find a suitable solution. Likewise, multi-task learning (MTL) in machine learning research is an approach that learns a problem together with other related problems at the same time, using shared knowledge. For the face recognition problems, which are popular topic in machine learning studies, there are many kinds of relevant multi-tasks such as classification of ages, genders, postures, and so on. Even though the targets of those tasks are all different, we usually use the common features because the input data for all of those are facial images. Therefore, multi-task learning has been applied to those facial information classification problems. A lot of recent researches based on deep learning are successfully applied for face recognition tasks [1–4], especially based on Convolutional Neural Networks (CNN), which has shown the state-of-the-art recognition performance in this research field. There are some studies for multi-task learning using the CNN in a supervised manner. Zhang and Zhang [2] used CNN to detect faces in the multi-view images and to

© Springer International Publishing AG 2016
A. Hirose et al. (Eds.): ICONIP 2016, Part III, LNCS 9949, pp. 547–554, 2016.
DOI: 10.1007/978-3-319-46675-0_60

recognize pose and landmarks. Li, Liu and Chan [5] used the CNN to detect body parts such as head, left upper arm, right lower arm, and so on. Those approaches only use the supervised learning methods for all the sub-tasks together. However, those supervised learning models are applicable to the specific tasks only. It is difficult to expand the trained features obtained by the supervised learning to other multi-tasks which are not considered in the task specific supervised learning. Therefore, we try to investigate the efficiency of unsupervised learning in the CNN architecture for multi-task classification to improve the overall multi-task classification performance related with facial information.

There exist several methods that have been tried to train meaningful features from unlabeled data for the CNN [6–8]. In the proposed method, we pre-train the CNN based on the auto-encoder framework [6, 7]. Since our purpose is to investigate the efficiency of unsupervised learning compared with the supervised learning only, we need to keep the original CNN structure in the pre-training process based on unsupervised learning. The auto-encoder is a good candidate, which can maintain the original structure of CNN model. When we use the auto-encoder for the CNN, it is carefully considered for the CNN not to be a just transformer for dimension reduction or data compression, and identity mapping [6]. So, to avoid that kind of trivial training results, we need to consider the sparsity constraints to obtain meaningful features. One of those kind of methods is Non-Maximum Suppression approaches (NMS) [6, 7]. The NMS can leave only trainable features for each convolutional kernel in order to reconstruct auto-encoder output as close to the input as possible, and the network should be trained so that each of the kernels learns different discriminative features [9]. Another useful constraint is robustness against noise [10]. Adding noises to each of the hidden node input is also used to prevent resulting in the identity mapping. To recover original input from the corrupted input, each convolutional kernel tries to keep general features from input rather than the meaningless errors.

There have been few studies to show the difference between the unsupervised and the supervised learning in training image features. Moreover, it is also not clear how much those unsupervised learning-based features are helpful for discriminative tasks in any kinds of classification problem. Therefore, in this paper, we try to evaluate the effect of these unsupervised learning-based features in multi-task classification. The performance of unsupervised learning is evaluated with several kinds of unlabeled face dataset, and characteristics of the obtained features are analyzed using a visualization and Linear Discriminant Analysis (LDA) [11]. We compare the characteristics of unsupervised features with those of supervised features in VGG Net [12] in multi-task classification problems.

2   Unsupervised Feature Learning Method

The Convolutional Auto Encoder (CAE) [6] consisted of only convolutional layers because of the irreversibility of pooling layers. However, we propose a new method to include pooling layers of the CNN to maintain data compression ability of the CNN.

2.1 Stacked Convolutional Auto-Encoder

The CNN model adopted in this paper is VGG Net of 16 layers [12]. This network consists of 6 blobs - one is fully connected layer and other 5 blobs are convolutional layers. The reference of this blob separation is pooling layers because of its irreversibility as mentioned above. Therefore, for the unsupervised training we consider a stacked convolutional auto-encoder, which can be trained in stacks. Figure 1 shows the first stack of CAE for the first blob of VGG Net structure.

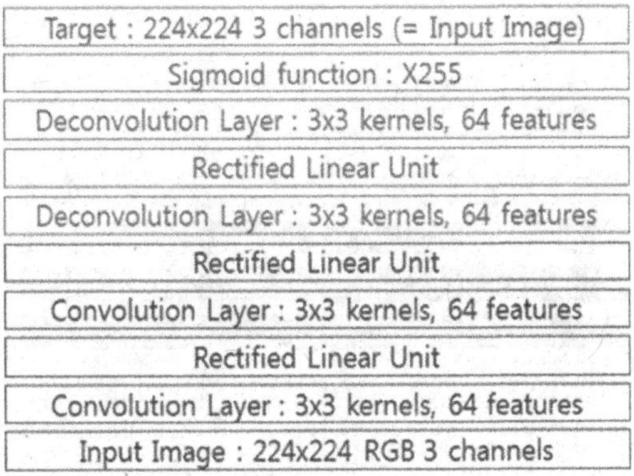

| Target : 224x224 3 channels (= Input Image) |
| Sigmoid function : X255 |
| Deconvolution Layer : 3x3 kernels, 64 features |
| Rectified Linear Unit |
| Deconvolution Layer : 3x3 kernels, 64 features |
| Rectified Linear Unit |
| Convolution Layer : 3x3 kernels, 64 features |
| Rectified Linear Unit |
| Convolution Layer : 3x3 kernels, 64 features |
| Input Image : 224x224 RGB 3 channels |

Fig. 1. A stack of convolutional auto-encoder for the first blob of VGG Net. The size of target input is 224×224 with RGB 3 color channels. The first blob consists of a couple of convolutional layers with 3×3 kernels for 64 output features each. Final output dimension is the same as input.

The lower half is the encoding unit composed of two convolutional layers with Rectified Linear Unit (ReLU) activation functions. The input image is 224×224 pixels, where each pixel has 3 values (RGB). The first and second convolutional layers use 64 kernels of size 3×3. The upper half is the decoding unit consisting of two deconvolutional layers of the same size as the encoding layers. However, the activation function at the last layer, yielding the auto-encoded output, is the sigmoid function. The final output is scaled by 0 to 255 [13].

When training the first blob is completed, the output is passed to the second blob. However, as shown in Fig. 2, the output of the final encoding layer of the first blob is connected to the input layer of the second blob, without using the decoded input images of the first blob. Note that when the weights of the second blob are trained, we fix the weights of the first blob, regarding the encoded layer as given observations but not hidden targets. In the pre-training, the first and the second blobs are trained independently to obtain better convergence of the unsupervised learning. In fine tuning of the entire auto-encoder network, all the weights are learned simultaneously to make a relevance among the blobs, similarly to the other auto-encoders. At the end of the second blob, a 2×2 pooling (subsampling) layer is applied. Since we cannot make perfect reconstruction of the input images if there exists a pooling layer, we use 2×2 interpolation to reverse the pooling of the same size, filling lost pixels with the neighboring ones. This entire auto-encoder structure for fine tuning is shown in Fig. 3. We add an interpolation (up-sampling) layer on top of the second blob, and also two new deconvolution layers are added, which is ignored in the first blob, to encode the original images from the output of the second blob. The final deconvolution layer weights are also based on the weights from first blob training, i.e. same deconvolution layers as shown in Fig. 1. We perform fine tuning of the combined network to obtain features for the subsequent steps.

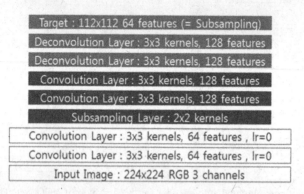

Fig. 2. Stacked convolutional auto-encoder for the second blob of VGG Net training. Target input is pooled 112×112 size of pooling layer output. The kernel size is 3×3 and number of output is 128 for each convolutional layers.

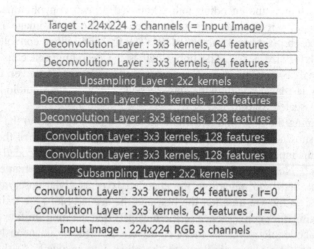

Fig. 3. Convolutional auto-encoder fine tuning structure for the first and second blob training. The up-sampling layer is resized from 112×112 size to 224×224 size.

2.2 Non-maximum Suppression and Dropout

Training generalized features through unsupervised learning is usually obtained by sparsity constraints. The sparsity constraint in the proposed model is the Non-maximum Suppression (NMS) in each sub region [6], not in the entire region [8]. We use 7×7 sub region with 1 stride. Output of this NMS filter is same as input size so it does not affect to auto-encoder structure.

Another important criterion is robustness to noise. This functionality is usually achieved by training augmented data by adding pepper-and-salt noise or Gaussian noise to the data [10]. In this study, the noise addition is derived from dropout technique [14]. Dropout is usually used in neural networks training for the network regularization purpose. If we use 0.5 as the dropout probability, we turn off each of hidden nodes with 0.5 probability. If the concept is applied to convolution layers, the result is the output feature map with 50% missing values. This is the same data corruption with noise addition. Also the missing value will not be trained and can be considered as sparsity constraint.

3 Experimental Results

The model we used in this paper is VGG Net [12] with 16 layers, which is designed to recognize ILSVRC2014 object dataset. The sizes of input images are all 224×224. The unsupervised trained model in this study has similar structure, and we used the same 224×224 input image size.

The experiment was performed on 668,264 unlabeled facial images. This dataset consists of several kinds of known face databases as shown in Table 1.

Table 1. Details of our dataset.

Name of dataset	Number of images
FM facepack	82,909
CASIA webface [3]	494,414
LFW [4]	13,244
Facescrub [15]	77,697
Total	668,264

The trained network was tested on another part of Facescrub dataset, which consists of 5,150 test images out of total 82,847 images of whole Facescrub database.

The Football Manager (FM) facepack (http://www.footballmanager.com/) consists of face images of football players. The original images from those face datasets mostly include upper body and the background. So we used a common face detection algorithm to collect images to extract facial regions only. The result of supervised learning is based on VGG Net trained with ImageNet dataset. We fine-tuned this with labeled Facescrub dataset in a supervised manner. The multi-task learning is done on the following 5 particular criteria: gender, posture, glasses, hair and age.

3.1   Evaluation with t-SNE

The Stochastic Neighbor Embedding (SNE) is a dimension reduction technique for the data visualization. The key concept in this method is modeling Euclidean distance between each pair of data as conditional probability and modeling low dimensional data distribution as same as the original. The t-SNE [16] uses t-distribution to model the distance distribution for better feature representation.

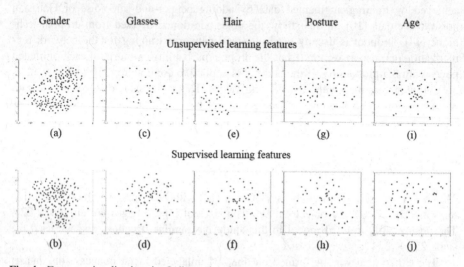

Fig. 4. Feature visualization in 2-dimensional space reduced by t-SNE. Top row: features obtained by unsupervised learning. Bottom row: features obtained by supervised learning.

Figure 4 shows how those features are distributed for the 5 kinds of multiple tasks for unsupervised and supervised training methods. The experiment was performed with the test set of Facescrub dataset. For the gender task, we used 100 male and 100 female samples that are labeled on the test images. For other tasks, we used only 60 samples for each criterion. In the visualized features, hair and glasses features look relatively well discriminated in unsupervised training, and posture looks well discriminated in supervised training. From this point, we can observe that supervised learning features ignores glasses or hair features because these are not important in discriminating faces. However, they have significant role in describing faces, so unsupervised learning tried to make clear distinction on those tasks.

3.2   Evaluation with LDA

The Table 2 indicates trained feature sample distance ratio based on Linear Discriminant Analysis (LDA) result for the test samples. A higher variance ratio of "Between class distance" and "Within class distance" indicates that it is a relatively more discriminative feature.

Table 2. Comparision of trained features for supervised and unsupervised learning using LDA.

Multi-tasks	Supervised learning			Unsupervised learning		
	Between distance	Within distance	Between/within	Between distance	Within distance	Between/within
Gender	0.127	0.022	5.80	2.1×10^{-8}	9.7×10^{-18}	2.2×10^{9}
Glass	0.391	0.015	26.67	2.3×10^{-8}	4.9×10^{-17}	4.6×10^{8}
Hair	0.346	0.043	8.03	1.9×10^{-8}	9.7×10^{-18}	1.9×10^{9}
Posture	0.701	0.020	35.55	4.8×10^{-8}	5.0×10^{-17}	9.6×10^{8}
Age	0.204	0.030	6.73	2.5×10^{-8}	1.4×10^{-16}	1.9×10^{8}

The ratio of between class distance with respect to the within class distance of the unsupervised learning is much more greater than that of the supervised learning. It means that the unsupervised learning features are more effective for multi-task classification than the supervised learning features.

4 Conclusion

In this paper, we tried to analyze the characteristics of the unsupervised feature learning and its effectiveness for the multi-task problems. We used stacked convolutional auto-encoder to maintain original structure as supervised learning model based on VGG Net including pooling layers. To learn more general and effective features from unlabeled dataset, we used non-maximum suppression constraint, and dropout to make the network robust against noise. For the experiment, we used 668,264 unlabeled facial images for the unsupervised learning. The analysis results show that the trained features using unsupervised learning are not that much distributed but the difference of "Between class distance" and "Within class distance" is significantly higher than that of supervised learning. Therefore, based on our results we argue that if we could design the supervised fine-tuning phase of unsupervised learning model more carefully, we would have the chance to carry the improved discriminative advantages of unsupervised trained features for multi-task learning problems.

In our future work, we are constructing the multi-task classification CNN model by supervised learning, in which the initial weights of the CNN are obtained by the proposed unsupervised learning.

Acknowledgement. This work was supported by the National Research Foundation of Korea (NRF) grant funded by the Korea government (MSIP) (No. NRF-2016R1E1A2020559) (50%) and Government Fund from Korea Copyright Commission. [2015-related-9500: Development of predictive detection technology for the search for the related works and the prevention of copyright infringement] (50%).

References

1. Wolf, L.: Deepface: closing the gap to human-level performance in face verification. In: 2014 IEEE Conference on Computer Vision and Pattern Recognition (CVPR). IEEE (2014)
2. Zhang, C., Zhang, Z.: Improving multiview face detection with multi-task deep convolutional neural networks. In: 2014 IEEE Winter Conference on Applications of Computer Vision (WACV). IEEE (2014)
3. Yi, D., Lei, Z., Liao, S., Li, S.Z.: Learning face representation from scratch (2014). arXiv preprint arXiv:1411.7923
4. Huang, G.B., Ramesh, M., Berg, T., Learned-Miller, E.: Labeled faces in the wild: a database for studying face recognition in unconstrained environments. Technical Report 07-49. University of Massachusetts, Amherst (2007)
5. Li, S., Liu, Z.-Q., Chan, A.: Heterogeneous multi-task learning for human pose estimation with deep convolutional neural network. In: Proceedings of the IEEE Conference on Computer Vision and Pattern Recognition Workshops (2014)
6. Masci, J., Meier, U., Cireşan, D., Schmidhuber, J.: Stacked convolutional auto-encoders for hierarchical feature extraction. In: Honkela, T. (ed.) ICANN 2011, Part I. LNCS, vol. 6791, pp. 52–59. Springer, Heidelberg (2011)
7. Makhzani, A., Frey, B.J.: Winner-take-all autoencoders. In: Advances in Neural Information Processing Systems (2015)
8. Paulin, M., Douze, M., Harchaoui, Z., Mairal, J., Perronin, F., Schmid, C.: Local convolutional features with unsupervised training for image retrieval. In: Proceedings of the IEEE International Conference on Computer Vision (2015)
9. Kavukcuoglu, K., Sermanet, P., Boureau, Y.-L., Gregor, K., Mathieu, M., Cun, Y.L.: Learning convolutional feature hierarchies for visual recognition. In: Advances in Neural Information Processing Systems (2010)
10. Vincent, P., Larochelle, H., Bengio, Y., Manzagol, P.-A.: Extracting and composing robust features with denoising autoencoders. In: Proceedings of the 25th International Conference on Machine Learning. ACM (2008)
11. Scholkopft, B., Mullert, K.-R.: Fisher discriminant analysis with kernels. Neural Networks Sig. Process. IX $\mathbf{1}$(1), 1 (1999)
12. Simonyan, K., Zisserman, A.: Very deep convolutional networks for large-scale image recognition (2014). arXiv preprint arXiv:1409.1556
13. Gülçehre, Ç., Bengio, Y.: Knowledge matters: importance of prior information for optimization (2013). arXiv preprint arXiv:1301.4083
14. Srivastava, N., Hinton, G., Krizhevsky, A., Sutskever, I., Salakhutdinov, R.: Dropout: a simple way to prevent neural networks from overfitting. J. Mach. Learn. Res. $\mathbf{15}$(1), 1929–1958 (2014)
15. Ng, H.-W., Winkler, S.: A data-driven approach to cleaning large face datasets. In: 2014 IEEE International Conference on Image Processing (ICIP). IEEE (2014)
16. Van der Maaten, L., Hinton, G.: Visualizing data using t-SNE. J. Mach. Learn. Res. $\mathbf{9}$(2579–2605), 85 (2008)

Sparse Auto-encoder with Smoothed l_1 Regularization

Li Zhang$^{(\boxtimes)}$, Yaping Lu, Zhao Zhang, Bangjun Wang, and Fanzhang Li

School of Computer Science and Technology and Joint International Research Laboratory of Machine Learning and Neuromorphic Computing, Soochow University, Suzhou 215006, Jiangsu, China
{zhangliml,cszzhang,wangbangjun,lfzh}@suda.edu.cn, yplu1990@163.com

Abstract. To obtain a satisfying deep network, it is important to improve the performance on data representation of an auto-encoder. One of the strategies to enhance the performance is to incorporate sparsity into an auto-encoder. Fortunately, sparsity for the auto-encoder has been achieved by adding a Kullback-Leibler (KL) divergence term to the risk functional. In compressive sensing and machine learning, it is well known that the l_1 regularization is a widely used technique which can induce sparsity. Thus, this paper introduces a smoothed l_1 regularization instead of the mostly used KL divergence to enforce sparsity for auto-encoders. Experimental results show that the smoothed l_1 regularization works better than the KL divergence.

Keywords: Auto-encoder · Sparsity · KL divergence · Smoothed l_1 regularization · Data representation

1 Introduction

Although the theoretical benefits of deep networks have been appreciated for many decades, the training procedure appears to often get stuck in poor solutions [1]. Fortunately, Hinton et al. introduced a greedy layer-wise unsupervised learning method [2,3], which may hold great promise as a principle to help address the issue of training deep networks. One of the particularly important aspects for this strategy is to pre-train the feature detector model of restricted boltzmann machines (RBMs) [4] as building blocks to achieve the initialization of deep networks. Soon after, Bengio et al. had verified that the layer-wise greedy unsupervised pre-training principle could be applied when using an auto-encoder [5] instead of a RBM as a layer building block [6]. Therefore, it is reasonable that improving the performance of auto-encoder is very important for quickly obtaining an efficient deep network.

Since the mammalian vision system uses sparse representations in early visual areas [7,8], some researchers try to improve the performance of auto-encoder by incorporating sparsity into it. Sparse coding [9,10] is the first model that can learn sparse representations for data by adding a sparsity penalty term to the

© Springer International Publishing AG 2016
A. Hirose et al. (Eds.): ICONIP 2016, Part III, LNCS 9949, pp. 555–563, 2016.
DOI: 10.1007/978-3-319-46675-0_61

risk functional. Inspired by this, sparsity has been successfully introduced into RBM [11,12] and auto-encoder [13]. Specifically, sparse auto-encoder [13–15] tries to learn sparse representations for data by adding a Kullback-Leibler (KL) divergence term to the risk functional so as to penalize those code units that are active.

The l_1 regularization based on sparse representation theory is known to induce sparse coefficients [16], and has been widely used in compressive sensing [17,18] and machine learning [19]. The goal of this paper is to replace the KL divergence with l_1 regularization. However, it is difficult to directly introduce l_1 regularization into auto-encoder since l_1 regularization is non-differentiable. A feasible approach is to utilize some smoothing technique like Moreau proximal smoothing [20] or Nesterov's smoothing [21] to get a smoothed l_1 regularization, an approximation to l_1 regularization. In [22], Beck and Teboulle presented a smoothed l_1 regularization via the inf-conv smoothing technique.

In this paper, we employ the smoothed l_1 regularization instead of the mostly used KL divergence to enforce sparsity for auto-encoder. Experiments are conducted to verify that the auto-encoder with the smoothed l_1 regularization is indeed able to learn sparse representations, and then to illustrate that the auto-encoder with the smoothed l_1 regularization works better than the one with the KL divergence.

2   Sparse Auto-encoder with the Smoothed l_1 Regularization

2.1   Smoothed l_1 Regularization

The famous technique inducing sparsity in machine learning is the l_1 regularization [16]. So far, however, no reference reports the auto-encoder with the l_1 regularization. The main reason is that the l_1 regularization is non-differentiable which results in the difficulty of optimization. In [22], Beck and Teboulle indicated that non-differentiable convex functions can be approximated by smooth functions by various techniques, such as Moreau proximal smoothing [20] and Nesterov's smoothing [21]. They also proposed a new smoothing technique called inf-conv smoothing technique, and as an application of the new technique, a smoothed l_1 regularization was given by

$$g_\mu(t) = \begin{cases} \frac{t^2}{2\mu} & |t| \leq \mu \\ |t| - \frac{\mu}{2} & |t| > \mu \end{cases}, \tag{1}$$

where $\mu > 0$ is a hyper-parameter that controls the similarity between the l_1 regularization and the smoothed l_1 regularization. Specifically, when $\mu \to 0$, the smoothed l_1 regularization tends to the l_1 regularization.

2.2   Sparse Auto-encoder with the Smoothed l_1 Regularization

Given a set of training samples $\{\mathbf{x}^{(i)}\}_{i=1}^m$, where $\mathbf{x}^{(i)} \in \mathbb{R}^d$, d is the dimensionality and m is the number of samples, the sparse auto-encoder aims to find a hypothesis $h_{\mathbf{W},\mathbf{b}}(\mathbf{x}^{(i)})$ with the parameters of weight $\mathbf{W}$ and bias $\mathbf{b}$ so as to get sparse representations for the training samples. To achieve this, we can minimize the following risk functional of the sparse auto-encoder

$$\frac{1}{m}\sum_{i=1}^{m}\left\|h_{\mathbf{W},\mathbf{b}}(\mathbf{x}^{(i)}) - \mathbf{x}^{(i)}\right\|_2^2 + \lambda\|\mathbf{W}\|_2^2 + \beta\sum_{j=1}^{n}S\left(\frac{1}{m}\sum_{i=1}^{m}a_j^{(i)}\right) \qquad (2)$$

where $\lambda > 0$ and $\beta > 0$ are hyper-parameters, n is the number of hidden units, $a_j^{(i)}$ represents the output of the jth hidden unit of the ith input sample, and $S(\cdot)$ is the sparse penalty term.

Now, what we need to do is to consider how to employ the smoothed l_1 regularization, instead of the mostly used KL divergence, as the sparsity penalty term of (2). However, if we select $\frac{1}{1+e^{-x}}$ as the activation function, $a_j^{(i)}$ only takes value from 0 to 1, which causes the average activation of each hidden unit j on the whole training set $\frac{1}{m}\sum_{i=1}^{m}a_j^{(i)}$ to be between 0 and 1 as well. That is to say, if we directly use (1) as the sparsity penalty term where $t = \frac{1}{m}\sum_{i=1}^{m}a_j^{(i)}$, only the situation of $t > 0$ needs to be considered. In other words, what we do is meaningless since there is no need to smooth the l_1 regularization at all in the situation of $t > 0$. Inspired by the KL divergence, we modify (1) to the right with γ units so as to avoid the problem discussed above. That is to say, in the paper, our sparsity penalty term of (2) is given by

$$S(t) = \begin{cases} \frac{(t-\gamma)^2}{2\mu}, & if \ |t-\gamma| \le \mu \\ |t-\gamma| - \frac{\mu}{2}, & if \ |t-\gamma| > \mu \end{cases} \qquad (3)$$

where $t = \frac{1}{m}\sum_{i=1}^{m}a_j^{(i)}$, γ and μ are hyper-parameters. Here, to ensure the efficiency, we should particularly pay attention to the range of values of γ and μ. Specifically, the parameter $0 < \gamma < 1$ controls the sparsity level, and also often sets to be a small value close to 0. As for the parameter μ, we have

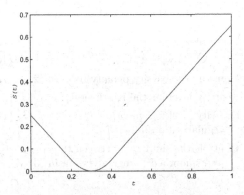

Fig. 1. Curve of the smoothed l_1 regularization modified, and the hyper-parameters of γ and μ are set to be 0.3 and 0.1, respectively.

$0 < \mu < max\{\gamma, 1 - \gamma\}$ so as to avoid the situation that our sparsity penalty term (3) may degenerate into l_2 regularization since the value of t is between 0 and 1.

Figure 1 shows the curve of the smoothed l_1 regularization modified with γ units. As we can see, this regularization has the same property with the KL divergence. Namely, $S(t) = 0$ when $t = \gamma$, and otherwise it increases monotonically as t diverges from γ. Therefore, we have reasons to believe that the smoothed l_1 regularization is able to introduce sparsity for auto-encoder, and what's more, in view of [16–19], we expect it to work better than the KL divergence.

3   Experiments

To verify that the smoothed regularizer can indeed induce sparsity for auto-encoders, we first conduct experiments on the Natural Images [23]. Then experiments are performed on the MNIST [24] dataset, aiming to quantitatively show that the sparse representations learned by the smoothed regularizer is more discriminative than the ones learned by the KL divergence. Finally, to make a further comparison, the two sparse auto-encoders are used as building blocks to achieve the trainings of deep networks on the datasets of MNIST and COIL [25].

To increase the credibility of comparisons, we employ the grid search method to select values for the hyper-parameters of λ, β, ρ, γ and μ. Since auto-encoder is an unsupervised model, hyper-parameters are determined according to the reconstruction error on the validation set. Table 1 shows the optional values of each hyper-parameter.

Additionally, as we have discussed in Sect. 3, when $\mu \geq max\{\gamma, 1 - \gamma\}$ and $\frac{1}{1+e^{-x}}$ is selected as an activation function, it is equivalent to employ the l_2 regularization as the sparsity penalty term since the value of $\frac{1}{m}\sum_{i=1}^{m} a_j^{(i)}$ is between 0 and 1. As a comparison, we shall also present the experimental results of this situation.

Table 1. List of the optional values of each hyper-parameter for grid search

Hyper-parameter	Description	Optional values
λ	Constant of weight decay term in (2)	$\{10^{-2}, 10^{-3}, 10^{-4}\}$
β	Constant of sparsity penalty term in (2)	$\{1, 3, 5\}$
ρ	Sparsity level for the KL divergence	$\{10^{-3}, 10^{-2}, 10^{-1}\}$
γ	Sparsity level for smoothed l_1 regularization in (3)	$\{10^{-3}, 10^{-2}, 10^{-1}\}$
μ	Controls the similarity between the l_1 and smoothed l_1 regularization in (3)	$\{0.001, 0.1, 0.5, 0.9\}$

3.1   Natural Images

We first train the auto-encoder with the smoothed l_1 regularizer on 100,000 gray-level patches of size 12×12 extracted from ten 512×512 natural images. Pre-processing of image patches consists of subtracting the mean pixel value, dividing the result by 255, and squashing the result to $[0.1, 0.9]$. The number of hidden units is set to be 196.

The learned bases are shown in Fig. 2. Specifically, Fig. 2(c) and (d) indicate that the auto-encoder with the KL divergence results in similar bases to the one with the smoothed l_1 regularization. They are both spatially localized, Gabor-like, and have different orientations, frequencies and scales, resembling the receptive fields of V1 neurons. This is consistent with the results obtained by applying other algorithms to learn sparse representations of this data (e.g., [11,12,16]). In a nutshell, the smoothed l_1 regularization can indeed induce sparsity in auto-encoders. Additionally, we can also see that the auto-encoder with l_2 regularization can only induce poor sparsity as shown in Fig. 2(b), and without sparsity penalty term, the visualizations of learned bases are meaningless images as given in Fig. 2(a).

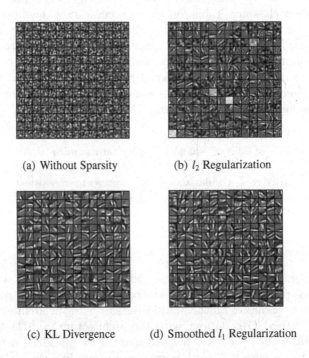

(a) Without Sparsity   (b) l_2 Regularization

(c) KL Divergence   (d) Smoothed l_1 Regularization

Fig. 2. Visualizations of 196 hidden unit bases, respectively learned by the auto-encoder (a) without sparsity term, (b) with l_2 regularization, (c) with the KL divergence, and (d) with the smoothed l_1 regularization.

3.2 MNIST Dataset

To further compare the performance of the smoothed l_1 regularization and the KL divergence, we use them to train the sparse auto-encoders on the MNIST dataset, and then employ the learned features to train the Softmax classifier.

The MNIST dataset has a total of 60,000 training images and 10,000 testing images, each 28×28, for handwritten digits 0 through 9, and here was rescaled to [0, 1] by dividing the data by 255. We apply a strategy called mini-batches [26], originally used to train RBMs [2], to train the sparse auto-encoders. The whole training images are randomly divided into 600 mini-batches with 100 images for each, and the number of hidden units is also set to be 196.

Figure 3 shows the visualizations of 196 hidden unit bases. Most of bases found by the auto-encoders with the KL divergence and the smoothed l_1 regularization (see Fig. 3(c) and (d)) roughly represent different "strokes" of which handwritten digits are comprised. This is consistent with the results obtained in [11,12]. Here, the l_2 regularization shows a better sparsity on the MNIST dataset than on the Natural Images as shown in Fig. 3(b).

To quantitatively show the effects of different sparsity penalty terms, the learned features corresponding to the whole training images are employed to train the Softmax classifier. Table 2 gives the accuracies of the different sparsity penalty terms on the whole testing images. It is obvious that we can obtain the more discriminative features by introducing the sparsity into auto-encoder. In addition, we can further improve the performance of auto-encoders when using the smoothed l_1 regularization, instead of the KL divergence, as the sparsity penalty term. In other words, the smoothed l_1 regularization is able to produce better sparsity for auto-encoder than the KL divergence.

Table 2. Accuracies of different sparsity penalty terms on the MNIST dataset (%)

No sparsity	l_2 regularization	KL divergence	Smoothed l_1
93.03	96.88	96.96	97.21

3.3 Train Deep Networks

To further present the effects of the two different sparsity terms, we use them as the building blocks to train deep networks with Softmax as output layer. Experiments are conducted on the MNIST dataset with the architecture of 784-500-500-10 and the COIL dataset with the architecture of 1024-400-400-100. It is notable that when pre-train the two sparse auto-encoders, decreasing the sparsity level ρ or γ with the depth increasing can produce surprising results.

The COIL dataset has a total of 7,200 gray-scale images, each 32×32, with 100 categories, 72 images for each class. We randomly pick up 50 images for each category, totally 5,000 images, as training set, then divide it into 50 mini-batches. The remaining images are used as the testing set.

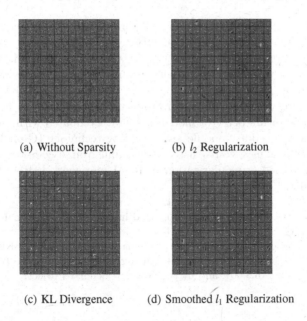

    (a) Without Sparsity   (b) l_2 Regularization

    (c) KL Divergence   (d) Smoothed l_1 Regularization

Fig. 3. Visualizations of 196 hidden unit bases, respectively learned by the auto-encoder (a) without sparsity term, (b) with l_2 regularization, (c) with the KL divergence, and (d) with the smoothed l_1 regularization.

Table 3 presents the results of the deep networks on the MNIST and COIL datasets. As we have expected, the deep network with the smoothed regularizer as building blocks works better than the one with the KL divergence. Particularly, in [2] Hinton et al. employed the RBM without sparsity as building blocks to train a deep network of 784-500-500-2000-10, which achieves a error rate of 1.2 %. That is to say, a deep network with a simple architecture that uses the auto-encoder with the smoothed regularizer as building blocks is able to achieve the same performance with the one with a complex architecture that uses the RBM without sparsity as building blocks.

Table 3. Accuracies of the deep networks on the MNIST and COIL datasets (%)

Dataset	KL divergence	Smoothed l_1 regularization
MNIST	98.72	98.80
COIL	99.45	99.59

4 Conclusions

In this paper, we introduce a smoothed l_1 regularization, an approximation of the l_1 regularization that is widely used in compressive sensing and machine learning, into the feature detector model auto-encoder instead of the mostly used KL divergence as the sparsity penalty term. Experimental results on the Natural Images, MNIST dataset, and COIL dataset show that the smoothed l_1 regularization can indeed induce sparsity in auto-encoders, and what's more, it can produce better sparsity with comparison to the mostly used KL divergence.

For the future work, we plan to introduce the smoothed l_1 regularization into the another feature detector model or RBM. Compared with auto-encoder, RBM only has an encoder module and sampling technique (Gibbs), instead of a decoder module, is used to reconstruct the input from a code vector. To some extent, the smoothed l_1 regularization in RBM may work better than that in the auto-encoder.

Acknowledgments. This work was supported in part by the National Natural Science Foundation of China under Grant Nos. 61373093, and 61402310, by the Natural Science Foundation of Jiangsu Province of China under Grant No. BK20140008, by the Natural Science Foundation of the Jiangsu Higher Education Institutions of China under Grant No.13KJA520001, and by the Soochow Scholar Project.

References

1. Bengio, Y.: Learning deep architectures for AI. Found. Trends Mach. Learn. **2**(1), 1–127 (2009)
2. Hinton, G.E., Salakhutdinov, R.R.: Reducing the dimensionality of data with neural networks. Science **313**(5786), 504–507 (2006)
3. Hinton, G.E., Osindero, S., Teh, Y.: A fast learning algorithm for deep belief nets. Neural Comput. **18**(7), 1527–1554 (2006)
4. Fischer, A., Igel, C.: An Introduction to restricted Boltzmann machines. In: Progress in Pattern Recognition, Image Analysis, Computer Vision, and Applications, pp. 14–36 (2012)
5. Hinton, G.E., Zemel, R.S.: Autoencoders, minimum description length and Helmholtz free energy. Adv. Neural Inf. Process. Syst. **6**, 3–10 (1993)
6. Bengio, Y., Lamblin, P., Popovici, D., Larochelle, H.: Greedy layer-wise training of deep networks. In: Conference on Neural Information Processing Systems, pp. 153–160 (2006)
7. Lennie, P.: The cost of cortical computation. Current Biol. **13**, 493–497 (2003)
8. Simoncelli, E.P.: Statistical Modeling of Photographic Images, 2nd edn. Academic Press, San Diego (2005)
9. Olshausen, B.A., Field, D.J.: Emergence of simple-cell receptive field properties by learning a sparse code for natural images. Nature **381**(6583), 607–609 (1996)
10. Olshausen, B.A., Field, D.J.: Sparse coding with an overcomplete basis set: a strategy employed by V1? Vis. Res. **37**(33), 3311–3325 (1997)
11. Lee, H., Ekanadham, C., Ng, A.Y.: Sparse deep belief net model for visual area V2. In: Conference on Neural Information Processing Systems, pp. 873–880 (2007)
12. Luo, H., Shen, R., Niu, C., Ullrich, C.: Sparse group restricted Boltzmann machines. In: AAAI Conference on Artificial Intelligence, pp. 429–434 (2011)
13. Ng, A.Y.: Sparse autoencoder. CS294A Lecture, Stanford University (2011). http://web.stanford.edu/class/cs294a/sparseAutoencoder_2011new.pdf

14. Le, Q.V., Ngiam, J., Coates, A., Lahiri, A., Prochnow, B., Ng, A.Y.: On optimization methods for deep learning. In: International Conference on Machine Learning, pp. 265–272 (2011)
15. Deng, J., Zhang, Z.X., Marchi, E., Schuller, B.: Sparse autoencoder-based feature transfer learning for speech emotion recognition. In: Humaine Association Conference on Affective Computing and Intelligent Interaction, pp. 511–516 (2013)
16. Lee, H., Battle, A., Raina, R., Ng, A.Y.: Efficient sparse coding algorithms. In: Conference on Neural Information Processing Systems, pp. 801–808 (2006)
17. Candes, E., Tao, T.: Decoding by linear programming. IEEE Trans. Inf. Theory 15(12), 4203–4215 (2005)
18. Donoho, D.L.: Compressed sensing. IEEE Trans. Inf. Theory 52(4), 1289–1306 (2006)
19. Ng, A.Y.: Feature selection, L_1 vs. L_2 regularization, and rotational invariance. In: International Conference on Machine Learning (2004)
20. Moreau, J.J.: Proximite et Dualite dans un espace Hilbertien. Bulletin de la Society Math matique de France 93, 273–299 (1965)
21. Nesterov, Y.: Smooth minimization of non-smooth functions. Math. Program. 103(1), 127–152 (2005)
22. Bech, A., Teboulle, M.: Smoothing and first order methods: a unified framework. SIAM J. Optimization 22(2), 557–580 (2012)
23. Ng, A.Y., Ngiam, J., Foo, C.Y., Mai, Y., Susen, C.: Ufldl Tutorial (2012). http://ufldl.stanford.edu/wiki/resources/sparseae_exercise.zip
24. LeCun, Y., Bottou, L., Bengio, Y., Haffner, P.: Gradient-based learning applied to document recognition. Proc. IEEE 86(11), 2278–2324 (1998)
25. Nene, S.A., Nayar, S.K., Murase, H.: Columbia Object Image Library (COIL-100). Technical Report, CUCS-006-96, Department of Computer Science, Columbia University (1996)
26. Hinton, G.E.: A practical guide to training restricted Boltzmann machines. Neural Netw. Tricks Trade 7700, 599–619 (2010)

Encoding Multi-resolution Two-Stream CNNs for Action Recognition

Weichen Xue, Haohua Zhao, and Liqing Zhang[✉]

Key Laboratory of Shanghai Education Commission for Intelligent Interaction and
Cognitive Engineering, Department of Computer Science and Engineering,
Shanghai Jiao Tong University, Shanghai, China
{xueweuchen,haoh.zhao,lqzhang}@sjtu.edu.cn

Abstract. This paper deals with automatic human action recognition
in videos. Rather than considering traditional hand-craft features such as
HOG, HOF and MBH, we explore how to learn both static and motion
features from CNNs trained on large-scale datasets such as ImagNet and
UCF101. We propose a novel method named multi-resolution latent concept descriptor (mLCD) to encode two-stream CNNs. Entensive experiments are conducted to demonstrate the performance of the proposed
model. By combining our mLCD features with the improved dense trajectory features, we can achieve comparable performance with state-of-
the-art algorithms on both Hollywood2 and Olympic Sports datasets.

Keywords: Deep learning · CNN · Action recognition

1 Introduction

Automatic action recognition is an important problem in computer vision and
surveillance systems and has drawn significant attention in recent years. Recent
research focuses on realistic datasets from movies, web videos such as Hollywood2
[6], UCF101 [11], and Olympic Sports [7]. The state-of-the-art performance of
action recognition is given by a bag-of-words (BoW) representation of local features like HOG, HOF and MBH [12]. Recently, Convolutional neural networks
(CNNs) are also introduced into action recognition task [10]. In some challenging
datasets like UCF101, CNNs [10] have reported better performance than traditional local features [12]. However, CNNs require a huge amount of annotated
training data. For some small size datasets such as Hollywood2 and Olympic
Sports, we lack of sufficient training samples to train CNNs adequately. Therefore, there are a large number of works [5,13] exploring how to utilize CNNs
trained on ImageNet to extract visual features.

In this work, we propose a novel algorithm to better employ the ImageNet
trained CNNs. Motivated by the popularity of spatial pyramids in image classification [9], we propose multi-resolution latent concept descriptor (mLCD) features. By encoding the LCD features from multiple scales, the final video feature is able to give a better representation. On the other hand, when transfer

A. Hirose et al. (Eds.): ICONIP 2016, Part III, LNCS 9949, pp. 564–571, 2016.
DOI: 10.1007/978-3-319-46675-0_62

ImageNet trained CNNs, mLCD is only used to encode last convolution layer features of spatial networks. We extend our mLCD to temporal networks to capture motion features. The main contributions of this paper are summarized as follows:

1. We propose a multi-resolution extension to LCD [13] named mLCD, which extracts visual features from video frames at multiple scales.
2. We combine our mLCD with the two-stream CNNs [10], that is, using mLCD to encode features from temporal networks. As we know, this is the first work which encodes the features from last convolution layer of temporal networks.
3. We combine our mLCD features with the traditional improved dense trajectory [12] features, and conduct experiments on Hollywood2 and Olympic Sports datasets. The experimental results of the proposed algorithm achieves state-of-the-art performance.

2 Method

In this section, we first provide a description of our action recognition framework. Then the details of the proposed mLCD method are further elaborated. A discussion of the Fisher Vector and VLAD is given at the end of the section.

2.1 Action Recognition Pipeline

Our action recognition framework is shown in Fig. 1, it mainly consists of three parts: feature extraction, feature encoding and classification. Our proposed framework combines the hand-craft local features and the learned deep local features, encoding them with different methods and finally combining them with late fusion.

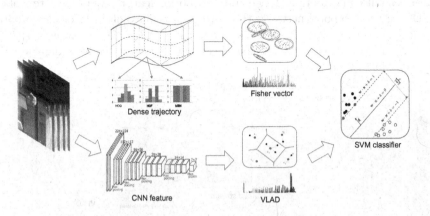

Fig. 1. Action recognition pipeline

For the hand-craft local descriptors, we adopt the improved dense trajectory features [12]. The densely sampled corner points are tracked to form dense trajectories. For each trajectory, different low-level features are computed in its spatial-temporal volume. In our framework, we compute histogram-based features including HOG, HOF and MBH. Normalization and PCA are applied to the local descriptors as mentioned in [12].

For the learned deep features, to capture both the semantic high-level features and motion features, we utilize the two-stream CNNs [10] to extract video features. In our framework, we combine LCD [13] (latent concept descriptor) and two-stream CNNs. We also extend LCD to its multi-resolution version, which is described in Sect. 2.2.

Once the improved dense trajectory features and multi-resolution LCD features are extracted, we encode the improved dense trajectory features with Fisher Vector and encode mLCD features with VLAD. The choice of encoding methods is discussed in Sects. 2.3 and 3.3. The two kinds of encoded features are combined using late fusion, and the concatenated video-level features are feed to SVM classifier to obtain the final classification result.

2.2 Multi-resolution LCD

LCD [13] (latent concept descriptor) is a method to encode the CNN extracted features by traditional BoW methods. LCD extracts the pooling layer feature rather than the full-connected layers, which contains spatial information. Specifically, for VGG16 architecture, the dimension of last convolution layer is $7 \times 7 \times 512$, which can be viewed as 49 local features of dimension 512. These local descriptors can be encoded by any BoW methods including Fisher Vector and VLAD. We propose two major improvements of the original LCD method.

Firstly, we combine LCD with the two-stream CNNs. Traditional LCD only utilizes the feature from the spatial network, thus, it can merely capture the static semantic information. To capture the motion feature, we propose to embed LCD into the temporal network, that is, the last convolution layer of temporal network is also viewed as local features and encoded by the same way as spatial network. The local descriptors from the two networks are encoded independently and combined with late fusion.

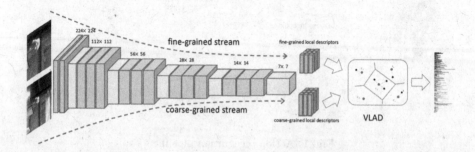

Fig. 2. Multi-resolution latent concept descriptor

The second important improvement of LCD is that we extend the LCD to its multi-resolution version, which is named mLCD. Our intuition is motivated by the success of spatial pyramid in both image classification [9] and action recognition [12]. For both the spatial network and temporal network, as shown in Fig. 2, we provide two kinds of input to the networks. The entire images (or optical flow) are feed into the network to gain the coarse-grained local descriptors. Fine-grained local descriptors are obtained by feeding the central crop of image into the same network. We call the two procedure coarse-grained stream and fine-grained stream respectively. We encoding the coarse-grained local features and fine-grained local features together to generate the feature of a video.

2.3 Fisher Vector or VLAD

Once the multi-resolution local descriptors are extracted by our networks, we can employ Fisher Vector [9] or VLAD [1] to encode the local descriptors into a video descriptor. Either Fisher Vector or VLAD can be viewed as an alternative of bag-of-words encoding, but both of them have shown better performance than traditional BoW encoding methods in image classification [9] and action recognition [8].

Fisher Vector [9] encoding, derived from Fisher Kernel, is the gradient of the log-likelihood with respect to a parameter. Generally, we fit the data with a Gaussian Mixture Model (GMM) with diagonal covariance matrix, so given a single local descriptor x, the gradient vector of log-likelihood respect to the model parameter is as follows:

$$\mathcal{G}_{\mu,k}^x = \frac{1}{\sqrt{\pi_k}}\gamma_k\left(\frac{x - \mu_k}{\sigma_k}\right) \tag{1}$$

$$\mathcal{G}_{\sigma,k}^x = \frac{1}{2\sqrt{\pi_k}}\gamma_k\left[\left(\frac{x - \mu_k}{\sigma_k}\right)^2 - 1\right] \tag{2}$$

where γ_k is the weight of local descriptor x to the k-th gaussian component, and is calculated by $\gamma_k = \frac{\pi_k \mathcal{N}(x;\mu_k,\Sigma_k)}{\sum_{k=1}^K \pi_k \mathcal{N}(x;\mu_k,\Sigma_k)}$

The Fisher Vector of one descriptor x is the concatenation of these gradients, $S_x = [\mathcal{G}_{\mu,1}^x, \mathcal{G}_{\sigma,1}^x, ..., \mathcal{G}_{\sigma,K}^x, \mathcal{G}_{\sigma,K}^x]$. The final Fisher Vector of one video is the sum of the Fisher Vectors of local features, $S = \sum S_x$.

VLAD [1] can be viewed as a simplified version of Fisher Vector. VLAD employs k-means algorithm to obtain the codebook rather than fit data with GMM. Once the codebook $\{d_i : i = 1, 2, ..., K\}$ is calculated, for each local descriptor x, its VLAD encoding can be calculated as $S_x = [\omega_1(x - d_1), \omega_2(x - d_2), ..., \omega_K(x - d_k)]$.

For our mLCD local descriptors, we make some quantitative analyses to decide which encoding method should be adopted. In Fig. 3, we plot the value distribution of Fisher Vector and VLAD, both of which have been normalized. It is shown that for Fisher Vector, most of the values are positive but the values of VLAD are distributed more uniform. Therefore, it is natural to assume VLAD is a better choice, and the experimental result in Sect. 3 validates our assumption.

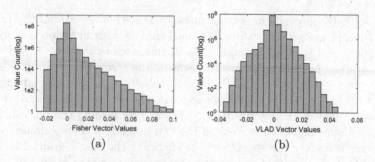

Fig. 3. The value distribution of Fisher Vector and VLAD

3 Experiments

In this section, we first introduce the datasets used in our experiments. Then we present the implement details of our algorithms. Some quantitative analyses are made to show the effectiveness of our method. Finally, a comparison with the state-of-the-art methods is given.

3.1 Datasets

Hollywood2. The Hollywood2 dataset [6] has been collected from 69 different Hollywood movies and includes 12 action classes. It contains 1,707 videos split into a training set (823 videos) and a test set (884 videos). The performance is measured by mean average precision (mAP) over all classes, as in [6].

Olympic Sports. The Olympic Sports dataset [7] consists of athletes practicing different sports collecting from YouTube. There are totally 16 sports actions (such as clean and jerk, bowling, basketball lay-up, discus throw), represented by a total of 783 video sequences. We use 649 sequences for training and 134 sequences for testing as recommended. mAP over all classes is reported as in [7].

3.2 Implement Details

In our experiments, HOG, HOG and MBH descriptors form a 396-dimension vector (96+108+96+96), dimension of which is reduced to half with PCA. The dense trajectory feature are encoded by Fisher Vector and square root normalization and L2 normalization are both applied. For the learned deep feature, we adopt VGG16 for both spatial network and temporal network. The spatial network is trained on the ImageNet and the temporal network is pre-trained on ImageNet and is finetuned on UCF101. The mLCD features are encoded by VLAD. For each local feature, we search 5 nearest neighbors in the codebook to encode it. Square root and L2 normalization is also applied at last. For multi-class SVM, we adopt the one-vs-all method. The hyperparameters of each SVM is decided via cross validation.

3.3   Quantitative Analysis

We conduct rich quantitative analyses to demonstrate the effectiveness of our algorithm. In this section, all the experiments are conducted on the Hollywood2 dataset. First, we make an analysis of the choice of some hyperparameters in our method. In Fig. 4(a), we encoding mLCD local descriptors with Fisher Vector and VLAD separately. It is showed that VLAD can gain a slightly better performance than Fisher Vector, which is consistent with the analyses in Sect. 2.3.

Figure 4(b) shows the impact of PCA dimension reduction to our method. In this experiment, we fix the encoding method as VLAD, and apply PCA to mLCD and LCD local space-time features to show the impact of PCA. As presented in Fig. 4(b), the performance of both mLCD and LCD is largely damaged by PCA dimension reduction. It is also clear that our mLCD method is more sensitive to PCA. For the temporal network, when the local discriptors keep the original dimension (512), mLCD reports a mAP of 45.41 %, while LCD gains 45.32 %. When PCA dimension is smaller, LCD obtains a better performance compared with mLCD. Therefore, in the following experiments, we do not apply PCA to mLCD local features.

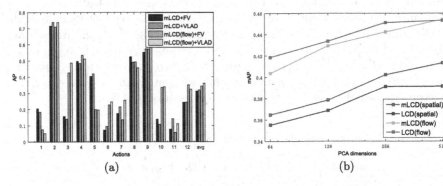

Fig. 4. Quantitative analysis of hyperparameters choice. (a) The mAP of mLCD when encoded by Fisher Vector and VLAD. (b) The mAP of LCD and mLCD when applied PCA dimension reduction

Table 1 compares our mLCD with LCD [13] under different settings. We can draw mainly two conclusions from this table. First, it is clear that mLCD outperforms LCD on both Hollywood2 and Olympic Sports under different configurations. Second, spatial network performs better than temporal network on both datasets. There are several possible reasons leading to this result: it may be decided by the scenes of two datasets, which indicate that on Hollywood2 dataset and Olympic Sports dataset, motion feature is less import than static feature. Another reason is that temporal network is trained on UCF101, which is relative small compared with ImageNet, therefore, temporal network tends to be overfitting on UCF101.

Table 1. The mAP of our method with different configurations

Methods	Hollywood2	Olympic sports
LCD	0.3915	0.7739
mLCD	0.4133	0.8003
LCD (flow)	0.4532	0.7429
mLCD (flow)	0.4541	0.7554
LCD + LCD (flow)	0.5101	0.8296
mLCD + mLCD (flow)	0.5163	0.8530

3.4   Comparison with the State of the Art

Table 2 compares our method with the most recent results reported in literature of th two datasets. On Hollywood2 dataset, trajectory based methods [2,4,12] achieve great success. Wang et al. [12] report 64.3 % by combining dense trajectories, motion features and human detectors. Jain et al. [3] report 66.6 % by introduce a large scale of concept detector. Our method improves the state-of-the-art result by around 0.1 % by combining shallow and deep features.

Olympic Sports is a collection of sports videos. This dataset contains rich structure information and significant camera motion. Therefore, traditional trajectory based methods [2,4] does not perform well on this dataset. Wang et al. [12] introduce human detectors to remove the background trajectories and gain a mAP of 91.1 %. Our experiments show that without computational expensive detectors, we can also obtain a slightly better result of 91.4 %.

Table 2. The mAP of our method with different settings

Methods	Hollywood2	Olympic Sports
Jiang et al. [4]	0.595	0.806
Manan Jain et al. [2]	0.625	0.832
Wang et al. [12]	0.643	0.911
Mihir Jain et al. [3]	0.666	–
IDT + mLCD (Ours)	0.669	0.914

4   Conclusion

In this paper, we explore how to effectively utilize CNNs trained on ImageNet and UCF101 to improve the performance of action recognition. We introduce multi-resolution latent concept descriptors (mLCD) to encode both spatial and temporal network, and conduct experiment on Hollywood2 and Olympic Sports datasets. We report a better result compared with the current state-of-the-art methods.

Acknowledgements. The work was supported by the National Natural Science Foundation of China (61272251), the Key Basic Research Program of Shanghai Municipality, China (15JC1400103) and the National Basic Research Program of China (2015CB856004).

References

1. Arandjelovic, R., Zisserman, A.: All about VLAD. In: CVPR. pp, 1578–1585. IEEE (2013)
2. Jain, M., Jégou, H., Bouthemy, P.: Better exploiting motion for better action recognition. In: CVPR, pp. 2555–2562. IEEE (2013)
3. Jain, M., van Gemert, J.C., Snoek, C.G.: What do 15,000 object categories tell us about classifying and localizing actions? In: CVPR, pp. 46–55 (2015)
4. Jiang, Y.-G., Dai, Q., Xue, X., Liu, W., Ngo, C.-W.: Trajectory-based modeling of human actions with motion reference points. In: Fitzgibbon, A., Lazebnik, S., Perona, P., Sato, Y., Schmid, C. (eds.) ECCV 2012, Part V. LNCS, vol. 7576, pp. 425–438. Springer, Heidelberg (2012)
5. Karpathy, A., Toderici, G., Shetty, S., Leung, T., Sukthankar, R., Fei-Fei, L.: Large-scale video classification with convolutional neural networks. In: CVPR, pp. 1725–1732. IEEE (2014)
6. Marszalek, M., Laptev, I., Schmid, C.: Actions in context. In: CVPR, pp. 2929–2936. IEEE (2009)
7. Niebles, J.C., Chen, C.-W., Fei-Fei, L.: Modeling temporal structure of decomposable motion segments for activity classification. In: Daniilidis, K., Maragos, P., Paragios, N. (eds.) ECCV 2010, Part II. LNCS, vol. 6312, pp. 392–405. Springer, Heidelberg (2010)
8. Peng, X., Wang, L., Wang, X., Qiao, Y.: Bag of visual words and fusion methods for action recognition: comprehensive study and good practice. arXiv preprint (2014). arXiv:1405.4506
9. Sánchez, J., Perronnin, F., Mensink, T., Verbeek, J.: Image classification with the fisher vector: theory and practice. IJCV **105**(3), 222–245 (2013)
10. Simonyan, K., Zisserman, A.: Two-stream convolutional networks for action recognition in videos. In: NIPS, pp. 568–576 (2014)
11. Soomro, K., Zamir, A.R., Shah, M.: Ucf101: a dataset of 101 human actions classes from videos in the wild. arXiv preprint (2012). arXiv:1212.0402
12. Wang, H., Schmid, C.: Action recognition with improved trajectories. In: ICCV, pp. 3551–3558. IEEE (2013)
13. Xu, Z., Yang, Y., Hauptmann, A.G.: A discriminative CNN video representation for event detection. In: CVPR, pp. 1798–1807 (2015)

Improving Neural Network Generalization by Combining Parallel Circuits with Dropout

Kien Tuong Phan[1(⊠)], Tomas Henrique Maul[1], Tuong Thuy Vu[1], and Weng Kin Lai[2]

[1] University of Nottingham Malaysia Campus,
43500 Semenyih, Selangor, Malaysia
khyx3pko@nottingham.edu.my
[2] Tunku Abdul Rahman University College,
Kuala Lumpur, Malaysia

Abstract. In an attempt to solve the lengthy training times of neural networks, we proposed Parallel Circuits (PCs), a biologically inspired architecture. Previous work has shown that this approach fails to maintain generalization performance in spite of achieving sharp speed gains. To address this issue, and motivated by the way Dropout prevents node co-adaption, in this paper, we suggest an improvement by extending Dropout to the PC architecture. The paper provides multiple insights into this combination, including a variety of fusion approaches. Experiments show promising results in which improved error rates are achieved in most cases, whilst maintaining the speed advantage of the PC approach.

Keywords: Deep learning · Parallel circuit · Dropout · DropCircuit

1 Introduction

Deep learning has repeatedly made significant improvements to the generalization capability of neural networks through several core strategies including: (i) increasing model size, (ii) undergoing longer training periods as well as (iii) adopting larger datasets. Unfortunately, this approach creates a tremendous overhead in terms of both time and computational resources. Deep learning is increasingly becoming more feasible with the support of specialized hardware systems (e.g. multicore CPUs, graphics processing units (GPUs), and high-performance computing (HPC)). However, these solutions tend to be costly and require careful tailoring to derive maximal benefits.

We attempt to apply deep learning to a remote sensing problem, within the constraints of an online platform with limited computational power. In order to address the feasibility concerns mentioned above, we decided to tackle the computational problem at the algorithmic level, as a pure hardware approach would be too expensive to solve the problem alone [1]. Within the scope of our previous paper [2], we have proposed Parallel Circuits (PCs), a biology-inspired Artificial Neural Network (ANN) architecture, as an attempt to reduce heavy computational loads without harming performance. Our preliminary experiments showed that PC architectures could decrease training time up to 40 % under some constraints. On the contrary, the impact on classification

A. Hirose et al. (Eds.): ICONIP 2016, Part III, LNCS 9949, pp. 572–580, 2016.
DOI: 10.1007/978-3-319-46675-0_63

accuracy was still debatable, as the proposed network exhibited unstable performance across configurations, especially when the size of the original ANN was small.

Dropout is a recent technique for regularization that has been boosting neural network accuracy in many applications. By randomly dropping nodes in the network, dropout successfully prevents the co-adaption of nodes [3]. Thanks to Dropout, over-reliance on specific input nodes is reduced [4]. Taking advantage of this valuable property, in this paper we propose a modification of Dropout whereby we scale it up to the level of Parallel Circuits (i.e. DropCircuit). In this paper, we hypothesize that Dropout would help circuits work more independently (achieving sparser perspectives of the problem domain) and would thus help harvest the benefits of PC modularity. The paper reports on the performance of different types of Dropout-PC combinations.

The paper is structured as follows: Sects. 2 and 3 provide brief reviews on Parallel Circuits and Dropout respectively; Sect. 4 explains the proposed approach; Sect. 5 describes and discusses the experimental results, comparing several Dropout implementations; and Sect. 6 concludes the paper.

2 Parallel Circuits

Our proposed architecture, Parallel Circuits, is inspired from one of nature's solutions to heavy workload, i.e.: parallelism. In implementing PCs, a standard fully-connected Multi-Layer Perceptron (MLP) is divided vertically, forming a series of independent sub-networks called circuits. It's worth mentioning that these circuits share the same input and output layers (i.e. the division is applied only to hidden layers). Thus, compared with MLPs having the same number of nodes, PC architectures reduce the number of connections by a factor of k, where k is the number of equal-sized circuits being used. This modification also defines a crucial assumption for PCs that the network should have at least two hidden layers.

It is not unreasonable to expect speed gains from the fact that PCs use fewer connections. On the other hand, it is probably more controversial to hypothesize that PC architectures are advantageous in terms generalization. In [5], an ANN training protocol was reported characterized by problem decomposition (repeatedly switching between multiple goals each with several sub-goals), leading to the emergence of modularity. Provided that modularity is already achieved through the independence of parallel circuits, we reversely hypothesize that this architecture should exhibit the property of automatic problem decomposition. Moreover, and in accordance with the divide and conquer principle, this automatic problem decomposition can further be hypothesized to improve generalization [6].

3 Dropout

As mentioned above, Dropout is a regularization technique that has been shown to be effective for a broad variety of neural networks and datasets. Some studies, such as [4, 7], have interpreted Dropout as an ensemble method similar to bagging. By randomly dropping nodes, a large number of thinned networks with shared weights are implicitly trained, which when recombined during classification with appropriate scaling, approximate an ensemble of averaged thinned networks [8]. Moreover, in the situation where Dropout is also applied to the input layer, it can be seen as a form of data augmentation whereby noise is added to the input patterns.

Probably due to both its effectiveness and ease of implementation, Dropout is attracting significant attention from researchers in the field. Multiple modifications have been developed for either specialized or general-purpose models. DropConnect is a successful descendent of Dropout, showing even better performance on a range of datasets [9]. In the context of convolutional neural networks (CNNs), one of the most state-of-the-art ANNs for vision-related problems, Dropout has been investigated in different parts of the model. For example, Hinton pioneered the trend of implementing the technique in the final fully connected layers [7], whereas Wu and Xu extended the application of Dropout to pooling layers [3].

4   Methodology

4.1   Parallel Circuit

According to this definition of Parallel Circuits, for circuit k, the input sum for the hidden node i over the layer l might be computed as follows:

$$z_{ki}^l = \sum_j^{n_k^{l-1}} y_{kj}^{l-1} * \omega_{ki\,j}^{l\,l-1} + b_{ki}^l \tag{1}$$

where n_k^{l-1} is the number of hidden nodes on layer $l-1$ of circuit k, y_{kj}^{l-1} represents the output of node j, in circuit k and layer $l-1$, $\omega_{ki\,j}^{l\,l-1}$ represents the weight of the connection from node j in layer $l-1$, to node i in layer l, in circuit k, and b_{ki}^l refers to the bias of node i in layer l, in circuit k. The first hidden layer is actually a special case, as all circuits shared the same input layer ($l = 0$). Therefore,

$$y_{1j}^0 = y_{2j}^0 = \ldots = y_{kj}^0 = x_j \tag{2}$$

On the other hand, the input for the output layer is equivalent to a hidden layer that concatenates the last hidden layers of all circuits (penultimate layer– PL), resulting in:

$$z_i^{out} = \sum_k \sum_j^{n_k^{PL}} y_{kj}^{PL} * \omega_{ki\,j}^{out\,PL} + b_i^{out} \tag{3}$$

4.2   Node Dropout

Node Dropout (ND) is an implementation of the standard dropout technique, where dropping is applied at the level of nodes. In the PC context, it is worth recalling that we treat each circuit independently. For any single circuit, whenever a training sample is introduced, a distinct mask is generated according to its predefined probability. This favors our attempt to enhance sparsity across subnetworks. Thus, for the PC case, the input sum (1) might be rewritten as follows:

$$m_{kj}^{l-1} = Bernoulli\left(p_k^{l-1}\right) \tag{4}$$

$$y'_{kj}^{l-1} = y_{kj}^{l-1} * m_{kj}^{l-1} \tag{5}$$

$$z_{ki}^l = \sum_j^{n_k^{l-1}} y'_{kj}^{l-1} * \omega_{ki\,j}^{ll-1} + b_{ki}^l \tag{6}$$

In which, m_{kj}^{l-1} stands for component j of the mask, corresponding to circuit k and layer $l-1$, and the remaining elements are defined as in Eq. (1).

4.3 DropCircuit

Our PC concept assumes that each circuit should produce a unique perspective of the problem. Therefore, it is critical to prevent circuits from adapting together. When scaling the standard Dropout technique from nodes to circuits, we could consider DropCircuit as a type of "uniform" Node Dropout, where particular nodes stick together for the entire training process. Once one of them is dropped, all the nodes tied to it will also be dropped. To implement this, we create a mask m_k describing the circuit k's dropping status. Every node belonging to the circuit inherits the mask and standard Node Dropout is applied accordingly.

$$m_k = Bernoulli(p) \tag{7}$$

$$y'_{kj}^{l-1} = y_{kj}^{l-1} * m_k \tag{8}$$

In this paper we report on two variants of circuit dropout, i.e.: fixed and non-fixed (mask). In Fixed DropCircuit (FD) we attempt to further specialize circuits by creating a fixed association between sets of patterns and circuits. In this case, a random mask m_k is generated initially for each training instance, and is maintained throughout the training process. On the contrary, Non-Fixed DropCircuit (NFD) works similar to standard Dropout but on the level of circuits, whereby masks are continually regenerated. In other words, Eq. (7) will be computed whenever a training sample is introduced, regardless of the epoch.

4.4 Experiment Setup

The purpose of the experiments was (i) to prove that Dropout-aided PC can improve generalization performance compared with its counterpart Dropout-aided SC (i.e. Single Circuit - fully connected MLP), and (ii) to determine the most efficient Dropout type for PCs.

The datasets adopted include MNIST and 4 smaller ones obtained from the UCI Machine Learning repository (e.g. Wisconsin Breast Cancer, Breast Tissue, Glass and Leaf) [10]. Two network structures were considered, one with 100 hidden nodes per layer and the other with 1000 nodes; both with two hidden layers. For each network, one SC and 3 PC implementations were examined (corresponding to ND, NFD and FD). The PC versions were further divided into 3 sub categories of 2, 5 and 10 circuits. The models were trained with 0.1 learning rate, 0.4 momentum, 0.0001 L2 sparsity penalty, and the chosen activation function was tanh. In all Dropout-aided conditions, the retaining probability was 0.5. For the four smaller datasets, each condition was run for 100 trials, whereas for MNIST, 10 trials per condition were used. This large number of trials explains our choice of 100 training epochs per training session. The experiments were conducted based on Theano library.

5   Experimental Results & Discussion

5.1   Parallel Circuit Versus Single Circuit

In this paper, we focus on investigating the ability of PCs fueled with Dropout. To have a fair comparison, the SC architecture was implemented with standard Dropout. Regarding total training times, the problem of PC inefficiencies on small networks is still unsolved. As mentioned in [2], the complexity reduction in small networks is not significant enough to compensate for the computational overhead involved in separating circuits. Therefore, in these networks, PCs, in contrast, increase the total training time to some extent. On the other hand, considering networks with 1000 nodes, we found that implementing PCs gives at least 30 % reduction in training time for all cases (Table 1). As mentioned above in Sect. 1, since one of the successful deep learning strategies is to improve performance by enlarging models, we believe that the reduction is significant. In small network cases, FD is always the approach closest to the speed of SCs while in larger models, both DropCircuit representatives (i.e. FD and NFD) occupy the first and second best positions respectively.

Table 1. Error rate and Training time (dark grey for best, light grey for second best)

		Error Rate				Training Time			
		SC	ND	NFD	FD	SC	ND	NFD	FD
100 nodes	BC	2.710	2.655	1.842	2.375	2.293	4.723	4.957	2.819
	BT	60	57.917	57.167	60.158	0.386	0.826	0.857	0.506
	GL	38.9375	38.546	30.579	34.267	0.814	1.707	1.766	1.037
	LF	93.908	93.464	91.208	93.494	1.490	2.938	3.111	1.794
	MN	9.178	8.650	6.881	7.442	362.914	424.498	412.547	372.583
1000 nodes	BC	2.575	2.235	1.917	2.668	89.073	29.930	29.606	25.557
	BT	60	58.292	58.375	60.133	16.605	5.258	5.020	4.399
	GL	35.575	29.492	30.258	33.588	34.542	10.539	10.346	8.865
	LF	89.033	92.489	86.064	92.481	53.541	18.392	18.316	15.796
	MN	9.356	8.317	7.827	8.058	6475.711	4060.208	3958.506	3595.458

Figure 1 illustrates the test errors for both model sizes, on different datasets, and across Dropout-aided conditions. It's clear that the statistics favor PCs more than SCs with generally lower test errors (39.336 % against 40.127 %). In fact, PCs always have at least one of its implementations perform better than SCs and its average testing error is lower in most of the cases. Regarding average PC vs. SC performance, this holds true except for only one case in which we used large networks (1000 nodes) to classify the Leaf dataset. For this particular configuration, the average error of all PC implementations (AVG) is higher than that of SCs (92.485 % and 89.033 % respectively). However, the best individual condition for this case (i.e. 1000 nodes and Leaf dataset) is still a PC condition (i.e. NFD). Significantly, in this context of generalization performance, PCs completely overshadow SCs for small sized networks (100 nodes), contrary to our earlier training time results. Finally, the lowest test errors for each dataset and model size combination were always obtained by a PC architecture.

5.2   Node Dropout Versus Drop Circuit

From Fig. 1, we can see that PCs on average (AVG) obtained a slight advantage over SCs (usually less than 1 % in difference). The Glass dataset revealed the largest gaps between SC and AVG, especially with 1000 node networks achieving more than 4 % in difference (35.575 % versus 31.540 % respectively). However, the argument that there is no significant improvement does not hold true. By closer observation, one can see that the performance of the 3 PC-Dropout conditions varied by a large extent. ND, which implements standard Node Dropout, achieved quite similar performance with SC, with a minimal lower average error rate (39.206 % versus 40.123 %). It exhibited better accuracy than SCs in most of the cases, except for the faulty case (LF + 1000 node family). According to Fig. 2, all the median, mean or whiskers of both SC and ND are nearly identical.

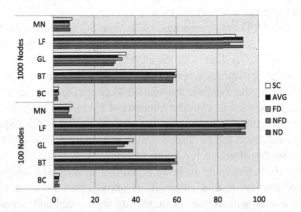

Fig. 1. Mean test errors for different conditions and datasets (MN-MNIST, LF-Leaf, GL-Glass, BT-Breast Tissue, BC-Breast Cancer, AVG-PC Average)

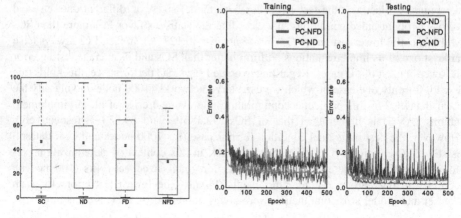

Fig. 2. Average test error of the 4 Drop-out-aided conditions

Fig. 3. Training and Testing error across epochs

NFD and FD are our proposed modification of Dropout for the PC approach (i.e. DropCircuit). NFD, without any hesitation, could be claimed as the champion of the experiments reported here. In every setup (even in the problematic Leaf case), NFD always outperformed SCs (37.212 % versus 40.127 % on average) and displayed up to 8.5 % in error difference for the Glass dataset. The median (32.5 %), mean (30.61 %) and both whiskers (0.5 %, 74.99 % respectively) are much lower compared with its counterparts. Recall that in our earlier experiments focusing on training speed [2], the original PC approach was completely beaten by SCs for small scale networks. This time, with NFD, the same setup (small network + PC) achieved the best generalization performance for all 5 datasets. For the Breast Cancer dataset, it scored around 2.375 % in error rate, compared with 3.66 % in our preliminary test. Thus, we can conclude that NFD is a strong candidate for balancing the speed/generalization trade off in our approach by boosting the accuracy of PCs.

On the other hand, FD only performed better than SC for half of the cases and beat ND 4 times (Table 1). After averaging, FD is still slightly better than SCs (39.466 % versus 40.127 %) but the median is far lower than that of SC (i.e. the major part in 100 trials achieved better performance compared with SC). One of the possible explanations might be due to the random association between specific data instances and circuits. In future work we will consider versions of FD, which adopt a more informed approach (e.g. via clustering) for associating sample instances with specific circuits. Since FD always achieved the best speed gains and considerably good accuracy, we believe this is a worthy direction for future research.

Finally, apart from providing insights pertaining to different Dropout-aided conditions, the experimental results summarized above, also make it clear that PCs themselves are a useful architecture for improving both speed and generalization. The fact that PCs outperform SCs, for conditions that adopt comparable architectures (e.g. number of nodes) and training techniques (e.g. Dropout), make it clear that this is an architecture that warrants further investigation in the context of Machine Learning and Neural Computation. Moreover, the results possibly shed additional light onto the rationale for parallel circuits in biological neural networks (BNNs). It is quite possible that parallel circuits in BNNs are partly motivated by the implementation of implicit ensembles aided by neurophysiological mechanisms related to Dropout.

5.3 MNIST

In Sect. 5.2, we showed that the PC approach could reduce training times. To test whether this result scaled to larger datasets, we compared conditions using MNIST, but this time extended training to 500 epochs using a [1000-500] network structure with a much higher learning rate (i.e. 1) to favor Dropout. Other parameters were kept unchanged. Figure 3 points out that both of the PC approaches converge faster, reaching a plateau around epoch 20 (especially PC-ND) compared to epoch 60 of SCs. Significantly, in this period, PC approaches reached lower plateaus than that of SC. Both of the PC versions exhibited the normal pattern of Dropout implementations, with error fluctuations throughout training. Especially in PC-NFD, the degree of fluctuation was even higher than PC-ND. In the end, PC-ND achieved slightly lower error rate than SC (best at 4.19 % and 4.38 % respectively) while PC-NFD reached 2.22 %. This revealed that (i) with smaller epoch numbers, PC approaches can achieve test errors at least similar to SCs and (ii) PC-NFD performs much better than NDs.

6 Conclusion

In this paper, we proposed an improvement of the parallel circuit approach by implementing Dropout, specifically targeting generalization performance. The experiments showed that combining Parallel Circuits with Dropout not only reduces training times but also enhances generalization performance in most cases. Our work provides multiple insights pertaining to this combination, and includes a benchmark comparing different variants. We found that Non-Fixed DropCircuit leads to the best improvements in generalization performance.

References

1. Suresh, S., Omkar, S.N., Mani, V.: Parallel implementation of back-propagation algorithm in networks of workstations. IEEE Trans. Parallel Distrib. Syst. **16**(1), 24–34 (2005). doi:10.1109/tpds.2005.11
2. Phan, K.T., Maul, T.H., Vu, T.T.: A parallel circuit approach for improving the speed and generalization properties of neural networks. In: Paper presented at the 11th International Conference on Natural Computation (ICNC), Zhangjiajie, 15–17 August 2015
3. Wu, H., Gu, X.: Towards dropout training for convolutional neural networks. Neural Netw. **71**, 1–10 (2015). doi:10.1016/j.neunet.2015.07.007
4. Baldi, P., Sadowski, P.: The dropout learning algorithm. Artif. Intell. **210**, 78–122 (2014). http://www.sciencedirect.com/science/article/pii/S0004370214000216

5. Matsugu, M., Mori, K., Mitari, Y., Kaneda, Y.: Subject independent facial expression recognition with robust face detection using a convolutional neural network. Neural Netw. Official J. Int. Neural Netw. Soc. 16(5–6), 555–559 (2003). doi:10.1016/s0893-6080(03)00115-1
6. Kashtan, N., Alon, U.: Spontaneous evolution of modularity and network motifs. Proc. Natl. Acad. Sci. U.S. A. 102(39), 13773–13778 (2005). doi:10.1073/pnas.0503610102
7. Hinton, G.E., Srivastava, N., Krizhevsky, A., Sutskever, I., Salakhutdinov, R.R.: Improving neural networks by preventing co-adaptation of feature detectors, pp. 1–18 (2012). doi:arXiv:1207.0580
8. Srivastava, N., Hinton, G.E., Krizhevsky, A., Sutskever, I., Salakhutdinov, R.: Dropout: a simple way to prevent neural networks from overfitting. J. Mach. Learn. Res. (JMLR) 15, 1929–1958 (2014). doi:10.1214/12-aos1000
9. Wan, L., Zeiler, M.: Regularization of neural networks using dropconnect. In: Proceedings of the 30th International Conference on Machine Learning (ICML-13)(1), pp. 109–111 (2013)
10. Lichman, M.: UCI Machine Learning Repository (2013). http://archive.ics.uci.edu/ml

Predicting Multiple Pregrasping Poses by Combining Deep Convolutional Neural Networks with Mixture Density Networks

Sungphill Moon, Youngbin Park, and Il Hong Suh[✉]

Department of Electronics and Computer Engineering, Hanyang University,
17 Haengdang-dong, Sungdong-gu, Seoul, Korea
sp9103@incrol.hanyang.ac.kr, {pa9301,ihsuh}@hanyang.ac.kr

Abstract. In this paper, we propose a deep neural network to predict the pregrasp poses of a three-dimensional (3D) object. Specifically, a single RGB-D image is used to determine multiple pregrasp position of three fingers of the robotic hand for various poses of known or unknown objects. Multiple pregrasping pose prediction typically involves the use of complex multi-valued functions where standard regression models fail. To this end, we propose a deep neural network containing a variant of the traditional deep convolutional neural network as well as a mixture density network. Furthermore, in order to overcome the difficulty of learning with insufficient data in the first part of the proposed network, we develop a supervised learning technique to pretrain the variant of the convolutional neural network.

Keywords: Grasping pose prediction · Deep convolutional neural network · Mixture density network

1 Introduction

According to the review by Sahbani et al. [1], grasping methods can be broadly categorized as analytic and data driven. Most analytic methods assume the availability of complete knowledge of the objects to be grasped, such as a complete three-dimensional (3D) model. These methods then involve the construction of a suitable grasp pose based on certain criteria, such as force closure or stability. Grasp synthesis is hence usually formulated as a constrained optimization problem over such criteria. Therefore, analytic methods typically require understanding the precise 3D shape of the object as well as massive amounts of computation to solving the optimization problem.

Data-driven approaches, on the other hand, investigate ways to avoid such disadvantages by imitating the human grasping. These methods select an appropriate grasp by building a direct mapping from vision to action. A majority of these methods have thus focused more on the use of vision-based features obtained from RGB or RGB-D images to predict grasp locations. The learning of visual features based on machine learning algorithms has enabled the easy

© Springer International Publishing AG 2016
A. Hirose et al. (Eds.): ICONIP 2016, Part III, LNCS 9949, pp. 581–590, 2016.
DOI: 10.1007/978-3-319-46675-0_64

generalization of grasp pose estimation to novel objects often encountered in uncontrolled environments. Methods that can capture the mapping from vision to action through a deep learning model [3,4] have recently garnered considerable attention, owing to the recent success of deep learning in a wide variety of tasks, including robotic grasping and manipulation [5,6] object recognition [7], semantic segmentation [8], and caption generation [9]. However, the main difficulty in deep learning is that training deep neural networks requires large amounts of data.

Fig. 1. The first three images to the left and the last three to the right show two qualitatively different pregrasp poses. In the former, the thumb appears in the upper part of the images, whereas it is in the lower part of the latter set of images. The three pregrasp poses on either side seem identical, but are actually slightly different.

Figure 1 shows a few pregrasp poses. We call a pregrasp pose a configuration where closing the fingers until resistance is encountered leads to a proper grasp pose. In this paper, we address the problem of multiple pregrasp pose regression of a 3D object using deep neural networks. Specifically, a single RGB-D image is used to determine multiple 3D positions of three fingers of the hand that can provide suitable pregrasp positions for various poses of a known or unknown object. To this end, we first create a large set of manually annotated pregrasp data. In the dataset, an image containing only a specific pose of an object is used as input, and the corresponding 150~300 human-supervised pregrasp poses are used for a set of labels. At least two qualitatively different finger configurations are used here for the set of labels.

In this case, it is challenging to collect a large amount of training data using traditional kinesthetic teaching procedures, where a human teacher moves a robotic arm to make the robot perform pregrasping action. To address this problem, we detached the robotic hand from the arm and attached optical markers to the tips of three of its fingers in order to track the 3D positions of the fingers using an optical motion capture system. The human teacher then grabbed the robotic hand and demonstrated possible pregrasp poses for some time, varying poses with small continuous movements. At this time, various 3D positions of the three fingers were recoded via the motion capture system. Once again, a qualitatively different finger configuration for the object's pose was provided by the human teacher, and the demonstration procedure was repeated. Figure 1 shows our data acquisition procedure. It should be noted that none of the images in Fig. 1 were used as training data. Only the 3D positions of the three figures were recorded for the targets. Optical markers have been omitted in the figure in order to clearly display the robotic hand.

The advantage of this data collection procedure is that the effort needed is considerably lesser than in traditional kinesthetic teaching. In this paper, we investigate pregrasp pose prediction instead of estimating grasp poses to exploit data collection scenario mentioned before. The rationale is that an accurate pregrasp leads to a high probability of successful robotic grasping.

We built our model based on the traditional deep convolutional neural network (DCNN). However, DCNN by itself is insufficient to model robotic pregrasp, especially in cases where the training dataset contains data for multiple possible pregrasps for a specific pose of an object. To address this limitation, we investigate a model that combines DCNN and the mixture density network (MDN) [10]. MDN has been shown to be successful for complex multi-valued functions where standard regression models fail. The combination of DCNN and MDN was trained using the manually annotated pregrasp dataset obtained using the data collection procedure proposed above. It should be noted that our dataset contained a large number of pregrasp pose labels, but a relatively small number of images containing objects. In this case, DCNN cannot learn rich visual features to predict suitable pregrasp locations. Therefore, we initialized the DCNN using a proposed supervised pretraining method.

2   Related Work

In this section, we mainly review several studies on robotic grasping using deep learning to predict the grasp location. For a comprehensive survey of robotic grasping, we refer the reader to recent work on the subject [1,2].

A robot's own trial-and-error experience is one way to collect training data for grasping tasks [11,12]. However, performing more than a few hundred trial-and-error runs using a physical robot is usually difficult. On the contrary, manually annotated benchmark datasets are used to train deep neural networks in a supervised manner [13]. The Cornell grasping dataset used in their study contains 1,035 images of 280 graspable objects. Each image is annotated by several ground-truth positive and negative grasping rectangles. Each of these patches is fed to the network to calculate a grasp quality score. Therefore, several feed-forward computations are performed to determine the best grasp pose.

Self-supervised learning of grasp poses is another way to train deep neural networks. Pinto and Gupta proposed a self-supervised data collection method without human supervision using a heuristic grasping system based on object proposals [3]. During training phase, they used the most recently learned model to gather data in order to make the procedure more efficient. The dataset contained more than 50,000 data points, collected in over 700 h of trial-and-error experiments using the Baxter robot with two grippers. Following training, given an image patch, an 18-dimensional likelihood vector, where each dimension represented the likelihood of whether the center of the patch is graspable at $0°$, $10°$, ..., $170°$, was estimated.

Levine et al. [4] proposed a self-supervised learning method similar to the one proposed by Pinto and Gupta. However, the training dataset in this case consisted of over 800,000 grasp attempts on a large variety of objects. To obtain a dataset of this size took them two months, using between six and 14 robotic manipulators at any given time. A deep convolutional neural network called grasp success predictor was trained to determine how likely a given motion is to produce a successful grasp. The performance of this method yielded state-of-the-art results for robotic grasping, but the data collection procedure was time consuming. Moreover, as in the work in [13], several feed-forward computations were required to determine the next movement following any given movement.

The deep neural network structure most closely related to the one proposed here is the one by Levine et al. [5]. They investigated a novel DCNN architecture. Specifically, the first half of the network contains three convolutional layers, followed by a spatial softmax and an expected position layer that converts pixelwise features to feature points. The expected position layer provides accurate spatial reasoning and reduces the number of parameters to avoid overfitting. The other half of the network was developed to generate motor torques of the robot arm for various object manipulation tasks. In this paper, we adapt the idea of converting pixel-wise features to feature points to build our proposed deep neural network.

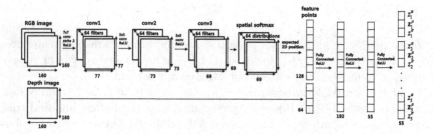

Fig. 2. Proposed neural network architecture.

3   The Proposed Neural Network Architecture

The proposed neural network architecture is shown in Fig. 2. The input of the network is a 160×160 RGB image that shows an object. The first part of the network contains three convolutional layers, followed by a spatial softmax and a feature-point layer that converts pixel-wise features into expected positions. The spatial softmax layer provides lateral inhibition, which suppresses low activations and retains only strong activations more likely to be accurate. The feature-point layer computes the expected position (x, y) of each channel in the softmax layer. As shown in Fig. 2, the depth of each expected position is added to the feature-point layer. The number of dimensions of the feature-point layer is thus three times the number of channels in the softmax layer. Adding depth information does not play an important role in pregrasp prediction and leads to only a slight improvement in performance. As mentioned above, this part of the network, except for the addition of depth data, is analogous to the network architecture proposed by Levine et al. [5]. In the following, we refer to the first part of the neural network as DCNN$^+$ for simplicity. We found that the proposed architecture did better to predict pregrasp poses than the architecture where DCNN$^+$ was replaced by the traditional DCNN. The spatial softmax and expected position layers have been minutely described in [14]. The second part of the network consists of a mixture density network (MDN), which is composed of three fully connected layers. In the following, we present the details of our two contributions: the MDN and the pretraining of DCNN$^+$.

3.1   Mixture Density Network

An MDN combines a mixture model with an artificial neural network. This paper employs a Gaussian mixture (GMM)-based MDN to predict multiple pregrasp poses. As shown in Fig. 2, a set of 3D feature points is the input for the MDN used in the proposed neural network architecture. The activations in the output layer are in turn transformed into the parameters of a GMM. These parameters can be derived from the MDN as

$$\alpha_i = \frac{exp(z_i^\alpha)}{\sum_{j=1}^{m} exp(z_j^\alpha)}, \quad \mu_{ik} = z_{ik}^\mu, \quad \sigma_i = exp(z_i^\sigma). \tag{1}$$

Here, the parameters for the i-th Gaussian, mixture weight, mean, and variance are denoted by α_i, μ_i, and σ_i, respectively. m denotes the number of kernels in the GMM; in our experiments, we used five Gaussians. In case of a multivariate GMM, $\boldsymbol{\mu}_i$ is the vector with component μ_{ik}. $\boldsymbol{\mu}_i$ in this work is a 9D vector because the pregrasp pose in this work is represented using 3D positions of three fingers of the hand. Although $\mathbf{z}_i^\mu$ shown in Fig. 2 is illustrated as scalar for visualization purposes, it is also a 9D vector. The covariance matrix for the i-th Gaussian is denoted by $\boldsymbol{\Sigma}_i$ and is equal to $\mathbf{I}\sigma_i$. The rationale underlying this is that the components of $\boldsymbol{\mu}_i$ can be assumed to be statistically independent and described by common variance σ_i. The details of this discussion are described in [10]. The full probability density function of a pregrasp pose $\mathbf{t}$, conditioned on a set of 3D feature points $\mathbf{x}$, is given as

$$p(\mathbf{t}|\mathbf{x}) = \sum_{i=1}^{m} \alpha_i(\mathbf{x})\phi_i(\mathbf{t}|\mathbf{x}) \tag{2}$$

where

$$\phi_i(\mathbf{t}|\mathbf{x}) = \frac{1}{(2\pi)^{c/2}\sigma_i(\mathbf{x})^c} exp\{-\frac{\|\mathbf{t} - \boldsymbol{\mu}_i(\mathbf{x})\|^2}{2\sigma_i(\mathbf{x})^2}\}. \tag{3}$$

Here, c represents the number of dimensions of the input vector. The loss of training of the MDN is intended to minimize the negative log likelihood, given the training data, as follows:

$$\ell = \sum_{j=1}^{n} \left[-ln\{\sum_{i=1}^{m} \alpha_i(\mathbf{x}^j)\phi_i(\mathbf{t}^j|\mathbf{x}^j)\} \right] \tag{4}$$

where $(\mathbf{x}^j, \mathbf{t}^j)$ is the j-th input/target pair and n is the number of training data items.

3.2 Pretraining

Our training dataset, collected using the method described in the Introduction, contained more than 100,000 pregrasp poses, but only 550 images containing object poses. In order to help the network learn rich and valuable visual features that can predict suitable pregrasp poses, we thus needed to pretrain DCNN$^+$ using more training data that included various objects and a large number of object poses. Finn et al. [14] proposed an unsupervised learning algorithm to pretrain identical network structures, where the loss function minimized reconstruction error. However, we found that this unsupervised learning often generated more than half the feature points from the background. To overcome this limitation, we developed a supervised pretraining algorithm. The output of the network for pretraining is the center (cx, cy) of the object in the image. We formed an additional dataset for pretraining. The center of the object was estimated using the simple background subtraction algorithm. For supervised training, 19 objects were used and 54,000 images were captured. To collect such a large dataset easily, we threaded objects and pulled them in several different directions while the RGB-D camera continued to record the scene.

4 Experiment

4.1 Experimental Setup

The dataset contained eight categories of objects: *cup, cellphone, pen, doll, lotion, can, small cylinder*, and *toy block*. Three different objects were used for each category in the training dataset. Two more objects were included for each category in the test dataset. Figure 3 shows all objects used for our training and test data. The objects in the right table are training data and those in the left table are objects not shown in the training phase. The objects in the right table were employed for testing as well but they are shown in different poses from the ones in the training phase. The size of the workspace was approximately $1\,\mathrm{m} \times 1\,\mathrm{m}$. In the training phase, an object was placed in five positions and rotated 3~8 times at each position. Between 150~300 human-supervised pregrasp poses were labeled for inputs. Therefore, the number of input images, where each included only one object, was 550, and the number of target pregrasp poses for the inputs was 119.243. For the test dataset, a known object appeared in 6~7 poses, and an unknown object was placed in five poses. There were hence 20 and 10 test data for each known and unknown object, respectively. The total number of test data for known and unknown objects were 160 and 80, in that order. Between 100~150 pregrasp poses were provided for the ground truth of each input in the test dataset. For an input image, two qualitatively different pregrasp poses were demonstrated by a human teacher in both the training and the test datasets.

Fig. 3. All objects used for our training and test data.

We implemented three methods for pregrasp pose regression for the sake of comparison: (1) DCNN$^+$ + SP (Pretraining DCNN$^+$ using the proposed supervised learning method), (2) DCNN$^+$ + USP + MDN (Pretraining DCNN$^+$ using the unsupervised learning method proposed in [14], and combining DCNN$^+$ and MDN), and (3) DCNN$^+$ + SP + MDN (Pretraining DCNN$^+$ using the proposed supervised learning method, and combining DCNN$^+$ and MDN). DCNN$^+$ + SP + MDN was the full implementation of our proposed methods.

We implemented the three methods on the Caffe deep learning framework. The number of epochs for all three methods in training phase was 35,000. All experiments were conducted using a robotic hand with three fingers, and Microsoft Kinect v2 was used for the RGB-D camera.

4.2 Quantitative Results

We measured the performance of the methods for pregrasp pose prediction using average error (AVE). While DCNN$^+$ + SP predicted only one pregrasp pose, the two methods that combined DCNN$^+$ and MDN produced multiple pregrasp poses. As illustrated in Sect. 3, the maximum number of the pregrasp poses produced by MDN is five. We can consider the i-th pregrasp pose reliable if the corresponding α_i in Eq. 1 is larger than a certain threshold γ: γ was set to 0.3 in this experiment. Then, to compute AVE, we selected one pregrasp pose from among the reliable predictions.

As mentioned above, the test data had 100~150 true pregrasp poses. In order to compute AVE, we selected one ground truth with the minimum Euclidean distance relative to the prediction. The Euclidean distance for each finger was then computed between the prediction and the selected ground truth. The average of the three Euclidean distances was determined to be the error for the test data.

Table 1. Average pregrasp pose prediction error. $DCNN^+ + USP + MDN$ and $DCNN^+ + SP + MDN$ selected the mean of the largest Gaussian as their predictions

	$DCNN^+ + SP$	$DCNN^+ + USP + MDN$	$DCNN^+ + SP + MDN$
AVE(known)	6.13 cm	5.85 cm	1.69 cm
AVE(unknown)	6.52 cm	5.79 cm	2.53 cm

Table 2. Average pregrasp pose prediction error. $DCNN^+ + SP + MDN$ selected the mean of the second-largest Gaussian as their prediction

AVE(known)	1.8 cm	AVE(unknown)	2.65 cm

Table 1 shows the AVEs of the three methods, where the μ_i of the highest α_i was determined as the prediction of $DCNN^+ + USP + MDN$ and $DCNN^+ + SP + MDN$. We observed that error in the two methods compared to ours significantly increased. This showed the effectiveness of the proposed methods: the use of the MDN and the supervised pretraining for $DCNN^+$:

Table 2 lists the advantages of combining $DCNN^+$ and MDN. In this experiment, the μ_i of the second-highest α_i was used for the prediction of $DCNN^+ + SP + MDN$. It was notable that the second reliable predictions produced by the networks yielded lower performances than the AVEs obtained by the first set of reliable predictions, but the reductions were small. This experiment showed that the proposed network, which combines $DCNN^+$ and MDN, can produce suitable multiple pregrasp poses. In the case of $DCNN^+ + USP + MDN$, the second-highest α_i was never higher than 0.3 for all test data. This supports the hypotheses that if poor visual features are learned in $DCNN^+$, it becomes difficult to successfully train the MDN.

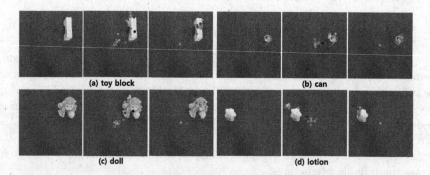

(a) toy block   (b) can

(c) doll   (d) lotion

Fig. 4. Each image to the left in (a-d) is the input image. Each image in the center of (a-d) shows values at the feature-point layer in Fig. 2, where $DCNN^+$ is pretrained using unsupervised learning. Each image to the right in (a-d) was generated by pretraining $DCNN^+$ using the proposed supervised learning method. Each feature is shown in a different color

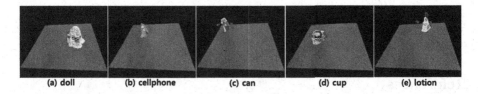

Fig. 5. (a-c) Known objects. (d-e) Unknown objects. Red dots represent ground truths, the blue dots are the predictions by DCNN$^+$ + SP + MDN, and the yellow and sky-blue dots represent predictions by DCNN$^+$+SP and DCNN$^+$ + USP + MDN (Color figure online)

4.3 Qualitative Results

Figure 4 shows the advantage of the proposed supervised pretraining. We observed that most feature points obtained from the DCNN$^+$, which was pretrained using our proposed method, were located on the object, whereas more than half of the feature points generated from DCNN$^+$, which was pretrained using unsupervised learning, were located in the background. It seems that such a small number of feature points from the object can lead to poor performance by DCNN$^+$ + USP + MDN in quantitative evaluations.

Figure 5 illustrates the advantages of the use of MDN. The 3D positions of the three fingers predicted by DCNN$^+$ + SP are located at the center of the objects, whereas the grasp poses predicted by DCNN$^+$ + SP + MDN are similar to the ground truths. DCNN$^+$ + SP nearly predicted the 3D positions of the three fingers at the center of the object because in the training phase, such regression can minimize error between multiple grasp poses and a prediction.

5   Conclusions

In this paper, we proposed a deep neural network architecture to predict multiple 3D positions of three fingers of the hand that provide suitable pregrasps for a known or unknown object in various poses. To this end, we proposed a deep neural network that combines a variant of the traditional deep convolutional neural network and a mixture density network. Moreover, to overcome the difficulty of learning with insufficient data for the variant of the convolutional neural network, we developed a supervised learning technique to pretrain the network.

We evaluated the performance of the proposed deep neural network against two methods. The results established the effectiveness of our method. Specifically, the use of MDN made it possible to predict multiple pregrasp poses, and the supervised learning pretraining provided rich visual features to predict suitable pregrasp poses to be learned.

Acknowledgement. This work was supported by the <Technology Innovation Industrial Program> funded by the Ministry of Trade, (MI, South Korea) [10048320,

Technology Innovation Program], by the National Research Foundation of Korea grant funded by the Korea Government (MEST) (NRF-MIAXA003- 2010-0029744). All correspondences should be addressed to I.H. Suh.

References

1. Sahbani, A., El-Khoury, S., Bidaud, P.: An overview of 3D object grasp synthesis algorithms. Robot. Auton. Syst. **60**(3), 326–336 (2012)
2. Bohg, J., Morales, A., Asfour, T., Kragic, D.: Data-driven grasp synthesis — a survey. IEEE Trans. Robot. **30**(2), 289–309 (2014)
3. Pinto, L., Gupta, A., Supersizing self-supervision: learning to grasp from 50k tries and 700 robot hours. arXiv:1509.06825 (2015)
4. Levine, S., Peter, P., Alex, K., Deirdre, Q.: Learning hand-eye coordination for robotic grasping with deep learning, large-scale data collection. arXiv:1603.02199 (2016)
5. Levine, S., Finn, C., Darrell, T., Abbeel, P.: End-to-end training of deep visuomotor policies. JMLR **17**(39), 1–40 (2016)
6. Han, W., Levine, S., Abbeel, P.: Learning compound multi-step controllers under unknown dynamics. In: Intelligent Robots and Systems (IROS), pp. 6435–6442 (2015)
7. Krizhevsky, A., Sutskever, I., Hinton, G.: ImageNet classification with deep convolutional neural networks. In: NIPS, pp. 1097–1105 (2012)
8. Noh, H., Hong, S., Han, B.: Learning deconvolution network for semantic segmentation. In: Proceedings of the IEEE International Conference on Computer Vision, pp. 1520–1528 (2015)
9. Xu, K., Ba, J., Kiros, R., Courville, A., Salakhutdinov, R., Zemel, R., Bengio, Y. Show Attend, tell: Neural image caption generation with visual attention. ICML, vol. 37 of JMLR Proceedings, pp. 2048–2057, 2015
10. Bishop, C.: Mixture density networks. Technical Report NCRG/94/004, Neural Computing Research Group, Aston University (1994)
11. Detry, R., Baseski, E., Popovic, M., Touati, Y., Kruger, N., Kroemer, O., Piater, J.: Learning object-specific grasp affordance densities. In: IEEE International Conference on Development and Learning (ICDL), pp. 1–7 (2009)
12. Paolini, R., Rodriguez, A., Srinivasa, S.S., Mason, M.T.: A data-driven statistical framework for post-grasp manipulation. Int. J. Robot. Res. **33**(4), 600–615 (2014)
13. Lenz, I., Lee, H., Saxena, A.: Deep learning for detecting robotic grasps. Int. J. Robot. Res. **34**(4–5), 705–724 (2015)
14. Finn, C., Tan, X. Y., Duan, Y., Darrell, T., Levine, S., Abbeel, P.: Deep spatial autoencoders for visuomotor learning. In: IEEE International Conference on Robotics and Automation (2016)

Recurrent Neural Networks for Adaptive Feature Acquisition

Gabriella Contardo[1]([✉]), Ludovic Denoyer[1], and Thierry Artières[2]

[1] Sorbonne Universités, UPMC Univ Paris 06, UMR 7606, LIP6, 75005 Paris, France
`gabriella.contardo@lip6.fr`
[2] Aix Marseille Univ, CNRS, Centrale Marseille, LIF, Marseille, France

Abstract. We propose to tackle the cost-sensitive learning problem, where each feature is associated to a particular acquisition cost. We propose a new model with the following key properties: (i) it acquires features in an adaptive way, (ii) features can be acquired *per block* (several at a time) so that this model can deal with high dimensional data, and (iii) it relies on representation-learning ideas. The effectiveness of this approach is demonstrated on several experiments considering a variety of datasets and with different cost settings.

1 Introduction

The development of attention models [1,13,14] is a recent trend in the neural network (NN) community. It usually consists in adding an attention module to classical NN architectures which goal is to select relevant information to use for predicting instead of using the whole input. These models have been mainly developed for particular types of data i.e. images and text and are specific to the nature of the inputs. More generally, the objective of selecting relevant information is not new and different models have been proposed in the Machine Learning domain during the last decades e.g. L1 regularization (e.g. [3]) or dimensionality reduction techniques [10]. These works are mainly motivated by the need to not only select relevant input information – as it is the case for attention models – but also to limit the inference cost in applications where the acquisition or computation of input features is expensive. Applications like medical diagnosis or personalized predictive tasks are intuitive examples of such setting, where some input features can be very expensive (e.g. fMRI exams). One can also think of many today applications such as spam detection ([17]), web-search ([6,24]), where one wants to answer huge numbers of prediction per second, per minute or per day. In that cases, limiting the number of input features used for prediction is a key factor for an algorithm, in order to constraint the "cost" of the information used, while keeping robust prediction ability. An optimal strategy to limit this cost should rely on an adaptive feature acquisition process (as in attention models), i.e. selecting features according to what has been currently observed of the input, since it is quite likely that not all inputs require the knowledge of the same subset of features to perform an accurate prediction.

© Springer International Publishing AG 2016
A. Hirose et al. (Eds.): ICONIP 2016, Part III, LNCS 9949, pp. 591–599, 2016.
DOI: 10.1007/978-3-319-46675-0_65

We propose a new sequential model based on a recurrent neural network architecture to tackle this problem of cost-sensitive features selection. At each time-step, the model chooses which features to acquire and builds a representation of the partially observed input based on all the acquired information. This learned representation is then used to both drive the future acquisition steps, but also to compute a final prediction at the end of the acquisition process. Our algorithm is thus an adaptive one which is able to select different subsets of features depending on the input and is learned based on the objective to find a good trade-off between the average acquisition cost and the quality of the prediction. At the opposite to recent NN-based techniques, our model is not specific to a particular nature of the data, and the attention part of the model is guided by the cost of the different features. Moreover, our algorithm is able to acquire multiple features at each timestep making it scalable for dealing with high dimensional data. These key aspects – adaptiveness, ability to handle different cost for different features, scalability, and representation learning based approach – are, to the best of our knowledge, novel in regard of the state of the art (see Sect. 4).

The paper is organized as follows: Sect. 2 details the cost-sensitive acquisition problem and details our RNN model and experimental results are provided in Sect. 3. Section 4 situates our work with respect to state of the art.

2   Cost-Sensitive Recurrent Neural Network

We consider the generic problem of computing a prediction $y \in \mathbb{R}^Y$ based on an input $x \in \mathbb{R}^n$ where n is the dimensionality of the input space, y is an output vector, and Y is the dimension of the output space. x_i (resp y_i) denotes the i-th features of x (resp. y). We particularly focus on the classification task where Y is the number of possible categories, and $y_i = 1$ if the input belongs to category i and $y_i = -1$ elsewhere.

2.1   Recurrent ADaptive AcquisitIon Network (RADIN[1])

The generic principle of adaptive feature acquisition may be resumed as follows. A model starts by acquiring a first subset of features from an input x. Then, new features are iteratively acquired at each timestep based on what has already been observed, we note T the number of steps made by the model. The final prediction is then performed based on the set of acquired features. Many models can be cast in this formalism. Non-adaptive feature selection approach stands for one step models ($T = 1$), while a decision tree may be thought as starting in the root node and acquiring a new feature one at a time that depends on the node in the tree.

We propose in this work to instantiate this general framework with a Recurrent Neural Network architecture (see Fig. 1). At each timestep, the acquired

---

[1] In French, "radin" Means "skinflint".

information enrich a latent representation of the input, and further decisions (acquisition of new features and final prediction) are made based on this representation. The internal state of the RNN (i.e. a continuous vector in $z \subset R^p$, p being the dimensionality of the latent space) is used to encode the information gathered on an input sample x through a subset of observed features. It is initialized as the null vector $z_0 = 0^p$. It is enriched all along the acquisition process, yielding a series of representations, z_1, z_2... up to a final iteration T and a final representation z_T. The use of multiple steps enables data dependent feature acquisition. The final representation of x, z_T, is used to perform prediction. It is worth noting that z_T is built from a partial view of x, i.e. only a subset of its features have been observed when performing prediction.

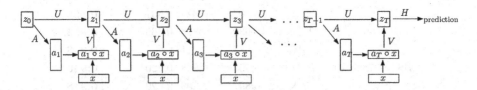

Fig. 1. Architecture of our recurrent acquisition network.

We discuss now the RNN architecture and how z_t's are updated. The underlying mechanism involve both an *attention layer* in charge of choosing which features to acquire, and an *aggregation layer* in charge of aggregating the newly acquired information to the previously collected features.

Attention Layer: While in classical RNN, the input at time t is usually a predetermined piece of the input (an element of an input sequence for example), in our case, this input is chosen by the model as a function of the previous state z_{t-1} in the following way: A specific *attention layer* computes a vector $a_t = f(A \times z_{t-1}) \in [0,1]^n$ whose component i denoted $a_{t,i}$ stands for the usefulness of feature i of the input denoted x_i. a_t is an attention vector that aims at selecting the features to acquire i.e. the features i such that $a_{t,i} > 0$. This vector is computed based on the previous representation z_{t-1} and different inputs will thus produce different values of the attention layer resulting in an adaptive acquisition model. f is typically a non-linear activation function and $A \in \mathbb{R}^{n \times p}$ corresponds to the parameters of the attention layer. In order to compute the input of the hidden layer, the attention vector is then "mixed" with the original input x by using the Hadamard product[2], the attention layer acting as a filter on the features of x. This input is denoted $x[a_t] = a_t \circ x$ in the following. Note that in the particular case where a_t would be a binary vector, this stands for a copy of x where features that should not be acquired are set to 0. This vector $x[a_t]$ is an additional input that is used to update the internal state, i.e. to compute z_t.

---

[2] Note that the Hadamard product is used during training since the training inputs are fully known. During inference on new inputs, the value of the Hadamard product is directly computed by only acquiring the chosen features.

Table 1. Results of the different models w.r.t percentage of features used on different datasets.

Corpus name	Nb. Feat.	Nb. Cat.	Feature used (%)	Model			
				SVM L_1	DT	GreedyMiser	RADIN
Cardio	21	10	90 %	0.683	0.775	**0.827**	0.824
			75 %	0.580	0.775	**0.825**	**0.825**
			50 %	0.496	0.775	0.751	**0.837**
			25 %	0.338	0.771	0.508	**0.775**
			10 %	0.259	0.643	0.325	**0.662**
Statlog	60	3	90 %	0.775	0.823	0.851	**0.859**
			75 %	0.741	0.823	0.846	**0.858**
			50 %	0.703	0.823	0.831	**0.858**
			25 %	0.630	0.823	0.765	**0.852**
			10 %	0.587	0.821	0.605	**0.833**
MNIST	780	10	90 %	0.897	0.808	0.920	**0.950**
			75 %	0.897	0.808	0.920	**0.948**
			50 %	0.882	0.808	0.903	**0.926**
			25 %	0.704	0.808	0.846	**0.920**
			10 %	0.577	0.808	0.776	**0.859**
gisette	5000	2	25 %	**0.970**	0.919	0.884	0.957
			10 %	**0.968**	0.919	0.884	0.957
			5 %	**0.963**	0.919	0.867	0.957
			1 %	0.910	0.919	0.785	**0.947**
r8	6224	8	25 %	**0.969**	0.901	0.948	0.962
			10 %	**0.968**	0.901	0.947	0.961
			5 %	0.951	0.901	0.945	**0.961**
			1 %	0.913	0.901	0.939	**0.959**
webkb	5388	4	25 %	0.891	0.793	0.861	**0.962**
			10 %	0.887	0.793	0.864	**0.961**
			5 %	0.859	0.793	0.857	**0.865**
			1 %	0.717	0.793	0.828	**0.831**

Aggregation Layer: Once newly features have been acquired, the internal state z_t is updated according to $z_t = f(U \times z_{t-1} + V \times x[a_t])$ (with U and V two weight matrices of size $p \times p$ and $p \times n$) as in classical RNN cells[3]. The internal state layer z_t is thus an aggregation of the information gathered from all previous acquisition steps up to step t.

---

[3] We also tested Gated Recurrent Unit ([8]).

Decision Layer: The final representation z_T, which is obtained after the T-acquisition step, is used to perform classification $o(x) = g(H \times z_T) \in R^Y$ with g a non linear function and H a weight matrix of size $Y \times p$, z_T being the representation of the input x at the end of the acquisition process.

Noting c_i the acquisition cost for feature i, $c_i \geq 0$ and $c \in \mathbb{R}^n$ the vector of all feature costs, the quantity $\sum_{t=1}^{T} a_t^\intercal.c$ stands for the actual acquisition cost provided that $a_{t,i}$ are actually binary values and that a feature cannot be acquired twice. In order to train the RNN to learn to acquire efficiently information from the inputs before classifying it we propose to optimize the weight parameters A, U, V, H to minimize the following empirical loss on a set of N training samples $(x^k, y^k)_{k=1..N}$:

$$\mathcal{J}^{emp}(A, U, V, H) = \sum_{k=1...N} \left[\Delta(o(x^k), y^k) + \lambda \sum_{t=1}^{T} a_t^\intercal.c \right] \qquad (1)$$

where the first term of the loss is a data fit term that measures how well prediction is performed on training samples and the second term is related to the constraint on the feature acquisition *budget*. In practice however, dealing with binary attention vectors a_t leads to a difficult optimization problem so that we use for our model continuous values $a_{t,i} \in [0,1]$ with a similar meaning i.e. x_i is acquired only if $a_{t,i} > 0$. In that case, the regularization term is an approximation of the *budget term* that acts as a penalty term that drives $a_{t,i}$ towards 0. This continuous relaxation is close to what it is usually done when using L_1-regularized models instead of L_0 ones.

It is worth mentioning here that the architecture we present allows feature acquisition to be performed **per block**, i.e. many features at a time (in one step), as a_t can have several non-null values. This is an interesting, and quite novel property with regard to state of the art methods for (cost-sensitive) problems, as it allows this model to scale well to data with a large number of features, reaching high accuracy while keeping a reasonable computational complexity of the process.

3   Experiments

This section provides results of various experiments on feature-acquisition problems with different cost-settings. We study the ability of our approach on several mono-label classification datasets[4]. Let us first describe our **experimental protocol** for validation (as our goal is both to optimize the accuracy as well as the acquisition cost, usual cross-validation protocol cannot be conducted here). Each dataset is split in training, validation and testing sets[5]. We then learn several

---

[4] Note that our approach also handles other problems such as multi-label classification, regression or ranking as long as the loss function Δ is differentiable.

[5] One third of the examples for each set, except for MNIST, where the split corresponds to 15%, 5%, 80% of the data.

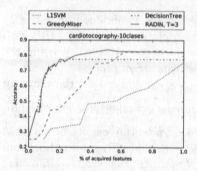

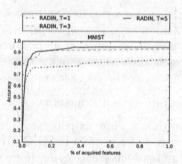

Fig. 2. Accuracy/Cost on *cardio*.

Fig. 3. RADIN with different T values on *MNIST*.

models with various hyper-parameters settings on the *training set*. Each learned model yields a two dimension point (accuracy, cost) on the *validation set*. The Pareto Front of this set of points is then computed to select the best models. At last, the selected models are evaluated on the *test set* on which results are reported. We used the following specifications for our model: a linear function for prediction o, GRU or RNN cells for the aggregation layer, and a hard logistic activation function for the attention layer. Δ is a mean-square error. The code used to conduct these experiments is available at http://github.com/ludc/radin. We compare our model with three different approaches: (i) a L_1-regularized linear **SVM**, (ii) a **Decision Tree**, (iii) a cost-sensitive method that constraints locally and globally the cost of a set of weak classifiers -decision trees- **GreedyMiser** ([22]). Note that the first two methods can't handle *cost-sensitive* problems. Due to a lack of space, we do not present here all the results obtained on many different datasets, and just focus on the more representative performance.

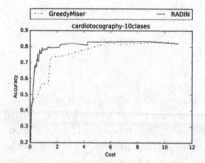

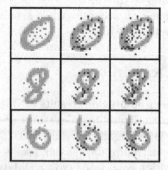

Fig. 4. Cost-sensitive setting on *cardio* dataset

Fig. 5. Illustration of the adaptive behavior of RADIN on three different MNIST inputs

Let us first focus on experiments with **uniform cost**, i.e. $\forall i, c_i = 1$. In this case, the acquisition cost is directly the number of features gathered. The cost is thus expressed as the percentage of features acquired w.r.t the total number of available features. Figure 2 illustrates the overall accuracy-cost curves for the dataset *cardio*. One can see for example that the GreedyMiser approach yields an accuracy of about 68 % for a cost of 0.4, i.e. acquiring 40 % of the features, while our model RADIN obtains approximately 82 % of accuracy for the same amount of acquired features. The results in Table 1 show the ability of our method to give competitive or better results on all datasets, including larger scale datasets, particularly in this case when the percentage of acquired features drops substantially.

The Fig. 3, which plots the accuracy/cost curves obtained with different number of acquisition steps T illustrates the adaptive behavior of our model and its ability to choose relevant features depending on the input. Moreover, one can see on Fig. 5 which features are acquired considering three particular inputs of the MNIST dataset. Each color corresponds to a particular acquisition step. If the features acquired at time $T = 1$ are the same, RADIN exhibits a different behavior for the following steps depending on the acquired information.

At last, we have considered a different cost-sensitive setting and show the performance obtained by RADIN and GreedyMiser on an artificial cost-sensitive dataset constructed from the *cardio* dataset, where the cost of feature i is defined as $c_i = \frac{i}{n}$. One can observe (Fig. 4) that our model yields better accuracy results than Greedy Miser for all the cost range, which indicates its ability to not only acquire the relevant features but also to integrate in the process their different costs c_i.

4   Related Work

Many methods have been proposed under a *static* features selection framework, i.e. with only one step of acquisition. A good overview of existing methods is provided in [10] which describe different approaches like *wrapper* methods [12] or *Embedded* methods [3,19] with l_1 and l_0-norm. Block feature selection has also been proposed but feature blocks have to be known beforehand [23]. Note that these approaches generally cannot handle non-uniform costs. Another family of algorithm proposes to tackle the features selection problem by estimating the information gain of the features [4] For example, [5] presents two greedy strategies to learn a test-cost sensitive naive Bayes classifier, while [18] propose to use reinforcement-learning to learn a value-function of the information gain. In the adaptive features selection literature, decision trees are naturally good candidates and they are used for example in [20,22] as several weak constraint classifiers. Another type of approach relies on learning *cascade* of classifiers, as in [7,16], the classifier being used depending on the input. More recently, [21] presented a method to learn a tree of classifiers which can be extended to cascade architecture inducing the possibility of *early-stopping*, which is an interesting aspect of adaptive prediction behavior. Feature acquisition can also be seen as a sequential decision process, and it has been studied under the MDP and Reinforcement Learning framework, as in [11], which models the problem as a partially observable MDP, or in [2,9,15] using classical RL algorithms. Here, these models usually suffer when the number of features is too large. At last, new deep learning models have recently emerged [1,13,14] and are closely related to our work. They consist in adding an attention mechanism to classical architecture. They have been mainly developed for images and text with the goal to increase the quality of the model. They thus don't consider a particular budget or cost-sensitive setting.

Regarding these various methods, our work differs on several aspects. It is, to the best of our knowledge, the first approach that tackles the adaptive cost-sensitive acquisition problem in a generic way with a RNN-like architecture.

5   Conclusion

We presented a recurrent neural network architecture to tackle the problem of adaptive cost-sensitive acquisition. Our approach can acquire the features per block and can be learned using efficient gradient descent algorithm. We showed that our model performs well on different problem settings and is able to acquire information resulting in a good cost/accuracy trade-off in an adaptive way.

Acknowledgements. This article has been supported within the Labex SMART supported by French state funds managed by the ANR within the Investissements d'Avenir programme under reference ANR-11-LABX-65. Part of this work has benefited from a grant from program DGA-RAPID, project LuxidX.

References

1. Ba, J., Mnih, V., Kavukcuoglu, K.: Multiple object recognition with visual attention. arXiv preprint arXiv:1412.7755 (2014)
2. Benbouzid, D., Busa-Fekete, R., Kégl, B.: Fast classification using sparse decision DAGs. In: ICML (2012)
3. Bi, J., Bennett, K., Embrechts, M., Breneman, C., Song, M.: Dimensionality reduction via sparse support vector machines. JMLR **3**, 1229–1243 (2003)
4. Bilgic, M., Getoor, L.: VOILA: efficient feature-value acquisition for classification. In: Proceedings of the National Conference on Artificial Intelligence (2007)
5. Chai, X., Deng, L., Yang, Q., Ling, C.X.: Test-cost sensitive naive Bayes classification. In: Data Mining, ICDM 2004 (2004)
6. Chapelle, O., Shivaswamy, P., Vadrevu, S., Weinberger, K., Zhang, Y., Tseng, B.: Boosted multi-task learning. Mach. Learn. **85**(1–2), 149–173 (2011)
7. Chen, M., Weinberger, K.Q., Chapelle, O., Kedem, D., Xu, Z.: Classifier cascade for minimizing feature evaluation cost. In: AISTATS, pp. 218–226 (2012)
8. Cho, K., van Merriënboer, B., Bahdanau, D., Bengio, Y.: On the properties of neural machine translation: encoder-decoder approaches. arXiv preprint arXiv:1409.1259 (2014)
9. Dulac-Arnold, G., Denoyer, L., Preux, P., Gallinari, P.: Sequential approaches for learning datum-wise sparse representations. Mach. Learn. **89**, 87–122 (2012)

10. Guyon, I., Elisseeff, A.: An introduction to variable and feature selection. JMLR **3**, 1157–1182 (2003)
11. Ji, S., Carin, L.: Cost-sensitive feature acquisition and classification. Pattern Recogn. **40**(5), 1474–1485 (2007)
12. Kohavi, R., John, G.H.: Wrappers for feature subset selection. Artif. Intell. **97**(1), 273–324 (1997)
13. Mnih, V., Heess, N., Graves, A., et al.: Recurrent models of visual attention. In: NIPS (2014)
14. Sermanet, P., Frome, A., Real, E.: Attention for fine-grained categorization. arXiv preprint arXiv:1412.7054 (2014)
15. Trapeznikov, K., Saligrama, V.: Supervised sequential classification under budget constraints. In: AISTATS (2013)
16. Viola, P., Jones, M.: Robust real-time object detection. Int. J. Comput. Vis. **4**, 51–52 (2001)
17. Weinberger, K., Dasgupta, A., Langford, J., Smola, A., Attenberg, J.: Feature hashing for large scale multitask learning. In: ICML. ACM (2009)
18. Weiss, D.J., Taskar, B.: Learning adaptive value of information for structured prediction. In: NIPS (2013)
19. Weston, J., Mukherjee, S., Chapelle, O., Pontil, M., Poggio, T., Vapnik, V.: Feature selection for SVMs. In: NIPS (2000)
20. Xu, Z., Huang, G., Weinberger, K.Q., Zheng, A.X.: Gradient boosted feature selection. In: ACM SIGKDD (2014)
21. Xu, Z., Kusner, M.J., Weinberger, K.Q., Chen, M., Chapelle, O.: Classifier cascades and trees for minimizing feature evaluation cost. JMLR **15**, 2113–2144 (2014)
22. Xu, Z., Weinberger, K., Chapelle, O.: The greedy miser: learning under test-time budgets. arXiv preprint arXiv:1206.6451 (2012)
23. Yuan, M., Lin, Y.: Efficient empirical Bayes variable selection and estimation in linear models. J. Am. Stat. Assoc. **100**, 100–1215 (2005)
24. Zheng, Z., Zha, H., Zhang, T., Chapelle, O., Chen, K., Sun, G.: A general boosting method and its application to learning ranking functions for web search. In: NIPS (2008)

Stacked Robust Autoencoder for Classification

Janki Mehta, Kavya Gupta, Anupriya Gogna, Angshul Majumdar[(✉)],
and Saket Anand

Indraprastha Institute of Information Technology, Delhi, India
{mehta1485,kavya1482,anupriyag,
angshul,anands}@iiitd.ac.in

Abstract. In this work we propose an l_p-norm data fidelity constraint for
training the autoencoder. Usually the Euclidean distance is used for this purpose;
we generalize the l_2-norm to the l_p-norm; smaller values of p make the problem
robust to outliers. The ensuing optimization problem is solved using the Augmented Lagrangian approach. The proposed l_p -norm Autoencoder has been
tested on benchmark deep learning datasets – MNIST, CIFAR-10 and SVHN.
We have seen that the proposed robust autoencoder yields better results than the
standard autoencoder (l_2-norm) and deep belief network for all of these
problems.

Keywords: Autoencoder · Deep learning · Classification · Robust estimation

1 Introduction

An autoencoder learns the analysis and the synthesis weights by minimizing the
l_2-norm between the input (training samples) and the output (training samples /corrupted
training samples). The l_2-norm is perhaps the most widely used data fidelity constraint in
signal processing and machine learning. It arises from the Gaussian /Normal assumption
of the distribution which fits a large class of problems in practice. But the practical
reason behind popularity of the l_2-norm stems from the fact that it is easy to solve; it is
smooth and convex and has a closed form solution (for linear problems).

The l_2-norm minimization works when the deviations are small – approximately
Normally distributed; but fail when there are large outliers. In statistics there is a large
body of literature on robust estimation. The Huber function [1] has been in use for more
than half a century in this respect. The Huber function is an approximation of the more
recent absolute distance based measures (l_1-norm). Recent studies in robust estimation
prefer minimizing the l_1-norm instead of the Huber function [2,3,4]. The l_1-norm does
not bloat the distance between the estimate and the outliers and hence is robust.

The problem with minimizing the l_1-norm is computational. However, over the
years various techniques have been developed. The earliest known method is based on
Simplex [5]; Iterative Reweighted Least Squares [6] used to be another simple yet
approximate technique. Other approaches include descent based method introduced by
[7] and Maximum Likelihood approach [8].

In this work, we propose a generalized l_p -norm autoencoder, for values of p between
0 and 1, l_p -norm fidelity is robust to outliers; l_p -norm is quasi-convex. Unfortunately

© Springer International Publishing AG 2016
A. Hirose et al. (Eds.): ICONIP 2016, Part III, LNCS 9949, pp. 600–607, 2016.
DOI: 10.1007/978-3-319-46675-0_66

this makes the problem non-differentiable; hence the standard gradient descent based techniques (e.g. backpropagation) cannot be applied here. One needs to solve it using sub-gradients. We invoke a state-of-the-art optimization approach to solve the ensuing problem; this is called the variable splitting augmented Lagrangian. This reduces our problem to a few simpler sub-problems; one of which needs to be solved using sub-gradients while the rest have an analytic solution. We test our proposed approach with standard autoencoder and deep belief network for benchmark problems in classification; we show that our results are indeed better.

The rest of the paper is organized into several sections. Section 2 describes our proposed approach. The experimental results are shown in Sect. 3. The conclusion of this work is discussed in Sect. 4.

2   Proposed Robust Autoencoder

An autoencoder consists of two parts (as seen in Fig. 1) – the encoder maps the input to a latent space, and the decoder maps the latent representation to the data [9, 10]. For a given input vector (including the bias term) x, the latent space is expressed as:

$$h = \phi(Wx) \qquad (1)$$

Here the rows of W are the link weights from all the input nodes to the corresponding latent node. The activation function is usually non-linear (sigmoid /tanh).

The decoder portion reverse maps the latent variables to the data space.

$$x = W'\phi(Wx) \qquad (2)$$

Since the data space is assumed to be the space of real numbers, there is no sigmoidal function here.

During training the problem is to learn the encoding and decoding weights – W and W'. In terms of signal processing lingo, W is the analysis operator and W' is the synthesis operator. These are learnt by minimizing the l_2-norm data fidelity constraint:

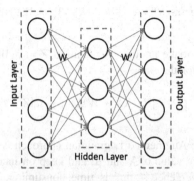

Fig. 1. Basic autoencoder

$$\arg\min_{W,W'} \|X - W'\phi(WX)\|_F^2 \tag{3}$$

Here $X = [x_1|\ldots|x_N]$ consists all the training sampled stacked as columns. The problem (4) is clearly non-convex. But can be solved by gradient descent techniques since the usual activation functions are smooth and continuously differentiable.

We do not change the autoencoder architecture. We only change the data fidelity constraint from l_2-norm to l_p-norm. This follows from our discussion on robust estimation. The p-norm is more generic and for values of p between 0 and 1; the estimation is more robust. There is a prior study on denoising autoencoders [11] which add noise to samples and then learn a denoising autoencoder; the goal is to learn robust encoding and decoding weights. Although intuitive, this study is at best heuristic. The robustness arising from our proposed formulation is mathematically and statistically optimal. The formulation we propose is:

$$\arg\min_{W,W'} \|X - W'\phi(WX)\|_p^p \tag{4}$$

The l_p-norm is not differentiable everywhere. Hence gradient based techniques cannot be applied. One need to compute sub-gradient. In this work, we propose to solve this problem using the Augmented Lagrangian approach.

First we substitute, $P = X - W'\phi(WX)$; thus converting (4) to the following,

$$\arg\min_{P,W,W'} \|P\|_p^p \text{ such that } P = X - W'\phi(WX) \tag{5}$$

The unconstrained Lagrangian is given by,

$$\arg\min_{P,W,W'} \|P\|_1 + L^T(P - (X - W'\phi(WX))) \tag{6}$$

The Lagrangian imposes equality at every step; this is too stringent a requirement in practice. One can relax the equality constraint initially and enforce it only during convergence. This is the Augmented Lagrangian formulation (7),

$$\arg\min_{P,W,W'} \|P\|_1 + \lambda\|P - (X - W'\phi(WX))\|_F^2 \tag{7}$$

In the next step, we make another substitution $Z = \phi(WX)$ and write down the Augmented Lagrangian for the same.

$$\arg\min_{P,W,W',Z} \|P\|_1 + \lambda\|P - (X - W'Z)\|_F^2 + \mu\|Z - \phi(WX)\|_F^2 \tag{8}$$

The problem with the Augmented Lagrangian approach is that, one needs to solve the full problem for every value of λ and μ; and keep on increasing them in order to enforce equality at convergence – this is time consuming. Besides, increasing the values of these hyper-parameters is heuristic. A more elegant approach is to introduce Bregman relaxation variables B_1 and B_2 [12].

$$\arg\min_{P,W,W',Z}\|P\|_1 + \lambda\|P - (X - W'Z) - B_1\|_F^2 + \mu\|Z - \phi(WX) - B_2\|_F^2 \qquad (9)$$

Although this problem is not completely separable, we can segregate (9) into alternate minimization of the following subproblems.

$$P1: \arg\min_{P}\|P\|_1 + \lambda\|P - (X - W'Z) - B_1\|_F^2 \qquad (10)$$

$$P2: \arg\min_{W}\|Z - \phi(WX) - B_2\|_F^2 \equiv \arg\min_{W}\|\phi^{-1}(Z - B_2) - WX\|_F^2 \qquad (11)$$

$$P3: \arg\min_{W'}\|P - (X - W'Z) - B_1\|_F^2 \qquad (12)$$

$$P4: \arg\min_{Z}\|Z - \phi(WX) - B_2\|_F^2 \qquad (13)$$

Subproblems P2-P4 are simple linear least squares problems. They have analytic solutions in the form of pseudo-inverse. Subproblem P1 is an l_p-minimization problem. This too has a closed form solution in the form of modified soft thresholding [13], given by.

$$P = signum(X + B_1 - W'Z)\max\left(0, |X + B_1 - W'Z| - \frac{\mu}{2}p|P|^{p-1}\right) \qquad (14)$$

The last step is to update the Bregman relaxation variables:

$$B_1 \leftarrow P - (X - W'Z) - B_1 \qquad (15)$$

$$B_2 \leftarrow Z - \phi(WX) - B_2 \qquad (16)$$

The problem is non-convex thus there is no guarantee of reaching a global optimum. In this case, we continue the iterations till the objective function does not change significantly in subsequent iterations. We also have a cap on the maximum number of iterations; we have kept it to be 50.

3   Experimental Evaluation

To test our formulation we used three datasets, MNIST, Street View House Numbers (SVHN) and CIFAR-10. The MNIST dataset is a handwriting recognition dataset developed by Y. LeCun et al. using the larger NIST dataset. It has 60,000 images of handwritten digits, which were used as training images and 10,000 images were used as test images. The SVHN dataset is obtained from Google Street View Images dataset.

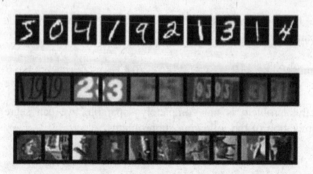

Fig. 2. Samples from datasets. Top – MNIST, Middle – SVHN, Bottom – CIFAR

It also involves recognition of digits, like the MNIST, however it is significantly harder to do so because of clustering of nearby digits and variety of backgrounds. It is a real world problem of recognizing the digits from natural scene images. It is a colored images database, with 73,257 images for training, and 26,032 images for test. There are also 531,131 simpler training images; however we do not use them. We use format 2 of the dataset, which is like the MNIST dataset. The CIFAR-10 dataset was compiled by Alex Krizhevsky et al. from the 80 million tiny images dataset. This dataset contains 50,000 32×32 training images with 10 classes which are mutually exclusive. CIFAR-10 contains images from various categories such as ship, frog, truck and more. This dataset contains 10,000 test images (Fig. 2).

Preprocessing

SVHN. We contrast normalize the Y channel of the YUV images of the dataset and use only the Y channel for training and classification. The Y channel is locally contrast normalized using a Gaussian neighborhood, with a 7×7 window. This made the images look more like the MNIST database. The resultant images reside in a $\mathbf{R}^{1024}$ space. From Fig. 3 we see that the Y channel contains the shape information in a clear and precise manner as compared to the U and V channels. Figure 3c, shows the preprocessed Y channels of the SVHN dataset. We only use the Y channel for training. The same preprocessing is applied to the test set before the classification step.

CIFAR-10. From each pixel we subtracted the mean of the image for all images in the dataset. This suppressed the brightness variation in the image. The resulting image is converted to greyscale. This is a very challenging problem as mentioned in [14].

MNIST. No preprocessing was required on this dataset.

Results

We show that by using the l_p-autoencoder, we can improve upon the (SAE) standard autoencoder (l_2-norm). The stacked autoencoders (both l_p-norm and l_2-norm) are of three levels; the number of nodes are halved in every successive level. The value of p is kept at 1 for the layers. Other combinations of p might yield better results, but we did not have time to test these. For the sake of comparison we also employ the deep

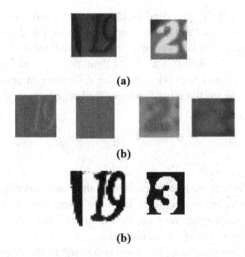

(a)

(b)

(b)

Fig. 3. (a) Y channels of SVHN, (b) U and V channels, (c) SVHN preprocessed images

belief network (DBN) for feature extraction; here also the number of nodes is halved in every successive.

We choose to use two non-parametric classifiers KNN and Sparse Representation based Classifier (SRC) [15], and a parametric classifier – SVM with RBF kernel. The SVM was tuned (via grid search) to yield the best results for proposed l_p-autoencoder, SAE and DBN (Tables 1, 2, and 3).

Table 1. Classification with KNN (K = 1)

Dataset	Proposed	SAE	DBN
MNIST	**97.44**	97.33	97.05
CIFAR-10	**50.01**	45.02	48.49
SVHN	**67.23**	63.93	65.70

Table 2. Classification with SRC

Dataset	Proposed	SAE	DBN
MNIST	**98.36**	98.33	88.43
CIFAR-10	**52.37**	45.11	46.38
SVHN	**69.90**	65.70	66.82

Table 3. Classification with SVM

Dataset	Proposed	SAE	DBN
MNIST	**98.64**	97.05	88.44
CIFAR-10	**53.29**	46.78	48.04
SVHN	**71.19**	66.42	68.01

The results show that the proposed stacked l_p-autoencoder always yields the best results. MNIST is a simple dataset, so the improvement on this dataset is not much. But for other datasets the difference between the proposed robust autoencoder and the non-robust version is significant. This is evident from the results between stacked autoencoder and deep belief network – here the difference in accuracy is marginal – between 1 % and 1.5 %; on the other hand our method improves over these by 4 % or more.

4   Conclusion

In this work we make a fundamental change to the basic autoencoder cost function. Instead of using the popular Euclidean norm to learn the encoding and decoding weights, we propose employing the l_p-norm. For small values of p (less than 1) – this makes the autoencoder more robust to outliers.

Minimizing the l_p-norm is more involved compared to the l_2-norm; this is because unlike the later, l_p-norm is not differentiable everywhere and hence gradient based techniques cannot be applied directly. We solve it using variable splitting and Augmented Lagrangian. This segregates the problem into several sub-problems; one of which needs to be solved using sub-gradient based techniques while the others have simple least squares solutions.

We carry out experiments on three benchmark deep learning datasets – MNIST, CIFAR-10 and SVHN. Three classifiers (KNN, SRC and SVM) were tested upon. In all three cases the proposed method yields the best results compared to the standard l_2-autoencoder and the deep belief network (DBN).

One may get better results by incorporating convolutional techniques into autoencoder [16] and DBN [17]; that is an entirely different direction of research and beating those results is not the goal of this work. Rather, our goal is to show that by moving from a non-robust Euclidean norm to a robust l_p-norm, one can achieve significant improvement in classification accuracy. Our technique does not bar incorporating convolutional techniques in our proposed robust framework; we plan to work on this topic in the future and expect to achieve further improvement in results.

References

1. Huber, P.J.: Robust estimation of a location parameter. Ann. Math. Stat. **35**(1), 73–101 (1964)
2. Branham Jr., R.L.: Alternatives to least squares. Astron. J. **87**, 928–937 (1982)
3. Shi, M., Lukas, M.A.: An L1 estimation algorithm with degeneracy and linear constraints. Comput. Stat. Data Anal. **39**(1), 35–55 (2002)
4. Wang, L., Gordon, M.D., Zhu, J.: Regularized least absolute deviations regression and an efficient algorithm for parameter tuning. In: IEEE ICDM, pp. 690–700 (2006)
5. Barrodale, I., Roberts, F.D.K.: An improved algorithm for discrete L1 linear approximation. SIAM J. Numer. Anal. **10**(5), 839–848 (1973)

6. Schlossmacher, E.J.: An iterative technique for absolute deviations curve fitting. J. Am. Stat. Assoc. **68**(344), 857–859 (1973)
7. Wesolowsky, G.O.: A new descent algorithm for the least absolute value regression problem. Commun. Stat. Simul. Comput. **B10**(5), 479–491 (1981)
8. Li, Y., Arce, G.R.: A maximum likelihood approach to least absolute deviation regression. EURASIP J. Appl. Sig. Process. **12**, 1762–1769 (2004)
9. Rumelhart, D.E., Hinton, G.E., Williams, R.J.: Learning representations by back-propagating errors. Nature **323**, 533–536 (1986)
10. Baldi, P., Hornik, K.: Neural networks and principal component analysis: learning from examples without local minima. Neural Netw. **2**, 53–58 (1989)
11. Vincent, P., Larochelle, H., Lajoie, I., Bengio, Y., Manzagol, P.A.: Stacked denoising autoen coders: learning useful representations in a deep network with a local denoising criterion. J. Mach. Learn. Res. **11**, 3371–3408 (2010)
12. Chartrand, R.: Nonconvex splitting for regularized low-rank + sparse decomposition. IEEE Trans. Sig. Process. **60**, 5810–5819 (2012)
13. Majumdar, A., Ward, R.K.: On the choice of compressed sensing priors: an experimental study. Sig. Process. Image Commun. **27**(9), 1035–1048 (2012)
14. Rifai, S., Vincent, P., Muller, X., Glorot, X., Bengio, Y.: Contractive auto-encoders: explicit invariance during feature extraction. In: ICML (2011)
15. Wright, J., Yang, A.Y., Ganesh, A., Sastry, S.S., Ma, Y.: Robust face recognition via sparse representation. IEEE Trans. Pattern Anal. Mach. Intell. **31**, 210–227 (2009)
16. Masci, J., Meier, U., Cireşan, D., Schmidhuber, J.: Stacked convolutional auto-encoders for hierarchical feature extraction. In: Honkela, T. (ed.) ICANN 2011, Part I. LNCS, vol. 6791, pp. 52–59. Springer, Heidelberg (2011)
17. Lee, H., Grosse, R., Ng, A.: Convolutional deep belief networks for scalable unsupervised learning of hierarchical representations. In: ICML (2009)

Pedestrian Detection Using Deep Channel Features in Monocular Image Sequences

Zhao Liu[1,2], Yang He[1], Yi Xie[1,2], Hongyan Gu[1], Chao Liu[1],
and Mingtao Pei[1(✉)]

[1] Beijing Lab of Intelligent Information, School of Computer Science,
Beijing Institute of Technology, Beijing 100081, People's Republic of China
{zhao.liu,heyang614,yxie.lhi,guhongyan,liuchao,peimt}@bit.edu.cn
[2] People's Public Security University of China,
Beijing 100038, People's Republic of China

Abstract. In this paper, we propose the Deep Channel Features as an extension to Channel Features for pedestrian detection. Instead of using hand-crafted features, our method automatically learns deep channel features as a mid-level feature by using a convolutional neural network. The network is pretrained by the unsupervised sparse filtering and a group of filters is learned for each channel. Combining the learned deep channel features with other low-level channel features (i.e. LUV channels, gradient magnitude channel and histogram of gradient channels) as the final feature, a boosting classifier with depth-2 decision tree as the weak classifier is learned. Our method achieves a significant detection performance on public datasets (i.e. INRIA, ETH, TUD, and CalTech).

Keywords: Pedestrian detection · Deep learning · Mid-level features · Deep channel features

1 Introduction

Pedestrian detection is a key problem in computer vision with many potential applications. There are many pedestrian detection methods such as deformable part based models [1], feature fusion models [2–4] and deep neural network models [5–7]. Features are the key elements to the detection task. Channel features [8] provide a possible representation of object identities for fast computation. Among all possible channels, LUV, HOG and gradient histogram channels achieve high performance in pedestrian detection [8]. However, channel features are low-level features, and lack semantic information of the objects to be detected. To improve detection performance, mid-level or even high-level features, which contain more task-oriented information, need to be combined with the low-level channel features.

In this paper, we propose the Deep Channel Features as an extension to Channel Features for pedestrian detection. Instead of using hand-crafted visual features, our method automatically learns deep channel features as mid-level

© Springer International Publishing AG 2016
A. Hirose et al. (Eds.): ICONIP 2016, Part III, LNCS 9949, pp. 608–615, 2016.
DOI: 10.1007/978-3-319-46675-0_67

feature by using a convolutional neural network. The network is pre-trained by the unsupervised sparse filtering and a group of filters is learned for each channel. Combining the learned deep channel features with other low-level channel features as the final feature, a boosting classifier with depth-2 decision tree as the weak classifiers is learned for pedestrian detection.

2   Feature Learning

2.1   Convolutional Neural Networks

CNNs were first promoted in hand-writing recognition by LeCun Yann [9]. In CNNS, a node is only related to its neighbor nodes, and the input signal is convoluted to obtain the output signal. Therefore the connection number is reduced greatly and the computation speed is improved. The basic operations in CNNs include convolution and pooling. The convolutional layer computes convolutions of the input with a series of filters. Many patches are extracted from the input to learn the filters. The learnt filters mainly preserve the edge, point and junction information. The pooling layer exacts one output from a small patch of the feature maps. Pooling reduces the feature dimensionality and represses the local distortion.

Besides convolution and pooling, some non-linear projections are proved to be helpful to improve the performance of CNNs, such as absolute value rectification and local contrast normalization.

The overall network structure is a multilayer multistage connected network. The output of each layer is fed as the input of the next layer. The output of each stage is used as the input of next stage. To utilize the features in multistage, the outputs of each stage are combined to form the final output. As show in Fig. 1, we use a two-stages network, which generate low-level local and high-level global features respectively.

2.2   Sparse Filtering

Sparse Filtering is an unsupervised feature learning algorithm which optimizes for the feature distribution. We use sparse filtering to lean the filters for the CNN.

Define a data matrix $X = \{x_1, x_2, ..., x_n\} \in R^{d \times n}$, where each column is a data point. Define $F = R^{t \times n}$ a feature distribution matrix over X, where each column is a sample and each row is a feature. F_{ij} is the response of the i-th feature on the j-th sample. The goal is to obtain a $S = [s_1, s_2, ..., s_n] \in R^{d \times t}$ which satisfies $F = S^T X$. Thus $F_{ij} = S_i^T X_j$. Then, each column of S can be regarded as a filter.

Let $f_{i,\Delta} \in R^{1 \times n}(i = 1, 2, ..., t)$ represents the i-th row of F, $f_{\Delta,j} \in R^{t \times 1}(j = 1, 2, ..., n)$ represents the j-th column of F. The objective function of the sparse filtering method is computed as:

1. Normalizing F by rows. Each feature is divided by its L2 norm across all examples: $\tilde{f}_{i,\Delta} = f_{i,\Delta}/\|f_{i,\Delta}\|_2$

2. Normalizing F by columns. Each sample is divided by its L2 norm across all features: $\overline{f}_{\Delta,i} = \tilde{f}_{\Delta,j}/\left\|\tilde{f}_{\Delta,j}\right\|_2$.

3. Sum up all entries in $\hat{F}$

The objective function of sparse filtering is

$$\min_s \left\|\hat{F}\right\| = \sum_{i=1}^{t} \sum_{j=1}^{n} \left|\hat{F}_{ij}\right| \tag{1}$$

which equals to

$$\min_x \sum_{j=1}^{n} \left\|\overline{f}_{\Delta,j}\right\|_1 = \sum_{j=1}^{n} \left\|\frac{\tilde{f}_{\Delta,j}}{\left\|\tilde{f}_{\Delta,j}\right\|_2}\right\|_1 \tag{2}$$

The minimization is achieved by L-BFGS [10].

2.3 Deep Channel Features

Aggregated Channel Feature (ACF) [16] is one of the channel features. ACF can be computed by one filtering and one subsampling, and is more efficient than integral image. The basic layers in a CNN include: filtering, non-linear projection, pooling and subsampling. Filtering and nonlinear projection can be regarded as a more complex nonlinear projection, which contains certain characteristic of the image. Thus, the computation structures of ACF and CNN share certain degree of similarity. Both contain non-linear projection and subsampling stages. However, ACF uses a hand-crafted projection, while the CNNs learns the filters from the input data.

We define a Deep Channel of the input image as one or more further transformations on a certain image channel. The depth of deep channel can be arbitrary. Figure 2 shows the deep channel features with depth of 2. To reduce the computational cost and memory requirement, we use two stages deep channel features in this paper. The first stage contains 10 channels (LUV, gradient magnitude and HOG channels). These channels are the inputs to the second stage. We use sparse filtering to learn the filters for the CNN. During the training process, a large amount of image fragments in each channel of the first stage is selected to learn the filters by LBFGS optimization. After processing all the 10 channels, the CNN learns a filter bank of 80 filters.

3 Detection

The 80 deep channels, 3 LUV channels, 1 gradient magnitude channel and 6 HOG channels are combined as the final feature. Then a boosting classifier with 2048 weak classifiers is learned. Each weak classifier is a depth-2 decision tree. Incremental and cascade training strategy is employed to train the classifier.

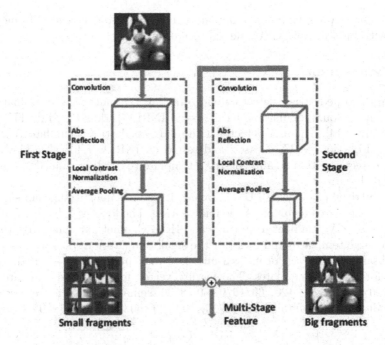

Fig. 1. The architecture of our two-stages convolutional neural network. The output of first stage is used as the input of the next stage to compute a higher lever features.

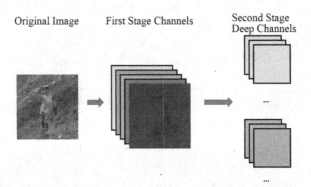

Fig. 2. Deep channel features with depth of 2. The deep features are extract from the first stage channels.

False positives of previous stage are added to the training set of current stage. The number of weak classifiers is increased gradually from 32, 128, 512 to 2048, 2048, 2048 in the 6 stages C0 to C5.

We use non-maximal suppression on the positive response windows for the final detection. When the overlap ratio of two detection windows is less than

0.6, the lowerscored window is eliminated. The algorithm achieves 2 frame/s on a PC with Intel i5 2.5G CPU and 4G memory.

4   Experiments

We train two pedestrian detectors with different template size to evaluate our algorithm on four popular public datasets: INRIA [11], ETH [12], TUD [13] and CalTech [14]. The first detector with a large pedestrian template (120×50) is trained on the INRIA dataset, and tested on INRIA, ETH and TUD. The second detector with a small pedestrian template (64×32) is trained and tested on CalTech dataset.

The 80 deep channels, 3 LUV channels, 1 gradient magnitude channel and 6 HOG channels are combined as the final feature. The low level channel features, namely the LUV, gradient magnitude and HOG channels, are blurred with 3×3 (for the large template) and 2×2 (for the small template) average windows, and are subsampled by step of 4. 3×3 and 2×2 max pulling is used in the CNN model for the two templates. The feature vector for the large template contains $120 \times 60/16 \times 10 + 120 \times 60/16/9 \times 80 = 8500$ elements. The feature vector for the small template contains $64 \times 32/16 \times 10 + 120 \times 60/16/4 \times 80 = 1950$ elements.

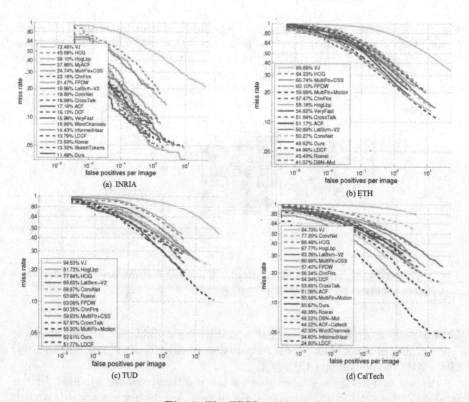

Fig. 3. The FPPI curves.

INRIA

ETH

TUD

CalTech

Fig. 4. Some detection results of our detector on the four public datasets.

Table 1. Compare results of miss rate.

Algorithm	Classifier	INRIA [%]	ETH [%]	TUD [%]	CalTech [%]
ChnFtrs	Boosting	22.18	57.47	60.35	56.34
FPDW	Boosting	21.47	60.10	63.08	57.40
Crosstalk	Boosting	18.98	51.94	57.97	53.88
ACF	Boosting	17.16	51.17	/	44.22
VeryFast	Boosting	15.96	54.82	/	/
ConvNet	Logistic regression	19.89	50.27	68.87	77.20
Ours	Boosting	11.48	48.62	52.61	50.67

We compare our algorithm with many classic and state of-the-art algorithms: VJ [15], HOG [11], HogLbp [2], ChnFtrs [8], ACF [16], ACF-CalTech [16], ConvNet [17], WordChannels [18], InformedHaar [19], MultiFtr+CSS [20], MultiFtr+Motion [20], LatSvm-v2 [1], SkethTokens [21], Roerei [22], LDCF [23], DBN-Mut [24], FPDW [25], Crosstalk [26], and VeryFast [27]. The evaluation method in the pedestrian benchmark [14] is used to compare the performance.

The threshold of overlap ratio is set as 0.5 for a true positive detection. Figure 3 shows the False Positive Per Image (FPPI) curve. Our method achieves outstanding performance on both INRIA and TUD datasets, and comparable result on ETH and CalTech datasets. Figure 4 shows some detection results of our pedestrian detector.

By employing deep channel features, our algorithm significantly outperforms the algorithms using channel feature solely, e.g. ChnFtrs, FPDW, Crosstalk, ACF, VeryFast and ConvNet. Deep features extend the representation capacity and improve the robust of the model. The miss rate of the above algorithms are listed in Table 1.

5   Conclusion

In this paper, we propose a novel feature representation called deep channel feature which processes the high level information in pedestrian detection in monocular image sequences without any depth and motion information. We successfully apply the deep features in pedestrian detection. By including the deep channel features, our method outperforms the methods with channel features alone. The proposed pedestrian detectors also outperform many current state-of-the-art algorithms in both effectiveness and efficiency on public datasets.

Acknowledgments. This work was supported in part by the Natural Science Foundation of China (NSFC) under Grant No. 61472038 and No. 61375044.

References

1. Felzenszwalb, P.F., Girshick, R.B., McAllester, D., Ramanan, D.: Object detection with discriminative trained part based models. IEEE Trans. Pattern Anal. Mach. Intell. **32**(9), 1627–1645 (2010)
2. Wang, X., Han, T.X., Yan, S.: An HOG-LBP human detector with partial occlusion handling. In: 2009 IEEE 12th International Conference on Computer Visionn, ICCV, pp. 32–39 (2009)
3. Zhang, S., Benenson, R., Schiele, B.: Filtered channel features for pedestrian detection. arXiv preprint arXiv:1501.05759 (2015)
4. Liu, W., Yu, B., Duan, C., et al.: A pedestrian-detection method based on heterogeneous features and ensemble of multi-view? Pose Parts. IEEE Trans. Intell. Transp. Syst. **16**(2), 813–824 (2015)
5. Luo, P., Tian, Y., Wang, X., Tang, X.: Switchable deep network for pedestrian detection. In: Conference on Computer Vision and Pattern Recognition, pp. 899–906 (2014)
6. Angelova, A., Krizhevsky, A., Vanhoucke, V.: Pedestrian detection with a large-field-of-view deep network. In: 2015 IEEE International Conference on Robotics and Automation (ICRA), pp. 704–711. IEEE (2015)
7. Cai, Z., Saberian, M., Vasconcelos, N.: Learning complexity-aware cascades for deep pedestrian detection. In: Proceedings of the IEEE International Conference on Computer Vision, pp. 3361–3369 (2015)

8. Dollár, P., Tu, Z., Perona, P., Belongie, S.: Integral channel features. In: BMVC 2009, London, England, pp. 1–11 (2009)
9. LeCun, Y., Bottou, L., Bengio, Y., Haffner, P.: Gradientbased learning applied to document recognition. Proc. IEEE **86**(11), 2278–2323 (1998)
10. Schmidt, M.: minFunc (2005). http://www.cs.ubc.ca//schmidtm/Software/minFunc.html
11. Dalal, N., Triggs, B.: Histograms of oriented gradients for human detection. In: CVPR Proceedings of 2005 IEEE Computer Society Conference on Computer Vision and Pattern Recognition, vol. 1, pp. 886–893 (2005)
12. Ess, A., Leibe, B., Schindler, K., Van Gool, L.: A mobile vision system for robust multi-person tracking. In: IEEE Conference on Computer Vision and Pattern Recognition, CVPR 2008, pp. 1–8 (2008)
13. Wojek, C., Walk, S., Schiele, B.: Multi-cue onboard pedestrian detection. In: IEEE Computer Society Conference on Computer Vision and Pattern Recognition Workshops, pp. 794–801 (2009)
14. Dollár, P., Wojek, C., Schiele, B., Perona, P.: Pedestrian detection: an evaluation of the state of the art. IEEE Trans. Pattern Anal. Mach. Intell. **34**(4), 743–761 (2012)
15. Viola, P., Jones, M.J., Snow, D.: Detecting pedestrians using patterns of motion and appearance. Int. J. Comput. Vis. **63**(2), 153–161 (2005)
16. Dollar, P., Appel, R., Belongie, S., Perona, P.: Fast feature pyramids for object detection. IEEE Trans. Pattern Anal. Mach. Intell. **36**(8), 1532–1545 (2014)
17. Sermanet, P., Kavukcuoglu, K., Chintala, S., Lecun, Y.: Pedestrian detection with unsupervised multi-stage feature learning. In: 2013 IEEE Conference on Computer Vision and Pattern Recognition, pp. 3626–3633 (2013)
18. Costea, A.D.: Word channel based multiscale pedestrian detection without image resizing and using only one classifier. In: CVPR, pp. 4321–4328 (2014)
19. Zhang, S., Bauckhage, C., Cremers, A.B.: Informed haar-like features improve pedestrian detection. In: CVPR 2014, pp. 947–954 (2014)
20. Walk, S., Majer, N., Schindler, K., Schiele, B.: New features and insights for pedestrian detection. In: Proceedings of IEEE Computer Society Conference on Computer Vision and Pattern Recognition, pp. 1030–1037 (2010)
21. Lim, J.J., Zitnick, C.L., Dollar, P.: Sketch tokens: a learned mid-level representation for contour and object detection. In: Proceedings of IEEE Computer Society Conference on Computer Vision and Pattern Recognition, pp. 3158–3165 (2013)
22. Benenson, R., Mathias, M., Tuytelaars, T., Van Gool, L.: Seeking the strongest rigid detector. In: Proceedings of the IEEE Computer Society Conference on Computer Vision and Pattern Recognition, pp. 3666–3673 (2013)
23. Nam, W., Dollár, P., Han, J.H.: Local decorrelation for improved detection. In: NIPS, pp. 1–9 (2014)
24. Ouyang, W., Zeng, X., Wang, X.: Modeling mutual visibility relationship in pedestrian detection. In: Proceedings of IEEE Computer Society Conference on Computer Vision and Pattern Recognition, vol. 1, pp. 3222–3229 (2013)
25. Dollar, P., Belongie, S., Perona, P.: The fastest pedestrian detector in the west. In: Proceedings of British Machine Vision Conference 2010, pp. 1–68 (2010)
26. Dollár, P., Appel, R., Kienzle, W.: Crosstalk cascades for frame-rate pedestrian detection. In: Fitzgibbon, A., Lazebnik, S., Perona, P., Sato, Y., Schmid, C. (eds.) ECCV 2012, Part II. LNCS, vol. 7573, pp. 645–659. Springer, Heidelberg (2012)
27. Benenson, R., Mathias, M., Timofte, R., Van Gool, L.: Pedestrian detection at 100 frames per second. In: Proceedings of IEEE Computer Society Conference on Computer Vision and Pattern Recognition, pp. 2903–2910 (2012)

Heterogeneous Multi-task Learning on Non-overlapping Datasets for Facial Landmark Detection

Takayuki Semitsu$^{(\boxtimes)}$, Xiongxin Zhao, and Wataru Matsumoto

Information Technology R&D Center, Mitsubishi Electric Corporation,
5-1-1, Ofuna, Kamakura, Kanagawa, Japan
Semitsu.Takayuki@dw.MitsubishiElectric.co.jp

Abstract. We propose a heterogeneous multi-task learning framework on non-overlapping datasets, where each sample has only part of the labels and the size of each dataset is different. In particular, we propose two batch sampling strategies for stochastic gradient descent to learn shared CNN representation. First one sets same number of iteration on each dataset while the latter sets same batch size ratio of one task to another. We evaluate the proposed framework by learning the facial expression recognition task and facial landmark detection task. The learned network is memory efficient and able to carry out multiple tasks for one feed forward with the shared CNN. In addition, we show that the learned network achieve more robust facial landmark detection under large variation which appears in the heterogeneous dataset, though the dataset does not include landmark labels. We also investigate the effect of weights on each cost function and batch size ratio of one task to another.

Keywords: Multi-task learning · Convolutional neural network · Facial expression · Facial landmark detection

1 Introduction

Facial landmark detection is an important research topic in many face analysis tasks such as facial attribute inference and face recognition. Though great progress has been made in this field, detection under variation in pose or occlusion remains difficult.

In Facial Image Analysis, deep neural networks have achieved success [1]. Convolutional neural networks (CNN) are one of the most popular architectures, which convolution calculation is suitable for recognizing 2D image structure. In addition, convolution can be computed efficiently using GPU. However, because of the large parameters of a deep neural network, it is hard to train a generalized network with limited data.

Some researchers tackle facial expression recognition and facial landmark detection tasks simultaneously using multi-task learning framework. Though these

© Springer International Publishing AG 2016
A. Hirose et al. (Eds.): ICONIP 2016, Part III, LNCS 9949, pp. 616–625, 2016.
DOI: 10.1007/978-3-319-46675-0_68

tasks are heterogeneous in terms of cost function, shared convolutional neural network and its joint global cost function are reported to work well [2].

These researches assume all labels are available for all training images. However, it is often the case that only part of the label is available for each image. This study aims to investigate the possibility to learn a task (e.g. facial landmark detection) with another related task (e.g. facial expression) using non-overlapping datasets (see Figs. 1, 2). As reported in [3], each non-overlapping dataset has its bias. Thus, convergence rates may differ greatly. To cope with this problem, some parameters in learning procedure have to be controlled. Since the stochastic gradient descent is popular for optimizing parameters in CNN, mini-batch sampling strategy has to be considered in order to cope with non-overlapping datasets. We propose two batch sampling strategy for non-overlapping datasets. We evaluate the effectiveness of our method by comparing it with single task learning.

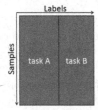

Fig. 1. Complete dataset. Each sample has labels for all tasks.

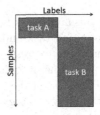

Fig. 2. Non-overlapping dataset. Each sample has only part of all tasks.

2 Related Work

Facial Landmark Detection. Conventional facial landmark detection methods can be divided into two categories, namely regression-based method and optimization-based method. A regression-based method estimates landmark locations explicitly by regression using image features. One popular regression-based approach is cascaded regression [4–6]. Cascaded fern regression with pixel-difference features are proposed [4,5]. Since most of these methods refine an initial guess of the landmark location iteratively, the first guess/initialization is critical. In contrast, our deep model takes raw image pixels as input without the need of any facial landmark initialization.

An optimization-based method builds face templates. Landmark detection is done by fitting template to input images using optimization [7,8]. Optimization-based method has recently been used for face fitting [9–11]. Zhu and Ramanan [11] show that face detection, facial landmark detection, and pose estimation can be jointly addressed. Our method differs in that we do not limit the learning of specific tasks, our method are learned on multiple dataset for heterogeneous tasks, thus potentially applicable to more general tasks.

Facial Expression Recognition. In facial expression analysis, popular approach is to divide face into set of Action Units (AUs) [12–15]. AUs are the fundamental actions of individual muscles or groups of muscles. Some researches utilize AU detections and facial landmark detections in successive expression recognition phase [16].

Multi-task Learning for Facial Analysis. Multi-task learning is typically applied when multiple tasks resemble each other and training data for each task is limited [17–19].

In [20], a heterogeneous multi-task model is trained by sharing the feature layers, which results in learning representation good for both tasks. Body-part detection and joint point regression task is learned by shared CNN [20]. Our method differs in that we tackles to train on non-overlapping multiple datasets, of which images have only part of the labels.

3   Proposed Method

Our heterogeneous multi-task framework consists of two types of tasks: (1) facial expression recognition task and (2) facial landmark detection task. The goal of (1) is to estimate the expression for given input image, and the goal of (2) is to predict the locations of facial landmarks in an image. Since we tackles the problem where the sizes of the training datasets differs for each task, we propose several batch sampling strategy for non-overlapping datasets in Sect. 3.5.

3.1   Facial Expression Recognition

For the facial expression recognition task, the goal is to classify the input image to several expression labels (e.g. pleasure, angry, sad, ...).

For expression recognition task, we minimize the cross-entropy error function,

$$E_c(\hat{Y}_{c,l}, Y_{c,l}) = -\sum_j^{N_c} y_{c,l}^j \log \hat{y}_{c,l}^j \tag{1}$$

where $Y_{c,l} = \{y_{c,l}^j | j = 1...N_c\}$ is the ground-truth label, and $\hat{Y}_{c,l} = \{\hat{y}_{c,l}^j | j = 1...N_c\}$ is the corresponding probability (i.e. softmax output) for the expression label l.

3.2   Facial Landmark Detection

For the facial landmark detection task, the goal is to estimate the location of each landmark. The coordinates of each facial landmark are taken as the target values. We use the squared-error as the cost function for our regression task,

$$E_r(\hat{Y}_{r,l}, Y_{r,l}) = \sum_i^{N_r} \|\hat{y}_{r,l}^i - y_{r,l}^i\|^2 \tag{2}$$

where $Y_{r,l} = \{y_{r,l}^i | i = 1...N_r\}$ is the ground truth landmark position, and $\hat{Y}_{r,l} = \{\hat{y}_{r,l}^i | i = 1...N_r\}$ is the estimated landmark position for landmark l.

3.3   Global Cost Function

Using cost functions defined in Sects. 3.1 and 3.2, our global cost function is the linear combination of the regression cost function for all facial landmarks and the classification cost function for all expression labels,

$$\Phi_{global} = \lambda_r \sum_{l_r}^{L_r} E_r(\hat{Y}_{r,l_r}, Y_{r,l_r}) + \lambda_c \sum_{l_c}^{L_c} E_c(\hat{Y}_{c,l_c}, Y_{c,l_c}) \tag{3}$$

where λ_r and λ_c are the weights for regression and classification tasks, respectively, N_r and N_c are the number of training images for each task, $l_r = (1...L_r)$ are labels for each keypoint and $l_c = (1...L_c)$ are labels for each expression label.

3.4   Network Structure

Figure 3 shows our network. It consists of shared layers and task-specific layers.

Shared Layers. For facial expression recognition and facial landmark detection, important points to see are likely to be similar to each other. So we share three layers of the network, which are encouraged to have generalized feature representation useful for both tasks. Each layer consists of convolutional layer and max-pooling layer.

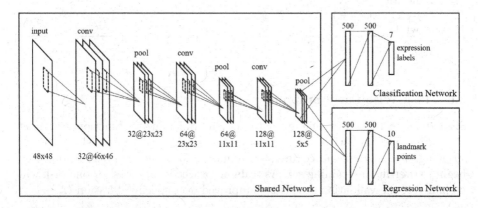

Fig. 3. Network structure of our proposed method. The input layer is 48 × 48 grayscale image. Shared Network consists of 3 convolutional layers and 3 max-pooling layers. Shared Network is then connected to Classification and Regression Network separately. Classification and Regression Network consists of 2 full-connection networks respectively

3.5 Mini-Batch Optimization

Since we apply Stochastic Gradient Descent as in [21], we have to choose batch size for each task respectively in order to deal with non-overlapping datasets. Suppose we have N_r samples for regression dataset and N_c for classification dataset. First strategy is to set batch sizes (B_r, B_c) so that iteration for each dataset is same $(N_r/B_r = N_c/B_c)$. We name it Same Iteration Multi-task SGD. See Algorithm 1 for detailed learning procedures. Second is to set specific batch sizes for each dataset so that the batch sizes holds specific ratio $\lambda_{batch}(B_r = \lambda_{batch} * B_c)$. In this case, iterations for each dataset differs. We name it Task Specific Batch Multi-task SGD. See Algorithm 2 for detailed learning procedures.

Algorithm 1. Same Iteration Multi-task SGD

1: $w_c \leftarrow$ initial vector of parameters for expression recognition(classification) task.
2: $w_r \leftarrow$ initial vector of parameters for landmark detection(regression) task.
3: $epoch \leftarrow 0$
4: **while** $epoch < N_{epoch}$ **do**
5:     Shuffle training sets(all tasks).
6:     **for** $i = 1...N_b$ **do**
7:         $batch_c^i \leftarrow$ one batch(size:B_c) from training set(classification).
8:         $batch_r^i \leftarrow$ one batch(size:B_r) from training set(regression).
9:         $update(w_c, batch_c^i)$
10:         $update(w_r, batch_r^i)$
11:     **end for**
12:     $epoch = epoch + 1$
13: **end while**

3.6 Task-Wise Early Stopping

Since different tasks and different datasets have different loss functions, learning difficulties, and thus with different convergence rates, we apply task-wise early stopping. The learning procedure stop weight updates for one task before they begin to over-fit the training set. As training proceeds, updates for one task will be halted if the validation cost is not improved over specific iteration (N_p).

4 Experimental Results

We present experiments using our method.

4.1 Training Data

We collect facial expression (classification) data from [22], facial landmark detection (regression) data from [1] respectively.

Algorithm 2. Task Specific Batch Multi-task SGD

1: w_c ← initial vector of parameters for expression recognition(classification) task.
2: w_r ← initial vector of parameters for landmark detection(regression) task.
3: $epoch_c$ ← 0
4: $epoch_r$ ← 0
5: Shuffle training sets(all tasks).
6: **while** $epoch_c < N_{epoch_c}$ **do**
7:      $batch_c^i$ ← one batch(size:B_c) from training set(classification).
8:      $batch_r^i$ ← one batch(size:B_r) from training set(regression).
9:      $update(w_c, batch_c^i)$
10:     $update(w_r, batch_r^i)$
11:     **if** all images in training set(classification) are sampled. **then**
12:         $epoch_c = epoch_c + 1$
13:         Shuffle training sets(classification tasks).
14:     **end if**
15:     **if** all images in training set(regression) are sampled. **then**
16:         $epoch_r = epoch_r + 1$
17:         Shuffle training sets(regression tasks).
18:     **end if**
19: **end while**

---

[22] includes 28709 images for training, 3589 images for testing. For facial landmark detection, [1] has 10000 images for training and 3466 for validation.

In order to construct shared CNN layers, which has one input image shape, we crop bounding boxes of the images of landmark detection dataset and resize them to (48 × 48), of which original image size is (250 × 250).

For the facial expression recognition task, the network predicts 7 labels of expressions for given images. For the facial landmark detection task, the network predicts 5 facial landmarks (centers of the eyes, nose, corners of the mouth). We train and evaluate our network on a PC with GTX Titan X. Training the network takes 1 to 2 days, while the evaluation for 3589 images takes 3–4 s.

4.2   Effect of the Weights on the Cost Functions

There are two parameters to set for our algorithm: weights for the cost functions of each task and batch sizes.

First we evaluate the effect of the weights for the classification and regression cost functions. We set different values for the weights. Figure 4 shows the training and testing error for Task Specific Batch Multi-task SGD. Batch sizes are both set to 100 so that the batch size ratio is 1 ($\lambda_{batch} = 1$). All the other parameters are kept the same.

Figure 5 shows the training and testing error for Same Iteration Multi-task SGD. Batch size of the facial landmark detection task is set to 51, while the batch size of the facial expression recognition task is set to 147. In this case the number of the iteration is same.

LambdaC/LambdaR	0	0.25	0.5	1	2	4	8	∞		
cls training	-	1.1772	1.0585	0.9704	0.9267	0.9410	1.0524	0.7903	MAX	MIN
cls test	-	1.1602	1.1154	1.1090	1.1075	1.0955	1.1615	1.0729	MAX	MIN
reg training	0.00618	0.00629	0.00629	0.00628	0.00627	0.00627	0.00628	-	MAX	MIN
reg test	0.00413	0.00409	0.00410	0.00410	0.00405	0.00405	0.00408	-	MAX	MIN

Fig. 4. Effect of changing the weights on each task for Task Specific Batch Multi-task SGD. Batch sizes for each task are same ($\lambda_{batch} = 1.0$). Cls (upper row) shows classification results while reg (lower row) shows regression results. The training and testing errors are for epoch 500.

LambdaC/LambdaR	0	0.0625	0.125	0.25	0.5	1	2	4	∞		
cls training	-	1.2519	1.1054	0.9081	0.7572	0.7822	0.7384	0.7336	0.2879	MAX	MIN
cls test	-	1.2212	1.1242	1.0780	1.0943	1.1149	1.1113	1.1044	1.3531	MAX	MIN
reg training	0.00488	0.00496	0.00495	0.00494	0.00495	0.00494	0.00494	0.00495	-	MAX	MIN
reg test	0.00326	0.00331	0.00333	0.00329	0.00332	0.00331	0.00331	0.00330	-	MAX	MIN

Fig. 5. Effect of changing the weights on each task for Same Iteration Multi-task SGD. The number of iterations for each task is the same. Cls (upper row) shows classification results while reg (lower row) shows regression results. The training and testing errors are for epoch 500.

As you can see in upper row of Fig. 4, the network has better generalization ability, i.e. better training error to testing error ratio on classification task.

Figure 5 shows that the classification gets overfit for single task learning, while multi-task learning shows better performance on classification task and competitive performance on regression task.

4.3 Effect of the Batch Sizes

Next we study the effect of the batch size ratio for Task Specific Batch Multi-task SGD. Weights for cost functions are kept $\frac{\lambda_C}{\lambda_R} = 1.0$. All the other parameters are kept the same. We show training and testing error in Fig. 6. We observed that best parameter for batch size ratio is 2 or 4.

Bc		50	100	200	400		
Br		100	100	100	100		
Bc/Br		0.5	1	2	4		
cls	training	1.0813	0.9704	0.8742	0.7881	MAX	MIN
cls	test	1.1148	1.1090	1.0926	1.0973	MAX	MIN
reg	training	0.00625	0.00628	0.00624	0.00634	MAX	MIN
reg	test	0.00410	0.00410	0.00409	0.00407	MAX	MIN

Fig. 6. Effect of changing the batch size on each task - training and testing errors are for epoch 500.

4.4   Evaluation

Finally we evaluate the effectivity of our proposed method, i.e., the joint learning of the facial expression recognition (classification) and facial landmark detection (regression) tasks using our method. For our proposed method, we use the Same Iteration Multi-task SGD (Sect. 3.5) with parameter setting of $\frac{\lambda_C}{\lambda_R} = 0.25$. Table 1 shows accuracy for the classification task and mean squared error for the regression task.

Table 1. Evaluation of multi-task learning. Left column shows the result of single-task learning networks trained on respective datasets. Right column shows the result of our proposed method. Note that in left columns, each row is generated by different network, while the row columns are generated by one network

	Single task (ER/LD)	Multi-task (ER+LD)
Expression recognition (Accuracy: %)	61.70 %	58.60 %
Landmark detection (MSE: pixel)	3.46	3.71

Table 1 shows that the learned network by our proposed framework has achieved competitive result to those of the networks trained on single task datasets, while our network is memory efficient and able to carry out multiple tasks at one feed forward of the shared CNN.

In addition, the learned network is more robust under large variation. Figure 7 shows the improvements of the landmark detection results on the heterogeneous dataset i.e. facial expression dataset. Though the dataset does not include landmark labels, the network can learn helpful feature representation for landmark detection at the shared CNN.

Fig. 7. Visualizes the improvement on heterogeneous dataset (facial expression validation dataset). Upper row depicts the result of the network trained on single task (landmark), while the lower row shows the result of our proposed framework. The dataset is for facial expression, so landmark labels are not included in the dataset. Results show that our network is more robust under hand occlusion (left 5 columns), rotation (column 5, 6), viewpoint variation (column 3, 4, 7, 8)

5   Conclusion

In this paper, we have proposed a heterogeneous multi-task learning framework on non-overlapping datasets. Two batch sampling strategies for stochastic gradient descent are introduced to learn shared CNN representation. Experimental results shows that it can achieve competitive results compared to single task, while the proposed network is memory efficient and able to carry out multiple tasks for one feed forward with shared CNN. We also investigate the effect of weights on each cost function and batch size ratio of one task to another. In addition, we show that the learned network achieve more robust facial landmark detection under large variation which appears in the heterogeneous dataset, though the dataset does not include landmark labels for training. Future work includes applying this method to advanced facial image analysis such as dense facial landmarks, and to other area such as human behavior analysis.

References

1. Sun, Y., Wang, X., Tang, X.: Deep convolutional network cascade for facial point detection. In: Proceedings of the IEEE Conference on Computer Vision and Pattern Recognition, pp. 3476–3483 (2013)
2. Zhang, Z., Luo, P., Loy, C.C., Tang, X.: Facial landmark detection by deep multi-task learning. In: Fleet, D., Pajdla, T., Schiele, B., Tuytelaars, T. (eds.) ECCV 2014, Part VI. LNCS, vol. 8694, pp. 94–108. Springer, Heidelberg (2014)
3. Torralba, A., Efros, A.A.: Unbiased look at dataset bias. In: 2011 IEEE Conference on Computer Vision and Pattern Recognition (CVPR), pp. 1521–1528, June 2011
4. Cao, X., Wei, Y., Wen, F., Sun, J.: Face alignment by explicit shape regression. Int. J. Comput. Vision **107**(2), 177–190 (2014)
5. Burgos-Artizzu, X., Perona, P., Dollár, P.: Robust face landmark estimation under occlusion. In: Proceedings of the IEEE International Conference on Computer Vision, pp. 1513–1520 (2013)
6. Wu, Y., Ji, Q.: Robust facial landmark detection under significant head poses and occlusion. In: Proceedings of the IEEE International Conference on Computer Vision, pp. 3658–3666 (2015)
7. Cootes, T.F., Edwards, G.J., Taylor, C.J.: Active appearance models. IEEE Trans. Pattern Anal. Mach. Intell. **6**, 681–685 (2001)
8. Liu, X.: Generic face alignment using boosted appearance model. In: IEEE Conference on Computer Vision and Pattern Recognition, CVPR 2007, pp. 1–8. IEEE (2007)
9. Asthana, A., Zafeiriou, S., Cheng, S., Pantic, M.: Robust discriminative response map fitting with constrained local models. In: Proceedings of the IEEE Conference on Computer Vision and Pattern Recognition, pp. 3444–3451 (2013)
10. Yu, X., Huang, J., Zhang, S., Yan, W., Metaxas, D.: Pose-free facial landmark fitting via optimized part mixtures and cascaded deformable shape model. In: Proceedings of the IEEE International Conference on Computer Vision, pp. 1944–1951 (2013)
11. Zhu, X., Ramanan, D.: Face detection, pose estimation, and landmark localization in the wild. In: 2012 IEEE Conference on Computer Vision and Pattern Recognition (CVPR), pp. 2879–2886. IEEE (2012)

12. Ekman, P., Friesen, W.V.: Manual for the Facial Action Coding System. Consulting Psychologists Press, Palo Alto (1978)
13. Zhao, K., Chu, W.-S., De la Torre, F., Cohn, J.F., Zhang, H.: Joint patch and multi-label learning for facial action unit detection. In: Proceedings of the IEEE Conference on Computer Vision and Pattern Recognition, pp. 2207–2216 (2015)
14. Zeng, J., Chu, W.-S., De la Torre, F., Cohn, J.F., Xiong, Z.: Confidence preserving machine for facial action unit detection. In: Proceedings of the IEEE International Conference on Computer Vision, pp. 3622–3630 (2015)
15. Rudovic, O., Pavlovic, V., Pantic, M.: Context-sensitive dynamic ordinal regression for intensity estimation of facial action units. IEEE Trans. Pattern Anal. Mach. Intell. **37**(5), 944–958 (2015)
16. Liu, P., Han, S., Meng, Z., Tong, Y.: Facial expression recognition via a boosted deep belief network. In: Proceedings of the IEEE Conference on Computer Vision and Pattern Recognition, pp. 1805–1812 (2014)
17. Evgeniou, T., Micchelli, C.A., Pontil, M.: Learning multiple tasks with kernel methods. J. Mach. Learn. Res. **6**, 615–637 (2005)
18. Yu, K., Tresp, V., Schwaighofer, A.: Learning gaussian processes from multiple tasks. In: Proceedings of the 22nd International Conference on Machine Learning, ICML 2005, pp. 1012–1019. ACM, New York (2005)
19. Yang, X., Kim, S., Xing, E.P.: Heterogeneous multitask learning with joint sparsity constraints. In: Advances in neural information processing systems, pp. 2151–2159 (2009)
20. Li, S., Liu, Z.-Q., Chan, A.: Heterogeneous multi-task learning for human pose estimation with deep convolutional neural network. In: Proceedings of the IEEE Conference on Computer Vision and Pattern Recognition Workshops, pp. 482–489 (2014)
21. Krizhevsky, A., Sutskever, I., Hinton, G.E.: Imagenet classification with deep convolutional neural networks. In: Advances in neural information processing systems, pp. 1097–1105 (2012)
22. Goodfellow, I.J., et al.: Challenges in representation learning: a report on three machine learning contests. In: Lee, M., Hirose, A., Hou, Z.-G., Kil, R.M. (eds.) ICONIP 2013. LNCS, vol. 8228, pp. 117–124. Springer, Heidelberg (2013). doi:10. 1007/978-3-642-42051-1_16

Fuzzy String Matching Using Sentence Embedding Algorithms

Yu Rong and Xiaolin Hu[✉]

Tsinghua National Laboratory for Information Science and Technology (TNList),
Department of Computer Science and Technology, Tsinghua University,
Beijing, China
rongyu9124@hotmail.com, xlhu@tsinghua.edu.cn

Abstract. Fuzzy string matching has many applications. Traditional approaches mainly use the appearance information of characters or words but do not use their semantic meanings. We postulate that the latter information may also be important for this task. To validate this hypothesis, we build a pipeline in which approximate string matching is used to pre-select some candidates and sentence embedding algorithms are used to select the final results from these candidates. The aim of sentence embedding is to represent semantic meaning of the words. Two sentence embedding algorithms are tested, convolutional neural network (CNN) and averaging word2vec. Experiments show that the proposed pipeline can significantly improve the accuracy and averaging word2vec works slightly better than CNN.

Keywords: Sentence embedding · Word2vec · CNN

1 Introduction

Fuzzy string matching has many applications, such as spelling checking. Automatic spelling check in a word processing system is such a problem because the input by users may contain errors but it is required to match the correct word and make correction suggestion. We are facing a similar problem which comes from a real world project. We have an Optical Character Recognition (OCR) system which extracts names of goods from images of receipts. Usually, OCR system are not perfect and their outputs may contain errors. For example, a Chinese name 可口可乐 *500* ml (Coca-Cola 500 ml in English) may be recognised as 可日可乐 *500* ml (口 is recognised as 日) and 可口可乐 *500* ml is also likely to be recognised as 可可乐 *500* ml (口 is missed). Our task is to retrieve the names of the goods in a large database (say one million) of names which are guaranteed to be correct. As our queries contain errors, we are facing a fuzzy string matching problem.

The most straightforward way is to compute similarity between the query string and the strings in the database one by one and return the most similar one. Traditional similarity metrics can be edit distance or some other metrics [1]

© Springer International Publishing AG 2016
A. Hirose et al. (Eds.): ICONIP 2016, Part III, LNCS 9949, pp. 626–633, 2016.
DOI: 10.1007/978-3-319-46675-0_69

that only depend on the appearance information of characters or words. However, simply using those approximate string matching algorithms can not get the best result due to the following reasons.

First, the approximate string matching can not handle the potential errors in string. For example, 青柠味 may be recognised as 清宁味 (青 and 柠 are wrongly recognised as 清 and 宁 respectively). From character level, 清宁味 is more similar to 清香味 than 青柠味. Approximate string matching only considers 青 not equals to 清 and 柠 not equals to 宁, it does not know 青 is easy to be wrongly recognised as 清 and 柠 is easy to wrongly recognised as 宁, so it can not make the right decision. However, if judged by human, it is easy to find the wrongly recognised characters and make the correct choice.

Second, approximate string matching only considers character level information and neglects the semantic information. For example, 洗面奶 and 洁面乳 are the same thing (facial cleanser), but from the character level, they only share one character. So if we only use appearance information, 洗面奶 and 洁面乳 may be judged as two different things.

Using sentence embedding algorithms [2], we can solve these problems. We can get error patterns of our OCR system and add them to training data during sentence embedding. In that case, our sentence embedding algorithm can learn these patterns and make right decisions while encountering recognition errors. As for semantic information, we can also extract it by using algorithms like word2vec [3,4]. Our task is retrieving the names in a large database with given recognised name of goods, so we mainly focus on fuzzy string matching and do not concern about how OCR system works.

2 Pipeline and Approximate String Matching

In this section, we will describe the pipeline of our algorithm first, then we will briefly introduce approximate string matching which is the first step of our pipeline.

2.1 Pipeline

Given the query string (string represents name of goods, the same below), we first use approximate string matching to find some candidates. These candidates are strings whose similarity with the query is beyond the pre-set threshold. After getting these candidates, we use sentence embedding algorithms to represent these strings (candidates and query string both) in vectors. Then we compute cosine distance between the vectors of the candidates and the query. The candidate that has the smallest cosine distance with the query is the final result we find. So our algorithm can be generalized as using approximate string matching to pre-select and sentence embedding to get the final result. The whole pipeline is depicted in Fig. 1.

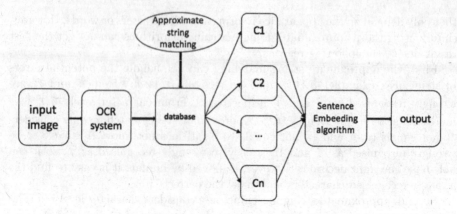

Fig. 1. The pipeline of the fuzzy string matching system

2.2 Approximate String Matching

We mainly use edit distance as our similarity function. Edit distance [5] is used to compute similarity between two strings by counting the minimum basic operations used to transform one string into another. There are three kinds of basic operations: insertion, deletion, substitution. Insertion means insert a single character (*kat* → *kate*, inserts *e* after *t*). Deletion means delete a single character (*kate* → *kat*, deletes *e* after *t*). Substitution means substitute one single character with another (*kate* → *kade*, substitutes *t* with *d*). Edit distance can be easily computed with dynamic programming, the detailed solving process can be found in this website [5]. Edit distance can be applied to Chinese strings in which a word means a single character.

3   Sentence Embedding

We focus on two kinds of sentence embedding methods. One method is based on convolutional neural network (CNN) [2] and word2vec and the other one is based on word2vec only. t is seen that both methods are based on word2vec, so we will describe the word2vec first. Word2vec is a two-layer shallow neural network that can generate word embedding which contains semantic information. During training, the network is given a word and its work is to guess which words is adjacent to this given word. The model after training can map the given word into a fixed length vector. Word2vec has two different models, one is skip-grams and another is continuous bag of words (CBOW).

Word2vec is an unsupervised learning algorithm. The algorithm is trained with those 100 million real names of goods. We randomly substitute and delete some characters according to error patterns of our OCR system and split the sentence into single characters. These split sentences are then used as our training data for word2vec. Word embedding we use in CNN and averaging word2vec are both trained in this way.

3.1   CNN

Recently, convolutional neural network has drawn remarkable success in many fields such as computer vision [6,9], natural language processing [7,10] and so on.

We train a classification convolutional neural network and make the output of the second last fully connected layer as our sentence embedding result. 50,000 names of goods are randomly chosen from our database and for each string we synthesis 10 strings as training data while making the original one as the testing data. In this way, we get a dataset of 50,000 classes, each original name of goods and its synthesis result form one class. Synthesis methods include adding meaningless words, randomly deleting or substituting one or two characters, disturbing the order of the words. Using the data, we train our classification network. Our task is not classification, category message is only used as supervisory information to help model the sentences.

Firstly, we get word embedding from word2vec. If we assume the length of our sentence is t, and the dimensionality of the word vectors is d, then the shape of the sentence matrix is $t \times d$. This sentence matrix can be viewed as an image and convolution operation can be performed on it.

During convolution operations, we must set the width of the filters to be the same as the dimension of the word embedding and leave the filters' height variable. Suppose the filters' width is d (dimension of the word embedding) and its height is h, then we can denote our filters by $w \in R^{h \times d}$ and the sentence matrix by $s \in R^{t \times d}$. The output is

$$O = f(w * s + b) \tag{1}$$

where $*$ represents convolution operation, b represents bias term, and f represents non-linear function. We choose ReLU [6] as our non-linear function. The output of convolution is a vector of length $t - h + 1$. The length of feature map generated by filters changes with the h and t, so we do 1-max pooling [2] to map the vector into a scalar. If n filters are used in convolution phase, then we get output matrix of shape $n \times (t - h + 1)$ after convolution. We do 1-max pooling on output matrix and get a vector of length n.

In the last step, the vectors are sent into two fully connected layers to represent it again and get a new vector of length n_1 (n_1 is the number of hidden units of the second fully connected layer). Finally, this n_1 length vector is fed into the last fully connected layer and softmax function to be classified.

After training, we feed the sentence matrix into the model and extract the output of the second fully connect layer as its vector representation. The structure of the convolutional neural network is depicted in Fig. 2.

3.2   Averaging Word2vec

In the previous section, the output of word2vec is used as the input of the convolutional neural network. In this section, it will be used in a more direct

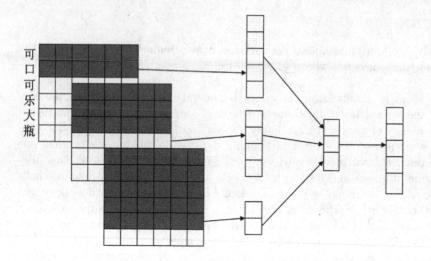

convolution with kernel 1,3,5   feature maps   1-max pooling   fully connected

Fig. 2. The structure of CNN

way. After getting the word embedding of every word in the sentence, we can get the sentence embedding by averaging these word embeddings. Suppose we have t words in the sentence and their word vectors are $v_1, v_2, \ldots v_t$, then we can get sentence embedding v_s by averaging them.

$$v_s = \frac{\sum_{i=1}^{t} v_i}{t} \tag{2}$$

This method looks simple, but it may work in practice.

We also try to use high frequency n-grams to replace single words. Assume a sentence can be split into $t_{(n)}$ words according to statistics results. $w_n^{(1)}, w_n^{(2)}, \ldots w_n^{(t_n)} s = w_n^{(1)} w_n^{(2)} \ldots w_n^{(t_n)}$. The length of $w_n^{(i)}$ may vary from 1 to n. We train word2vec models using these n-grams, then get word vectors, $v_1^{(n)}, v_2^{(n)}, v_{t_n}^{(n)}$. Again, we get sentence embedding v_s by averaging them.

$$v_s = \frac{\sum_{i=1}^{t_n} v_n^{(i)}}{t_n} \tag{3}$$

4   Experiments

4.1   Data

We are facing a real world problem: retrieving relevant name of goods in a large database using strings extracted from receipts by an OCR system which may contain errors. Each sample in the dataset is composed of the following parts:

one query string, one true result (one of the candidate string). This dataset is used for testing only and we compute testing accuracy on the dataset then take it as our criterion to judge the algorithms.

4.2   Baseline

In our paper, approximate string matching is used as baseline. During testing, we choose the string with the smallest edit distance as the final result. In our testing set, the accuracy of this method is 52.9 %.

4.3   Averaging Word2vec

Our model is trained with Google's implementation of word2vec[1]. Word2vec is an unsupervised algorithm, so we use real goods names in the database to train the model. We mainly focus on the effect of output vector size and window size and hold other parameters unchanged. The values of other parameters are in Table 1.

Table 1. Unchanged parameters in averaging word2vec.

Negative sampling	Model	Iteration number
No	skip-grams	5

First, we do experiments on the vector size. We hold the window size to be 5 and changed output vector size. Detailed results are presented in Table 2. From Table 2, we can see that the accuracy reaches highest when the vector size is 1000. Then the vector size is held to be 1000 and the window size changes, Detailed results can be seen in Table 3. From Table 3, we can find that the accuracy reaches highest when window size is 5, increasing or decreasing the window size will both have bad influence on performance.

Table 2. Experiment results with different vector size.

Vector size	200	500	1000	2000
Test accuracy(%)	80.8	83.4	**84.9**	80.8

In the n-gram experiments, we analyse all the n-grams in the real goods names. One n-gram is viewed as valid only when its occurrence time is beyond some preset threshold. Parameter settings of n-gram experiments can be viewed in Table 4. The accuracy of Averaging word2vec with n-gram is 82.1 % which is lower than the accuracy of averaging word2vec with single character.

---

[1] http://word2vec.googlecode.com/svn/trunk/.

Table 3. Experiment results with different window size.

Window size	3	5	10
Test accuracy(%)	80.5	**84.9**	80.8

Table 4. Parameter settings of averaging word2vec with n-grams.

Parameter	Iteration number	Window size	Vector size	n	Threshold
Value	10	10	1000	4	300

4.4   CNN

We use Caffe [8] to build the CNN. During training and testing phases, the length of sentence is held to be 45 ($t = 45$). The shorter sentences are padded with zeros while the longer sentences are truncated. Again, we use Google's implementation of word2vec to get the word embedding. During word2vec training, window size and length of output vector are set to be 5 and 200, respectively. Other parameters are in Table 1.

We have two fully connected layers. The number of their hidden units are 1500, 50000, respectively. We also try to add one more fully connected layers or increase the number of hidden units, but do not get better results. Since our sentences are short, we use a small convolution kernel height, which is 3. We have tried different kernel height of 5 and 7, the results remains the same. We set the number of feature map to be 300 which is the best in our experiments.

In the experiments below, we set the convolution kernel height to be 3 and the highest test accuracy we get is 83.1 %. We also try to use multiple convolution kernel heights which are 1, 3, 5. The number of output feature maps of each kind of convolution kernel is held to be 100, thus, we get 300 feature maps after convolution operation. Other parameters remain the same as before. With this setting, test accuracy is improved to 84.4 %.

4.5   Comparison

From the experiment, we find that averaging word2vec performs slightly better than CNN. The reason may be that name of goods we deal with is different with normal sentences. For a name of goods, it is what words it contains instead of the order of the words that matters. For example, 可口可乐 *500* ml and *500* ml 可口可乐 are exactly the same thing. Averaging word2vec neglects the order information and only considers the word information, so it performs better. We also find CNN with multiple convolution kernel size performs better than CNN with single one. The reason is that these multiple kernels can catch both fine and coarse information.

5   Conclusion

In this study, we use approximate string matching and sentence embedding to solve a real world problem. Through experiments, we show that approximate string matching is not good enough to solve this problem while sentence embedding can improve the results. We try two different sentence embedding methods: CNN and averaging word2vec. The results show that averaging word2vec performs slightly better than CNN. The proposed pipeline is not restricted to this particular application, and may be adapted to solve other similar string matching problems.

Acknowledgements. This work was supported in part by the National Basic Research Program (973 Program) of China under Grant 2012CB316301 and Grant 2013CB329403, and in part by the National Natural Science Foundation of China under Grant 61273023, Grant 91420201, and Grant 61332007.

References

1. Navarro, G.: A guided tour to approximate string matching. ACM Comput. Surv. (CSUR) **33**(1), 31–88 (2001)
2. Ye, Z., Byron, C.W.: A sensitivity analysis of (and practitioners guide to) convolutional neural networks for sentence classification. arXiv preprint arXiv:1510.03820 (2015)
3. Mikolov, T., Sutskever, I., Chen, K., Corrado, G.S., Dean, J.: Distributed representations of words and phrases and their compositionality. In: Proceedings of Advances in Neural Information Processing Systems, South Lake Tahoe, pp. 3111–3119 (2013)
4. Mikolov, T., Chen, K., Corrado, G., Dean, J.: Efficient estimation of word representations in vector space. arXiv preprint arXiv:1301.3781 (2013)
5. Edit Distance. https://en.wikipedia.org/wiki/Edit_distance
6. Krizhevsky, A., Sutskever, I., Hinton, G.E.: Imagenet classification with deep convolutional neural networks. In: Proceedings of Advances in Neural Information Processing Systems, South Lake Tahoe, pp. 1097–1105 (2012)
7. Sutskever, I., Vinyals, O., Le, Q.V.: Sequence to sequence learning with neural networks. In: Proceedings of Advances in Neural Information Processing Systems, Montréal, Quebec, Canada, pp. 3104–3112 (2014)
8. Jia, Y., Shelhamer, E., Donahue, J., Karayev, S., Long, J., Girshick, R., Guadarrama, S., Darrell, T.: Caffe: convolutional architecture for fast feature embedding. In: Proceedings of the ACM International Conference on Multimedia, pp. 675–678. ACM (2014)
9. Szegedy, C., Liu, W., Jia, Y., Sermanet, P., Reed, S., Anguelov, D., Erhan, D., Vanhoucke, V., Rabinovich, A.: Going deeper with convolutions. In: Proceedings of the IEEE Conference on Computer Vision and Pattern Recognition, Boston, pp. 1–9 (2015)
10. Zhang, X., Zhao, J., LeCun, Y.: Character-level convolutional networks for text classification. In: Proceedings of Advances in Neural Information Processing Systems, Montréal, Quebec, Canada, pp. 649–657 (2015)

Initializing Deep Learning
Based on Latent Dirichlet Allocation
for Document Classification

Hyung-Bae Jeon[1,2] and Soo-Young Lee[3(✉)]

[1] Department of Bio and Brain Engineering,
Korea Advanced Institute of Science and Technology,
Daejeon, Republic of Korea
hbjeon@etri.re.kr
[2] Electronics and Telecommunications Research Institute, Daejeon, Korea
[3] School of Electrical Engineering,
Korea Advanced Institute of Science and Technology,
Daejeon, Republic of Korea
sylee@kaist.ac.kr

Abstract. The gradient-descent learning of deep neural networks is subject to local minima, and good initialization may depend on the tasks. In contrast, for document classification tasks, latent Dirichlet allocation (LDA) was quite successful in extracting topic representations, but its performance was limited by its shallow architecture. In this study, LDA was adopted for efficient layer-by-layer pre-training of deep neural networks for a document classification task. Two-layer feedforward networks were added at the end of the process, and trained using a supervised learning algorithm. With 10 different random initializations, the LDA-based initialization generated a much lower mean and standard deviation for false recognition rates than other state-of-the-art initialization methods. This might demonstrate that the multi-layer expansion of probabilistic generative LDA model is capable of extracting efficient hierarchical topic representations for document classification.

Keywords: Document classification · Deep learning · Latent dirichlet allocation · Good initialization

1 Introduction

With the recent abundance of electronic documents, it has become very important to classify and search text documents with latent semantics, context, and topics. Until recently, latent Dirichlet allocation (LDA) was the most popular approach for developing generative topic models from bag-of-words document representations [1]. On the

This work was supported by the ICT R&D program of MSIP/IITP, Republic of Korea (R0126-15-1117, Core technology development of spontaneous speech dialogue processing for the language learning).

© Springer International Publishing AG 2016
A. Hirose et al. (Eds.): ICONIP 2016, Part III, LNCS 9949, pp. 634–641, 2016.
DOI: 10.1007/978-3-319-46675-0_70

other hand, '*deep learning*' has recently attracted significant attention owing to several notable successful cases involving speech recognition and image classification tasks [2–4]. However, open issues still exist, such as those related to *good* initialization based on various pre-training algorithms [5, 6]. Although it may be possible to obtain *reasonable* performance by simply utilizing big data and fast computing systems, gradient-descent algorithms converge to a local minimum and the learning process may not be optimal. Unless stochastic optimization is incorporated, it is necessary to start with a good initialization, which must be close to the global minimum and therefore depends on the task.

The recent renaissance of artificial neural networks owes heavily to the good initialization of deep neural networks (DNNs) by restricted Boltzmann machine (RBM). However, stochastic RBM-based initialization requires significant computing power, and another popular initialization based on an auto-encoder (AE) demonstrated only marginal improvement over the '*big data and fast computer*' approach. In this Letter we propose a new *layer-by-layer* initialization method based on LDA for document classification tasks. Similar to RBM, LDA is a probabilistic generative model; however, its learning process is far more efficient than that of RBM [1]. Furthermore, LDA demonstrated learning excellent topic-based representation of documents.

2 Deep Learning with LDA-Based Initialization

Figure 1 shows the flow diagram of a DNN and three learning steps. Following the recent trend, rectified linear unit (ReLU) was used for the nonlinear functions of hidden-layers, while the SoftMax function was adopted at the output layer. At the first step, several lower layers were pre-trained layer-by-layer using LDA. At the second step, the remaining higher layers were trained by standard error backpropagation. Starting from the learned synaptic weights at the first and second steps, the third and final step provided the fine-tuning of synaptic weights for all layers.

Based on the *bag-of-words* representation, the input to the classifier is determined as a word-frequency vector W_n, of which the mth element is defined as $W_{mn} = c_{mn}/C_n$ with $1 \leq n \leq N$ and $1 \leq m \leq M$. Here, c_{mn} is the number of times the mth word w_m occurs in the nth document d_n, and C_n is the total number of words present in the nth document [7]. In addition, M and N denote the number of words in the vocabulary and the number of documents in the training database, respectively. $\bar{W} = [W_1 \cdots W_N]$ constitutes a training word-document matrix.

The uniqueness of the proposed learning algorithm resides in the LDA-based unsupervised pre-training of lower layers. By maximizing the likelihood of the word-document matrix $\bar{W}$ obtained from the training data, LDA learns word probability distributions of the kth topic, i.e., $\beta_k = [\beta_{k1},..,\beta_{kM}]^T$ with $\beta_{km} = p(w_m|t_k)$, and topic probability distributions of the nth document, i.e., $\theta_n = [\theta_{n1},..,\theta_{nK}]^T$ with $\theta_{nk} = p(t_k|d_n)$, for $1 \leq k \leq K$, $1 \leq m \leq M$, and $1 \leq n \leq N$. Here, t_k is the kth topic and K is the number of topics. Parameters α and η control the Dirichlet distributions of θ_n and β_k, respectively. Once the LDA parameters are learned, the clustering or classification of a document d_n may be performed from θ_n.

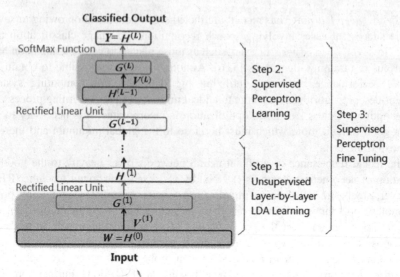

Fig. 1. Deep neural networks architecture and learning steps. Layer-by-layer LDA-based unsupervised learning is incorporated at Step 1.

In general, the number of topics K is much smaller than the number of words M, and θ_n may be regarded as a dimensionality-reduced representation of document d_n. Therefore, the layer-by-layer LDA learning may start from the document vector W_n as the input, and the first hidden-layer neuronal activations $H_n^{(1)} = f(G_n^{(1)})$ as the output with $G_n^{(1)} \equiv \theta_n$ and the ReLU nonlinear function $f(.)$. Then, for $l \geq 1$, the $(l+1)$th layer may be added with $H_n^{(l)}$ as the input.

However, LDA is a probabilistic generative model and, even with a pre-trained $B = [\beta_1,.., \beta_K]$, the estimation of θ_n from a new W_n for testing is nontrivial. To approximate this mapping deterministically using a single-layer neural network, one noticed that

$$p(w_m|d_n) = \sum\nolimits_{k=1}^{K} p(w_m|t_k)p(t_k|d_n) = \sum\nolimits_{k=1}^{K} \theta_{nk}\beta_{km} \tag{1}$$

$$E[W_{mn}] \cong \sum\nolimits_{k=1}^{K} \theta_{nk}\beta_{km}, W_n \cong B\theta_n. \tag{2}$$

Here, $E[.]$ denotes an expectation operator, and one incidence of normalized word-frequency W_n is assumed to approximate word probability $p(w_m|d_n)$. Then, an approximate θ_n may be obtained as $\hat{\theta}_n \cong B^+ W_n$. Here, superscript '+' denotes pseudo-inverse of a rectangular matrix. Therefore, the lth layer is defined as

$$G^{(l)} = (B^{(l)})^+ H^{(l-1)}, H^{(l)} = ReLU(G^{(l)}). \tag{3}$$

Here, $B^{(l)}$ denotes learned LDA parameter B at the lth layer. Even though all elements of B are non-negative, B^+ and therefore $G^{(l)}$ may have negative elements. Therefore, the ReLU nonlinearity is adopted at each hidden layer.

3 Experimental Database and Networks

The proposed initialization scheme was applied to the deep learning of a document classification task using the Reuters-21578 corpus. The modified Apte (ModApte) split consists of 9603 training documents and 3299 test documents. In our experiment, all documents with multiple categories were removed, and only documents related to the top 20 categories were selected. As a result, 6201 and 2771 documents were selected for the training and test sets, respectively. Functional words and numerals were removed, and the top 10,000 words were selected. The average number of words in each document is approximately 130, and the *bag-of-words* representation of documents is quite sparse.

Although recurrent neural networks, such as Long Short-Term Memory (LSTM) model, become quite popular for the learning of sentence representation with huge training data, the *bag-of-words* representation is still advantageous for small training data and is less sensitive to syntax and grammar.

In the experiments the DNN consisted of four layers, of which first two layers were initialized based on LDA. The numbers of neurons at the input and output layers were set to 10,000 for 10,000 words and 20 for 20 classes, respectively. Based on previous works [8], the number of neurons at all the hidden layers was set to 500, i.e., $K = 500$. Therefore, the LDA learning assumed 500 topics and 500 meta-topics.

4 Performance of LDA-Based Deep Learning

Because θ_n will be provided as the input to the next layer, the characteristics of the topic probabilities for each document θ_n are very important. Figure 2 presents histograms of $B = [\beta_1.. \beta_K]$ and $\hat{\theta}_n$. Although the word probabilities per topic are sparser for smaller α values, $\hat{\theta}_n$ distribution is insensitive to α values.

Therefore, as shown in Fig. 3, the cross entropy error for the training database converged well during the fine-tuning stage, and the learning convergence on the training data was not sensitive to α values. However, false recognition rates (FRRs) on the test data show differences. It agrees with the common understanding that different initializations may converge to local minima and result in different generalization performance.

In Table 1, the final FRRs from ten different initializations were summarized. The mean test FRRs were about 4 %, and the small standard deviations showed insensitivity to α values and random initializations.

In Figs. 4 and 5, the learning performance of the LDA-based initialization method for the test data was compared with other popular initialization methods such as random, auto-encoder (AE), and supervised learning with one hidden layer neural

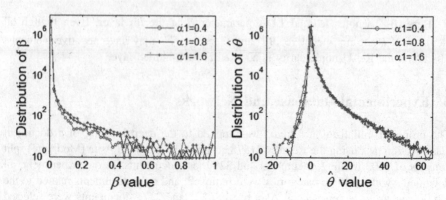

Fig. 2. Distributions of data representations at the LDA-trained first layer. (a) Histograms of β_{km} values, i.e., word probabilities for each topic; (b) Histogram of $\hat{\theta}_n$ values, i.e., topic probabilities of documents.

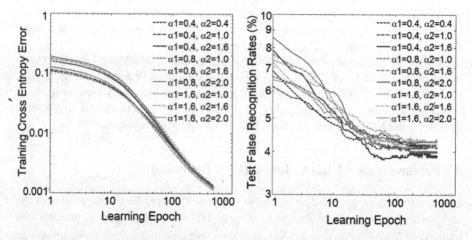

Fig. 3. Learning convergence of four-layer networks during fine-tuning stage with LDA-based initialization of the first two-layers. (a) Cross entropy error of training database; (b) False recognition rates (%) for the test database. Parameters α_1 and α_2 control the Dirichlet distributions of θ_n at the first and second layer, respectively.

Table 1. Mean and standard deviation of false recognition rates (%) evaluated from ten LDA-based initializations

	False Recognition Rates (%)				FRR Standard Deviation (%)			
	$\alpha_2 = 1.0$	$\alpha_2 = 1.6$	$\alpha_2 = 2.0$	$\alpha_2 = 2.5$	$\alpha_2 = 1.0$	$\alpha_2 = 1.6$	$\alpha_2 = 2.0$	$\alpha_2 = 2.5$
$\alpha_1 = 0.4$	4.00	**3.97**	4.17	4.01	0.12	**0.19**	0.09	0.13
$\alpha_1 = 0.8$	4.16	4.20	**4.19**	4.22	0.11	0.17	**0.08**	0.07
$\alpha_1 = 1.6$	4.39	4.25	**3.97**	4.04	0.14	0.07	**0.14**	0.17

networks (denoted as '1-L NN' hereafter). RBM was not compared owing to its slow convergence with only 'not-so-significant' improvements over AE. Here, the same DNN architecture with two pre-trained layers and two additional layers was used.

As shown in Fig. 4, the cross entropy error for the training database converged well during the fine-tuning stage for all initialization methods. However, the converged false recognition rates showed large differences. Actually, while the training errors were not

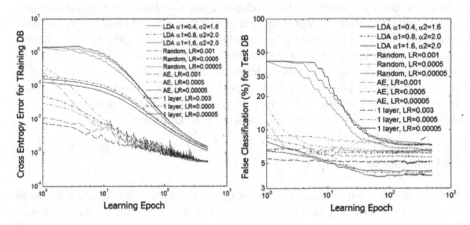

Fig. 4. Comparison with other initialization methods. (a) Cross entropy errors for the training database versus fine-tuning the learning epoch. (b) False recognition rate (%) for the test database versus fine-tuning the learning epoch.

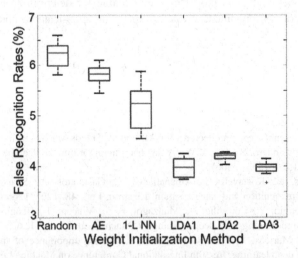

Fig. 5. False recognition rates (%) generated by the LDA-based initialization and other initialization methods from 10 different random initializations. 'Random', 'AE', and '1-L NN' denote random, auto-encode, and 1 hidden-layer supervised learning initialization, respectively. LDA1/2/3 denote proposed LDA-based initialization with 3 best parameters shown as bold fonts in Table 1.

the lowest, the three LDA-based initializations resulted in much lower false recognition rates (FRRs) for the test data. Owing to the supervised learning at the layer-by-layer stacking stage, the '1 layer' methods exhibited very good test performance even at the start; however, the fine-tuning did not result in significant improvements.

Figure 5 shows the converged FRRs from 10 different random initializations. Here, LDA1, LDA2, and LDA3 denote three top LDA-based initializations in Table 1 with $(\alpha_1 = 0.4, \alpha_2 = 1.6)$, $(0.8, 2.0)$, and $(1.6, 2.0)$, respectively. Compared with other popular initialization methods, the developed LDA-based initialization method generated much lower FRRs as well as much lower FRR standard deviations. Owing to the supervised learning at the layer-by-layer stacking stage, the '1-L NN' method exhibited good test performance even at the start of the fine tuning stage; however, the fine-tuning did not result in significant improvements. Actually, although not shown in Fig. 5, even the LDA-based initializations with three-layer networks (two LDA-based layers and one additional layer) outperform all other conventional initialization methods with a 4.21 % average FRR and a 0.13 % standard deviation.

5 Conclusion

In this paper, we demonstrated that the proposed LDA-based initialization results in much better classification performance, in terms of both the mean accuracy and standard deviation from randomness. The reduction of the FRR from 5.81 % (AE) and 5.20 % (1 hidden-layer supervised learning) to 3.97 % is significant, and the performance is robust on the choice of LDA parameter α. This excellent performance may result from the match between the task and the initialization algorithm. In fact, the probabilistic generative LDA model may be suitable for determining hieratical topic representations of text documents while not incurring significant computing costs.

References

1. Blei, D.M., Ng, A.Y., Jordan, M.I.: Latent dirichlet allocation. J. Mach. Learn. Res. **3**, 993–1022 (2003)
2. Bengio, Y.: Learning deep architectures for AI. Found. Trends Mach. Learn. **2**, 1–127 (2007)
3. Hinton, G.E., Osindero, S., Teh, Y.W.: A fast learning algorithm for deep belief nets. Neural Comput. **18**, 1527–1554 (2006)
4. Charalampous, K., Kostavelis, I., Amanatiadis, A., Gasteratos, A.: Sparse deep-learning algorithm for recognition and categorisation. Electron. Lett. **48**, 1265–1266 (2012)
5. Erhan, D., Bengio, Y., Courville, A., Manzagol, P.-A., Vincent, P., Bengio, S.: Why does unsupervised pre-training help deep learning? J. Mach. Learn. Res. **11**, 625–660 (2010)
6. Sutskever, I., Martens, J., Dahl, G., Hinton, G.: On the importance of initialization and momentum in deep learning. In: 30th International Conference on Machine Learning, Atlanta, USA, pp. 1139–1147, June 2013

7. Haidar, M.A., O'Shaughnessy, D.: Unsupervised language model adaptation using LDA-based mixture models and latent semantic marginals. Comput. Speech Lang. **29**, 20–31 (2015)
8. Song, H.A., Kim, B.K., Xuan, T.L., Lee, S.Y.: Hierachical feature extraction by multi-layer non-negative matrix factorization network for classification task. Neurocomputing **165**, 63–74 (2015)

Author Index

Printed in the United States
By Bookmasters